Frequency Stability: Fundamentals and Measurement

OTHER IEEE PRESS BOOKS

Electronic Displays, *Edited by H. I. Refioglu*
Spread-Spectrum Communications, *Edited by C. E. Cook, F. W. Ellersick, L. B. Milstein, and D. L. Schilling*
Color Television, *Edited by T. Rzeszewski*
Advanced Microprocessors, *Edited by A. Gupta and H. D. Toong*
Biological Effects of Electromagnetic Radiation, *Edited by J. M. Osepchuk*
Engineering Contributions to Biophysical Electrocardiography, *Edited by T. C. Pilkington and R. Plonsey*
The World of Large Scale Systems, *Edited by J. D. Palmer and R. Saeks*
Electronic Switching: Digital Central Systems of the World, *Edited by A. E. Joel, Jr.*
A Guide for Writing Better Technical Papers, *Edited by C. Harkins and D. L. Plung*
Low-Noise Microwave Transistors and Amplifiers, *Edited by H. Fukui*
Digital MOS Integrated Circuits, *Edited by M. I. Elmasry*
Geometric Theory of Diffraction, *Edited by R. C. Hansen*
Modern Active Filter Design, *Edited by R. Schaumann, M. A. Soderstrand, and K. R. Laker*
Adjustable Speed AC Drive Systems, *Edited by B. K. Bose*
Optical Fiber Technology, II, *Edited by C. K. Kao*
Protective Relaying for Power Systems, *Edited by S. H. Horowitz*
Analog MOS Integrated Circuits, *Edited by P. R. Gray, D. A. Hodges, and R. W. Brodersen*
Interference Analysis of Communication Systems, *Edited by P. Stavroulakis*
Integrated Injection Logic, *Edited by J. E. Smith*
Sensory Aids for the Hearing Impaired, *Edited by H. Levitt, J. M. Pickett, and R. A. Houde*
Data Conversion Integrated Circuits, *Edited by D. J. Dooley*
Semiconductor Injection Lasers, *Edited by J. K. Butler*
Satellite Communications, *Edited by H. L. Van Trees*
Frequency-Response Methods in Control Systems, *Edited by A. G. J. MacFarlane*
Programs for Digital Signal Processing, *Edited by the Digital Signal Processing Committee, IEEE*
Automatic Speech & Speaker Recognition, *Edited by N. R. Dixon and T. B. Martin*
Speech Analysis, *Edited by R. W. Schafer and J. D. Markel*
The Engineer in Transition to Management, *I. Gray*
Multidimensional Systems: Theory & Applications, *Edited by N. K. Bose*
Analog Integrated Circuits, *Edited by A. B. Grebene*
Integrated-Circuit Operational Amplifiers, *Edited by R. G. Meyer*
Modern Spectrum Analysis, *Edited by D. G. Childers*
Digital Image Processing for Remote Sensing, *Edited by R. Bernstein*
Reflector Antennas, *Edited by A. W. Love*
Phase-Locked Loops & Their Application, *Edited by W. C. Lindsey and M. K. Simon*
Digital Signal Computers and Processors, *Edited by A. C. Salazar*
Systems Engineering: Methodology and Applications, *Edited by A. P. Sage*
Modern Crystal and Mechanical Filters, *Edited by D. F. Sheahan and R. A. Johnson*
Electrical Noise: Fundamentals and Sources, *Edited by M. S. Gupta*
Computer Methods in Image Analysis, *Edited by J. K. Aggarwal, R. O. Duda, and A. Rosenfeld*
Microprocessors: Fundamentals and Applications, *Edited by W. C. Lin*
Machine Recognition of Patterns, *Edited by A. K. Agrawala*
Turning Points in American Electrical History, *Edited by J. E. Brittain*
Charge-Coupled Devices: Technology and Applications, *Edited by R. Melen and D. Buss*
Spread Spectrum Techniques, *Edited by R. C. Dixon*
Electronic Switching: Central Office Systems of the World, *Edited by A. E. Joel, Jr.*
Electromagnetic Horn Antennas, *Edited by A. W. Love*
Waveform Quantization and Coding, *Edited by N. S. Jayant*
Communication Satellite Systems: An Overview of the Technology, *Edited by R. G. Gould and Y. F. Lum*
Literature Survey of Communication Satellite Systems and Technology, *Edited by J. H. W. Unger*
Solar Cells, *Edited by C. E. Backus*
Computer Networking, *Edited by R. P. Blanc and I. W. Cotton*
Communications Channels: Characterization and Behavior, *Edited by B. Goldberg*
Large-Scale Networks: Theory and Design: *Edited by F. T. Boesch*
Optical Fiber Technology, *Edited by D. Gloge*

Frequency Stability: Fundamentals and Measurement

Edited by
Věnceslav F. Kroupa
Institute of Radio Engineering and Electronics
Czechoslovak Academy of Sciences

A volume in the IEEE PRESS Selected Reprint Series,
prepared under the sponsorship of the IEEE
Instrumentation and Measurement Society.

IEEE
PRESS

The Institute of Electrical and Electronics Engineers, Inc., New York

1884 1984
A CENTURY OF ELECTRICAL PROGRESS

PRINTED IN THE UNITED STATES OF AMERICA

IEEE Order Number: PC01644

Library of Congress Cataloging in Publication Data
Main entry under title:

Frequency stability.

Includes index.
1. Frequency stability—Addresses, essays, lectures.
I. Kroupa, Věnceslav F., 1923-
TK381.F74 1983 621.37'47 83-18336
ISBN 0-87942-171-1

Contents

To the memory of my mother and
the hundredth anniversary of my father's birthday

Preface

WITH the advent of atomic clocks and the space age more than 25 years ago arose the need for precise measurements of frequency and frequency stability on one hand and of the common "language" between different laboratories on the other hand. After feverish research in the early 1960's, an IEEE and NASA Symposium on Short Term Frequency Stability was held in 1964, and a special issue of the PROCEEDINGS OF THE IEEE devoted to the problem was published in February, 1966. After these two events the art of frequency stability characterization, both its theory and practical measurements, was perfected by many workers in America as well as in Europe, and many papers were published.

However, there is no book or monograph dealing exclusively with this subject. To fill the gap, we proposed that the IEEE Press publish in its Selected Reprint Series this volume, a collection of 38 representative papers. Unfortunately, as a consequence of the inevitable limit in size of a reprint book, we have been forced to leave out many valuable articles and treatises. To provide at least a partial remedy, we have mentioned some of them in introductions to individual parts.

As to the organization of the present book, we have divided the material into six parts, starting with the problem of flicker noise, an omnipresent physical phenomenon which limits frequency stability in many oscillators and requires a special statistical treatment.

Part II is devoted to the theory of noise in free-running and synchronized oscillators.

In the third part, the reader is acquainted with the evolution of the mathematical theory of frequency stability and the present state of the art, which may be considered to be classical with respect to the time domain and frequency domain definitions and to their mutual conversions.

The papers dealing with measurement systems, procedures, and devices are collected in Parts IV and V, the latter being devoted to measurements of frequency stability in microwave ranges.

The last part provides information about noise behavior of frequency generators—oscillators and frequency synthesizers—and measurement devices used in determining the frequency stability.

We hope that this reprint book may be used with advantage by both practicing engineers as a reference book and by the students of the art as a textbook.

As to the required background knowledge readers of this book will need, we have to assume familiarity with the Laplace and Fourier transforms, the fundamentals of probability, random variables, and stochastic processes, and some aquaintance with PLL and frequency synthesis techniques.

Finally, the editor would like to thank the IEEE Press for including the present volume in its Selected Reprint Series and the IEEE Instrumentation and Measurement Society for its backing of this enterprise. My thanks are also due to the Institute of Radio Engineering and Electronics of the Czechoslovak Academy of Sciences, and especially to the Institute's Director, Dr. V. Zima, for his generous support of the present work.

VĚNCESLAV F. KROUPA
January 1983

Part I
1/f (Flicker) Noise and Its Statistical Treatment

THE omnipresent fluctuations or "noise" in physical systems limit the eventual precision with which any quantity or parameter can be determined. Of particular concern are noises of which the power spectral density increases with lowering Fourier frequency, f.

There are three common noises found in nature: white noise with $S(f) \sim$ constant, flicker noise with $S(f) \sim 1/f$, and the random walk process with $S(f) \sim 1/f^2$ (e.g., the position of a particle undergoing one-dimensional Brownian motion). The latter two fall into the category where no improvement can be achieved by utilizing longer averaging times.

Since the discovery of electronic noise more than 60 years ago [I-1], a large number of papers and books have been devoted to the problem. In this connection, we call the readers' attention to the IEEE Press book *Electrical Noise: Fundamentals & Sources* [I-2] which contains, in addition, a copious bibliography.

Reverting to the matter of measurement of frequency stability, we find flicker noise to be one of the major limiting factors, since there is some evidence that the random walk of frequency is caused mainly by fluctuations of the environmental temperature and as a consequence can often be effectively reduced, as is testified by the first of the reprint papers and many others [I-3], [I-4], [IV-9].

In contradistinction, the $1/f$ noise or flicker noise is ubiquitous and seems to be a general fluctuation phenomenon mostly connected with the collective motion of particles. A good brief review is provided in the article by Voss [I-5] and in the second reprint paper, by Wolf, which deals with the state of the art in electronic systems. For further information, the interested reader is referred to the *Proceedings of the Symposium on 1/f Fluctuations,* held in Tokyo in 1977 [I-6].

The common difficulty with low-frequency divergent noises, with spectral density $|f|^\alpha$ where $\alpha \leq -1$, is the nonstationarity of the respective random processes. The consequence is that the autocorrelation function is time dependent; e.g., both Radeka and Keshner [I-7], [I-8] arrived at the following result for $1/f$ noise:

$$R(t, \tau) \sim \lg (4t) - \lg \tau. \qquad \text{(I-1)}$$

Another difficulty is to find a practical model. From the theory of random processes it is known that the derivation in the time domain is represented in the frequency domain by multiplication of the original power spectral density by $S(\omega)$ by ω^2 [I-9]. Similarly, one may assume that the one-order integration is modeled by dividing $S(\omega)$ by ω^2. Proceeding further this way, one could arrive at $1/f$ noise behavior by half-order integration of the white noise process, as suggested by Barnes and Allan in their "Statistical Model of Flicker Noise" [I-10]. A practical physical model proposed both by Radeka and Keshner [I-7], [I-8], is provided by an infinitely distributed RC-line driven from a white noise current source. Keshner proposed another linear system yielding $1/f$ noise and consisting of resistor-capacitor sections separated and isolated by unity gain amplifiers. The result may seem to be somewhat surprising: approximately one properly designed section is required (one state variable—the voltage on the capacitor) per decade of frequency to fit a $1/f$ power spectral density (cf. the paper by van Vliet [I-11]). On the contrary, Brownian motion requires only one state variable.

The above mentioned models for generation of $1/f$ noise are too restrictive since some studies, concerned with solid-state devices, suggest that it can be caused by perturbations occurring randomly in time and having certain specific distributions of lifetimes. The situations being such, Halford has proposed, in the third reprint paper, a general model in which a broad class of "reasonable" perturbations are found to satisfy the criteria for generation of $1/f$ noise. The advantage of the Halford's approach is that noise spectral density of any desired law or combination of laws can be accurately and efficiently generated by a simple computer program.

At the end of this part we reprint two historical papers which are still quoted today. The leading ideas were to find dependable and repeatable procedures to evaluate the behavior of atomic frequency standards and clocks even in ranges governed by divergent noises. In the first of these papers, Barnes introduces the concept of finite differences of time (or phase) and arrives both experimentally and theoretically to the conclusion that

$$\langle \Delta^2 \varphi_n \rangle = 6R_\varphi(0) - 8R_\varphi(\tau) + 2R_\varphi(2\tau) \qquad \text{(I-2a)}$$

$$\sim K\tau^{-\alpha-3} \qquad \text{(I-2b)}$$

where α is the exponent of the considered frequency divergent noise (in instances where secular terms, aging etc., have been excluded). In this way, he succeeded in eliminating the problem of nonstationarity, i.e., the time-dependent term of the autocorrelation function $R_\varphi(t,\tau)$.

In the second paper, Allan introduces as a measure of the noise behavior of generators the standard deviation

$$\sigma^2 = \frac{2}{T^2}\left[R_\varphi(0) - R_\varphi(\tau)\right]. \quad \text{(I-3a)}$$

$$= \frac{1}{T^2} U(\tau) \quad \text{(I-3b)}$$

and investigates its properties with respect to the measurement time T, the sampling time τ, and the number of samples N. He again finds a similar dependence on α as Barnes did, but he deplores the fact that the standard deviation diverges with N. This problem was later overcome by equating N to 2 (see Part III, paper 3).

References

[I-1] W. Schottky, "Über spontane Stromschwankungen in vershiedenen Elektrizitätsleitern," *Ann. Phys.* vol. 57, pp. 541-67, 1918.

[I-2] M. S. Gupta, Ed., *Electrical Noise: Fundamentals & Sources.* New York: IEEE Press, 1977.

[I-3] R. Burgoon and R. L. Wilson, "Performance results of an oscillator using the SC cut crystal," in *Proc. 33rd Annu. Symp. Frequency Contr.*, Atlantic City, NJ, May–June 1979, pp. 406–410.

[I-4] Y. Noguchi, Y. Teramachi, and T. Musha, "1/*f* frequency fluctuation of a quartz crystal oscillator and temperature fluctuation," in *Proc. 35th Annu. Symp. Frequency Contr.*, Philadelphia, PA, May 1981, pp. 484–491.

[I-5] R. F. Voss, "1/*f* (flicker) noise: A brief review," in *Proc. 33rd Annu. Symp. Frequency Contr.,* Atlantic City, NJ, May–June 1979, pp. 40–46.

[I-6] *Proc. Symp. 1/f fluctuations*, Tokyo, Japan, 1977.

[I-7] V. Radeka, "1/*f* noise in physical measurements," *IEEE Trans. Nucl. Sci.*, vol. NS-16, pp. 17–35, Oct. 1969; also in M. S. Gupta, Ed., *Electrical Noise: Fundamentals & Sources.* New York: IEEE Press, 1977, pp. 178–196.

[I-8] M. S. Keshner, "1/*f* noise," *Proc. IEEE,* vol. 70, pp. 212–218, Mar. 1982.

[I-9] A. Papoulis, *Probability, Random Variables and Stochastic Processes*. New York: McGraw-Hill, 1965, p. 339.

[I-10] J. A. Barnes and D. W. Allan, "A statistical model of flicker noise," *Proc. IEEE*, vol. 54, pp. 176–178, Feb. 1966.

[I-11] K. M. van Vliet, "Noise in semiconductors and photoconductors," *Proc. IRE*, vol. 46, pp. 1004–1018, June 1958; also in M. S. Gupta, Ed., *Electrical Noise: Fundamentals & Sources.* New York: IEEE Press, 1977, pp. 148–162.

Spectrum Analysis of Extremely Low Frequency Variations of Quartz Oscillators*

The rather numerous discussions of the spectrum of oscillator frequency fluctuations prompted the application of the Tukey[1] technique of spectrum analysis to five years of daily frequency comparisons between an atomic frequency standard and free-running quartz crystal oscillators at the National Bureau of Standards. Three of the four spectra shown in Fig. 1 were computed from observations made on oscillators 5a and 35 which are 100-kc quartz oscillators utilizing vacuum tubes in amplifiers and automatic level control circuits. The aging rate of these oscillators had stabilized several years before these measurements were begun.

The fourth spectrum of Fig. 1 is an estimate of the contribution to the oscillator spectra due to instrumentation noise. From a series of random numbers this estimate was computed assuming that the daily frequency measurements were in error by as much as ± 1 part in 10^{10} and that the measurement error on any given day was independent of the errors made on neighboring days. Before undertaking the spectrum analyses a detrending procedure described by Norton, *et al.*,[2] was used to remove frequency drifts and variations at frequencies lower than the lowest frequency for which the spectrum was obtained. The spectrum of $\Delta f/f$, the relative or fractional frequency deviations of the oscillator, is shown with normalization such that integration from 0 to ∞ with respect to the Fourier frequency F in cycles per second would formally yield the total relative mean-square fluctuation. Fig. 1 shows spectral densities $G_{\Delta f/f}(F)$ which appear to increase without limit as one considers lower and lower Fourier frequencies. This behavior is similar to that of klystron and LC oscillators at Fourier frequencies about 10^8 times greater than those of Fig. 1.[3,4] A least-squares-type fit to the lowest oscillator curve of Fig. 1 is $G(F)=3.6 \times 10^{-23}/F^{1.4}$ (fractional frequency deviation)^2 per cycles per second.

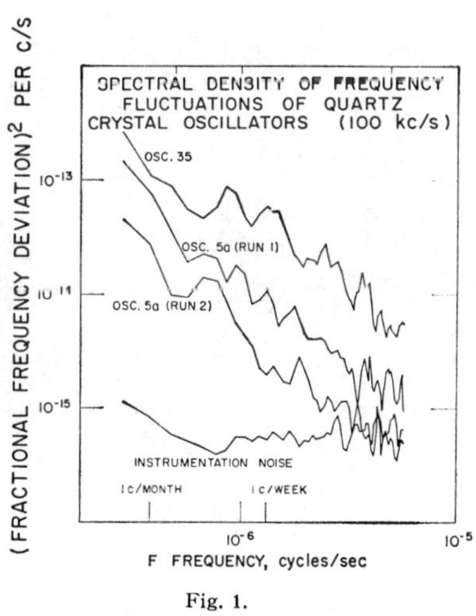

Fig. 1.

Other spectrum measurements were made in a higher fluctuation, or Fourier, frequency range on more recently constructed transistorized 5-Mc quartz oscillators. The equipment consisted of an NH_3 maser, frequency multipliers, a mixer, an electronic frequency meter, and a tunable mean-square indicator. Analysis of these data yielded $G(F)=(5\times 10^{-23})/0.9$(fractional frequency deviation)^2 per cycles per second in the range $0.02<F<2$ cps. This result is comparable to that for the lower frequency range. Above $F=2$ cps the spectral density was observed to increase with increasing Fourier frequency F.

For both types of crystal oscillators studied, the spectral behavior observed below $F=2$ cps is similar to that of current fluctuations in current carrying semiconductors. Current noise possessing a spectrum of this type occurs widely in nature and has been referred to as excess noise, flicker noise, or simply $1F$ noise.[5,6] Excess noise has been observed at frequencies as low as 6×10^{-5} cps.[7] In many theoretical treatments[8–10] of oscillator stability it has been customary to consider only the direct effects of tube shot noise and thermal noise on frequency stability with resulting theoretical spectra which disagree with the actually observed spectra reported here. The effect of flicker noise on oscillator stability has been treated theoretically by Troitsky.[11] His analysis resulted in a spectrum similar to the oscillator spectra of Fig. 1; however, experiments of Driagin[4] failed to verify certain aspects of this theory.

A difficulty associated with the oscillator spectra of the type shown in Fig. 1 is in determining how to extrapolate curves to lower frequencies in a meaningful way. It is desirable to compute for an oscillator such quantities as the power spectrum width, the mean-square frequency variation, and the expected time error of a clock connected to the oscillator. Formal relationships may be easily derived expressing these quantities as integrals over frequency of the frequency fluctuation spectrum times some appropriate filter function which weights the spectrum in accordance with the application under consideration. However, these integrals sometimes diverge at the lower limit if the spectrum is assumed to behave at arbitrarily low frequencies as a straight line extrapolation of Fig. 1 might suggest. The convergence difficulty was discussed by Malakhov[6] with reference to the expressions that lead to the spectral distribution of energy of an oscillator randomly frequency modulated with flicker noise. After disposing of the problem associated with the upper limit, Malakhov proposed proceeding according to the customary formalism that is valid for stationary processes, with the exception that the lower limit $F=0$ was to be replaced by $F=1/T$ where T is characteristic of the duration of an experiment which would determine the width of the power spectrum. This replacement seems reasonable and could be applied in the computation of the fluctuational aspects of various experiments involving oscillators. However, the exact relationship between T and the duration of an observation or experiment is somewhat vague, though this in itself would be of little consequence were it not for the fact that in some cases examined, the pertinent integral was found to be rather dependent on the exact value of the lower limit. Closer analysis of the whole problem might show that the weighting function, rather than the lower limit, could, and should, incorporate the duration of the experiment in such a way as to avoid the difficulty.

The authors wish to thank V. E. Heaton, who made and recorded the frequency measurements, A. W. Kirkpatrick, who supplied the program for performing the detrending and spectrum analysis, and J. A. Barnes, who made available the ammonia maser.

W. R. Attkinson
L. Fey
J. Newman
Radio Standards Lab.
National Bureau of Standards
Boulder, Colo.

* Received August 17, 1962.
1 R. B. Blackman and J. W. Tukey, "The Measurement of Power Spectra," Dover Publications, Inc., New York, N. Y.; 1959.
2 K. A. Norton, J. W. Herbstreit, H. B. Janes, K. O. Hornberg, C. F. Peterson, A. F. Barghaussen, W. E. Johnson, P. D. Wells, M. C. Thompson, Jr., M. J. Vetter, and A. W. Kirkpatrick, "An Experimental Study of Phase Variations in Line of Sight Microwave Transmissions," NBS Mono. No. 33; November 1, 1961.
3 G. A. Espersen, "Noise studies on two-cavity CW klystrons," IRE Trans. on Microwave Theory and Techniques, vol. MTT-8, pp. 474–477; September, 1960.
4 Iu. A. Driagin "Investigating technical frequency drifts in oscillators " 93–97; December, 1958.
5 D. A. Bell, "Electrical Noise: Fundamentals and Physical Mechanism," D. Van Nostrand Co., Inc., Princeton, N. Y.; 1960.
6 A. N. Malakhov "Spectra of flicker noise " *Radiotekh. Elektron.*, vol. 4 pp. 54–62, January, 1959.
7 T. E. Firl and H. Winston, "Noise measurements in semiconductors at very low frequencies " *J. Appl. Phys.* vol. 26 pp. 716–718; June, 1955.
8 W. A. Edson "Noise in oscillators," Proc. IRE, vol. 48, pp. 1454–1466; August, 1960.
9 J. A. Mullen, "Background noise in nonlinear oscillators," Proc. IRE, vol. 48, pp. 1467–1473; August, 1960.
10 M. J. Golay, "Monochromaticity and noise in a regenerative electrical oscillator," Proc. IRE, vol. 48, pp. 1473–1477; August, 1960.
11 V. S. Troitsky, "Certain problems of the theory of fluctuations in oscillators " *Radiofizika*, vol. 1, pp. 20–23, February, 1958.

Reprinted from *Proc. IEEE*, vol. 51, p. 379, Feb. 1963.

1/f-Noise

Dietrich Wolf

Institut für Angewandte Physik der Universität
D-6000 Frankfurt a.M., Fed. Rep. of Germany

1. Introduction

Low-frequency fluctuations showing a power spectral density inversely proportional to the frequency are observed in various physical, technical, biological, and economic systems. This peculiar phenomenon which is called 1/f-noise or flicker-noise has stimulated the research efforts of numerous scientists since the early work of JOHNSON [1] and SCHOTTKY [2]. In the past an impressive number of papers have been devoted to the experimental and theoretical aspects of 1/f-noise. Today this subject seems to be more fascinating than ever before. A remarkable sign of growing interest has been the Tokyo Symposium in 1977 [3], which solely dealt with 1/f-fluctuations. Table 1 reviews important systems in which 1/f-fluctuations have been observed. The observed quantities x are given together with some references.

Though many results have been found and ingenious theoretical models have been developed the basic problem of the origin of 1/f-noise in general remains unsolved up to now. The exciting question whether 1/f-noise is a general fluctuation phenomenon inherent in the collective motion of particles or is a property of a specific system is still a challenge to us.

Recently VAN DER ZIEL [32] and HOOGE [33] have presented excellent surveys on the state of the art of 1/f-noise in physical systems. Therefore, in this article - without the claim for completeness - some special problems will be discussed which illuminate the present situation on the 1/f-scene.

2. General Properties

The 1/f-fluctuation phenomena in physical systems generally can be characterized by their Gaussian amplitude distribution density and by a power spectral density $S_x(f)$ which follows a $f^{-\alpha}$ law in a wide range of frequency f with an exponent α close to unity. x denotes the fluctuating quantity. Thus 1/f-noise may be considered to be a stationary Gaussian random process. Another characteristic feature attributed to 1/f-noise is the typical dependence $S_i(f) \sim I^{\gamma}$ of the power spectral density $S_i(f)$ of the current fluctuation i on the d.c. current I with $1 < \gamma \leq 2$. At last 1/f-noise is commonly regarded as to be independent on temperature.

In real physical systems more or less significant deviations from ideal properties are observed and in a set of equivalent systems the actual values are statistically distributed. This situation has given rise to permanent - sometimes academic - discussions and controversies about the basic characteristics of 1/f-noise.

At the beginning some critical comments and remarks on the basic features shall be given.

1) Amplitude Distribution Density

Reprinted with permission from *Proc. 5th Int. Conf. Noise*, Bad Nauheim, West Germany, Mar. 13–16, 1978, pp. 122–133.

Table 1 Systems showing 1/f-noise

System	x	Ref.
Carbon film	Current	4
Metal film	Current	5,6
Semiconductor	Current	5
Metal Contact	Current	5
Semiconductor Contact	Current	5,7
Ionic Solution Contact	Current	7
Superconductor	Flux Flow	8,9,10
Vacuum Tube	Current	11,12,13
Junction Diode	Current	14
Schottky Diode	Current	15,16,17
Zener Diode	Current	18
Bipolar Transistor	Current	14
Field Effect Transistor	Current	14
Thermocell	Thermovoltage	19,20
Electrolytic Concentration Cell	Voltage	7
Quartz Oscillator	Frequency	21,22,23
Earth (5 days mean of rotation)	Frequency	24,25,26
Sound and Speech Sources	Loudness	27,28
Nerve Membrane	Potential	29,30
Highway Traffic	Current	31

Frequently the one-dimensional amplitude distribution density has been measured and found to be Gaussian [34,35]. Appearing deviations could be explained by the interference with additional low frequency noise components, in particular burst noise. In higher order distributions the Gaussian nature has never been verified, but the relation

$$\langle i^2(t)\cdot i^2(t+\tau)\rangle = 2\langle i(t)i(t+\tau)\rangle^2 + \langle i^2(t)\rangle^2$$

has found to be valid which follws from the second order Gaussian distribution density [36]. This may be considered to be an indirect confirmation of second order Gaussianity.

2) Stationarity

Several authors analysing the fluctuations of the short-time variance - the so-called variance noise - came to the conclusion [38] that 1/f-noise might be a non-stationary random process and proposed specific models for non-stationary random processes [39]. Own measurements on the variance noise of 1/f-noise in bipolar transistors as well as in carbon resistors have indicated clearly that there exists no discrepancy to the assumption of stationarity [36].

3) Power Spectral Density

The shape of the power spectral density is of the f^{-a}-type strictly only by idealizing. Often the spectrum is structured and spectra of equivalent systems are not identical. The "constant" a frequently is not a constant in the whole range of f. But, in general, values of a are found between 0.8 and 1.4. Thus, the values 0.5 and 1.5 which are characteristic for some diffusion controlled processes and 2.0 which is well known from a single time constant relaxation process are excluded distinctly. Fig.1 shows typical 1/f-noise spectra of several equivalent

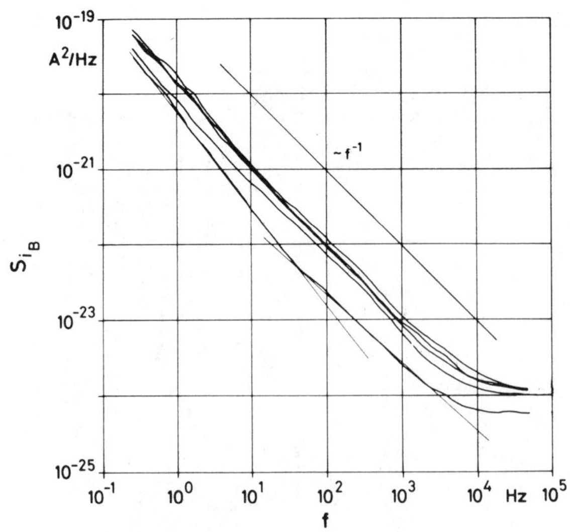

Fig.1 Examples of 1/f-noise spectra

bipolar transistors taken from the same wafer and differing in shape and magnitude. As is demonstrated some spectra are characterized by a single constant a in the whole frequency range from 10^{-1} Hz up to 10^{4} Hz, while one sample spectrum requires two values of a. Measuring a spectrum with higher resolution and precision even the uniform spectra often reveal humps and hollows along their shapes. Fig.2 gives examples of high resolution spectra, each of them being constituted of 1600 measured values. It also shows that 1/f-noise cannot be observed at frequencies higher than about 10^{5} Hz when white shot noise with the power spectral density $2qI_B$ becomes predominant. Looking into very low frequencies one finds a monotonously increasing noise power spectral density proportional to 1/f without visible limit. Further measurements in electrical systems down to 10^{-6} Hz have not yet revealed any lower frequency limit in the 1/f-spectrum. The theoretical problem still remains unsolved if there exists such a limit at all.

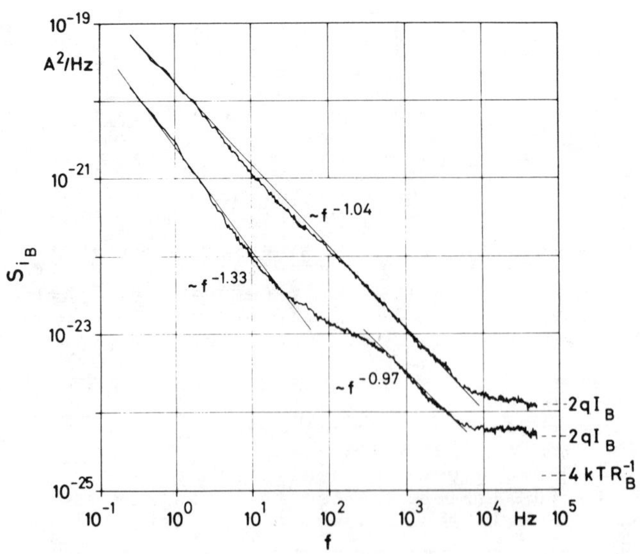

Fig.2 High resolution spectra of 1/f-noise. For comparison the power spectral density $2qI_B$ of white shot noise of the base current I_B as well as the power spectral density $4kTR_B^{-1}$ of the thermal noise producted by the base resistor R_B is indicated.

4) Current Dependence

In homogeneous conducting materials a strict I^2-dependence of 1/f-noise has been verified, especially considering recent investigations. These observations have led to the conclusion that 1/f-noise originates in resistance fluctuations. In junction diodes and bipolar transistors the current spectral density $S_i(f)$ in general is observed to be proportional to I^γ with γ between 2.0 and 1.0 [40,41]. Our recent measurements including three kinds of equivalent bipolar transistors which were processed under different conditions indicate that the exponent γ might depend on the processing and configuration of the transistor. This is illustrated by Fig.3 showing the spectrum of base current fluctuation S_{iB} vs. base current I_B. The three different symbols correspond to the three kinds of transistors analysed. Assuming this 1/f-noise to be caused by fluctuations of the recombination velocity VAN DER ZIEL has evaluated γ to be in the range between 1.0 and 2.0 [42].

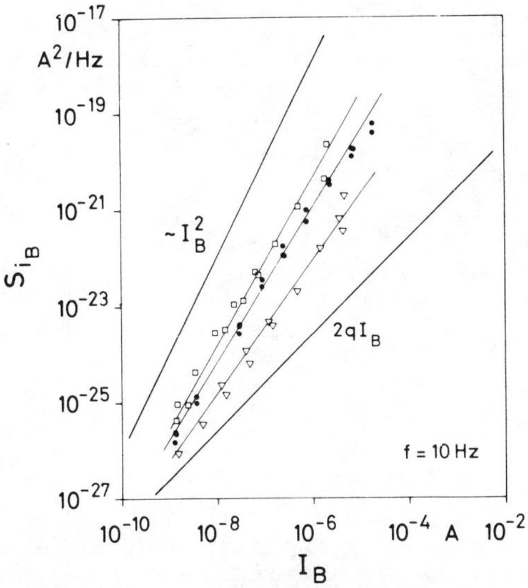

Fig.3 Influence of different processing methods on the power spectral density S_{iB} of base current fluctuations in transistors versus base current I_B

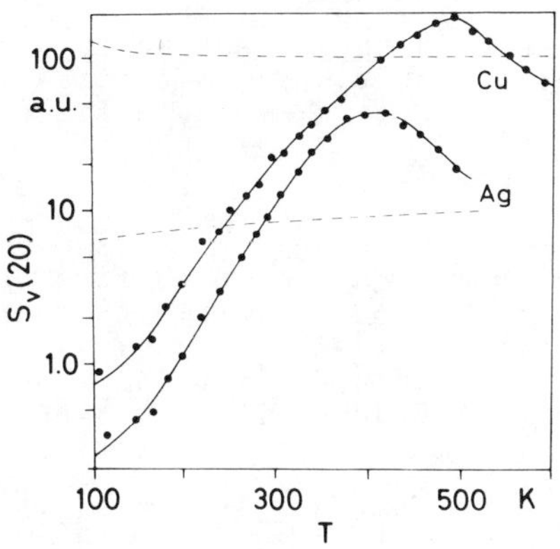

Fig.4 Temperature dependence of the noise power spectral density $S_v(f)$ measured at f = 20 Hz for thin metal films
—•—•— experimental (after[44])
----- theoretical (after [6])

5) Temperature Dependence

1/f-noise depends on temperature as the current depends on temperature. Besides this indirect temperature influence earlier studies concluded that no other direct temperature dependence exists [43]. That conclusion was based on experimental results which included considerable uncertainties. Today this conclusion must be rejected. In the last year EBERHARD and HORN [44] presented new remarkable results on 1/f-noise in metal films which show the distinct temperature dependence as given in Fig.4. Similar results have been obtained by MAY and HORWOOD [45]. This feature cannot be explained by any of the present theoretical models. Further investigations on the temperature dependence may provide us with a better insight into the mechanism of 1/f-noise.

3. Experimental Facts and Theoretical Interpretation

1) Hooge's Formula

In 1969 HOOGE [46] established his famous phenomenological relation of the power spectral density $S_i(f)$ of current fluctuations - or $S_{\Delta R}(f)$ of resistance fluctuations -

$$\frac{S_i(f)}{I^2} = \frac{S_{\Delta R}(f)}{R^2} = C \cdot \frac{1}{f} \qquad (1)$$

which well describes the observed 1/f-noise in homogeneous metals and semiconductors. I and R are the means of current and resistance, resp.

In systems with homogeneous current density the intensity factor

$$C = \frac{\alpha}{nV} \qquad (2)$$

only depends on the total number $n \cdot V$ of the free charge carriers where n is the carrier density and V the sample volume. The dimensionless quantity α empirically has been found to be about $2 \cdot 10^{-3}$. However, up to now no theoretical evidence has been given that α should be an universal constant.

Eqs. (1) and (2) may be generalized for situations in homogeneous systems, when the current density becomes inhomogeneous, e.g. in the case of point contacts [47] or fourpoint configurations [48,49]. Applying (1) to subvolumes of the sample in which the current density may be considered to be homogeneous one finds by linear independent superposition of the subvolume noise contributions a total noise power spectral density which is now determined by a factor

$$C = \frac{\alpha}{nV^*} \; . \qquad (3)$$

V^* represents an effective volume in which the 1/f-noise source seems to be concentrated. In general, V^* is a complicated expression that has to be calculated numerically. In special cases V^* has been derived in closed form. For point contacts HOOGE and HOPPENBROUWERS [47] obtained

$$V^* = \frac{20}{\pi^2} (2\pi r)^3 = \frac{20}{\pi^2} \left(\frac{\rho}{R}\right)^3 . \qquad (4)$$

r is the contact radius and α the resistivity.

HOOGE's formula has been examined and verified by many authors. The investigations dealt with point contacts of Ag, Al, Au, Brass, Cr, Cu, Pb, Pt, Sn, Steel [50] and Manganin [51], metal films of Ag, Au, Bi, Cu, and Sn [52] with various semiconductor materials like n- and p-type Ge, Si [40], GaAs, and InSb, n-GaP [53], and Li-doped MnO [54], with plastically deformed p- and n-type Si [55]. The results throughout are in accordance with (1) yielding values of α in the range of $1 \cdot 10^{-3}$ to $9 \cdot 10^{-3}$. Recently HOOGE et al. applied (1) to liquid Hg and Ga and measured $\alpha \simeq 3 \cdot 10^{-3}$ [56].

Furthermore it has been tried to employ (1) with more complicated structures such as ion-implanted resistors [57] or junction (J) or MOS FETs [5]. While the 1/f-noise in the resistors could to some extent be interpreted by (1) the results of noise analysis in MOST's are 10^3 to 10^4 times smaller than predicted by (1). These results suggest that in general inhomogeneous materials do not obey (1). Further investigations are desirable.

Summarizing one can state that with HOOGE's relation (1) a valuable tool has been given for practical calculations of 1/f-noise intensity in homogeneous metals as well as in semiconductor materials.

So far resistance fluctuations have been taken for the origin of 1/f-noise. Consequently the problem of the origin of the resistance fluctuations arises. Since the resistance is determined by density and mobility of the carriers both of these quantities could cause the observed resistance fluctuations. The investigations of KLEINPENNING [20] who analysed thermovoltage fluctuations clearly indicate the thermovoltage fluctuations to be an effect of mobility fluctuations. Experiments with concentration cells led to the same conclusion [50].

Supposing that similar results could be achieved with other physical systems showing 1/f-noise, the basic question still remains: What kind of mechanism originates the mobility fluctuations?

2) Approach of Clark and Voss

CLARK and VOSS [58][6] suggested to interpret the resistance fluctuations as a result of thermal equilibrium temperature fluctuations which modulate the temperature dependent resistance. This approach was supported by their observation that manganin characterized by a negligible temperature coefficient β of the resistance does not show 1/f-noise as compared to other metals with normal values of β. They assumed that spatially and temporally uncorrelated temperature perturbances δT with the variance

$$<(\delta T)^2> = kT^2/c_V \tag{5}$$

randomly occur which by heat conduction propagate within the sample. In (5) c_V is the specific heat and k Boltzmann's constant. A straightforward treatment of this model does not yet yield a 1/f-shaped spectrum. But introducing certain spatial correlations between the perturbances one finds the spectrum fitting to a 1/f-shape in the analysed frequency range. Adequate normalization to the variance then leads to the result

$$\frac{S_{\Delta R}(f)}{R^2} = \frac{(\beta T)^2}{3(3 + 2 \ln Q)} \frac{1}{f} = \frac{\alpha}{N} \frac{1}{f} \tag{6}$$

which corresponds to (1) with $\alpha = 3.6 \cdot 10^{-3}$ and agrees with the experiments. Q denotes the length-width ratio of the sample, N the number of free carriers.

The hypothesis of spatial correlations could be confirmed by VOSS and CLARK [52], too, by perturbing the temperature of the sample and then analysing the response.

The thermal equilibrium fluctuations of the resistance should lead to corresponding fluctuations in thermal noise intensity. This has been verified by VOSS and CLARK and in addition by BECK and SPRUIT [59].

The model of VOSS and CLARK has given rise to several disputes. VANDAMME [51] could not confirm the extreme low 1/f-noise intensity in manganin but found full agreement with (1). Analysing temperature fluctuations in metals and semiconductors KLEINPENNING [59] came to the conclusion that common 1/f-noise and temperature induced resistance fluctuations have to be well distinguished. He found the temperature induced resistance fluctuations to be negligible in semiconductors while in metals this effect can predominate at frequencies close to the inverse thermal relaxation time. Furthermore the temperature dependence of 1/f-noise measured by EBERHARD and HORN [44] cannot be explained by the thermal equilibrium fluctuation model, which yields the dashed curves in Fig.4.

3) McWhorter's Surface Noise Model

Another famous approach for the explanation of 1/f-noise is MCWHORTER's surface model [14], [60]. This model of the origin of 1/f-noise in semiconductors is based on the idea that the charge carriers in the material by tunneling interact with traps distributed in the surface oxide layer close to the semiconductor-oxide interface. The fluctuating occupancy of the traps modulates the surface potential and thus the carrier density in the region close to the surface. Noise is detected when a current passes this region of the semiconductor. Thus, 1/f-noise appears to be the superposition of processes with Lorentzian-type spectra

$$\frac{\tau}{1+(2\pi f\tau)^2}$$

where the time constants τ are widely distributed due to the tunnel probability distribution. Assuming a homogeneous distribution of the traps in the oxide at a distance between 0 and 20 Å of the semiconductor-oxide interface and an appropriate tunneling model the time constant τ varies over many orders of magnitude. This results in a 1/f-spectrum extended over a corresponding number of frequency decades.

MCWHORTER's Model is fully accepted for the interpretation of 1/f-noise in MOST's or in other devices where the charge transport in regions near the surface predominates. In junction diodes and, especially, in bipolar transistors 1/f-noise may be explained by MCWHORTER's model as long as the current is controlled by the surface potential.

4) Other Models

For the explanation of 1/f-noise further specific models have been proposed. Some of them are based upon fundamental physical principles such as thermodynamical equilibrium [61], transport of particles, or interaction of moving charge carriers with potential barriers [62].

Others make use of very special - sometimes artificial looking - assumptions. They may be helpful to some extent but they are restricted to particular situations [63, 64].

A third group of models is established under pure mathematical and signal theoretical aspects [65]. There are authors who treat 1/f-noise as a random sequence of random pulses [66, 67, 68, 69], or as a queuing problem [70, 4], or as a result of nonlinear transformation of white noise [71, 72, 73, 74, 75], or as a non-stationary stochastic process [10, 76, 77].

One outstanding recent approach is the quantumtheoretical treatment of 1/f-noise by HANDEL [62]. He takes into account that bremsstrahlung is emitted while moving charges are scattered. In this theory bremsstrahlung means emission or absorption of any massless particles which lead to infrared divergence in lowest significant order of the perturbation theory, such as photons, phonons, electron-hole pairs on the Fermi surface of metals, etc. Two types of 1/f-noise are obtained: the first is a self-interference effect while the second represents transition probability oscillations. Both types act as mobility fluctuations. For metals HANDEL finds HOOGE's constant to be approximately $2 \cdot 10^{-3}$ in the low temperature limit.

4. Conclusion

Looking at the interesting material on 1/f-noise to both experimental and theoretical aspects presented the last years, we can state some remarkable progress in the understanding of this subject. Some results indicate 1/f-noise to be closely related to the collective motion of particles. From another point of view one can ask whether 1/f-noise is an inherent non-equilibrium phenomenon or is an effect also present in thermal equilibrium.

References

1. J.B. Johnson: Phys. Rev. 26, 71-85 (1925)
2. W. Schottky: Phys. Rev. 28, 74-103 (1926)
3. Proc. Symposium on 1/f Fluctuations, Tokyo (1977)
4. D.A. Bell: *Electrical Noise* (van Nostrand, London 1960)
5. F.N. Hooge: Physica 83 B+C, 14-23 (1976)
6. J. Clark and R.F. Voss: Phys. Rev.Letters 33, 24-27 (1974)
7. F.N. Hooge and J.L. Gaal: Philips Res. Repts. 26, 77-90 (1971)
8. D.J. van Ooijen and G.J. van Gurp: Philips Res. Repts. 21, 343-367 (1966)
9. J.R. Clem: Phys. Rev. B1, 2140-2155 (1970)
10. H.M. Choe and A. van der Ziel: Physica 81B, 237-244 (1976)
11. H. Rothe und W. Kleen: *Elektronenröhren als Anfangsstufenverstärker* (Akademische Verlagsges., Leipzig 1948), Kap. XV
12. A. van der Ziel: in: *Noise in Electronic Devices*, ed. by L.D. Smullin and H.A. Haus (Wiley, New York 1959) Chapter 2
13. H. Pfeiffer: *Elektronisches Rauschen* (J.A. Mayer V., Aachen 1962)
14. A. van der Ziel: Proc. IEEE, 58, 1178-1206 (1970)
15. S.T. Hsu: IEEE Trans. Electr. Devices, ED-17, 496-506 (1970)
16. S.T. Hsu: IEEE Trans. Electr. Devices, ED-18, 882-887 (1971)
17. E.L. Wall: Solid State Electronics 19, 389-396 (1976)

18. T.J. Boehm, H.R.Bilger and J.L. Tandon: Proc. Symposium on 1/f-Fluctuations, Tokyo (1977) pp. 62-67
19. F.N. Hooge and J.L.M. Gaal: Philips Res. Repts. 26, 345-358 (1971)
20. Th.G.M. Kleinpenning: Physica 77, 78-98 (1974)
21. J.A. Barnes et al.: IEEE Trans. on Instr. and Measurement, IM-20, 105-120 (1971)
22. D.W. Allen: Proc. Symposium on 1/f-Fluctuations, Tokyo (1977) pp. 158-172
23. J.J. Gagnepain and J. Uebersfeld: Proc. Symposium on 1/f-Fluctuations, Tokyo (1977) pp. 173-181
24. W.H. Munk: *The rotation of the Earth* (Cambridge Univ. Press, New York 1960) p. 192
25. J.A. Barnes and D.W. Allan: Frequency 5, 15 (1967)
26. N. Matsunami and K. Nakajima: Proc. Symposium on 1/f-Fluctuations, Tokyo (1977) pp. 192-198
27. R.F. Voss and J. Clarke: Nature 258, 317-318 (1975)
28. R.F. Voss: Proc. Symposium on 1/f-Fluctuations, Tokyo (1977) pp. 199-205
29. A.A. Verveen and H.E. Derksen: Proc. IEEE 56, 906-916 (1968)
30. A.A. Verveen et al.: Proc. Symposium on 1/f-Fluctuations, Tokyo (1977) pp. 97-105
31. T. Musha and H. Higuchi: Jap.J.Appl.Phys. 15, 1271-1275 (1976)
32. A. van der Ziel: Proc. Symposium on 1/f-Fluctuations, Tokyo (1977) pp. 1-6
33. F.N. Hooge: Proc. Symposium on 1/f-Fluctuations, Tokyo (1977) pp. 88-96
34. F.N. Hooge and A.M.H. Hoppenbrouwers: Physica 42, 331-339 (1969)
35. J.J. Brophy: Phys. Rev. 166, 827-831 (1968)
36. M. Stoisiek and D. Wolf: J. Appl. Phys. 47, 362-364 (1976)
37. J.J. Brophy: J. Appl. Phys. 40, 3551-3553 (1969), and 41, 2913-2916 (1970)
38. J.K.E. Tunaley: J. Appl. Phys. 43, 3851-3855 (1972)
39. J.L. Tandon and H.R. Bilger: J. Appl. Phys. 47, 1697-1701 (1976)
40. J.M. Ribeyrol:Thèse, Université de Toulouse, Toulouse (1969)
41. J.C. Martin, D. Esteve, G. Blasquez, and J.M. Ribeyrol: Electronic Letters 6, 128-130 (1970)
42. A. van der Ziel: Physica 48, 242-246 (1970)
43. F.N. Hooge and A.M.H. Hoppenbrouwers: Physica 45, 386-392 (1969)
44. J.W. Eberhard and P.M. Horn: Phys. Rev. Letters 39, 643-646 (1977)
45. E.J.P. May and J.M.K. Horwood: Proc. Symposium on 1/f-Fluctuations, Tokyo (1977) pp. 124-127
46. F.N. Hooge: Physics Letters 29 A, 139-140 (1969)
47. F.N. Hooge and A.M.H. Hoppenbrouwers: Physics Letters 29 A, 642-643 (1969)
48. L.K.J. Vandamme and W.M.G. van Bokhoven: Appl. Phys. 14, 205-215 (1977)
49. T.G.M. Kleinpenning: J. Appl. Phys. 48, 2946-2949 (1977)
50. F.N. Hooge: Physica 60, 130-144 (1972)
51. F.N. Hooge: private communication
52. R.F. Voss and J. Clark: Phys. Rev. B 13, 556-573 (1976)
53. L.K.J. Vandamme: Physics Letters 49 A, 233-234 (1974)
54. T.G.M. Kleinpenning: J. Phys.Chem. Solids 37, 925-928 (1976)
55. J.T.M. Stroeken and T.G.M. Kleinpenning: J. Appl. Phys. 47, 4691-4692 (1976)
56. F.N. Hooge: private communication
57. H.G.E. Beck: Solid State Electronics 20, 951-954 (1977)
58. H.G.E. Beck and W.P. Spruit: private communication
59. T.G.M. Kleinpenning: Physica 84 B, 353-361 (1976)
60. A. McWhorter: Ph.D. thesis, Massachusetts Institute of Technology, Cambridge, Massachusetts (1955)
61. M. Mikulinski and S. Fishman: Proc. Symposium on 1/f-Fluctuations, Tokyo (1977) pp. 7-11
62. P.H. Handel: Physics Letters 53 A, 438-440 (1975);
In *Linear and Nonlinear Electronic Transport in Solids*, Proc. NATO Adv. Study Inst. Antwerpen 1975, ed. by J.T. Devreese and V.van Doren, (Plenum Press, New York 1976) p. 515;
Proc. Symposium on 1/f-Fluctuations, Tokyo (1977) pp. 12-18
63. O. Jäntsch and I. Feigt: Solid State Electronics 16, 1517-1520 (1973)

64. O. Müller: IEEE Trans. Electron. Devices ED-21, 539-540 (1974)
65. B. Mandelbrot: IEEE Trans. Inform. Theory IT-13, 289-298 (1967)
66. D. Halford: Proc. IEEE 56, 251-258 (1968)
67. T.H. Bell jr.: J. Appl. Phys. 45, 1902-1904 (1974)
68. A. Ambrozy: Electronic Letters 13, 137-138 (1977)
69. H. Schönfeld: Zs. Naturforschung 10a, 291-300 (1955)
70. D.A. Bell: Proc. Royal Soc. 72, 27-32 (1958)
71. S. Putterman: Phys. Rev. Letters 39, 585-587 (1977)
72. V.V. Losev, Y.A. Oksman: Sovj. Phys. Semicond. 6, 442-443 (1972)
73. T. Kawakubo: Proc. Symposium on 1/f-Fluctuations, Tokyo (1977) pp. 36-40
74. J.A. Barnes and D.W. Allen: Proc. IEEE 54, 176-178 (1966)
75. S. Teitler and M.E.M. Osborne: J. Appl. Phys. 41, 3274-3276 (1970)
76. J.L. Tandon: Ph.D. thesis, Oklahoma State University, Stillwater, Oklahoma (1976)
77. M. Agu and T. Kinoshita: Proc. Symposium on 1/f-Fluctuations, Tokyo (1977) pp. 41-48

A General Mechanical Model for $|f|^{\alpha}$ Spectral Density Random Noise with Special Reference to Flicker Noise $1/|f|$

DONALD HALFORD

***Abstract*—Any class of reasonable time-dependent perturbations occurring at random, under certain internal constraints, generates random noise having a spectral density varying as $|f|^{\alpha}$ over an arbitrarily large range of spectral frequency f only for $-2 \leq \alpha \leq 0$. A class is the set of all perturbations which are equivalent under some individual independent scaling of amplitude, scaling of time, and translation of time. A subclass is characterized by $P(\tau)$ and $A^2(\tau)$. $P(\tau)$ is the lifetime probability density. $A^2(\tau)$ is a mean square amplitude of perturbations having lifetime τ. For a given class, $|f|^{\alpha_\infty}$ and $|f|^{\alpha_0}$ are the frequency smoothed laws in the limits of infinite and zero frequencies, respectively. Any reasonable perturbation has $\alpha_\infty \leq -2$ and $\alpha_0 \geq 0$. To generate random noise having an $|f|^{\alpha}$ law over an arbitrarily large range of f from a subclass chosen from any class characterized by α_∞ and α_0, it is necessary that $\alpha_\infty \leq \alpha \leq \alpha_0$. For $\alpha_\infty < \alpha < \alpha_0$, it is necessary and sufficient that such subclasses satisfy the condition, $P(\tau)A^2(\tau) \approx B\tau^{-\alpha-3}$ with B constant, over a suitable range of τ, and that $P(\tau)A^2(\tau)$ not be larger than $B\tau^{-\alpha-3}$ outside the range. This general mechanical model is of immediate value in the formulation and criticism of specific physical models of $|f|^{\alpha}$ noise, including flicker noise, and in computer simulation of $|f|^{\alpha}$ noise.**

I. Introduction

A. The Importance of Flicker Noise in Precise Experiments

IN THE DESIGN of a precision measurement or of a standard for measurement, one of the aims is to minimize the random noise of the system. It is often found that the random fluctuations are dominated by the lowest frequency portion of the noise modulation spectrum. The spectral density of this noise is observed to increase without limit as its frequency decreases. A typical behavior is as $|f|^{\alpha}$, with α constant and equal to -1.0, but α varies from system to system, and it is not always constant; values of -0.8 to -1.3 are common. In many cases this behavior is found to hold over many decades of frequency.[1]–[8] In this paper the term "flicker noise" will be used in a strict sense for the prototype law $|f|^{-1}$ and in a loose sense for the general behavior described above. When it was discovered in vacuum tubes[9],[10] this curious phenomenon was called "flicker effect." Different names found in the literature for similar spectral behavior in other devices are "$1/f$ noise," "excess noise," "semiconductor noise," "low-frequency noise," "contact noise," and "pink noise."

A measurement with results which are limited by random noise with an $|f|^{-1}$ law has a fixed precision which cannot be improved by increasing the length of each run.[11],[12] In contrast, for a measurement limited by white noise (spectral density law of f^0) any desired precision may be attained by increasing the duration τ_e of the experiments; the improvement accrues as $\sqrt{\tau_e}$, as, for example, when thermal agitation is the limiting noise. The spot noise factor[13] of a device displaying flicker noise is necessarily appreciably greater than unity below some particular modulation frequency and becomes progressively worse toward lower frequencies.

For an example, consider the stability of quartz crystal oscillators, which are widely used as secondary frequency standards. Even the best quartz crystal oscillators are

Manuscript received July 13, 1966; revised January 8, 1968. The original version of this paper was presented at the Conference on Precision Electromagnetic Measurements at the National Bureau of Standards, Boulder, Colo., June 1966.

The author is with the Atomic Frequency and Time Standards Section, Radio Standards Laboratory, National Bureau of Standards, Boulder, Colo. 80302

Reprinted from *Proc. IEEE*, vol. 56, pp. 251–258, Mar. 1968.

limited by their flicker of frequency fluctuations to have an imprecision of about three parts in 10^{13} or worse.[14]–[18] No improvement beyond the three parts in 10^{13} is possible by utilizing longer averaging times. If it were not for the flicker noise, the precision of ten-second-interval measurements of frequency would be better than one part in 10^{14}. Such precision *is needed* by the author in a slave oscillator to be used with the atomic hydrogen maser frequency standard.

Flicker noise is ubiquitous. It is found not only in quartz crystal oscillators, but also in such things as vacuum tubes, field effect and bipolar transistors, semiconductor diodes, resistors, thermistors, carbon microphones, thin films, and light sources.[1]–[9],[11],[14] The fluctuations of a membrane potential in a biological system have recently been reported to have a flicker noise spectrum.[19],[20] No electronic low-frequency amplifier has been found to be free of flicker noise. Yet in most devices its cause is not understood. It may be presumed that if the cause were understood, then the flicker noise could be reduced. It is hoped that the model presented here will help in the understanding and ultimate control of flicker noise.

Although the case of $\alpha \approx -1$ is common, there are some interesting processes found in nature where α is much more negative than -1. For example, the fluctuations of the frequency of rotation of the earth[21] are described by $\alpha \approx -2$, and the power spectral density of galactic radiation noise[22] is described by $\alpha \approx -2.7$. The consequences of the low-frequency divergence of fluctuation amplitude when $\alpha \leq -1$ are often not easily comprehended by someone accustomed only to white noise spectral density. The mechanical model presented here can help one comprehend the divergence and its consequences. The model is applicable to the noise law $|f|^\alpha$ for any value of α.

B. The Problem of Generating $|f|^\alpha$ Noise

Since flicker noise or other low-frequency divergent noise is often encountered in measurements, it is useful to be able to fashion models which can generate noise having the appropriate spectral density. The models should be physically plausible, and preferably they should be chosen to correspond as closely as possible to the physical process being measured. This paper describes a rather general model which can generate noise with a spectral density of $|f|^\alpha$, any α, for a very broad range of physically plausible processes.

We consider a system in which some process of interest is occurring, and in which the process is suffering many time-dependent perturbations in a random manner due to some unspecified agent or agents. The uniform portion of the process constitutes a signal, and the random fluctuations are the associated noise. We wish to answer at least two questions.

1) Can such a superposition of time-dependent perturbations of a specified shape, occurring at random, give rise to an $|f|^\alpha$ dependence of the noise spectral density with α constant over an arbitrarily large range of spectral frequency, f?

2) What conditions must be satisfied to obtain the law $|f|^\alpha$?

To answer these questions, the discussion needs to be concerned only with the mathematical description of the time-dependent perturbations. Hence there is no reference to specific *physical models*. To avoid the possibility of the model thereby synthesized being nonphysical in principle, some constraints are imposed, e.g., certain infinities are excluded from consideration. We use the term *mechanical model* for a model which is physical in principle, although not necessarily encountered in reality.

II. Synthesis of a Mechanical Model of $|f|^\alpha$ Noise

A. Definitions

Let H be the representation in the time domain (t) of a perturbation of the class $\mathscr{H}$. The representation $H(a, \tau)$ has lifetime τ and squared amplitude a^2. These quantities will be defined in this part of Section II. A class is the set of all perturbations which are equivalent under some individual independent scaling of amplitude a, scaling of time t, and translation of time. Special consideration is given to reasonable perturbations. By definition a *reasonable perturbation* 1) is everywhere finite, 2) has a finite time integral, 3) has finite, nonzero energy, 4) has finite, nonzero lifetime, and 5) is a real function of time t. These may be stated symbolically, as follows.

$$-\infty < H < \infty, \quad \text{all } t, \tag{1a}$$

$$-\infty < \int_{-\infty}^{+\infty} H dt < \infty, \tag{1b}$$

$$0 < \int_{-\infty}^{+\infty} H^2 dt < \infty, \tag{1c}$$

$$0 < \tau < \infty, \tag{1d}$$

and

$$H = H^* \equiv \text{complex conjugate of } H. \tag{1e}$$

Throughout the following, the time t need not appear explicitly as an argument of H, and it will not be shown. Under scaling of amplitude, the representation has the property

$$a_0 H(a, \tau) = a H(a_0, \tau). \tag{2}$$

Let $F(\omega, \tau)$ denote the Fourier transform[23] of $a^{-1}H(a, \tau)$,

$$F(\omega, \tau) \equiv a^{-1} \int_{-\infty}^{+\infty} H(a, \tau) e^{-i\omega t} dt = F^*(-\omega, \tau), \tag{3}$$

$$\omega \equiv 2\pi f, \tag{4}$$

where f is cyclic frequency. Under scaling of frequency and time, the Fourier transform has the property

$$\tau_0 F(\omega, \tau) = \tau F(\omega\tau/\tau_0, \tau_0). \tag{5}$$

The energy q of a perturbation is defined as

$$q(a, \tau) \equiv \int_{-\infty}^{+\infty} H^2(a, \tau) dt = a^2 \int_{-\infty}^{+\infty} |F(\omega, \tau)|^2 df. \tag{6}$$

Under scaling, q has the property

$$a_0^2\tau_0 q(a,\tau) = a^2\tau_0 q(a_0,\tau) = a^2\tau q(a_0,\tau_0). \quad (7)$$

The lifetime τ of a perturbation is defined implicitly by

$$\int_{-1/(2\pi\tau)}^{+1/(2\pi\tau)} |F(\omega,\tau)|^2 df \equiv \frac{1}{2}\int_{-\infty}^{+\infty} |F(\omega,\tau)|^2 df. \quad (8)$$

This definition of τ is chosen so that $|\omega|_m = 2\pi|f|_m = \tau^{-1}$, where $|f|_m$ is the median absolute frequency of the energy distribution. The utility of this definition may be seen later (in Section II–B–C) when we consider the frequency range over which a particular noise law is valid.

The square of the amplitude is defined as

$$a^2 \equiv \frac{q(a,\tau)}{\tau}. \quad (9)$$

For simplicity we set a_0^2 and τ_0 equal to unity. Hence $q(a_0, \tau_0)$ equals unity, and we note

$$q(a,\tau) = a^2\tau, \quad (10)$$

$$\int_{-\infty}^{+\infty} |F(x,\tau_0)|^2 dx = 2\pi\tau_0 = 2\pi. \quad (11)$$

Suppose a process is perturbed by a large number of time-dependent perturbations which form a subclass of the class $\mathscr{H}$. We define $A^2(\tau)$ as the mean square amplitude of perturbations in the subclass having the same lifetime τ, and we define $P(\tau)$ as the normalized probability of occurrence in the subclass of perturbations of lifetime τ. We further suppose the perturbations occur randomly (as a Poisson process) with the finite average total rate of R perturbations per unit interval of time. Hence the rate at which perturbations of lifetime τ occur is given by the product $RP(\tau)$.

The definitions given above establish a mathematical description of a general mechanical model for the generation of random noise. Using these concepts in the situation described in the preceding paragraph, the equation for the two-sided spectral density $S(f)$ of the noise is defined as

$$S(f) \equiv R\int_0^{+\infty} P(\tau)A^2(\tau)|F(\omega,\tau)|^2 d\tau. \quad (12)$$

This definition of the spectral density is consistent with that of Blackman and Tukey.[23] The dimensionality of $S(f)$ is the same as the dimensionality of the ratio H^2/f. For pure real perturbations, $S(f)$ is symmetrical about zero frequency.

Appendix I illustrates by mathematical example some of the above concepts. Fig. 1 shows two members of the specific class of reasonable perturbations discussed in Appendix I. Fig. 2 shows the corresponding squared Fourier transforms.

B. Generation of Flicker Noise, $\alpha = -1$

Equation (12) for the spectral density is put in the form

$$S(f) = \frac{R}{|\omega|}\int_0^{+\infty} [P(\tau)A^2(\tau)][\tau/\tau_0]^2 |F(\omega\tau/\tau_0, \tau_0)|^2 d(|\omega|\tau). \quad (13)$$

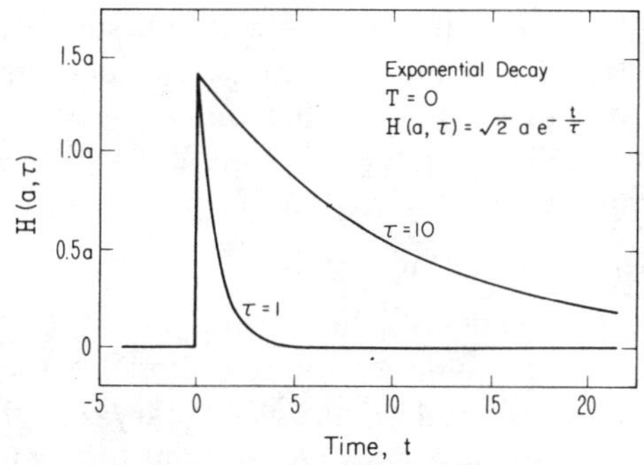

Fig. 1. Two reasonable perturbations from the class having, in the time domain, a step change followed by exponential decay back to the origin. Two different lifetimes are shown (time domain). See Appendix I.

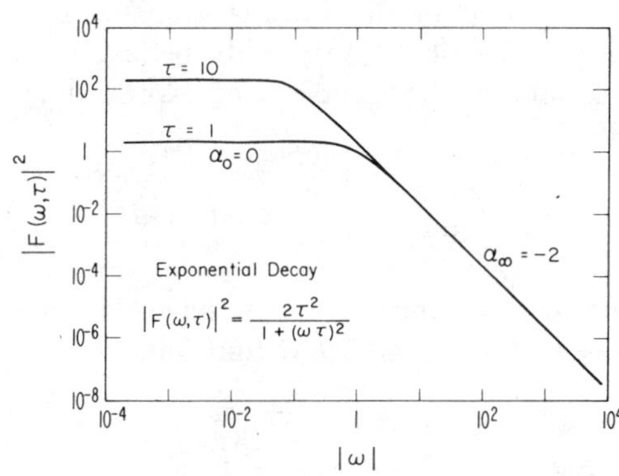

Fig. 2. The square of the normalized Fourier transform of the two reasonable perturbations of Fig. 1 are plotted against ω (frequency domain). See Appendix I.

The dependence on τ of $P(\tau)A^2(\tau)$ is expressed as

$$P(\tau)A^2(\tau) \approx B\tau^{-\delta-3}. \quad (14)$$

In general, B and δ may be functions of τ, but our interest will focus on those situations in which B and δ are independent of τ over some finite range. Consider the case where, in the range $\varepsilon \le \tau \le \zeta$, $\varepsilon > 0$,

$$\delta = -1,$$
$$B = \text{constant},$$
$$P(\tau)A^2(\tau) \approx B\tau^{-2},$$

and $P(\tau)A^2(\tau) \le B\tau^{-2}$ for τ outside of the range ε to ζ. For the spectral range $\zeta^{-1} \ll |\omega| \ll \varepsilon^{-1}$,

$$S(f) \approx \frac{RB}{|\omega|}\int_{|\omega|\varepsilon}^{|\omega|\zeta} |F(\omega\tau, 1)|^2 d(|\omega|\tau), \quad (15)$$

$$\int_{|\omega|\varepsilon}^{|\omega|\zeta} |F(x,1)|^2 dx \approx \int_0^{+\infty} |F(x,1)|^2 dx = \pi. \quad (16)$$

Hence,

$$S(f) \approx \tfrac{1}{2}RB|f|^{-1} = \tfrac{1}{2}RP(1)A^2(1)|f|^{-1}, \quad (17)$$

for frequencies in the arbitrarily large but finite range $(2\pi\zeta)^{-1} \ll |f| \ll (2\pi\varepsilon)^{-1}$.

This result shows that a spectral density law of $|f|^{-1}$ can be obtained over an arbitrarily large frequency range from a subclass *chosen from any class of reasonable perturbations*. The *sufficient* condition is that such subclass have the property $P(\tau)A^2(\tau) \approx (\text{const.})\tau^{-2}$ within a suitable range of τ, with $P(\tau)A^2(\tau) \leq (\text{const.})\tau^{-2}$ outside that range. It will be concluded in Section III that this is also the *necessary* condition for obtaining the $|f|^{-1}$ noise law from a subclass chosen from *any class* of reasonable perturbations.

C. Extension to $|f|^\alpha$ Noise, Any α

In the case of general α the starting point is (12), which is exact. Some approximations follow. Possibly the approximations are more crude than is necessary, but the author believes that they introduce no error into the conclusions. It is hoped that the simplicity of the treatment allows each conclusion and its significance to be quickly understood.

Let G be a frequency-smoothed representation of the magnitude squared of the Fourier transform of a time-dependent perturbation. Consider the behavior of G at very high and at very low frequencies, and express it as

$$\lim_{|f|\to\infty} G = (\text{const.})|f|^{\alpha_\infty}, \tag{18}$$

$$\lim_{|f|\to 0} G = (\text{const.}')|f|^{\alpha_0}. \tag{19}$$

All perturbations of the same class have the same α_∞ and the same α_0. For perturbations of finite, nonzero energy q,

$$\alpha_\infty < -1, \tag{20a}$$

$$\alpha_0 > -1. \tag{20b}$$

It has been pointed out by van der Ziel[24] that these inequalities apply to the net noise spectral density for any physical process having a bounded continuous autocovariance function. However, for any process suffering from reasonable perturbations, the exponents must satisfy the stronger constraints

$$\alpha_\infty \leq -2, \tag{21a}$$

$$\alpha_0 \geq 0. \tag{21b}$$

Appendix II has a brief discussion of α_∞ and α_0, and it includes a demonstration of the above constraints.

Consider all subclasses of reasonable perturbations for which α_∞ and α_0 are the limiting exponents as defined above. The noise spectral density of each is given by (12). We make a substitution using (14),

$$P(\tau)A^2(\tau) \approx B\tau^{-\delta-3},$$

and further limit the consideration to all such subclasses for which B and δ are approximately independent of τ over the finite range $\varepsilon \leq \tau \leq \zeta$, $\varepsilon > 0$, with the additional constraint that $P(\tau)A^2(\tau) \leq B\tau^{-\delta-3}$ for all τ outside the range ε to ζ. For the range

$$\zeta^{-1} \ll |\omega| \ll \varepsilon^{-1}, \tag{22}$$

$$S(f) \approx \frac{R}{|\omega|}\int_{|\omega|\varepsilon}^{|\omega|\zeta} B\tau^{-\delta-1}|F(\omega\tau, 1)|^2 d(|\omega|\tau), \tag{23}$$

$$S(f) \approx RB|\omega|^\delta \int_{|\omega|\varepsilon}^{|\omega|\zeta} |x|^{-\delta-1}|F(x, 1)|^2 dx. \tag{24}$$

The definite integral in (24) has the following properties:

$$\int_{|\omega|\varepsilon}^{|\omega|\zeta} x^{-\delta-1}|F(x,1)|^2 dx \approx \begin{cases} C|\omega|^{\alpha_0-\delta}, \text{ for } \delta > \alpha_0; & \text{(25a)} \\ C'|\omega|^0 = C', \text{ for } \alpha_\infty < \delta < \alpha_0; & \text{(25b)} \\ C''|\omega|^{\alpha_\infty-\delta}, \text{ for } \delta < \alpha_\infty; & \text{(25c)} \end{cases}$$

where C, C', and C'' are constants, independent of ω. Hence the frequency dependence of the noise spectral density is

$$S(f) \approx \begin{cases} (\text{const.})|f|^{\alpha_0}, & \text{for } \delta > \alpha_0; & \text{(26a)} \\ (\text{const.}')|f|^{\delta}, & \text{for } \alpha_\infty < \delta < \alpha_0; & \text{(26b)} \\ (\text{const.}'')|f|^{\alpha_\infty}, & \text{for } \delta < \alpha_\infty. & \text{(26c)} \end{cases}$$

Equations (25a–c) are obtained in Appendix III, where it is also shown that

$$C' \approx \pi \frac{(\alpha_0 + 1)(\alpha_\infty + 1)}{(\alpha_0 - \delta)(\alpha_\infty - \delta)}. \tag{27}$$

Hence, for $\alpha_\infty < \delta < \alpha_0$, the two-sided noise spectral density is

$$S(f) \approx RB\pi \frac{(\alpha_0 + 1)(\alpha_\infty + 1)}{(\alpha_0 - \delta)(\alpha_\infty - \delta)}|\omega|^\delta. \tag{28}$$

Note that for the case of $\delta = -1$, (28) is the same result as (17).

Equations (26a–c) give the interesting result that, for any smooth distribution of lifetimes of the perturbations, the exponent α of the frequency dependence of the noise spectral density cannot lie outside of the interval α_∞ to α_0 over an arbitrarily large range of frequency for all subclasses characterized by α_∞ and α_0. Equation (26b) shows that under certain conditions, which are of interest, the exponent α is equal to δ.

III. Results

We have found some interesting and useful answers to our questions. The concepts α_∞ and α_0 are important when considering the general problem of creating noise with $|f|^\alpha$ spectral density. Not all classes of reasonable perturbations characterized by α_∞ and α_0 can generate $|f|^\alpha$ noise over an arbitrarily large range of f unless there is a sufficiently broad, smooth distribution of lifetimes in the chosen subclasses. An arbitrarily smooth distribution which is arbitrarily broad gives rise to an arbitrarily smooth spectrum in which local peaks and valleys no longer exist even though they may be present in the individual $|aF(\omega, \tau)|^2$ spectra which make up the superposition. In such a superposition, at sufficiently high frequencies the spectrum varies as $|f|^{\alpha_\infty}$, and at sufficiently low frequencies the spectrum varies as $|f|^{\alpha_0}$. For any arbitrarily large range of intermediate frequencies, as the operative value of δ is varied over its domain from an arbitrarily small value to an arbitrarily large value, the only values of α which arise from such distributions lie in the interval $\alpha_\infty \leq \alpha \leq \alpha_0$. For the interval $\alpha_\infty < \alpha < \alpha_0$, α and δ have the one-to-one correspondence $\alpha = \delta$.

Some conclusions are summarized in the following statements.

1) To generate random noise having an $|f|^\alpha$ spectral density over an arbitrarily large range of f from a subclass of reasonable perturbations occurring at random, chosen from

any class characterized by α_∞ *and* α_0, it is necessary that $\alpha_\infty \le \alpha \le \alpha_0$.

2) To generate random noise having an $|f|^\alpha$ spectral density over the arbitrarily large range

$$(2\pi\zeta)^{-1} \ll |f| \ll (2\pi\varepsilon)^{-1} \tag{29}$$

from a subclass of reasonable perturbations occurring at random, chosen from *any class characterized by* α_∞ *and* α_0, for α in the more restricted interval $\alpha_\infty < \alpha < \alpha_0$, it is *necessary and sufficient* that such subclass satisfy the condition

$$P(\tau)A^2(\tau) \approx B\tau^{-\alpha-3}, \tag{30}$$

with B constant, over the range $\varepsilon \le \tau \le \zeta$, and that $P(\tau)A^2(\tau) \le B\tau^{-\alpha-3}$ for τ outside the range ε to ζ.

3) Any reasonable perturbation has $\alpha_\infty \le -2$ and $\alpha_0 \ge 0$. At least one class of reasonable perturbations exists which has $\alpha_\infty = -2$ and $\alpha_0 = 0$ (see Appendix I). Indeed, an unlimited number of such classes exist.

4) It follows from statements 1), 2), and 3) that the only values of α for which $|f|^\alpha$ noise, over an arbitrarily large range of f, can be generated by a subclass chosen from *any class* of reasonable perturbations are in the interval $-2 \le \alpha \le 0$. *This interval of* α *is centered on the flicker law,* $\alpha = -1$.

5) For the smoothed random noise spectral density due to a subclass of reasonable perturbations occurring at random, chosen from any class characterized by α_∞ and α_0, the law $|f|^{\alpha_\infty}$ holds at sufficiently high frequencies. The law $|f|^{\alpha_\infty}$ also holds for the spectral density in the frequency range $(2\pi\zeta)^{-1} \ll |f| \ll (2\pi\varepsilon)^{-1}$ if, for $\varepsilon \le \tau \le \zeta$,

$$P(\tau)A^2(\tau) \approx B\tau^{-\delta-3},$$
$$B = \text{constant},$$
$$\delta < \alpha_\infty,$$

and also if, for $\tau < \varepsilon$, $P(\tau)A^2(\tau) \le B\tau^{-\alpha_\infty-3}$. Similarly, for such a subclass, the smoothed law $|f|^{\alpha_0}$ holds at sufficiently low frequencies. The law $|f|^{\alpha_0}$ also holds for the spectral density in the frequency range $(2\pi\zeta)^{-1} \ll |f| \ll (2\pi\varepsilon)^{-1}$ if, for $\varepsilon \le \tau \le \zeta$,

$$P(\tau)A^2(\tau) \approx B\tau^{-\delta-3},$$
$$B = \text{constant},$$
$$\delta > \alpha_0,$$

and also if, for $\tau > \zeta$, $P(\tau)A^2(\tau) \le B\tau^{-\alpha_0-3}$.

It is helpful to chart the logarithm of the spectral density versus the logarithm of the frequency when considering the above results.

IV. Discussion

This general method of generating an $|f|^\alpha$ spectral density gives special attention to the factor $P(\tau)A^2(\tau)$ of (12). A second method exists, in which the special attention is given instead to the *shape* of each of the time-dependent perturbations.[25],[26] This second method is severely restricted in its choice of classes of perturbations which can generate $|f|^\alpha$ noise for any specific value of α. In this second method, a physical explanation of a flicker noise spectrum extending over more than a couple of decades of frequency requires postulating the existence of specially shaped, time-dependent perturbations which are rarely, or never, found. On the other hand, the random perturbations found to be affecting physical processes often do have broad distributions of their lifetime square-amplitude product, one of the necessary conditions for the proper application of the first method. The diffusion process is a prime example of a physical process which gives rise to broad distributions of lifetimes.[27]–[32] The first person to point out that a distribution of τ might be important appears to have been Bernamont[33] in 1937. Since then, many investigators have invoked particular distributions of τ in attempts to explain flicker noise.[1]–[3],[24],[27],[30]–[32],[34]–[37]

The effects of noise on the action of an instrument can be simulated on a computer. With this general mechanical model, noise spectral density of any desired law, or combination of laws, can be accurately and efficiently generated by a simple computer program.[38] This mechanical model, besides being general, is more efficient than a mathematical model proposed recently by Barnes and Allan[26] for simulating flicker noise with a computer.

The model as presented deals with certain portions of the total spectrum. To give an approximate description of the entire spectral domain, the model can be applied successively to each portion of the spectrum and to each operative class, as necessary.

It is apparent that the experimental observation, per se, of a noise law of $|f|^\alpha$, $-2 \le \alpha \le 0$, gives no information concerning the identity of the class (or classes) of reasonable perturbations which is operative. The corresponding statement is not true for other values of α, and the further from this range is the observed α, the more restricted is the set of classes of reasonable perturbations which might be operative.

From a physical viewpoint, the restrictions which have been placed on reasonable perturbations are not severe. Many random noise processes of physical importance may be represented by reasonable perturbations, and to them the results of this paper are applicable without modification. In some cases it is helpful to consider a time derivative or a running time integral of a process, instead of the process itself, if the perturbations involved are thereby converted into reasonable perturbations. The present results are also applicable to many random physical processes where each member of the ensemble of perturbing agents is most conveniently regarded as being of infinite liefetime,[3] provided that the principle of ergodicity can be used.

The present results are of immediate value in allowing one to evaluate and criticize proposed physical models of $|f|^\alpha$ noise. The present results also can give insight into the problem of devising models which may explain noise spectra observed in specific physical situations. For example, a new mechanism, hypothetical and untested but physically plausible, for the flicker of frequency of quartz crystals has been devised.[39] It is emphasized that, to develop a physical model for flicker noise, the crucial problem is to find the physical circumstances which cause the product

$P(\tau)A^2(\tau)$ to vary approximately as $\tau^{-\alpha-3}$; the shape or class of the perturbation is probably irrelevant. That it is the product $P(\tau)A^2(\tau)$ which is the pertinent factor seems not to have been appreciated in previous discussions of flicker noise.[1]–[3],[10],[24],[27],[33]–[37]

Appendix I

An example of a specific class will help to clarify the H representation and some of the concepts associated with it. Consider the class of all perturbations each of which is represented in the time domain by a step change followed by exponential decay back to the origin (see Figs. 1 and 2).

$$H(a, \tau) = 0, t \le T.$$
$$H(a, \tau) = \sqrt{2}ae^{-(t-T)/\tau}, t \ge T.$$
$$H(1, \tau) = \sqrt{2}e^{-(t-T)/\tau}, t \ge T.$$
$$F(\omega, \tau) = \sqrt{2}[\tau^{-1} + i\omega]^{-1}e^{-i\omega T}.$$
$$F(\omega, 1) = \sqrt{2}[1 + i\omega]^{-1}e^{-i\omega T}.$$
$$G = (\text{const.})[\tau^{-2} + \omega^2]^{-1}.$$
$$\alpha_0 = 0.$$
$$\alpha_\infty = -2.$$

Appendix II

For reasonable perturbations,

$$\lim_{\omega \to 0} |F(\omega, \tau)|^2 = |F(0, \tau)|^2 = \left| \int_{-\infty}^{+\infty} H(1, \tau) dt \right|^2 < \infty. \quad (31)$$

This requires the frequency-smoothed behavior, (19), which defines α_0, to have the constraint

$$\lim_{f \to 0} G = \lim_{f \to 0} (\text{const.})|f|^{\alpha_0} < \infty. \quad (32)$$

Hence, $\alpha_0 \ge 0$ for reasonable perturbations.

It follows from the square-integrability of reasonable perturbations (1c) that α_∞ must be less than -1 for all classes, while Appendix I shows that α_∞ equals -2 for at least one class. We will now show that there are no reasonable perturbations with α_∞ greater than -2.

By analogy with the Wiener-Khinchin theorem,[40] $|F(\omega, 1)|^2$ is the Fourier transform of the autocovariance function for $H(1, 1)$. If $H(1, 1)$ is a reasonable perturbation, then the autocovariance function is everywhere continuous, and the first derivative with respect to time of the autocovariance function is everywhere finite. Hence

$$\left| \frac{d}{dT} \int_{-\infty}^{+\infty} \frac{1}{2\pi} \left| F(\omega, 1) \right|^2 e^{i\omega T} d\omega \right| < \infty, \quad (33)$$

$$\left| \int_0^\infty \left| F(\omega, 1) \right|^2 \omega \sin \omega T d\omega \right| < \infty. \quad (34)$$

Equation (34) must hold for all T, including in the limit as T goes to zero.

We approximate the magnitude squared of the normalized Fourier transform of a reasonable perturbation with

$$|F(x, 1)|^2 \approx K|x|^{\alpha_0} \quad (35)$$

for $|x| \le 1$, and for $|x| \ge 1$ we use the approximation

$$|F(x, 1)|^2 \approx K|x|^{\alpha_\infty}, \quad (36)$$

where K is a constant, and $0 < K < \infty$. Equations (35) and (36) are used in the integral in (34), which is considered in three parts, and each part is evaluated in the limit as T goes to zero.

$$\lim_{T \to 0} \left| \int_0^1 \omega^{\alpha_0 + 2} T d\omega + \int_1^{1/T} \omega^{\alpha_\infty + 2} T d\omega + \int_{\omega T = 1}^{\omega T = \infty} \frac{(\omega T)^{\alpha_\infty + 1} \sin \omega T d(\omega T)}{T^{\alpha_\infty + 2}} \right| < \infty, \quad (37)$$

$$\lim_{T \to 0} \left| \left(\frac{1}{\alpha_0 + 3} - \frac{1}{\alpha_\infty + 3} \right) T + \left(\frac{1}{\alpha_\infty + 3} + \int_1^\infty x^{\alpha_\infty + 1} \sin x dx \right) \left(\frac{1}{T} \right)^{\alpha_\infty + 2} \right| < \infty, \quad (38)$$

$$\lim_{T \to 0} \left| \left(\frac{1}{T} \right)^{\alpha_\infty + 2} \right| < \infty. \quad (39)$$

By (39) we see that $\alpha_\infty \le -2$ for reasonable perturbations.

It is interesting to note that $\alpha_0 = 0$ for a reasonable perturbation H, if and only if

$$\int_{-\infty}^{+\infty} H(a, \tau) dt \ne 0,$$

a condition which describes a shot noise perturbation, for example. For a reasonable perturbation H, $\alpha_\infty = -2$ if and only if H has at least one finite discontinuity. Arbitrarily large positive values of α_0 may be realized by interposing an arbitrarily large number of high-pass filters between the perturbing agents and the process of interest. Independently, arbitrarily large negative values of α_∞ may be realized by interposing an arbitrarily large number of low-pass filters between the perturbing agents and the process of interest.

Appendix III

As in Appendix II, we approximate the magnitude squared of the normalized Fourier transform of a reasonable perturbation with

$$|F(x, 1)|^2 \approx K|x|^{\alpha_0} \quad (40)$$

for $|x| \le 1$, and for $|x| \ge 1$ we use the approximation

$$|F(x, 1)|^2 \approx K|x|^{\alpha_\infty}. \quad (41)$$

The common coefficient K is a constant, and it is evaluated by equating

$$\int_0^1 K|x|^{\alpha_0} dx + \int_1^\infty K|x|^{\alpha_\infty} dx = \int_0^\infty |F(x, 1)|^2 dx = \pi. \quad (42)$$

Hence,

$$K = \pi \frac{(\alpha_0 + 1)(\alpha_\infty + 1)}{(\alpha_\infty - \alpha_0)}. \quad (43)$$

Using (24), with (22) valid,

$$S(f) \approx RB|\omega|^{\delta} I(\omega), \tag{44}$$

where

$$I(\omega) \equiv \int_{|\omega|\varepsilon}^{|\omega|\zeta} |x|^{-\delta-1}|F(x,1)|^2 dx. \tag{45}$$

Using (40) and (41),

$$I(\omega) \approx K\left[\int_{|\omega|\varepsilon}^{1} |x|^{\alpha_0-\delta-1}dx + \int_{1}^{|\omega|\zeta} |x|^{\alpha_\infty-\delta-1}dx\right]. \tag{46}$$

For $\delta \neq \alpha_0$ and $\delta \neq \alpha_\infty$,

$$I(\omega) \approx K\left[\frac{1-(|\omega|\varepsilon)^{\alpha_0-\delta}}{(\alpha_0-\delta)} + \frac{(|\omega|\zeta)^{\alpha_\infty-\delta}-1}{(\alpha_\infty-\delta)}\right]. \tag{47}$$

Because we are considering only reasonable perturbations for which $\alpha_\infty < \alpha_0$, then, for the range of ω given by (22), and for $\alpha_\infty < \delta < \alpha_0$, $I(\omega)$ can be approximated by

$$\begin{aligned} I(\omega) &\approx K\left[\frac{1}{(\alpha_0-\delta)} - \frac{1}{(\alpha_\infty-\delta)}\right] \\ &\approx \pi \frac{(\alpha_0+1)(\alpha_\infty+1)}{(\alpha_0-\delta)(\alpha_\infty-\delta)}. \end{aligned} \tag{48}$$

Similarly, for $\delta > \alpha_0$,

$$\begin{aligned} I(\omega) &\approx K\left[-\frac{(|\omega|\varepsilon)^{\alpha_0-\delta}}{(\alpha_0-\delta)}\right] \\ &\approx -\pi \frac{(\alpha_0+1)(\alpha_\infty+1)\varepsilon^{\alpha_0-\delta}}{(\alpha_0-\delta)(\alpha_\infty-\alpha_0)} |\omega|^{\alpha_0-\delta}. \end{aligned} \tag{49}$$

Similarly, for $\delta < \alpha_\infty$,

$$\begin{aligned} I(\omega) &\approx K\left[\frac{(|\omega|\zeta)^{\alpha_\infty-\delta}}{(\alpha_\infty-\delta)}\right] \\ &\approx \pi \frac{(\alpha_0+1)(\alpha_\infty+1)\zeta^{\alpha_\infty-\delta}}{(\alpha_\infty-\delta)(\alpha_\infty-\alpha_0)} |\omega|^{\alpha_\infty-\delta}. \end{aligned} \tag{50}$$

For $\delta = \alpha_0$, by (46),

$$I(\omega) \approx K\left[\ln\frac{1}{|\omega|\varepsilon} + \frac{(|\omega|\zeta^{\alpha_\infty-\alpha_0}-1}{(\alpha_\infty-\alpha_0)}\right], \tag{51}$$

$$\begin{aligned} I(\omega) &\approx K \ln\frac{1}{|\omega|\varepsilon} \\ &\approx -\pi \frac{(\alpha_0+1)(\alpha_\infty+1)(\ln|\omega|\varepsilon)}{(\alpha_\infty-\alpha_0)}. \end{aligned} \tag{52}$$

For $\delta = \alpha_\infty$, by (46),

$$I(\omega) \approx K\left[\frac{1-(|\omega|\varepsilon)^{\alpha_0-\alpha_\infty}}{(\alpha_0-\alpha_\infty)} + \ln|\omega|\zeta\right], \tag{53}$$

$$\begin{aligned} I(\omega) &\approx K \ln|\omega|\zeta \\ &\approx +\pi \frac{(\alpha_0+1)(\alpha_\infty+1)(\ln|\omega|\zeta)}{(\alpha_\infty-\alpha_0)}. \end{aligned} \tag{54}$$

The logarithmic dependence on $|\omega|$ which appears in (52) and (54) is expected. It is not a spurious consequence of the approximations.

Acknowledgment

The author appreciates many discussions with associates at the National Bureau of Standards and the encouragement and invaluable help of Dr. J. A. Barnes, in particular.

References

[1] A. van der Ziel, *Noise*. Englewood Cliffs, N. J.: Prentice-Hall, 1954.

[2] ——, *Fluctuation Phenomena in Semiconductors*. New York: Academic Press, 1959.

[3] A. L. McWhorter, "1/f noise and related surface effects in germanium," M.I.T. Lincoln Lab., Lexington, Mass., Tech. Rept. 80, May 1955 (identical to M.I.T. Sc.D. dissertation of same title). See also "1/f noise and germanium surface properties," in *Semiconductor Surface Physics*, R. H. Kingston, Ed. Philadelphia, Pa.: University of Pennsylvania Press, 1957, pp. 207–228.

[4] R. E. Burgess *et al.*, *Valve and Circuit Noise. A Survey of Existing Knowledge and Outstanding Problems*, Radio Research Special Rept. 20. London: His Majesty's Stationery Office, 1951.

[5] B. G. Bosch and W. A. Gambling, "Noise in reflex klystrons and backward-wave oscillators," *J. Brit. IRE*, vol. 24, pp. 389–403, November 1962.

[6] S. T. Eng, "A new low 1/f noise mixer diode: Experiments, theory and performance," *Solid-State Electronics*, vol. 8, pp. 59–77, January 1965.

[7] A. H. Hoffait and R. D. Thornton, "Limitations of transistor dc amplifiers," *Proc. IEEE*, vol. 52, pp. 179–184, February 1964.

[8] H. C. J. van Burik and A. C. E. Wessels, "Heterodyne spectral noise analyzer," *Rev. Sci. Instr.*, vol. 34, pp. 798–801, July 1963.

[9] J. B. Johnson, "The Schottky effect in low frequency circuits," *Phys. Rev.*, vol. 26, pp. 71–85, July 1925.

[10] W. Schottky, "Small-shot effect and flicker effect," *Phys. Rev.*, vol. 28, pp. 75–103, July 1926.

[11] M. P. Klein and G. W. Barton, Jr., "Enhancement of signal-to-noise ratio by continuous averaging: Application to magnetic resonance," *Rev. Sci. Instr.*, vol. 34, pp. 754–759, July 1963.

[12] R. H. Dicke, "The measurement of thermal radiation at microwave frequencies," *Rev. Sci. Instr.*, vol. 17, pp. 268–275, July 1946.

[13] H. A. Haus, chairman, "IRE standards on methods of measuring noise in linear twoports, 1959," *Proc. IRE*, vol. 48, pp. 60–68, January 1960; and "Representation of noise in linear twoports," *ibid.*, pp. 69–74.

[14] W. R. Attkinson, L. Fey, and J. Newman, "Spectrum analysis of extremely low frequency variations of quartz oscillators," *Proc. IEEE* (*Correspondence*), vol. 51, p. 379, February 1963.

[15] L. S. Cutler and C. L. Searle, "Some aspects of the theory and measurement of frequency fluctuations in frequency standards," *Proc. IEEE*, vol. 54, pp. 136–154, February 1966.

[16] J. A. Barnes, "Atomic timekeeping and the statistics of precision signal generators," *Proc. IEEE*, vol. 54, pp. 207–220, February 1966.

[17] D. W. Allan, "Statistics of atomic frequency standards," *Proc. IEEE*, vol. 54, pp. 221–230, February 1966.

[18] J. A. Barnes, private communication.

[19] H. E. Derksen and A. A. Verveen, "Fluctuations of resting neural membrane potential," *Science*, vol. 151, pp. 1388–1399, March 18, 1966.

[20] H. E. Derksen, "Axon membrane voltage fluctuations," *Acta Physiol. Pharmacol. Neerl.*, vol. 13, pp. 373–466, 1965.

[21] D. Brouwer, "A study of the changes in the rate of rotation of the earth," *Astron. J.*, vol. 57, pp. 125–146, September 1952. See also J. A. Barnes and D. W. Allan, "An approach to the prediction of coordinated universal time," *Frequency*, vol. 5, pp. 15–20, November–December 1967.

[22] H. V. Cottony and J. R. Johler, "Cosmic radio noise intensities in the VHF band," *Proc. IRE*, vol. 40, pp. 1053–1060, September 1952.

[23] R. B. Blackman and J. W. Tukey, *The Measurement of Power Spectra*. New York: Dover, 1959.

[24] A. van der Ziel, "On the noise spectra of semi-conductor noise and of flicker-effect," *Physica*, vol. 16, pp. 359–372, April 1950.

[25] H. Schönfeld, "Beitrag zum 1/f-Gesetz beim Rauschen von Halbleitern," *Zeit. für Naturforsch.*, vol. 10a, pp. 291–300, April 1955.

[26] J. A. Barnes and D. W. Allan, "A statistical model of flicker

noise," *Proc. IEEE*, vol. 54, pp. 176–178, February 1966.

[27] J. M. Richardson, "The linear theory of fluctuations arising from diffusional mechanisms—An attempt at a theory of contact noise," *Bell Sys. Tech. J.*, vol. 29, pp. 117–141, January 1950.

[28] G. G. MacFarlane, "A theory of contact noise in semiconductors," *Proc. Phys. Soc.* (London), vol. 63B, pp. 807–814, October 1950.

[29] R. E. Burgess, "Contact noise in semiconductors," *Proc. Phys. Soc.* (London), vol. 66B, pp. 334–335, April 1953.

[30] K. M. van Vliet and A. van der Ziel, "On the noise generated by diffusion mechanisms," *Physica*, vol. 24, pp. 415–421, 1958.

[31] M. Lax and P. Mengert, "Influence of trapping, diffusion and recombination on carrier concentration fluctuations," *J. Phys. Chem. Solids*, vol. 14, pp. 248–267, July 1960.

[32] K. M. van Vliet and J. R. Fassett, "Fluctuations due to electronic transitions and transport in solids," in *Fluctuation Phenomena in Solids*, R. E. Burgess, Ed. New York: Academic Press, 1965, pp. 267–354.

[33] J. Bernamont, "Fluctuations in the resistance of thin films," *Proc. Phys. Soc.*, vol. 49 (Suppl.), pp. 138–139, July 1937.

[34] F. K. du Pre, "A suggestion regarding the spectral density of flicker noise," *Phys. Rev. Lett.*, vol. 78, p. 615, June 1950.

[35] D. A. Bell, *Electrical Noise*. Princeton, N. J.: Van Nostrand, 1960.

[36] W. R. Bennett, *Electrical Noise*. New York: McGraw-Hill, 1960.

[37] B. Mandelbrot, "Some noises with $1/f$ spectrum, a bridge between direct current and white noise," *IEEE Trans. Instrumentation and Measurement*, vol. IT-13, pp. 289–298, April 1967.

[38] L. Fey, J. A. Barnes, and D. W. Allan, "An analysis of a low information rate time control unit," *Proc. 20th Ann. Symp. on Frequency Control*, pp. 629–635, 1966.

[39] D. Halford, to be published.

[40] W. B. Davenport, Jr., and W. L. Root, *An Introduction to the Theory of Random Signals and Noise*. New York: McGraw-Hill, 1958, p. 106.

Atomic Timekeeping and the Statistics of Precision Signal Generators

JAMES A. BARNES

Abstract—Since most systems that generate atomic time employ quartz crystal oscillators to improve reliability, it is essential to determine the effect on the precision of time measurements that these oscillators introduce. A detailed analysis of the calibration procedure shows that the third finite difference of the phase is closely related to the clock errors. It was also found, in agreement with others, that quartz crystal oscillators exhibit a "flicker" or $|\omega|^{-1}$ type of noise modulating the frequency of the oscillator.

The method of finite differences of the phase is shown to be a powerful means of classifying the statistical fluctuations of the phase and frequency for signal generators in general. By employing finite differences it is possible to avoid divergences normally associated with flicker noise spectra. Analysis of several cesium beam frequency standards have shown a complete lack of the $|\omega|^{-1}$ type of noise modulation.

Introduction

AN ORDINARY clock consists of two basic systems: a periodic phenomenon (pendulum), and a counter (gears, clock face, etc.) to count the periodic events. An atomic clock differs from this only in that the frequency of the periodic phenomenon is, in some sense, controlled by an atomic transition (atomic frequency standard). Since microwave spectroscopic techniques allow frequencies to be measured with a relative precision far better than any other quantity, the desirability of extending this precision to the domain of time measurement has long been recognized [1].

From the standpoint of precision, it would be desirable to run the clock (counter) directly from the atomic frequency standard. However, atomic frequency standards in general are sufficiently complex that reliable operation over very extended periods becomes somewhat doubtful (to say nothing of the cost involved). For this reason, a quartz crystal oscillator is often used as the source of the "periodic" events to run a synchronous clock (or its electronic equivalent). The frequency of this oscillator is then regularly checked by the atomic frequency standard and corrections are made.

These corrections can usually take on any of three forms: 1) correction of the oscillator frequency, 2) correction of the indicated time, or 3) an accumulating

Manuscript received September 8, 1965; revised November 6, 1965.

The author is with the National Bureau of Standards, Boulder, Colo.

Reprinted from *Proc. IEEE*, vol. 54, pp. 207–220, Feb. 1966.

record of the difference from atomic time of the apparent or indicated time shown by the clock. Both methods 1) and 2) require a calculation of the time difference, and it is sufficient to consider only the last method and the errors inherent in it.

A careful consideration of the calibration procedure leads to the development of certain functionals of the phase which have a very important property—existence of the variance even in the presence of a flicker $(1/|\omega|)$ type of frequency noise. The simplest of these functionals, the second and third finite differences of the phase, turn out to be stationary, random variables whose auto-covariance function is sufficiently peaked to insure rapid convergence of the variance of a finite sample toward the true (infinite sample) variance. These functionals of the phase have the added features of being closely related to the errors of a clock run from the oscillator as well as being a useful measure of oscillator stability.

With the aid of these functionals, it is possible to classify the statistical fluctuations observed in various signal sources. In agreement with work of others [2]–[5], a flicker noise frequency modulation was observed for all quartz crystal oscillators tested. Similar studies on several commercial rubidium gas cells gave uniform indications of flicker noise modulation of levels comparable to those of the better quartz crystal oscillators.

In Section I, the effects on the precision of a time scale due entirely to the calibration procedure of the quartz crystal oscillator and the oscillator's inherent frequency instability are considered.

In Section II, the experimental results of Section I are used as the basis for a theoretical model of oscillator frequency fluctuations, and the results are compared to those of other experimenters.

In Section III, the statistics of an atomic frequency standard of the passive type (e.g., Cs-beam or Rb-gas cell) are considered, and the composite Clock system is treated. Section IV is devoted to a brief discussion of stability measures for signal sources.

I. Quartz Crystal Oscillator Phase Fluctuations

Typical Gross Behavior

Figure 1 shows a typical aging curve for a fairly good quartz crystal oscillator. The oscillator had been operating for a few months prior to the date shown in the graph. On May 1, 1963, the frequency of the oscillator was reset in order to maintain relatively small corrections.

Least square fits of straight lines to the two parts of Fig. 1 yield aging rates of 0.536×10^{-10} per day and 0.515×10^{-10} per day, respectively. This difference in aging rates could be explained by an acceleration of the frequency of about -9×10^{-15} per day per day. This acceleration of the frequency is sufficiently small over periods of a few days when compared to other sources of error that it can be safely ignored. Thus, the frequency of conventional quartz oscillators can be written in the form

$$\Omega(t) = \Omega_0[1 + \alpha t + \dot{\epsilon}(t)] \tag{1}$$

where α is the aging rate, $\dot{\epsilon}(t)$ is a variation of the frequency probably caused by noise processes in the oscillator itself, and t can be considered to be some rather gross measure of the time (since α and $\dot{\epsilon}$ are quite small corrections).

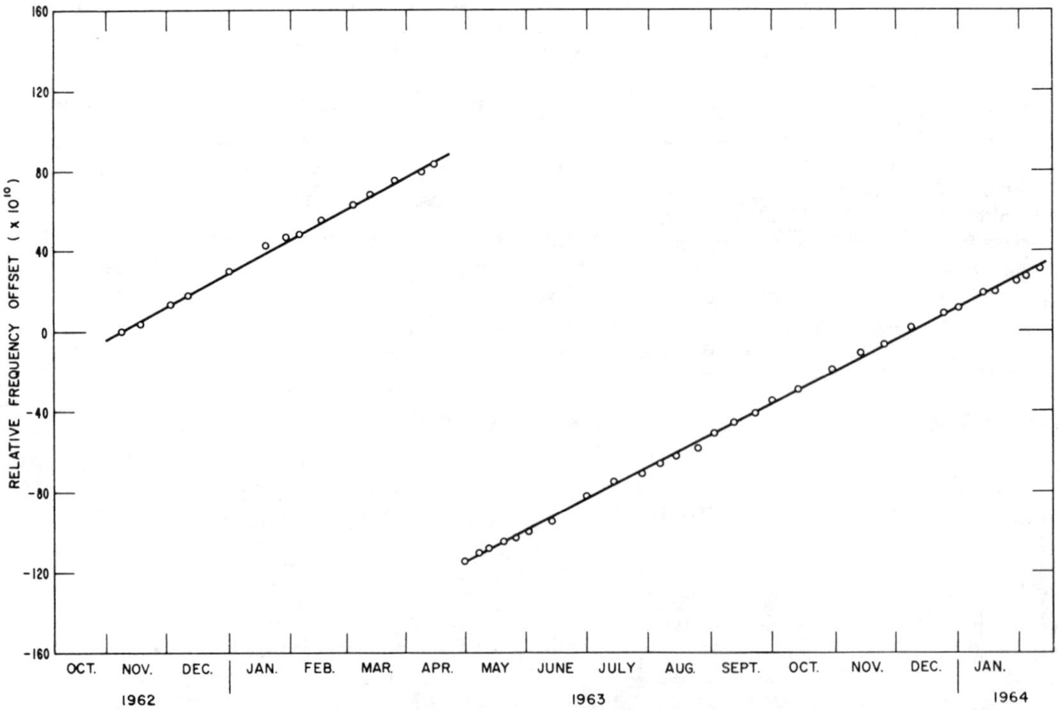

Fig. 1. Frequency aging of a good quartz crystal oscillator.

Actual Calibration Procedure of a Clock System

The fundamental equation for atomic timekeeping is [6]

$$\Omega = \frac{d\phi}{dt}, \tag{2}$$

where Ω is the instantaneous frequency of the oscillator as measured by an atomic frequency standard, $d\phi$ is the differential phase change, and dt is an increment of time as generated by this clock system. Since *time* is to be generated by this system, and ϕ and Ω are the directly measured quantities, it is of convenience to assume that $\Omega = \Omega(\phi)$ and to write the solution of (2) in the form,

$$\Delta t = \int_{\phi_1}^{\phi_2} \frac{d\phi}{\Omega(\phi)}. \tag{3}$$

If one divides the output phase of the oscillator by Ω_0, and defines the apparent or indicated time t_A to be

$$t_A \equiv \frac{\phi}{\Omega_0}, \tag{4}$$

(1), (3), and (4) can be combined to give

$$\Delta t = \int_{t_{A1}}^{t_{A2}} \frac{dt_A}{1 + \alpha t_A + \dot{\epsilon}(t_A)}. \tag{5}$$

As it is indicated in Fig. 1, it is possible to maintain the magnitude of the relative frequency offset $|\alpha t_A + \dot{\epsilon}|$ within fixed bounds of 10^{-8}. Expanding (5) to first order in this relative frequency offset yields

$$\Delta t \simeq \Delta t_A \quad \int_{t_A}^{t_{A2}} [\alpha t_A + \dot{\epsilon}(t_A)]\, dt_A. \tag{6}$$

Equation (6) should then be valid to about one part in 10^{16}.

Normally the frequency of the oscillator is measured over some period of time (usually a few minutes) at regular intervals (usually a few days). At this point, it is desirable to restrict the discussion to the case where the calibration is periodic (i.e., period T, determined by t_A) and then generalize to other situations later. One period of the calibration is as follows:

t_A start of calibration interval
$t_A + \frac{1}{2}(T-\tau)$ start of frequency measurement $(\tau < T)$
$t_A + \frac{1}{2}(T+\tau)$ end of frequency measurement
$t_A + T$ end of calibration interval,

where τ is the frequency measurement interval. If $\dot{\epsilon}$ were constant in time, the frequency measured during the interval $t_A + \frac{1}{2}(T-\tau)$ to $t_A + \frac{1}{2}(T+\tau)$ would be just the average frequency during the complete measurement interval T since the oscillator would have an exactly linear drift in frequency. Also, if $\dot{\epsilon}$ were constant, (6) could be written as

$$\Delta t = \Delta t_A - T\left\langle \frac{\delta\Omega}{\Omega_0} \right\rangle_\tau, \tag{7}$$

where $\langle \delta\Omega/\Omega_0 \rangle_\tau$ is the average relative frequency offset during the interval $t_A + \frac{1}{2}(T-\tau)$ to $t_A + \frac{1}{2}(T+\tau)$.

Even though, in general, $\dot{\epsilon}$ is not constant, ϵ and $\dot{\epsilon}$ are not knowable, and thus one is usually reduced to using (7) anyway. The problem, then, is to determine how much error is introduced by using (7).

The time error δt accumulated over an interval T committed by using (7) can be expressed in the form,

$$\delta t = \int_{t_A}^{t_A+T} [\alpha t_A + \dot{\epsilon}]\, dt_A - T\left\langle \frac{\delta\Omega}{\Omega_0} \right\rangle, \tag{8}$$

where the quantity $\langle \delta\Omega/\Omega_0 \rangle_\tau$ is given by

$$\left\langle \frac{\delta\Omega}{\Omega_0} \right\rangle_\tau = \frac{1}{\tau} \int_{t_A+1/2(T-\tau)}^{t_A+1/2(T+\tau)} [\alpha t_A + \dot{\epsilon}]\, dt_A. \tag{9}$$

Equations (8) and (9) can be combined to give

$$\delta t = \epsilon(t_A + T) - \epsilon(t_A) - \frac{T}{\tau}\left[\epsilon\left(t_A + \frac{T+\tau}{2}\right) - \epsilon\left(t_A + \frac{T-\tau}{2}\right)\right]. \tag{10}$$

It is this equation which relates the random phase fluctuations with the corresponding errors in the time determination.

Meaningful Quantities

It is again of value to further restrict the discussion to a particular situation and generalize at a later point. In particular, let $T = 3\tau$, then (10) becomes

$$\delta t = \epsilon(t_A + 3\tau) - \epsilon(t_A) - 3[\epsilon(t_A + 2\tau) - \epsilon(t_A + \tau)]. \tag{11}$$

It is now possible to define the discrete variable ϵ_n by the relation

$$\epsilon_n \equiv \epsilon(t_A + n\tau), \qquad n = 0, 1, 2, \cdots$$

and rewrite (11) in the simpler form (see Table I)

$$\delta t \equiv \Delta^3 \epsilon_n, \tag{12}$$

where $\Delta^3\epsilon_n$ is the third finite difference of the discrete variable ϵ_n.

TABLE I
FINITE DIFFERENCES
$\epsilon_n \equiv \epsilon(t_A + n\tau)$

Discrete variable		Definition
ϵ_n	ϵ_n	$\epsilon(t_A+n\tau)$
$\Delta\epsilon_n$	$\epsilon_{n+1}-\epsilon_n$	$\epsilon(t_A+\tau)-\epsilon(t_A)$
$\Delta^2\epsilon_n$	$\Delta\epsilon_{n+1}-\Delta\epsilon_n$	$\epsilon(t_A+2\tau)-2\epsilon(t_A+\tau)+\epsilon(t_A)$
$\Delta^3\epsilon_n$	$\Delta^2\epsilon_{n+1}-\Delta^2\epsilon_n$	$\epsilon(t_A+3\tau)-3[\epsilon(t_A+2\tau)-\epsilon(t_A+\tau)]-\epsilon(t_A)$
$\Delta^4\epsilon_n$	$\Delta^3\epsilon_{n+1}-\Delta^3\epsilon_n$	$\epsilon(t_A+4\tau)-4[\epsilon(t_A+3\tau)+\epsilon(t_A+\tau)]$ $+6\epsilon(t_A+2\tau)+\epsilon(t_A)$

Similarly, (2) may be integrated directly using (1) to obtain

$$\phi(t_A) = \Omega_0\left[t_A + \frac{\alpha}{2}t_A{}^2 + \epsilon(t_A)\right] + \phi(0). \qquad (13)$$

Thus, by defining another discrete variable ϕ_n, the third difference of (13) yields the relation

$$\Delta^3\phi_n = \Omega_0\Delta^3\epsilon_n, \qquad (14)$$

or equivalently,

$$\delta t = \left(\frac{1}{\Omega_0}\right)\Delta^3\phi_n. \qquad (15)$$

It is now possible to set up a table of meaningful quantities for time measurement (see Table II).

If three oscillators are used, it is possible to independently measure the three quantities σ_{12}, σ_{13}, and σ_{23}. Thus there exist three independent equations:

$$\left.\begin{aligned}\sigma_{12}{}^2 &= \sigma_1{}^2 + \sigma_2{}^2\\ \sigma_{13}{}^2 &= \sigma_1{}^2 + \sigma_3{}^2\\ \sigma_{23}{}^2 &= \sigma_2{}^2 + \sigma_3{}^2\end{aligned}\right\}. \qquad (18)$$

While the three equations,

$$\delta t_{ij} = \Delta^3\epsilon_n{}^{(i)} - \Delta^3\epsilon_n{}^{(j)} \qquad \begin{cases} i, j = 1, 2, 3 \\ i < j \end{cases},$$

are not linearly independent, the standard deviations $\sigma_{ij}{}^2$ given by (18), in fact, form linearly independent equations (subject only to certain conditions analogous to the triangle inequalities). Thus, the systems of (18) are solvable for the three quantities $\sigma_i{}^2 = \langle(\delta t_i)^2\rangle$. It is thus possible to estimate the statistical behavior of each individual oscillator.

TABLE II

MEANINGFUL QUANTITIES

Quantity		Discussion
Phase	$\}\phi_n$	Must exist for all times of interest and be measurable.
First difference of the phase	$\}\Delta\phi_n$	Proportional to the average frequency in the interval τ. Must exist for all time.
Second difference of the phase	$\}\Delta^2\phi_n$	Related to the drift of the oscillator frequency. Must exist for all time.
Variance of the second difference	$\}\langle[\Delta^2\phi_n - \langle\Delta^2\phi_n\rangle]^2\rangle$	(It is possible to construct a time scale even if this does not exist. However, experiment indicates that it is probably finite.)
Third difference of the phase	$\}\Delta^3\phi_n$	Proportional to the clock error in the time interval $T=3\tau$. Must exist if clock is to be of value.
Mean square third difference	$\}\langle(\Delta^3\phi_n)\rangle^2$	Proportional to the precision of time interval measurements. Must exist if clock is to be of value.

Where all averages are defined by the relation:

$$\langle f(t)\rangle = \lim_{T\to\infty}\frac{1}{T}\int_{-T/2}^{T/2} f(t)dt.$$

Experimental Determination of Phase Fluctuations

If one measures the phase difference between two oscillators, (15) applies to both, and hence the difference phase $\theta_n = \phi_n{}^{(1)} - \phi_n{}^{(2)}$ is related to the difference time $\delta t_{12} = \delta t_1 - \delta t_2$ by the relations

$$\delta t_{12} = \frac{1}{\Omega_0}\Delta^3\theta_n = \Delta^3\epsilon_n{}^{(1)} - \Delta^3\epsilon_n{}^{(2)}. \qquad (16)$$

Provided that the cross-correlation coefficient $\langle(\Delta^3\epsilon_n{}^{(1)})(\Delta^3\epsilon_n{}^{(2)})\rangle$ is zero (i.e., the $\epsilon_n{}^{(i)}$ are noncorrelated), the variance of δt_{12} becomes

$$\sigma_{12}{}^2 \equiv \langle(\delta t_{12})^2\rangle = \langle(\Delta^3\epsilon_n{}^{(1)})^2\rangle + \langle(\Delta^3\epsilon_n{}^{(2)})^2\rangle, \qquad (17)$$

since it is assumed that $\langle\Delta^3\epsilon_n\rangle = 0$. The assumption that the cross-correlation coefficients vanish is equivalent to postulating an absence of linear coupling between the oscillators either electrically or through their environment. It is of value to develop a scheme which is capable of classifying individual oscillators rather than treating ensembles of assumed identical members. Thus, this development is restricted to time averages of individual oscillators rather than ensemble averages.

Apparatus

A phase meter was used similar to one described in Cutler and Searle [5]. The basic system was aligned with an electrical-to-mechanical angle tolerance of about ± 0.25 percent of one complete cycle, and since the phase meter is operated at 10 Mc/s, this implies the possibility of measuring the time difference to ± 0.25 nanoseconds. The output shaft, in turn, drives a digital encoder with one hundred counts per revolution and a total accumulation (before starting over) of one million. Thus the digital information is accurate to within one ns and accumulates up to one ms. Since measurements are made relatively often, it is easy for a computer to spot when the digital encoder has passed one ms on the data, and this restriction on the data format in no way hampers the total length of the data handled.

Since the oscillator is assumed to have a linear drift in frequency with time, the variance (about the mean) of the second difference of the phase should depend only on the $\epsilon_n^{(i)}$. For the evaluation of clocks, it is the mean square time error which is important and, thus, the mean square (rather than variance) of the third difference of the phase is the quantity of importance. Calculations of these quantities from the phase θ_n were accomplished on a digital computer. Normally the phase difference is printed every hour for at least 200 hours and the computer program computes the $\Delta^2\theta_n$ and $\Delta^3\theta_n$ for τ equal to 2, 4, 6, 8, etc., hours. Thus it is possible to plot the square root of the variance of the second difference and the root mean square third difference of the phase as a function of the variable τ. A typical plot of these two quantities appears in Fig. 2.

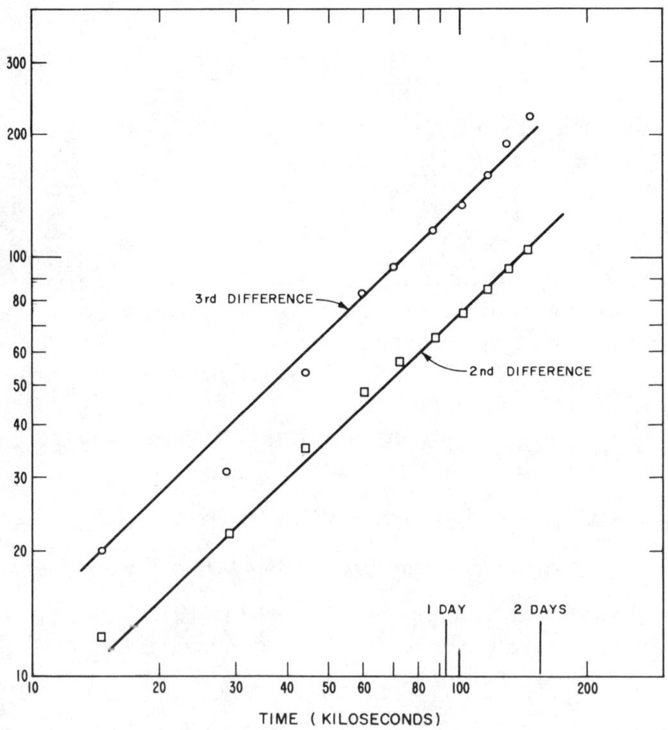

Fig. 2. Variance of second and third difference as a function of τ.

Since many revolutions of the phase meter are normally encountered between data points, the combined effects of nonperfect electrical-to-mechanical phase and rounding errors of the digital encoder can be combined into one stationary, statistical quantity γ, that can be assumed to have delta-function auto-correlation. Let χ represent the rounding errors of the digital encoder defined by the difference between the encoded number and the actual angular position of the shaft. Then these rounding errors form a rectangular distribution from -0.5 ns to $+0.5$ ns, and, thus, contribute an amount

$$\frac{\int_{-1/2}^{1/2} \chi^2\, d\chi}{\int_{-1/2}^{1/2} d\chi} = 0.08 \text{ ns}^2$$

to $\langle\gamma^2\rangle$. The nonperfect electrical-to-mechanical phase conversion is sinusoidal in nature and, thus, contributes an amount equal to

$$\tfrac{1}{2}(0.25)^2 \simeq 0.03 \text{ ns}^2.$$

The final value for $\langle\gamma^2\rangle$ should then be about 0.11 ns^2.

This was checked by taking two very good oscillators operating at a rather large difference frequency ($\sim 5\times10^{-9}$), and printing the phase difference every 10 seconds for several minutes. Since frequency fluctuations of the oscillators are small compared to 1×10^{-10}, the resultant scatter can be attributed to the measuring system. The results of this experiment gave the value 0.19 ns^2 for $\langle\gamma^2\rangle$.

Since γ is assumed to have a delta-function auto-correlation, reference to Table I shows that $\langle(\Delta^3\gamma)^2\rangle = 20\langle\gamma^2\rangle$, and thus (18) may be more precisely written,

$$\left.\begin{aligned} \sigma_{12}^2 &= \sigma_1^2 + \sigma_2^2 + 20\langle\gamma^2\rangle \\ \sigma_{13} &= \sigma_1^2 + \sigma_3^2 + 20\langle\gamma^2\rangle \\ \sigma_{23} &= \sigma_2^2 + \sigma_3^2 + 20\langle\gamma^2\rangle \end{aligned}\right\}. \tag{18'}$$

For the best oscillators tested, σ_{ij}^2 became about $2\langle(\Delta^3\gamma)^2\rangle$ for $\tau = 10^3$ seconds, and, therefore, measurements were limited on the lower end to 20 minutes or 1200 seconds. The longest run made lasted a little over a month or about one thousand hours. Thus the largest value of τ which might have reasonable averaging is about one hundred hours or 3.6×10^5 seconds. This limits the results to about two orders of magnitude variation on τ.

While it is possible to build appropriate frequency multipliers and mixers to improve the resolution of the phase meter and reduce the lower limit on τ, it was considered that the longer time intervals are of greater interest because T $(=3\tau)$ is normally between one day and one week.

II. Theoretical Development

Introductory Remarks

In the development which follows, certain basic assumptions are made. It is assumed that the coefficients Ω_0 and α of (1) may be so chosen that the average value of $\epsilon(t)$ is zero; i.e.,

$$\langle\epsilon(t)\rangle = 0.$$

It is also assumed that a translation in the time axis, $t_A \to t_A + \zeta$, (stationarity) causes no change in the value of the auto-covariance function,

$$\begin{aligned} R_\epsilon(\tau) &= \langle\epsilon(t)\cdot\epsilon(t+\tau)\rangle \\ &\equiv \langle\epsilon(t+\xi)\cdot\epsilon(t+\xi+\tau)\rangle. \end{aligned} \tag{19}$$

The only justification of this assumption lies in the fact that the results of the analysis agree well with experiment and the results of others. In the development which follows, one *cannot* assume that

$$R_\epsilon(0) = \langle[\epsilon(t)]^2\rangle$$

exists (i.e., is finite) and, hence, Wiener [7] cannot guarantee that (19) is valid. While it may be that $R_\epsilon(\tau)$ does not exist, quantities such as

$$U(\tau) = 2[R_\epsilon(0) - R_\epsilon(\tau)] \tag{20}$$

may exist and be meaningful if limits are approached properly. It is, thus, assumed that relations such as (20) have meaning and may be handled by conventional means.

Development of Experimental Results

All oscillator pairs tested, which exhibited a stable drift rate as indicated in Fig. 1 and did not have obvious diurnal fluctuations in frequency, showed a definite, very nearly linear dependence on τ for both the standard deviation of the second difference and the root mean square third difference of the phase. It was observed that if an oscillator were disturbed accidentally during a measurement, the plot would have a more nearly $\tau^{3/2}$ dependence. This is probably because the assumption of a negligible quadratic dependence of frequency with time is not valid when the oscillator is disturbed. All oscillators tested, therefore, were shock mounted and all load changes and physical conditions were changed as little as possible.

A least square fit of all reliable data to an equation of the form

$$\sqrt{\langle(\Delta^m \epsilon_n)^2\rangle} = \sqrt{k_m}\tau^\mu \qquad m = 2, 3, 4 \tag{21}$$

gave a value of 1.09 for the average of the μ's. The values of μ ranged from about 0.90 to 1.15 (and even to 1.5 when the oscillators were disturbed during the measurement). It is of interest to postulate that, for an "ideal," undisturbed quartz crystal oscillator, the value of μ is exactly the integer one, and to investigate the consequences of this assumption.

Because certain difficulties arise at the value $\mu = 1$, it is essential to calculate with a general μ and then pass to the limits $\mu \to 1^{(-)}$ for the quantities of interest. This is equivalent to considering a sequence of processes which approach, as a limit, the case of the "ideal" crystal oscillator. Thus (21) may be rewritten for $m = 2$ in the form

$$\langle(\Delta^2 \epsilon_n)^2\rangle = k_2 |\tau|^{2\mu}. \tag{22}$$

Using Table I, (22) may then be rewritten as

$$6\langle[\epsilon(t_A)]^2\rangle - 8\langle\epsilon(t_A)\cdot\epsilon(t_A+\tau)\rangle + 2\langle\epsilon(t_A)\cdot\epsilon(t_A+2\tau)\rangle = k_2|\tau|^{2\mu}$$

or, equivalently,

$$6R_\epsilon(0) - 8R_\epsilon(\tau) + 2R_\epsilon(2\tau) = k_2|\tau|^{2\mu}. \tag{23}$$

As mentioned above, the function $U(\tau)$ is defined by the relation

$$\begin{aligned} U(\tau) &\equiv \langle[\epsilon(t_A) - \epsilon(t_A + \tau)]^2\rangle \\ &= 2[R_\epsilon(0) - R_\epsilon(\tau)] \end{aligned} \tag{24}$$

and is assumed to exist. Equations (23) and (24) may be combined to give

$$4U(\tau) - U(2\tau) = k_2|\tau|^{2\mu}. \tag{25}$$

If a trial solution of the form

$$U(\tau) = A(\beta)\cdot|\tau|^\beta$$

is used, one obtains

$$A(\beta)\cdot|\tau|^\beta[4 - 2^\beta] = k_2|\tau|^{2\mu}$$

from which one concludes that

$$\beta = 2\mu \le 2,$$

$$(4 - 2^\beta)A(\beta) = k_2,$$

and

$$U(\tau) = \frac{2}{4 - 2^{2\mu}}|\tau|^{2\mu}, \tag{26}$$

where $\mu \le 1$ since $U(\tau)$ and k_2 are non-negative. Equation (21) may, thus, be satisfied if

$$\lim_{\mu \to 1^{(-)}} (4 - 2^{2\mu})A(2\mu) = k_2, \tag{27}$$

which implies that $A(2)$ is infinite. It is for this reason that the limiting process must be employed.

It is not necessary, however, to assume a particular form for $A(2\mu)$ because (27) is sufficient for the purposes of this development.

It is now possible to determine the mean square third difference of ϵ_n; i.e., from Table I,

$$\begin{aligned} \langle(\Delta^3\epsilon_n)^2\rangle &= 20\langle[\epsilon(t_A)]^2\rangle - 30\langle[\epsilon(t_A)\cdot\epsilon(t_A + \tau)]\rangle \\ &\quad + 12\langle[\epsilon(t_A)\cdot\epsilon(t_A + 2\tau)]\rangle - 2\langle[\epsilon(t_A)\cdot\epsilon(t_A + 3\tau)]\rangle, \end{aligned} \tag{28}$$

where use has again been made of (19). Equation (28) may equivalently be written in the form

$$\langle(\Delta^3\epsilon_n)^2\rangle = 15U(\tau) - 6U(2\tau) + U(3\tau), \tag{29}$$

which may be combined with (26) to yield

$$\langle(\Delta^3\epsilon_n)^2\rangle = \frac{k_2|\tau|^{2\mu}}{(4 - 2^{2\mu})}[15 + 6(2^{2\mu}) + 3^{2\mu}]. \tag{30}$$

If one now passes to the limit $\mu \to 1^{(-)}$, (30) becomes

$$\langle(\Delta^3\epsilon_n)^2\rangle = \frac{k_2\tau^2(24 \ln 2 - 9 \ln 3)}{4 \ln 2}. \tag{31}$$

Thus, a quadratic dependence of the variance of the second difference of the phase implies a quadratic dependence of the mean square third difference, and the ratio

$$\sqrt{\frac{\langle(\Delta^3\epsilon_n)^2\rangle}{\langle(\Delta^2\epsilon_n)^2\rangle}} = 1.5601 \cdots \tag{32}$$

is independent of τ. The average ratio of the points plotted in Fig. 2 is 1.65. Values ranging from 1.4 to 1.7 were observed for various runs on different oscillator pairs. Average value for all reliable data taken is 1.52.

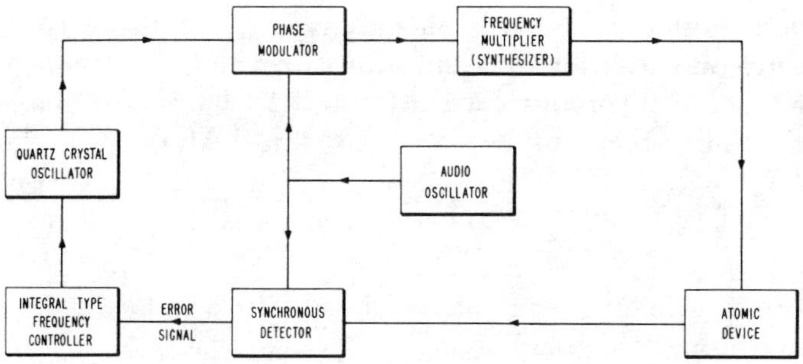

Fig. 3. Typical control loop for an atomically controlled oscillator.

Generalization of the Time Error Problem

The average frequency of an oscillator over an interval of time is just the total elapsed phase in the interval divided by the time interval. Since errors of the frequency standard are not presently being considered, the calibration interval of Section I gives rise to an error time δt which could be expressed as a sum

$$\delta t = \frac{1}{\Omega_0} \sum_{n=0}^{m} a_n \phi(t_A + b_n \tau), \tag{33}$$

where $m+1$ is the total number of terms and the set $\{a_n, b_n\}$ are chosen to fit the particular calibration procedure. Indeed, any calibration procedure must give rise to an error time which is expressible in the form of (33).

There are, however, certain restrictions on the $\{a_n, b_n\}$ which are of importance. First, it is a matter of convenience to require that $b_l > b_n$ for $l > n$. Also, if the oscillator were absolutely perfect, and $\epsilon(t_A)$ were identically zero, one should logically require that the error time δt be identically zero, independent of t_A, the drift rate α, and the basic time interval τ. That is, from (13)

$$\delta t = \sum_{n=0}^{m} a_n \cdot (t_A + b_n \tau) + \frac{\alpha}{2} \sum_{n=0}^{m} a_n \cdot (t_A + b_n \tau)^2 \equiv 0 \tag{34}$$

for all t_A, α, and τ. One is thus led to the three conditions:

$$\sum_{n=0}^{m} a_n = 0, \tag{35}$$

$$\sum_{n=0}^{m} a_n b_n = 0, \tag{36}$$

$$\sum_{n=0}^{m} a_n b_n^2 = 0. \tag{37}$$

It is of interest to form the quantity

$$\left(\sum_{n=0}^{m} a_n b_n^2\right)\left(\sum_{n=0}^{m} a_l\right) = \sum_{n=0}^{m} a_n^2 b_n^2 + \sum_{n<l}^{m} a_n a_l b_n^2 + \sum_{l<n}^{m} a_n a_l b_n^2, \tag{38}$$

and it is now possible to interchange the subscripts n and l in the last term of (38) and write the equation in the form

$$\sum_{n=0}^{m} a_n^2 b_n^2 + \sum_{n<l}^{m} a_n a_l (b_n^2 + b_l^2) = 0. \tag{39}$$

Subtracting the square of (36),

$$\left(\sum_{n=0}^{m} a_n b_n\right)\left(\sum_{l=0}^{m} a_l b_l\right) = \sum_{n=0}^{m} a_n^2 b_n^2 + 2 \sum_{n<l}^{m} a_n a_l b_n b_l,$$

one obtains

$$\sum_{n<l}^{m} a_n a_l (b_l - b_n)^2 = 0 \tag{40}$$

as another condition on the $\{a_n, b_n\}$. Equation (40) is, of course, not independent of (35)–(37).

For the actual situation where $\epsilon(t_A)$ is not identically zero, one is reduced, as before [see (7)], to using conditions (35)–(37) since $\epsilon(t_A)$ is not knowable. The time error then becomes

$$\delta t = \sum_{n=0}^{m} a_n \cdot \epsilon(t_A + b_n \tau) \tag{41}$$

where use has been made of (13), (33), and the restrictions (35)–(37). Note that (41) is the generalization of (10). The square of (41) can be written in the form

$$(\delta t)^2 = \sum_{n=0}^{m} a_n^2 \epsilon^2(t_A + b_n \tau) + 2 \sum_{n<l}^{m} a_n a_l [\epsilon(t_A + b_n \tau) \cdot \epsilon(t_A + b_l \tau)], \tag{42}$$

which may be averaged over t_A to yield

$$\langle(\delta t)^2\rangle = \sum_{n=0}^{m} a_n^2 \langle(\epsilon(t_A))^2\rangle + 2 \sum_{n<l}^{m} a_n a_l \langle[\epsilon(t_A) \cdot \epsilon(t_A + (b_l - b_n)\tau)]\rangle, \tag{43}$$

where use has again been made of (19).

The square of (35) may be written as

$$\sum_{n=0}^{m} a_n^2 + 2 \sum_{n<l}^{m} a_n a_l = 0,$$

and (43) then becomes

$$\langle(\delta t)^2\rangle = - \sum_{n<l}^{m} a_n a_l \{\langle \epsilon^2(t_A) \} \{ 2\langle \epsilon^2(t_A) \rangle - 2\langle[\epsilon(t_A) \cdot \epsilon(t_A + (b_l - b_n)\tau)]\rangle\}. \tag{44}$$

Combining (44) with (21) one obtains

$$\langle(\delta t)^2\rangle = - \sum_{n<l} a_n a_l U((b_l - b_n)\tau). \tag{45}$$

Substitution of (26) in (45) yields

$$\langle(\delta t)^2\rangle = -k_2|\tau|^{2\mu}\frac{\sum_{n<l}^{m} a_n a_l (b_l - b_n)^{2\mu}}{4 - 2^{2\mu}}, \quad (46)$$

which is indeterminate of the form (0/0) for $\mu=1$ because of (40). Making use of L'Hospital's rule, the limit of this expression as $\mu \to 1^{(-)}$ is

$$\langle(\delta t)^2\rangle = \frac{k_2\tau^2}{4 \ln 2}\sum_{n<l}^{m} a_n a_l (b_l - b_n)^2 \ln (b_l - b_n). \quad (47)$$

Equation (47) is thus the generalization of (31). Since many terms appear in the summation in (47), a computer program was written for evaluation with particular sets of $\{a_n, b_n\}$.

Comparison With Others

If one were to assume that (20) could be solved for $R_\epsilon(\tau)$, one would obtain

$$R_\epsilon(\tau) = R_\epsilon(0) - \tfrac{1}{2}U(\tau) \quad (48)$$

which expresses the auto-covariance function in terms of $U(\tau)$. If one assumes still further that the Wiener-Khinchin theorem applies to (48), the power spectral density of $\epsilon(t)$ is then the Fourier transform of $R_\epsilon(\tau)$; i.e.,

$$S^2(\omega) = \text{F.T.}\left[R_\epsilon(0) - \frac{k_2|\tau|^{2\mu}}{2(4 - 2^{2\mu})}\right].$$

Fortunately, Fourier transforms of functions like $|\tau|^{2\mu}$ have been worked out in Lighthill [8]. The result is

$$S_\epsilon(\omega) = \frac{R_\epsilon(0)}{2\pi}\delta(\omega) - \frac{k_2\left[\cos\frac{\pi(2\mu+1)}{2}\right][(2\mu)!]|\omega|^{-2\mu-1}}{2\pi(4 - 2^{2\mu})}. \quad (49)$$

The factor of $(1/2\pi)$ occurs because $S_\epsilon(\omega)$ is assumed to be a density relative to an angular frequency ω, rather than a cycle frequency $(f=\omega/2\pi)$ as is used in [8]. The first term on the right of (49) indicates an infinite density of power at zero frequency, i.e., a nonzero average. The zero frequency components are not measurable experimentally, and hence this term will be dropped as not being significant to these discussions. The second term on the right of (49) is indeterminate for $\mu=1$. As before, the limit as $\mu \to 1^{(-)}$ may be obtained, yielding

$$S_\epsilon(\omega) = \frac{k_2}{8 \ln 2}|\omega|^{-3} \quad (50)$$

for the final result.

The assumptions needed to arrive at (50) are not wholly satisfying, and it is of value to show that (50) implies that $\langle(\Delta^2\epsilon_n)^2\rangle$ is indeed given by (21) for $m=2$. Only one additional assumption is needed, the Wiener-Khinchin theorem. Since $R_\epsilon(\tau)$ and $S_\epsilon(\omega)$ are real quantities, this theorem may be written in the form

$$R_\epsilon(\tau) = 2\int_0^\infty S_\epsilon(\omega)\cos \omega\tau \, d\omega. \quad (51)$$

From Table I, one may obtain (after squaring and averaging) the expression

$$\langle(\Delta^2\epsilon_n)^2\rangle = 6R_\epsilon(0) - 8R_\epsilon(\tau) + 2R_\epsilon(2\tau). \quad (52)$$

Substitution of (51) in (52) yields

$$\langle(\Delta^2\epsilon_n)^2\rangle = 4\int_0^\infty S_\epsilon(\omega)[3 - 4\cos(\omega\tau) + \cos(2\omega\tau)]\, d\omega. \quad (53)$$

Using (50) for $S_\epsilon(\omega)$ is a lengthy but straightforward process to evaluate the integral in (53). The result is, in fact,

$$\langle(\Delta^2\epsilon_n)^2\rangle = k_2\tau^2,$$

in agreement with (21).

If $g_\epsilon(\omega)$ is the Fourier transform of $\epsilon(t)$, it is shown, for example, in Lighthill ([8] p. 20), that

$$g_{\dot\epsilon}(\omega) = i\omega g_\epsilon(\omega)$$

is the Fourier transform of $\dot\epsilon(t)$. Thus, the power spectral density of $\dot\epsilon(t)$ is related to the power spectral density of $\epsilon(t)$ by the familiar relation

$$S_{\dot\epsilon}(\omega) = \omega^2 S_\epsilon(\omega).$$

Thus, the power spectral density of the *frequency* fluctuations of an "ideal" oscillator may be given as

$$S_{\dot\epsilon}(\omega) = \frac{k_2}{8 \ln 2}|\omega|^{-1},$$

that is, flicker noise frequency modulation. The existence of this type of noise modulating the frequency of good quartz crystal oscillators has been reported by several others [2]–[5], [9].

Comments on the "Ideal" Oscillator

It has thus been shown that the assumptions of stationarity and "ideal" behavior form a basis for a mathematical model of a quartz crystal oscillator which is in quite good agreement with several experiments and experimenters. On the basis of this model, it is now possible to predict the behavior of systems employing "nearly ideal" oscillators with the hope of committing no great errors. There are compelling reasons to believe that $U(\tau)$ actually exists (see Section IV) for real oscillators in spite of (26). Such conditions require that the $|\omega|^{-1}$ behavior for the power spectral density of the frequency fluctuations cut off at some small, nonzero frequency. From some experiments [2] conducted, however, this cutoff frequency is probably much smaller than one cycle per year. Such small differences from zero frequency are of essentially academic interest to the manufacturer and user of oscillators. Quantities which may be expressed in the form of (41), however, where the coefficients satisfy conditions (35)–(37), have finite averages even in the limit as flicker noise behavior approaches zero modulation frequency. Such quantities are called cutoff *in*dependent in contrast to quantities like $U(\tau)$ which will exist only if the flicker noise cuts off at some nonzero frequency.

III. Atomic Frequency Standards

Passive Devices

There are in use today two general types of atomic frequency standards: 1) the active device such as a maser whose atoms actually generate a coherent signal whose frequency *is* the standard, and 2) the passive type such as a cesium beam or rubidium gas cell. In the passive type, a microwave signal irradiates the atoms and some means is employed to detect any change in the atom's energy state. This paper is restricted to the passive type of frequency standard. Some experiments are in progress, however, to determine the statistical behavior of a maser type oscillator.

It is first of value to discuss in what way a "standard" can have fluctuations or errors. Consider the cesium beam. *Ideally* the standard would be the exact frequency of the photons emitted or absorbed at zero magnetic field in the $(F=4,\ m_f=0)\leftrightarrow(F=3,\ m_f=0)$ transition of cesium^{133} in the ground electronic state for an infinite interaction time. This is, of course, impossible. This means that the standard is at least less than ideal and one is, thus, led to speak, in some sense of the word, about "errors" or even "fluctuations" of the standard.

Figure 3 shows a block diagram of a typical standard of the passive type. An equivalent diagram of this frequency-lock servo is shown in Fig. 4, where $V_1(\omega)$ is the Fourier Transform of the noise generated in the detectors, associated demodulating circuitry, and the frequency multipliers of Fig. 3. $V_2(\omega)$ is an equivalent noise voltage driving the reactance tube in the oscillator to produce the $\dot{\epsilon}(t_A)$ term in the unlocked oscillator, such that flicker noise FM results. The power spectrum of $V_2(\omega)$, then, is given by

$$C^2 S_{V_2} = \frac{h_s}{|\omega|}, \tag{54}$$

where $h_s/|\omega|$ is the power spectral density of the *frequency* fluctuations of the unlocked oscillator.

It is easiest to treat the servo equations by the use of the variable $\dot{\gamma}$, defined to be the difference between the output frequency of the multiplier and the "ideal" frequency of the atomic transition (the output of the "atomic device," as shown in Fig. 4, is then assumed to be the constant zero). In order to preserve the dimensions of voltage for the addition networks, it is convenient to assume that the output of the subtraction network is $-\beta\dot{\gamma}$ where β has the dimensions of volt-seconds. Thus, the equation governing the operation of the servo can be expressed, in the frequency domain, as

$$\left[V_2(\omega) + \frac{V_1(\omega) - \beta\dot{\gamma}(\omega)}{j\omega\tau_1}\right]CN = \dot{\gamma}(\omega). \tag{55}$$

This leads to a power spectral density for $\dot{\gamma}$ given by

$$S_{\dot{\gamma}}(\omega) = \frac{(NC)^2 S_{V_1}(\omega) + N^2\tau_1^2 h_s|\omega|}{\omega^2\tau_1^2 + (\beta NC)^2} \tag{56}$$

where use has been made of (54).

Normally, $\omega\tau_1$ becomes of the order of (βNC) for ω of the order of 10 s^{-1}. Thus, for small ω, (56) becomes

$$S_{\dot{\gamma}}(\omega) \approx \frac{1}{\beta^2} S_{V1}(\omega) \tag{57}$$

for $|\omega| < 1\ s^{-1}$.

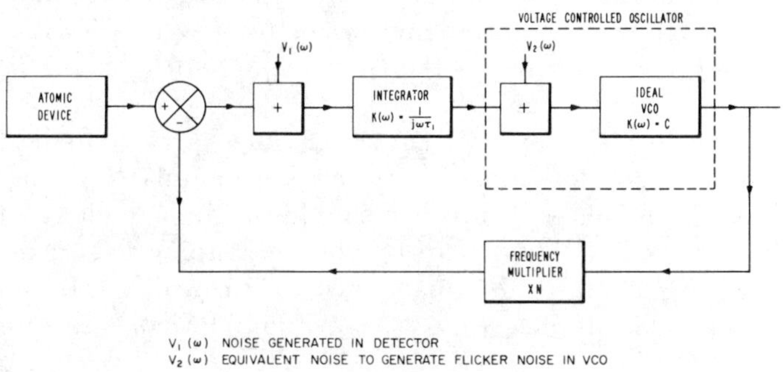

Fig. 4. Equivalent servo diagram of passive type frequency standard.

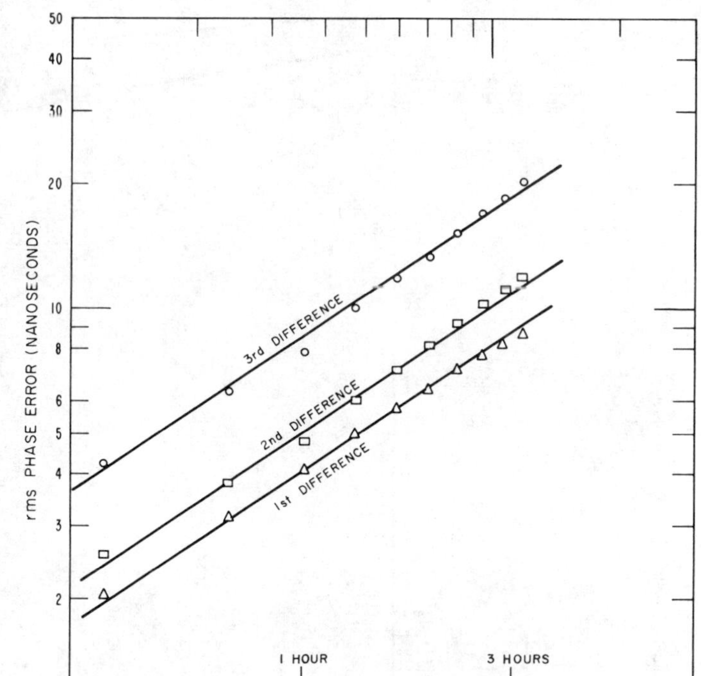

Fig. 5. Variance of differences of the relative phase difference between two atomic standards.

By applying the techniques of Section I to the difference phase of two cesium beams, the curves shown in Fig. 5 were obtained. Through least square fits to the data and comparing the ratios of the variances of differences (see Appendix), the result was obtained that the data fit curves of the form

$$\langle(\Delta^m \gamma_n)^2\rangle = B_m |\tau|^\eta$$

for $\eta = 1.34$, with an uncertainty (standard deviation) of about ± 0.04. Again using [8] as before, this leads to the result that

$$S_{\dot{\gamma}}(\omega) = \frac{g}{|\omega|^\mu}, \qquad g \approx B_1 \frac{\sqrt{3}}{4\pi} \Gamma\left(\frac{7}{3}\right) \tag{58}$$

where $\mu = 0.34 \pm 0.04$. One is, thus, led to the very strange conclusion that the spectral distribution of the detector and multiplier noise varies as $\sim |\omega|^{-1/3}$. The source of this noise was later traced to a faulty preamplifier. With a proper amplifier used, the spectral distribution appears white as other papers [5], [10], [11] indicate should be the case.

Ideally, for measurements over times large compared to the servo time constant, the error accumulated during one measurement interval should be independent of errors accumulated during nonoverlapping intervals, i.e., mathematically analogous to Brownian Motion [12]. This implies that $S_{\dot{\gamma}}(\omega)$ should be constant for $|\omega| < 1\ \text{s}^{-1}$.

The Composite Clock System

If an oscillator's frequency is measured by an atomic standard, the error in measurement of the frequency is given by

$$\delta f = \frac{1}{2\pi\tau} \left[\gamma(t_A + \tfrac{1}{2}(T + \tau)) - \gamma(t_A + \tfrac{1}{2}(T - \tau))\right] \tag{59}$$

for a calibration interval as given in Section I. Also, if f_s is the defined output frequency of the frequency standard, the total error time for the entire calibration interval T, due to the Standard, is given by $T\delta f/f_s$, and if the clocks contribute an additional error (uncorrelated to the standard's error), of δt, the total error δT is given by

$$\delta T = \delta t + \frac{T\delta f}{f_s}. \tag{60}$$

It is of value to explore the dependence of $\langle(\delta t)^2\rangle$ on the ratio of T/τ. In particular, let $T = (2n+1)\tau$ for (10). This can be put in the form of (41) with the following assignments:

n	a_n	b_n
0	-1	0
1	$2n+1$	n
2	$-2n-1$	$n+1$
3	1	$2n+1$

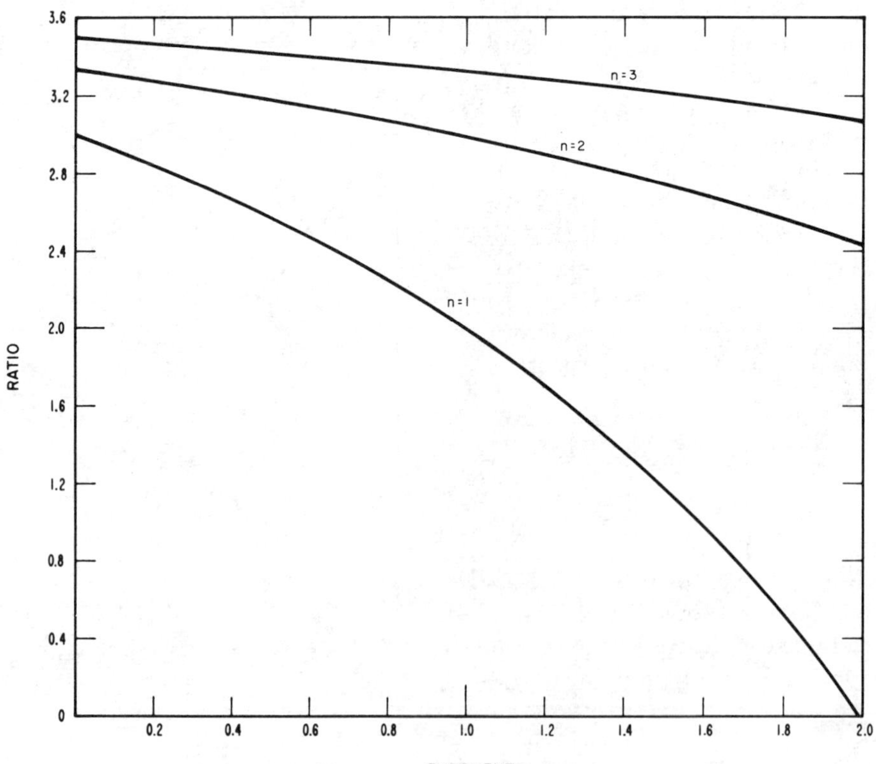

Fig. 6. Ratio of variances, $\langle(\Delta^{n+1}\epsilon m)^2\rangle, / \langle(\Delta^n \epsilon m)^2\rangle$, as a function of the exponent η.

It is easy to show that these coefficients satisfy (35)–(37).

Substitution of these coefficients into (47) yields

$$\langle(\delta t)^2\rangle = 2h\tau^2\{-2n^2(2n+1)\ln(n) + 2(n+1)^2(2n+1)\ln(n+1) - (2n+1)^2\ln(2n+1)\}, \quad (61)$$

where $h = k_2/8 \ln 2$.

Since τ is normally small compared to T, it is reasonable to approximate (61) for large n. That is, the relative time error is approximately given by

$$\frac{\sqrt{\langle(\delta t)^2\rangle}}{T} \approx \sqrt{2h\ln(n)}, \quad (62)$$

where the approximations

$$n \gg 1, \ln(n+1) \approx \ln(n) + \frac{1}{n}$$

have been used. It is apparent from (62) that for a given interval T, the errors accumulated by the clock are not *critically* dependent on the measuring time τ, since $\sqrt{\ln(n)}$ is a very slowly changing quantity.

The mean square error associated with the standard, however, can be written in the form

$$T^2\langle(\delta f)^2\rangle = \left(\frac{T}{\tau}\right)^2\langle(\Delta\gamma_n)^2\rangle, \quad (63)$$

since (59) may be written as the first difference of γ_n. From (57) it is apparent that the relative time error is then given by

$$\frac{T^2\langle(\delta f)^2\rangle}{T^2 f_s^2} = \frac{B_1}{(2\pi\tau f_s)^2}\tau^\eta, \quad (64)$$

where $\eta = 1$ ("white" noise). Or, combining (59), (62), and (64),

$$\frac{\sqrt{\langle(\delta T)^2\rangle}}{T} \approx \sqrt{2h\ln(n) + \frac{B_1}{(2\pi f_s)^2\tau}}. \quad (65)$$

As one should expect, for a given T, the errors get less for larger τ but not rapidly. In the limit of $\tau = T$ [not using the approximate (65)], the errors are those of the standard alone, i.e., $\sqrt{B_1\tau}/2\pi f_s$. Also, as $\tau \to 0$, the clock errors are unbounded.

Compounding Time Errors

Equation (65) represents a reasonable approximation to the time errors of a clock system after one calibration interval. The next question is: how do the errors of many calibrations compound to give a total error $\delta(NT)$ after N calibrations? Returning to (59), one may write

$$\delta(NT) = \sum_{n=1}^{N} \delta T_n, \quad (66)$$

where δT_n is the time error associated with the nth calibration interval. There have been papers published [13] that assume that the errors of one calibration interval are not correlated to the errors of any other calibration interval. It is now possible to investigate this assumption more precisely.

In particular, the clock errors (not including the frequency standard errors) δt_n compound to give $\delta(Nt)$, given by

$$\delta(Nt) = \sum_{n=1}^{N} \delta t_n, \quad (67)$$

which can obviously be put in the form of (41). For N equal to any number larger than one, the algebra becomes much too lengthy for actual calculation by hand and it is desirable to make use of a digital computer. Table III shows the results of this calculation for compounding several third-difference-type calibrations. It is interesting to note that, in fact, the total mean square error after N calibrations is very nearly equal to N-times the mean square error after one calibration. This is in quite good agreement with DePrins [13]. Thus, the rms relative eror may be approximated by the relation

$$\sqrt{\frac{\langle(\delta(NT))^2}{NT}} \approx \sqrt{\frac{2h\ln(n) + B_1\tau}{N}}, \quad (68)$$

where h, n, B_1, f_s, and τ have the same meanings as in (65).

TABLE III
THEORETICAL DETERMINATION OF TIME ERROR PROPAGATION

N	$\dfrac{\left\langle\left(\sum_{n=1}^{N}\delta t_n\right)^2\right\rangle}{\langle(\delta t_n^2)\rangle}$
2	2.054
3	3.109
4	4.165
5	5.220
6	6.276
7	7.332

The conclusions which can now be drawn are that: 1) the total rms time error of this clock system from an "ideal" atomic clock is unbounded as time increases, and 2) the relative rms error time to total elapsed time (68) approaches zero about as fast as $N^{-1/2}$. It should be mentioned here that systematic errors in the atomic standard have not been considered. While this is a very important problem, it has been treated rather thoroughly elsewhere [10], [11], [14]–[16].

IV. Measures of Frequency Stability

General Restrictions

It is of value to consider the problem of establishing a stability measure in a very general sense. Consider some functional of the phase

$$\chi = \chi(\phi(t)),$$

from which the stability measure Ψ is obtained according to the relation

$$\Psi^2 = \lim_{T\to\infty} \frac{1}{T}\int_{-T/2}^{T/2} |\chi|^2\, dt,$$

provided this limit exists.

In practice it is not possible to pass to the limit $T\to\infty$, and, thus, one measures for some fixed time, T; i.e.,

$$\Psi_T^2 = \frac{1}{T}\int_{-T/2}^{T/2} |\chi|^2\, dt.$$

Under favorable conditions, Ψ_T may be a reasonable approximation to Ψ. Unfortunately, this may not always be the case.

The frequency emitted by any physically realizable device must be bounded by some upper bound, say B. The following inequalities must, then, be valid:

$$|\Omega(t)| \le B, \quad \text{for some } B > 0;$$

$$\lim_{T\to\infty} \frac{1}{T}\int_{-T/2}^{T/2} (\Omega(t))^2\, dt \le B^2.$$

With $S_{\dot\phi}(\omega)$ being the power spectral density of $\Omega(t)$, it follows from the definition of power spectra that

$$\lim_{T\to\infty} \frac{1}{T}\int_{-T/2}^{T/2} (\Omega(t))^2\, dt = 2\int_0^\infty S_{\dot\phi}(\omega)\, d\omega$$

for real $\Omega(t)$, and thus

$$2\int_0^\infty S_{\dot\phi}(\omega)\, d\omega \le B^2. \tag{69}$$

If $S_\phi(\omega)$ has a flicker noise spectrum for small ω, it is apparent that this $1/|\omega|$ type of noise cannot persist to absolute zero frequency or the inequality (69) would be violated. It is, thus, reasonable to postulate the existence of a lower cutoff frequency ω_L for the flicker noise modulation.

It is apparent from the preceding considerations that stability measures may exist for which Ψ_T begins to approach Ψ only after T is several times larger than $1/\omega_L$. From some of the experiments on crystal oscillators [2], this may require T to exceed several years in duration. This is quite inconvenient from a manufacturing or experimental standpoint. The logical conclusion is to consider only those stability measures Ψ which are "cutoff independent," that is, those measures of stability which would be valid even in the limit $\omega_L \to 0^+$.

Finite Differences

It was shown in Section II that an expression of the form

$$\delta t = \frac{1}{\Omega_0}\sum_{n=0}^{m} a_n \phi(t + b_n\tau) \tag{70}$$

will have a finite variance if the $\{a_n, b_n\}$ satisfy conditions (35)–(37). It is easy to show that the first difference of the phase (i.e., frequency) cannot be put in the form of (70) with the coefficients satisfying conditions (35)–(37). Indeed, the limit

$$\lim_{\mu\to 1(-)} U(\tau) = \lim_{\mu\to 1(-)} \frac{k_2|\tau|^{2\mu}}{4 - 2^{2\mu}}$$

does not exist, and, hence, $U(\tau)$ is not a good measure of stability.

The variances of the second and third finite differences, however, are convergent. It is of interest to note that the first line of Table III may be expressed in the form

$$\frac{\langle |\Delta^3\phi_{n+3} + \Delta^3\phi_n|\rangle}{\langle(\Delta^3\phi_n)^2\rangle} = 2.054,$$

which may be simplified to the form

$$\frac{\langle(\Delta^3\phi_{n+3})(\Delta^3\phi_n)\rangle}{\langle(\Delta^3\phi_n)^2\rangle} = 0.027.$$

This equation expresses the fact that a third difference has a very small correlation (~3 percent) to an adjacent, nonoverlapping third difference. By extending this procedure with the other values given in Table III, it is found that the correlation of one third difference with a nonoverlapping third difference becomes small very rapidly as the interval between these differences becomes large. This is sufficient to insure that the variance of a finite sample of third differences will approach the "true" variance (infinite average) in a well-behaved and reasonable fashion as the sample gets larger.

Variance of Frequency Fluctuations for Finite Averaging Times

Even though the variance of the first difference of the phase does not satisfy the condition of being cutoff independent, it is possible (by specifying both the sample time and the total averaging time) to construct a cutoff independent measure of the frequency fluctuations. Instead of Ψ_T, the variance of N adjacent samples of the frequency will be denoted by $\sigma^2(\tau, N)$ where τ is the sample time for each of the N measurements of frequency. The variance is given by the conventional formula

$$\sigma^2(\tau, N) = \frac{1}{N-1}\sum_{i=1}^{N}\left[\frac{\Delta\phi_i}{\tau} - \frac{\phi_{N+1} - \phi_1}{N\tau}\right]^2.$$

If one neglects the drift rate α, which is essentially equivalent to obtaining the standard deviation around a linear drift, one obtains

$$\langle\sigma^2(\tau, N)\rangle = \frac{1}{(N-1)\tau^2}\left\{N\langle[\epsilon(t+\tau)-\epsilon(t)]^2\rangle - \frac{1}{N}\langle[\epsilon(t+N\tau)-\epsilon(t)]^2\rangle\right\}. \quad (71)$$

For the case of an "ideal" crystal oscillator, (19), (24), and (26) allow (71) to be simplified (after passing to the limit $\mu \to 1^{(-)}$ to the form

$$\langle\sigma^2(\tau, N)\rangle = \frac{2hN \ln (N)}{N-1}.$$

Thus, as N increases, the expected value of $\sigma^2(\tau, N)$ increases without bound (at least until $N\tau \sim 1/\omega_L$). It is interesting to note that for a given oscillator, $\langle\sigma^2(\tau, N)\rangle$ has a minimum value for $N=2$. Obviously one would have to average several experimental determinations of $\sigma^2(\tau, 2)$ in order to have a reasonable approximation to $\langle\sigma^2(\tau, 2)\rangle$.

While $\sigma(\tau, N)$ is a cutoff independent measure of frequency stability, it has the significant disadvantage of being a function of two variables. Indeed, in order to compare the stability of two oscillators, both the τ's and the N's should have nearly corresponding values.

Delayed Frequency Comparison

In radar work, often the frequency of a signal is compared to the frequency of the same source after it has been delayed in traversing some distance—often a very great distance. One might, thus, be interested in defining a stability measure in an analogous fashion:

$$\Psi^2(\tau, T) = \left\langle\left[\frac{\phi(t+T+\tau)-\phi(t+T)}{\tau} - \frac{\phi(t+\tau)-\phi(t)}{\tau}\right]^2\right\rangle. \quad (72)$$

Again neglecting the drift rate α of the oscillator, (24) and (72) combine to yield

$$\Psi^2(\tau, T) = \frac{1}{\tau^2}[2U(\tau)-U(T-\tau)+2U(T)-U(T+\tau)]. \quad (73)$$

After substitution of (26) into (73) the equation can be rearranged to give (again passing to the limit, $\mu \to 1^{(-)}$)

$$\Psi^2(\tau, T) = -2h[\rho^2 \ln \rho - (1+\rho)^2 \ln (1+\rho) - (\rho-1)^2 \ln (\rho-1)],$$

where $\rho \equiv T/\tau$. Although this is a rather complicated expression, it may be simplified with the approximation $\rho \equiv T/\tau \gg 1$. The result is

$$\Psi^2(\tau, T) = 4h\left(2 + \ln\frac{T}{\tau}\right), \quad \text{for } \frac{T}{\tau} \gg 1 \quad (74)$$

for an "ideal" oscillator. It is interesting to note here that even when considering only $1/|\omega|$ type of noise, one cannot pass to the limit $\tau=0$ for this problem. In the limit as $\tau \to 0^+$, the expression

$$\lim_{\tau\to 0^+}\left[\frac{\phi(t+\tau)-\phi(t)}{\tau}\right] \equiv \dot{\phi}(t) \equiv \Omega(t)$$

and thus

$$\lim_{\tau\to 0^+} \Psi^2(\tau, T) = \langle[\Omega(t+T)-\Omega(t)]^2\rangle \to \infty$$

from (74), even though $T \ll 1/\omega_L$. The source of this difficulty is the high-frequency divergence of the flicker noise spectrum. If the system is limited at the high frequency by ω_H, then one should pass to the limit $\tau \to 1/\omega_H$.

Again, $\psi(\tau, T)$ is a function of two variables with all of the associated annoyances. It may, however, be useful in certain applications.

Conclusion

The assumptions of stationarity and "ideal" behavior for a quartz crystal oscillator lead to a statistical model which agrees well with many different experiments. One finds, however, that certain quantities are unbounded as averaging times are extended and it is important to consider only those quantities which have reasonable hope of converging toward a good value in reasonable time. Thus, the concepts of cutoff dependent and cutoff independent measures of frequency stability form a natural classification for all possible frequency stability measures.

On the basis of "ideal" behavior, it has been shown that the errors of a clock, run from a quartz crystal oscillator and periodically referenced to an atomic frequency standard, accumulate error at a probable rate proportional to the square root of the number of calibrations. That is, the errors of one calibration interval are essentially uncorrelated to errors of nonoverlapping intervals in spite of the fact that "ideal" behavior is highly correlated for long periods of time.

It has also been shown that the method of finite differences can be a useful method of determining spectral distributions of noise, as well as being a possible measure of frequency stability. By using higher order finite differences, phase fluctuations with even a higher order pole at zero-modulation frequency can similarly be treated. The need for higher than second or third differences, however, has not yet been demonstrated.

It should be noted that the existence of higher-frequency modulation noise of different origin also has significant affect on stability measures. In general, the factors which limit the system to a finite bandpass are sufficient to insure convergence of the stability measures as $\omega \to \infty$. If it is primarily the measuring system which limits the system bandpass, however, the results may be significantly altered by the measuring system itself.

Appendix

Ratio of Variances

Let $\epsilon(t)$ be a real generalized function such that $\langle\epsilon(t)\rangle = 0$ and define the discrete variable ϵ_m by the relation

$$\epsilon_m \equiv \epsilon(t + m\tau). \tag{75}$$

Also, let the auto-covariance function of $\epsilon(t)$ be, as before, independent of a simple time transition. One may now write (see Table I)

$$\langle(\Delta\epsilon_m)^2\rangle = 2[\langle(\epsilon_n)^2\rangle - \langle[\epsilon(t + \tau)\epsilon(t)]\rangle] \tag{76}$$

and assume that

$$\langle(\Delta\epsilon_m)^2\rangle = k_1\tau^\eta, \tag{77}$$

where k_1 is a constant for a given η. It is also possible to obtain the variance of the second difference:

$$\langle(\Delta^2\epsilon_m)^2\rangle = 6\langle(\epsilon_m)^2\rangle - 8\langle[\epsilon(t + \tau)\cdot\epsilon(t)]\rangle + \langle[\epsilon(t + 2\tau)\epsilon(t)]\rangle. \tag{78}$$

Using (77) and (78), one may obtain

$$\frac{\langle(\Delta^2\epsilon_m)^2\rangle}{\langle(\Delta\epsilon_m)^2\rangle} = \frac{4k_1\tau^\eta - k_1(2\tau)^\eta}{k_1\tau^\eta} = 4-(2)^\eta. \tag{79}$$

Since (79) must be non-negative (ϵ is real), the exponent is restricted to the range $\eta \leq 2$.

Similarly, one may obtain the variance of the third difference

$$\langle(\Delta^3\epsilon_m)^2\rangle = 15k_1\tau^\eta - 6k_1(2\tau)^\eta + k_1(3\tau)^\eta, \tag{80}$$

and hence the ratio

$$\frac{\langle(\Delta^3\epsilon_m)^2\rangle}{\langle(\Delta^2\epsilon_m)^2\rangle} = \frac{15 - 6\cdot(2)^\eta + 3^\eta}{4 - (2)^\eta}. \tag{81}$$

Similarly,

$$\frac{\langle(\Delta^4\epsilon_m)^2\rangle}{\langle(\Delta^3\epsilon_m)^2\rangle} = \frac{56 - 28\cdot(2)^\eta + 8\cdot(3)^\eta - (4)^\eta}{15 - 6\cdot(2)^\eta + (3)^\eta}. \tag{82}$$

Equations (79), (81), and (82) are plotted in Fig. 6 as a function of the exponent η.

For $\eta = 4/3$, as in Fig. 6, the theoretical ratios,

$$\frac{\langle(\Delta^{n+1}\epsilon_m)^2\rangle}{\langle(\Delta^n\epsilon_m)^2\rangle}$$

for $n = 1$ and 2 are 1.48 and 2.84, respectively. The straight lines drawn in Fig. 5 were made to have these ratios and slope 2/3 (the square root of the variances).

Acknowledgment

The author is sincerely indebted to Drs. P. Wacker and E. Crow of the National Bureau of Standards for some very valuable suggestions and references. The author also wishes to acknowledge the assistance of D. Allan for his aid in data acquisition and analysis, R. E. Beehler for making the cesium beams of the NBS available, and R. L. Fey for some helpful criticism.

References

[1] H. Lyons, "The atomic clock," *Am. Scholar*, vol. 19, pp. 159–168, Spring 1950.
[2] W. R. Arkinson, L. Fey, and J. Newman, "Spectrum analysis of extremely low frequency variations of quartz oscillators," *Proc. IEEE (Correspondence)*, vol. 51, p. 379, February 1963.
[3] L. Fey, W. R. Atkinson, and J. Newman, "Obscurities of oscillator noise," *Proc. IEEE (Correspondence)*, vol. 52, pp. 104–105, January 1964.
[4] V. Troitsky, "Certain problems of the theory of fluctuations in oscillators. The influence of flicker noise," *Izv. Vysshikh Uchebn. Zavedenii, Radiofiz.*, vol. 1, pp. 20–33, 1958.
[5] L. Cutler and C. Searle, "Some aspects of the theory and measurement of frequency fluctuations in frequency standards," this issue, page 136.
[6] J. A. Barnes and R. C. Mockler, "The power spectrum and its importance in precise frequency measurements," *IRE Trans. on Instrumentation*, vol. I–9, pp. 149–155, September 1960.
[7] N. Wiener, *Extrapolation, Interpolation, and Smoothing of Stationary Time Series*. Cambridge, Mass.: Technology Press of Mass. Inst. of Tech., and New York: Wiley, January 1957.
[8] M. Lighthill, *Introduction to Fourier Analysis and Generalized Functions*. New York: Cambridge, 1962.
[9] L. R. Malling, "Phase-stable oscillators for space communications, including the relationship between the phase noise, the spectrum, the short-term stability, and the Q of the oscillator," *Proc. IRE*, vol. 50, pp. 1656–1664, July 1962 (see also [3]).
[10] P. Kartaschoff, "Influence de l'effet de grenaille sur la frequence d'un oscillateur asservi a un etal on a jet atomique," Laboratoire Suisse de Recherches Horlogeres, Neuchatel, Switzerland, No. 08-64-01.
[11] P. Kartaschoff, "Etude d'un etalon de frequence a jet atomique de cesium," Laboratoire Suisse de Recherchess Horlogeres, Neuchatel, Switzerland, November 1963.
[12] S. Basri, "Time standards and statistics," to be published.
[13] J. DePrins, "Applications des masers A $N^{15}H_3$ a la mesure et a la definition du temps," unpublished Doctor of Sciences dissertation, Université Libre de Bruxelles, Belgium, 1961.
[14] R. E. Beehler, W. R. Atkinson, L. E. Heim, and C. S. Snider, "A comparison of direct and servo methods for utilizing cesium beam resonators as frequency standards," *IRE Trans. on Instrumentation*, vol. I–11, pp. 231–238, December 1962.
[15] R. E. Beehler and D. J. Glaze, "The performance and capability of cesium beam frequency standards at the National Bureau of Standards," to be published.
[16] R. E. Beehler and D. J. Glaze, "Evaluation of a thallium atomic beam frequency standard at the National Bureau of Standards," to be published.

Statistics of Atomic Frequency Standards

DAVID W. ALLAN

Abstract—A theoretical development is presented which results in a relationship between the expectation value of the standard deviation of the frequency fluctuations for any finite number of data samples and the infinite time average value of the standard deviation, which provides an invariant measure of an important quality factor of a frequency standard. A practical and straightforward method of determining the power spectral density of the frequency fluctuations from the variance of the frequency fluctuations, the sampling time, the number of samples taken, and the dependence on system bandwidth is also developed. Additional insight is also given into some of the problems that arise from the presence of "flicker noise" (spectrum proportional to $|\omega|^{-1}$) modulation of the frequency of an oscillator.

The theory is applied in classifying the types of noise on the signals of frequency standards made available at NBS, Boulder Laboratories, such as: masers (both H and $N^{15}H_3$), the cesium beam frequency standard employed as the U. S. Frequency Standard, and rubidium gas cells.

"Flicker noise" frequency modulation was not observed on the signals of masers for sampling times ranging from 0.1 second to 4 hours. In a comparison between the NBS hydrogen maser and the NBS III cesium beam, uncorrelated random noise was observed on the frequency fluctuations for sampling times extending to 4 hours; the fractional standard deviations of the frequency fluctuations were as low as 5 parts in 10^{14}.

I. Introduction

As ATOMIC TIMEKEEPING has come of age, it has become increasingly important to identify quality in an atomic frequency standard. Some of the most important quality factors are directly related to the inherent noise of a quantum device and its associated electronics. For example, a proper measurement and statistical classification [1] of this inherent noise makes it possible to determine the probable rate of time divergence of two independent atomic time systems, as well as giving insight concerning the precision and accuracy obtainable from an atomic frequency standard.

In the realm of precise frequency measurements, the properties of noise again play an important role. The relative precision obtainable with atomic frequency standards is unsurpassed in any field, and the precision limitations in this field are largely due to inherent noise in the atomic device and the associated electronic equipment. The standard deviation of the frequency fluctuations can be shown to be directly dependent on the type of noise in the system, the number of samples taken, and the dead-time between samples.

A very common and convenient way of making measurements of the noise components on a signal from a frequency standard is to compare two such standards by measuring the period of the beat frequency between the two standards. It is again the intent of the author to show a practical and easy way of classifying the statistics, i.e., of determining the power spectral density of the frequency fluctuations using this type of measuring system.

An analysis has already been made of the noise present in passive atomic frequency standards [1], such as cesium beams, but a classification of the types of noise exhibited by the maser type of quantum-mechanical oscillator has not been made in the long term area, i.e., for low frequency fluctuations. Though this paper is far from exhaustive, the intent is to give additional information on the noise characteristics of masers. Because a maser's output frequency is more critically parameter dependent than a passive atomic device, it has been suggested [2] that the output frequency might appear to be "flicker noise" modulated, where "flicker noise" is defined as a type of power spectral density which is inversely proportional to the spectral frequency $\omega/2\pi$. It has been shown that if "flicker noise" frequency modulation is present on a signal from a standard, some significant problems arise, such as the logarithmic divergence of the standard deviation of the frequency fluctuations as the number of samples taken increases, and also the inability to define precisely the time average frequency. It thus becomes of special interest to determine whether "flicker noise" is or is not present on the signal from a maser so that one might better evaluate its quality as a frequency standard.

Throughout the paper, the paramount mathematical concern is the functional form of the equations with the hope of maintaining simplicity and of providing better understanding of the material to be covered.

II. Methods Employed to Measure Noise

A. Power Spectrum and Variance Relationship

The average angular frequency $\Omega_\tau(t)$ of an oscillator (to distinguish it from spectral frequency ω) over a time interval τ can be written

$$\Omega_\tau(t) = \frac{1}{\tau}[\phi(t+\tau) - \phi(t)], \tag{1}$$

where ϕ is the phase angle in radians. Now the variance of the frequency deviations is the square of the standard deviation σ. Define the time average of a function as

$$\langle f(t)\rangle = \lim_{T\to\infty}\frac{1}{T}\int_{-T/2}^{T/2} f(t)\,dt. \tag{2}$$

Manuscript received September 17, 1965; revised December 7, 1965.

The author is with the National Bureau of Standards, Boulder, Colo.

Reprinted from *Proc. IEEE*, vol. 54, pp. 221–230, Feb. 1966.

One may, therefore, write the square of the standard deviation as follows:

$$\sigma^2 = \langle \Omega_\tau(t)^2 \rangle - \langle \Omega_\tau(t) \rangle^2. \tag{3}$$

One may assume with no loss of generality that the second term in (3) can be set equal to zero by a proper translation, since it is the square of the time average frequency. Therefore, $\Omega_\tau(t)$ is now the frequency deviation from the average value, and $\phi(t)$ the integrally related phase deviation.

Substituting (1) into (3) gives

$$\sigma^2 = \frac{1}{\tau^2} \left[\langle \phi(t+\tau)^2 \rangle - 2\langle \phi(t+\tau) \cdot \phi(t) \rangle + \langle \phi^2(t) \rangle \right]. \tag{4}$$

The time average of $\phi(t+\tau) \cdot \phi(t)$ is the autocovariance function of the phase—denoted $R_\phi(\tau)$. One is justified in assuming that a time translation has no effect on the autocovariance function [1], therefore [4],

$$\sigma^2 = \frac{2}{\tau^2} \left[R_\phi(0) - R_\phi(\tau) \right]. \tag{5}$$

It is now possible to relate the variance of the squared frequency deviations to the power spectral density by use of the Wiener-Khinchin Theorem, which states that the autocovariance function of the phase is equal to the Fourier transform (F.T.) of the power spectral density of the phase $S_\phi(\omega)$. The power spectral density of the frequency is related to this by the useful equation

$$S_{\dot\phi}(\omega) = \omega^2 S_\phi(\omega).$$

Most of the discussion that follows is based on the restriction

$$S_\phi(\omega) = h \left| \omega \right|^\alpha. \tag{6}$$

That a singular type of power spectrum predominates over a reasonable range of ω has been verified experimentally. The region of interest for α is $-3 \leq \alpha \leq -1$, and $\alpha = 0$. This covers white noise phase modulation ($S_\phi(\omega) = h$), "flicker noise" frequency modulation ($S_{\dot\phi}(\omega) = h\left|\omega\right|^{-1}$), and includes, of course, white noise frequency modulation ($S_{\dot\phi}(\omega) = h$). Fortunately, the Fourier transforms of functions of the above form [3] have been tabulated, and the following transforms can be established:

$$\text{F.T.} \left| \omega \right|^\alpha = a'(\alpha) \cdot \left| \tau \right|^{-\alpha-1} \quad \text{for } \alpha \neq 0 \text{ or not an integer}$$

$$\text{F.T.} \left| \omega \right|^{-0} = \delta(\tau)$$

$$\text{F.T.} \left| \omega \right|^{-2} = a'(-2) \cdot \left| \tau \right| \tag{7}$$

where a' is an α dependent coefficient. A useful substitution is the following [1]:

$$U(\tau) = 2\left[R_\phi(0) - R_\phi(\tau) \right]. \tag{8}$$

Because of finite system bandwidths, ω_B, a better representation of $U(\tau)$ is obtained by replacing $R_\phi(0)$ with $R_\phi(1/\omega_B)$ (see Appendix). If the sample time τ is large compared to the reciprocal system bandwidth $1/\omega_B$, then $R_\phi(1/\omega_B)$ is negligible compared to $R_\phi(\tau)$ in the region where $-2 < \mu < 0 (-1 > \alpha > -3)$. $R_\phi(1/\omega_B)$ becomes the larger term, however, in the region of $-1 < \alpha < 0$, and if one assumes $(\omega_B \tau)^{\alpha+1} \gg 1$, then $R_\phi(\tau)$ is neglectable (see Appendix). The following equations for $U(\tau)$ may, therefore, be written:

$$U(\tau) = \begin{cases} \left| a(\alpha) \right| \left| \tau \right|^{-\alpha-1}; & -3 < \alpha < -1 \\ a(\alpha) \left| \omega_B \right|^{\alpha+1}; & -1 < \alpha \leq 0. \end{cases} \tag{9}$$

The standard deviation squared may, therefore, be written:

$$\sigma^2 = \begin{cases} a(\mu) \cdot \left| \tau \right|^\mu; & -3 < \alpha < -1 \\ a(\alpha) \left| \omega_B \right|^{\alpha+1} \left| \tau \right|^{-2}; & -1 < \alpha \leq 0 \end{cases} \tag{10}$$

where $\mu = -\alpha - 3$.

$A(\mu)$ has a small dependence on ω_B, implicit within the previous assumptions. Considering the results of (10), the Appendix, and that to be discussed in Section II-D on "flicker noise" modulation ($\alpha = -1, -3$), an informative graph of μ into α may be established as illustrated below.

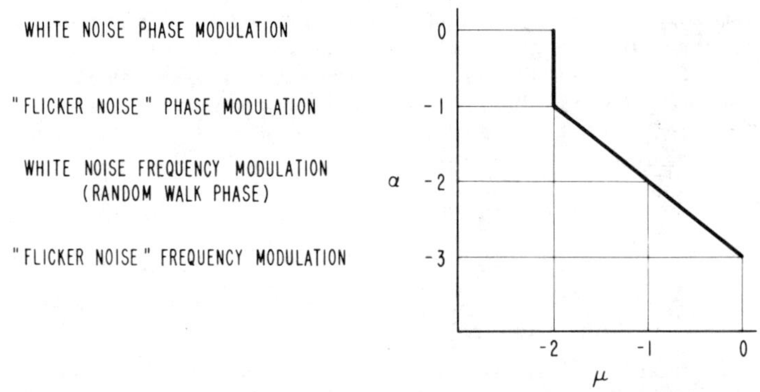

Using (6) and (10) along with the above graph, one may, therefore, deduce the power spectral density from the dependence of the standard deviation of the frequency fluctuations on the sampling time, with restrictions on the experimental parameters as will be shown in the following.

B. Adjacent Sampling of Data

In actual practice, of course, the number of frequency or phase samples must be finite. The case to be considered now is one for which the phase or the frequency is monitored on a continuous basis. Two of the techniques used by the author to accomplish this were as follows: A device, described elsewhere [1], was used to monitor the phase of the beat frequency between two oscillators at prescribed time intervals; the second technique was to measure the period of the beat frequency between two oscillators with two counters so that the dead-time of one counter corresponded to the counting time of the other.

A very powerful and meaningful method of analysis of data taken by a phase monitoring technique has been developed by J. A. Barnes [1] at the National Bureau of Standards, Boulder, Colo. The method employs the use of finite differences and is especialy useful in analyzing "flicker noise" and long-term frequency fluctuations in general. On the other hand, if one uses the period counting technique for data acquisition, the following form of analysis is useful. It also may be cast in a form where one may use the finite difference technique.

Data are often obtained with one counter measuring the frequency or the period of the beat note between two oscillators with a dead-time between counts.[1] The concern of Section II is with the dead-time being zero, but it is convenient to develop the general case for which the dead-time is nonzero for use in Section III, and specialize this to the continuous sampling case for which the dead-time is zero, which is the case of interest in this section.

Let T be the period of sampling, τ the sample time, and N the number of samples. The standard deviation, $\sigma(N, T, \tau)$, of the frequency fluctuations[2] may, therefore, be written as:

$$\sigma^2(N, T, \tau) = \frac{1}{N-1}\left\{\sum_{n=0}^{N-1}\left[\frac{\phi(nT+\tau)-\phi(nT)}{\tau}\right]^2 - \frac{1}{N}\left[\sum_{n=0}^{N-1}\frac{\phi(nT+\tau)-\phi(nT)}{\tau}\right]^2\right\}. \quad (11)$$

Taking the expectation value of $\sigma^2(N, T, \tau)$ and making the substitution given in (8) yields

$$\langle\sigma^2(N, T, \tau)\rangle = \frac{1}{\tau^2}\left\{U(\tau) + \frac{1}{N(N-1)} \sum_{n=1}^{N-1}(N-n)[2U(nT)-U(nT+\tau)-U(nT-\tau)]\right\}. \quad (12)$$

If the dead-time were zero, then $T=\tau$, and (12) becomes [1]

$$\langle\sigma^2(N, \tau)\rangle = \frac{1}{(N-1)\tau^2}\left[NU(\tau) - \frac{1}{N}U(N\tau)\right]. \quad (13)$$

Remembering that $\mu=-\alpha-3$ and substituting (9) into (13) gives

$$\langle\sigma^2(N, \tau)\rangle = \frac{a(\mu)N\,|\,\tau\,|^\mu}{N-1}[1-N^\mu]; \qquad -2<\mu<0 \quad (14)$$

which establishes the interesting result of the dependence of the expectation value of the standard deviation of the frequency fluctuations on the number of samples, the sample time, and the power spectral density. It will be noted that

$$\langle\sigma^2(\infty, \tau)\rangle = a(\mu)\,|\,\tau\,|^\mu; \qquad -2<\mu<0 \quad (15)$$

in agreement with (10).

[1] See L. S. Cutler and C. L. Searle, "Some aspects of the theory and measurement of frequency fluctuations in frequency standards," this issue, page 136. This paper shows that the fluctuations $\Delta\tau$ in the period τ are a good approximation to the frequency fluctuations if $\Delta\tau \gg \tau$.

[2] Note that this $\sigma(N, T, \tau)$ is not the same as the σ in (3)–(5), and (10). $\sigma(N, T, \tau)$ is over a finite number of data samples N and, to avoid confusion, the variable N will always be used with σ in the finite sampling case as in (11).

By keeping N constant and assuming μ to be constant over several different values of τ, it may be seen that the value of μ is the slope on a log-log plot of $\langle\sigma^2(N, \tau)\rangle$ vs. τ. This provides a means of determining the power spectral density simply by varying the sample time over the region of interest [4].

It is informative to look at the family of curves obtained from a plot of the dependence of $\langle\sigma^2(N, \tau)\rangle$ as a function of N for various pertinent values of μ to see how it approaches $\langle\sigma^2(\infty, \tau)\rangle$. The family of curves is shown in Fig. 1. One may notice that the convergence is much faster in the region between white noise frequency modulation and white noise phase modulation than between white noise frequency modulation and "flicker noise" frequency modulation.[3] In fact, as $\mu \to 0$, the ratio approaches zero, and one would conjecture that the

$$\lim_{N\to\infty}\langle\sigma^2(N, \tau)\rangle$$

is infinite in the presence of "flicker noise" frequency modulation—a result proven by J. A. Barnes et al. [1].

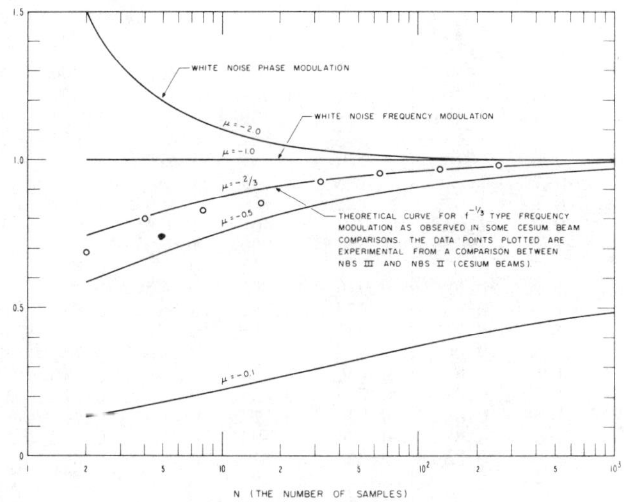

Fig. 1. A plot showing the dependence of the standard deviation of the frequency fluctuation on the number of samples and the type of noise present.

The data points plotted in Fig. 1 were extracted from a cesium beam-cesium beam comparison analyzed elsewhere [1], and exhibit in this new formulation a type of frequency modulation proportional to $|\omega|^{-1/3}$, giving confirmation to this strange type of power spectral density.

It is possible to utilize the dependence of the standard deviation on the number of samples to determine a value of μ by considering a function which takes into account the extreme values of N obtainable from a finite set of data, namely,[4]

[3] To see that $\mu=0$ corresponds to "flicker noise" see Section II-D.

[4] It will be noted that the τ dependence cancels in the expression for χ and hence it is N and μ dependent only. In the table and graphs, the τ dependence is not shown and is, therefore, suppressed since the μ dependence is the thing emphasized.

$$\chi(N, \mu) = \frac{\langle \sigma^2(N, \tau) \rangle}{\langle \sigma^2(2, \tau) \rangle}. \tag{16}$$

Tabulated values of this function are given in Table I, and a plot of $\chi(N, \mu)$ as a function of μ for various values of N is given in Fig. 2. It may be noted that the function is most sensitive in the region between "flicker noise" frequency modulation $\mu = 0$ and white noise frequency modulation $\mu = -1$—one of basic interest. In practice, the table and graph have proven very useful. $\chi(\infty, \mu)$ is plotted for comparison and computational purposes.

Note that $\chi(\infty, 0) = \infty$. In the development thus far, no consideration has been given to the experimental fact that there must exist a lower cutoff frequency that keeps the functions considered from going to infinity, corresponding to certain types of noise, such as "flicker noise." The value of the cutoff frequency is not important other than to say that the functions considered are valid for times up to the order of $1/(\omega \text{ cutoff})$. This time is apparently more than a year for quartz crystal oscillators [5]. If "flicker noise" is present in some atomic frequency standards, the value of $1/(\omega \text{ cutoff})$ is probably less than for quartz crystal oscillators for reasons discussed later, and if "flicker noise" is not present, infinities do not occur in the functions considered for most other types of pertinent noise and hence there is no concern.

TABLE I

THE DEPENDENCE OF $\langle \sigma(N, \tau) \rangle$ ON THE NUMBER OF SAMPLES FOR A FIXED SAMPLE TIME τ, THUS DETERMINING THE VALUE OF μ AND THE STATISTICS. THE VALUES LISTED ARE OF

$$\chi(N, \mu) = \frac{\langle \sigma^2(N, \tau) \rangle}{\langle \sigma^2(2, \tau) \rangle}.$$

μ	N (Number of Samples)					
	4	16	64	256	1024	∞
0.0	1.337	2.133	3.048	4.016	5.004	∞
−0.1	1.288	1.928	2.580	3.190	3.736	7.464
−0.2	1.247	1.753	2.215	2.598	2.899	3.855
−0.3	1.208	1.604	1.928	2.167	2.332	2.660
−0.4	1.171	1.475	1.700	1.847	1.937	2.062
−0.5	1.138	1.365	1.517	1.606	1.655	1.705
−0.6	1.106	1.270	1.369	1.422	1.447	1.467
−0.7	1.077	1.188	1.249	1.278	1.291	1.299
−0.8	1.049	1.116	1.150	1.165	1.171	1.174
−0.9	1.023	1.054	1.068	1.074	1.076	1.076
−1.0	1.000	1.000	1.000	1.000	1.000	1.000
−1.1	0.977	0.952	0.942	0.938	0.937	0.937
−1.2	0.956	0.910	0.893	0.887	0.886	0.886
−1.3	0.937	0.873	0.851	0.844	0.842	0.842
−1.4	0.919	0.841	0.815	0.807	0.805	0.804
−1.5	0.902	0.812	0.784	0.776	0.774	0.773
−1.6	0.886	0.786	0.756	0.748	0.746	0.745
−1.7	0.871	0.763	0.733	0.725	0.723	0.722
−1.8	0.858	0.743	0.712	0.704	0.702	0.701
−1.9	0.845	0.724	0.693	0.685	0.683	0.682
−2.0	0.833	0.708	0.677	0.669	0.667	0.667

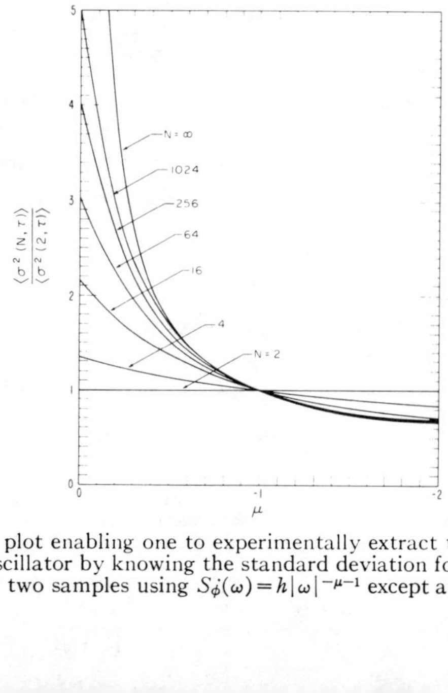

Fig. 2. A plot enabling one to experimentally extract the statistics of an oscillator by knowing the standard deviation for N samples and for two samples using $S_{\dot\phi}(\omega) = h|\omega|^{-\mu-1}$ except at $\mu = -2$.

C. Non-Adjacent Sampling of Data

The next consideration is to determine the effect of counter dead-time on one's ability to deduce the statistics of an oscillator using the techniques developed in Section II-B. This form of data acquisition is one of the most common, and hence merits attention.

It was shown earlier that (12) is applicable to the present case, and if (9) is substituted into (12), with the assignment that $r = T/\tau$ (the ratio of the period of sampling to the sample time), then

$$\langle \sigma^2(N, T, \tau) \rangle = a(\mu) \,|\, \tau \,|^{\mu} \left\{ 1 + \sum_{n=1}^{N-1} \frac{(N-n)(nr)^{\mu+2}}{N(N-1)} \left[2 - \left(1 + \frac{1}{nr}\right)^{\mu+2} - \left(1 - \frac{1}{nr}\right)^{\mu+2} \right] \right\}; \qquad -2 < \mu < 0. \tag{17}$$

An important result from (17) is that if N and r are held constant, it is still possible to determine the value of μ by varying τ. Therefore, the relationship between the power spectral density and the standard deviation has the same form as for the continuous sampling case. Additional insight may be obtained by considering some special cases. If $\mu = -1$, $(S_{\dot\phi}(\omega) = h)$ [6], the series in (17) goes to zero for all possible values of r and hence

$$\langle \sigma^2(N, T, \tau) \rangle = \frac{a(-1)}{|\tau|} \tag{18}$$

for white noise frequency modulation; this is the same result obtained in the continuous data sampling case. One notices that (18) is independent of N as would be espected since the frequency fluctuations are uncorrelated.

If $\mu = -2$, the series in (17) again goes to zero for all values of $r > 1$. The value of α is degenerate except that one may say $-1 < \alpha$. In this domain, the frequency fluctuations appear to be uncorrelated as long as the measurement dead-time is nonzero. Using the proper form of $U(\tau)$ from (9) and substituting into (13) for the case where $r = 1$ (zero dead-time), gives:

$$\langle \sigma^2(N, \tau) \rangle = \left[a(\alpha) \,|\, \omega_B \,|^{\alpha+1} (N+1)/N\tau^2 \right].$$

One, therefore, has the unusual result that

$$\langle\sigma^2(N, T, \tau)\rangle = \frac{N + \delta(r-1)}{N} \cdot \frac{a(\alpha)\,|\,\omega_B\,|^{\alpha+1}}{\tau^2} \tag{19}$$

where

$$\delta(r-1) = \begin{cases} 1, & r = 1 \\ 0, & r \neq 1, \end{cases} \quad \text{for } -1 < \alpha \text{ and } (\omega_B \tau)^{\alpha+1} \gg 1.$$

A slight N dependence then appears only in the continuous sampling case. Experimentally, one can show that the Kronecker δ-function is replaced by $R_\phi(r-1)/R_\phi(0)$ because of the finite bandwidths involved.

It will be recalled that the curves in Figs. 1 and 2 are for $r=1$ (the dead-time equal zero). The results of (19) show a character change in the curves for $r>1$ and $-2<\mu<-1$, for now the curve for $\mu=-2$ is coincident with the curve for $\mu=-1$ in Fig. 1 and $\chi(N, -2) = \chi(N, -1) = 1.0$ in Fig. 2. No profound character change occurs for $-1<\mu<0$.

It is possible to show that the series in (17) approaches zero as N approaches infinity for all values of $r>1$ and $\mu<0$, hence $\langle\sigma^2(N, T, \tau)\rangle$ has the same asymptotic value as for the continuous sampling technique, independent of the counter dead-time.

If a binomial expansion is made of the second two terms in the series expression of (17), and fourth-order terms and higher in $1/nr$ are neglected, the following simplification occurs:

$$\langle\sigma^2(N, T, \tau)\rangle = a(\mu)\,|\,\tau\,|^\mu\left[1 - \frac{(\mu+2)(\mu+1)}{N(N-1)} r^\mu \sum_{n=1}^{N} (N-n)n^\mu\right]. \tag{20}$$

On the first observation of (20), one may notice that as r becomes large, the standard deviation approaches its asymptotic value. This occurs when one is taking samples much shorter than the capable reset time of the counter. The dependence on N is, therefore, reduced as r^μ. In fact, it has been determined by a computer analysis of (17) that the net effect of increasing the dead-time is to collapse the curves in Figs. 1 and 2 towards the unit axis.

D. The "Flicker Noise" Problem

The existence of "flicker noise" frequency modulation on the signal of quartz crystal oscillators has caused difficulty in handling such quantities as the autocovariance function of the phase and the standard deviation of the frequency fluctuations. As mentioned previously, the development by J. A. Barnes [1] makes it possible to classify "flicker noise" frequency modulation without any divergence difficulties or dependence on the value of the low-frequency cutoff.

Some of the other difficulties associated with the presence of "flicker noise" frequency modulation, as Barnes has shown, are illustrated in the following equations:

$$\langle\sigma^2(N, \tau)\rangle = \frac{2hN \ln N}{N-1}, \tag{21}$$

and

$$\lim_{\tau\to 0^+} \left\langle\left[\frac{\phi(t+\tau) - \phi(t)}{\tau}\right]\right\rangle = \langle\Omega(t)\rangle. \tag{22}$$

The fact that the standard deviation diverges with N, as shown in (21), is an annoyance, aside from the fact that it becomes more difficult to write specifications on the frequency fluctuations of an oscillator. The limit expressed in (22) does not exist or is dependent on the low-frequency cutoff; this, of course, affords difficulty in defining a frequency, and would be an unfortunate property to be present in a frequency standard. As has been stated previously for all values of μ considered thus far, $-2<\mu<0$, the limit as N approaches infinity of $\langle\sigma^2(N, \tau)\rangle$ exists; and as the sampling time increases, $\langle\sigma^2(N, \tau)\rangle$ converges toward zero or perfect precision.

Data have been analyzed that indicate the presence of "flicker noise" frequency modulation on the signals from rubidium gas cell frequency standards. It is of concern to determine if this type of noise is present on the signals of other atomic frequency standards, and specifically masers. It is not the intent to determine the source or sources of "flicker noise," as this is a ponderous problem in and of itself, but perhaps only infer where such noise might arise.

Consider, now, ways to establish the presence of "flicker noise." If the data were on a continuous basis, the method of finite differences [1] developed by Barnes would be very useful. One may also notice from (21) that if N is held constant, the value of μ is zero for values of τ in the "flicker noise" frequency modulation region. If data were taken by the common technique of non-adjacent samples, the following considerations are of value.

For "flicker noise" frequency modulation, $U(\tau)$ takes on a different form as a result of the divergence of the autocovariance function of the phase [1],

$$U(\tau) = \lim_{\mu\to 0^-} \frac{k\,|\,\tau\,|^{\mu+2}}{4 - 2^{\mu+2}}. \tag{23}$$

The constant k is dependent on the quality of the oscillator. If (23) is substituted into (12), letting $\mu\to 0$ ($\mu=0$ corresponds to "flicker noise"), an indeterminate form results for $\langle\sigma^2(N, T, \tau)\rangle$. Applying L'Hospital's Rule, and then passing to the limit, gives the following equation:

$$\langle\sigma^2(N, T, \tau)\rangle = \frac{k}{N(N-1)} \sum_{n=1}^{N-1} (N-n)\Big[-2(nr)^2 \ln(nr) + (nr+1)^2 \ln(nr+1) + (nr-1)^2 \ln(nr-1)\Big]. \tag{24}$$

Hence, for a fixed ratio $r=T/\tau$, and a determined number of samples N, the standard deviation is constant, independent of τ. So for both adjacent and nonadjacent data sampling, the value of μ is zero. One may, therefore, determine if "flicker noise" frequency modulation is present on a signal sampled in a nonadjacent fashion, subject to the above constraints.

A consideration of interest at this point is "flicker noise" phase modulation ($S_\phi(\omega) = h|\omega|^{-1}$). One can show from the work of J. A. Barnes[5] that the value of the function $U(\tau)$ for $\alpha = -1$ is

$$U(\tau) = 4h(2 + \ln \tau\omega_B), \qquad \tau\omega_B \gg 1, \tag{25}$$

where ω_B is the system bandwidth. Substituting (25) into (13) yields the following:

$$\langle \sigma^2(N, \tau) \rangle = \frac{(N+1)4h}{N\tau^2}\left[2 + \ln \tau\omega_B - \frac{\ln N}{N^2 - 1}\right] \qquad \alpha = -1, \quad \tau\omega_B \gg 1. \tag{26}$$

One thus obtains the somewhat unfortunate result that experimentally it would be unlikely that one could distinguish between "flicker noise" phase modulation and any other noise with $-1 < \alpha$ up to and including white noise frequency modulation using the dependence of $\langle \sigma^2(N, T, \tau) \rangle$ on τ, since $\mu \simeq -2$ in this range. However, one might be able to infer from the experimental setup which of the two types of noise was being observed. If this were not possible, one would be forced to determine the type of noise present by some other technique, such as the one employed by Vessot in which he varies ω_B.[6]

If the number of samples N, and the ratio of the period of sampling to the sample time r, are held constant, the following general equation can be written for both adjacent and nonadjacent sampling of data:

$$\langle \sigma^2(N, T, \tau) \rangle = K(N, r)|\tau|^\mu. \tag{27}$$

Using (27) and (6), ($S_\phi(\omega) = h|\omega|^\alpha$), coupled with the results established thus far, one is now able to extract the power spectral density from the dependence of the standard deviation on sample time for values of μ ranging from $-2 < \mu \leq 0$. All of the results have therefore been obtained to make the previous mapping of μ into α for finite data sampling.

III. Atomic Frequency Standards

A. *Passive Atomic Standards*

Devices such as atomic beam machines and rubidium gas cells have been made to generate impressively stable frequencies [7], [8]. In October, 1964, the appropriately authorized International Committee of Weights and Measures adopted as a provisional definition for the measurement of time, the transition between the $F=4$, $m_F=0$ and $F=3$, $m_F=0$ hyperfine levels of the ground state $^2S_{1/2}$ of the atom of cesium 133, unperturbed by external fields, with the assigned frequency for the transition of 9 192 631 770 Hz. The cesium beam at NBS has also been established as the United States Frequency Standard. The analysis of the theory of operation of these quantum devices has been covered in many publications [7], [8], along with the methods of slave-locking an oscillator to a given transition [8]. An analysis of the noise that should be present in an oscillator servoed to an atomic transition has been made by Kartaschoff [9], Cutler [6], and others. The author desires to reiterate at this point some of the results of these analyses: the frequency should appear white noise modulated; therefore, the phase fluctuations will go as the random walk phenomena, and hence the mean square time error in a clock running from one of these frequency standards would be proportional to the running time. Barnes [1] has also shown that

$$\langle (\Delta^3\phi)^2 \rangle \sim \langle (\delta t)^2 \rangle, \tag{28}$$

where Δ^3 denotes the 3rd finite difference and δt the clock error time for a running time t, and also that

$$\langle (\Delta^3\phi)^2 \rangle \sim |\tau|^{\mu+2} \qquad -2 < \mu \leq 0. \tag{29}$$

The results from combining (28) and (29) are obvious, but still very important, i.e., if either the power spectrum or the dependence of $\langle \sigma^2 \rangle$ on the sample time is known, one may then determine the rate of time divergence and conversely [see (7) and 10)],

$$\langle (\delta t)^2 \rangle \sim |\tau|^{\mu+2}; \qquad -2 < \mu \leq 0. \tag{30}$$

An example of the above is illustrated by the following: if in fact "flicker noise" frequency modulation is present on the signal of rubidium gas cells, and if one assumes the rms time errors were equal on clocks driven by a rubidium gas cell and a cesium beam of theoretical form at $\frac{1}{3}$ of a day—say, for example, 0.1 microsecond—then the accumulated rms time error after 1 year would be of the order of 100 microseconds for the rubidium cell and 3 microseconds for the cesium beam.

B. *Masers*

Masers, in contrast to the passive atomic devices discussed in Section III-A, may be used as quantum mechanical oscillators [10]. Since the sources of "flicker noise" frequency modulation could arise from many different mechanisms, the suggestion exists that perhaps the active character of masers could be influenced by one of these mechanisms. Most of the basic experimental research associated with this paper and performed by the author has been to determine if the above suggestion is valid or not, and specifically to determine the type or types of noise present on the signals from masers.

One $N^{15}H_3$ maser has been in operation at NBS for

[5] See Barnes [1], equation (74), page 219.

[6] R. Vessot, L. Mueller, and J. Vanier, "The specification of oscillator characteristics from measurements made in the frequency domain," *this issue*, page 199. (Also see Appendix.)

several years. It is desirable to compare similar atomic devices in looking at noise, so another $N^{15}H_3$ maser was put into operation for direct comparison purposes. Worth considering at this time is the basic operation of a maser so that arguments made later on will be understandable.

Three basic elements of a maser are the source, the energy-state selector, and the resonant cavity. A needle valve and a collimating nozzle are coupled to the source to provide a highly directed beam of $N^{15}H_3$ down the axis of the four-pole focusers. The four-pole focusers are the energy-state selectors and consist of four electrodes with alternate high voltages so that on-axis (in line with the resonant cavity) the electric field is zero, but slightly off-axis the field is large and increases as the distance off axis. Because ammonia has an induced electric dipole moment, an interaction occurs as it enters this electric field region, and as it has been shown [10] the low energy states of the inversion levels of ammonia are defocused while the high energy states are focused on axis and into the resonant cavity. Therefore, if a noise component of the proper frequency is present and the ammonia beam flux is sufficient, a regenerative process will take place in which a high energy-state molecule will decay and radiate a quantum of energy "$h\nu$" causing the field to increase in the cavity and hence inducing other molecules to undergo the transition, etc. The radiation may be coupled off with a waveguide and then, with an appropriate detection scheme, one may observe the maser oscillations at a frequency ν where, neglecting any frequency pulling effects, ν is approximately 22 789 421 700 Hz for $N^{15}H_3$.

There are many parameters that affect the output frequency of an ammonia maser. It is from these parameters that one may expect correlation of frequency fluctuations over long times such as exist for "flicker noise" frequency modulation. To see the mechanism, consider the fundamental pulling equation [10]

$$\nu_0 - \nu \doteq \frac{\Delta\nu_1}{\Delta\nu_c}(\nu_0 - \nu_c) + B, \qquad (31)$$

where ν_0 is the unperturbed transition frequency, ν the maser output frequency, $\Delta\nu_1$ the transition line width, $\Delta\nu_c$ the cavity bandwidth, ν_c the cavity's frequency, and B is a term involving the basic parameters such as ammonia beam flux, the voltage on the state selectors, and the magnetic and electric field intensities inside the cavity. Most of the quantities in (31) are dependent on temperature either directly or through the coupling electronics. Some of the quantities in (31) are dependent on the maser's alignment affording another mechanism for correlation of the long-term frequency fluctuations.

As described in detail elsewhere [11], it is possible to construct an electronic servo that will tune the resonant frequency of the cavity to that of the ammonia transition frequency ($\nu_0 = \nu_c$). The tuning is not perfect because of the noise present. The effect this has, as can be seen from (31), is profound, greatly reducing the effect of the basic parameters and almost entirely eliminating cavity dimensional instability. The correlation now existing in the long term frequency fluctuations may well be expected to be masked out by other noise in the system.

Since the servomechanism is of the frequency lock type, and if the noise in the servo is white over some finite bandwidth, then the fluctuations on the output frequency would be white inasmuch as they were caused by the servo. This is the same as for passive atomic devices—a result not altogether unexpected.

IV. Experiments Performed

A. $N^{15}H_3$ Maser Comparison

Two $N^{15}H_3$ masers were compared by the following technique. Each maser has coupled onto its output waveguide a balanced crystal detector and onto each detector a 30 MHz IF amplifier. Since the frequency of the inversion transition in $N^{15}H_3$ is about 22 790 MHz, a local oscillator signal at 22 760 MHz is inserted into each remaining leg of the balanced crystal detectors. The 30 MHz beat frequencies resulting from the output of the IF amplifiers are then compared in a balanced mixer and its audio output is analyzed; the bandwidth of this mixer is about 20 kHz. The frequency of the audio signal can be adjusted to a reasonable value by offset tuning the cavity of one of the masers. The period of this audio signal is determined with a counter; the data are punched on paper tape for computer analysis. The computer determines the average value of the fractional standard deviation σ, along with its confidence limit for pertinent discrete values of the sampling time τ.

In all of the maser data taken it was necessary to subtract out a systematic but well-understood linear drift—due to a change of the ammonia beam flux in the $N^{15}H_3$ maser and due to cavity dimensional drift from changing temperature in the case of the H maser. This was accomplished by the method of least squares [12]. The ability to change the number of samples N was also built into the computer program.

A plot of the computer output for some of the data analyzed in the $N^{15}H_3$ maser comparison is shown in Fig. 4. The output frequencies of the masers were not servo-controlled for these data. The slope is very nearly that of white noise frequency modulation. It will be noted that for a time of one second, the fractional standard deviation of the frequency fluctuations is less than 2×10^{-12}.

B. Cesium Beam, Maser, Quartz Oscillator Intercomparisons

Many comparisons were made between all the different types of atomic devices made available in the Atomic Frequency and Time Standards Section of NBS. Most of the comparisons were made at 5 MHz by measuring the period of the beat frequency with a counter, and the resulting data were processed by a computer.

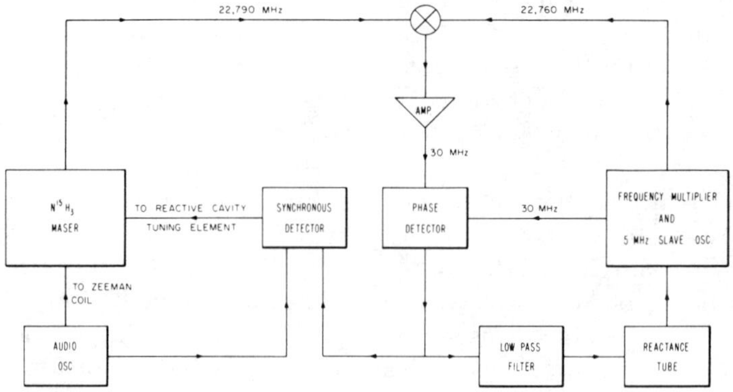

Fig. 3. Block diagram of maser cavity servo system.

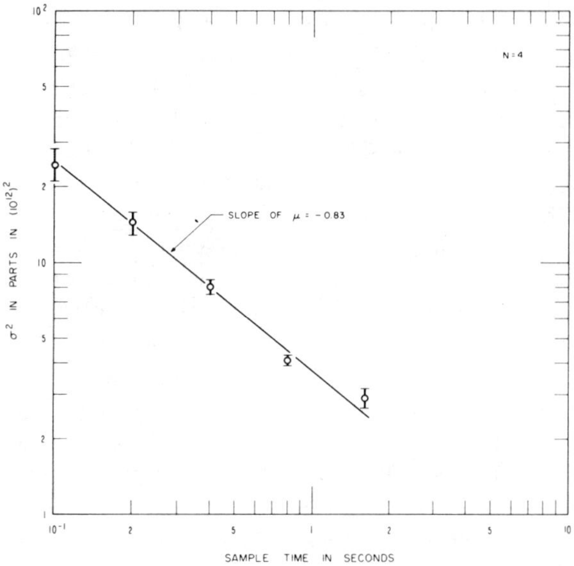

Fig. 4. Comparison of two $N^{15}H_3$ masers; standard deviation of the frequency fluctuations as a function of sampling time.

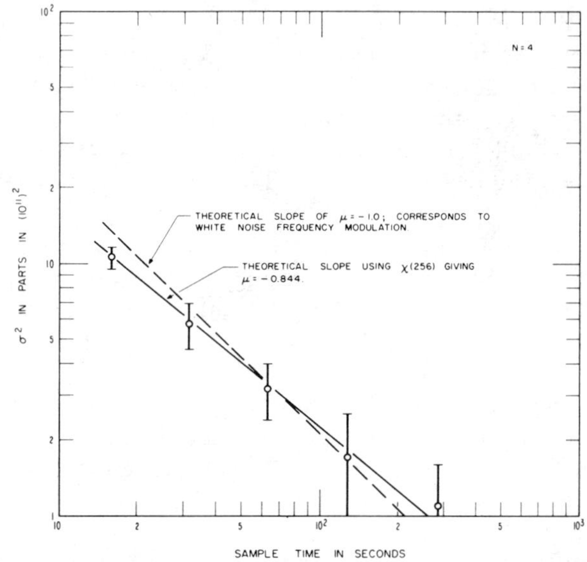

Fig. 5. Cesium beam (NBS III), $N^{15}H_3$ maser (2) comparison; standard deviation of the frequency fluctuations as a function of sampling time.

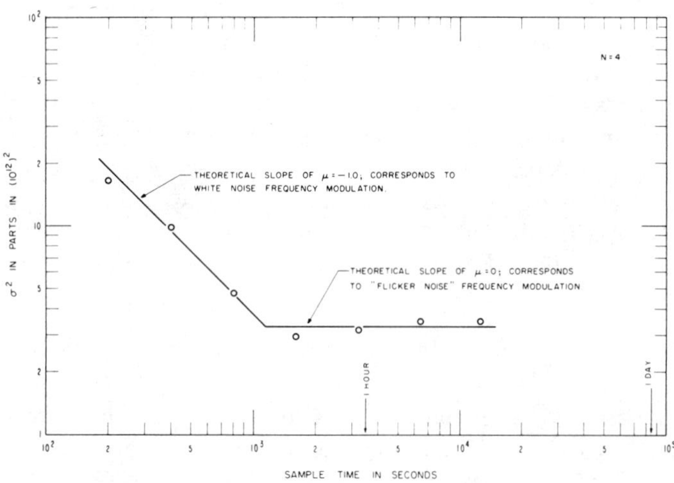

Fig. 6. Cesium beam (NBS III), quartz crystal oscillator comparison; standard deviation of the frequency fluctuations as a function of sampling time.

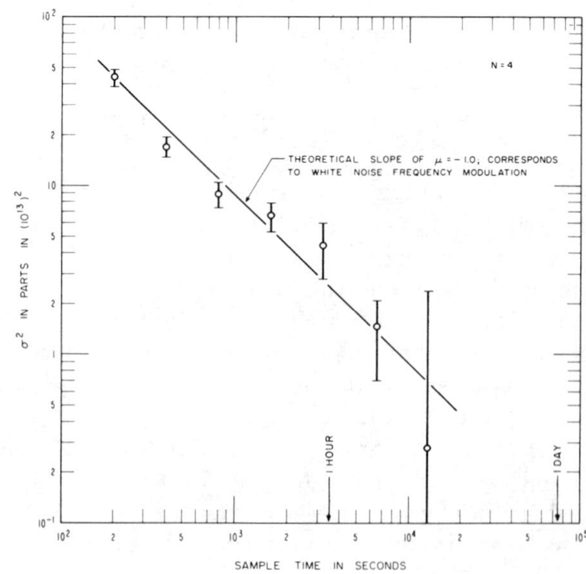

Fig. 7. Cesium beam (NBS III), hydrogen maser comparison; standard deviation of the frequency fluctuations as a function of sampling time.

The bandwidth of the 5 MHz mixer that produces the beat frequency was about 33 Hz.

A comparison was made between the double beam $N^{15}H_3$ maser with the cavity servo in operation as illustrated in Fig. 3, and the NBS III cesium beam. A 5 MHz quartz crystal oscillator of good spectral purity was phase locked by a double heterodyne technique to the output frequency of the maser, and this oscillator was compared with a high quality quartz crystal oscillator that was frequency locked to NBS III. A plot of the data is shown in Fig. 5. The noise is undoubtedly that of the maser cavity servo system, and though it exhibits very nearly its theoretical value of $\mu = -1$ for white noise frequency modulation, the noise level will be seen to be well over an order of magnitude higher than for the free-running maser.

The free-running, single beam, $N^{15}H_3$ maser was compared with NBS III on a longer time basis and indicated the presence of "flicker noise" frequency modulation. There was indicated a fair amount of uncertainty to the data, however.

In Fig. 6 one sees a very interesting plot in the comparison of the NBS III cesium beam with a very high quality quartz crystal oscillator. The comparison was again made at 5 MHz as indicated above. The first part of the curve shows the white noise frequency modulation in the cesium beam servo system, and then the curve changes slope indicating the presence of "flicker noise" frequency modulation as is typically exhibited by quartz crystal oscillators. This shows experimentally the theoretical result discussed previously, that one may have quite a high level of white noise in a system and eventually the system will be better than one with "flicker noise" frequency modulation.

An extended comparison was made (58 hours) between the H maser, NBS III cesium beam, a rubidium gas cell, and a quartz crystal oscillator. Some of the data are still to be analyzed, but some interesting and impressive results have been obtained thus far. Figure 7 shows the analysis of some of the data obtained from the H maser, cesium beam (NBS III) comparison. White noise frequency modulation is exhibited for sampling times extending to 4 hours with the very impressive standard deviation of $\sim 6\times 10^{-14}$ for $\tau = 4$ hours. "Flicker noise" frequency modulation was observed on both the rubidium gas cell and on the quartz crystal oscillator in this comparison.

V. Conclusions

An invariant quality factor of a frequency standard having definite esthetic value would be the infinite time average, standard deviation of the frequency fluctuations. From the theoretical development a least biased estimate of this factor may be established in terms of the expectation value of the standard deviation for any finite number of samples:

$$\sigma^2 = \frac{(N-1)\langle\sigma^2(N,\tau)\rangle}{N(1-N^{\mu})},$$

$$-2 \le \mu \le 0 \text{ and } \left|\tau\omega_B\right|^{|\alpha+1|} \gg 1. \quad (32)$$

σ^2 also has the desirable feature of being dependent on only one variable, τ, except for $\mu = -2$, where it is also necessary to specify ω_B, the system bandwidth. Equation (32) is valid for all types of noise between and including "flicker noise" frequency modulation and white noise phase modulation ($-3 \le \alpha \le 0$). However, if $\alpha = -3$, some other means of specification need be employed, such as Barnes' [1] finite difference technique or $\langle\sigma^2(2, \tau)\rangle$ which, interestingly, is equal to $\langle(\Delta^2\phi)^2\rangle/2\tau^2$ (Δ^2 denotes the second finite difference).

"Flicker noise" frequency modulation was not observed on the frequency fluctuations of the maser type of atomic frequency standard for any of the reliable data obtained—the time coverage here being from about 0.1 second to 4 hours. The presence of "flicker noise" was indicated for free-running masers (both H and $N^{15}H_3$) in the long term range, though the data were somewhat unreliable. This type of noise was eliminated in the case of the $N^{15}H_3$ maser by use of a cavity servo mechanism, and one might infer from the similarities in the masers that the same would hold true for the H maser.

It is acknowledged that if "flicker noise" is present on the frequency fluctuations of a free-running maser for longer times, the lower cutoff frequency might be greater than for quartz crystal oscillators—for indeed the maser operator is himself a part of a cavity servo when he retunes the cavity or perhaps recoats the quartz bulb.

The existence of "flicker noise" frequency modulation on the frequency fluctuations of rubidium gas cells is just another indication that they would not at present make a reliable primary frequency standard. One cannot conclude from the data analyzed that "flicker noise" is not present on masers—even if a cavity servo is employed. It is possible, however, to determine an upper limit for the level of "flicker noise." If $G_\phi(\omega) = h\left|\omega\right|^{-3}$, then using (22), the upper limit of h can be determined from the minimum value of $\sigma^2(N, \tau)$. The value of h for $N^{15}H_3$ (determined from Fig. 4) is $\sim 7\times 10^{-25}$ seconds^{-2}. It is of interest to compare this with the value of h determined for the quartz crystal oscillator shown in Fig. 6—$\sim 8\times 10^{-25}$ seconds^{-2}. Though comparable, it should be stated that this oscillator is one of the best that has been analyzed at NBS. The upper limit on h for the H maser, cesium beam (NBS III) comparison shown in Fig. 7 is $\sim 2\times 10^{-27}$ seconds^{-2}—being a factor of 350 and 400 times better than the $N^{15}H_3$ maser and quartz crystal oscillator, respectively.

There are many areas where additional data analysis would be informative. It would be very interesting to look at the behavior of the cesium beam, hydrogen maser comparison, shown in Fig. 7, for extended times. Another analysis of interest would be that of a cavity-servoed H maser. Such a servo system is under development for the NBS H maser. To then compare the cavity-servoed H maser with a cesium beam and its associated servo system would afford great insight into determining which of these competitive atomic standards should be primary.

Appendix

The general expression for the autocovariance function for infinite bandwidth is:

$$R_\varphi(\tau) = a'(\alpha) \mid \tau \mid^{-\alpha-1}, \tag{33}$$

$\alpha \neq 0$, and not an integer, where

$$a'(\alpha) = 2\left[\cos\left(\frac{\pi(\alpha+1)}{2}\right)\right] \cdot \Gamma(\alpha+1).$$

Physically, $R_\varphi(0)$ is kept from diverging in the region $-1<\alpha$ because of finite system bandwidths, ω_B; i.e., for any $\epsilon < 1/\omega_B$, $R_\varphi(\epsilon) \approx R_\varphi(1/\omega_B)$. A good approximation to $R_\varphi(0)$ in this region is, therefore, $R_\varphi(1/\omega_B)$. One may notice that the variance of the frequency fluctuations diverges [see (5)] with increasing ω_B. The variance of the frequency fluctuations may, therefore, be written:

$$\sigma^2 = \frac{a(\alpha)}{\tau^2}\left[\mid \omega_B \mid^{\alpha+1} - \mid \tau \mid^{-\alpha-1}\right]. \tag{34}$$

In the region $-3<\alpha<-1$, $R_\varphi(0)$ is approximately zero; and if $\left|\tau\omega_B\right|^{|\alpha+1|} \gg 1$, then the equation for $U(\tau)$ are as written in (9).

Consider now a finite number of data samples N in the domain where $-1<\alpha<0$, and $\left|\tau\omega_B\right|^{\alpha+1} \gg 1$. Using the appropriate form of $U(\tau)$ from (9) and substituting into (13) gives:

$$\langle\sigma^2(N,\tau)\rangle = \frac{N+1}{N} a(\alpha) \mid \tau \mid^{\mu} \mid \tau\omega_B \mid^{\alpha+1}, \tag{35}$$

where

$$\mu = -\alpha - 3.$$

Hence, by keeping N and $\tau\omega_B$ constant, one can determine the value of μ (of α) on a log $\langle\sigma^2(N,\tau)\rangle$ vs. log τ plot, thus eliminating the degeneracy in this region. Equation (35) may also be written:

$$\langle\sigma^2(N,\tau)\rangle = \frac{(N+1)\cdot a(\alpha) \cdot \mid \omega_B \mid^{\alpha+1}}{N\tau^2}. \tag{36}$$

Hence, if N and τ are held constant and ω_B is varied, the slope on a log-log plot of $\langle\sigma^2(N,\tau)\rangle$ vs. ω_B would be $\alpha+1$—affording another technique of determining the spectral density in this domain.

It is of interest to note that the most sensitive parameters to use in arriving at the value of α are: the bandwidth, ω_B, if $-1<\alpha<0$; the sample time τ, if $-2<\alpha<-1$; and the number of samples N, if $-3<\alpha<-2$.

Acknowledgment

The author wishes to acknowledge the help of many of the personnel in the Atomic Frequency and Time Standards Section of NBS, Boulder, Colo. The author is sincerely indebted to J. A. Barnes for many informative discussions and for much aid in data processing; he has also contributed significantly in making manuscript criticisms for which the author is also very grateful. The help rendered by R. E. Beehler and C. S. Snider in making available the cesium beam frequency standards is greatly appreciated. Appreciation is also expressed to Dr. F. R. Petersen and to Dr. D. Halford for the hydrogen maser availability, and to V. Heaton for his help in data processing.

References

[1] J. A. Barnes, "Atomic timekeeping and the statistics of precision signal generators," this issue, page 207.
[2] J. A. Barnes and R. Vessot, private communication.
[3] M. Lighthill, *Introduction to Fourier Analysis and Generalized Functions*. Cambridge, England: University Press, 1962.
[4] C. L. Searle, R. D. Posner, R. S. Badessa, and V. J. Bates, "Computer-aided calculation of frequency stability," *1964 Proc. Symp. on the Definition and Measurement of Short-Term Frequency Stability*. Washington, D. C.: U. S. Govt. Printing Office, NASA SP-80, pp. 273–277.
[5] W. R. Atkinson, R. L. Fey, and J. Newman, "Spectrum analysis of extremely low frequency variations of quartz oscillators," *Proc. IEEE (Correspondence)*, vol. 51, p. 379, February 1963.
[6] L. S. Cutler, "Some aspects of the theory and measurement of frequency fluctuations in frequency standards," *1964 Proc. Symp. on the Definition and Measurement of Short-Term Frequency Stability*, pp. 89–100.
[7] R. C. Mockler, "Atomic beam frequency standards," *Advan. Electron. and Electron. Physics*, vol. 15, 1961.
[8] R. E. Beehler, W. R. Atkinson, L. E. Heim, and C. S. Snider, "A comparison of direct and servo methods for utilizing cesium beam resonators as frequency standards," *IRE Trans. on Instrumentation*, vol. I-11, pp. 231–238, December 1962.
[9] P. Kartaschoff, "Influence de l'Effect de Grenaille sur la Frequence d'un Oscillateur Asservi A un Etalon a Jet Atomique," Laboratoire Suisse de Recherches Horlogères, Neuchatel, Switzerland, Rapport L.S.R.H. 08-64-01. See also, P. Kartaschoff, "Shot-effect influence on the frequency of an oscillator locked to an atomic beam resonator," *1964 Proc. Symp. on the Definition and Measurement of Short-Term Frequency Stability*, pp. 303–308.
[10] A. A. Vuylsteke, *Elements of Maser Theory*. Princeton, N. J.: Van Nostrand.
[11] J. A. Barnes, D. W. Allan, and A. E. Wainwright, "The ammonia beam maser as a standard of frequency," *IRE Trans. on Instrumentation*, vol. I-11, pp. 26–30, June 1962.
[12] C. R. Wylie, *Advanced Engineering Mathematics*. New York: McGraw-Hill, p. 527, 1951.

Part II
Theoretical Backgrounds of Oscillator Noise

THE theory of noise in free-running and synchronized oscillators has been dealt with by many authors for nearly half a century. However, it seems that some important questions still remain to be answered. Particularly, there is no satisfactory theory which would explain actual phase noise characteristics in the whole range of useful Fourier frequencies. The difficulty is caused by the excess noises (the frequency and phase flicker noise contributions), the origins of which are generally unknown or only guessed. As a consequence, authors limit themselves only to the investigation of the effects of thermal or shot noises.

Another interesting feature encountered is that application of simple mathematics, by considering the oscillator as a linear noise filter with a narrow bandwidth, reveals results which generally agree with those arrived at by far more sophisticated techniques. The situation being such, we have decided to reprint only four papers which we consider to be representative of different approaches on one hand and are not too difficult to understand on the other hand.

The theory of noise in free-running oscillators was first established by Berstein in 1938 [II-1]. Since then, the same subject has been discussed by many authors both in Europe [II-2]–[II-5] and in the U.S. [II-6]–[II-10]. Among all of these papers, the paper by Hafner [II-10] is one of the most fundamental.

However, for the introduction into the problem of the oscillator noise, we have chosen the paper by Edson who investigates both the initiation of oscillations and the amplitude and phase behavior at sustained oscillations by using the noise buildup approach. On the other hand, Kurokawa, in the second reprint paper, starts by deriving the fundamental differential equations for the phase and amplitude of the oscillator current in the presence of small perturbations. Both authors effectively arrived at the same results with respect to the autocorrelation function of the voltage or current and the respective power spectral densities. Kurokawa's equations (15) and (25) lead to the phase noise power spectral density, close to the carrier, to be

$$S_\varphi(f) = \frac{1}{\omega^2} \frac{\omega_0^2}{4Q_{ext}^2} \left[F_0 \frac{kT}{P_0} + \frac{S_{\Delta X}(f)}{R_0^2} \right] \quad \text{(II-1)}$$

where P_0 is the power dissipated in the load R_0, F_0 is the noise figure, and $S_{\Delta X}(f)$ is the power spectral density of the reactance fluctuations ($X = [\omega_0 L - 1/\omega_0 C]$).

In addition, Kurokawa investigated properties of synchronized oscillators. He arrived at the conclusion, found earlier by Adler [II-11], that the system behaves as a phase-locked loop of the first order with all the consequences to the phase noise (see, e.g., Part VI, paper 9); i.e., in the PLL "passband" the synchronizing oscillator noise predominates and in the "stopband" the noise of the synchronized oscillator predominates. This problem is of importance in VHF and microwave frequency ranges, and the interested reader can find further information in [II-12], [II-13], and [II-14].

However, with the advent of the transferred electron (Gunn) and transit time (impatt diode) microwave oscillators, the approximation of the nonlinear characteristic by a first-order Taylor series expansion around the operating point was inadequate and a large signal noise theory had to be derived. One of the approaches was via the "describing function" [II-15], [II-16] known from the study of nonlinear control systems. As a representative paper for the application of this theory, we have chosen one by Schuenemann and only mention some others [II-17]–[II-19].

For the benefit of engineers who are not interested in background theory, but rather in actual noise characteristics, we reprint the short paper by Leeson. The underlying idea is that of the feedback system with a zero phase around the loop; this condition requires compensation of spurious phase fluctuations $\varphi(t)$ by frequency fluctuations $\omega(t)$, i.e.,

$$\varphi(t) + \int \omega(t)dt = 0 \quad \text{(II-2)}$$

where

$$\omega(t) \approx \frac{\varphi(t)}{\tau} \quad \text{(II-3)}$$

and the time delay is defined as

$$\tau = \frac{\Delta\varphi}{\Delta\omega} \approx \frac{2Q}{\omega_0}. \quad \text{(II-4)}$$

With the assistance of the Laplace transform and (II-1) we obtain a very useful relation

$$S_{\varphi,osc}(f) \approx \left[1 + \left(\frac{f_0}{2Q_{ext} f}\right)^2\right]\left[\frac{a_{-1}}{f} + a_0\right] \quad \text{(II-5)}$$

where a_{-1} and a_0 are flicker and white noise constants; their approximate magnitudes, found experimentally, are given in Part VI, paper 9. Additional analysis is to be found in [II-20], [II-21], and Part VI, paper 7.

References

[II-1] I. Berstein, "On fluctuations in the neighbourhood of periodic motion of an auto-oscillating system," *C.R. Acad. Sci. USSR*, vol. XX, no. 1, pp. 11–16, 1938.

[II-2] A. Spälti, "Der Einfluss des thermischen Widerstandsrauschens and des Schrotteffektes auf die Störmodulation von oscillatoren," *Bull. Schweizer. Electrotech. Ver.*, vol. 39, pp. 419–427, June 1948.

[II-3] A. Blaquière, "Effet du bruit de fond sur la fréquence des auto-oscillateurs à lampes," *Ann. Radioelec.*, vol. 8, pp. 36–80, Jan. 1953; and ——, "Spectre de puissance d'un oscillateur non-linéaire perturbé par le bruit," *Ann. Radioelect.*, vol. 8, pp. 153–179, Aug. 1953.

[II-4] P. I. Kuznetsov, R. L. Stratonovich, and V. I. Tikhonov, "The effects of electrical fluctuations on a vacuum tube oscillator," *J. Exp. Theor. Phys. USSR*, vol. 1, pp. 510–519, 1955; also (in Russian), *Zh. Eksp. Teor. Fiz.*, vol. 28, pp. 509–523, 1955.

[II-5] S. M. Rytov, "Fluctuations in oscillating systems of the Thomson type," *Sov. Phys.*, vol. 2, pp. 217–235, 1956; also (in Russian), *Zh. Eksp. Teor. Fiz.* USSR, vol. 29, pp. 304–333, 1955.

[II-6] R. M. Lerner, "The effects of noise on the frequency stability of a linear oscillator," in *Proc. Nat. Electro. Conf.*, vol. 7, 1951, pp. 275–280.

[II-7] W. A. Edson, *Vacuum Tube Oscillators.* New York: Wiley, 1953.

[II-8] J. A. Mullen, "Background noise in nonlinear oscillators," *Proc. IRE*, vol. 48, pp. 1467–1473, Aug. 1960.

[II-9] L. R. Malling, "Phase-stable oscillators for space communications, including the relationship between the phase noise, the spectrum, the short-term stability, and the Q of the oscillator," *Proc. IRE*, pp. 1656–1664, July 1962.

[II-10] E. Hafner, "The effects of noise in oscillators," *Proc. IEEE*, vol. 54, pp. 179–198, Feb. 1966.

[II-11] R. Adler, "A study of locking phenomena in oscillators," in *Proc. IRE Waves Electrons*, vol. 34, June 1946, pp. 351–357.

[II-12] L. W. Couch, "A study of a driven oscillator with FM feedback by use of a phase-lock-loop model," *IEEE Trans. Microwave Theory Tech.*, vol. MTT-19, pp. 357–366, Apr. 1971.

[II-13] R. L. Kuvås, "Noise in single-frequency oscillators and amplifiers," *IEEE Trans. Microwave Theory Tech.*, vol. MTT-21, pp. 127–134, Mar. 1973.

[II-14] K. Kurokawa, "Injection locking of microwave solid-state oscillators," *Proc. IEEE*, vol. 61, pp. 1386–1410, Oct. 1973.

[II-15] A. Gelb and W. E. van der Velde, *Multiple-Input Describing Functions and Nonlinear System Design.* New York: McGraw-Hill, 1968.

[II-16] L. Gustafsson, G. H. B. Hansson, and K. I. Lundström, "On the use of describing functions in the study of nonlinear active microwave circuits," *IEEE Trans. Microwave Theory Tech.*, vol. MTT-20, pp. 402–409, June 1972.

[II-17] R. Knöchel and K. Schünemann, "Noise and transfer properties of harmonically synchronized oscillators," *IEEE Trans. Microwave Theory Tech.*, vol. MTT-26, pp. 939–944, Dec. 1978.

[II-18] A. Takaoka and K. Ura, "Noise analysis of nonlinear feedback oscillator with AM-PM conversion coefficient," *IEEE Trans. Microwave Theory Tech.*, vol. MTT-28, pp. 654–662, June 1980.

[II-19] R. Knöchel, "Unified large-signal stability and noise theory for synchronized oscillators," *Proc. Inst. Elec. Eng.*, vol. 128, Pt. H, pp. 137–145, June 1981.

[II-20] G. Sauvage, "Phase noise in oscillators: A mathematical analysis of Leeson's model," *IEEE Trans. Instrum. Meas.*, vol. IM-26, pp. 408–410, Dec. 1977.

[II-21] D. J. Healey and S. Y. Kwan, "SC-cut quartz crystal units in low-noise oscillator application at VHF," in *Proc. 35th Annu. Frequency Contr. Symp.*, Philadelphia, PA, May 1981, pp. 440–454.

Noise in Oscillators*

W. A. EDSON†, FELLOW, IRE

***Summary*—Noise affects the behavior of oscillators in at least two important ways. During sustained oscillation, noise creates undesired perturbations or modulation in both the amplitude and the phase of the wave. The amplitude perturbations produce a continuous spectrum which in typical situations is quite weak and broader than the bandwidth of the resonator. The phase perturbations disperse the nominal frequency into a continuous distribution which is of the same form but much stronger and narrower than for the amplitude perturbations.**

During the initiation of oscillation, noise constitutes the starting voltage and therefore affects the time required for the wave to reach some pre-established amplitude. The resulting jitter in the starting time of pulsed oscillators is objectionable because it degrades the signal-to-noise ratio in systems employing super-regenerative receivers or pulse-time modulation. The time and spectral distributions of noise effects in typical oscillators are derived and discussed in the following sections.

List of Symbols

A = amplitude of oscillation
C = capacitance
F = frequency
G = conductance
I = a current
K = constant of integration
L = inductance
P = power
Q = selectivity
R = the autocorrelation
T = temperature °K
U = mean noise amplitude
V = a voltage
ω = angular velocity
σ = probable error
ρ = an amplitude function
ϕ = a phase function
δ = logarithmic decrement
τ = a time interval
α = expansion coefficient
β = a variable related to decrement
γ = a dimensionless multiplier
π = the number 3.14159
μ = a dimensionless multiplier
a = a variable
b = a variable
d = the differential operator
e = the number 2.71828
f = frequency
h = a variable
i = a variable
j = a variable
k = Boltzmann's constant 1.38×10^{-23}
n = number of half-cycles
p = a variable, also probability
q = a variable, also cumulative probability
r = a variable
s = the saturation coefficient
t = time
u = an amplitude coeffcient
v = a voltage
x = an amplitude coefficient

Introduction

Of THE many publications concerned with noise, [2], [13], [15], a considerable number, [4], [7], [14], concern themselves with the more special problem of noise in oscillators. A variety of approaches has been taken, and results of undoubted value have been obtained. However, it appears that many important questions remain to be answered. In particular, it is believed that no previous treatment adequately describes either the distribution during build-up of oscillation or the spectrum during sustained oscillation. Conventional noise formulas are directly applicable only to steady-state situations in linear systems. Unfortunately, the build-up problem involves the transient state, and sustained oscillation involves nonlinearity. Thus, normal methods are inapplicable, and it has been necessary to return to basic considerations.

Only white noise of thermal or electronic origin is considered in this treatment. While microphonics, element drift and other effects contribute to the instability of oscillators, their action can be calculated by these or other methods and no specific treatment is given here. The analysis indicates that low frequency abnormalities such as the flicker effect may affect the amplitude distribution of the output but should have only a second order effect upon the frequency.

Narrow Band Noise

Oscillators are characterized by relatively selective circuits and the fact that all the effects of interest are concentrated in a relatively narrow band of frequencies. Therefore, we may restrict our attention to noise within fractionally narrow bandwidths. The voltage which results when white noise is passed through any relatively narrow band-pass filter may be described by

$$v(t) = x_1(t) \cos \omega_0 t + x_2(t) \sin \omega_0 t \qquad (1)$$

where ω_0 is the midband angular frequency and $x_1(t)$

* Original manuscript received by the IRE, June 9, 1959; revised manuscript received, Janaury 6, 1960.
† General Electric Microwave Lab., Palo Alto, Calif.

Reprinted from *Proc. IRE*, vol. 48, pp. 1454–1466, Aug. 1960.

and $x_2(t)$ are *uncorrelated* functions of time which vary slowly and randomly about zero in a manner described by the normal or Gaussian probability function shown as Fig. 1. That is, the most probable value is zero and during half the time, the function lies within the range of probable error $\pm 0.675\sigma$, where σ, the *standard deviation*, is the rms value. The numerical value of σ depends upon the temperature, impedance, and bandwidth of the circuit and is discussed later.

Although zero is the most probable value for $x_1(t)$ and $x_2(t)$ separately, it is very improbable that the two coefficients are simultaneously zero. That this is true may be seen from the expression equivalent to (1), also given by Davenport and Root, [5],

$$v(t) = \rho(t) \cos [\omega_0 t + \phi(t)]. \qquad (2)$$

Again ω_0 is the midband radian frequency and $\rho(t)$ and $\phi(t)$ are slowly varying functions. However, entirely new statistical properties are involved. The phase varies randomly and with equal probabilities in all quadrants: therefore, $\phi(t)$ has a rectangular distribution, uniform throughout the interval $\pm\pi$ radians. The amplitude $\rho(t)$ is now a magnitude or envelope function and has the Rayleigh distribution shown in Fig. 2. The standard deviation has the same significance as before, and during half the total time the envelope amplitude equals or exceeds 1.177 of the rms value. The fact that the same physical phenomenon is correctly described by both Gaussian and Rayleigh distributions is important to all the following discussions.

I. Initiation of Oscillation

In the interval which follows the instant that an oscillator is first turned on, the amplitude builds up from a small initial value until it becomes large enough to produce limiting at some point in the system. During this period, the behavior of typical oscillators is substantially linear. Therefore, it is possible to treat this part of the problem without recourse to nonlinear mathematics.

In low-frequency oscillators, it is very difficult to achieve the turn-on process without simultaneously generating a transient which, within the bandwidth of interest, is large compared with background noise. Under such circumstances, the oscillation starts from this injected transient rather than from noise, and the phase of oscillation is uniquely related to the turn-on signal. Such oscillations are referred to as coherent and are desirable in many situations. However, they are beyond the scope of the present treatment, which is limited to situations in which the starting transient is negligible and oscillation builds up from the white noise signal which is inherent in all physical circuits. Because pure noise has no correlation with respect to the turn-on signal, there is no phase relationship between the phase of the growing oscillation and the turn-on signal. Such oscillations are referred to as incoherent. Phase coherence is a convenient practical test for the existence and relative magnitude of a starting transient.

At best, the problem to be analyzed is complicated and difficult. For this reason the following treatment is divided into several steps of increasing complexity, each introducing some additional consideration and, therefore, being a closer approximation to a physical system.

A. Noise Build-Up in a Passive Circuit

Referring to Fig. 3, assume that L, C, and G are positive fixed elements and that the switch has been closed for a long time. Noise effects identifiable with the conductance G are represented by the current I. The sys-

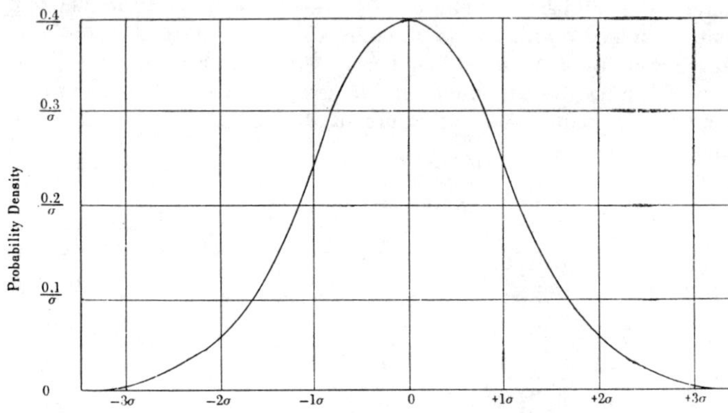

Fig. 1—The Gaussian probability density function

$$p(x) = \frac{1}{\sigma\sqrt{2\pi}} e^{-x^2/2\sigma^2}.$$

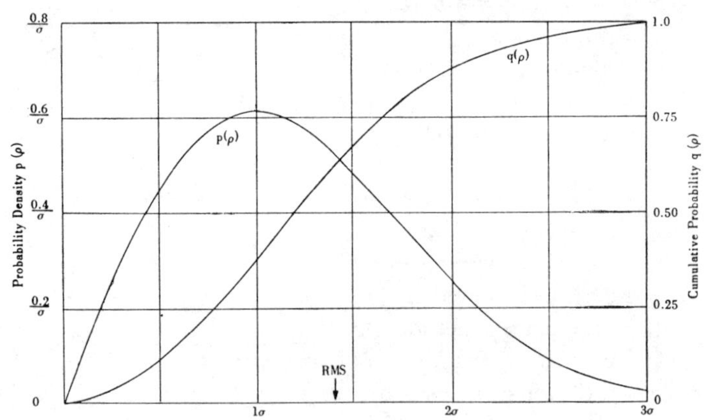

Fig. 2—The Rayleigh probability density function

$$p(\rho) = \frac{\rho}{\sigma^2} e^{-\rho^2/2\sigma^2}.$$

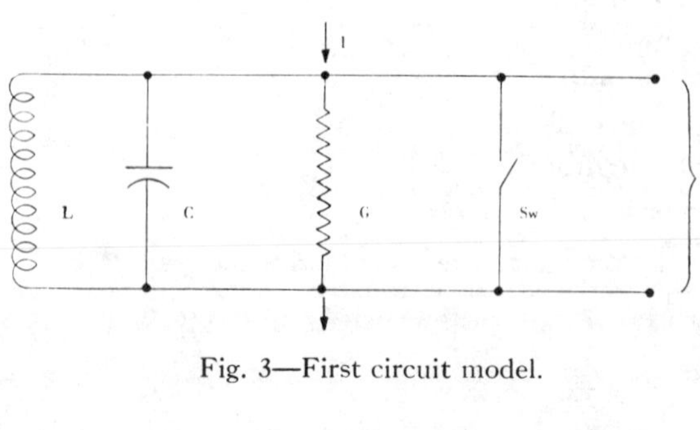

Fig. 3—First circuit model.

tem, which is initially at rest, is driven by the current I in the interval which follows the opening of the switch at time $t=0$.

Normal tube or thermal noise consists of a great many small impulses, randomly distributed in time. However, as already shown, the energy storage in a tuned circuit reduces the rapidity of response which is possible. Therefore, no loss of generality results if the impulses which occur within some finite period are collected or grouped together. For the present purpose, it is convenient to use one-quarter cycle as the collecting interval. On page 31 of Woodyard [17], it is shown that this procedure is correct and that it meets the requirements of sampling theory; this discussion also accounts for the sine and cosine components of (1).

In the present section, it is assumed that the sampling period is perfectly uniform. However, in the final section on phase perturbations this period is modified to conform with the existing phase of the wave as perturbed by preceding impulses. No appreciable error is introduced because the perturbation per cycle is very small.

The situation is best visualized by imagining that a very large number of identical oscillatory systems are subjected simultaneously to the same experiment. It is now possible to make meaningful statements as to the statistical behavior of this ensemble. In the following paragraphs the amplitudes u and v are to be interpreted in this sense.

During the first quarter-cycle after the switch is opened, the noise current will excite in the system a voltage of sine phase with an amplitude identifiable with some mean-square value u^2. In the ensuing quarter-cycle, the noise current excites a voltage of cosine phase and the same mean square amplitude. The initially excited sine term is not affected because sine and cosine terms are orthogonal. Since the sine and cosine components experience similar influences, it is sufficient to explore either. The cosine term is selected because it offers minor simplification of notation.

During the last quarter of the initial cycle, two opposing effects take place. The cosine oscillation excited during the second quarter cycle is decreased in amplitude by a factor β associated with the positive damping of the passive circuit. However, this effect is more than compensated by the additional impulse received from the driving current I. Thus, at the end of the second half-cycle, the amplitude of the cosine component is representable by

$$v_2^2 = u^2(1 + \beta) \quad (3)$$

where β is related to the logarithmic decrement δ by

$$\beta = e^{-\delta} \quad (4)$$

derived from

$$\beta^2 = e^{-2\delta} \quad (5)$$

which more closely represents the actual physical situation. For future reference it is noted that, because δ is small in the cases of interest, (4) may be approximated by

$$1 - \beta = \delta. \quad (6)$$

Consistent with ordinary usage, δ is taken as positive for a passive circuit.

The validity of (3) depends upon the fact that the noise impulses are statistically independent so that the square of a deviation is equal to the sum of the squares of all component deviations.

Iteration of the above reasoning shows that the mean square of the coefficient of the cosine voltage at the end of n half-cycles is represented by the finite sum

$$v_n^2 = u^2(1 + \beta + \beta^2 + \beta^3 + \cdots \beta^{n-1}). \quad (7)$$

The stepwise derivative of this series with respect to n is obtained by noting that, because $dn=1$,

$$v_{n+1}^2 - v_n^2 = dv_n^2 = u^2\beta^n dn = u^2 e^{-n\delta} dn. \quad (8)$$

Integration gives

$$v_n^2 = K - u^2(e^{-n\delta})/\delta. \quad (9)$$

The constant of integration K is evaluated by fitting the initial condition,

$$v_n = 0 \quad \text{when } n = 0 \text{ with the result that}$$

$$v_n^2 = u^2(1 - e^{-n\delta})/\delta = u^2(1 - \beta^n)/\delta.[1] \quad (10)$$

The noise build-up represented by (7) or (10) is plotted as curve (A) in Fig. 4. It is seen to reach the asymptotic value

$$U^2 = u^2/\delta. \quad (11)$$

However, Pierce, [12], has shown that this equilibrium amplitude is given by

$$U^2 = kT/C \quad (12)$$

where $k=1.38\times10^{-23}$ is Boltzmann's constant, and T is the effective absolute temperature of G. The coefficient in (12) is correct because U corresponds to the peak value of the cosine wave, which in turn is equal to the

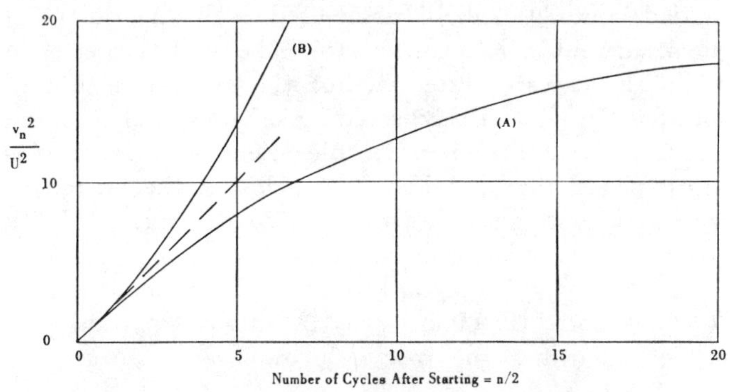

Fig. 4—Growth of the cosine component of voltage: (A) Passive circuit, $Q=+20\pi$; $\delta=+0.05$. (B) Active circuit, $Q=-20\pi$; $\delta=-0.05$.

[1] An alternate derivation of this result is given in Appendix I.

rms value of the total wave. This expression can be put in several equivalent forms by use of the following well-known relationships:

$$\omega_0 = 1/\sqrt{LC}, \tag{13}$$

$$Q = \omega_0 C/G, \tag{14}$$

and

$$\delta = \pi/Q = \pi G/\omega_0 C. \tag{15}$$

The magnitude of the mean-square noise impulse delivered per quarter-cycle is given by eliminating U^2 between (11) and (12) to obtain

$$u^2 = \pi kT/CQ = \pi kTG/\omega_0 C^2 \quad \text{peak volts}^2. \tag{16}$$

This relation is basic to all that follows, and may be taken as the principal result of this section.[2]

B. Noise Build-Up in an Active Circuit

Now assume that G is negative rather than positive but that no other change is made in the system previously described. Because numerical relationships between voltage and current are independent of whether G is positive or negative, it follows that the equations already developed can be used. The essential new fact is that disturbances caused by noise or other influences increase rather than decrease with the passage of time.

Specifically, δ is now inherently negative and β is larger (rather than smaller) than unity. Eqs. (7) and (10) are valid, but now diverge rather than converge. The situation corresponding to the negative selectivity, $Q=-20\pi$, is plotted as curve (B) in Fig. 4. As before, this curve represents the mean square value of the coefficient of the cosine component of voltage, and deviations from this value are given by the probability distribution of Fig. 2.

C. Change of Circuit Parameters

Let us now extend the analysis to include a change of circuit parameters as shown in Fig. 5. It is assumed that L, C and G are all positive and that an initial statistical equilibrium has been reached with respect to the noise current I identified with G at temperature T. The added conductance γG associated with an additional noise current μI is negative and numerically larger than G in the cases of interest, but the analysis is not so limited. In view of the preceding sections, and recalling that a conductance is incapable of energy storage, it is anticipated that subsequent to closing the switch at $t=0$, the voltage across the system will increase with time from some finite initial value.

Evidently, U_0^2, the value of the cosine voltage at time $t=0$, is correctly given by (12), and the parameter β which identifies the rate of growth is given by

$$\beta = e^{-\pi(1+\gamma)/Q}. \tag{17}$$

[2] Here and throughout the remainder of this paper, noise power is identified with positive frequencies, and no use is made of the negative-frequency concept.

The total noise impulse received per quarter-cycle is, from (16),

$$u^2 = (\pi kT/CQ)(1+\mu^2). \tag{18}$$

Therefore, the cosine component of voltage which exists at the end of n half-cycles after the switch is closed is

$$v_n^2 = U_0^2\beta^n + u^2(1+\beta+\beta^2+\cdots+\beta^{n-1}). \tag{19}$$

This equation may be converted with use of (10) to

$$v_n^2 = U_0^2\beta^n + \frac{u^2}{\delta}(1-\beta^n) = \frac{u^2}{\delta} + \left(U_0^2 - \frac{u^2}{\delta}\right)\beta^n \tag{20}$$

where β, u, and δ have the values given in (17) and (18).

To illustrate the application of these results, assume that in Fig. 5, $\gamma=-2$ and $\mu=1$. Further assuming that the passive circuit has a selectivity of $Q=20\pi$, it can be determined from (17) that $\delta=-0.05$ and $\beta=1.0514$. Consistent with (18), $u^2=0.1\,U_0^2$. The result of substituting these values in (20) yields the expression $v_n^2/U_0^2=-2+3(1.0514)^n$, shown as curve (A) in Fig. 6. The slope is relatively steep for small values of n because noise impulses add to the exponential growth due to the negative conductance. After about ten cycles, the effect of additional noise impulses becomes negligible and the plot becomes essentially straight on the semilog coordinates used.

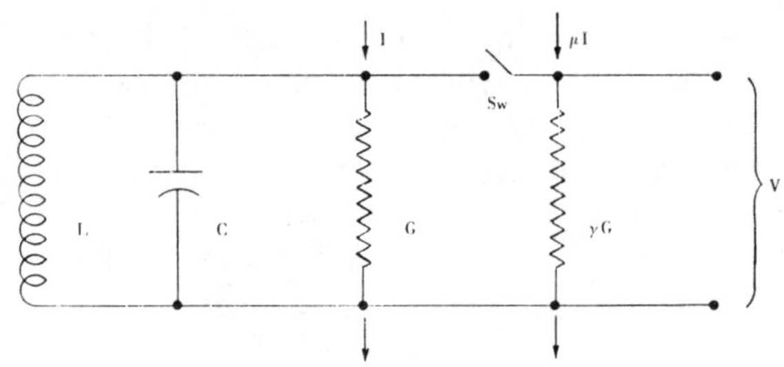

Fig. 5—Second circuit model.

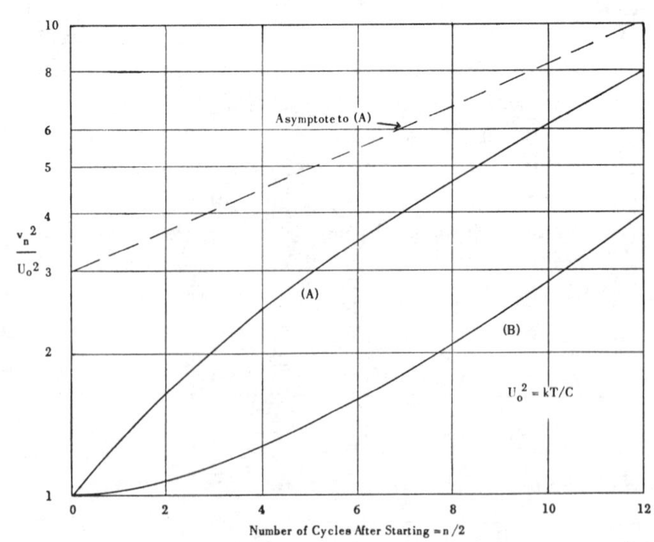

Fig. 6—Growth of the cosine component of voltage: (A) Parameters changed discontinuously. (B) Parameters changed continuously.

D. *Build-Up with Continuously Variable Parameters*

The analyses of the foregoing sections have in common the underlying assumption that the parameters of the system are changed discontinuously, that is, in a time which is short compared with one cycle of the oscillation. In practice, such a sudden change would be associated with transients which are large compared with the level of noise. Moreover, no means is known for producing switching rapid enough to meet this criterion at microwave frequencies. Therefore, the concern here is with systems in which the total conductance is changed in some orderly manner from a positive to a negative value. Because such a change is ordinarily made by means of an electronic device and because appreciable noise contributions are associated with such devices, it is quite desirable to obtain an analysis which permits time variation of the noise current.

The system under consideration is shown in Fig. 7 which differs from Fig. 3 only in that G and I are both functions of time. At time $t=0$, the voltage has an initial equilibrium value U_0 identifiable with the system comprised of L, C, G_0 and I_0, where G_0 and I_0 refer to the values of these parameters during all negative values of time. Starting at time $t=0$, both G and I change, ordinarily in such a way as to increase I and to decrease G from a positive to a negative value. No serious restriction of generality results from the assumption that neither G nor I changes appreciably during any one cycle of the oscillation. Therefore, the convolution integral and methods of linear, time-variable, systems may be used. The analysis of superregeneration by Wheeler [16], and others has much in common with the following discussion.

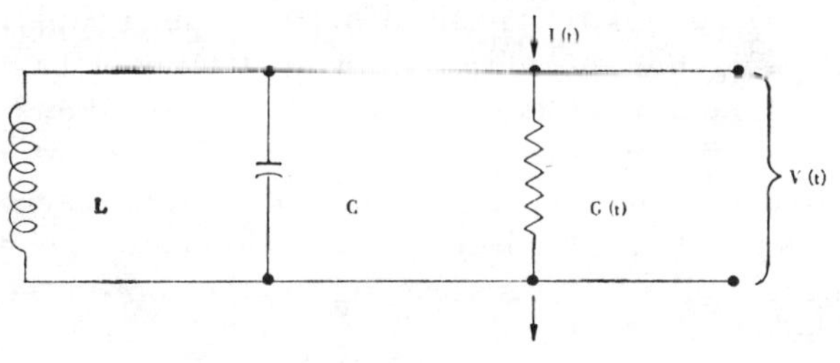

Fig. 7—Third circuit model.

Drawing on the results derived in preceding sections, the following expression can be written for the cosine component of noise in Fig. 7 at the end of n half-cycles:

$$\begin{aligned} v_n^2 = U_0^2(\beta_1\cdot\beta_2\cdot\beta_3\cdot\beta_4 \cdots \beta_n) \\ + u_1^2(\beta_2\cdot\beta_3\cdot\beta_4 \cdots \beta_n) \\ + u_2^2(\beta_3\cdot\beta_4 \cdots \beta_n) \\ + u_3^2(\beta_4 \cdots \beta_n) \\ + \cdots\cdots\cdot \\ + \cdots\cdot \\ + u_n^2 \end{aligned} \qquad (21)$$

where u_n and β_n refer to the noise impulse and the expansion factor applicable to the nth half cycle.

This expression may be written more compactly as

$$v_n^2 = \sum_{i=0}^{i=n} u_i^2 \prod_{j=i+1}^{j=n} \beta_j \qquad (22)$$

where the initial condition is represented by the special convention $u_0^2 = U_0^2$.

Computation is facilitated by using an expression equivalent to (21) which has the form

$$v_n^2 = \beta_n v_{n-1}^2 + u_n^2. \qquad (23)$$

The net change between v_n^2 and v_{n-1}^2 is

$$\frac{dv_n^2}{dn} = (1 - \beta_n)v_{n-1}^2 + u_n^2. \qquad (24)$$

Because this equation is not readily integrated in its general form, a typical numerical example will be used. Assume that the initial conditions are the same as those in Fig. 6 and that the noise and negative conductance both increase (almost) linearly with time so that ten full cycles are required to carry out the transition which is achieved stepwise by switching in Fig. 5.

Consistent with the above,

$$u_n^2 = 0.05U_0^2\left(1 + \frac{n}{20}\right) \qquad (25)$$

and

$$\beta_n = 0.95 + n/200. \qquad (26)$$

By definition,

$$v_0^2 = U_0^2. \qquad (27)$$

Therefore, iterative use of (23), (25) and (26) permits evaluation of v_n^2 for any desired value of n. The results obtained by this method are plotted as curve (B) in Fig. 6. It is noted that the slope of curve (B) at the end of ten cycles is the same as that of curve (A) for the same ordinate, and barring further changes in u and β, the two are simply displaced in time for all larger amplitudes.

E. *Generalizations on Build-Up*

Each of the foregoing analyses has led to a Gaussian distribution for the separate sine and cosine components of voltage. Hence, we conclude that for all oscillators of this type and at all times during the build-up interval, the distribution of the separate components is Gaussian and the distribution of the total amplitude is Rayleigh. This result, which follows from the fact that successive impulses are statistically independent, is quite general, and is later used to determine a distribution function which describes the jitter in the starting time of an ensemble of oscillators.

An additional feature of oscillation build-up is worthy of mention. Prior to turn-on, the noise voltage of a given oscillator remains small and changes randomly throughout a range comparable with its greatest magnitude. Likewise, the phase is subject to large random excursions. After turn-on, the level increases in a generally exponential manner to a relatively large value. Because the perturbing effects of noise remain nearly constant, as growth occurs it becomes increasingly less probable that the phase will be greatly shifted or that an oscillator which started with a smaller-than-normal amplitude will be able to advance into a larger-than-normal amplitude class. Explicitly, in terms of Fig. 6, increase of time makes it less and less likely that a given oscillator can move from one group to another. The situation might be paraphrased as "the social strata become more firmly fixed as their general level rises."

To obtain a numerical measure of this decreasing mobility of level, a digression is in order to introduce the autocorrelation function, which is a widely used measure of predictability. The formal definition for the autocorrelation of a stationary statistical function for the time interval τ is

$$R(\tau) = \lim_{T\to\infty} \frac{1}{T}\int_0^T v(t)v(t+\tau)dt. \tag{28}$$

For steady-state noise in a passive (rectangular) narrow band circuit, (156) of Bennett, [1], may be converted to the form

$$R(\tau) = \frac{2U^2GQ}{\omega_0\tau}\sin\frac{\omega_0\tau}{2Q}\cos\omega_0\tau. \tag{29}$$

For the present purpose, it is convenient to consider the period of one full cycle and to introduce as the *autocorrelation per cycle*,

$$R_1 = (U^2GQ/\pi)(\sin\pi/Q). \tag{30}$$

This is recognized as the product of a (sine x)/x function and the mean-square power in the circuit. Interest centers in the extent to which the (sine x)/x function differs from unity. Eliminating Q with the use of (11) and (15), this expression can be converted to

$$R_1 = \frac{U^4G}{u^2}\sin\frac{u^2}{U^2}\cdot \tag{31}$$

Specifically, this is the predictability during the period of one cycle of a function having a Rayleigh distribution with mean-square magnitude U^2 and perturbed by noise impulses of magnitude u^2.

Returning to a consideration of the build-up process in an active circuit, it is seen that here, too, are found a Rayleigh amplitude distribution, a mean square amplitude which changes relatively little during any one cycle, and noise impulses measured by u^2. Therefore, it is concluded that (31) also represents the autocorrelation per cycle of an expanding wave, provided U^2 is interpreted as the mean-square amplitude of the total wave over the cycle in question, and u^2 represents the mean-square value of the impulses received during that cycle.

The result just derived does not distinguish between amplitude and phase perturbations, nor is it particularly easy to interpret. Fortunately, relevant statistics for both amplitude and phase distributions have been worked out and are tabulated. The essential fact is that, during any one cycle, a sinusoidal oscillation representable as a unique phasor is modified by the addition of another phasor which is randomly distributed. The situation is illustrated by Fig. 8, which shows in an exaggerated form the effect produced upon the oscillation phasor by noise impulses received during a single cycle. The extensive curves published by K. A. Norton, *et al.*, [9] [10], in another connection are directly applicable to the present problem.

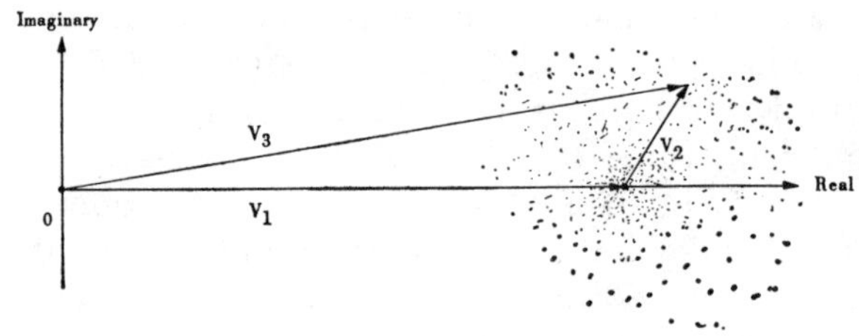

Fig. 8—Cosine wave plus random noise impulse.

F. *Jitter in Starting Time*

The starting time of an oscillator may be defined as the time after turn-on required for the voltage to build up to some preassigned reference value v_r which is below saturation but large compared to the level of noise. The starting time of a given oscillator for any single turn-on is a perfectly definite and exact quantity. However, this time varies in a statistical manner if the same oscillator is repeatedly tested, or if an ensemble of many oscillators is tested once.

To determine this distribution, it is noted that in the region of interest the level of a given oscillator increases exponentially with time. Therefore, the concern here is with amplitude distributions expressed in terms of a logarithmic (neper) scale rather than a linear scale.

The situation is illustrated by Fig. 9, which shows the growth of voltage toward the reference level. Like Fig. 6, it is plotted to semilog scales. However, several lines representing the cumulative probability distribution replace the single mean-square line of build-up. The superimposed curve at the left, representing the cumulative probability distribution, is replotted from Fig. 2. The shape of this curve is markedly altered by the change from a linear to a logarithmic abscissa. However, no mathematical subtleties are involved in this transformation of variable.

The construction of Fig. 9 makes clear the fact that the cumulative probability distribution of starting times (the upper superimposed curve) has the same shape as that for the logarithm of the voltages. Therefore, the cumulative Rayleigh distribution plotted against a suitable logarithmic abscissa represents the universal distribution of jitter in starting time. This curve is plotted upright and on a general basis as curve (B) of Fig. 10. The derivative of this curve, which represents the relative probability density distribution, is plotted as curve (A).

The reversal of the time and voltage scales of Fig. 10 arises from the fact that oscillators with relatively large amplitudes have shorter starting times than those with smaller amplitudes. Moreover, the effect of the logarithmic voltage scale is to inhibit short build-up times and to increase the number of long build-ups.

The numerical factor between the two abscissa scales of Fig. 10 derives from the exponential build-up equation. Setting[3]

$$\ln (v/v_r) = n\delta/2 = 1, \tag{32}$$

$$\tau = n/2F, \qquad = n\pi/\omega_0 \tag{33}$$

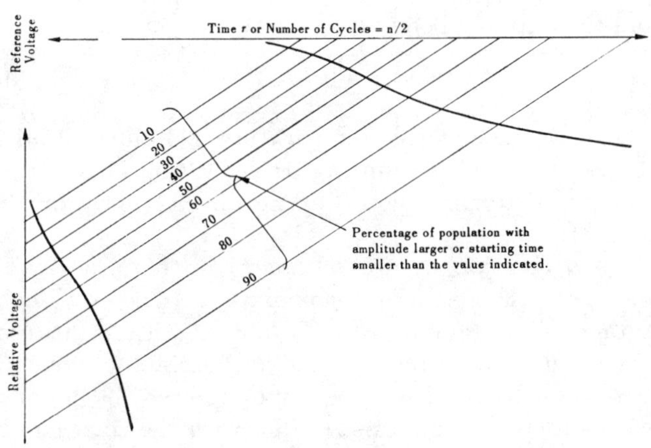

Fig. 9—Jitter in starting time caused by Rayleigh amplitude distribution.

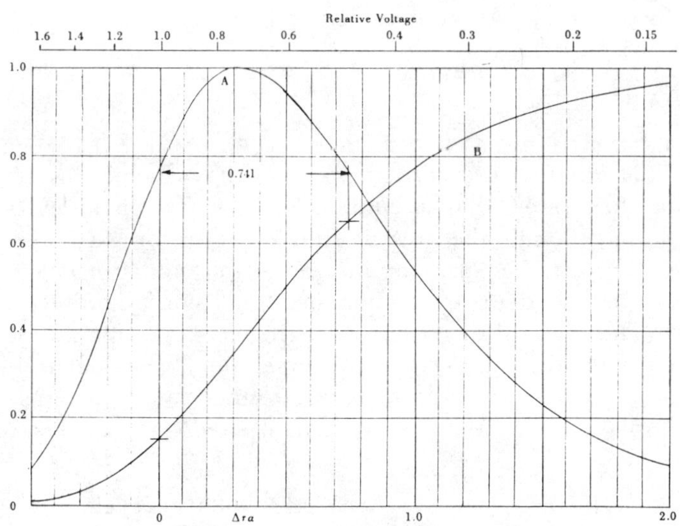

Fig. 10—Distribution of starting time jitter. (A) Relative probability density. (B) Cumulative probability.

[3] The symbol σ which would normally be used in (34) has already been used for the standard derivation and is therefore replaced by α.

and

$$\alpha = -F\delta, \tag{34}$$

one sees that a voltage ratio of one neper corresponds to a time interval

$$\Delta\tau = 1/\alpha. \tag{35}$$

Because the density distribution curve (A) of Fig. 10 is skewed, it is necessary to treat it with some caution. Exploring the steepest portion of curve (B), we see that 50 per cent of the total population (15 per cent to 65 per cent) is included in a voltage ratio of 2.09 corresponding to 0.741 neper. The associated time jitter, which is the principal result of this section is

$$\tau_j = 0.741/\alpha \text{ seconds.} \tag{36}$$

Ignoring the question of skew distribution, this value is twice the probable error of an observed starting time.

The important conclusion is that *the amount of jitter in the starting time of an oscillator varies inversely with the expansion coefficient* α *and is independent of all other influences.* In particular, reduction of the noise level of the oscillator affords no relief at all. Except for a slight numerical refinement, this result is in agreement with one previously derived, [6], by a much less rigorous argument.

In certain situations the magnitude of the starting transient produced by switching on an oscillator is comparable to the noise level. Under these conditions, the amount of jitter and the randomness of starting phase are reduced. It is clear that the foregoing methods, together with the results of Norton, *et al.*, [9], [10], are sufficient to calculate the behavior which will occur under such circumstances.

Finally, for the sake of emphasis, it is repeated that none of the foregoing results predict the behavior of a particular oscillator when turned on once; they correctly represent the statistical behavior of a given oscillator subjected to many repetitive tests or to the behavior of a great many identical oscillators tested simultaneously. The behavior of a given oscillator is thoroughly random and disorderly in the interval preceding and immediately following turn-on. However, as build-up occurs, the noise disturbances become relatively less significant, and the growth is more and more accurately exponential. The growth period comes to an end when saturation sets in and brings about the equilibrium condition commonly referred to as sustained oscillation which is treated in the following section. The essential new fact is that nonlinearity now couples the sine and cosine components so that they are no longer statistically independent. Fortunately, amplitude and phase components of an envelope function *are independent*, even in a nonlinear system. The results obtained in the following section are based upon this fact.

II. Sustained Oscillation

A. Amplitude Perturbation

In the absence of noise, an oscillator would generate a succession of nearly sinusoidal cycles identical in amplitude, period and shape. The presence of noise perturbs this situation slightly, making both period and amplitude vary randomly about the previous undisturbed value. To evaluate the magnitude of this disturbance, it should be noted that the amplitude of any oscillator changes slightly when a locking or synchronizing signal is injected. If the injected signal is small and has exactly the same frequency and phase as that of the spontaneous oscillation, then the amplitude change is directly proportioned to the input amplitude. This condition is representable by

$$sG = I/\Delta V \qquad (37)$$

where ΔV is the increase in amplitude which results from injection of a synchronizing current I. Here G is the normal circuit conductance and s is a dimensionless parameter which depends upon the degree of limiting or saturation and is of the order of ten in typical oscillators. The product sG represents an equivalent conductance closely related to one of the compliance coefficients of Huntoon and Weiss, [8]. Additional insight into the situation is gained by noting that if it is increased or decreased slightly from its normal value, the amplitude of a free-running oscillator relaxes exponentially toward its normal value at a rate corresponding to the time constant C/sG where C is as usual the capacitance of the oscillatory circuit. The introduction of the parameter s represents one form of the process of equivalent linearization. Because this step is basic to what follows, it is well to examine its validity. Experience shows that within a narrow range of amplitude, the deviation *is* accurately proportional to the perturbing influence and that the deviations of interest *are* very small. Therefore, the procedure *is* valid. The value of s depends upon the passive components of the system, the characteristics of the tube or other nonlinear elements, and the magnitude of the applied voltages. Thus, while a unique value of s exists for each condition of oscillation, it may be hard to evaluate in practical situations.

The treatment is directly applicable to conventional self-limiting oscillators. Oscillators employing automatic output control or other sophisticated forms of limiting will relax toward the normal level of oscillation in a somewhat more complicated manner. Therefore, the amplitude perturbations calculated by the following method are only approximate for unconventional forms of oscillators.

Let us now assume that a noise-free system is in the condition of sustained oscillation described by

$$v = A \cos \omega t, \qquad (38)$$

and that starting at time $t = 0$, noise consisting of a series of impulses is introduced. Consistent with the methods developed in Section I on the Initiation of Oscillation, consider first only those impulses occurring at odd quarter-cycles and generating cosine-phase voltages.

In view of the foregoing discussion, it can be seen that the result of the noise is the addition to the original voltage wave of another which has the statistical properties associated with the given noise impulses operating on a passive equivalent circuit consisting of L, C, and sG. That is, the cosine-phase noise impulses act to amplitude-modulate the original wave.

Consistent with (11), the mean-square value of the envelope deviation reached in the steady state is

$$U^2 = u^2/s\delta, \qquad (39)$$

where U^2 and δ have the values given in (16) and (15), respectively.

Elimination of undesired terms yields, as the mean-square voltage deviation of the oscillation envelope from the unperturbed value,

$$V^2 = kT/sC \qquad (40)$$

where T is the effective temperature including all noise sources, and s is the limiting or saturation factor. It is seen that this expression differs from (12) only by the factor s.

The derivation of (40) shows that the situation is exactly the same as if noise currents produced by a conductance G at temperature T acted upon an antiresonant circuit consisting of L, C and a conductance sG. Therefore, the resulting spectrum is lower and broader by a factor s than that associated with the corresponding passive circuit. Taking this factor into account, together with the quantity 2 arising from the fact that only the cosine component is involved, an equation from Pierce, [12], can be modified to read

$$V_1^2(\omega) = \frac{2kTG}{s^2G^2 + 4C^2(\omega - \omega_0)^2} \text{ effective volts}^2 \text{ per cycle.} \qquad (41)$$

This equation is regarded as one of the principal results of the present paper, and, as will be proven later, is quite general. The present simple derivation gives correct results because amplitude and phase disturbances are independent, even though the system is nonlinear.

The spectral distribution of the envelope function is often of interest. Because the upper and lower sidebands are coherent and add directly, voltage must be superimposed rather than power. The desired expression is

$$V_e^2 = \frac{8kTG}{s^2G^2 + 4\omega_e^2C^2} \text{ effective volts}^2 \text{ per cycle,} \qquad (42)$$

where $\omega_e = (\omega - \omega_0)$ is the envelope angular frequency.

In anticipation of future use, consider the autocorrelation function corresponding to the envelope distribution of (42). The needed partial autocorrelation function is given by the definition.

$$R_1(\tau) = \frac{1}{2\pi}\int_0^\infty v^2(\omega)\cos(\omega\tau)d\omega. \tag{43}$$

Eq. (42), with special consideration of the origin, results in

$$R_1(\tau) = \frac{A^2}{2} + \frac{4}{\pi}\int_0^\infty \frac{kTG}{s^2G^2 + 4\omega_e^2C^2}\cos(\omega_e\tau)d\omega_e \tag{44}$$

$$= \frac{A^2}{2} + \frac{4kT}{\pi s^2 G}\int_0^\infty \frac{\cos(\omega_e\tau)d\omega_e}{1 + 4\omega_e^2C^2/s^2G^2}. \tag{45}$$

Using Peirce, [11], formula 490, letting

$$x = 2\omega_e C/sG, \tag{46}$$

and normalizing to $A^2/2$ yields

$$R_1(\tau) = 1 + \frac{2kT}{A^2 sC}e^{-(sG\tau/2C)}. \tag{47}$$

In all typical situations the first term, specifying the nominal power output, is very large compared with the second term which describes amplitude fluctuations.

B. Phase Perturbation

To evaluate the phase perturbations of the oscillator output we consider the sine-phase noise produced by impulses occurring in even-numbered quarter cycles. The situation is best visualized in terms of Fig. 8 showing a phasor, which, in a noise-free system, would have constant length and a constant angular velocity. Length perturbations are counteracted by the inherent limiting action, and produce the random amplitude modulation described in the previous section.

In contrast, no mechanism exists to counteract phase perturbations. The situation corresponds to a well-regulated clock; if once set forward it continues to read fast until it is reset. The sine-phase noise impulses consitute a series of random clock-settings, which produce much the same effect as a random deviation of rate.

Using (16) we obtain, as the standard deviation or rms value of the phase error accumulated in the period of n half-cycles, the value

$$\phi_n = \sqrt{n}u/A = \sqrt{\pi nkTG/\omega_0 C^2 A^2}\ \text{rms radians}. \tag{48}$$

It is clear that this result is valid for small values of n because $u \ll A$ and the angle ϕ is closely equal to its tangent. For large values of n the accumulated phase error also becomes large, and it might be thought that the method would fail. Actually (48) is valid for all values of n. The situation is saved by the fact that noise impulses continually and naturally resolve themselves parallel and perpendicular to the instantaneous direction of the phasor. This corresponds to the small and gradual readjustment of the sampling interval mentioned in Section I-A.

To obtain a basis for converting (48) to other more useful forms we note that the total generated power is given by

$$P = A^2G/2. \tag{49}$$

Eqs. (14) and (33) may be used with (49) to convert (48) to the convenient form

$$\phi_n = \sqrt{\pi n\omega_0 kT/2PQ^2} = \sqrt{\omega_0^2 kT\tau/2PQ^2}\ \text{rms radians}. \tag{50}$$

Consistent with (7), the standard deviation of the cumulative phase error increases with the square root of the number of cycles (or half-cycles) which elapse. Thus, while the probable *phase error* increases without limit, the probable *frequency error* approaches zero over a long period of time because it varies in the manner $\sqrt{n}/n = 1/\sqrt{n}$.

In an interval of n half-cycles, the total average phase will advance by $n\pi$ radians. Thus, the standard deviation or rms value of the fractional phase or frequency in an interval τ is

$$\phi_n/\phi = \sqrt{\omega_0 kT/2n\pi PQ^2} = \sqrt{kT/2\tau PQ^2}. \tag{51}$$

It is significant and perhaps surprising that ω_0 does not appear in this expression; (51) is important because it represents the ultimate limit of the accuracy with which the frequency of an oscillator may be observed in a given period τ.

A useful measure of frequency indeterminacy may be defined as the *period of coherence*. This is the time interval τ_c required for the standard deviation of the phase to increase to a full radian. Setting (50) equal to one radian, the period of coherence may be stated

$$\tau_c = 2PQ^2/kT\omega_0^2. \tag{52}$$

This period, which is quite long for typical oscillators, is the interval during which the rms phase deviation between an actual and an ideal oscillator would reach one radian, and represents approximately the longest period over which a useful measurement may be made by the zero-beat method.

The autocorrelation is established as the most meaningful measure of coherence. Referring only to phase perturbations,

$$R_2(\tau) = \frac{A^2}{2}\overline{\cos\{\omega_0 t + \phi(t)\}\cos\{\omega_0[t+\tau] + \phi(t+\tau)\}} \tag{53}$$

where the bar indicates averaging over the ensemble for the time interval τ. This expression is equivalent to

$$R_2(\tau) = \frac{A^2}{2}\overline{\cos\phi(t)\cos[\omega_0\tau + \phi(t+\tau)] - \sin\phi(t)\sin[\omega_0\tau + \phi(t+\tau)]}, \tag{54}$$

$$\frac{A^2}{2}\overline{\cos[\omega_0\tau + \phi(t+\tau) - \phi(t)]} \tag{55}$$

$$= \frac{A^2}{2}\cos\omega_0\tau\,\overline{\cos[\phi(t+\tau) - \phi(t)]} - \sin\omega_0\tau\,\overline{\sin[\phi(t+\tau) - \phi(t)]}. \tag{56}$$

It has already been shown that $[\phi(t+\tau)-\phi(t)]$ is a random variable with a standard deviation given by (50). Since the argument has a symmetrical distribution, the term $\overline{\sin[\phi(t+\tau)-\phi(t)]}$ is equal to zero, and the desired autocorrelation simplifies to

$$R_2(\tau) = \frac{A^2}{2}\cos(\omega_0\tau)\cdot\overline{\cos[\phi(t+\tau) - \phi(t)]}. \tag{57}$$

To evaluate the last term, a cosine weighting function is applied to the Gaussian phase distribution. The probability distribution of the phase deviation ϕ accumulated in a time interval τ may be written in the form

$$p = be^{-a^2\phi^2}, \tag{58}$$

where, from (50) and the definition of the normal distribution,

$$b = \sqrt{PQ^2/\pi\omega_0^2\tau kT} \tag{59}$$

and

$$a^2 = PQ^2/\omega_0^2\tau kT. \tag{60}$$

Thus,

$$\overline{\cos[\phi(t+\tau) - \phi(t)]} = b\int_{-\infty}^{\infty} e^{-a^2\phi^2}\cos\phi\, d\phi. \tag{61}$$

Use of B. O. Peirce, [11], formula 508 reduces this expression to

$$\overline{\cos[\phi(t+\tau) - \phi(t)]} = \frac{b\sqrt{\pi}}{a}e^{-(1/4a^2)}. \tag{62}$$

As a result, the autocorrelation with respect to phase fluctuations is

$$R_2(\tau) = \frac{A^2}{2}\cos(\omega_0\tau)e^{-\omega_0^2\tau kT/4PQ^2}. \tag{63}$$

To obtain the corresponding spectral distribution, the following identity is used:

$$v^2(\omega) = 4\int_0^{\infty} R(\tau)\cos(\omega\tau)d\tau. \tag{64}$$

Substitution of (63) in this expression yields

$$v_2^2(\omega) = 2A^2\int_0^{\infty}\cos(\omega_0\tau)\cos(\omega\tau)e^{-\omega_0^2\tau kT/4PQ^2}d\tau. \tag{65}$$

This formidable expression is evaluated by the use of (7), section 262 of Bierens de Haan, [3], which gives

$$\int_0^{\infty} e^{-px}\cos(qx)\cos(rx)dx$$

$$= \frac{p^2 + q^2 + r^2}{[p^2 + (q-r)^2][p^2 + (q+r)^2]}, \tag{66}$$

where

$$r = \omega, \qquad q = \omega_0 \quad \text{and} \quad p = \omega_0^2 kT/4PQ^2.$$

In typical situations p is very small compared to q and r. Moreover, since the interest here is in the narrow-band situation, ω may be replaced by ω_0 except where the difference occurs. With these simplifications (66) reduces to

$$v_2^2(\omega) = \frac{pA^2}{[p^2 + (q-r)^2]}\cdot \tag{67}$$

By substituting and using (14) and (49),

$$v_2^2(\omega) = \frac{2kTG}{(\omega_0^2 CkT/2PQ^2)^2 + 4C^2(\omega_0 - \omega)^2}$$

effective volts² per cycle. (68)

From this expression, by doubling the denominator term, the full noise bandwidth[4] is obtained:

$$\delta\omega = \omega_0^2 kT/2PQ^2 \text{ (radians per second)}, \tag{69}$$

which is the reciprocal of the period of coherence τ_c given by (52).

Finally, the fractional bandwidth of phase noise is

$$\delta\omega/\omega_0 = \omega_0 kT/2PQ^2. \tag{70}$$

These three equations constitute the principal result of the present section. Because the numerators and second denominator terms of (41) and (68) are identical, it follows that, at frequencies remote from the carrier, the disturbances caused by amplitude and phase fluctuations are equal.

[4] This result is exactly half as large as that given as (15.21) in [6], which was derived in a much simpler manner. The discrepancy is attributed to the fact that the earlier treatment included equally the effects of amplitude and phase fluctuations.

C. Complete Spectrum

The product of (47) and (63) gives the following equation as the autocorrelation function for the complete wave.

$$R(\tau) = \frac{A^2}{2}\cos(\omega_0\tau)e^{-(\omega_0{}^2\tau kT/4PQ^2)} + \frac{kT}{sC}\,e^{-(sG\tau/2C+\omega_0{}^2\tau kT/4PQ^2)}. \tag{71}$$

As noted previously, in all practical situations the coefficient of the first term is extremely large compared with the coefficient of the second term. Likewise, the first term of the exponent in the second term is large compared with its associated quantity. In the light of these statements, the operation indicated by (71) for determining the complete spectrum simply leads to the sum of (41) and (68). Therefore, it is concluded that the total spectrum of an oscillator represents the sum of two expressions already derived. These are similar in form but have very different proportions. The amplitude spectrum which is very weak is also extremely broad, whereas the phase perturbed spectrum contains nearly all the energy of the wave in a very narrow spectrum. Because the derivation of the phase-perturbed spectrum does not involve nonlinearity, the results are applicable to all kinds of oscillators, including those using automatic output control.

III. Harmful Effects of Noise

A. General Discussion

The superheterodyne method is used in nearly all kinds of receivers to simplify the problem of obtaining the needed amplification and selectivity.

In typical situations, the local oscillator supplies a voltage large compared with that to be received, and the desired difference frequency is produced in some sort of nonlinear element, usually a rectifier. Such nonlinear elements preserve phase relationships, so that any irregularity in the phase of the local oscillator appears at the output, indistinguishable from phase irregularities in the input signal. That is, phase noise of the local oscillator is fully preserved.

Receivers for frequency-modulated waves usually employ clipping in the intermediate-frequency amplifier. Such a clipper removes amplitude fluctuations of all kinds, whether caused by the local oscillator or otherwise. In view of these facts, it is concluded that *for receivers of frequency modulated signals, the amplitude fluctuations of the local oscillator have no significance and any noise contributed to the output is caused by phase perturbation and is calculable by use of (50).*

Just the opposite situation exists in receivers for amplitude modulated signals. To show this, assume that no signal is present and note that for small variations of the local oscillator level the rectified output varies in a linear manner with changes of the oscillator level. Stated in a different way, amplitude perturbations in the output of the local oscillator are indistinguishable from the desired signal. Phase perturbations have substantially no effect because the rectification process (in the absence of a signal) is insensitive to phase. Thus, the threshold sensitivity of a receiver for amplitude modulated signals is affected only by amplitude perturbations described by (42). In particular, it is noted that knowledge of the complete noise spectrum is neither necessary nor desirable for the calculation of local oscillator noise. The reason is that correlation exists between noise components in the upper and lower sidebands, both of which are ordinarily applied to the nonlinear element. Therefore, the net result differs from what would be anticipated on the basis of simple spectrum considerations.

B. Numerical Examples

To appraise the implications of the foregoing analysis, some numerical values should be obtained. As two rather extreme examples, consider a K-band reflex klystron of the variety commonly used for local oscillators and a low-frequency precision oscillator using a very excellent quartz crystal resonator. Numerical values typical of such oscillators are given in Table I. The very low value of power attributed to the low-frequency oscillator is identified with the fact that precision crystals are adversely affected by higher powers. The capacitance value is an equivalent value appropriate for the present purposes.

Substitution in various formulas yields the results presented in Table II.

Suppose that a radar receiver uses a simple superheterodyne arrangement, and that the first mixer may be taken as an ideal noise-free diode. Thus, assume that in the absence of a received signal the noise voltage delivered to the input of the IF amplifier is simply the envelope of the wave developed by the local oscillator. Substituting in (42) the values given in Table I for the klystron and choosing a nominal intermediate frequency of 30 mc,

$$v_s{}^2 \approx 3 \times 10^{-16} \text{ volts}^2 \text{ per cycle.} \tag{72}$$

This value changes relatively slowly in the region of 30 mc. Therefore, in a 5-mc band centered on 30,

$$\bar{v}^2 = 1.4 \times 10^{-9} \text{ volts}^2 \quad \text{or} \quad \bar{v} = 3.8 \times 10^{-5} \text{ volts.} \tag{73}$$

Because in typical cases the level of oscillation is about five times higher than that applied to the mixing rectifier, 6×10^{-6} volts may be taken as approximating the threshold of sensitivity of the receiver. Assuming that the impedance level of the signal source is 100 ohms, the limiting power sensitivity is 4×10^{-13} watts or -94

TABLE I
OSCILLATOR PARAMETERS

	Klystron	Crystal
T degrees K	2×10^{4}	4×10^{2}
C farads	2×10^{-12}	4×10^{-5}
G mhos	2×10^{-4}	2×10^{-4}
s	5	5
ω_0 radians per second	10^{11}	10^{7}
Q	10^{3}	2×10^{6}
P watts	5×10^{-2}	10^{-5}

TABLE II
OSCILLATOR NOISE CHARACTERISTICS

	Klystron	Quartz
Bandwidth envelope noise, radians per second	5×10^{8}	25
Bandwidth phase noise, radians per second	2.76×10^{-2}	6.9×10^{-15}
Envelope deviation, rms volts	1.66×10^{-4}	5.25×10^{-9}
Standard deviation of frequency per second	1.66×10^{-12}	8.3×10^{-15}
Period of coherence, seconds	36.2	1.45×10^{14}
Fractional phase noise bandwidth	2.76×10^{-13}	6.9×10^{-22}

dbm, a plausible figure. The sensitivity of a practical receiver using such a local oscillator will be inferior to this figure because of signal loss and noise in the converter. Conversely, as is well known, nearly all the noise caused by the local oscillator may be eliminated by use of a balanced mixer. Therefore, other factors limit the sensitivity of typical practical radar receivers.

IV. Effect of an Auxiliary Filter

It is clear that any kind of imperfect spectrum can be improved by passage through a sufficiently narrow filter. This is true because the energy storage of the filter elements tends to remove existing perturbations in both amplitude and phase. Because the primary spectrum of practical oscillators is very narrow, one might think that no substantial improvement could be obtained with filters of conventional selectivity. This pessimism is unjustified because the most troublesome effects come from perturbations which occur in relatively short time intervals and therefore correspond to relatively broad spectra.

To explore this situation, refer to Fig. 11 which differs from Fig. 3 by the addition of the load circuit ($L'C'$, G'), which is tuned to ω_0 and presumed to have such a high impedance that the noise characteristics of the voltage V are unaffected. Interest centers in the spectral purity of I' as compared with that of V. Evidently, an important parameter is the selectivity of the tuned load circuit in relationship to that of the basic oscillator. It is readily seen that rapid variations of amplitude or phase are smoothed out by the tuned load circuit but that slow changes are little affected. Stated another way, the added filtering removes remote spectral components but transmits nearby ones.

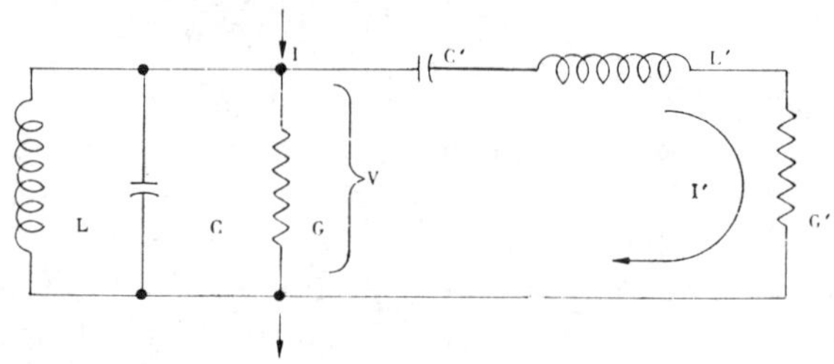

Fig. 11—Effect of tuned load circuit.

The foregoing analysis has shown that at frequencies far from the nominal frequency of oscillation, the spectrum of an oscillator is correctly represented by calculating the filtering effect of the resonator upon the smooth noise spectrum of the driving current I. This idea is readily extended to cover the present situation. The relative suppression of noise in I' with respect to that in V is obtained directly from the selectivity of the added elements.

The use of auxiliary filtering in low-frequency oscillators is rarely necessary. However, at microwave frequencies it is considerably more attractive. This is true for two principal reasons. Microwave oscillators such as magnetrons and reflex klystrons are inherently quite noisy, and mechanical construction problems often preclude the attainment of a large value of Q in the basic resonator. Under these circumstances, it is often possible to obtain substantial benefits from the addition of an auxiliary high-Q resonator. Ordinarily, the added cavity is tightly coupled to the other so that a single resonance results. In effect, large values of inductive and capacitive susceptance are added to the basic circuit without appreciably affecting the inherent impedance level. This arrangement increases the long-term stability of the system to the extent that the added susceptances are more stable than those of the original internal resonator. However, superior performance as a local oscillator is to be expected if the auxiliary cavity is loosely coupled. This is true because the attenuation of remote spectral components produced by the two resonators is compounded in nepers or db in this situation, whereas the tightly coupled arrangement produces a fixed additive figure which is only 6 db if the effect of the added resonator is to double the working Q. Thus, the choice of loose vs tight coupling depends upon whether interest centers in slow or fast perturbations of the output amplitude and phase.

Appendix I[5]

Let $I(t)$ represent the typical member of the noise current generator ensemble and $g(t)$ the impulse response for $I(t)$ applied at time $t=0$ is given by

$$V(t) = \int_0^t v(x)g(t-x)dx.$$

The ensemble average of the squared voltage at fixed time t is then

$$\begin{aligned} \text{av } V^2(t) &= \int_0^t v(x)g(t-x)dx \int_0^t v(y)g(t-y)dy \\ &= \int_0^t \int_0^t g(t-x)g(t-y) \text{ av } [v(x)v(y)]dxdy \\ &= \int_0^t \int_0^t R_I(x-y)g(t-x)g(t-y)dxdy \end{aligned}$$

where $R_I(\tau)$ is the autocovariance function of the I ensemble. If the current generator is white thermal noise,

$$R_I(\tau) = 2GkT\delta(\tau)$$

and

$$\text{av } V^2(t) = 2GkT \int_0^t g^2(t)dt.$$

In the circuit of Fig. 3,

$$g(t) = \frac{\exp(-\alpha t)}{\omega_1 C}(\omega_1 \cos \omega_1 t - \alpha \sin \omega_1 t),$$

$$\omega_1^2 = \frac{1}{LC} - \alpha^2, \qquad \alpha = \frac{G}{2C},$$

$$\text{av } V^2(t) = \frac{kT}{C}\left[1 - e^{-2\alpha t}\left(1 + \frac{\alpha^2}{\omega_1^2} - \frac{\alpha}{\omega_1} \sin 2\omega_1 t - \frac{\alpha^2}{\omega_1^2} \cos 2\omega_1 t\right)\right].$$

Setting $t = n\pi/\omega_1$, where n is an integer, gives

[5] The author gratefully acknowledges this rigorous analysis provided by an anonymous reviewer.

$$\begin{aligned} \text{av } V^2(n\pi/\omega_1) &= \frac{kT}{C}(1 - e^{-2n\pi\alpha/\omega_1}) \\ &= \frac{kT}{C}(1 - \beta^n) \text{ where } \beta = e^{-\pi G/\omega_1 C}. \end{aligned}$$

This checks (10), except that the text sets $\omega_1 = \omega_0$, which is valid when $\alpha^2 \ll 1/LC$ but represents an unnecessary restriction.

Acknowledgment

The author is indebted to Dr. W. W. Harman for reading the manuscript and contributing to the evaluation of the spectrum. Thanks are also due to Dr. L. C. Hale for suggesting the application of the noise impulse method to this problem. Acknowledgment is also made to W. R. Bennett and to S. O. Rice for their publication of very helpful papers on the subject.

Bibliography

[1] W. R. Bennett, "Methods of solving noise problems," Proc. IRE, vol. 44, pp. 609–639; May, 1956.
[2] W. R. Bennett, "Characteristics of noise," *Electronics*, vol. 29, pp. 154–160; March, 1956.
[3] D. Bierens de Haan, "Nouvelles Tables d'Intégrales Définies," (Leyden 1867), G. E. Stechert and Co., New York, N. Y.; 1939.
[4] A. Blaquiére, "Effect of background noise on the frequency of vacuum-tube oscillators, ultimate precision of electronic clocks," *Annales de Radioélectricité Compagnies Françaises Associées de T.S.F.*; March, 1953.
[5] W. B. Davenport, Jr. and W. L. Root, "An Introduction to the Theory of Random Signals and Noise," McGraw-Hill Book Co. Inc., New York, N. Y.; 1958.
[6] W. A. Edson, "Vacuum Tube Oscillators," John Wiley & Sons, Inc., New York, N. Y., ch. 15; 1953.
[7] M. A. Garstens, "Noise in nonlinear oscillators," *J. Appl. Phys.*, vol. 28, pp. 352–356; March, 1957.
[8] R. D. Huntoon and A. Weiss, "Synchronization of oscillators," Proc. IRE, vol. 35, pp. 1415–1423; December, 1947.
[9] K. A. Norton, L. E. Vogler, W. V. Mansfield and P. J. Short, "The probability distribution of the amplitude of a constant vector plus a Rayleigh distributed vector," Proc. IRE, vol. 43, pp. 1354–1361; October, 1955.
[10] K. A. Norton, E. L. Schultz and H. Yarbrough, "The probability distribution of the resultant vector sum of a constant vector plus a Rayleigh distributed vector," *J. Appl. Phys.*, vol. 23, pp. 137–141; January, 1952.
[11] B. O. Peirce, "A Short Table of Integrals," Ginn & Company, Boston, Mass.; 1910.
[12] J. R. Pierce, "Physical sources of noise," Proc. IRE, vol. 44, pp. 601–608; May, 1956.
[13] S. O. Rice, "Mathematical analysis of random noise," *Bell Sys. Tech. J.*, vol. 23, pp. 282–332, July, 1944; and vol. 24, pp. 46–156, January, 1945.
[14] J. L. Stewart, "Frequency modulation noise in oscillators," Proc. IRE, vol. 44, pp. 372–376; March, 1956.
[15] A. Van der Ziel, "Noise," Prentice-Hall, Inc., Englewood Cliffs, N. J.; 1954.
[16] H. A. Wheeler, "A Simple Theory and Design Formulas for Super-Regenerative Receivers," Wheeler Mono., No. 3; 1953.
[17] P. M. Woodward, "Probability and Information Theory, With Applications to Radar," McGraw-Hill Book Co., Inc., New York, N. Y.; 1955.

Noise in Synchronized Oscillators

KANEYUKI KUROKAWA, MEMBER, IEEE

Abstract—**Noise in synchronized oscillators is analyzed using a simple oscillator model. It is shown that the FM noise can be considerably improved by synchronization while the AM noise degrades slightly. It is also shown how the noise spectrum becomes unsymmetrical with respect to the synchronizing frequency when it differs from the free-running frequency.**

I. Introduction

THE THEORY OF NOISE in free-running oscillators was established first by Berstein[1] in the Soviet Union in 1938. Since then, the same subject has been discussed by many authors[2] including Edson[3] and Mullen[4] who provided clear descriptions of oscillator output spectra.

The theory of oscillator phase locking has also been well known for some time.[5] The pioneering work was done by van der Pol[6] in the 1920s, followed by Krylov and Bogoliubov,[7] Huntoon and Weiss,[8] and many others who either improved or generalized his results. The same theory is also presented in Slater's celebrated book *Microwave Electronics* in slightly different form.

By combining these two theories, this paper shows how the noise of oscillators may be improved by synchronization. To this end, the fundamental differential equations for the phase and amplitude of the oscillator current are first derived in Section II, assuming the simplest model of oscillators. In Section III, important relations among the various parameters of a synchronized oscillator are obtained for later use. Although the current always increases under the synchronized condition compared with the free-running case, the output power does not necessarily increase. Therefore, the condition for the increase or decrease is established. In Section IV, the noise in free-running oscillators is discussed to provide again a background for later use. The noise in synchronized oscillators is discussed in detail in Section V. It is shown that by injecting a synchronizing signal the FM noise can be greatly improved while the AM noise degrades slightly. It is also shown how the noise spectrum becomes unsymmetrical with respect to the synchronizing frequency when it differs from the free-running frequency. The case of a noisy synchronizing signal is discussed briefly in Section VI, and Section VII contains the discussion of $1/f$ noise in oscillators.

The objective of this paper is to show how a simple oscillator model can be used to give a good explanation for the observed noise reduction[9] in synchronized microwave oscillators, including those using IMPATT and Gunn diodes. No attempt is made to generalize the discussion so as to cover more complicated oscillators[10] for which the simple model is not applicable. Although the effect of noise in synchronized oscillators has been discussed previously[11] it was not in a convenient form for explaining microwave oscillator performance.

Manuscript received November 3, 1966; revised July 15, 1967.
The author is with Bell Telephone Laboratories, Inc., Murray Hill, N. J.

II. Equations for Phase and Amplitude of Oscillating Current

As the simplest model of oscillators, let us consider the circuit shown in Fig. 1. The negative resistance $-R$ is a function of the current of the present as well as of the past (as in the case of transit time device). R_i is the internal resistance and R_0 the load resistance which generally represents a waveguide or coaxial impedance. The equation for the current $i(t)$ is given by

$$L\frac{di}{dt} + (R_i + R_0 - R)i + \frac{1}{C}\int idt = e(t). \qquad (1)$$

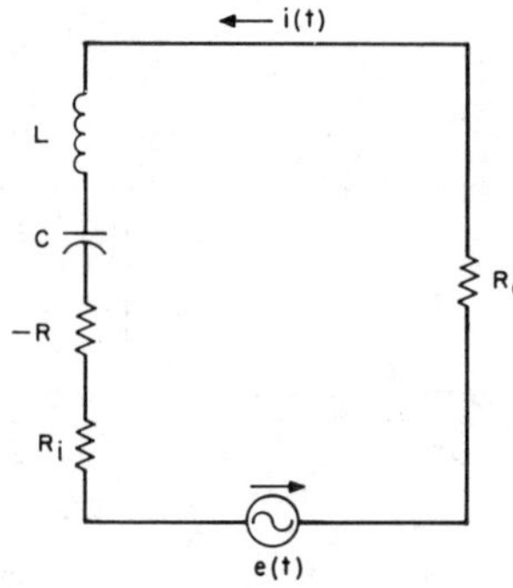

Fig. 1. Equivalent circuit of oscillators.

In the case of free-running noiseless oscillators $e(t)$ is equal to zero. Under this condition, the current is expected to be a periodic function of time which may be expressed in the form

$$i(t) = A_1 \cos(\omega t + \phi_1) + A_2 \cos(2\omega t + \phi_2) + A_3 \cos(3\omega t + \phi_3) + \cdots.$$

Since, in most cases, the higher harmonics in the output of oscillators are small, let us assume $A_2, A_3, \ldots$ are small compared to A_1. Now suppose that a small perturbation is given to the system by applying a small $e(t)$. Then $i(t)$ may no longer be a periodic function of time. However, in many cases $i(t)$ is similar to the original waveform. We restrict ourselves to these cases only and assume that $i(t)$ can be represented by

Reprinted from *IEEE Trans. Microwave Theory Tech.*, vol. MTT-16, pp. 234–240, Apr. 1968.

$$i(t) = A_1(t)\cos[\omega t + \phi_1(t)] + A_2(t)\cos[2\omega t + \phi_2(t)] + \cdots$$

where $A_n(t)$, $\phi_n(t)$, etc., do not change appreciably over the period of $2\pi/n\omega$, that is, over one cycle of each corresponding harmonic. Under this condition di/dt and $\int i dt$ can be approximated by

$$\frac{di}{dt} = -A_1\left(\omega + \frac{d\phi_1}{dt}\right)\sin(\omega t + \phi_1) + \frac{dA_1}{dt}\cos(\omega t + \phi_1)$$
$$+ \text{(slowly varying function)}\sin(2\omega t + \phi_2)$$
$$+ \text{(slowly varying function)}\cos(2\omega t + \phi_2) + \cdots$$

and

$$\int i dt = \left(\frac{A_1}{\omega} - \frac{A_1}{\omega^2}\frac{d\phi_1}{dt}\right)\sin(\omega t + \phi_1)$$
$$+ \frac{1}{\omega^2}\frac{dA_1}{dt}\cos(\omega t + \phi_1)$$
$$+ \text{(slowly varying function)}\sin(2\omega t + \phi_2)$$
$$+ \text{(slowly varying function)}\cos(2\omega t + \phi_2) + \cdots$$

where the two lowest order terms in an expansion in $1/\omega$ are retained. Substituting these results into (1), multiplying by $\sin(\omega t+\phi_1)$ or $\cos(\omega t+\phi_1)$, and integrating with respect to t from $t-T_0$ to t where T_0 is a period of the oscillation, because of the orthogonal relations between sine and cosine functions, we obtain an approximate differential equation of ϕ_1 or that of A_1 only.[7] The integration with respect to time is necessary to eliminate other variables since A_2, $d\phi_2/dt$, etc., are small, but $e(t)$ is also small and their effect may be of the same order of magnitude in the original equation. Since we are not interested in the harmonics any more, the subscript 1 is henceforth dropped. Then the preceding equations become

$$\left(-\omega L + \frac{1}{\omega C}\right) - \left(L + \frac{1}{\omega^2 C}\right)\frac{d\phi}{dt} = \frac{2}{AT_0}\int_{t-T_0}^{t} e(t)\sin(\omega t + \phi)dt \quad (2)$$

$$\left(L + \frac{1}{\omega^2 C}\right)\frac{dA}{dt} + (R_i + R_0 - \overline{R})A = \frac{2}{T_0}\int_{t-T_0}^{t} e(t)\cos(\omega t + \phi)dt \quad (3)$$

where

$$\overline{R} = \frac{2}{AT_0}\int_{t-T_0}^{t} RA\cos^2(\omega t + \phi)dt.$$

$\overline{R}$ is an average resistance over one cycle. In general, $\overline{R}$ is a function of A and may look something like Fig. 2. For steady-state free-running oscillation, since $dA/dt=0$ and $e(t)=0$, the amplitude of the current is determined from (3) as the intersection of $\overline{R}$ and R_i+R_0. Let the amplitude be A_0. Draw a tangential line to $\overline{R}$ at this point and let the intersection with the vertical axis be KA_0 above R_i+R_0 as shown in the figure. Then for a small variation ΔA from A_0 we have

$$R_i + R_0 - \overline{R} = K\Delta A. \quad (4)$$

Suppose ΔA is not equal to zero at $t=0$. If $K>0$, $|\Delta A|$ decreases exponentially with time. Therefore, A_0 represents a stable operating point. On the other hand, if $\overline{R}$ intersects R_i+R_0 from the other side, $K<0$ and $|\Delta A|$ grows indefinitely with time. This means that such a point does not provide stable operation.

The free-running frequency ω_0 is given by $1/\sqrt{LC}$ from (2).

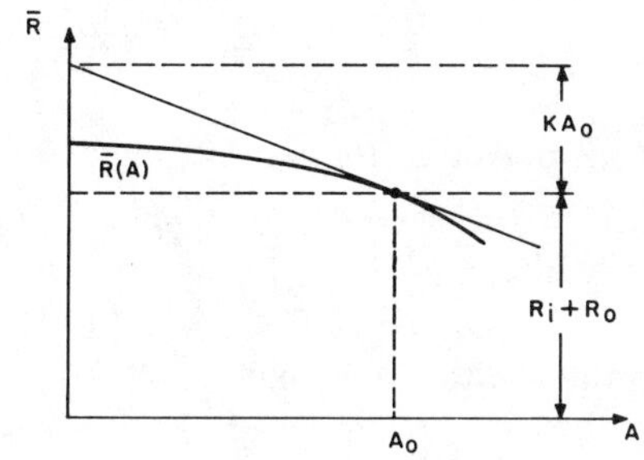

Fig. 2. $\overline{R}$ versus A.

III. Synchronized Oscillators

To study the phase-locking mechanism[5] of oscillators, let $e(t)$ be $a_0\cos\omega_s t$ in (2), where a_0 is assumed to be small. When the oscillator is synchronized, the oscillating frequency must be the same as ω_s. The right-hand side of (2) is given by $(a_0/A)\sin\phi$. Strictly speaking, A may not be the same as A_0. However, since a_0 is small the effect of the difference ΔA_0 between A and A_0 can be neglected. Thus, the right-hand side of (2) becomes $(a_0/A_0)\sin\phi$. For the steady state, $d\phi/dt=0$ and ϕ must be a constant. Let this constant be ϕ_0 and ω_s be $\omega_0+\Delta\omega_0$ where $\Delta\omega_0$ is the difference between the synchronizing and free-running frequencies. Then, from (2), one obtains

$$\Delta\omega_0 = \frac{-a_0}{2LA_0}\sin\phi_0. \quad (5)$$

Since $\sin\phi_0$ must be less than 1,

$$|\Delta\omega_0| \leq \frac{a_0}{2LA_0}. \quad (6)$$

This determines the synchronizing range $|\Delta\omega_0|_{max}$ beyond which no synchronization can take place. For stable locking, writing $\phi=\phi_0+\Delta\phi$, $|\Delta\phi|$ has to decrease with time if $\Delta\phi$ is not equal to zero at $t=0$. Thus,

$$\frac{a_0}{2LA_0}\cos\phi_0 > 0 \quad (7)$$

is required where $L+1/\omega_s^2 C\approx 2L$ is used. Equations (5) and (7) determine ϕ_0 uniquely.

The right-hand side of (3) is given by $a_0 \cos \phi$. When the oscillation is in the steady state $dA/dt=0$ and (3) gives

$$\Delta A_0 = \frac{a_0 \cos \phi_0}{K A_0} \cdot \tag{8}$$

Because of (7), ΔA_0 is always positive, i.e., the current increases.

The preceding discussions are in terms of voltage and current. However, in the microwave range, power is a more meaningful quantity. Therefore, let us try to express, for instance, the synchronizing range in terms of powers. The available power of the synchronizing source is given by

$$P_s = \frac{\left(\frac{a_0}{\sqrt{2}}\right)^2}{4R_0} = \frac{a_0^2}{8R_0}$$

and the output power of the free-running oscillator by

$$P_0 = R_0\left(\frac{A_0}{\sqrt{2}}\right)^2 = \frac{R_0 A_0^2}{2} \cdot$$

Substituting these into (6) we have

$$\left(\frac{\Delta\omega_0}{\omega_0}\right)^2 \leq \frac{1}{Q_{\text{ext}}^2} \frac{P_s}{P_0} \tag{9}$$

where the external Q is defined by

$$Q_{\text{ext}} = \frac{\omega_0 L}{R_0} \cdot$$

The injection of a synchronizing signal as well as the measurement of the output power of an oscillator are generally done through a circulator in the microwave range. In this case, the output power is not given by $R_0(A/2^2)$. However it must be the power emerging from the oscillator toward the circulator. Using the power wave concept,[12] this is given by

$$P = \frac{|V - R_0 I|^2}{4R_0}$$

where V is the voltage between the terminals of R_0 given by

$$V = E - R_0 I = \frac{a_0}{\sqrt{2}} - R_0 \frac{A_0 + \Delta A_0}{\sqrt{2}} e^{i\phi_0}.$$

Thus, to a first-order approximation, one has

$$P = P_0 + A_0 a_0 \cos \phi_0 \left(\frac{R_0}{K A_0} - \frac{1}{2}\right). \tag{10}$$

This equation tells us that P becomes larger or smaller than P_0, depending on whether $2R_0 > KA_0$ or $2R_0 < KA_0$, respectively. Also it shows that the difference between P and P_0 should decrease as $\cos \phi_0$ approaches zero, i.e., as the synchronizing frequency moves away from the free-running frequency. When the circuit is adjusted for maximum power, K is equal to $2R_0/A_0$ and P does not change by the injection of synchronizing signal as a first-order approximation.

IV. Noise in Free-Running Oscillators

For the discussion of noise, the integrals on the right-hand sides of (2) and (3) have to be evaluated first for the noise voltage $e(t)$. Since the equations are linear, the principle of superposition holds. Keeping this in mind, let us decompose $e(t)$ into elementary pulses and first consider the effect of just one pulse. The pulse is represented by $\epsilon\delta(t-t_0)$ where ϵ gives the strength of the pulse and t_0 the time of its occurrence. Both ϵ and t_0 are independent random variables from one pulse to the next. However, the time average of the square of the pulses over a period of time must be equal to $\overline{e(t)^2}$, i.e.,

$$\frac{1}{2T}\int_{-T}^{T} (\Sigma\epsilon\delta(t - t_0))^2 dt = \overline{e(t)^2}\cdot \tag{11}$$

For one pulse represented by $\epsilon\delta(t-t_0)$, the integral $2/T_0\int_{t-T_0}^{t} e(t) \sin(\omega t+\phi)dt$ becomes a rectangular pulse with constant height $(2/T_0)\epsilon \sin(\omega t_0+\phi)$ and the length T_0 as shown in Fig. 3. Since T_0 is considered short for any variation in ϕ or A, this pulse can be considered as another pulse in the form of a δ function with strength $2\epsilon \sin(\omega t_0+\phi)$ and being located at $t=t_0$ for all practical purposes of calculating the effect using (2) or (3). If $n_1(t)$ indicates the total effect of $2/T_0\int_{t-T_0}^{t} e(t) \sin(\omega t+\phi)dt$, $n_1(t)$ consists of many such pulses. Since the time average of the square of these pulses is given by

$$\frac{1}{2T}\int_{-T}^{T} (\Sigma 2\epsilon \sin(\omega t_0 + \phi)\delta(t - t_0))^2 dt$$

$$= \frac{1}{T}\int_{-T}^{T} (\Sigma\epsilon\delta(t - t_0))^2 dt$$

from (11), we have

$$\overline{n_1(t)^2} = \overline{2e(t)^2}. \tag{12}$$

Similarly, indicating the effect of $2/T_0\int_{t-T_0}^{t} e(t)\cos(\omega t+\phi)dt$ by $n_2(t)$, one obtains

$$\overline{n_2(t)^2} = \overline{2e(t)^2}. \tag{13}$$

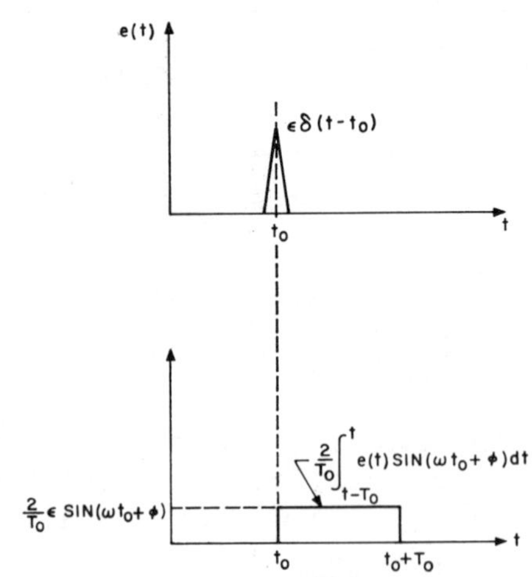

Fig. 3. The relation between $2/T_0\int_{t-T_0}^{t} e(t) \sin(\omega t+\phi)dt$ and $e(t)$ for a single pulse in $e(t)$.

Further, because of the orthogonality between the sine and cosine functions, we have

$$\overline{n_1(t)n_2(t)} = 0. \tag{14}$$

The same result can be obtained by using a narrowband noise expression of the form

$$e(t) = e_1(t)\sin(\omega_0 t + \phi) - e_2(t)\cos(\omega_0 t + \phi)$$

where $e_1(t)$ and $e_2(t)$ are slowly varying functions of time.

We are now in a position to calculate ϕ and A. For free-running oscillators $\omega = \omega_0$ and (2) becomes

$$-2L\frac{d\phi}{dt} = \frac{1}{A_0}n_1(t).$$

A may not be equal to A_0 as shall shortly be discussed. However, the difference is assumed to be small and in the denominator on the right-hand side of the above equation A has been replaced by A_0. From this, we obtain

$$|\phi(f)|^2 = \frac{|n_1(f)|^2}{4\omega^2L^2A_0^2} = \frac{2|e|^2}{4\omega^2L^2A_0^2} \tag{15}$$

where $|n_1(f)|^2$ and $|e|^2$ have the dimensions of (effective voltage)^2 per unit bandwidth and f varies from $-\infty$ to $+\infty$.

Next, for the amplitude, writing $\Delta A = A - A_0$, we have from (3)

$$2L\frac{d\Delta A}{dt} + KA_0\Delta A = n_2(t)$$

which gives

$$|\Delta A(f)|^2 = \frac{|n_2(f)|^2}{4\omega^2L^2 + (KA_0)^2} = \frac{2|e|^2}{4\omega^2L^2 + (KA_0)^2}. \tag{16}$$

An ordinary spectrum analyzer shows the spectrum density of the output waveform directly. Therefore, let us calculate it. To do so, the best way may be to calculate the autocorrelation function of the waveform and transform the result to the spectrum density function by a Fourier transformation. Since $A(t)$ and $\phi(t)$ have no correlation as is easily seen by (14), the autocorrelation function $R_i(\tau)$ of the current is easily calculated to be

$$R_i(\tau) = \overline{A(t)A(t+\tau)\cos(\omega_0 t + \phi(t))\cos(\omega_0 t + \omega_0\tau + \phi(t+\tau))}$$
$$= \tfrac{1}{2}(A_0^2 + R_{\Delta A}(\tau))\cos\omega_0\tau\,\overline{\cos(\phi(t+\tau) - \phi(t))}.$$

from which one obtains (3)

$$R_i(\tau) = \frac{1}{2}\left(A_0^2 + \frac{|e|^2}{2LKA_0}\exp\left[-\frac{KA_0}{2L}|\tau|\right]\right)\cdot\exp\left[-\frac{|e|^2}{4L^2A_0^2}|\tau|\right]\cos\omega_0\tau.$$

Thus, noting that $KA_0/2L \gg |e|^2/4L^2A_0^2$, we have

$$|i(f)|^2 = \int_{-\infty}^{\infty} R_i(\tau)e^{-i\omega\tau}d\tau = \frac{|e|^2}{8L^2}\left\{\frac{1}{(\omega-\omega_0)^2 + \left(\frac{|e|^2}{4L^2A_0^2}\right)^2} + \frac{1}{(\omega+\omega_0)^2 + \left(\frac{|e|^2}{4L^2A_0^2}\right)^2}\right\} + \frac{|e|^2}{8L^2}\left\{\frac{1}{(\omega-\omega_0)^2 + \left(\frac{KA_0}{2L}\right)^2} + \frac{1}{(\omega+\omega_0)^2 + \left(\frac{KA_0}{2L}\right)^2}\right\}. \tag{17}$$

This result is originally due to Edson. The first term on the right-hand side is called the FM noise, since a spectrum analyzer shows approximately this spectrum if the output of the oscillator is displayed after eliminating the amplitude fluctuation by a limiter. The second term is called the AM noise. Since $KA_0/2L \gg |e|^2/4L^2A_0^2$ the first term predominates over the second when ω is close to ω_0. However, for large $\omega - \omega_0$, both terms give equal contributions.

Neglecting the contribution from the noise source in R_0, $2R_0|i(f)|^2$ gives the output power spectrum density function $P(f)$ $(f>0)$ of the oscillator. For simplicity, let us neglect R_i. Then, since $\omega - \omega_0 \ll \omega + \omega_0$,

$$P(f) = \frac{\omega_0^2}{2Q_{ext}^2}N\left\{\frac{1}{(\omega-\omega_0)^2 + \left(\frac{\omega_0^2}{4Q_{ext}^2}\right)^2\frac{N^2}{P_0^2}} + \frac{1}{(\omega-\omega_0)^2 + \left(\frac{\omega_0}{Q_{ext}}\right)^2\left(\frac{KA_0}{2R_0}\right)^2}\right\} \tag{18}$$

where N is given by

$$N = \frac{|e|^2}{2\bar{R}}.$$

It is interesting to note that if an amplifier were built using the same negative resistance with the same noise source, the optimum noise measure obtainable would be given by

$$M_{opt} = \frac{N}{k \times 290°}$$

where k is Boltzmann's constant.

V. Noise in Synchronized Oscillators

In the case of free-running oscillators ϕ could take almost any value. However, when synchronized, ϕ should stay in the vicinity of ϕ_0 because of the restoring force due to the synchronizing signal. Writing $\phi=\phi_0+\Delta\phi$ and $A=A_0+\Delta A_0+\Delta A$ and assuming $\Delta\phi$ and ΔA are small, (2) and (3) become, respectively,

$$-2L\frac{d\Delta\phi}{dt} - \frac{a_0}{A_0}\Delta\phi\cos\phi_0 = \frac{1}{A_0}n_1(t)$$

$$2L\frac{d\Delta A}{dt} + KA_0\Delta A + a_0\Delta\phi\sin\phi_0 = n_2(t).$$

From these equations, we obtain

$$|\Delta\phi(f)|^2 = \frac{|n_1(f)|^2}{4\omega^2L^2A_0^2 + a_0^2\cos^2\phi_0} = \frac{2|e|^2}{4\omega^2L^2A_0^2 + a_0^2\cos^2\phi_0} \tag{19}$$

$$|\Delta A(f)|^2 = \frac{|n_2(f)|^2}{4\omega^2L^2 + (KA_0)^2} + \frac{a_0^2\sin^2\phi_0}{4\omega^2L^2 + (KA_0)^2}\,\frac{|n_1(f)|^2}{4\omega^2L^2A_0^2 + a_0^2\cos^2\phi_0} = \frac{2|e|^2}{4\omega^2L^2 + (KA_0)^2}\,\frac{4\omega^2L^2A_0^2 + a_0^2}{4\omega^2L^2A_0^2 + a_0^2\cos^2\phi_0}. \tag{20}$$

It follows from this that the envelope fluctuation becomes slightly larger by synchronization but the phase fluctuation is considerably reduced in the frequency range $|\omega| < (a_0\cos\phi_0/2LA_0)$.

In order to calculate the spectrum density function of i first calculate $R_i(\tau)$. Noting that both $\Delta\phi$ and ΔA are small, we have

$$R_i(\tau) = \tfrac{1}{2}(A_0 + \Delta A_0)^2(1 - \overline{\Delta\phi^2})\cos\omega_s\tau + \tfrac{1}{2}R_{\Delta A}(\tau)\cos\omega_s\tau + \tfrac{1}{2}A_0^2R_{\Delta\phi}(\tau)\cos\omega_s\tau + \tfrac{1}{2}A_0\sin\omega_s\tau(\overline{\Delta A(\tau)\Delta\phi(0)} - \overline{\Delta A(0)\Delta\phi(\tau)}).$$

From this, one can obtain the spectrum density function of i by a standard method

$$|i(f)|^2 = \tfrac{1}{4}(A_0 + \Delta A_0)^2\left(1 - \frac{|e|^2}{2LA_0}\frac{1}{a_0\cos\phi_0}\right)\{\delta(f-f_s) + \delta(f+f_s)\}$$
$$+ \frac{|e|^2}{8L^2}\left\{\frac{1}{(\omega-\omega_s)^2 + \dfrac{a_0^2\cos^2\phi_0}{4L^2A_0^2}} + \text{similar term with } (\omega+\omega_s)\right\}$$
$$+ \frac{|e|^2}{8L^2}\left\{\frac{1}{(\omega-\omega_s)^2 + \left(\dfrac{KA_0}{2L}\right)^2}\;\frac{(\omega-\omega_s)^2 + \dfrac{a_0^2}{4L^2A_0^2}}{(\omega-\omega_s)^2 + \dfrac{a_0^2\cos^2\phi_0}{4L^2A_0^2}} + \text{similar term with } (\omega+\omega_s)\right\}$$
$$+ \frac{|e|^2}{8L^2}\left\{\frac{\omega-\omega_s}{(\omega-\omega_s)^2 + \left(\dfrac{KA_0}{2L}\right)^2}\;\frac{\dfrac{a_0\sin\phi_0}{LA_0}}{(\omega-\omega_s)^2 + \dfrac{a_0^2\cos^2\phi_0}{4L^2A_0^2}} - \text{similar term with } (\omega+\omega_s)\right\}. \tag{21}$$

The first term is a pure spectrum and the second term represents the FM noise. The third term corresponds to the AM noise and the last term gives the interaction between the FM and AM noises. As we can easily see by comparison, the FM noise becomes of the same order of magnitude as in the free-running case when $(\omega-\omega_s)^2$ is large. However, as ω approaches ω_s a considerable improvement is obtainable by synchronization. On the other hand, the AM noise is not improved by synchronization. As a matter of fact, since $\cos^2\phi_0$ is smaller than 1, it degrades slightly. If $\cos^2\phi_0$ becomes small, i.e., if the synchronizing frequency is near the edge of the synchronizing range, both FM and AM noise become extremely large. In the limit of $\cos^2\phi_0$ approaching zero, however, the assumption of $\Delta\phi$ and ΔA being small is no longer valid and the preceding result should not be applied. The interaction term shifts the noise from one side of ω_s to the other if ω_s and ω_0 do not coincide. As a result, the noise spectrum becomes unsymmetrical with respect to ω_s and the sign of this interaction term is such that the ω_0 side of ω_s always becomes noisier.

The noise power density function $P_n(f)$ ($f>0$) corresponding to the last three terms on the right-hand side of (21) is given by

$$P_n(f) = \frac{\omega_0^2}{2Q_{\text{ext}}^2} N \left\{ \frac{1}{(\omega - \omega_s)^2 + |\Delta\omega_0|_{\max}^2 \cos^2\phi_0} + \frac{1}{(\omega - \omega_s)^2 + \left(\frac{\omega_0}{Q_{\text{ext}}}\right)^2 \left(\frac{KA_0}{2R_0}\right)^2} \frac{(\omega - \omega_s)^2 + |\Delta\omega_0|_{\max}^2}{(\omega - \omega_s)^2 + |\Delta\omega_0|_{\max}^2 \cos^2\phi_0} + \frac{\omega - \omega_s}{(\omega - \omega_s)^2 + \left(\frac{\omega_0}{Q_{\text{ext}}}\right)^2 \left(\frac{KA_0}{2R_0}\right)^2} \frac{2|\Delta\omega_0|_{\max}\sin\phi_0}{(\omega - \omega_s)^2 + |\Delta\omega_0|_{\max}^2 \cos^2\phi_0} \right\}. \tag{22}$$

This should be compared with (18).

VI. Noisy Synchronizing Signals

In practice, the synchronizing signal itself originates in another oscillator. The signal is, therefore, expected to be noisy also. The noise spectrum has already been one discussed in Section IV. Thus, if $(a_0+\Delta a)\cos(\omega_s t+\psi)$ represents the synchronizing signal, ψ has a spectrum density function of the form given by (15) and, disregarding the difference in dimension and magnitude, Δa has that given by (16). To distinguish the parameters of the synchronizing source from those of the oscillator to be synchronized, we shall use the subscript s for the former. Equation (2) becomes

$$-2L\Delta\omega_0 - 2L\frac{d\phi}{dt} = \frac{a_0 + \Delta a}{A}\sin(\phi - \psi) + \frac{n_1(t)}{A}.$$

Assuming that the noise of the synchronizing source is not too large compared to the noise of the oscillator to be synchronized, if one considers the fact that the synchronizing signal is generally fed in after a considerable attenuation, the contribution from Δa may be neglected. In this case, to the first-order approximation, the above equation is reduced to

$$-2L\frac{d\Delta\phi}{dt} - \frac{a_0}{A_0}\Delta\phi\cos\phi_0 = 2L\frac{d\psi}{dt} + \frac{n_1(t)}{A_0}$$

where $\phi=\psi+\phi_0+\Delta\phi$ and $(a_0/A_0)\sin\phi_0=-2L\Delta\omega_0$ are used. From this we have

$$(\phi - \phi_0)(\omega) = \psi(\omega) + \Delta\phi(\omega) = \frac{-n_{s1}(\omega)}{j\omega 2L_s A_s}\left(1 - \frac{2j\omega LA_0}{2j\omega LA_0 + a_0\cos\phi_0}\right) - \frac{n_1(\omega)}{2j\omega LA_0 + a_0\cos\phi_0}.$$

The spectrum density function is given by

$$|(\phi - \phi_0)(f)|^2 = \frac{1}{4\omega^2 L^2 A_0^2 + a_0^2\cos^2\phi_0} \cdot \left\{\frac{2a_0^2\cos^2\phi_0}{4\omega^2 L_s^2 A_s^2}|e_s|^2 + 2|e|^2\right\}. \tag{23}$$

As ω approaches zero, this becomes almost identical with $|\psi(f)|^2$. When ω becomes large, the spectrum density approaches that of (19), i.e., the case of a noiseless synchronizing signal.

From (3), the equation for the amplitude becomes

$$2L\frac{d\Delta A}{dt} + KA_0(\Delta A_0 + \Delta A) = a_0\cos(\phi - \psi) + n_2(t).$$

Using the relation $KA_0\Delta A_0 = a_0\cos\phi_0$ the above equation is reduced to

$$2L\frac{d\Delta A}{dt} + KA_0\Delta A = -a_0\Delta\phi\sin\phi_0 + n_2(t).$$

From this,

$$|\Delta A(f)|^2 = \frac{1}{4\omega^2L^2 + (KA_0)^2}\left\{\frac{a_0^2\sin^2\phi_0}{4\omega^2L^2A_0^2 + a_0^2\cos^2\phi_0} \cdot \left(\frac{LA_0}{L_sA_s}2|e_s|^2 + 2|e|^2\right) + 2|e|^2\right\}. \tag{24}$$

It is difficult to obtain the spectrum density function of i. However, for small $(\omega-\omega_s)$ the noise from ψ predominates over all other components. Therefore, the spectrum must be very similar to that of the synchronizing source for small $(\omega-\omega_s)$. For large $(\omega-\omega_s)$ the spectrum should approach that of Section V.

VII. 1/*f* Noise in Oscillators

So far, the contribution from $1/f$ noise has been completely neglected. In practice, since an oscillator contains either a vacuum tube or semiconductor device, the circuit parameters are expected to fluctuate slowly with time.[13] The spectrum of the fluctuations should be nearly proportional to $1/f$. This type of noise does not play any significant role at all in the case or ordinary high-frequency amplifiers. However, in the case of oscillators the situation is different. As shown in Section II the behavior of the phase and amplitude of an oscillating current is governed by (2) and (3). In these equations, L, C, and R or $X=[\omega L-1/\omega C]$ and $\overline{R}$ are expected to fluctuate with time.[1] Let us now concentrate

[1] An alternative mechanism for low-frequency oscillators was proposed by Grivet and Blaquière.[2]

on (2). If $[-\omega L+1/\omega C]=-X$ fluctuates, transfer the fluctuating part ΔX to the right-hand side. The coefficient of the second term may also fluctuate. However, its contribution can be neglected since $(1/\omega)/(d\phi/dt)$ is small. If $\Delta X(t) = n_x(t)/A$ is written, $n_x(t)$ has the dimension of voltage and the equation becomes

$$\left(-\omega L+\frac{1}{\omega C}\right)-\left(L+\frac{1}{\omega^2 C}\right)\frac{d\phi}{dt} = \frac{2}{AT_0}\int_{t-T_0}^{t} e(t)\sin(\omega t+\phi)dt+\frac{n_x(t)}{A}\cdot \quad (25)$$

Similarly, (3) becomes

$$\left(L+\frac{1}{\omega^2 C}\right)\frac{dA}{dt}+(R_i+R_0-\overline{R})A = \frac{2}{T_0}\int_{t-T_0}^{t} e(t)\cos(\omega t+\phi)dt+n_R(t) \quad (26)$$

where $n_R(t)=A\Delta\overline{R}(t)$.

Equations (25) and (26) are the fundamental equations for ϕ and A when $1/f$ noise is taken into account. $|n_x(f)|^2$ and $|n_R(f)|^2$ are both proportional to $1/f$. In general, $n_x(t)$ and $n_R(t)$ are independent of $n_1(t)$ and $n_2(t)$, the white noises we introduced in Section IV. However, $n_x(t)$ and $n_R(t)$ may be cross correlated to each other and this makes the calculation of $|i(f)|^2$ difficult. On the other hand, the spectrum calculation of phase and amplitude fluctuations remains straightforward. For instance, the results for $|\phi(f)|^2$ and $|\Delta A(f)|^2$ of free-running oscillators remains valid if $2|e|^2$ is simply replaced by $2|e|^2+|n_x(f)|^2$ and $2|e|^2+|n_R(f)|^2$, respectively. It is worth noting that the spectrum of frequency fluctuation $|d\phi/dt(f)|^2$ is no longer white. Depending on the relative magnitude of $2|e|^2$ and $|n_x(f)|^2$ the spectrum shows a certain slope with the steepest slope being about 3 dB/octave.

The calculated results of the phase fluctuation of synchronized oscillators also remains valid if $2|e|^2$ is replaced by $2|e|^2+|n_x(f)|^2$. Therefore, phase fluctuations can be greatly reduced by synchronization even if $1/f$ noise is taken into account. The amplitude fluctuation has a new interaction term between $n_x(t)$ and $n_R(t)$. Whether the amplitude fluctuation is reduced by synchronization or not depends on the correlation between the two. If they are independent the amplitude fluctuation will be increased by synchronization as before.

VIII. Conclusion

The noise of synchronized oscillators is analyzed using a simple model of oscillators. It is shown that the FM noise can be considerably improved by synchronization while the AM noise degrades slightly. It is also shown how the noise spectrum becomes unsymmetrical with respect to the synchronizing frequency when it differs from the free-running frequency.

Acknowledgment

The author wishes to thank R. S. Englebrecht for his valuable comments and discussions and M. R. Barber and J. G. Josenhans who made available many experimental facts on oscillator noise which are not yet published.

References

[1] I. L. Berstein, "On fluctuations in the neighborhood of periodic motion of an auto-oscillating system," *Doklad. Akad. Nauk.*, vol. 20, p. 11, January 1938. See also Blaquière.[2]

[2] A. Blaquière, *Nonlinear System Analysis*. New York: Academic, 1966.

[3] W. A. Edson, "Noise in oscillators," *Proc IRE*, vol 48, pp. 1454–1466, August 1960.

[4] J. A. Mullen, "Background noise in nonlinear oscillators," *Proc IRE*, vol. 48, pp. 1467–1473, August 1960.

[5] N. Minorsky, *Nonlinear Oscillations*. Princeton, N. J.: Van Nostrand, 1962. Also, J. J. Stoker, *Nonlinear Vibrations*, New York: Wiley, 1950.

[6] B. van der Pol, "Forced osciilations in a circuit with nonlinear resistance," *Phil. Mag.*, vol. 3, pp. 65–80, January 1927.

[7] N. Krylov and N. Bogoliubov, *Introduction to Nonlinear Mechanics*. Princeton, N. J.: Princeton University Press, 1943.

[8] R. D. Huntoon and A. Weiss, "Synchronization of oscillators," *Proc. IRE*, vol. 35, pp. 1415–1423, December 1947.

[9] J. Josenhans, "Noise spectra of Read diode and Gunn oscillators," *Proc. IEEE (Letters)*, vol. 54, pp. 1478–1479, October 1966.

[10] J. A. Mullen, "Background noise in oscillators with RF pushing," *J. Electronics and Controls*, vol. 10, pp. 127–138, February 1961. See also R. Esposito and J. A. Mullen, "Noise in oscillators with general tank circuits," *IRE Internat'l Conv. Rec.*, pt. 4, pp. 202–208, 1961.

[11] R. Spence and A. R. Boothroyd, "The discriminating properties of the synchronized oscillator," *Proc. IEE* (London), vol. 110, pp. 481–492, March 1963.

[12] K. Kurokawa, "Power waves and the scattering matrix," *IEEE Trans. Microwave Theory and Techniques*, vol. MTT-13, pp. 194–202, March 1965.

[13] P. I. Kuznetsov, R. L. Stratonovich, and V. I. Tikkonov, "The effect of electrical fluctuations on a vacuum tube oscillator," *Soviet Phys.—JETP*, vol. 1, pp. 510–514, November 1955.

Nonlinear Noise Theory for Synchronized Oscillators

KLAUS F. SCHÜNEMANN, MEMBER, IEEE, AND KARL BEHM

Abstract—**A nonlinear theory of noise in synchronized oscillators is outlined, thus extending Kurokawa's work [1] from small to arbitrary injection levels. The description is of phenomenological nature: it uses the describing function method of control theory for calculating the carrier waves, and the circuit theory of periodically driven nonlinear systems for an analysis of the noise sidebands. Simple expressions are derived for the various output noises and noise conversion factors in the case when the nonlinear characteristic of the active device can be described by a third-order (van-der-Pol) polynomial.**

I. Introduction

THE THEORY OF NOISE in free-running and synchronized oscillators of the negative resistance type has recently been established [1] and surveyed [2] by Kurokawa, who combined previous work in a convenient form for explaining microwave oscillator noise performance. Since the nonlinear characteristic of the active device (Gunn element, IMPATT diode, etc.) was approximated by a first-order Taylor series expansion around the operating point, every injection signal must be small compared to the free-running amplitude. Due to this restriction, the applicability of the theory is severely limited. Especially the following cases of practical importance are not met:

1) noise and transfer properties of synchronized oscillators for moderate injection levels ("moderate" means that the injected power is -20 dB or less below the output power of the free-running oscillator);
2) noise in tightly to moderately coupled oscillators for mutual synchronization in a power combining network, or in a mounting structure which contains several active elements, or in a phased array system;
3) noise in subharmonically synchronized oscillators;
4) noise in harmonically synchronized oscillators which are of importance as a frequency-divider circuit;
5) noise in multifrequency oscillators.

In cases 3), 4), and 5), Kurokawa's theory even fails for small injection levels, for these synchronization schemes are based upon the nonlinear interaction of the synchronizing and synchronized signals in the active device.

Manuscript received April 30, 1978.
The authors are with the Institut für Hochfrequenztechnik, Technische Universität, Postfach 3329, D-3300 Braunschweig, Germany.

A further restriction of the general noise theory of [1], [2] is the assumption that the open-circuit noise voltage source does not depend on the signal level. To overcome this, Hines [3], Fikart and Goud [4], and Goedbloed and Vlaardingerbroek [5] developed a large-signal noise theory for IMPATT diodes in free-running oscillators, by applying the theory of parametric systems to the solution of the nonlinear Read equation. Thus the system equations have to be solved twice: for the noise-free carrier amplitudes and, subsequently, for the small perturbation signals at the sideband frequencies, which have been assumed to model the noise behavior. The method can, consequently, be extended to cover the case of a synchronized oscillator, which has been done by Goedbloed and Vlaardingerbroek in [6]. Their theory gives a good quantitative description of noise and transfer properties of locked amplifiers with IMPATT diodes. It takes into account both the bias level and the RF-amplitude dependence of the equivalent noise sources as well as the contribution of upconverted bias noise to the output noise spectrum. The theory has been demonstrated to adequately describe synchronization and noise performance of realized IMPATT diode oscillators.

The work to be reported here deals with the same subject as that in [6] but from a different point of view. It is our aim to outline a nonlinear noise theory for synchronized oscillators by directly extending Kurokawa's work to the case of arbitrary injection levels. The description of the active device is thus phenomenological rather than physical: the nonlinearity has been modeled by a third-order (van-der-Pol-type) polynomial, which leads to lucid and tractable calculations. The method is, hence, suited to be applied to more complex systems as, e.g., mutually coupled oscillators and (sub)harmonically synchronized oscillators. In comparison to the noise theory of [6], one can state: our theory is inferior if reliable, quantitative calculations have to be made concerning a realized oscillator with IMPATT diode. It turns out to be superior: 1) as far as lucidity and tractability of the calculations and as physical insight into the phenomena involved are concerned, 2) if complex synchronization schemes shall be investigated, and 3) if the noise performance of a synchronized oscillator with Gunn element shall be analyzed, because a van-der-Pol-type current–voltage characteristic in parallel to a nearly constant electronic capacitance have been shown to adequately model the nonlinear behavior of a Gunn element [7]. Furthermore, the equivalent

Reprinted from *IEEE Trans. Microwave Theory Tech.*, vol. MTT-27, pp. 452–458, May 1979.

open-circuit noise voltage source (or the short-circuit noise current source) may be assumed to be RF-amplitude-independent in this case.

Our first-order nonlinear noise theory, hence, presents a compromise between Kurokawa's work, which arrives at simple analytical expressions for the oscillator output spectrum, and that of Goedbloed and Vlaardingerbroek, which needs time-consuming computations and a complicated matrix formulation. It will be shown that many of the interesting quantities can be derived in an explicit form, although the nonlinear effect of a large injection signal on the oscillator amplitude and phase has been taken into account.

II. Carrier Wave Analysis

The oscillator will be described by a single-tuned circuit as in [1], [6]. It is shown in Fig. 1. N_A is the nonlinear element, and i_i is an injection current source. This equivalent circuit will later be extended to cover the case of a circulator-coupled oscillator. First of all, the noise-free case shall be regarded. For analyzing the "carrier waves" in the equivalent circuit of Fig. 1, the active device must be represented by its nonlinear characteristic, e.g., by its current–voltage characteristic. Then the describing function method of control theory, which has been introduced into the microwave field in [8], is an adequate tool for an analysis, provided that the filtering effect of the linear part of the network is such that the voltage waveform across the active device can be guessed. It may consist of a finite number of sinusoids, of a bias signal, and of Gaussian noise. This condition normally holds in the case of a microwave oscillator. Then the analysis proceeds in the following way: the voltage waveform being known, the current waveform is calculated from the current–voltage characteristic by Fourier analysis. Then the fundamental components of current and voltage are related by the so-called describing function N_A, which means the effective admittance of the nonlinear device. It is amplitude- and, for dynamic nonlinearities, frequency-dependent. With the nonlinear element being replaced by the quasi-linear operator N_A, the circuit of Fig. 1 can be analyzed by standard techniques yielding amplitude and frequency of the oscillation. For further details and generalizations, the reader is referrred to [8].

Although the shape of the nonlinearity is arbitrary, we will model it by a third-order polynomial, because a van-der-Pol-type characteristic is thought to be a consistent extension of Kurokawa's work towards an inclusion of nonlinear effects. The current–voltage characteristic is hence assumed to be

$$i = -a_1 v + a_2 v^2 + a_3 v^3. \tag{1}$$

The voltage waveform across the active device is given by

$$v = v_0 + \hat{v}\cdot\cos(\omega t). \tag{2}$$

Following [9], the various quantities are normalized

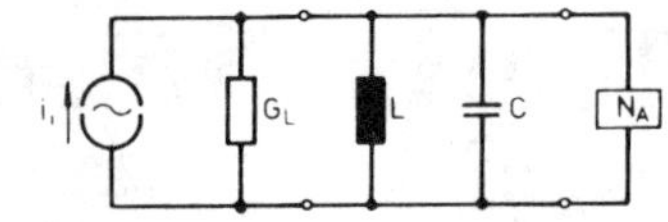

Fig. 1. Simple equivalent circuit of synchronized oscillator.

according to

$$x = v/V_n \qquad y = i/I_n \qquad g_L = G_L/G_n \qquad p = P/P_n$$

with

$$V_n = \sqrt{a_1/a_3} \qquad I_n = a_1 V_n$$
$$G_n = a_1 \qquad P_n = V_n I_n. \tag{3}$$

Equations (1) and (2) can then be rewritten

$$y = -x + ax^2 + x^3 = f(x), \qquad \text{with } a = \sqrt{a_2/(a_1 a_3)} \tag{1a}$$

$$x = x_0 + \hat{x}\cdot\cos(\omega t). \tag{2a}$$

The describing function is defined via

$$n_A = N_A/G_n = \frac{1}{\pi\hat{x}}\int_{-\pi}^{\pi} f(x)\cdot\cos(\omega t)\,d(\omega t)$$
$$= -1 + 2ax_0 + 3x_0^2 + 3/4\hat{x}^2. \tag{4}$$

Maximizing the power $p_L = P_L/P_n$ in the load conductance with respect to x_0 and g_L yields for the free-running case $y_i = 0$:

$$x_0 = -a/3 \qquad g_L = (1 + a^2/3)/2$$
$$\hat{x}^2 = 2(1 + a^2/3)/3 \qquad p_L = (1 + a^2/3)^2/6. \tag{5}$$

The free-running oscillation frequency is

$$\omega/\omega_0 = 1, \qquad \text{with } \omega_0 = 1/\sqrt{LC}\,. \tag{6}$$

In the case of a nonzero injection current

$$y_i = \hat{y}_i\cdot\cos(\omega_i t + \theta) \tag{7}$$

yields the oscillator node equation

$$[n_A + g_L(1 + j\beta)]\hat{x} = \hat{y}_i\cdot e^{j\theta},$$
$$\text{with } \beta = Q_L(\nu - 1/\nu), \quad \nu = \omega_i/\omega_0, \quad Q_L = \frac{\omega_0 C}{G_L} \tag{8}$$

a third-order polynomial in $\hat{x}^2$, from which the voltage amplitude can be calculated. The phase of the injection current is given by

$$\tan(\theta) = g_L\beta/(3\hat{x}^2/4 + g_L - 1 + 2ax_0 + 3x_0^2). \tag{9}$$

Equations (8) and (9) are valid provided that the oscillator has been locked to the injection signal. In order to investigate the stable synchronization range, the system is tested whether or not additionally injected incremental input signals will grow up or die away with time [8]. It turns out that the effect of the nonlinear element on the incremental signal can be described by a quasi-linear operator n_i which is closely related to n_A via

$$n_i = n_A + \frac{\hat{x}}{2}\frac{\partial n_A}{\partial \hat{x}}(1 + k\cdot e^{-2j\gamma}). \tag{10}$$

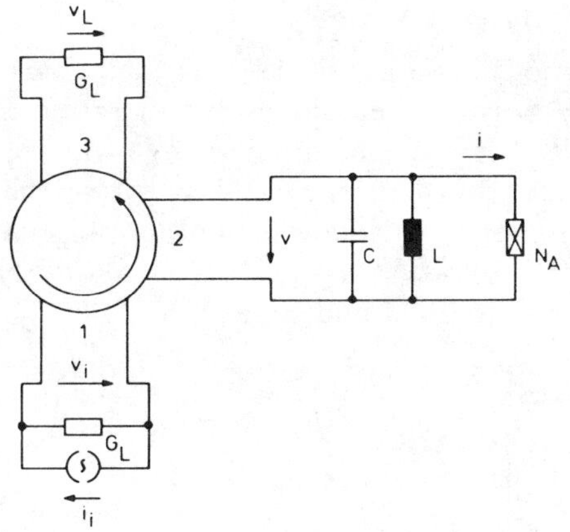

Fig. 2. Equivalent circuit of circulator-coupled locked amplifier.

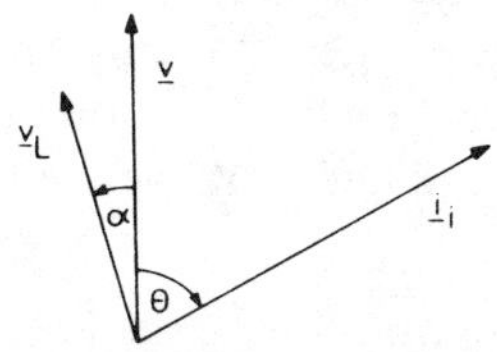

Fig. 3. Phasor diagram for the equivalent circuit of Fig. 2.

In (10) $k=1$ for a synchronous input signal and $k=0$ for a nonsynchronous one. The stability curves of the synchronized oscillator can then be calculated in the following way: The nonlinear element in the equivalent circuit of Fig. 1 is replaced by n_i. Setting $y_i=0$, the oscillator node equation then yields the boundary and locus curves of the oscillator [8]. If the actual voltage amplitude $\hat{x}$ of the synchronized oscillator lies above them, the oscillation is stable; otherwise it is not.

So far, the carrier wave analysis enables us to determine both the voltage amplitude of the oscillation and the stable synchronization range. We will now regard the more realistic equivalent circuit of Fig. 2, in which the injection source and the oscillator are coupled via an ideal circulator. The injected voltage wave $x_i=v_i/V_n$ and the current source y_i are related by

$$x_i=y_i/(2g_L). \tag{11}$$

As the voltage x across the active device is composed of the injected voltage wave x_i plus the reflected voltage wave $x_L=v_L/V_n$, which appears across the load conductance, the latter can be determined from

$$\hat{x}_L\cdot e^{j\alpha}=\hat{x}\cdot e^{j\theta}-\frac{\hat{y}_i}{2g_L}\cdot e^{j\theta}. \tag{12}$$

Here $\hat{x}_L$ means amplitude and α phase of x_L. The corresponding phasor diagram is shown in Fig. 3.

Equations (11) and (12) establish a connection between the equivalent circuits of Figs. 1 and 2. Thus the output power at port 3 of the circulator can be calculated with the injection amplitude $\hat{y}_i$ and the injection frequency ν as parameters.

III. Sideband Noise Analysis

We are in a position now to calculate the AM- and PM-output noise spectra of the synchronized oscillator, which are due to both the intrinsic noise sources and the injected noise from the current source y_i. With respect to the small noise signals, the system is periodically driven by the carriers. Provided that the perturbations are band-limited in a frequency band surrounding the carrier, and that the perturbations are small, the noise analysis should follow the guidelines of a "circuit theory of periodically driven nonlinear systems" which has been formulated by Penfield [10]. It is shown that when the perturbations are expressed in terms of slowly varying voltages and currents, the relations among these voltages and currents are linear and time-invariant. There do exist several ways of representing the perturbations in terms of slowly varying functions of time: the cosine–sine representation, the amplitude–phase representation, and the upper sideband representation have been proposed in [10], while a combination of the lower with the upper sideband has been used in [3]–[6]. The only difference between the well-known network theory for the carriers and that for the sideband signals is that any relation between a perturbation signal at one network port to that at another one is now a matrix equation between the corresponding pairs of sideband signals instead of a simple scalar equation as in the case of the carrier analysis. The elements of such matrices do, of course, depend on the carriers of the nonlinear device.

As the circuit theory for the small perturbation signals is linear, the principle of superposition holds. We will, hence, calculate the contributions to the output noise spectra from the intrinsic and from the injected noise separately.

A. Effect of Intrinsic Noise Sources

The equivalent circuit for the perturbation signals is as shown in Fig. 1. Here the current source i_i may represent the intrinsic short-circuit noise current source of the active device which is composed of a pair of pseudo-sinusoids with random amplitudes and phases at the lower sideband frequency $\omega_l=\omega-\Delta\omega$ and at the upper sideband frequency $\omega_u=\omega+\Delta\omega$. In the following, every voltage or current has to be represented by its lower sideband portion (index "l") and by its upper sideband portion (index "u").

In order to establish a relation between the noise voltage and current at the active device, we assume small perturbations Δy and Δx to be present according to

$$y+\Delta y=f(x+\Delta x), \qquad \Delta y=\frac{df}{dx}\Delta x=f'(x)\Delta x,$$
$$\text{with } f'(x)=-1+2ax+3x^2. \tag{13}$$

$f'(x)$ can be expanded into a Fourier series with coefficients $g_m=G_m/G_n$

$$f'(x)=\sum g_m e^{jm\omega t},$$
$$g_m=\frac{1}{2\pi}\int_{-\pi}^{\pi} f'(x)e^{-jm\omega t}d(\omega t). \tag{14}$$

In the case of a van-der-Pol-type nonlinearity, the g_m are

$$g_0=-1+2ax_0+3x_0^2+3\hat{x}^2/2 \qquad g_1=g_{-1}=a\hat{x}+3x_0\hat{x}$$

$$g_2=g_{-2}=3\hat{x}^2/4 \qquad g_m=0, \qquad \text{for } |m|>2. \tag{15}$$

If, in addition to the upper and lower sideband vectors, a baseband signal (index "b") at frequency $\Delta\omega$ is assumed too, (13)–(15) yield the desired matrix relation

$$\begin{vmatrix} y_b \\ y_l^* \\ y_u \end{vmatrix} = \begin{vmatrix} g_0 & g_1 & g_1 \\ g_1 & g_0 & g_2 \\ g_1 & g_2 & g_0 \end{vmatrix} \cdot \begin{vmatrix} x_b \\ x_l^* \\ x_u \end{vmatrix}. \tag{16}$$

The asterisk denotes the complex conjugate of the corresponding quantity.

We will in the following restrict ourselves to the case that upconversion from and downconversion to the baseband can be neglected, because this leads to simple analytic expressions for the output noise. Moreover, g_1 turns out to be zero if the bias is adjusted for maximum generated power, as will be shown in the next section. Hence, (16) may be replaced by

$$\begin{vmatrix} y_l^* \\ y_u \end{vmatrix} = \begin{vmatrix} g_0 & g_2 \\ g_2 & g_0 \end{vmatrix} \cdot \begin{vmatrix} x_l^* \\ x_u \end{vmatrix}. \tag{16a}$$

In order to complete the analysis, two node equations have to be added:

$$\begin{aligned} y_{il} &= y_l + x_l g_L(1+j\beta_l), \\ y_{iu} &= y_u + x_u g_L(1+j\beta_u), \\ \beta_l &= Q_L(\nu_l - 1/\nu_l), \qquad \nu_l = \omega_l/\omega_0 \\ \beta_u &= Q_L(\nu_u - 1/\nu_u), \qquad \nu_u = \omega_u/\omega_0. \end{aligned} \tag{17}$$

From (16a) and (17) the noise voltage vectors x_l and x_u can be calculated. Assuming a Gaussian amplitude distribution of the noise current sources y_{il} and y_{iu} according to a spectral density

$$\phi_i = \frac{2\varphi_i^2/\omega_c}{1+(\omega/\omega_c)^2} \tag{18}$$

with φ_i the variance and ω_c a characteristic frequency, the AM and PM portions of the spectral density of the noise voltage $x_l + x_u$ across the active device are given by

$$\begin{aligned} \phi_{AM} &= \langle \text{Re}^2\,(x_l + x_u)\rangle \phi_i \\ \phi_{PM} &= \langle \text{Im}^2\,(x_l + x_u)\rangle \phi_i. \end{aligned} \tag{19}$$

The brackets denote an average over the two independent phase angles of y_{il} and y_{iu}. The AM- and PM-noise power spectra follow from (19) according to

$$p_{AM} = \phi_{AM} \cdot g_L \cdot 1\ \text{Hz} \qquad p_{PM} = \phi_{PM} \cdot g_L \cdot 1\ \text{Hz}. \tag{20}$$

It will be shown in the next section that, for some special cases, the evaluation of (19) and (20) leads to simple analytic expressions for the AM- and PM-noise spectra.

In order to calculate the noise power spectra at port 3 of the circulator, the phase shift of the output carrier wave has to be taken into account. According to (12) and the phasor diagram of Fig. 3, the noise voltage $x_l + x_u$ in (19) has to be multiplied by $e^{-j\alpha}$ before the real or imaginary part is taken. With this modification the output noise can be calculated from (19) and (20).

B. Effect of Injected Noise

That part of the output noise spectrum, which is due to the injected noise, is determined by analyzing the equivalent circuit of Fig. 2. The input voltage wave x_i has now to be replaced by a pair x_{il} and x_{iu} of sideband noise vectors. In order to calculate the output noise vectors x_{Ll} and x_{Lu}, the following system of equations has to be solved: 1) the modified node equations

$$\begin{aligned} y_l + x_l g_L(1+j\beta_l) &= g_L(x_{il} - x_{Ll}) \\ y_u + x_u g_L(1+j\beta_u) &= g_L(x_{iu} - x_{Lu}) \end{aligned} \tag{21}$$

2) the equations corresponding to (12)

$$x_{Ll} + x_{il} = x_l \qquad x_{Lu} + x_{iu} = x_u \tag{22}$$

and 3) the matrix equation (16a).

It is useful to regard the AM and PM portions of the input noise separately. In the case of injected AM noise the stochastic phase angles are related to one another according to

$$\begin{aligned} x_{il} &= |x_i| e^{-j\epsilon} \cdot e^{j\theta} \\ x_{iu} &= |x_i| e^{j\epsilon} \cdot e^{j\theta}\ \text{(AM noise)} \end{aligned} \tag{23}$$

and in the case of PM noise according to

$$\begin{aligned} x_{il} &= -|x_i| e^{-j\epsilon} \cdot e^{j\theta} \\ x_{iu} &= |x_i| e^{j\epsilon} \cdot e^{j\theta}\ \text{(PM noise)}. \end{aligned} \tag{24}$$

The output noise power spectra due to injected AM or PM noise are then given by

$$\begin{aligned} p_{AM}(\text{AM}) \text{ or } p_{AM}(\text{PM}) &= \langle \text{Re}^2\,(x_{Ll} + x_{Lu})e^{-j\alpha}\rangle \phi_i' \cdot 1\ \text{Hz} \cdot g_L \\ p_{PM}(\text{AM}) \text{ or } p_{PM}(\text{PM}) &= \langle \text{Im}^2\,(x_{Ll} + x_{Lu})e^{-j\alpha}\rangle \phi_i' \cdot 1\ \text{Hz} \cdot g_L \end{aligned} \tag{25}$$

with ϕ_i' the spectral density of the injected noise voltage wave. The average has to be taken over the stochastic phase ϵ.

IV. Discussion of Results

Due to the special choice of the nonlinear characteristic, simple expressions can be derived for the noise spectra in several cases. In order to simplify the evaluation, we will first derive some general relations for the Fourier coefficients g_m.

1) If the bias is adjusted for maximum generated power p_g, g_1 turns out to be zero. This can be proven by regarding

$$p_g = \hat{x}^2 n_A/2 \tag{26}$$

and differentiating with respect to x_0

$$\frac{\partial p_g}{\partial x_0} \sim \frac{\partial n_A(\hat{x}, x_0)}{\partial x_0} = 0. \tag{27}$$

Inserting (4) into (27) yields

$$\frac{\partial n_A}{\partial x_0} = \frac{1}{\pi \hat{x}} \int_{-\pi}^{\pi} f'(x) \frac{dx}{dx_0} \cos(\omega t) d(\omega t). \tag{28}$$

Taking (2a) and (14) into account

$$\frac{\partial n_A}{\partial x_0} = \frac{2}{\hat{x}} g_1 \tag{29}$$

i.e., $g_1 = 0$ for optimum bias.

2) For the free-running oscillator, $g_0 = 0$ is valid. This can be proven by differentiating p_g with respect to x

$$\frac{\partial p_g}{\partial \hat{x}} = \frac{1}{2\pi}\int_{-\pi}^{\pi} f(x)\cos(\omega t)\, d(\omega t) + \frac{\hat{x}}{2\pi}\int_{-\pi}^{\pi} f'(x)\cos^2(\omega t)\, d(\omega t) = \frac{\hat{x}}{2}(n_A + g_0 + g_2). \tag{30}$$

In order to further simplify (30), we integrate the right-hand side of n_A from (4) by parts

$$n_A = \frac{1}{\pi \hat{x}}\left[f(x)\sin(\omega t)\Big|_{-\pi}^{\pi} - \int_{-\pi}^{\pi} f'(x)\frac{dx}{d(\omega t)}\sin(\omega t)\, d(\omega t)\right]. \tag{31}$$

Inserting (2a) yields

$$n_A = g_0 - g_2. \tag{32}$$

This relation holds in general for free-running and for synchronized oscillators.

By combining (30) with (32)

$$\frac{\partial p_g}{\partial \hat{x}} = \hat{x} g_0. \tag{33}$$

Hence, $g_0 = 0$ in the free-running state.

3) Last but not least, the incremental input describing function (10) for a synchronous input signal shall be related to the Fourier coefficients g_m. To this end we write

$$n_i = n_A + \frac{\hat{x}}{2}\frac{dn_A}{d\hat{x}}(1 + \cos(2\gamma) - j\sin(2\gamma)). \tag{34}$$

Using the Fourier integral (4) for n_A, we can verify that

$$\frac{\hat{x}}{2}\frac{dn_A}{d\hat{x}} = g_2. \tag{35}$$

Hence, taking (32) into account,

$$n_i = g_0 + g_2(\cos(2\gamma) - j\sin(2\gamma)). \tag{36}$$

It shall be emphasized that (29), (32), (33), and (36) hold for arbitrary current–voltage characteristics $f(x)$.

Making use of these rather general results, simple expressions can be derived for the AM- and PM-noise spectra. In the case of a free-running oscillator

$$p_{AM} = \phi_i \cdot 1\ \text{Hz} \cdot \frac{1}{2g_L}\frac{1}{1 + \Delta\nu^2 Q_L^2}$$
$$p_{PM} = \phi_i \cdot 1\ \text{Hz} \cdot \frac{1}{2g_L}\frac{1}{\Delta\nu^2 Q_L^2}, \qquad i_i = 0 \tag{37}$$

with $\Delta\nu = \Delta\omega/\omega_0$. Of course, the spectral dependence of both the AM and the PM noise corresponds to what has been derived in [1].

In the case where the oscillator is synchronized at its free-running frequency ($\nu = 1$), the near-carrier output noise ($\Delta\nu \to 0$) due to the intrinsic noise sources is given by

$$p_{AM} = \phi_i \cdot 1\ \text{Hz} \cdot 2g_L \frac{1}{(g_0 + g_2 + g_L)^2}$$
$$p_{PM} = \phi_i \cdot 1\ \text{Hz} \cdot 2g_L \frac{1}{(g_0 - g_2 + g_L)^2},$$
$$\text{for } \nu = 1, \quad \Delta\nu \to 0, \quad i_i \neq 0. \tag{38}$$

The second expression shows the stabilizing effect of the synchronizing signal on the PM noise. For $i_i = 0$ the denominator vanishes because of (32). This gives rise to a sharp PM-noise peak, which is considerably damped even for small injected signals $i_i \neq 0$ for then $n_A \neq g_L$. The AM noise is, on the other hand, only weakly influenced by the injected signal.

Another simple relation, which gives physical insight into the stabilization mechanism, can be obtained for $\nu \neq 1$ but $\Delta\nu \to 0$:

$$p_{AM} = \phi_i \cdot 1\ \text{Hz} \cdot 2g_L \frac{(g_0 - g_2 + g_L)^2 + g_L^2 Q_L^2(\nu - 1/\nu)^2}{((g_0 + g_L)^2 - g_2^2 + g_L^2 Q_L^2(\nu - 1/\nu)^2)^2}$$
$$p_{PM} = \phi_i \cdot 1\ \text{Hz} \cdot 2g_L \frac{(g_0 + g_2 + g_L)^2 + g_L^2 Q_L^2(\nu - 1/\nu)^2}{((g_0 + g_L)^2 - g_2^2 + g_L^2 Q_L^2(\nu - 1/\nu)^2)^2},$$
$$\text{for } \nu \neq 1, \quad \Delta\nu \to 0, \quad i_i \neq 0. \tag{39}$$

It is well known [1] that both AM and PM noise become infinite at the borders of the stable synchronization range. This can be derived from (39) as well if one utilizes (36) in order to calculate the stability condition for synchronous perturbations. It reads

$$n_i + g_L(1 + jQ_L(\nu - 1/\nu)) = 0. \tag{40}$$

Inserting (36) and eliminating γ shows that the left-hand side exactly equals the denominator of (39). The denominator hence coincides with the boundary curve for stable oscillations.

Simple expressions can also be given for the various output noise spectra due to the injected AM or PM noise in the limiting case of $\nu = 1$ and $\Delta\nu \to 0$. Then the AM noise due to AM injection is

$$p_{AM}(\text{AM}) = \phi_i' \cdot 1\ \text{Hz} \cdot 2g_L \frac{((g_L - g_2)^2 - g_0^2)^2}{((g_0 + g_L)^2 - g_2^2)^2},$$
$$\text{for } \nu = 1, \quad \Delta\nu \to 0 \tag{41}$$

while $p_{PM}(\text{AM}) = 0$ at $\nu = 1$. The PM noise due to PM injection is

$$p_{PM}(\text{PM}) = \phi_i' \cdot 1\ \text{Hz} \cdot 2g_L \frac{((g_L + g_2)^2 - g_0^2)^2}{((g_0 + g_L)^2 - g_2^2)^2},$$
$$\text{for } \nu = 1, \quad \Delta\nu \to 0 \tag{42}$$

while $p_{AM}(\text{PM}) = 0$ at $\nu = 1$.

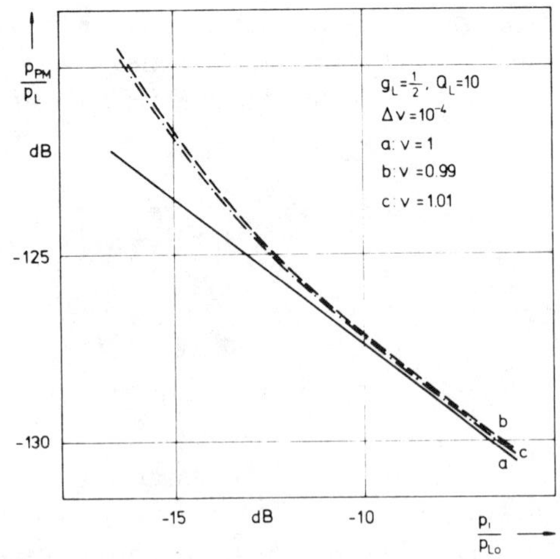

Fig. 4. PM-noise power at the oscillator output due to intrinsic noise versus the injection power (p_L means load power).

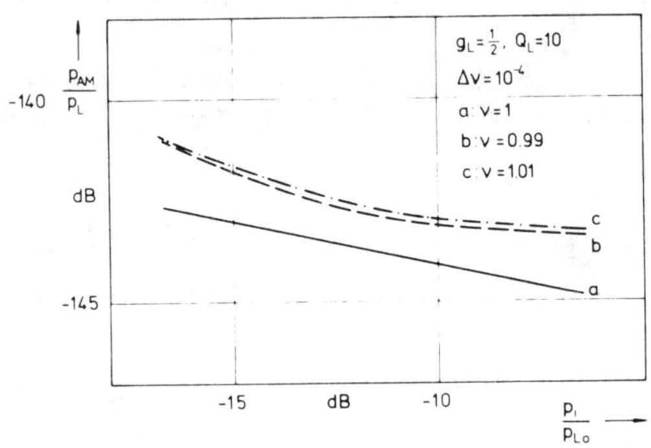

Fig. 5. AM-noise power at the oscillator output due to intrinsic noise versus the injection power (p_L means load power).

Concluding, some numerical examples shall be given. In Fig. 4 (Fig. 5), the intrinsic PM (AM) noise at the circulator output port is shown versus the injected power. The normalizing quantities are given by

$$p_L=\hat{x}_L^2 g_L/2 \qquad p_i=\hat{y}_i^2/(8g_L)$$
$$p_{L0}=p_L(y_i=0)=1/6. \tag{43}$$

ϕ_s has been set to $\phi_s=1/3\times 10^{-14}/\text{Hz}$. The frequency stabilization effect increases with increasing injection power. Both AM and PM noise at the circulator output port are typically 1–2 dB larger than at the device port.

The noise conversion factors PM PM, AM AM, and AM PM have been shown versus the injection power in Figs. 6–8. The quantities, which have been labeled by an additional index "0," denote the corresponding values of the free-running oscillator. Supplementary to (43)

$$p_{\text{AM0}}=2\phi_i'\cdot 1\ \text{Hz}\cdot g_L\cdot p_{L0}/p_i \tag{44}$$

is the double-sideband AM-noise power of a free-running oscillator which is coupled to port 1 of the circulator via an attenuator with transmission coefficient $p_i/p_{L0}<1$. A similar expression holds for p_{PM0} except that ϕ_i' now is different from ϕ_i' in (44). These spectral densities have been chosen so that $p_{\text{AM0}}/p_{L0}=-160$ dB and $p_{\text{PM0}}/p_{L0}=-128$ dB.

The PM→PM conversion factor (Fig. 6) is nearly independent from the injection power, while the AM→AM

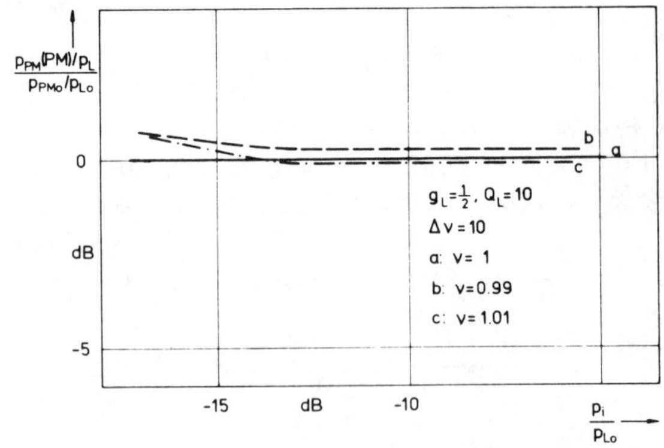

Fig. 6. PM→PM conversion factor of a synchronized oscillator versus the injection power.

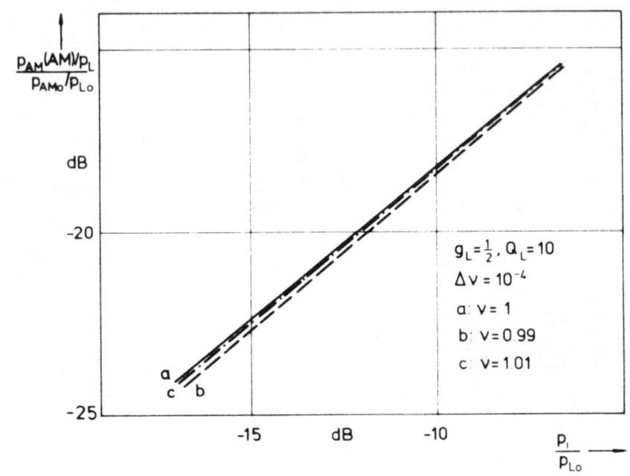

Fig. 7. AM→AM conversion factor of a synchronized oscillator versus the injection power.

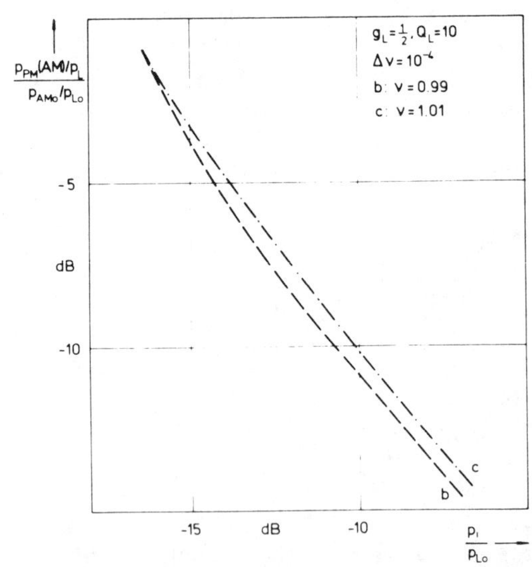

Fig. 8. AM→PM conversion factor of a synchronized oscillator versus the injection power.

conversion factor (Fig. 7) linearly increases with p_i. The AM compression is, however, always better than 10 dB. The AM→PM conversion factor (which is identical to the PM→AM conversion factor) strongly decreases with increasing injection power (Fig. 8). It is, of course, zero for $\nu=1$.

V. Conclusions

A nonlinear theory of noise in synchronized oscillators has been presented which extends Kurokawa's work [1] from small to arbitrary injection levels. The theoretical approach differs from that of Goedbloed and Vlaardingerbroek [6] in that a phenomenological model of the active device has been chosen which considerably simplifies the mathematical derivations. As a consequence, simple expressions could be obtained for various output noises and noise conversion factors. It is this simplicity that makes the model a suitable means, if more complex synchronization schemes (mutually coupled or (sub) harmonically synchronized oscillators, e.g.) shall be analyzed. Moreover, the model adequately describes the output noise of synchronized oscillators with Gunn elements. It could be extended to account for upconversion and downconversion effects as well as for an RF-amplitude-dependent intrinsic noise source.

References

[1] K. Kurokawa, "Noise in synchronized oscillators," *IEEE Trans. Microwave Theory Tech.*, vol. MTT-16, pp. 234–240, 1968.

[2] ——, "Injection locking of microwave solid-state oscillators," *Proc. IEEE*, vol. 61, pp. 1386–1410, 1973.

[3] M. E. Hines, "Large-signal noise, frequency conversion, and parametric instabilities in IMPATT diode networks," *Proc. IEEE*, vol. 60, pp. 1534–1548, 1972.

[4] J. L. Fikart and P. A. Goud, "A theory of oscillator noise and its application to IMPATT diodes," *J. Appl. Phys.*, vol. 44, pp. 2284–2296, 1973.

[5] J. J. Goedbloed and M. T. Vlaardingerbroek, "Noise in IMPATT diode oscillators at large signal levels," *IEEE Trans. Electron Devices*, vol. ED-21, pp. 342–351, 1974.

[6] ——, "Theory of noise and transfer properties of IMPATT diode amplifiers," *IEEE Trans. Microwave Theory Tech.*, vol. MTT-25, pp. 324–332, 1977.

[7] W. Freude, "Measurement of the statistics of a Gunn oscillator signal and comparison with a mathematical model," *Electron. Commun.*, vol. 30, pp. 209–218, 1976.

[8] L. Gustafsson, G. H. B. Hansson, and K. I. Lundstroem, "On the use of describing functions in the study of nonlinear active microwave circuits," *IEEE Trans. Microwave Theory Tech.*, vol. MTT-20, pp. 402–409, 1972.

[9] M. E. Hines, "Negative-resistance diode power amplification," *IEEE Trans. Electron Devices*, vol. ED-17, pp. 1–8, 1970.

[10] P. Penfield, Jr., "Circuit theory of periodically driven nonlinear systems," *Proc. IEEE*, vol. 54, pp. 266–280, 1966.

A Simple Model of Feedback Oscillator Noise Spectrum

Introduction

This letter contains brief thoughts on the following points.

1) The relationships among four commonly used spectral descriptions of oscillator short-term stability or noise behavior.
2) A heuristic derivation, presented without formal proof, of the expected spectrum of a feedback oscillator in terms of known oscillator parameters.
3) Some experimental results which illustrate the validity of the simple model.
4) Comments on the effect of nonlinearity, specific spectral requirements for several applications, choice of resonator frequency and active element, and expected spectrum characteristics of several oscillator types.

Spectral Models of Phase Variations

Consider a stable oscillator whose measurable output can be expressed as

$$v(t) = A \cos [\omega_0 t + \phi(t)].$$

It is common to treat $\phi(t)$ as a zero-mean stationary random process describing deviations of the phase from the ideal. The frequency domain information about phase or frequency variations is contained in the "power" spectral density $S_{\dot\phi}(\omega_m)$ of the phase $\phi(t)$ or, alternatively, in the "power" spectral density $S_\phi(\omega_m)$ of the frequency $\dot\phi$. By analogy to modulation theory, we use ω_m to mean the modulation, video, baseband, or offset frequency associated with the noise-like variations in $\phi(t)$. The units of $S_\phi(\omega_m)$ are radians^2/cps bandwidth or dB relative to 1 radians^2/cps BW; $S_{\dot\phi}(\omega_m)$ is expressed in (radians/sec)^2 per c/s BW [1] [2]. The two are related by $S_{\dot\phi}(\omega_m) = \omega_m{}^2 S_\phi(\omega_m)$.

$S_{\dot\phi}(\omega_m)$ can also be expressed in terms of the equivalent rms frequency deviation Δf_{rms} in a given video bandwidth. Further, subject to the limitations that $\overline{\phi^2} \ll 1$ (small total modulation indes) and that AM$\ll$FM components, the normalized RF power spectrum $G(\omega - \omega_0)$ is identical to the two-sided spectrum of the phase $S_\phi(\omega_m)$; i.e., RF sidebands relative to the carrier are down by $S_\phi(\omega_m)$ expressed in decibels relative to 1 radian^2/BW.

Relation to Oscillator Internal Noise

A basic requirement on an oscillator noise model is that it show clearly the relationship of the spectrum of the phase $S_\phi(\omega_m)$ to the known or expected noise and signal levels and resonator characteristics of the oscillator. A simple picture can be constructed using a model of a linear feedback oscillator. Minor corrections to the results are necessary to account for nonlinear effects which must be present in a physical oscillator. Assume a single resonator feedback network of fractional bandwidth $2B/\omega_0 = 1/Q$, where Q is the operating, or loaded, quality factor. For small phase deviations at video rates which fall within the feedback half-bandwidth $\omega_0/2Q$, a phase error at the oscillator input due to noise or parameter variations results in a frequency error determined by the phase-frequency relationship of the feedback network, $\Delta\theta = 2Q\dot\phi/\omega_0$. Thus, for modulation rates *less* than the half-bandwidth of the feedback loop, the spectrum of the *frequency* $S_{\dot\phi}(\omega)$ is identical (with a scale factor) to the spectrum of the uncertainty of the oscillator input phase due to noise and parameter variations. This uncertainty will be denoted $\Delta\theta(t)$, and its two-sided power spectral density $S_{\Delta\theta}(\omega_m)$.

For modulation rates *large* compared to the feedback bandwidth, a series feedback loop is out of the circuit. At these modulation rates, the power spectral density of the output *phase* $S_\phi(\omega)$ is identical to the spectrum of the oscillator input phase uncertainty $S_{\Delta\theta}(\omega_m)$.

For a physical oscillator the spectrum $S_{\Delta\theta}(\omega_m)$ of the input phase uncertainty $\Delta\theta(t)$ is expected to have two principal components. One component is due to phase uncertainties resulting from additive white noise at frequencies around the oscillator frequency, as well as noise at other frequencies mixed into the pass band of interest by nonlinearities. The second component is due to parameter variations at video frequencies which affect the phase (such as variations in the phase shift of a transistor due to carrier density fluctuations in the base resistance). The additive noise component of $S_{\Delta\theta}(\omega_m)$ is identical to the spectral density of the noise voltage squared relative to the mean square signal voltage. For white additive noise, this component is flat with frequency. For a feedback oscillator with an effective noise figure F, the two-sided $S_{\Delta\theta}(\omega) = 2\text{FKT}/P_S$; P_S is the signal level at oscillator active element input.

Manuscript received December 10, 1965; revised December 29, 1965

The video spectrum of parameter variations is found typically to have a power spectral density varying inversely with frequency (a $1/\omega_m$ or $1/f$ spectrum). The total power spectral density of oscillator input phase errors is of the form $S_{\Delta\theta}(\omega_m) = \alpha/\omega_m + \beta$, where α is a constant determined by the level of $1/f$ variations and β is $= 2\text{FKT}/P_S$ for two-sided spectra.

To find $S_\phi(\omega_m)$ or $S_{\dot\phi}(\omega_m)$, we use the fact that

$$\text{for} \quad \omega_m < \frac{\omega_0}{2Q} \qquad S_{\dot\phi}(\omega) = \left[\frac{\omega_0}{2Q}\right]^2 S_{\Delta\theta}(\omega_m)$$

$$\omega_m > \frac{\omega_0}{2Q} \qquad S_\phi(\omega) = S_{\Delta\theta}(\omega_m).$$

A suitable composite expression is

$$S_\phi(\omega_m) = S_{\Delta\theta}\left[1 + \left(\frac{\omega_0}{2Q\omega_m}\right)^2\right].$$

This yields an asymptotic model for $S_\phi(\omega)$ shown on log-log scales in Fig. 1.

The model can be summarized as follows.

$S_\phi(\omega_m)$ decreases with ω_m
- at 9 dB/octave up to the point where $1/f$ effects no longer predominate.
- at 6 dB/octave from that point up to the feedback loop half-bandwidth.
- at 0 dB/octave above that frequency up to a limit imposed by subsequent filtering.

$S_{\dot\phi}(\omega)$ decreases at 3 dB/octave up to the first breakpoint, is flat with frequency up to the feedback baseband bandwidth, and increases at 6 dB octave above that point.

The case where $1/f$ effects predominate only for frequencies small compared with the feedback loop bandwidth is shown here as an

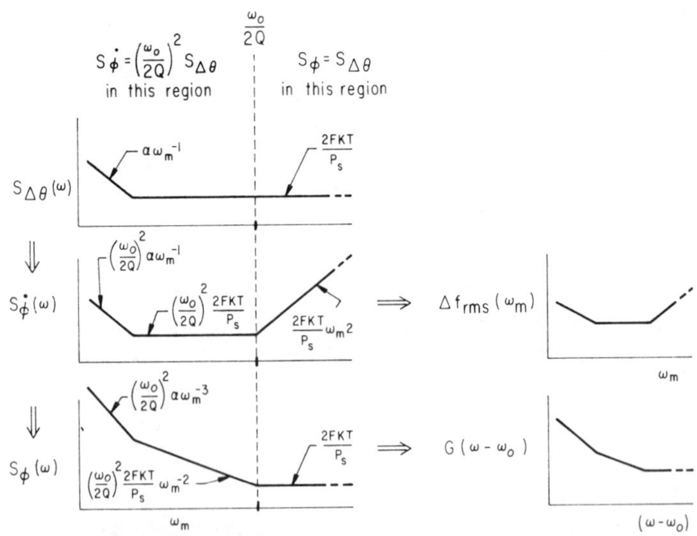

Fig. 1. Derivation of Oscillator Spectra. The logical sequence leading from oscillator parameters to spectrum characteristics is presented here. The power spectra of output phase or frequency are derived from the spectrum of input phase uncertainties and from the oscillator feedback bandwidth. The calculable constants of the oscillator are FKT, P_s, and $\omega_0/2Q$; the $1/f$ constant α is not accurately predictable but can be inferred from data. The amplitude spectrum of frequency deviation and the RF spectrum can be derived as shown, subject to limitations discussed in the text.

Reprinted from *Proc. IEEE*, vol. 54, pp. 329–330, Feb. 1966.

example. For a high-Q oscillator, $1/f$ effects in $S_{\Delta\theta}$ can predominate out to a modulation rate exceeding $\omega_0/2Q$; in this case there is no 6 dB per octave region in $S_{\phi}(\omega)$. A similar spectrum results where large additive noise in following amplifier stages or measuring equipment obscures the oscillator internal noise, except at very low modulation frequencies.

Note that there is a portion of the curve $S_{\phi}(\omega_m)$ which is proportional to $1/\omega_m^2$, leading to a $1/\omega_m$ or $1/f$ variation for rms phase deviation. This is often confused with the true $1/f$ effects associated with parameter variations leading to the $1/f$ portion of the curve for $S_{\Delta\theta}(\omega_m)$ and $S_{\dot{\phi}}(\omega_m)$. These two are not the same thing; "$1/f$" refers to a power spectral density rather than an amplitude spectrum.

In practice, the measurable $S_{\phi}(\omega_m)$ is always modified by subsequent bandlimiting filtering and by additive noise contributed by following amplifiers. It is conceivable that, for a two-terminal oscillator, the filtering action of the resonator eliminates the additive phase noise component for $\omega_m > (\omega_0/2Q)$.

Experimental Verification

Measurements were taken on a stable microwave signal source[1] designed to have a spectral purity limited only by the oscillator, which was a 100 M/s crystal oscillator. This unit employs two large-jump step recovery diode multipliers with amplification between them. The data are presented in Fig. 2 in comparison with a model derived from the following constants:

Feedback bandwidth	= 16 kc/s
P_S	$= -4$ dBm
F	= 9 dB
KT	$= -174$ dBm in 1 c/s BW
Multiplication ratio	= 100 = 40 dB
$N^2 2FKT/P_S$	$= +40+3+9-174+4 = -118$ dB.

This leads to an asymptotic value for $S_{\phi}(\omega_m)$ of -118 dB relative to 1 radian²/BW in 1 c/s bandwidth, i.e., a carrier-to-sideband ratio of 118 dB. The "$1/f$" region (9 dB/octave) constant α is estimated for best data fit.

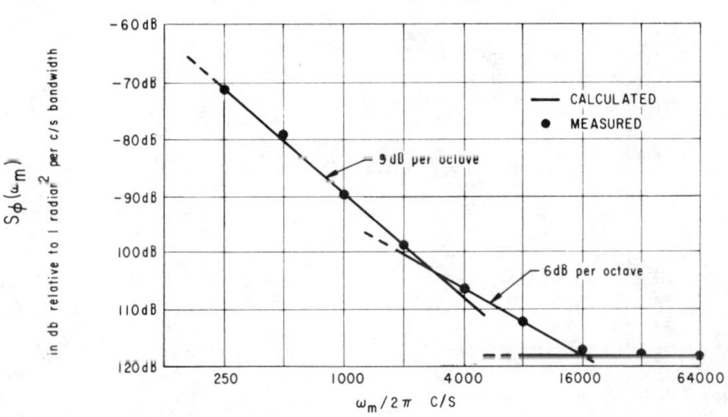

Fig. 2. $S_{\phi}(\omega_m)$ for Stable Microwave Signal Source. The data presented here is the average of two independent measurements which were in excellent agreement. These measurements were made at X Band on the multiplied output of a 100 Mc/s voltage controlled crystal oscillator having a 16 kc/s feedback half-bandwidth. Since this bandwidth can be reduced by a considerable factor without exceeding the present state of the art, the data is not intended to represent ultimate attainable levels, but rather serves as an illustrative example. The $1/f$ constant is chosen for best data fit. Slopes and other calculated parameters are derived from known oscillator characteristics.

Nonlinear Effects

The data was based on an estimated transistor noise figure of 9 dB. This was taken high to account for nonlinear mixing of noise at third harmonic and higher frequencies which is mixed into the pass band of interest by second harmonic periodic parameter variations caused by the nonlinearity. The excellent fit of the data implies that this degradation of effective noise figure may well be an adequate description of the effect of nonlinearity.

Video Frequency Range of Interest

A number of applications which have been dealt with in this issue of the Proceedings may be summarized in terms of the video frequency range of interest. Space systems and Doppler radar applications are of particular interest to the author. For these two, interest lies in the range of a few c/s up to 100 kc/s. Space applications typically concentrate on the range where, for a crystal oscillator, $S_{\dot{\phi}}(\omega_m)$ is proportional to $S_{\Delta\theta}(\omega_m)$ [3], while Doppler radar applications place additional emphasis on the region above the oscillator feedback loop bandwidth [4]. Both applications typically require microwave systems which employ multiplication from the oscillator frequency.

Choice of Oscillator Frequency for Crystal Oscillator-Multiplier

It is of interest to inspect the effect of oscillator frequency upon the output spectrum of an oscillator-multiplier system having a fixed output frequency. Two assumptions which aid the calculation are a) constant oscillator input signal-to-noise ratio, and b) resonator Q varying inversely with the oscillator frequency ω_0. Under these assumptions a comparison of two oscillator frequencies yields the following results.

1) For $\omega_m < (\omega_0/2Q)$, of the lower frequency oscillator, the multiplied output $S_{\phi}(\omega_m)$ is identical for either choice.
2) For $\omega_m \gg (\omega_0/2Q)$, the output $S_{\phi}(\omega_m)$ varies as the square of the multiplication ratio (i.e., inversely as the square of the oscillator frequency).

This can be verified by a simple graphical construction.

Choice of Active Element in a Transistor Oscillator

It is apparent that $1/f$ variations and nonlinearity can have significant deleterious effects on the attainable low levels of $S_{\phi}(\omega_m)$. In the light of suggestions by O. Mueller that microthermal effects [6] contribute to $1/f$ noise in transistors, it is suggested that AGC oscillators using large area transistors having high power capabilities may provide simultaneous improvements in $1/f$ level and in nonlinear effects.

Spectrum Characteristics of Microwave Solid State Sources

The spectrum model given here allows simple prediction of spectrum shape and level for microwave sources of the types discussed by Johnson et al [5]. Comparison with their data shows good agreement —their measurements for crystal oscillator units extend to $\omega_m \gg (\omega_0/2Q)$, while microwave oscillators are characterized by Q factors such that, for the measurements cited, $\omega_m < (\omega_0/2Q)$.

Acknowledgment

The author is pleased to acknowledge helpful discussions with members of IEEE Subcommittee 14.7, of which this letter may be considered a brief summary. Prepublication access to all of the papers contained in this issue is also freely acknowledged. The influence of L. S. Cutler, J. A. Mullen, and W. L. O. Smith has been of special value in the preparation of this correspondence.

D. B. Leeson
Applied Technology, Inc.
Palo Alto, Calif.

References

[1] L. S. Cutler and C. L. Searle, "Some aspects of the theory and measurement of frequency fluctuations in frequency standards," *this issue*, page 136.
[2] E. J. Baghdady, R. N. Lincoln, and B. D. Nelin, "Short-term frequency stability: characterization, theory, and measurement" *Proc. IEEE*, vol. 53, pp. 704–722, July 1965.
[3] R. L. Sydnor, J. J. Caldwell, and B. E. Rose, "Frequency stability requirements for space communications and tracking systems," *this issue*, page 231.
[4] D. B. Leeson and G. F. Johnson, "Short-term stability for a Doppler radar: requirements, measurements, and techniques," *this issue*, page 244.
[5] S. L. Johnson, B. H. Smith, and D. A. Calder, "Noise spectrum characteristics of low-noise microwave tubes and solid-state devices," *this issue*, page 258.
[6] O. Mueller, "Thermal feedback and $1/f$-flicker noise in semiconductor devices," *1965 Internat'l Solid State Circuits Conference, Digest*, p. 68 ff.

[1] 9.5 Gc/s Solid State Local Oscillator PN 31-007191, manufactured by Applied Technology, Inc., Palo Alto, Calif. Measurements are average of values measured by the author and D. J. Healey, III, Westinghouse Corp., Baltimore, Md., using Spectra Electronics SE-200 and Westinghouse proprietary noise test sets.

Part III
Mathematical Theory of Frequency Stability

THE technological progress of the 1940's and 50's resulted in the effort to establish a precise and widely acceptable definition of frequency stability, particularly of short-term frequency stability. To this end, IEEE and NASA held a symposium in November 1964 [III-1], and soon afterwards a special issue of the PROCEEDINGS OF THE IEEE (February 1966) was dedicated to this problem [III-2]. Here, in more than two hundred pages, foundations of the mathematical theory and measurement procedures of frequency stability have been laid, the major features of which remain valid to this day. From the papers published in the mentioned proceedings, we have decided to reprint in this part only the survey article by Cutler and Searle since it is quoted so often and, further, since it summarizes the basic definitions in such a lucid manner. Some practical aspects of short-term frequency stability are discussed by Baghdady *et al.* in the second reprint paper, particularly with respect to precision range-rate measurements.

Five years later, members of the Subcommittee on Frequency and Time of the IEEE Group on Instrumentation and Measurement wrote another basic paper dealing with "Characterization of Frequency Stability," which we reprint here as the third paper, its major contribution being the application of the piecewise linearized oscillator noise characteristics making possible the translation among frequency domain and time domain stability measurements. The difficulty with this paper is that some errors crept into the conversion tables (see errata). Since the reader will encounter slightly different conversion tables between the important two-sample variance (often called "Allan" variance) and the noise coefficients h_α in other papers on frequency stability, we reproduce here a correct one verified, once more, by the editor himself; see Table III-1.

Since the idea of the transfer functions relating the time domain and frequency domain measurements was established, viz.

$$\sigma^2(\tau) = \int_0^\infty S_y(f)\, |H(f)|^2\, df, \qquad \text{(III-1)}$$

$S_y(f)$ being the one-sided power spectral density of the normalized frequency

$$y = \Delta\, \nu(t)/\nu_0, \qquad \text{(III-2)}$$

the search for more convenient "equivalent filters" than that of the two-sample variance started. Baugh suggested in 1971 [III-3] a more-sample variance, the so-called Hadamard variance, which exhibits a larger effective Q on one hand, but a lot of spurious responses on the other hand. In the same paper, Baugh proposed to solve the problem of spurious responses with the assistance of binomial weighting. However, the other advantage of the binomial weighting, namely the removing of the long-term drifts, remained unrecognized until the introduction of the structure functions into the theory of oscillator instability by Lindsey and Chie (see the fourth reprint paper in this part and the ninth in the next part).

Up to now, we have been referring to American authors only; however, there is also an important European school that has been investigating short-term frequency stability for nearly 20 years. The theoretical work deals with the statistical properties of the oscillator noise and with suitable mathematical models [III-4]–[III-6].

From a more practical point of view, Sauvage and Rutman investigated the cosinusoidal weighting of the Hadamard variance with the intention of suppressing some harmonic responses [III-7], whereas a simultaneous binomial and cosinusoidal weighting, suggested by Lesage and Audoin [III-8], resulted in the removal of spurious responses in the neighborhood of the main passband as well.

Rutman turned the transfer function approach the other way and proposed to characterize frequency stability via an actual filtering of the phase noise [III-9], the major advantage being the simplification of the measurement systems (cf. the sixth and seventh reprint papers in Part IV).

The culmination of the work done by the European school is presented in an excellent survey paper by Rutman which we reprint here. The only difficulty with this treatise is the omission of FFT systems recently introduced into frequency stability measurements. The principles of the theory are contained in the monographic work by Howe *et al.* [III-10] in its Part X which we reprint here; note that it is based on the book by Brigham [III-11]. Additional information can be found in the paper by Babitch and Oliviero reprinted in the next part. A good introduction to DFT and FFT theory is provided in a booklet by Yuen and Fraser [III-12].

To give the reader a uniform approach to all digital systems used for frequency stability measurements, we repeat here the major results of an earlier study [III-13] dealing with the effective transfer function of FFT devices.

By choosing in the more-sample variance the weighting coefficients (cf. (6.1) or (6.4) of the reprint paper by Rutman)

$$a_k = e^{j\omega_i k\tau} \tag{III-3}$$

we actually get the DFT "spectral line" $Y^*(\omega_i)$.

The variance of the FFT output signal is

$$\sigma^2_{y,\ \omega i}(\tau) = \langle Y^*(\omega_i)\, Y(\omega_i) \rangle = \int_{-\infty}^{\infty} S_y(f)\,|H(f)|^2\, df \tag{III-4}$$

from which we have computed the effective transfer function to be

$$|H(\omega)|^2 = \left[\frac{1}{2N} \cdot \frac{\sin \omega\tau/2}{\omega\tau/2} \cdot \frac{\sin N(\omega_i - \omega)\tau}{\sin \dfrac{\omega_i - \omega}{2}\tau}\right]^2 \tag{III-5}$$

where $2N$ is the number of the samples used, τ is the measurement interval for evaluation of $\bar{y}_k$, $\omega_i = 2\pi i/2N\tau$ $(i = 1,2,\cdots,N)$, and $2N\tau$ is the total time duration of one measurement.

The major advantage of the measurement systems based on DFT or FFT is that at the same number of the samples $2N$ we can determine spectral densities at N different frequencies. However, with the increasing normalized frequency $x_i = \omega_i\tau$, the gap between it and the cutoff frequency $x_N = \pi$ of the antialiasing filter diminishes. For this reason we must be content with a smaller number of values $S_y(f_i)$, generally with some $N/2$ values.

In contradistinction, in systems based on two or more sample variances, spectral density at only one frequency is determined each time, and in addition, no means are being taken for removing harmonic responses. As a consequence, we cannot safely distinguish between the white and flicker phase noises. To cope with the difficulty, the cosinusoidal weighting of the samples $\bar{y}_k$ was suggested, but only systems with the dead time $T_D = \tau/2$ found wider applications, even on a commercial basis.

Recently, Allan and Barnes solved the problem by introducing a new so-called "modified Allan variance" [III-14], [III-15] defined as

$$\begin{aligned}\text{Mod}\ \sigma^2_y(\tau) &= \frac{1}{2\tau^2} \left\langle \left[\frac{1}{n}\sum_{i=1}^{n} (x_{i+2n} - 2x_{i+n} + x_i)\right]^2 \right\rangle \\ &= \frac{1}{2n^2} \left\langle \left[\sum_{i=1}^{n} (\bar{y}_{i+n} - \bar{y}_i)\right]^2 \right\rangle \end{aligned} \tag{III-6}$$

where

$$\begin{aligned}\bar{y}_{i+n} &= \frac{x_{i+2n} - x_{i+n}}{\tau} \\ &= \frac{\varphi(t + i\tau_0 + 2n\tau_0) - \varphi(t + i\tau_0 + n\tau_0)}{2\pi\nu_0\tau}\end{aligned} \tag{III-7}$$

and

$$\tau = n\tau_0. \tag{III-8}$$

By computing the effective transfer function, we arrive at

$$|H(\omega\tau_0)| = 2\,\frac{\sin^3(n\omega\tau_0/2)}{(n^2\omega\tau_0/2)\sin(\omega\tau_0/2)}. \tag{III-9}$$

After introducing the above relation into (III-1) we can compute conversion formulas between spectral densities $S_y(f)$ and the "modified Allan variance;" see Table III-1. The major achievement is the different slope of Mod $\sigma^2_y(\tau)$ for white and flicker noise processes.

The final task we face is the estimation of the confidence interval of the measured values. The central limit theorem teaches that with the increased number M of individual experiments, the expected error decreases in proportion to $\sqrt{M}$. However, in instances where the measurement time is large (tens of seconds, minutes, days), where the number of averaged results is limited (as with modern automatic measurement systems), or where a cross correlation between individual measurements is suspected, the calculation or at least estimation of the error variance should be performed. The problem, in respect to frequency stability measurements, is discussed in the last three reprint papers.

All of these authors investigated the uncertainty of the frequency stability measurements based on two-sample variances without and with a dead time based on the estimation

$$\sigma^2(\tau) = \frac{1}{M}\sum_{i=1}^{M} (y_i - y_{i+1})^2. \tag{III-10}$$

The sought value was the variance of $\hat{\sigma}^2(\tau)$ as a function of the number M of "events." For the benefit of our readers we summarize here the results in Tables III-2 and III-3 (note that Yoshimura's M equals $m - 1$ of Lesage and Audoin and $N - 2$ of Howe *et al.* [III-10]).

It was Yoshimura who suggested that the probability distribution of the sum in (III-10) might be governed by χ^2 statistics [III-16]. In the case of r degrees of freedom its normalized variance is given by

$$\frac{\text{Var}[\hat{\sigma}^2]}{\sigma^4} = \frac{2}{r}. \tag{III-11}$$

By comparing the above result to those in Tables III-2 and III-3, we may conclude that for white and flicker phase noises the number of degrees of freedom is about one half of M, whereas for flicker and random walk frequency noises the individual samples $(\bar{y}_i - \bar{y}_{i+1})^2$ are nearly independent of each other and the process retains the full number of degrees of freedom. According to Yoshimura, this is also true in instances where there is a large dead time between measurements (see Table III-3).

The idea of χ^2 statistics and the effective degrees of freedom instigated by workers in NBS caused them to show the benefits of a more efficient use of data of two-sample variances to tighten the uncertainty tolerance [III-10]. They recommend averaging all possi-

TABLE III-1

CONVERSION BETWEEN SPECTRAL DENSITIES, $S_y(f)$, TWO-SAMPLE VARIANCES $\sigma_y^2(\tau)$, AND "MODIFIED ALLAN VARIANCES" MOD $\sigma_y^2(\tau)$ FOR POWER-LAW SPECTRAL DENSITY MODEL

$S_y(f)$	$\sigma_y^2(\tau)$	Slope of the $\sigma_y^2(\tau)$ versus τ on a log-log plot	Mod $\sigma_y^2(\tau)$	Slope of the Mod $\sigma_y^2(\tau)$ versus on a log-log plot
$h_2 f^2$	$\frac{3h_2 f_H}{(2\pi\tau)^2}$ $(2\pi\tau f_H >> 1)$	−20 dB/decade	$\frac{3h_2 f_H}{n(2\pi n\tau_0)^2} \approx 0.076 \frac{\tau_0}{\tau_H} \cdot \frac{h_2}{(n\tau_0)^3}$	−30 dB/decade
$h_1 f$	$\frac{h_1}{(2\pi\tau)^2}\left[3\gamma + 3\lg(2\pi\tau f_H) - \lg 2\right] = \frac{h_1}{(2\pi\tau)^2}\left[1.038 + 3\lg(2\pi\tau f_H)\right]$	−20 dB/decade	$\approx h_1 \frac{3.3 + (3/n)\lg(\omega_H \tau_0)}{(2\pi n\tau_0)^2} \approx h_1 \frac{0.084}{(n\tau_0)^2}$ $n >> 1$	−20 dB/decade
h_0	$\frac{h_0}{2\tau}$	−10 dB/decade	$\approx \frac{h_0}{4(n\tau_0)} = h_0 \frac{0.25}{n\tau_0}$	−10 dB/decade
$\frac{h_{-1}}{f}$	$2h_{-1}\lg 2$	zero change	$\approx 0.675 h_{-1} 2\lg 2 \approx 0.94 h_{-1}$	zero change
$\frac{h_{-2}}{f^2}$	$\frac{(2\pi)^2 \tau h_{-2}}{6}$	+10 dB/decade	$\approx 0.824 h_{-2} \frac{(2\pi)^2 n\tau_0}{6} \approx 5.4 n\tau_0 h_{-2}$	+10 dB/decade
Sinusoidal FM modulation $\frac{1}{2}\left(\frac{\Delta\nu_0}{\nu_0}\right)^2 \delta(f - f_m)$	$\left(\frac{\Delta\nu_0}{\nu_0}\right)^2 \frac{\sin^4(\pi f_m \tau)}{(\pi f_m \tau)^2}$	periodic change	$\left(\frac{\Delta\nu_0}{\nu_0}\right)^2 \left[2 \frac{\sin^3(n f_m \tau_0)}{n f_m \tau_0 \sin(f_m \tau_0)}\right]^2$	periodic change
Linear drift (aging) of y $\dot{y} = D_y$	$\frac{(\tau D_y)^2}{2}$	+20 dB/decade	$\frac{(n\tau_0 D_y)^2}{2}$	+20 dB/decade

TABLE III-2

THE NORMALIZED VARIANCE VAR$[\hat{\sigma}^2(\tau)]/\sigma^4(\tau)$ COMPUTED BY DIFFERENT AUTHORS WITH ZERO DEAD TIME BETWEEN MEASUREMENTS

α	Lesage–Audoin	$M >> 1$	Yoshimura	$M >> 1$	Howe *et al.*	$M >> 1$
2	$\frac{1}{9M^2}(35M-18)$	$\frac{3.89}{M}$	$\frac{1}{9M^2}(35M-18)$	$\frac{3.89}{M}$	$\frac{4(M+1)}{M(M+3)}$	$\frac{4}{M}$
1	$\frac{1}{9M^2}(35M-18)$	$\frac{3.89}{M}$	$M=2$ $2.83/M$	$M=100$ $3.73/M$		
0	$\frac{1}{M^2}(3M-7)$	$\frac{3}{M}$	$\frac{3M-1}{M^2}$	$\frac{3}{M}$	$\frac{9(M+2)}{3M^2+7M+6}$	$\frac{3}{M}$
−1	$\frac{1}{M^2}(2.3M-0.3)$	$\frac{2.3}{M}$	$M=2$ $2.1/M$	$M=100$ $2.27/M$	$\frac{2.3M-0.3}{M^2}$	$\frac{2.3}{M}$
−2	$\frac{1}{4M^2}(9M-1)$	$\frac{2.25}{M}$	$\frac{2}{M}$	$\frac{2}{M}$	$\frac{2(M-1)^2}{M(M+1)^2 - 3(M+1) + 4}$	$\frac{2}{M}$

TABLE III-3

THE NORMALIZED VARIANCE VAR$[\hat{\sigma}^2(\tau)]/\sigma^4(\tau)$ COMPUTED BY DIFFERENT AUTHORS WITH A DEAD TIME COMPARABLE TO THE EVALUATION INTERVAL

α	Lesage–Audoin $T_m - \tau \approx \tau$ $M >> 1$	Yoshimura $T_m = 2\tau$ $M >> 1$
2	$\frac{3}{M}$	$\frac{2.1}{M}$
1	$\frac{3}{M}$	$\frac{2}{M}$
0	$\frac{3}{M}$	$\frac{2}{M}$
−1	$\frac{3}{M}$	$\frac{2}{M}$
−2	$\frac{2}{M}$	$\frac{2}{M}$

ble two-sample variances of a given $n \cdot \tau$ value that one can compute from the data taken at N time (or phase) measurements one τ apart.

For a deeper study of the problems of frequency instability, we would like to call to the attention of the readers a short paper by Percival [III-17] that is an essence of his Ph.D. dissertation.

To those readers interested in long-term stability (sample times of days and weeks) we recommend they consult the NBS Monograph 140 [IV-2] or the book by Kartaschoff [III-18].

REFERENCES

[III-1] IEEE-NASA Symp. Short-Term Frequency Stability, Goddard Space Flight Center, Greenbelt, MD, Nov. 23–24, 1964, Document NASA SP-80.

[III-2] Special Issue on Frequency Stability, *Proc. IEEE*, vol. 54, pp. 101–338, Feb. 1966.

[III-3] R. A. Baugh, "Frequency modulation analysis with the Hadamard variance," in *Proc. 25th Annu. Frequency Contr. Symp.*, Atlantic City, NJ, Apr. 1971, pp. 222–225.

[III-4] J. de Prins, G. Descornet, M. Gorski, and J. Tamine, "Frequency-domain interpretation of oscillator phase stability," *IEEE Trans. Instrum. Meas.*, vol. IM-18, pp. 251–261, Dec. 1969.

[III-5] E. Boileau and B. Picinbono, "Statistical study of phase fluctuations and oscillator stability," *IEEE Trans. Instrum. Meas.*, vol. IM-25, pp. 66–75, Mar. 1976.

[III-6] E. Boileau, "Improvements of the procedures used to study the fluctuations of oscillators," *IEEE Trans. Instrum. Meas.*, vol. IM-27, pp. 210–214, Sept. 1978.

[III-7] G. Sauvage and J. Rutman, "Analyse spectrale du bruit de fréquence des oscillateurs par la variance de Hadamard," *Ann. Telecommun.*, vol. 28, pp. 301–314, July–Aug. 1973.

[III-8] P. Lesage and C. Audoin, "A time domain method for measurement of the spectral density of frequency fluctuations at low Fourier frequencies," in *Proc. 29th Annu. Symp. Frequency Contr.*, May 28–30, 1975, pp. 394–403.

[III-9] J. Rutman, "Characterization of frequency stability: A transfer function approach and its application to measurement via filtering of phase noise," *IEEE Trans. Instrum. Meas.*, vol. IM-23, pp. 40–48, Mar. 1974.

[III-10] D. A. Howe, D. W. Allan, and J. A. Barnes, "Properties of signal sources and measurement methods," in *Proc. 35th Annu. Frequency Contr. Symp.*, Philadelphia, PA, May 1981, pp. A1–A47.

[III-11] E. O. Brigham, *The Fast Fourier Transform.* Englewood Cliffs, NJ: Prentice-Hall, 1974.

[III-12] C. K. Yuen and D. Fraser, *Digital Spectral Analysis*. East Melbourne, Australia: CSIRC; and London, England: Pitman, 1979.

[III-13] V. F. Kroupa, "Digital systems for frequency stability measurements," in *Proc. 6th Summer Symp. Circuit Theory 1982, Main Lectures,* Prague, Czechoslovakia, July 12–16, 1982, pp. 56–70.

[III-14] D. W. Allan and J. A. Barnes, "A modified 'Allan variance' with increased oscillator characterization ability," in *Proc. 35th Annu. Frequency Contr. Symp.*, Philadelphia, PA, May 1981, pp. 470–475.

[III-15] V. F. Kroupa, "The transfer function approach to the 'modified' Allan variance," unpublished.

[III-16] A. Papoulis, *Probability, Random Variables and Stochastic Processes.* New York: McGraw-Hill, 1965, p. 250.

[III-17] D. B. Percival, "Characterization of time and frequency instabilities: Prediction error variance," in *Conf. Precision Electromag. Meas. CPEM-82 Dig.*, pp. N7–N9.

[III-18] P. Kartaschoff, *Frequency and Time.* London: Academic, 1978.

Some Aspects of the Theory and Measurement of Frequency Fluctuations in Frequency Standards

L. S. CUTLER, MEMBER, IEEE, AND C. L. SEARLE, MEMBER, IEEE

Abstract—**Precision quartz oscillators have three main sources of noise contributing to frequency fluctuations: thermal noise in the oscillator, additive noise contributed by auxiliary circuitry such as AGC, etc., and fluctuations in the quartz frequency itself as well as in the reactive elements associated with the crystal, leading to an f^{-1} type of power spectral density in frequency fluctuations. Masers are influenced by the first two types of noise, and probably also by the third.**

The influence of these sources of noise on frequency fluctuation vs. averaging time measurements is discussed. The f^{-1}-spectral density leads to results that depend on the length of time over which the measurements are made. An analysis of the effects of finite observation time is given.

The characteristics of both passive and active atomic standards using a servo-controlled oscillator are discussed. The choice of servo time constant influences the frequency fluctuations observed as a function of averaging time and should be chosen for best performance with a given quartz oscillator and atomic reference.

The conventional methods of handling random signals, i.e., variances, autocorrelation, and spectral densities, are applied to the special case of frequency and phase fluctuations in oscillators, in order to obtain meaningful criteria for specifying oscillator frequency stability. The interrelations between these specifications are developed in the course of the paper.

I. Introduction

The purpose of this paper is to present some of the theoretical and practical aspects of frequency fluctuation measurements in frequency standards. In Section II the fundamental definitions of terms used in specifying frequency stability both in the time domain and the frequency domain are developed, and the interrelations between these terms are established. Section III describes briefly several of the commonly used techniques for measuring frequency stability and in each case relates the measurement to the appropriate mathematical description in Section II.

The three main sources of frequency fluctuation in oscillators are discussed in Section IV: thermal and shot noise in the oscillator, additive noise (in the amplifier, for example), and oscillator parameter changes giving rise to an f^{-1} power spectral density of frequency fluctuations. Finally, in Section V we consider the frequency fluctuations in atomic frequency standards.

Manuscript received October 11, 1965; revised December 1, 1965. This work was supported in part by the Joint Services Electronics Program under Contract DA 36-039-AMC-03200(E).

L. S. Cutler is with the Physics Research Division, Hewlett-Packard Company, Palo Alto, Calif.

C. L. Searle is with the Department of Electrical Engineering and the Research Laboratory of Electronics, Massachusetts Institute of Technology, Cambridge, Mass.

II. Specification of Frequency Stability

A. Definitions

The signal from an oscillator may be described by

$$v(t) = A(t) \cos [\omega_0 t + \phi(t)] \tag{1}$$

where $v(t)$ represents a voltage or current, $A(t)$ and $\phi(t)$ are slowly varying real functions of time, and ω_0 is a constant. $A(t)$ is the variable amplitude of the signal and does not contribute directly to frequency fluctuations. (Of course, in frequency multipliers and other nonlinear devices, conversion of amplitude modulation to phase modulation and vice versa can occur. Nonetheless, at any given point in the system, (1) is valid.) It is assumed that the time origin and ω_0 are chosen so that $\phi(t)$ has zero time average and

$$\lim_{T\to\infty} \frac{\phi(T/2) - \phi(-T/2)}{T} = 0.$$

These conditions simplify the mathematics (but will have to be relaxed later). The instantaneous angular frequency is

$$\omega(t) = \frac{d}{dt}[\omega_0 t + \phi(t)] = \omega_0 + \dot{\phi}(t). \tag{2}$$

In all that follows we will refer to angular frequency as frequency.

The average frequency of the signal $v(t)$ is by definition

$$\begin{aligned}\langle\omega(t)\rangle &= \lim_{T\to\infty} \frac{1}{T}\int_{-T/2}^{T/2} \omega(t)\, dt \\ &= \omega_0 + \lim_{T\to\infty} \frac{\phi(T/2) - \phi(-T/2)}{T} = \omega_0 \end{aligned} \tag{3}$$

where the symbol $\langle\ \rangle$ indicates time average over an infinite time. Therefore, we see from (1) and (2) that $\phi(t)$ is the instantaneous phase angle of the oscillator with respect to an ideal oscillator of frequency ω_0, and $\dot{\phi}(t)$ is the frequency departure away from ω_0.

Of particular importance in the specification of frequency stability is the frequency departure averaged over some finite time τ:

Reprinted from *Proc. IEEE*, vol. 54, pp. 136–154, Feb. 1966.

$$\langle\dot{\phi}\rangle_{t,\tau} = \frac{1}{\tau}\int_{t-\tau/2}^{t+\tau/2} \dot{\phi}(t')\,dt' \quad (4a)$$

$$= \frac{\phi(t+\tau/2) - \phi(t-\tau/2)}{\tau} \quad (4b)$$

where the symbol $\langle\ \rangle_{t,\tau}$ signifies the average over a finite time τ, centered at time t. Similarly, we can define the finite-time average of the phase

$$\langle\phi\rangle_{t,\tau} = \frac{1}{\tau}\int_{t-\tau/2}^{t+\tau/2} \phi(t')\,dt'. \quad (5)$$

For future calculations we shall need the accumulated phase over a time τ. From (4),

$$\Delta_\tau\phi(t) \equiv \phi(t+\tau/2) - \phi(t-\tau/2) = \tau\langle\dot{\phi}\rangle_{t,\tau}. \quad (6)$$

Equation (4) can be depicited in signal-processing terms as shown in Fig. 1. Equation (4a) indicates that $\langle\dot{\phi}\rangle_{t,\tau}$ is found by averaging $\dot{\phi}$ over a time τ. This averaging can be accomplished by passing the signal $\dot{\phi}(t)$ (in the form of a voltage, perhaps) through a filter with a square impulse response, as shown in Fig. 1. The required shape of this response is found by solving (4a), assuming that $\dot{\phi}(t)$ is a unit impulse at $t=0$.[1] Mathematically, this impulse response can be described by

$$h_a(t) = \frac{u(t) - u(t-\tau)}{\tau} = \frac{\text{rect}_\tau\,(t-\tau/2)}{\tau}. \quad (7)$$

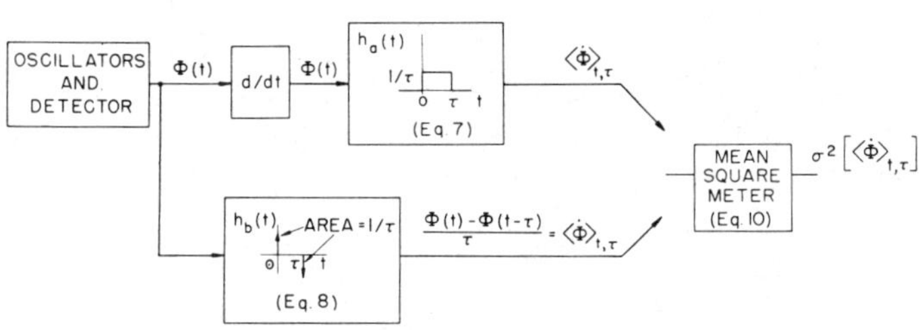

Fig. 1. Representation of (4a) and (4b) in signal-processing terms.

In a similar way, (4b) can be depicted in signal-processing terms as shown in the lower branch of Fig. 1. If we work directly with the phase signal $\phi(t)$, then by inspection the appropriate impulse response in this case contains two impulses, both of area $1/\tau$, as shown. (Again a time delay of $\tau/2$ seconds has been introduced.) This response is described by

$$h_b(t) = \frac{\delta(t) - \delta(t-\tau)}{\tau}. \quad (8)$$

Because both filters h_a and h_b are clearly difficult to realize in practice, systems to measure $\langle\dot{\phi}\rangle_{t,\tau}$ usually do not conform directly to either of these signal-processing diagrams. Practical systems for measuring $\langle\dot{\phi}\rangle_{t,\tau}$ are presented in Section III.

[1] For mathematical symmetry, the integrals have been written with limits symmetric about t. However, to insure physical realizability of the filter, the impulse response has been drawn from 0 to τ, implying that the lower and upper limits of integration have been changed to t and $t+\tau$, respectively. This corresponds to a time delay of $\tau/2$ seconds.

B. Variance

Because $\phi(t)$, and hence $\dot{\phi}(t)$ and $\langle\dot{\phi}\rangle_{t,\tau}$ are random variables, some measure of their dispersion is needed. For this purpose we often calculate the standard deviation σ or the variance σ^2, the latter being defined as

$$\sigma^2[X] = \overline{(X-\overline{X})^2} = \overline{X^2} - (\overline{X})^2 \quad (9)$$

where the bar signifies either time or statistical average. The variance of the frequency averaged over time τ is thus

$$\sigma^2[\langle\dot{\phi}\rangle_{t,\tau}] = \overline{\langle\dot{\phi}\rangle_{t,\tau}^2}. \quad (10)$$

Here the variance is identical to the mean-square value because the mean value is zero:

$$\overline{\langle\dot{\phi}\rangle_{t,\tau}} = \frac{1}{\tau}\int_{t-\tau/2}^{t+\tau/2} \overline{\dot{\phi}(t')}\,dt' = 0 \quad (11)$$

on the basis that, by definition, $\overline{\dot{\phi}(t')}$ equals zero. A block corresponding to (10) is included in Fig. 1.

C. Autocorrelation and Spectral Density

The frequency stability of an oscillator can also be specified in terms of autocorrelation functions (also called autocovariance—see, for example, Blackman and Tukey [1]) or spectral densities. Returning for a moment to (1), it is clear that four possible spectra might be of interest in describing the properties of the signal $v(t)$.

1) The complete spectrum of $v(t)$. This is often called the RF spectrum, because it includes the carrier and all sidebands.
2) The spectral density of $A(t)$, which could be found by passing $v(t)$ through an ideal AM detector, and measuring the spectrum of the detector output.
3) $S_\phi(\omega)$, the spectral density of $\phi(t)$, found by passing $v(t)$ through an ideal phase detector, and measuring the detector output spectrum.
4) $S_{\dot{\phi}}(\omega)$, the spectrum of $\dot{\phi}(t)$, obtained by spectral analysis of the output of an ideal FM detector.

The RF spectrum of $v(t)$ is of great importance in many applications, but for very stable oscillators in which the sidebands are orders of magnitude smaller than the carrier everywhere except close to the carrier, it is often difficult to measure directly. Thus, for stable oscillators the most useful spectral representation of the frequency stability is usually $S_\phi(\omega)$ or $S_{\dot{\phi}}(\omega)$. For the frequently encountered case where the AM power spectral density is negligible, and the mean-square value of the phase is much less than one rad², for frequencies greater than some minimum value, the RF power spectral density (two-sided) is

$$S_{\mathrm{RF}}(\omega) \approx \frac{P}{2}(S_\phi(\omega+\omega_0)+S_\phi(\omega-\omega_0))$$

where P is the carrier power, and the expression is valid except for the carrier frequencies, $\pm\omega_0$, and narrow bands surrounding them containing all the high modulation-index, low-frequency modulation sidebands. The low modulation-index portion of the power spectral density of the sidebands has the same shape as the phase power spectral density because only the first-order sidebands are of any consequence.

The spectral density of a random signal is defined as the Fourier transform of its autocorrelation function [2]. Thus, we first define the autocorrelation of the phase as

$$\begin{aligned} R_\phi(\tau) &= \overline{\phi(t+\tau/2)\phi(t-\tau/2)} \\ &= \lim_{T\to\infty}\frac{1}{T}\int_{-T/2}^{T/2}\phi(t+\tau/2)\phi(t-\tau/2)\,dt \qquad (12)\end{aligned}$$

where the bar signifies statistical average, and ergodicity is assumed. Similarily, $R_{\dot\phi}(\tau)$ is the autocorrelation function of the frequency departure

$$\begin{aligned} R_{\dot\phi}(\tau) &= \overline{\dot\phi(t+\tau/2)\dot\phi(t-\tau/2)} \\ &= \lim_{T\to\infty}\frac{1}{T}\int_{-T/2}^{T/2}\dot\phi(t+\tau/2)\dot\phi(t-\tau/2)\,dt. \qquad (13)\end{aligned}$$

Writing these both as functions of τ alone implies that ϕ and $\dot\phi$ are stationary in the wide sense [3].

As pointed out above, the spectral density of the phase (using two-sided spectra) is

$$S_\phi(\omega) = \int_{-\infty}^{\infty} R_\phi(\tau)e^{-j\omega\tau}\,d\tau = 2\int_0^{\infty} R_\phi(\tau)\cos\omega\tau\,d\tau. \qquad (14)$$

It also follows that $R_\phi(\tau)$ is the Fourier transform of $S_\phi(\omega)$:

$$R_\phi(\tau) = \frac{1}{2\pi}\int_{-\infty}^{\infty} S_\phi(\omega)e^{j\omega\tau}\,d\omega = \frac{1}{\pi}\int_0^{\infty} S_\phi(\omega)\cos\omega\tau\,d\omega. \qquad (15)$$

In a similar way, $S_{\dot\phi}(\omega)$ and $R_{\dot\phi}(\tau)$ also form a Fourier transform pair, identical in form to (14) and (15).

On the basis of our original definitions, $\dot\phi(t)$ was the time derivative of $\phi(t)$. Differentiation in the time domain corresponds to multiplication by $j\omega$ in the frequency domain and, hence, multiplication by ω^2 in terms of spectral densities. Thus

$$S_{\dot\phi}(\omega) = \omega^2 S_\phi(\omega).$$

In dealing with electrical signals it is customary to calculate the *power* spectral density (units of watts/hertz) for signals that exist for all time, and the *energy* spectral density (units of joules/hertz) for signals that exist only for some finite time (pulse signals). Equations (12), (14), and (15) imply that $S_\phi(\omega)$ is a "power" spectral density, even though there is no power involved (the units are actually radians² per hertz). In this paper we shall define all spectra in terms of "power" and not "energy." To emphasize this we shall use the somewhat inappropriate term "power spectral density." Also, in what follows we shall always use two-sided spectra.

There are, of course, other spectral representations of frequency stability that are closely related to $S_\phi(\omega)$ and $S_{\dot\phi}(\omega)$. As an example, often the FM noise of a microwave oscillator is specified in terms of the rms voltage measured in a 1-kc bandwidth at the output of a frequency discriminator [27]. Following the convention for frequency-modulated waves, this measurement is plotted as Δf, the rms frequency deviation measured in a 1-kc bandwidth, vs. frequency f_m (the modulating frequency in the sinusoidal case, but the center frequency of the 1-kc measurement passband).

This measurement can be related to $S_{\dot\phi}(\omega)$, the spectral density of the frequency, by noting from (13) that for $\tau=0$,

$$R_{\dot\phi}(0) = \overline{\dot\phi^2(t)}$$

and, because of the Fourier transform relationship between $R_{\dot\phi}(\tau)$ and $S_{\dot\phi}(\omega)$,

$$R_{\dot\phi}(0) = \frac{1}{2\pi}\int_{\infty}^{\infty} S_{\dot\phi}(\omega)\,d\omega.$$

Combining these two equations, we obtain

$$\overline{\dot\phi^2(t)} \equiv \frac{1}{2\pi}\int_{-\infty}^{\infty} S_{\dot\phi}(\omega)\,d\omega.$$

This result, which in general relates the mean-square value of a function to the total area under its spectral density curve, is a special case of Parseval's Theorem. For the present case, we want the mean-square value of the signal out of a 1-kc filter, and thus the appropriate relation is

$$\left.\overline{\dot\phi^2(t)}\right|_{1\ \mathrm{kc}} \equiv (\Delta f)^2 = \frac{1}{\pi}\int_{2\pi(f_m-500)}^{2\pi(f_m+500)} S_{\dot\phi}(\omega)\,d\omega$$

for a sharp-edged filter. In the case where $S_{\dot\phi}(\omega)$ is fairly constant over the 1-kc bandwidth, we find

$$(\Delta f)^2 \cong 2000 S_{\dot\phi}(\omega).$$

D. Relations Between Variance and Spectral Densities

The two major methods of specifying the frequency stability of an oscillator presented above, namely the variance of the frequency departure averaged over time τ and the power spectral density, are in fact closely related via the autocorrelation function [4], [5] (see Appendix I). Specifically, the variance of the average frequency departure can be written either in terms of $R_\phi(\tau)$ or $R_{\dot\phi}(\tau)$.

$$\sigma^2[\langle\dot{\phi}\rangle_{t,\tau}] = \frac{2}{\tau^2}\,[R_\phi(0) - R_\phi(\tau)] \tag{16a}$$

$$= \frac{2}{\tau}\int_0^\tau R_{\dot{\phi}}(\tau')\left(1 - \frac{\tau'}{\tau}\right) d\tau'. \tag{16b}$$

This variance is often normalized to yield the variance of the average *fractional* frequency departure by dividing both sides of (16) by ω_0^2.

By substituting (15) into (16a), we can eliminate R_ϕ, and, hence, express the variance directly in terms of the power spectral density of frequency or phase. The resulting expressions are

$$\sigma^2[\langle\dot{\phi}\rangle_{t,\tau}] = \frac{1}{2\pi}\int_{-\infty}^{\infty} S_{\dot{\phi}}(\omega)\,\frac{\sin^2 \omega\tau/2}{(\omega\tau/2)^2}\,d\omega \tag{17a}$$

$$= \frac{2}{\pi\tau^2}\int_{-\infty}^{\infty} S_\phi(\omega)\sin^2 \omega\tau/2\,d\omega. \tag{17b}$$

Although $\sigma^2[\langle\dot{\phi}\rangle_{t,\tau}]$ (or its normalized equivalent) is the variance most commonly used in specifying frequency stability, we include for convenience formulas for several other variances written in terms of autocorrelations and spectral densities:

$$\sigma^2[\langle\phi\rangle_{t,\tau}] = \frac{2}{\tau}\int_0^\tau R_\phi(\tau')\left(1 - \frac{\tau'}{\tau}\right) d\tau'$$

$$= \text{variance of average phase} \tag{18}$$

$$\sigma^2[\Delta_\tau\phi] = 2[R_\phi(0) - R_\phi(\tau)]$$

$$= \text{variance of accumulated phase} \tag{19}$$

$$\sigma^2[\phi(t)] = R_\phi(0) = \frac{1}{\pi}\int_0^\infty S_\phi(\omega)\,d\omega$$

$$= \text{variance of phase} \tag{20}$$

$$\sigma^2[\dot{\phi}(t)] = R_{\dot{\phi}}(0) = \frac{1}{\pi}\int_0^\infty S_{\dot{\phi}}(\omega)\,d\omega$$

$$= \text{variance of frequency} \tag{21}$$

$$= \frac{1}{\pi}\int_0^\infty \omega^2 S_\phi(\omega)\,d\omega.$$

The preceding formulas hold for wide-sense stationary random processes or for time functions that have stationary means and autocorrelation functions that depend only on the time difference.

E. Signal-Processing Summary

The interrelations between power spectral density and variance of average frequency departure, which have been stated mathematically above, are shown in signal-processing terms in Fig. 2. Thus, the figure serves as a summary of the preceding discussion. Practical methods of performing these data-processing steps are discussed in Section III.

At the top left of Fig. 2 is the system to be analyzed (for example, two crystal oscillators plus phase detector and nulling servo, see Section III-C). We assume that the output from the system is the phase difference, $\phi(t)$. The first row indicates the real-time data processing required to find $\langle\dot{\phi}\rangle_{t,\tau}$ and its variance, based on Fig. 1. The second row in Fig. 2 shows the data reduction required to specify the oscillator stability in terms of spectral densities, either $S_\phi(\omega)$ or $S_{\dot{\phi}}(\omega)$. In addition, Fig. 2 shows the steps necessary to find the variance from $S_{\dot{\phi}}(\omega)$, i.e., (17a). Here, however, we emphasize the relationship between the time-domain operations and the corresponding operations required in the frequency domain. Corresponding to convolution with $h_a(t)$, we have multiplication by $|H_a(j\omega)|^2$, where $H_a(j\omega)$ is the

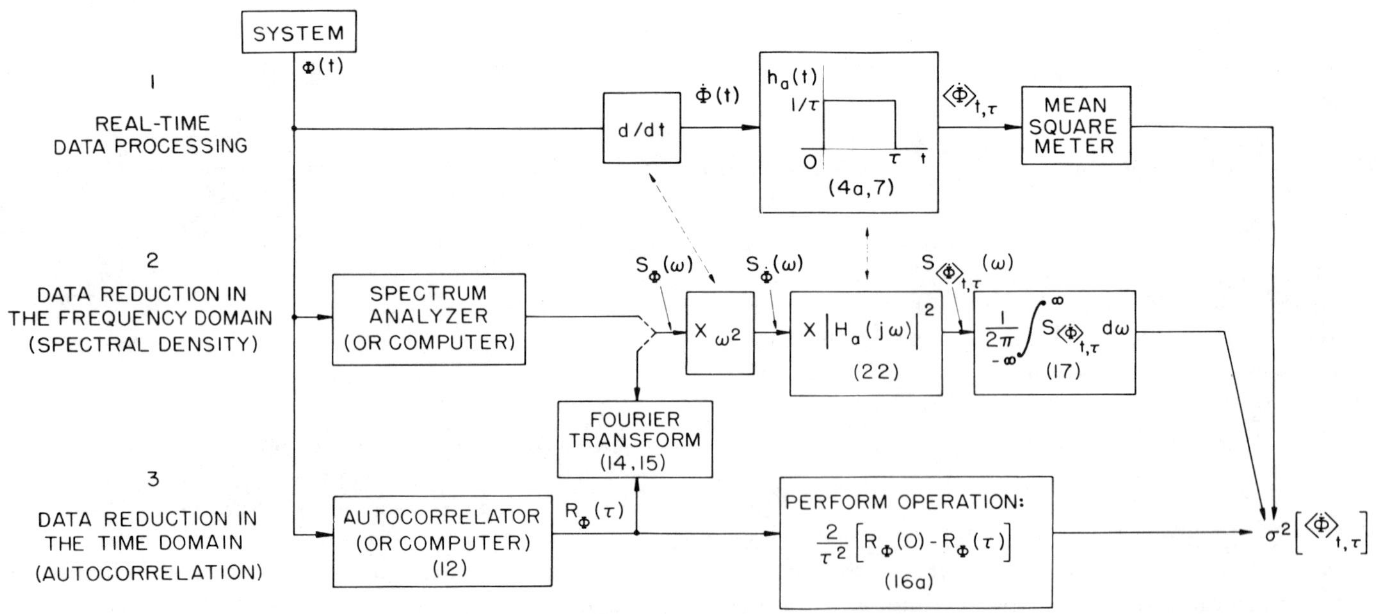

Fig. 2. Summary of methods of data processing. Numbers in brackets refer to equation numbers in the text. Double arrows (< - - - - - >) indicate Fourier transform relationship (time< - - - - - >power spectral density).

Fourier transform of $h_a(t)$. Specifically

$$|H_a(j\omega)|^2 = \frac{\sin^2 \omega\tau/2}{(\omega\tau/2)^2}. \tag{22}$$

Thus, we find that the spectral density of $\langle\dot{\phi}\rangle_{t,\tau}$ is

$$S_{\langle\dot{\phi}\rangle_{t,\tau}}(\omega) = S_{\dot{\phi}}(\omega)\,|H_a(j\omega)|^2.$$

Parseval's Theorem states that finding the mean-square value of $\langle\dot{\phi}\rangle_{t,\tau}$ is identical to finding the area under $S_{\langle\dot{\phi}\rangle_{t,\tau}}(\omega)$. The required integration of $S_{\langle\dot{\phi}\rangle_{t,\tau}}(\omega)$ is specified in Fig. 2.

The preceding discussion indicates that the first two rows of Fig. 2 are closely related in that each operation in one row has its counterpart in the other. Thus, the diagram can be traversed not only along the indicated solid lines, but also vertically between corresponding variables in the first two rows. Mathematically this vertical movement is accomplished by applying an appropriate Fourier relation (time ←- - - → spectral density); conceptually we merely move from the time domain to the frequency domain.

The third row in Fig. 2 shows the operations required to find the variance via the autocorrelation function, i.e., (16a).

Figure 2 indicates that there are alternate paths for calculating many of the quantities of interest here. Although for a particular problem one method will usually be much simpler than others, the alternate paths provide a useful check on computations. As an example of such a calculation, assume that $S_\phi(\omega)$ is band-limited white noise, as shown in Fig. 3(a). (This model for a phase power density spectrum is often encountered in frequency standards.) Let us first calculate $\sigma^2[\langle\dot{\phi}\rangle_{t,\tau}]$ following the method outlined in the second row of Fig. 2. First we find $S_{\dot{\phi}}(\omega)$. The result, shown in Fig. 3(b), indicates a predominance of high-frequency noise. The variance can now be found by multiplying $S_{\dot{\phi}}(\omega)$ by $|H_a|^2$ to yield $S_{\langle\dot{\phi}\rangle_{t,\tau}}(\omega)$ and then integrating. Of course, $S_{\langle\dot{\phi}\rangle_{t,\tau}}$ is a function of both ω and the averaging time τ; in this case

$$S_{\langle\dot{\phi}\rangle_{t,\tau}}(\omega) = \frac{4}{\tau^2}\sin^2$$

$$\left.\begin{aligned} \frac{\omega\tau}{2} &= \frac{2}{\tau^2} - \frac{2}{\tau^2}\cos\omega\tau, \quad -\omega_c < \omega < \omega_c \\ &= 0 \qquad \text{elsewhere} \end{aligned}\right].$$

From Fig. 2

$$\sigma^2[\langle\dot{\phi}\rangle_{t,\tau}] = \frac{1}{2\pi}\int_{-\omega_c}^{\omega_c}\left(\frac{2}{\tau^2} - \frac{2}{\tau^2}\cos\omega\tau\right)d\omega.$$

Without performing the integration, we can see that for large τ, the first integral will dominate, and thus

$$\sigma^2[\langle\dot{\phi}\rangle_{t,\tau}] = \frac{2\omega_c}{\pi\tau^2} \qquad (\omega_c\tau \gg 2\pi).$$

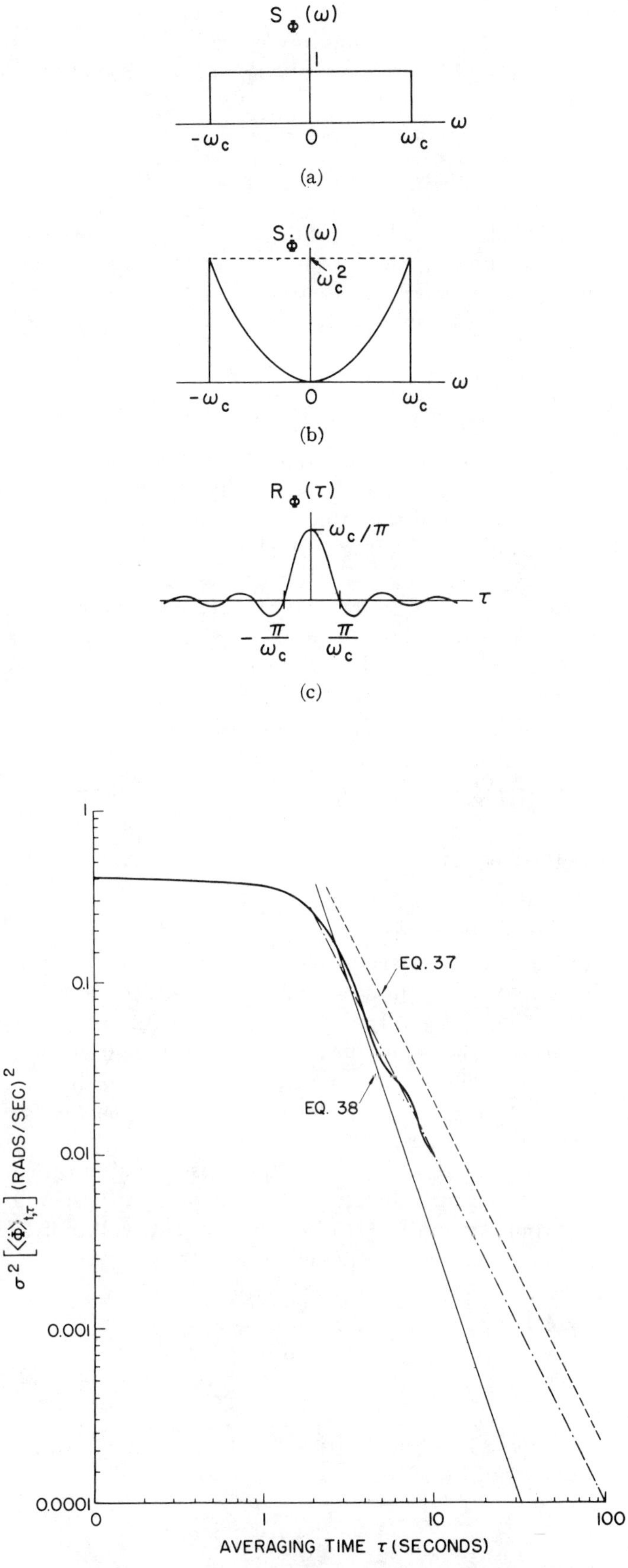

Fig. 3. Example: calculation of $\sigma^2[\langle\dot{\phi}\rangle_{t,\tau}]$ for band-limited white phase noise. (a) Power density spectrum of phase, rad²/hertz. (b) Power density spectrum of frequency, (rad/s)²/hertz. (c) Autocorrelation of phase, rad². (d) Variance of average frequency, (rad/s)².

Also, for τ very small, the $H_a(j\omega)$ function is approximately flat over the range $-\omega_c<\omega<\omega_c$, so multiplication by $|H_a|^2$ leaves $S_{\dot\phi}$ relatively unchanged. Thus, in this case

$$\sigma^2 = \frac{1}{2\pi}\int_{-\omega_c}^{\omega_c} \omega^2\, d\omega = \frac{\omega_c^3}{3\pi} \qquad (\omega_c\tau \ll 1).$$

The third row of Fig. 2 indicates that an alternative method of finding the variance is via the autocorrelation. Because $S_\phi(\omega)$ is a square pulse of unit height and length $2\omega_c$, we find from (15) that

$$R_\phi(\tau) = \frac{\omega_c}{\pi}\left[\frac{\sin \omega_c\tau}{\omega_c\tau}\right]$$

as shown in Fig. 3(c). Equation (16a) indicates that the variance can be found by multiplying the difference between $R_\phi(\tau)$ and its value at $\tau=0$ by the factor $2/\tau^2$. The result is shown in Fig. 3(d); here ω_c, the width of the noise spectrum, has been arbitrarily set equal to $\pi/2$ for ease of calculation. Note that the general shape of the variance can be sketched by inspection as a basic $1/\tau^2$ behavior perturbed by the autocorrelation difference term. As a check on previous calculations, we see that because $R_\phi(\tau)$ approaches zero for large τ,

$$\sigma^2[\langle\dot\phi\rangle_{t,\tau}] = \frac{2}{\tau^2}\,[R_\phi(0)] = \frac{2\omega_c}{\pi\tau^2} \qquad (\omega_c\tau \gg 2\pi)$$

as before.

F. *Finite Data Length*

The preceding derivations and discussion have all been based on the assumption that the functions $\phi(t)$ and $\dot\phi(t)$ exist for all time. In practice this is not so; measurements can only be made over some finite time T. Under these conditions we can only make *estimates* of σ and S. These estimates (denoted $\hat\sigma$, $\hat S$) are themselves statistical variables, with variances and all other properties associated with random processes. This problem is treated in Appendix II, and at greater length in Allan [6]. From Appendix II, (76), the average value of the variance of average frequency departure is

$$\overline{\hat\sigma^2[\langle\dot\phi\rangle_{t,\tau}]} = \frac{1}{2\pi}\int_{-\infty}^{\infty} S_{\langle\dot\phi\rangle_{t,\tau}}(\omega)\left(1 - \frac{\sin^2 \omega(T-\tau)/2}{[\omega(T-\tau)/2]^2}\right) d\omega. \qquad (23)$$

The corresponding relation for the case of continuous data is, from Fig. 2,

$$\sigma^2[\langle\dot\phi\rangle_{t,\tau}] = \frac{1}{2\pi}\int_{-\infty}^{\infty} S_{\langle\dot\phi\rangle_{t,\tau}}(\omega)\, d\omega.$$

Equation (23) differs from this expression, in that effectively the low-frequency portion of the spectrum of $S_{\langle\dot\phi\rangle_{t,\tau}}$ (i.e., below $\omega\cong 2/T$) is removed before the integration.

Clearly, the effect of the variance of measuring $\phi(t)$ for only a finite time T will depend on the shape of the spectral density. Specifically, if $S_{\langle\dot\phi\rangle_{t,\tau}}$ is relatively wideband, such that most of the power is at frequencies higher than $\omega\cong 2/T$, then the variance is not going to be appreciably affected by the finite measurement time T. On the other hand, if $S_{\langle\dot\phi\rangle_{t,\tau}}$ is relatively narrow-band, or is greatly peaked at low frequencies, then the apparent variance based on data over time T is going to be considerably smaller than the true variance. As an example of this, in the case of $1/f$-frequency noise (i.e., $S_{\dot\phi}(\omega) = k/\omega$), the true variance is finite, whereas a finite observation time always gives a finite result for $\overline{\hat\sigma^2}$ (see Appendix II).

We can assess from a signal-processing point of view the effect on the spectral densities S_ϕ and $S_{\dot\phi}$ of making measurements for a finite rather than infinite time. In the time domain, phase data acquisition for a length of time T can be represented by multiplying the signal $\phi(t)$ (assumed to exist for all time), by a pulse of unit height extending from $-T/2$ to $T/2$. The corresponding operation in the frequency domain is convolution with the function

$$|H_c(j\omega)|^2 = \frac{T\sin^2 \omega T/2}{(\omega T/2)^2} = Q_c(\omega) \qquad (24)$$

(see Appendix II).

For example, the average value of the estimate of the phase spectral density is

$$\overline{\hat S_\phi(\omega)} = S_\phi(\omega) \otimes |H_c(j\omega)|^2 \qquad (25)$$

where the symbol $\otimes$ indicates convolution. Thus, the effect of finite data length is to blur the spectrum somewhat, thereby making it impossible to resolve frequency components on a finer scale than $\Delta\omega\cong 2/T$ (see Appendix II).

III. Measurement Techniques

There are several well-known techniques for making measurements of some of the quantities described above. Some of these will now be considered.

A. *Multiple-Period Technique*

The general system of a multiple-period measurement is shown in Fig. 4. Two signal sources slightly offset in average frequency feed two identical channels through optional frequency multipliers to a mixer. (The multipliers may be used to increase the resolution and sensitivity of the measurement.) The counter then measures the period of the mixer output.

The output from the mixer contains all the desired frequency and phase information. That is, if the two signal sources have exactly the same statistics for $\phi_1(t)$ and $\phi_2(t)$ but are uncorrelated, then all the fluctuation may be assumed to be in one channel $\sqrt{2}$ times as large as that channel alone, while the other channel may be assumed to be perfect.

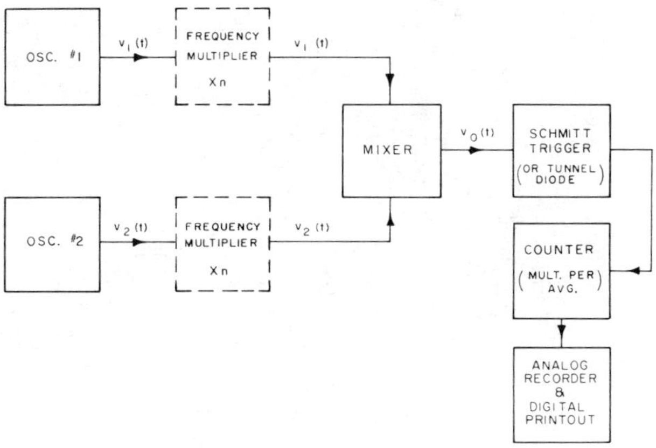

Fig. 4. Multiple-period measuring system.

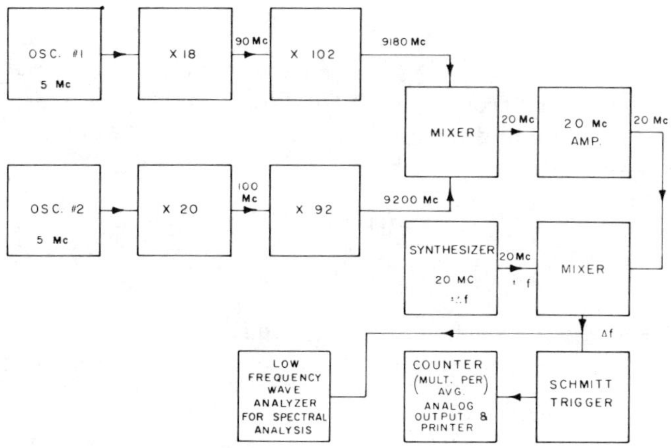

Fig. 5. Versatile multiple-period measuring system.

Thus, we can assume the mixer output signal to be of the form

$$v_0(t) = A \cos (\omega_0 t + \phi(t)) \tag{26}$$

where A, ω_0 and ϕ are as defined in (1), except that here ω_0 represents the constant frequency *difference* between the two oscillators, and ϕ is the phase *difference*. The period of the mixer output waveform is measured by using a counter to determine the time τ required for N_0 periods of $v_0(t)$. The relation between the counter output τ and the phase is

$$2\pi N_0 = \omega_0 \tau + \phi(\tau) - \phi(0). \tag{27}$$

Note that, in this method, the time τ is not constant. In fact, τ is the quantity that carries the information about the phase variations. Specifically, let

$$\tau_0 = \frac{2\pi N_0}{\omega_0} \tag{28}$$

and

$$\tau = \tau_0 - \Delta\tau. \tag{29}$$

Then (27) reduces to

$$\phi(\tau) - \phi(0) = \omega_0 \Delta\tau. \tag{30}$$

If $\omega_0 \Delta\tau \ll 1$ and $\dot{\phi}(0)\Delta\tau \ll 1$, then only very small error is caused by replacing $\phi(\tau)$ by $\phi(\tau_0)$. (The process of averaging over many measurements helps here.) The multiple-period technique thus measures essentially $\phi(t+\tau_0) - \phi(t)$. Dividing by τ_0 to reduce it to the form of (4), we thus obtain from (30)

$$\frac{\phi(t + \tau_0) - \phi(t)}{\tau_0} = \langle \dot{\phi} \rangle_{t,\tau_0} \cong \frac{\omega_0 \Delta\tau}{\tau_0} \cdot \tag{31}$$

Thus the mixer-counter combination does, in fact, perform the function described by (4) and Fig. 1. The standard deviation of $\langle \dot{\phi} \rangle_{t,\tau_0}$ can be found by making repeated measurements:

$$\sigma[\langle \dot{\phi} \rangle_{t,\tau_0}] = \frac{\omega_0 \sigma[\Delta\tau]}{\tau_0} \cdot \tag{32}$$

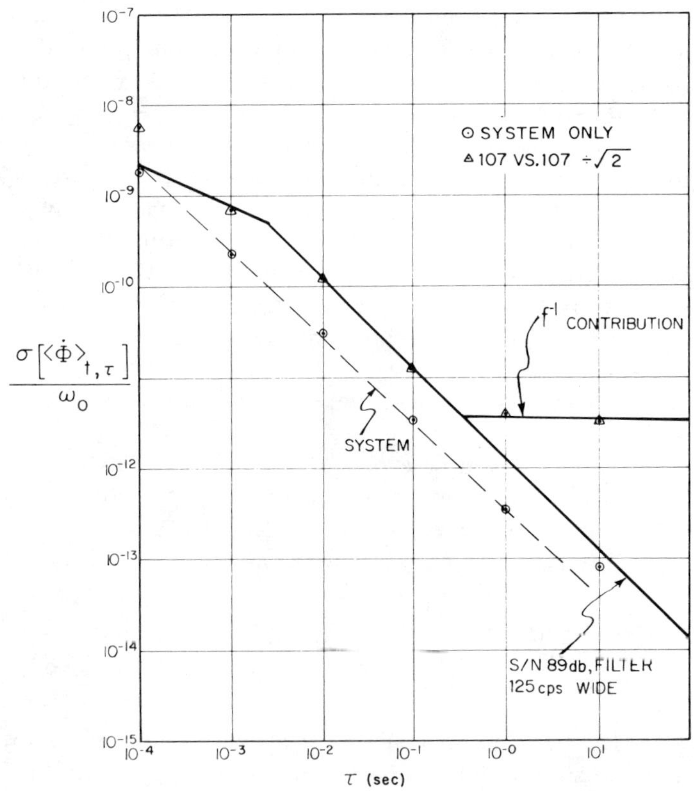

Fig. 6. Measurements with multiple-period system.

Figure 5 shows a block diagram of a versatile system that allows the two oscillators to have zero offset. This feature permits the system noise to be evaluated by feeding both channels from one source. The offset is obtained by the frequency synthesizer whose fluctuations do not greatly degrade the measurement, since the oscillator fluctuations have been multiplied by 1840 in the 20 Mc/s difference frequency before the comparison is made. Figure 6 shows some typical results obtained with this system.

From a practical standpoint the system gives good results for averaging times greater than 10^{-4} seconds, and is particularly good for times greater than 10^{-2} seconds.

B. Phase Detector Techniques [7], [8]

Figure 7 shows a typical phase detector or multiplier measurement technique. If the two signals are identical in frequency and are placed in quadrature they may be represented as

$$v_1(t) = A_1(t) \cos (\omega_0 t + \phi_1(t)),$$
$$v_2(t) = A_2(t) \sin (\omega_0 t + \phi_2(t)). \quad (33)$$

If the difference between $\phi_2(t)$ and $\phi_1(t)$ is small, the phase detector output will be

$$v_0(t) \cong \frac{A_1(t) A_2(t)}{2} \phi(t) \quad (34)$$

if we designate the phase difference $\phi_2(t) - \phi_1(t)$ as $\phi(t)$. If the variations in A_1 and A_2 are small, as is usually the case, they may be neglected and the phase detector output is essentially the difference between the instantaneous phases. Both signals may be heterodyned down to a convenient low frequency by means of two mixers and a common local oscillator. Because this phase detector is linear only for small angles, measurements can be taken only for time intervals during which the phase difference $\phi(t)$ is less than $\pi/6$. Within this restriction, the phase detector output may be analyzed by a low-frequency narrow-band wave analyzer to estimate the spectral density of the phase $S_\phi(\omega)$ directly. Alternatively, an rms voltmeter can be used to estimate $\sigma[\phi(t)]$.

Often it is desirable to place some form of network at the phase detector output in order to have the voltmeter read $\sigma[\langle\dot{\phi}\rangle_{t,\tau}]$ directly. To find $\sigma[\langle\dot{\phi}\rangle_{t,\tau}]$ exactly, Fig. 2 indicates that the required network is a differentiator followed by a filter with a response

$$|H_a(j\omega)|^2 = \frac{\sin^2 \omega\tau/2}{(\omega\tau/2)^2}.$$

Clearly, the $(\sin x)/x$ response is difficult to approximate with a simple RLC filter; thus it is appropriate to discuss two other filters that might be used as substitutes. The first is a simple single-pole low-pass filter, with a frequency response $H_d(j\omega)$ equal to

$$H_d(j\omega) = \frac{1}{1 + j\omega\tau/2}. \quad (35)$$

The second is a multiple-pole, sharp-cutoff filter with response approaching

$$\left.\begin{aligned} H_e(j\omega) &= 1, \quad -\omega_0 < \omega < \omega_0 \\ &= 0 \quad \text{elsewhere} \end{aligned}\right]. \quad (36)$$

Using H_d, the corresponding equations for the variances are as follows.

$$\sigma_d^2 = \frac{1}{2\pi} \int_{-\infty}^{\infty} S_{\dot{\phi}}(\omega) \frac{1}{1 + (\omega\tau/2)^2} d\omega \quad (37)$$

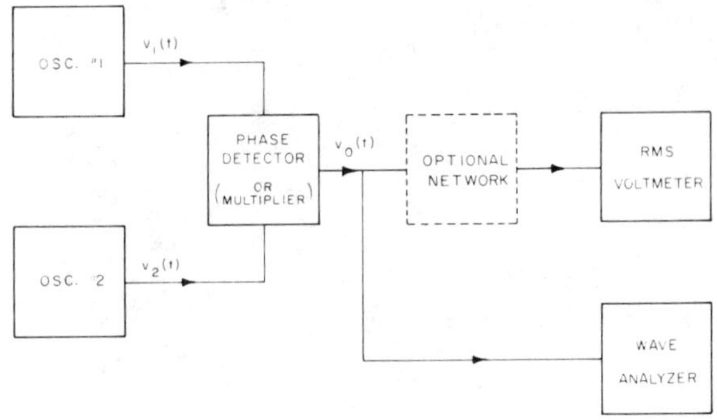

Fig. 7. Phase detector system.

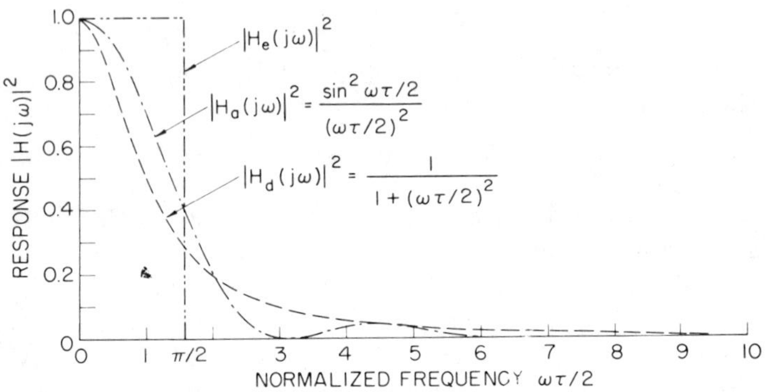

Fig. 8. Frequency response of three filters.

and using H_e

$$\sigma_e^2 = \frac{1}{2\pi} \int_{-\omega_0}^{\omega_0} S_{\dot{\phi}}(\omega) \, d\omega. \quad (38)$$

Because the filter characteristic of interest is $|H(j\omega)|^2$, the squared magnitude of response for each of the three filters in question is plotted in Fig. 8. Because σ^2 is proportional to the area under the power spectral density curve, the three characteristics in Fig. 8 have been normalized to have the same area. This requires that the square filter, (36), have a bandwidth $\omega_0 = \pi/\tau$.

Figure 8 can now be used as a basis for comparing σ_d^2 and σ_e^2 with the true variance $\sigma^2[\langle\dot{\phi}\rangle_{t,\tau}]$. First, all three functions have the same area, so σ_d^2 and σ_e^2 will be identical to the true variance if $S_{\dot{\phi}}(\omega)$ is flat for all frequencies. Second, all three functions have about the same ordinates for small $\omega\tau/2$. Thus, the variances measured for values of τ that make the filters much broader than the spectrum of $S_{\dot{\phi}}(\omega)$ will be about the same for the three filters. Specifically, if $S_{\dot{\phi}}(\omega)$ is band-limited to frequencies below ω_c, then the values of σ_d^2 and σ_e^2 measured for $\tau \ll 2/\omega_c$ will again agree closely with the true value.

Figure 8 indicates that significant divergence in results can be expected when $\omega\tau/2$ is large. In this range, $|H_d|^2$ still has the correct general behavior, i.e., it varies as $1/\omega^2$, but the area is about a factor of two larger than is required. On the other hand, the H_e filter bears no relation to the required $(\sin^2 x)/x^2$ behavior. Thus if

$S_{\dot{\phi}}(\omega)$ is peaked at high frequencies, we can expect σ_d^2 measured for large τ to be about a factor of two larger than the true value. Under the same conditions, however, σ_e^2 measured for large τ can be less than the true value by orders of magnitude, depending on the specific constants involved.

To illustrate some of these points, the example in Section II-E involving a flat band-limited phase spectrum has been reworked using the H_d and H_e filters as approximations to the desired H_a filter. The asymptotic results for σ^2 are as follows:

Filter	σ^2	
	$\tau \ll 2/\omega_c$	$\tau \gg 2\pi/\omega_c$
$H_a(j\omega)$ (true dependence)	$\frac{\omega_c^3}{3\pi}$	$\frac{2\omega_c}{\pi\tau^2}$
$H_d(j\omega)$	$\frac{\omega_c^3}{3\pi}$	$\frac{4\omega_c}{\pi\tau^2}$
$H_e(j\omega)$	$\frac{\omega_c^3}{3\pi}$	$\frac{\pi^2}{3\tau^3}$

These results are consistent with the preceding general discussion.

The large τ asymptotes for these cases have been added to Fig. 3(d) to facilitate the intercomparison among the filters. Note particularly in this example that for large τ, σ_d^2 is a factor of 2 large, whereas the calculation based on H_e yields a σ_e^2 which varies as τ^{-3} instead of τ^{-2}. Thus σ_e^2 is too small, i.e., *optimistic*, by a factor that increases linearly with averaging time τ.

We conclude, therefore, that for calculating $\sigma^2[\langle\dot{\phi}\rangle_{t,\tau}]$ the simple one-pole low-pass filter, (35) is a reasonable approximation to the required $(\sin x)/x$ filter, but serious errors may be introduced if a sharp-cutoff filter, (36), is used as the approximation. It is certainly perfectly valid to place a sharp filter at the output of the phase detector and measure the rms voltage appearing at the filter output as an indirect measurement of $S_{\dot{\phi}}(\omega)$ (see, for example, [7]). The error appears when one tries to identify this rms voltage with $\sigma[\langle\dot{\phi}\rangle_{t,\tau}]$.

C. *Elapsed Phase Difference Method*

Figure 9 shows a block diagram of a simple elapsed phase difference measurement system. The two signals are heterodyned down by means of a common local oscillator to a convenient low frequency suitable for the resolver phase shifter. The servo system maintains the output phase from the phase shifter in fixed relation to the phase in the other channel. Consequently, the phase shift introduced by the phase shifter is equal to the relative phase difference in the two channels.

Since the relative phase is preserved in the heterodyning process, it is equal to the relative phase of the two oscillators. The phase shift may be read electrically by an optical shaft encoder and thus fed directly to computation devices. If the phase is read at intervals of time τ, this technique gives an estimate of $\Delta_\tau\phi(t)$ and hence can be used as a basis of computation of $\langle\dot{\phi}\rangle_{t,\tau}$. If the oscillators being compared operate at 5 Mc/s and the resolver encoder can be read to 0.005 of one revolution, the sensitivity is 10^{-12} in $\langle\dot{\phi}\rangle_{t,\tau}/\omega_0$, for a τ of 1000 seconds.

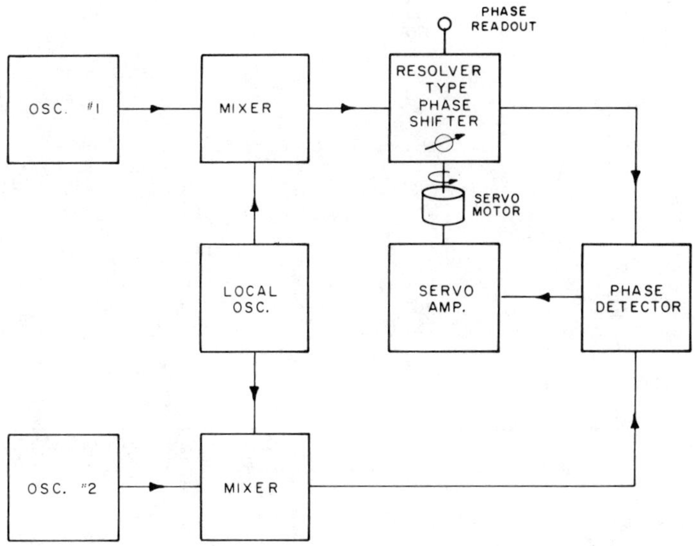

Fig. 9. Elapsed phase method.

Because of the slow speed of the servo, the high-frequency components of the spectrum do not appear at the encoder output. Some of the high-frequency components can be recovered by measuring the servo-error signal out of the phase detector, provided that a suitable filter is added to compensate for the response of the servo motor.

As pointed out above, the resolver system is readily adaptable to digital computation of oscillator stability. One such method is to calculate $R_\phi(\tau)$ from $\phi(t)$ as provided at the encoder output, and hence calculate $S_\phi(\omega)$ and $\sigma[\langle\dot{\phi}\rangle_{t,\tau}]$. Typical results are shown in Fig. 10 [9]. The phase difference between two 262-MHz signals was sampled and recorded digitally on punched tape every 10 seconds for about ten hours. A plot of the sampled phase is shown in Fig. 10(a). The computer-generated plot of the autocorrelation of the phase is shown in Fig. 10(b). Although the calculation was made for delay times as large as 10 000 seconds, the plot has significant error (because of the finite data length T) beyond $T/10 = 3500$ seconds.

To conserve computer time, the power spectral density of phase, Fig. 10(c), was calculated by forming the Fourier transform of a 1000 point sample of Fig. 10(a). A Hamming routine was used to smooth the resulting spectrum. (See Blackman and Tukey [1].) In spite of the finite data length, we assume that the spectrum thus obtained is approximately the true power spectral density within the limitations discussed in Section II-F.

The curve of standard deviation vs. averaging time, Fig. 10(d), was formed from Fig. 10(b) using (16a).

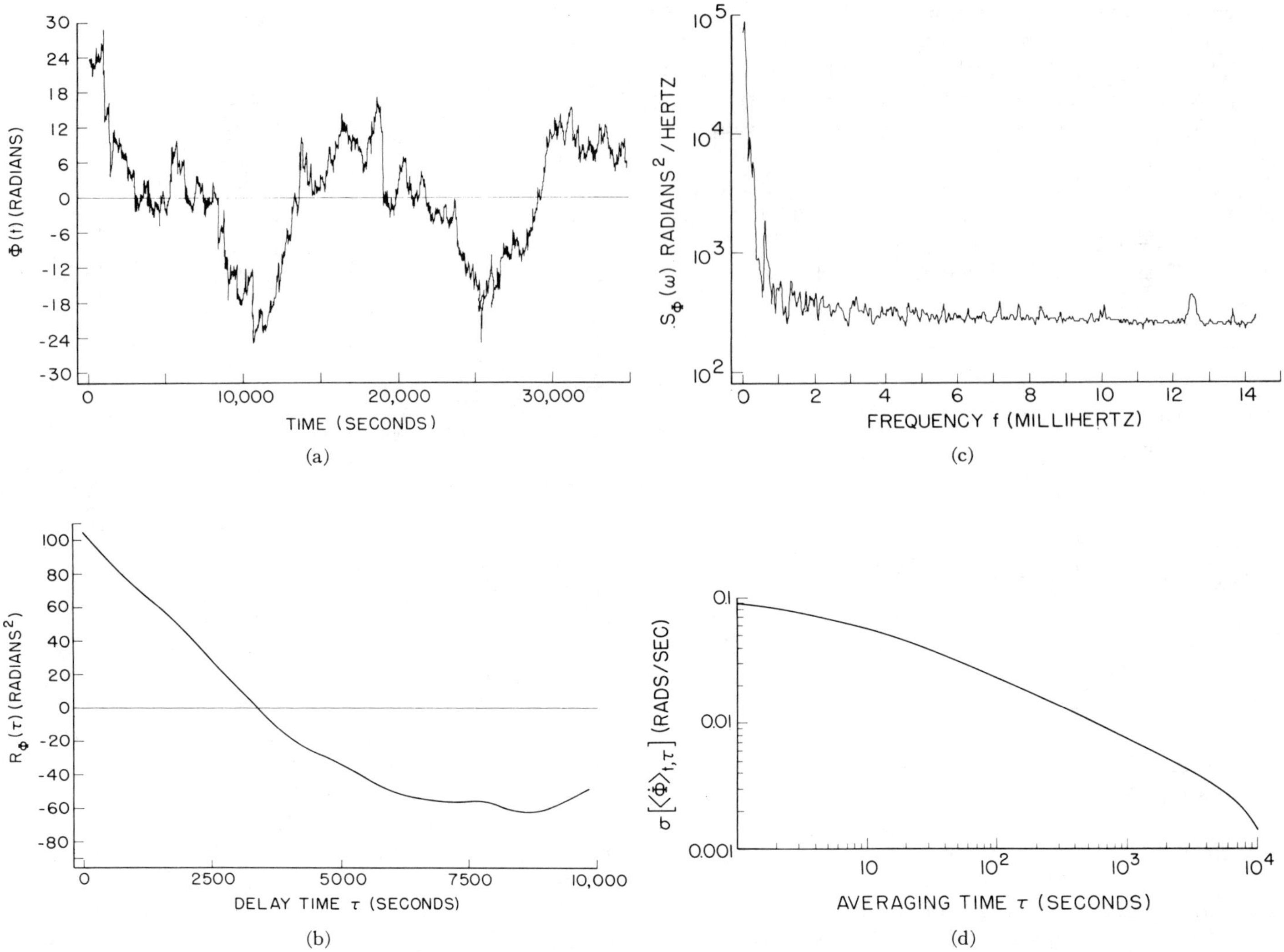

Fig. 10. Computer analysis of 10 hour length of phase data. (a) Phase difference measured at 262 MHz. (b) Autocorrelation of phase difference. (c) Power spectral density of phase. (d) Standard deviation of average frequency.

IV. Noise in Oscillators

Practical oscillators appear to have three main sources of noise contributing to frequency fluctuations. These are: A. additive noise associated with the oscillator and accessory circuits, such as AGC and amplifiers, which does not perturb the oscillation but is merely added to the signal; B. thermal and shot noise in the oscillator itself which actually perturbs the oscillation; and C. fluctuations in the resonator frequency either in the resonator itself or due to circuit parameter changes influencing the resonance frequency. The frequency fluctuations of an oscillator due to the last mentioned source appear to have an f^{-1} power spectral density [10].

A. Additive Noise

In oscillators used for quartz frequency standards, great care is taken to couple very lightly into the oscillating circuit, and because of the nonlinearity of the resonator [11] it is necessary to stabilize the oscillation at a very low power level (typically about 10^{-7}–10^{-6} watts). The nonlinearity causes a coefficient of about 1×10^{-9} per dB of drive at 50 μA in a 2.5-Mc/s fifth overtone crystal [12]. As a consequence, the noise of the amplifiers following the oscillator is the predominant factor for fluctuations involving times of the order of 0.1 second or less.

Assume that the additive noise is band limited by a narrow-band filter with transfer function

$$H(\omega) \simeq \frac{1}{1 + j\dfrac{\omega - \omega_0}{\omega_1}} \tag{39}$$

where ω_1 is the half bandwidth of the filter. As will be shown, this is often a desirable procedure. Such a transfer function, of course, is not realizable but is a good approximation for the narrow-band case. Assume also that the additive noise is white and has a power spectral density S_0. Then the power spectral density of the noise out of the filter will be

$$\frac{S_0}{1 + \left(\dfrac{\omega - \omega_0}{\omega_1}\right)^2}\,.$$

The total noise power will be

$$P_N = \frac{1}{2\pi}\int_{-\infty}^{\infty} \frac{S_0}{1+\left(\frac{\omega-\omega_0}{\omega_1}\right)^2}\, d\omega = \frac{\omega_1 S_0}{2}. \tag{40}$$

If the total signal power P_S is large in comparison to P_N so $P_N/P_S \ll 1$ then it is well known that half the noise power appears as amplitude modulation sidebands on the signal centered in the noise and the other half as phase modulation sidebands. The power spectral density of this effective phase modulation is then

$$S_\phi(\omega) = \frac{1}{\omega_1}\frac{P_N}{P_S}\frac{\omega_1^2}{\omega^2+\omega_1^2} \tag{41}$$

and the autocorrelation function is

$$R_\phi(\tau) = \frac{P_N}{P_S}\frac{e^{-\omega_1|\tau|}}{2}. \tag{42}$$

From this, using (16a),

$$\frac{\sigma(\langle\dot\phi\rangle_{t,\tau})}{\omega_0} = \frac{1}{\omega_0\tau}\left[\frac{P_N}{P_S}(1-e^{-\omega_1\tau})\right]^{1/2} \tag{43}$$

(for a single filter). A normalized plot of the asymptotes of this function is shown in Fig. 11. For $\omega_1\tau \gg 1$,

$$\frac{\sigma}{\omega_0} \simeq \frac{1}{\omega_0\tau}\left(\frac{P_N}{P_S}\right)^{1/2};$$

for $\omega_1\tau \ll 1$,

$$\frac{\sigma}{\omega_0} \simeq \frac{1}{\omega_0}\left(\frac{\omega_1}{\tau}\frac{P_N}{P_S}\right)^{1/2}.$$

For $\omega_1\tau \gg 1$, σ/ω_0 is proportional to $\omega_1^{1/2}$ for constant noise spectral density and for $\omega_1\tau \ll 1$ it is proportional to ω_1. Consequently, using a narrow-band filter can greatly improve short-term stability. For two cascaded filters each of half-bandwidth ω_1 the result is

$$\frac{\sigma(\langle\dot\phi\rangle_{t,\tau})}{\omega_0} = \frac{1}{\omega_0\tau}\left\{\frac{P_N}{P_S}(1-e^{-\omega_1\tau}(\omega_1\tau+1))\right\}^{1/2} \tag{44}$$

(for a double filter).

In this case σ is constant for $\omega_1\tau \ll 1$. For any shape of band-pass filter with white additive noise

$$\frac{\sigma}{\omega_0} \simeq \frac{1}{\omega_0\tau}\left(\frac{P_N}{P_S}\right)^{1/2}$$

when τ is much greater than the reciprocal half bandwidth.

As a practical example, consider a 5-Mc/s oscillator followed by a single filter of half bandwidth of 62.5 c/s and $P_N/P_S = -87$ dB. Then for $\tau = 1$ second, $\sigma(\langle\dot\phi\rangle_{t,\tau})/\omega_0 = 1.4\times10^{-12}$. For a comparison with experimental results see Figs. 6 and 12.

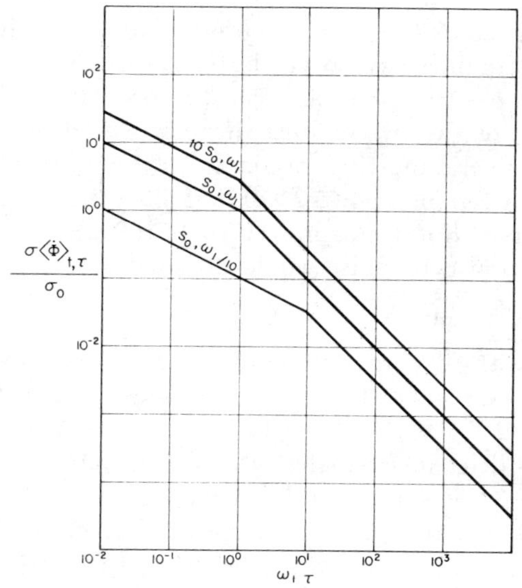

Fig. 11. Normalized additive noise contributions.

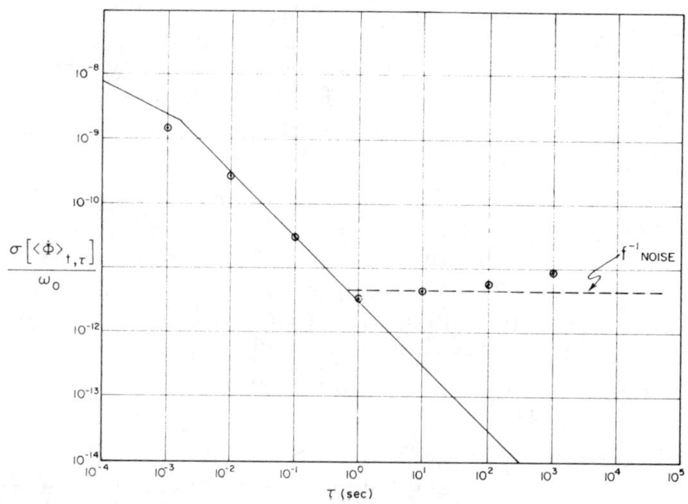

Fig. 12. Theoretical and experimental standard deviation of average fractional frequency departure for a 5.0-Mc/s precision quartz oscillator.

B. Noise that Perturbs the Oscillation

The perturbing effects of thermal and shot noise in oscillators are well known [13]. The phase does a random walk due to the perturbations. It has been shown that this leads to

$$\frac{\sigma(\langle\dot\phi\rangle_{t,\tau})}{\omega_0} = \left(\frac{kT}{2PQ^2\tau}\right)^{1/2} \tag{45}$$

where

- k = Boltzmann's constant
- T = the effective noise temperature
- P = the total power delivered to the resonator and the load, and
- Q = the loaded Q of the resonator.

As an example consider a 5-Mc/s oscillator with $Q = 2\times10^6$, $P = 10^{-7}$ watts, and $T = 10^{3\circ}$ K. For $\tau = 1$ second $\sigma(\langle\dot\phi\rangle_{t,\tau})/\omega_0 = 1.3\times10^{-13}$. From this example and the previous one, it is apparent that the additive noise will dominate over perturbation noise for short averaging times in precision crystal oscillators. Here the effects of the two sources of noise would become equal at an averaging time of about 100 seconds with a total fluctuation of about 2×10^{14} if no other sources were effective. For times longer than 100 seconds, the perturbation type noise would dominate.

C. f^{-1} Noise

The third main source of noise is that which has an f^{-1} power spectral density of frequency fluctuation, i.e., $S_{\dot\phi}(\omega)=K/|\omega|$, K a constant. In addition to this, oscillators drift in frequency (drift is usually accompanied by an f^{-1} power spectral density). Because of the drift in frequency, the phase is not a stationary process. Also, since an f^{-1} power spectral density for frequency fluctuations corresponds to $S_\phi(\omega)$ proportional to $|\omega|^{-3}$,

$$\sigma^2(\langle\dot\phi\rangle_{t,\tau}) = \frac{1}{\pi\tau^2}\int_{-\infty}^{\infty} S_\phi(\omega)(1-\cos\omega\tau)\,d\omega$$

[from (16a) with the autocorrelation functions written in terms of $S_\phi(\omega)$ using (15)] does not converge since $1-\cos\omega\tau$ only goes to zero as ω^2 for ω approaching zero. This is not surprising. If the f^{-1} spectrum persisted down to zero frequency we would see infinitely large fluctuations by observing over all time.

In the actual situation observations are made over a finite time T. If the frequency drift is removed during the time T by subtraction of the least squares fit of a straight line, this behaves like a high-pass filter acting on the low-frequency components of the phase (see Appendix II). The output of this filter goes to zero as ω^2 for ω approaching zero, and the filter starts to cut off at $\omega\simeq 2/T$. This gives finite results for $\sigma(\langle\dot\phi\rangle_{t,\tau})/\omega_0$ as shown in Appendix II. It is apparent from the functional form of the integral that if the ratio of observing time to averaging time T/τ is constant (corresponding to a fixed number of samples) then $\sigma(\langle\dot\phi\rangle_{t,\tau})/\omega_0$ is constant. Since there is no theory giving the strength of the f^{-1}-spectral density, no prediction can be made as to the constant value of σ. The dependence of σ^2 on T/τ is proportional to $1.04+\frac{1}{2}\log T/2\tau$.

Actual oscillators measured under the conditions of finite T and removal of drift exhibit the predicted behavior (see Figs. 6 and 12). For T/τ of about 100, $\sigma(\langle\dot\phi\rangle_{t,\tau})/\omega_0$ flattens out at a value somewhat greater than 1×10^{-12}. Since this is larger than the fluctuations due to the perturbing type of noise at the τ for which additive noise becomes dominant, it is apparent that the effects of the perturbing type of noise are not seen at all in many frequency standards. Figure 13 shows the effects due to the three sources of fluctuations.

J. Barnes [14] of the NBS Laboratories, Boulder, Colo., has recently shown that taking successive differences of the phase has interesting consequences. For example, taking second differences removes the linear frequency drift and also gives convergent results independent of observation time for an assumed f^{-1}-power spectrum. As shown in Appendix I

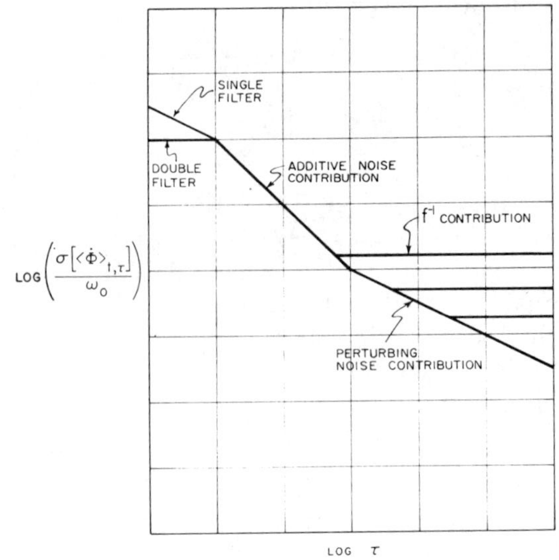

Fig. 13. Contributions to oscillator fluctuations.

$$\sigma\left(\frac{\Delta_\tau{}^2\phi(t)}{\omega_0\tau}\right) = \frac{1}{\omega_0\tau}\,[6R_\phi(0) - 8R_\phi(\tau) + 2R_\phi(2\tau)]^{1/2} \qquad (46)$$

where $\Delta_\tau{}^2\phi(t)\equiv\phi(t+\tau)-2\phi(t)+\phi(t-\tau)$ is the second phase difference. Using this on an f^{-1}-power spectral density of frequency fluctuations, $S_\phi(\omega)=K/|\omega|^3$, gives

$$\sigma = \frac{(8K\log 2)^{1/2}}{\omega_0}\,. \qquad (47)$$

Since this is independent of observation time T, it appears to be a good measure for the f^{-1} characteristic of oscillators.

D. Noise in Masers

It is interesting to compute the effects of the perturbing thermal noise and additive noise in maser oscillators. Following Kleppner et al. [15], [16], but solving the first-order perturbation exactly and correcting the thermal noise expression [17] gives a total spectral density of phase (two-sided) for a hydrogen maser.

$$S_\phi(\omega) = \frac{kT}{2P}\left(\frac{Q_e}{Q_{c,l}} + \frac{\omega_0^2}{\omega^2 Q_l^2}\,G_c(\omega)\right). \qquad (48)$$

Here

k = Boltzmann's constant
T = absolute temperature
P = power delivered by the beam to the loaded cavity
Q_e = external cavity Q
$Q_{c,l}$ = loaded cavity Q
Q_l = atomic line $Q=\omega_0/2\gamma$, and
$G_c(\omega)=\omega_1^2/(\omega^2+\omega_1^2)$, the equivalent low-pass magnitude squared impedance function of the cavity, where $\omega_1=\omega_0/(2Q_{c,l})$.

In deriving (48) it is assumed that the maser is coupled to its load through an isolator in thermal equilibrium with the cavity and consequently the available signal power is $P_0 = P(Q_{c,l}/Q_e)$ and the available noise power spectral density is just $(kT)/2$. The first term in (48) is just the additive noise (white phase noise) and the second term (low-pass filtered white frequency noise) arises from the perturbation of the oscillation. Note that for $\omega > (\omega_0/Q_l)(Q_{c,l}/Q_e)^{1/2}$, the additive noise predominates. It is interesting to note that a portion of the additive noise term is actually present in the solution to the maser signal in the cavity [17]. This is similar to results obtained by Hafner [18] for an oscillator with two coupled tuned circuits.

The corresponding result for the rubidium maser has not yet been derived, but the mechanism of operation is very similar so that the result is probably the same except that P should be replaced by the power delivered to the loaded cavity by the rubidium atoms. We will assume here that the results are valid for the rubidium maser.

Since the additive white phase noise term in (48) is very broadband, most applications involving the maser will require that the bandwidth of the output be restricted. If the system has an effective single tuned bandpass characteristic of width $2\omega_1'$ with $\omega_1' \ll \omega_1$, and a noise factor F (referred to the maser input port), we obtain

$$\frac{\sigma(\langle\dot{\phi}\rangle_{t,\tau})}{\omega_0} = \left\{\frac{kT}{2P}\left[\frac{F\omega_1' Q_e}{\omega_0^2 \tau^2 Q_{c,l}}(1 - e^{-\omega_1'\tau}) + \frac{1}{Q_l^2 \tau}\left(1 - \frac{(1 - e^{-\omega_1'\tau})}{\omega_1'\tau}\right)\right]\right\}^{1/2}. \quad (49)$$

For $\omega_1'\tau \gg 1$ this reduces to

$$\frac{\sigma(\langle\dot{\phi}\rangle_{t,\tau})}{\omega_0} \approx \left\{\frac{kT}{2P}\left[\frac{F\omega_1' Q_e}{\omega_0^2 \tau^2 Q_{c,l}} + \frac{1}{Q_l^2 \tau}\right]\right\}^{1/2} \quad (50)$$

which is similar in form to that obtained earlier by Vessot [19].

For $\omega_1'/\omega_1 \ll \omega_1'\tau \ll 1$,

$$\frac{\sigma(\langle\dot{\phi}\rangle_{t,\tau})}{\omega_0} \approx \left\{\frac{kT}{2P}\left[\frac{F\omega_1'^2 Q_e}{\omega_0^2 \tau Q_{c,l}} + \frac{\omega_1'}{2Q_l^2}\right]\right\}^{1/2}. \quad (51)$$

Note that the perturbing noise term in (50) is exactly the same for a maser as that for an ordinary oscillator [see (45)]. For $\tau < F\omega_1' Q_l^2 Q_e/\omega_0^2 Q_{c,l}$ (if $\omega_1'\tau \gg 1$) or $\tau < 2F\omega_1' Q_l^2 Q_e/\omega_0^2 Q_{c,l}$ (if $\omega_1'\tau \ll 1$) the fluctuations due to the additive noise dominate. For a hydrogen maser with $Q_l = 2\times10^9$, $Q_e/Q_{c,l} = 5$, $F = 2$, and $\omega_1' = 60$ the additive noise dominates for $\tau < 30$ seconds. For a rubidium maser with $Q_l = 1\times10^8$ [20] and all other factors the same, the additive noise dominates for $\tau < 6.5\times10^{-3}$ seconds (here $\omega_1'\tau < 1$). Figure 14 is a plot of the theoretical $\sigma(\langle\dot{\phi}\rangle_{t,\tau})/\omega_0$ vs. averaging time for a hydrogen and a rubidium maser. Note again the influence of the narrow filter on the results.

The factors in the two terms of $\sigma(\langle\dot{\phi}\rangle_{t,\tau})/\omega_0$ that depend only on the maser may be collected into figures of mert M_A and M_P, allowing σ to be written as

$$\frac{\sigma(\langle\dot{\phi}\rangle_{t,\tau})}{\omega_0} = \left\{\frac{kT}{2}\left[\frac{F\omega_1'}{M_A^2 \tau^2}(1 - e^{-\omega_1'\tau}) + \frac{1}{M_P^2 \tau}\left(1 - \frac{(1 - e^{-\omega_1'\tau})}{\omega_1'\tau}\right)\right]\right\}^{1/2} \quad (52)$$

where

$$M_A^2 = \frac{\omega_0^2 P Q_{c,l}}{Q_e} \quad \text{and} \quad M_P^2 = Q_l^2 P.$$

Some typical figures of merit are given in Table I. The figures for quartz assume fairly tight coupling to the load to obtain a large M_A. This is what would be needed

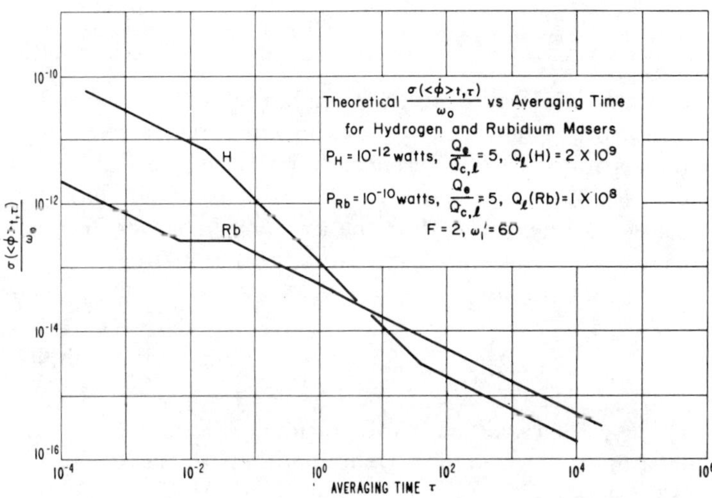

Fig. 14. Theoretical standard deviation of average fractional frequency departure for hydrogen and rubidium masers.

TABLE I

CHARACTERISTICS AND FIGURES OF MERIT OF VARIOUS OSCILLATORS

Type of Oscillator	ω_0—Seconds^{-1}	P—watts	Q_l	$Q_e/Q_{c,l}$	M_P—watts$^{1/2}$	M_A—watts$^{1/2}$ Seconds^{-1}
Ammonia	1.5×10^{10}	1×10^{-10}	5×10^6	5	50	6.7×10^5
Rubidium 87	4.3×10^{10}	1×10^{-10}	1×10^8	5	1000	1.9×10^5
Hydrogen	8.9×10^9	1×10^{-12}	2×10^9	5	2000	4×10^3
Quartz (5 Mc/s)	3.1×10^7	1×10^{-6}	2×10^6	5	2000	1.4×10^4

for an oscillator servoed by a short time constant loop to an atomic device but contrary to the requirements for good long-term stability and freedom from circuit influences in a free running oscillator.

It is very likely that masers also exhibit an f^{-1} power spectral density of frequency fluctuation. This is due to the fact that the frequency is dependent on many external factors that may have an f^{-1} spectral density such as cavity tuning (all masers), magnetic field (hydrogen and rubidium), buffer gas pressure and temperature and light intensity (rubidium), and wall shift (hydrogen). Allan [6] has reported one case where an f^{-1} spectral density was observed for an ammonia maser with some uncertainty in the data. Not many experiments have been performed as yet, partially because of the long times necessary to gather reliable data and also because the field is quite new. This is an interesting area for future research.

V. Fluctuations in Atomic Standards Using a Servo-Controlled Quartz Oscillator

A. Atomic Beam Device

Consider an oscillator compared against a reference such as an atomic beam device [21] and controlled in frequency by a servo actuated by the error signal. Figure 15 shows a system block diagram. The power spectral density of the fluctuations in the output frequency is

$$S_{\dot\phi}(\omega)_0 = S_{\dot\phi}(\omega)_{\text{Osc}} \frac{1}{|1 + G(\omega)|^2} + S_{\dot\phi}(\omega)_{\text{Ref}} \left| \frac{G(\omega)}{1 + G(\omega)} \right|^2 \quad (53)$$

where $S_{\dot\phi}(\omega)_{\text{Osc}}$ refers to the open loop fluctuations in the oscillator, $S_{\dot\phi}(\omega)_{\text{Ref}}$ refers to the equivalent frequency fluctuations in the reference, and $G(\omega)$ is the total loop gain $K_0 K_R g(\omega)$. In deriving (53) it is assumed that the system is linear and that the two noise sources are uncorrelated. The simplest useful form for $G(\omega)$ (also a very practical one) is

$$G(\omega) = \frac{\omega_c}{j\omega} \quad (54)$$

which is just gain and integration. This is a good approximation to what is done in practice. Substituting, (53) becomes

$$S_{\dot\phi}(\omega)_0 = S_{\dot\phi}(\omega)_{\text{Osc}} \frac{\omega^2}{\omega^2 + \omega_c^2} + S_{\dot\phi}(\omega)_{\text{Ref}} \frac{\omega_c^2}{\omega^2 + \omega_c^2} \quad (55)$$

so that the oscillator noise is high-pass filtered and the reference noise is low-pass filtered.

Consider first the fluctuations due to the oscillator only. If it is assumed to have characteristics similar to those previously discussed, namely additive noise filtered by a narrow-band filter of half width ω_1, and an f^{-1} behavior, then, if $\omega_1 \gg \omega_c$ (as is usually the case), the additive noise contribution is virtually unchanged by the servo loop. The f^{-1} portion gives, in the closed loop condition, [from (55) and (17a)]

$$\frac{\sigma(\langle\dot\phi\rangle_{t,\tau})}{\omega_0} = \frac{B}{\omega_0 \omega_c \tau} (1.16 + 2 \log \omega_c \tau - e^{-\omega_c \tau} \mathcal{E}i^*(\omega_c \tau) - e^{\omega_c \tau} \mathcal{E}i(-\omega_c \tau))^{1/2} \quad (56)$$

where B is a constant depending on the strength of the f^{-1} noise; $\gamma \approx 0.577$ is Euler's constant;

$$\mathcal{E}i(X) = \int_{-\infty}^{X} \frac{e^t}{t} dt, \qquad X < 0,$$

is the exponential integral; and $\mathcal{E}i^*(X)$ is the principal value of the integral for $X > 0$. This behaves like $(B/\omega_0)[5/2 - \gamma - \log (\omega_c \tau)]^{1/2}$ (very slowly varying) for $\omega_c \tau \ll 1$ and like $(B/(\omega_0 \omega_c \tau))[2 \log (\omega_c \tau) + 2\gamma]^{1/2}$ for $\omega_c \tau \gg 1$.

The contribution due to the reference noise, which is assumed to be white with power spectral density S_R, is [from (55) and (17a)]

$$\frac{\sigma(\langle\dot\phi\rangle_{t,\tau})}{\omega_0} = \frac{1}{\omega_0} \left\{ \frac{S_R}{\omega_c \tau^2} (\omega_c \tau + e^{-\omega_c \tau} - 1) \right\}^{1/2}. \quad (57)$$

This approaches a constant value of $(1/\omega_0)(S_R \omega_c/2)^{1/2}$ for $\omega_c \tau \ll 1$. For $\omega_c \tau \gg 1$, $\sigma = (1/\omega_0)(S_R/\tau)^{1/2}$ (this holds for any $G(\omega)$ such that $|G(\omega)/(1+G(\omega))|^2 \to 1$ as $\omega \to 0$ for $\omega_c \tau \gg 1$). Figure 16 shows the behavior of the various contributions to the overall fluctuations. It is apparent that there is an optimum choice of loop cutoff ω_c for a given oscillator and beam tube.

It is of interest to calculate the power spectral density

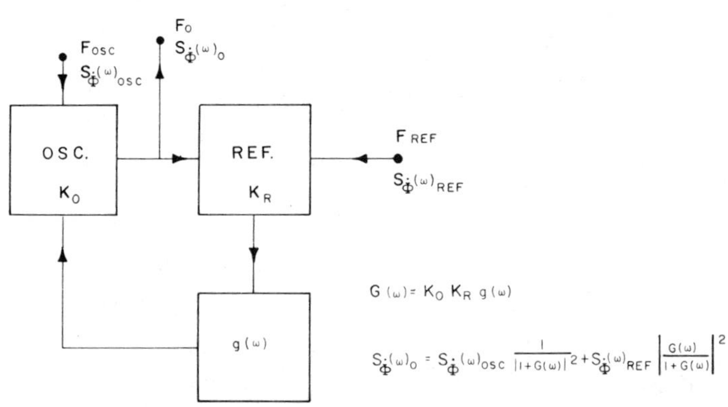

Fig. 15. Block diagram of passive atomic standard.

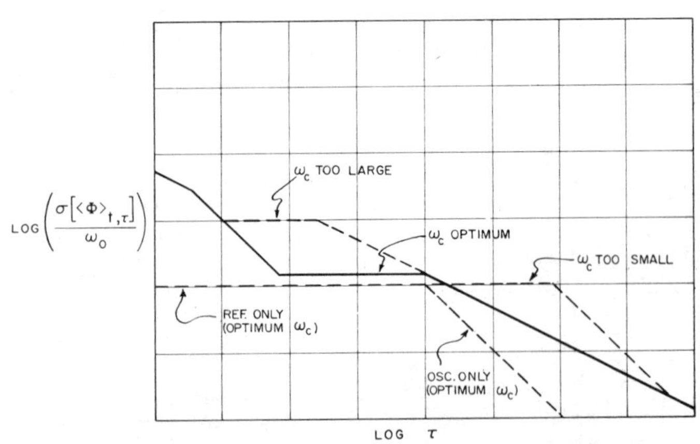

Fig. 16. Contributions to passive atomic standard fluctuations.

of the noise for the case of a cesium beam reference. For Ramsey excitation the response may be written for small departures from line center approximately as

$$I(t) = \frac{I_m}{2}\left\{1 + \cos \pi\left(\frac{\Delta\omega(t)}{\omega_l}\right)\right\} \tag{58}$$

where $I(t)$ is the output current which has a maximum value I_m, $\Delta\omega(t)$ is the departure from line center, and ω_l is the full line width. Let $\Delta\omega(t)/\omega_l = \epsilon + \alpha \cos \omega_m t$ to allow for the usual sinusoidal modulation, where ω_m is the modulation frequency assumed small in comparison with ω_l so that the dynamic behavior is essentially the same as the static behavior. The result of synchronous detection of this signal is

$$I_0 = \frac{I_m\omega_m}{2\pi}\int_{-\pi/\omega_m}^{\pi/\omega_m} \frac{1}{2}(1 + \cos \pi(\epsilon + \alpha \cos \omega_m t)) \cos \omega_m t \, dt$$

$$= -I_m J_1(\pi\alpha)\frac{\sin \pi\epsilon}{2} \approx \frac{-I_m J_1(\pi\alpha)}{2}\epsilon\pi \tag{59}$$

for ϵ small. For a given offset, ϵ, this has the maximum absolute value of $0.91\ \epsilon I_m$ with $\pi\alpha \approx 1.8$, the first maximum of the Bessel function. This represents the signal output for a given fractional mistuning ϵ.

In a beam tube, the noise is mainly due to shot noise and so the power spectral density is proportional to $I(t)^{1/2}$ and is independent of ω. Since $I(t)$ is a function of time, the noise output is not a stationary function. If $i_n(t)$ is the instantaneous noise current with no modulation and $\epsilon = 0$ then

$$I_n(t) = i_n(t)\left|\cos\left(\frac{\pi}{2}\alpha \cos \omega_m t\right)\right|. \tag{60}$$

The autocorrelation function after synchronous detection is desired. Performing the statistical and the time averages gives

$$R_I(\tau) = \delta(\tau)\frac{S_I}{4}(1 + J_0(\pi\alpha) - J_2(\pi\alpha)) \tag{61}$$

where S_I is the spectral density of the noise current. From this and (59) the spectral density S_R is

$$S_R = \frac{S_I[1 + J_0(\pi\alpha) - J_2(\pi\alpha)]\omega_l^2}{J_1^2(\pi\alpha)I_m^2\pi^2} \tag{62}$$

where S_I is the spectral density of the noise current, I_m is the peak signal current, and $\alpha = (\Delta\omega/\omega_l)_{\max}$ is the ratio of peak frequency swing due to the modulation to the full line width, ω_l. Using these results and assuming optimum modulation amplitude,

$$\frac{\sigma(\langle\dot\phi\rangle)_{t,\tau}}{\omega_0} = 0.386\frac{f_l}{f_0}\frac{I_n}{I_m}(\tau)^{1/2}$$

for averaging time τ long in comparison with the loop time constant, where I_n is the noise current from the beam tube in a 1-c/s noise bandwidth centered about the modulation frequency; $f_l/f_0 = \omega_l/\omega_0$.

B. Maser with Phase-Locked Oscillator

Essentially the same sort of analysis that was applied to an oscillator slaved to a passive atomic device may be used for an oscillator phase-locked to a maser. A block diagram of a simple system with the various important noise sources is shown in Fig. 17. Again we have

$$S_{\dot\phi}(\omega)_0 = S_{\dot\phi}(\omega)_{\text{Osc}}\frac{1}{|1 + G(\omega)|^2} + S_{\dot\phi}(\omega)_{\text{Ref}}\left|\frac{G(\omega)}{1 + G(\omega)}\right|^2 \tag{63}$$

where

$$S_{\dot\phi}(\omega)_{\text{Ref}} = \frac{S_{\dot\phi}(\omega)_{\text{mult}} + S_{\dot\phi}(\omega)_m + S_{\dot\phi}(\omega)_R + S_{\dot\phi}(\omega)_S}{N^2}$$

(see Fig. 17).

As before, the oscillator noise is high-pass filtered and the reference noise is low-pass filtered. Great care must be exercised in the design of the frequency multiplier, receiver, and synthesizer so that their frequency power spectral densities within the loop bandwidth are as small as possible. Solid-state frequency multipliers (and amplifiers) are often plagued with an f^{-1} power spectral density of phase which frequently can be due to f^{-1} noise in biasing or collector voltage supplies. For this type of application temperature control of the frequency multiplier is desirable to remove slow phase drifts with temperature change.

As an example of a possible design, consider locking a relatively high level, tightly coupled 10-Mc/s quartz oscillator to a hydrogen maser. Reasonable estimates for the various pertinent quantities are (hydrogen maser parameters same as in Table I—see also Fig. 17):

$$P_{\text{Osc}} = 10^{-5} \text{ watts},$$

$$\omega_{\text{Osc}} = 6.3 \times 10^7 \text{ second}^{-1},$$

$$N = 140,$$

$$S_{\dot\phi}(\omega)_{\text{Osc}} = 10^{-14}\omega^2 + \frac{2 \times 10^{-5}}{|\omega|}, \text{ and}$$

$$S_{\dot\phi}(\omega)_{\text{Ref}} = 2 \times 10^{-12}(\omega^2 + 1).$$

It is assumed that the oscillator has an f^{-1}-frequency spectrum plus white phase noise and that the f^{-1} portion leads to a $\sigma(\langle\dot\phi\rangle_{t,\tau})/\omega_0 \simeq 10^{-10}$ for $T/\tau \simeq 100$. The reference spectral density is assumed to be due mainly to the additive noise caused by the preamp and mixer (white phase noise) and the random walk term in the maser. The f^{-1} spectral density in the maser is neglected, since presently not much is known about its magnitude (assuming it exists).

In order to reduce the effects of oscillator drift and the f^{-1} spectral density, it is desirable to use a phase-lock loop with two poles at the origin and a zero at ω_2, i.e.,

$$G(\omega) = \frac{2_i\omega_2\omega + \omega_2^2}{-\omega^2}.$$

The phase detector provides one pole, and the other

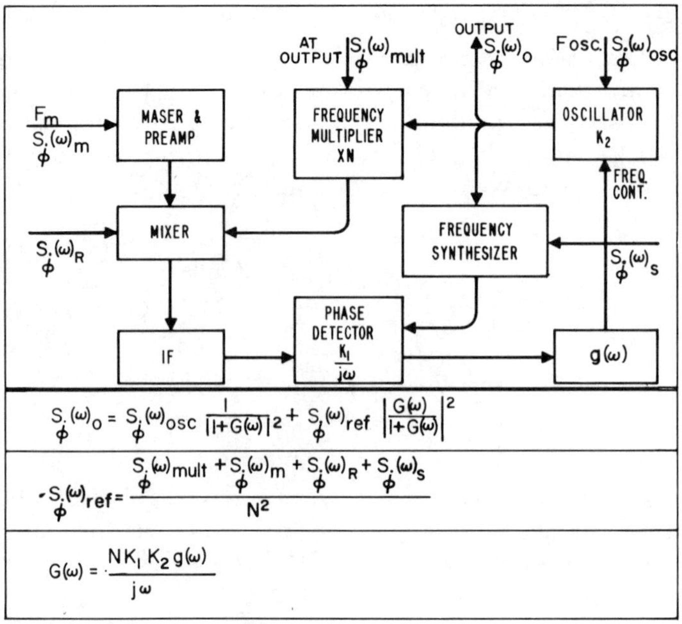

Fig. 17. Block diagram of an active atomic standard with phase-locked quartz oscillator.

pole and the zero may be approximated very well with an operational amplifier with feedback. The loop will be critically damped with this choice of $G(\omega)$, and ω_2 determines the loop cutoff frequency ω_c: i.e., $\omega_c = 2.06\ \omega_2$. With this loop, a constant drift rate in the oscillator produces no frequency error.

The loop cutoff frequency in this example is optimum at the point where the f^{-1} spectral density from the oscillator equals the white phase noise term from the reference. This occurs at $\omega \simeq 215$ seconds^{-1} or $f \simeq 34$ c/s.

The additive noise contribution from the oscillators must be smaller than that from the reference; otherwise, the optimum loop cutoff frequency would be very large. This is the reason for using a fairly high drive level, high-frequency oscillator. Again, careful use of narrow-band filtering of the oscillator can be of great value. Excessive narrow-band filtering may violate servo-stability requirements.

VI. Conclusions

Some of the considerations of the theory and measurement of fluctuations in frequency standards have been presented. It was shown that the fluctuations could be characterized by the autocorrelation function of either frequency or phase or the power spectral density of either frequency or phase. From these the many measures of stability may be computed. Measurement techniques were discussed as well as the techniques for estimating spectral densities, etc. Sources of fluctuations in quartz oscillators, passive atomic standards, and masers were treated and some of the theory presented. In some cases, details have been lightly treated or omitted due to the length of the material presented. It is hoped that the material covered will be useful in promoting understanding and guiding designs.

Appendix I

If $\phi(t)$ is wide-sense stationary and has zero mean

$$\sigma^2(\Delta_\tau\phi(t)) = \overline{(\phi(t+\tau/2) - \phi(t-\tau/2))^2}$$
$$= \overline{\phi(t+\tau/2)^2} - 2\overline{\phi(t+\tau/2)\phi(t-\tau/2)}$$
$$+ \overline{\phi(t-\tau/2)^2} = 2(R_\phi(0) - R_\phi(\tau)) \quad (64)$$

which leads to (16a) in the text, since $\sigma^2(\langle\dot\phi\rangle_{t,\tau}) = (1/\tau^2)\sigma^2(\Delta_\tau\phi(\tau))$.

$$\sigma^2(\langle\dot\phi\rangle_{t,\tau}) = \frac{1}{\tau^2}\int_{-\tau/2}^{\tau/2} dt' \int_{-\tau/2}^{\tau/2} dt''\, \overline{\dot\phi(t')\dot\phi(t'')}$$
$$= \frac{1}{\tau^2}\int_{-\tau/2}^{\tau/2} dt' \int_{-\tau/2}^{\tau/2} dt'' R_{\dot\phi}(t''-t'). \quad (65)$$

Let $(t''-t')=\tau'$. Using this substitution and interchanging the order of integration gives

$$\sigma^2(\langle\dot\phi\rangle_{t,\tau}) = \frac{2}{\tau}\int_0^\tau R_{\dot\phi}(\tau')\left(1 - \frac{\tau'}{\tau}\right) d\tau' \quad (66)$$

which gives (16b).

The second phase difference $\Delta_\tau^2\phi(t)$ is $[\phi(t+\tau)-\phi(t)] - [\phi(t)-\phi(t-\tau)]$ or $\phi(t+\tau)-2\phi(t)+\phi(t-\tau)$.

$$\sigma^2(\Delta_\tau^2\phi(t)) = \overline{(\phi(t+\tau) - 2\phi(t) + \phi(t-\tau))^2}$$
$$= 6R_\phi(0) - 8R_\phi(\tau) + 2R_\phi(2\tau) \quad (67)$$

leading to (46).

Appendix II

Assume that a signal $v(t)$ is observed for a time T, and it is desired to remove the mean and the linear drift in a least square departure sense. This requires that

$$\int_{T/2}^{T/2} (v(t) - (a+bt))^2\, dt = \text{minimum}. \quad (68)$$

For this to be satisfied

$$a = \frac{1}{T}\int_{-T/2}^{T/2} v(t)\, dt,$$
$$b = \frac{12}{T^3}\int_{-T/2}^{T/2} tv(t)\, dt. \quad (69)$$

Consider the Fourier transform of

$$v'(t) = v(t) - (a+bt), \qquad |t| < T/2$$
$$= 0, \qquad |t| > T/2$$
$$V'(\omega) = \int_{-T/2}^{T/2} v'(t)e^{-j\omega t}\, dt. \quad (70)$$

Expanding the kernel gives an approximation to $V'(\omega)$ for $\omega T/2 < 1$:

$$V'(\omega) = \int_{-T/2}^{T/2} v'(t)\left(1 - j\omega t - \frac{\omega^2 t^2}{2} + \cdots\right) dt$$
$$= \int_{-T/2}^{T/2} (v(t) - a - bt)\left(1 - j\omega t - \frac{\omega^2 t^2}{2} + \cdots\right) dt. \quad (71)$$

If the integrals that can be done are carried out and the values of a and b substituted, it will be seen that $V'(\omega)$ has neither a constant term nor a term linear in ω. Also the transform of $a+bt$ goes to zero rapidly for $\omega \gg 2/T$. Thus, sampling for time T and removing the mean and the linear drift is quite similar to passing $v(t)$ through a high-pass filter with two zeros, cutting off at about $\omega = 2/T$. Another way to show this is to consider the minimization of

$$\frac{1}{2\pi}\int_{-\infty}^{\infty} |V'(\omega)|^2 \, d\omega = \int_{-T/2}^{T/2} (v'(t))^2 \, dt \tag{72}$$

the equality holding because of Parseval's Theorem. Adjusting the parameters a and b for minimum total energy is really removing the maximum amount of energy possible in the band of frequencies occupied by the transform of $a+bt$. Since this transform has a half width of the order of $2/T$, the whole process is quite similar to the high-pass filter mentioned earlier. Subtracting a second-order and third-order term in t would about double the cutoff frequency.

We may obtain the corresponding general result for subtraction of an Nth order polynomial, $g_N(t)$, from $v(t)$. For

$$\int_{-T/2}^{T/2} (v(t) - g_N(t))^2 dt$$

to be a minimum we may let

$$g_N(t) = \sum_{n=0}^{N} a_n P_n(2t/T)$$

where $P_n(x)$ is the Legendre polynomial of the first kind and a_n is a constant to be determined. ($N=1$ is the linear regression case considered earlier.) This follows from the mean square approximation property of orthonormal sets of functions: $P_n(2t/T)$ is the orthonormal set of polynomials for the interval $|t| < T/2$ with weight function unity. In the usual way we find

$$a_n = \frac{2n+1}{T}\int_{-T/2}^{T/2} v(t) P_n(2t/T) dt$$
$$= \frac{2n+1}{T}\int_{-\infty}^{\infty} \hat{v}(t) P_n(2t/T) dt$$

where $\hat{v}(t) = v(t)$ for $|t| < T/2$ and $\hat{v}(t) = 0$ for $|t| > T/2$.

Consider the Fourier transform, $\hat{G}_N(w)$, of $\hat{g}_N(t)$ where $\hat{g}_N(t) = g_N(t)$ for $|t| < T/2$ and $\hat{g}_N(t) = 0$ for $|t| > T/2$.

$$\hat{G}_N(w) = \int_{-\infty}^{\infty} \hat{g}(t) e^{-j\omega t} dt$$
$$= \sum_{n=0}^{N} a_n \int_{-T/2}^{T/2} P_n(2t/T) e^{-j\omega t} dt$$
$$= T \sum_{n=0}^{N} a_n (-j)^n j_n(wT/2)$$

where $j_n(x)$ is the spherical Bessel function of the first kind. We may express the coefficient a_n in terms of $\hat{V}(w)$, the Fourier transform of $\hat{v}(t)$:

$$a_n = \frac{2n+1}{T}\int_{-\infty}^{\infty} \hat{v}(t) P_n(2t/T) dt$$
$$= \frac{2n+1}{T}\left[P_n\left((2j/T)\left(\frac{d}{dw}\right)\right)\hat{V}(w)\right]\Bigg|_{w=0}$$

where d/dw is the differential operator and the expression is evaluated at $w=0$ after the differentiations are performed. We find, using the expansions for the Bessel functions, that as

$$WT/2 \to 0$$

$$\hat{G}_0(w) \to \hat{V}(0),$$
$$\hat{G}_1(w) \to \hat{V}(0) + w\hat{V}'(0),$$
$$\hat{G}_2(w) \to \hat{V}(0) + w\hat{V}'(0) + \frac{w^2}{2}\left(\hat{V}''(0) + \frac{\hat{V}(0)T_2}{12}\right),$$
$$\left(\hat{V}'(0) = \left[\frac{d}{dw}\hat{V}(w)\right]\Bigg|_{w=0}\right)$$

and for $wT/2 \gg 1$, $\hat{G}_N(w) = 0$. Thus, if $\hat{g}_0(t)$ (the average value of $\hat{v}(t)$ is subtracted from $\hat{v}(t)$, the transform of the difference has a single zero at $w=0$, and subtraction of $\hat{g}_1(t)$ (linear regression) gives two zeros. Note that subtraction of higher order terms such as $\hat{g}_2(t)$ does not introduce more zeros. Also, since $\hat{G}_N(w) \to 0$ for $wT/2 \gg 1$, the transform of $\hat{v}(t) - \hat{g}_N(t)$ is just $\hat{V}(w)$ for large $WT/2$. This again demonstrates the high-pass filtering action.

The next consideration is to calculate the average of the estimates of the variance for a signal $v(t)$ observed for only a finite time T. It is assumed that $v(t)$ represents a stationary ergodic process and has zero mean. The variance is

$$\sigma^2(v) = \overline{v(t)^2} = \langle v^2 \rangle = R_v(0) = \frac{1}{2\pi}\int_{-\infty}^{\infty} S_v(\omega) \, d\omega. \tag{73}$$

The estimate of the variance over time T is

$$\hat{\sigma}^2(v) = \langle v^2 \rangle_{t,T} - \langle v \rangle_{t,T}{}^2$$
$$= \frac{1}{T}\int_{t-T/2}^{t+T/2} v^2(t') \, dt' - \frac{1}{T^2}\left(\int_{t-T/2}^{t+T/2} v(t') \, dt'\right)^2. \tag{74}$$

The average value of the estimate variance is

$$\overline{\hat{\sigma}^2(v)} = \frac{1}{T}\int_{t-T/2}^{t+T/2} \overline{v^2(t')} \, dt'$$
$$- \frac{1}{T^2}\int_{t-T/2}^{t+T/2} dt' \int_{t-T/2}^{t+T/2} dt'' \overline{(v(t')v(t''))}$$
$$= \sigma^2(v) - \frac{1}{T^2}\int_{t-T/2}^{t+T/2} dt' \int_{t-T/2}^{t+T/2} dt'' R_v(t'' - t')$$
$$= \frac{1}{2\pi}\int_{-\infty}^{\infty} d\omega S_v(\omega)\left(1 - \frac{\sin^2 \omega T/2}{(\omega T/2)^2}\right). \tag{75}$$

(The bar here means either statistical or time average.) Thus, the average of the estimated variance is the same as the true variance except for the factor

$$\left(1 - \frac{\sin^2 \omega T/2}{(\omega T/2)^2}\right)$$

in the integrand. For $\omega T/2 \ll 1$, this factor is approximately $1/3\ (\omega T/2)^2$ and for $\omega T/2 \gg 1$, the factor goes to 1. Here again is the high-pass filtering action, the filter cutting off at $\omega \simeq 2/T$.

If $v(t)$ is taken to be

$$\langle \dot\phi \rangle_{t,\tau} = \frac{1}{\tau} \int_{t-\tau/2}^{t+\tau/2} \dot\phi(t')dt' = \frac{\Delta_\tau \phi(t)}{\tau}$$

corresponding to *averaging* the frequency over time τ and *observing* it for time T,

$$\overline{\hat\sigma^2(\langle \dot\phi \rangle_{t,\tau})} = \frac{1}{2\pi} \int_{-\infty}^{\infty} S_{\dot\phi}(\omega) \left(\frac{\sin^2 \omega\tau/2}{(\omega\tau/2)^2}\right) \cdot 1 - \frac{\sin^2 \omega(T-\tau)/2}{(\omega(T-\tau)/2)^2} d\omega, \tag{76}$$

since

$$S_{\langle \dot\phi \rangle_{t,\tau}}(\omega) = S_{\dot\phi}(\omega) \frac{\sin^2 \omega\tau/2}{(\omega\tau/2)^2}$$

which may be shown either by direct calculation of the Fourier transform of the autocorrelation function of

$$\langle \dot\phi \rangle_{t,\tau} = (\phi[t + (\tau/2)] - \phi[t - (\tau/2)])/\tau$$

and using the relation $S_{\dot\phi}(\omega) = S_{\dot\phi}(\omega)/\omega^2$ or by the method used in the text following (22). $T - \tau$ appears in place of T in (76), since only $T - \tau$ is available for averaging. Note that since the factor

$$[1 - \sin^2 \omega(T-\tau)/2/(\omega(T-\tau)/2)^2]$$

goes to zero as ω^2, the integral will be convergent for an f^{-1} behavior of $S_{\dot\phi}(\omega)$. If $\langle \dot\phi \rangle_{t,\tau}$ is taken to be a discrete variable (as is usually done in the actual measurements) the average of the estimated variance will be [6], [14]

$$\overline{\hat\sigma^2_{N,\langle \dot\phi \rangle t,\tau}} = \frac{1}{2\pi} \int_{-\infty}^{\infty} S_{\dot\phi}(\omega) \frac{\sin^2 \omega\tau/2}{(\omega\tau/2)^2} - \frac{\sin^2 N\omega\tau/2}{(N\omega\tau/2)^2} d\omega \tag{77}$$

which is very similar to (76). Here, instead of using a continuous average over T to compute σ, the average is over N discrete consecutive samples of $\langle \dot\phi \rangle_{t,\tau}$ each spaced time τ apart so that $N\tau$ corresponds to T in the continuous case. For a detailed discussion of the discrete variable formulation see Allan [6] and Barnes [14].

The bar over $\hat\sigma^2$ in (75), (76), etc. implies averaging over a number of periods of time of length $(T - \tau)$ to obtain the average of $\hat\sigma^2$. This is necessary since $\hat\sigma$ is really a statistical variable and has its own variance [1].

Usually $T \gg \tau$ in (76) and the spectral density will be fairly smooth so that if (76) converges the high-pass filter action may be approximated by a low-frequency cutoff on the integral. Thus

$$\overline{\hat\sigma(\langle \dot\phi \rangle_{t,\tau})} \approx \frac{1}{\pi} \int_{2/T}^{\infty} S_{\dot\phi}(\omega) \left\{\frac{\sin^2 \frac{\omega\tau}{2}}{\left(\frac{\omega\tau}{2}\right)^2}\right\} d\omega. \tag{78}$$

Let us now calculate the fluctuation expected in a finite time of observation of an oscillator with an f^{-1} power spectral density of frequency. Assume

$$S_{\dot\phi}(\omega) = \frac{K}{|\omega|}$$

so that

$$S_\phi(\omega) = \frac{K}{|\omega|^3}.$$

If the oscillator were observed for an infinite time

$$\sigma^2(\Delta_\tau \phi(t)) = 2(R_\phi(0) - R_\phi(\tau)) = \frac{2K}{\pi} \int_0^{\infty} \frac{1 - \cos \omega\tau}{\omega^3} d\omega$$

which of course does not converge. If we cut off the integral at the lower limit $2/T$ corresponding to (78)

$$\sigma^2 = \frac{2K}{\pi} \int_{2/T}^{\infty} \frac{1 - \cos \omega\tau}{\omega^3} d\omega = \frac{2K\tau^2}{\pi} \int_{2\tau/T}^{\infty} \frac{1 - \cos x}{x^3} dx.$$

For constant T/τ the integral is a constant and $\sigma^2 = C^2\tau^2$, C a constant. Then, we have

$$\frac{\sigma(\langle \dot\phi \rangle_{t,\tau})}{\omega_0} = \frac{1}{\omega_0 \tau} \sigma(\Delta_\tau \phi(t)) = \text{constant}.$$

To get the dependence on T/τ we can estimate the integration. (The integral can be done exactly in terms of the cosine integral.)

$$\frac{\sigma^2(\langle \dot\phi \rangle_{t,\tau})}{\omega_0^2} = \frac{2K}{\pi\omega_0^2} \int_{2\tau/T}^{\infty} \frac{1 - \cos x}{x^3} dx = \frac{2K}{\pi\omega_0^2} \cdot \left(1.04 + \frac{1}{2} \log\left(\frac{T}{2\tau}\right) + O\left(\left(\frac{2\tau}{T}\right)^2\right)\right).$$

So for $T/2\tau \gg 1$ $\sigma^2(\langle \dot\phi \rangle_{t,\tau})/\omega_0^2$ is proportional to log $(T/2\tau)$.

The problem of spectral density estimation from data over a finite time is well covered in the literature [1], [22]. For the relatively high-frequency portions, say $\omega > 20$, analysis of the signal representing $\phi(t)$ or $\dot\phi(t)$ by the usual narrow-band and tunable filter is simple and useful if the resolution required is not too great. The techniques of carrier cancellation and sideband exaltation by using a dispersive bridge or a nondispersive delay line are also extremely useful [23], [24].[3]

[3] Several papers and correspondence items in this issue contain material pertaining to these types of measurements.

For lower frequencies it is usually necessary to do the spectral analysis from the time data. One technique is to compute an approximate autocorrelation function, modify it with a time window (Hanning, Hamming, or other [1]), $D(\tau)$, which is symmetric about $\tau = 0$, equals 1 for $\tau = 0$, and goes smoothly to zero and then stays zero for some $\tau \leq T$, where T is the total time duration of the data. Then the Fourier transform of this modified autocorrelation function is computed to get the estimate of the spectral density. One can then show that if $\hat{S}(\omega)$ is an estimate of the spectral density computed in this way then $\overline{\hat{S}(\omega)} = S(\omega) \otimes Q(\omega)$ where $S(\omega)$ is the true spectral density, $\otimes$ signifies convolution, the bar signifies the statistical average, and $Q(\omega)$ is the Fourier transform of the time window. The variance of $\hat{S}(\omega)$ depends on the shape of the time window as well as the ratio of the length (in time) of the time window to the total length in time of the data. For small ratios the variance is small, but the smearing of the spectrum due to the consequently wide $Q(\omega)$ is appreciable. For any given length of data, therefore, there is some variance and some smearing. The time window represented by (24) is

$$D_c(\tau) = 1 - \frac{|\tau|}{T}, \quad |\tau| \leq T; \qquad D_c(\tau) = 0, \quad |\tau| > T.$$

Since this time window is as wide as is possible from the data length, the variance of $\hat{S}(\omega)$ will be fairly large. As pointed out in Blackman and Tukey [1], it is wise to remove the mean and the linear drift before attempting spectral analysis via the autocorrelation function route.

Low-frequency analysis can also be done by observing the slope of $\sigma(\langle\phi\rangle_{t,\tau})/\omega_0$, or by successive differences of phase [14], or by observing

$$\sigma^2(\dot{\phi}'(\omega))_{\omega c} = \frac{1}{2\pi}\int_{-\infty}^{\infty} S_{\dot{\phi}}(\omega) \,|\, H(\omega, \omega_c) \,|^2 d\omega$$

as a function of the cutoff frequency ω_c of a low-pass filter $H(\omega, \omega_c)$ through which the signal representing $S_{\dot{\phi}}(\omega)$ is passed [25].

Acknowledgment

The authors would like to thank R. Vessot of Varian Associates, Beverly, Mass., J. Barnes of NBS, Boulder, Colo., and J. W. Graham, R. P. Rafuse, and J. G. King at M.I.T., Cambridge, Mass., for helpful discussions; also, L. Bodily and A. Foster of the Hewlett-Packard Co., Palo Alto, Calif., R. S. Badessa of M.I.T., and R. D. Posner, formerly of M.I.T., now with Aeronutronics, Newport Beach, Calif., for making some of the measurements.

References

[1] R. B. Blackman and J. W. Tukey, *The Measurement of Power Spectra.* New York: Dover, 1959.
[2] W. B. Davenport and W. L. Root, *Random Signals and Noise.* New York: McGraw-Hill, 1958, ch. 6.
[3] *Ibid.*, ch. 4, sec. 4.5.
[4] C. L. Searle and D. A. Brown, "Comparison of performance criteria of frequency standards," *Proc. 1962 16th Annual Symp. on Frequency Control*, pp. 259–266.
[5] J. Rarity, L. Saporta, and G. Weiss, "Study of short term stability of crystal oscillators," New York University, New York, N. Y., Tech. Rept., Contract DA-36-039-SE-87450, DA Project 3A-99-15-02, -02-02.
[6] D. W. Allan, "Statistics of atomic frequency standards," this issue, page 221.
[7] R. Vessot, L. Mueller, and J. Vanier, "The specification of oscillator characteristics from measurements made in the frequency domain," this issue, page 199. See also Vessot et al., "A cross-correlation technique for measuring the short-term properties of stable oscillators," *IEEE-NASA Symp. on Short-Term Frequency Stability*, Washington, D. C.: U. S. Govt. Printing Office, pp. 111–118, NASA SP-80.
[8] V. Van Duzer, "Short term stability measurements," *ibid.*, pp. 269–272, and private communication.
[9] R. D. Posner, "Digital computer aided calculation of frequency stability," M.S. thesis, M.I.T., Cambridge, Mass., June 1965.
[10] W. R. Atkinson, L. Fey, and J. Newman, "Spectrum analysis of extremely low frequency variations of quartz oscillators," *Proc. IEEE (Correspondence)*, vol. 51, p. 379, February 1963.
[11] D. Hammond, C. Adams, and L. Cutler, "Precision crystal units," *Proc. 1963 17th Annual Symp. on Frequency Control.*
[12] A. W. Warner, "Design and performance of ultraprecise 2.5 mc quartz crystal units," *Bell Sys. Tech. J.*, vol. 39, p. 1193, September 1960.
[13] W. A. Edson, "Noise in oscillators," *Proc. IRE*, vol. 48, pp. 1454–1466, August 1960. See also J. A. Mullen, "Background noise in oscillators," and M. Golay, "Monochromaticity and noise in a regenerative electrical oscillator" in the same issue.
[14] J. A. Barnes, private communication, and J. A. Barnes and D. W. Allan, "Effects of long term stability on the definition and measurement of short term stability," *Proc. 1964 Symp. on Definition and Measurement of Short-Term Frequency Stability.* Also, J. A. Barnes, "Atomic timekeeping and the statistics of precision signal generators," this issue, page 207.
[15] D. Kleppner, H. Goldenberg, and N. F. Ramsey, *Phys. Rev.*, vol. 126, p. 603, 1962.
[16] K. Shimoda, T. Wang, and C. H. Townes, *Phys. Rev.*, vol. 192, p. 1308, 1956.
[17] L. Cutler, to be published.
[18] E. Hafner, "The effects of noise in oscillators," this issue, page 179.
[19] R. F. C. Vessot and H. Peters, "Frequency beat experiments with hydrogen masers," *Proc. 1963 17th Annual Symp. on Frequency Control.*
[20] P. Davidovits and R. Novick, "The optically pumped rudidium maser," this issue, page 155.
[21] J. J. Bagnall, Jr., "The effect of noise on an oscillator controlled by a primary reference," *NEREM 1959 Record*, pp. 84–86.
[22] W. B. Davenport and W. L. Root [2], sec. 6-6.
[23] I. Berstein and G. Gorelik, "Frequency modulation noise in oscillators," *Proc. IRE (Correspondence)*, vol. 45, p. 94, January 1957.
[24] I. Berstein, "On the fluctuations of a vacuum tube oscillator," *Reports of the Academy of Science, USSR*, vol. 68, no. 3, pp. 469–472, 1949.
[25] R. F. C. Vessot, private communication. See also Vessot et al. [7].
[26] E. J. Baghdady, R. N. Lincoln, and B. D. Nelin, "Short term frequency stability characterization, theory, and measurement"; L. S. Cutler, "Some aspects of the theory and measurement of frequency fluctuations in frequency standards;" V. J. Bates, R. S. Badessa, R. D. Posner, and C. L. Searle, "Computer-aided calculation of frequency stability," all in *Proc. IEEE-NASA Symp. on Short Term Frequency Stability.* Washington D. C.: U. S. Govt. Printing Office, 1964, NASA SP-80, no. 20402. Baghdady et al. also appears in *Proc. IEEE*, vol. 53, pp. 704–722, July 1965.
[27] S. L. Johnson, B. H. Smith, and D. A. Calder, "Noise spectrum characteristics of low-noise microwave tubes and solid-state devices," this issue, page 258.

Short-Term Frequency Stability: Characterization, Theory, and Measurement

E. J. BAGHDADY, SENIOR MEMBER, IEEE, R. N. LINCOLN AND B. D. NELIN, MEMBER, IEEE

Abstract—**An analysis is first presented of the manner in which oscillator short-term instabilities limit performance in a number of applications. This analysis provides guidelines for theoretical formulations, the definition of "short-term stability" and the development of measurement techniques for characterizing short-term instabilities. The factors that affect short-term stability are then discussed, and basic models for theoretical analysis are formulated. The models are used in investigations of the characteristics of outputs of "stable" sources. Measurement techniques are proposed and compared with techniques employed by other investigators.**

I. Introduction

THIS PAPER presents a treatment of the problems of characterizing, analyzing and measuring short-term instabilities of ultrastable sinusoidal oscillation sources as well as of loop-controlled oscillations.

The term "short-term stability" appears with increasing frequency in discussions of timing and synchronization systems, frequency synthesis techniques, and moving-target tracking systems of all types. But this term continues to mean different things to different individuals even in the same applications. As of the time of this writing, and in almost all applications, there is

1) no common understanding of what "short-term" stability means,
2) no common agreement on why it is needed,
3) no common understanding of how to characterize, model, analyze and measure oscillation instabilities, and
4) no widely accepted accounts of how the instabilities will affect performance in applications, and how the requirement for short-term stability should be specified.

Stable oscillators are used extensively as

1) original sources of timing and synchronizing signals,

Manuscript received December 8, 1965; revised March 28, 1965. This paper is based on work performed for the National Aeronautics and Space Administration, Marshall Space Flight Center, Huntsville, Ala., under Contract No. NAS 8-11228.
The authors are with ADCOM, Inc., Cambridge, Mass.

Reprinted from *Proc. IEEE*, vol. 53, pp. 704–722, July 1965.

2) original sources of carriers or subcarriers for target tracking applications, communication signals and multiple-access utilization of aerospace vehicles as relay centers,
3) local independent reference signals and timing, phase, or frequency measurement, and
4) local oscillation of controllable phase and frequency which tracks the instantaneous phase or frequency of an incoming signal, thereby providing a continuous estimate of its instantaneous phase or frequency.

In some of these operations, the emphasis is placed on the unavoidable fluctuations in phase or frequency caused by sources of instability in others on the spectral structure or "spectral purity" of the oscillation. Thus, the characterization and measurement of oscillator instability may be pursued along the lines of

1) characterizing the fluctuations in the phase and frequency of the oscillator output, or
2) characterizing the structure of the spectrum of the oscillator output.

In general, one cannot concentrate on one and ignore the other of these two aspects of an oscillation completely, although either of them potentially can be made to provide a complete representation of the source output. Thus, exclusive concentration on the structure of the spectrum or the mean-square spectral density obscures the fact that not all of the spectral components will actually cause phase and frequency fluctuations—fluctuations in the amplitude of the oscillation also spread out its spectrum and may be completely unrelated to the phase and frequency fluctuations. Similarly, exclusive concentration on the phase or frequency fluctuations obscures the fact that a considerable source of the observed instability may actually be spurious filterable components that will be effectively reduced in the ultimate system. In the final analysis, the type of characterization sought must be the one most clearly related to the intended application of the oscillation.

In Section II we explore the manner in which oscillator short-term instabilities limit system performance in a number of applications. This analysis shows that the mean-square spectral density of the instantaneous frequency fluctuations of the oscillation frequently provides the most direct basis for the characterization, measurement, and specification of the short-term frequency and phase instabilities. Although the probability density functions of the instantaneous phase and frequency fluctuations may be of interest, often only an estimate of the peak factor of the fluctutations is sufficient. Guidelines are provided for user estimation of the shortest interval of time over which instability errors will become noticeable in a specified application. A user quantitative interpretation of "short-term" is thereby made possible. Situations in which the mean-square spectral density of the oscillation itself provides an adequate abstraction are also discussed.

In Section III a survey is made of mathematical models and associated characteristics of unstable oscillations. Two general approaches for modeling are presented, and several models in each category are discussed.

In Section IV attention is directed to the problem of measuring the characteristics of frequency unstable oscillations. The general properties of satisfactory measurement systems are developed from which the basic structure of these systems is obtained. The gross performance of the class of recommended systems is contrasted with that of previous systems for the measurement of oscillator instability. Finally, the areas of unsolved problems in measurement are discussed.

II. Effect of Oscillation Instability in Applications

The characterization, modeling, analysis, measurement, and specification of bounds on oscillation short-term, or rapid, instability must ultimately be made in the light of its effects on the performance of systems that rely on the oscillation. General guidelines will now be sought in terms of a number of specific applications.

A. Effect of Instability on Precision Range Measurement

An application of great current interest, and one whose methodology of treatment is representative of a number of other applications, is encountered in precision range (or radial distance to a vehicle) measurement systems. Such a system is illustrated in Fig. 1. In this system, a tone of suitable frequency is transmitted to the vehicle to be tracked and is returned by it to the tracking station. Range from the tracking station to the vehicle is then determined from a measurement of the phase of the returned tone relative to the phase of the original transmitted tone on the basis of a current segment of the continuing output of the source of the original tone. In the ensuing discussion we consider only the errors caused by the short-term instabilities of the source of the ranging tone which cause the current seg-

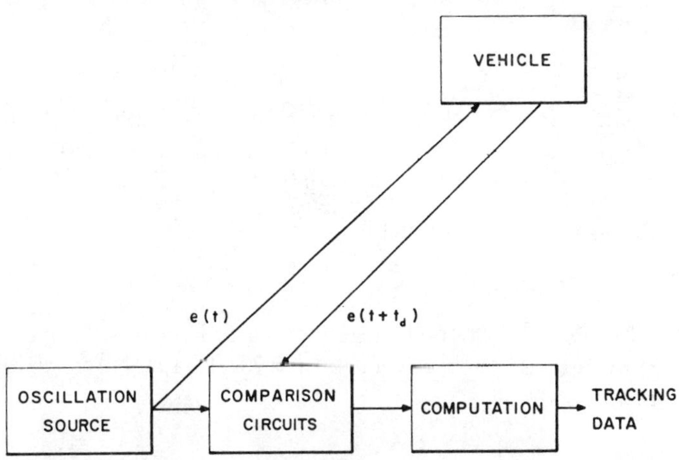

Fig. 1. Basic functions of a target-tracking system with sinusoidal ranging signal.

ment to differ from the original one (or a perfect replica of it) and ignore all other effects as extraneous to this discussion.

Thus, let t_d be the two-way signal propagation time between the tracking station and the vehicle. The radial distance to the vehicle is then

$$R = ct_d/2 \tag{1}$$

where c is the velocity of signal propagation in the intervening medium. If the original transmitted tone is expressed as

$$e_{\text{trans}}(t) = E(t) \cos \theta(t) \tag{2}$$

then, assuming ideal propagation conditions and no delay in turn-around at the vehicle, we have for the returned signal

$$e_{\text{ret}}(t) = E(t + t_d) \cos \theta(t + t_d). \tag{3}$$

The desired information is contained in

$$\begin{aligned}\Delta\theta(t, t_d) &= \theta(t + t_d) - \theta(t) \\ &= \int_t^{t+t_d} \dot{\theta}(\mu) d\mu \\ &= \dot{\theta}(t) \otimes p_{\text{rect},t_d}(t + t_d/2) \qquad (4)\end{aligned}$$

where $\dot{\theta}(t) = d\theta(t)/dt$, $\otimes$ denotes convolution and

$p_{\text{rect},t_d}(t)$ = rectangular pulse of unit height extending from $t = -t_d/2$ to $t = t_d/2$.

Now, the instabilities in the oscillation frequency may be accounted for by writing

$$\theta(t) = 2\pi f_0 t + \phi(t) \tag{5}$$

so that

$$\dot{\theta}(t) = 2\pi f_0 + \dot{\phi}(t)$$

where

f_0 = the value of the instantaneous frequency of the tone at the start of the t_d round-trip time to be measured,

and

$\phi(t)$ = instantaneous fluctuations in the phase of the oscillation about the phase $2\pi f_0 t$ of an absolutely stable source of frequency f_0.

Substitution in (4) yields

$$\Delta\theta(t, t_d) = 2\pi f_0 t_d + \dot{\phi}(t) \otimes p_{\text{rect},\, t_d}(t + t_d/2). \tag{6}$$

This shows that the frequency instability of the ranging tone introduces an uncertainty in the measured value of t_d given by

$$\Delta t_d = \frac{1}{2\pi f_0} \dot{\phi}(t) \otimes p_{\text{rect},t_d}(t + t_d/2). \tag{7}$$

In view of the random character of $\dot{\phi}(t)$, a convenient measure of the error it causes is usually the rms (or standard deviation). Thus, if σ_{t_d} denotes the rms value of Δt_d,

$$\sigma_{td} = \frac{1}{2\pi f_0} \sqrt{E[\{\dot{\phi}(t) \otimes p_{\text{rect},\, td}(t + t_d/2)\}^2]} \tag{8}$$

where $E[\ \ ]$ denotes "statistical average." Since the Fourier transform of the convolution of two functions equals the product of their Fourier transforms, we have

$$\sigma_{td}/t_d = \frac{1}{2\pi f_0} \sqrt{\frac{1}{2\pi} \int_{-\infty}^{\infty} S_{\dot{\phi}}(\omega) \left(\frac{\sin \omega t_d/2}{\omega t_d/2}\right)^2 d\omega} \tag{9}$$

where $S_{\dot{\phi}}(\omega)$ is the mean-square spectral density of $\dot{\phi}(t)$.

Equation (9) brings out the following important facts:

1) The fractional rms error caused by the frequency instability of the oscillator is determined by the *mean-square spectral density*, $S_{\dot{\phi}}(\omega)$, of the *instantaneous frequency perturbations*.
2) The spectral density $S_{\dot{\phi}}(\omega)$ is weighted by a filtering effect,

$$\left(\frac{\sin \omega t_d/2}{\omega t_d/2}\right)^2 \tag{10}$$

plotted in Fig. 2, which depends upon the interval of time t_d—peculiar to the application—over which the oscillator instabilities may cumulate their influence on the measurement.

It is clear from Fig. 2 that the weighting function favors only the part of the spectrum contained *within* $|\omega| \leq 2\pi/t_d$. As t_d increases, the edges of this interval shrink closer and closer to $\omega = 0$, indicating that the faster fluctuations in $\dot{\phi}(t)$ become less and less significant in determining the error in the measurement.

In practice, additional filtering is introduced to smooth out the effects of noise from various sources, the bandwidth being chosen to maximize the noise smoothing without intolerable conflict with requirements on speed of response imposed by range resolution specifications and the dynamics of the vehicle trajectory. The introduction of such a filter, characterized by $H_{lp}(j\omega)$, multiplies the integrand in (9) by $|H_{lp}(j\omega)|^2$. If the nominal passband of $H_{lp}(j\omega)$ is restricted to $|\omega| \leq 2\pi B_{lp}$, then for

$$B_{lp} \leq (1/t_d)/10$$

only the spectral components within $|\omega| \leq 2\pi B_{lp}$ will contribute instability errors. In the absence of this filter action, or when B_{lp} is of the order of 10 times $1/t_d$ or more, components with frequency up to B_{lp} c/s may contribute [depending upon $S_{\dot{\phi}}(\omega)$] significantly to the instability error, although components within $|\omega| \leq 2\pi/t_d$ fall on the more favorable part of the curve in Fig. 2.

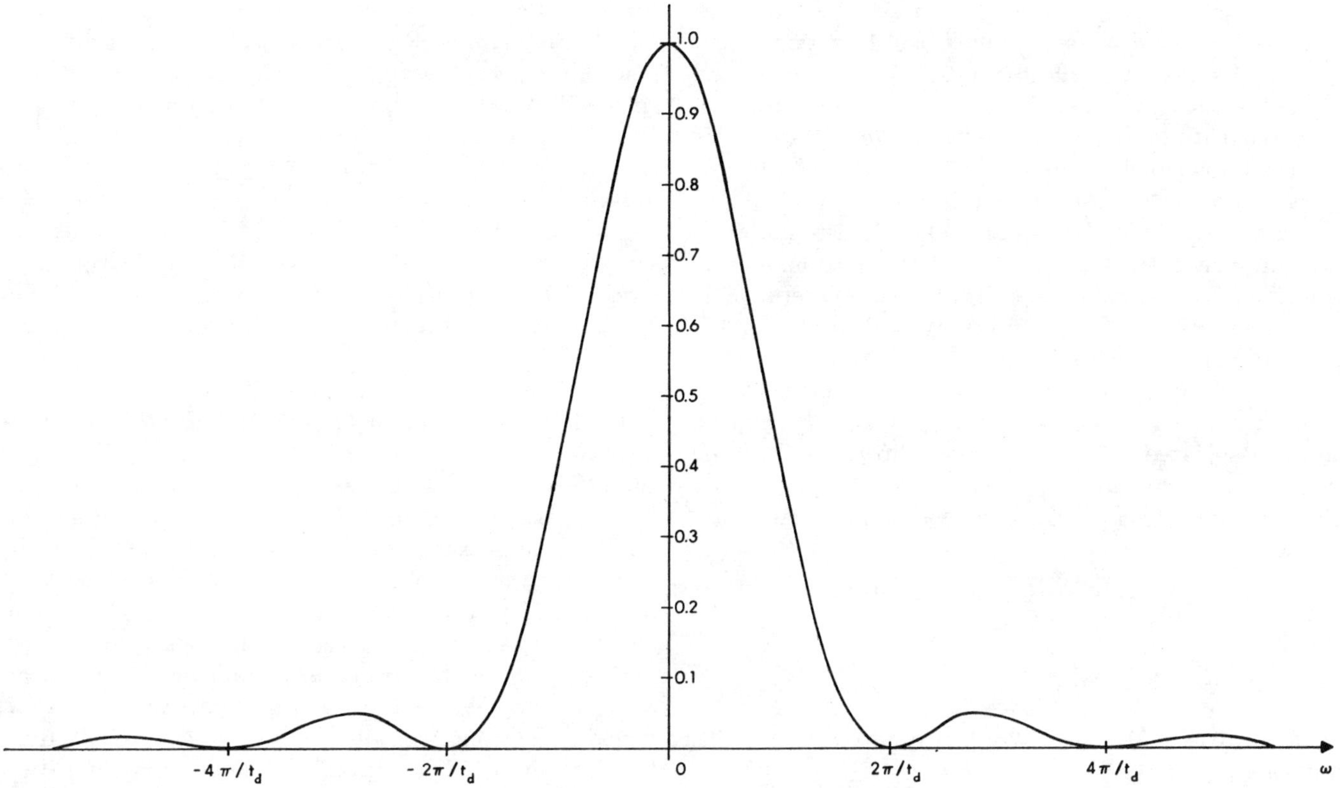

Fig. 2. Filtering effect of interval, t^7, over which the instabilities may cumulate their influence on the measurement.

Note that the range measurement application just considered is representative of all applications in which information is derived from a comparison of samples of the oscillation *phase* taken at different times. If the precision of the phase-shift measuring instrument is sufficient to resolve ρ_θ rad, then the effects of instability will be negligible provided that

$$(\text{phase uncertainty cumulated in } t_d \text{ sec}) < \rho_\theta/10. \quad (11)$$

The peak value of the error cumulated in t_d sec may be expressed as the product of the expected rms error σ_{t_d} and a peak factor, $(\text{p.f.})_\phi$, associated with the fluctuation caused by oscillator instability. Thus, if the peak error is to satisfy condition (11), we must have

$$\frac{(\text{p.f.})_\phi}{2\pi f_0} t_d \sqrt{\frac{1}{2\pi}\int_{-\infty}^{\infty} S_{\dot\phi}(\omega)\left(\frac{\sin \omega t_d/2}{\omega t_d/2}\right)^2 | H_{lp}(j\omega) |^2 d\omega} \leq \rho_\theta/10. \quad (12)$$

This shows that, as far as the rms error caused by instabilities of the oscillation source is concerned, each application is completely characterized by a smoothing filter, $H_{lp}(j\omega)$, the nominal ranging tone frequency, f_0, the value (or range of values) of the time separation, t_d, between the compared samples of tone phase, and the resolution of the phase-measuring instrument, ρ_θ. Of these, only t_d is independent of equipment. Therefore, for a given setup, t_d may be considered to characterize the application. The oscillator instabilities are completely characterized by $S_{\dot\phi}(\omega)$ and a nominal peak factor for conversion from rms to peak phase fluctuation

Close inspection of the left-hand member of (12) shows that the peak phase uncertainty caused by oscillator instability is an increasing function of the time separation t_d between the compared phase samples, all other factors (such as choice of B_{lp}) being the same. This is particularly evident when the smoothing filter bandwidth, B_{lp}, is much smaller than $1/t_d$, in which case the error will rise monotonically with increasing t_d. This observation, combined with the fact that of all the parameters in condition (12) only t_d is beyond the control of the designer, suggests that, in any specific application (specified equipment and oscillator), the equation

$$\frac{(\text{p.f.})_\phi}{2\pi f_0} t_d \cdot \sqrt{\frac{1}{2\pi}\int_{-\infty}^{\infty} S_{\dot\phi}(\omega)\left(\frac{\sin \omega t_d/2}{\omega t_d/2}\right)^2 | H_{lp}(j\omega) |^2 d\omega} = \rho_\theta \quad (13)$$

may be solved for a value of t_d for which the expected peak error will equal the phase resolution capability ρ_θ. If this value of t_d is denoted t_{dp}, then, from the viewpoint of the user, the phase discrimination errors caused by oscillator instability become noticeable only for values of t_d which are comparable with or in excess of t_{dp}. In this way, the term "short-term" in a specified application may be interpreted to mean "over intervals of duration t_d which are comparable with t_{dp}."

A word of caution is necessary here. The preceding interpretation of "short-term" is strictly intended to provide a criterion for determining, in a specified application, whether or not measurements based on a comparison of two samples of the oscillation phase, separated in time by t_d sec, will be subject to noticeable errors attributable to instability of the oscillation source. We assumed knowledge of the application—specifically, f_0, $H_{lp}(j\omega)$, and ρ_θ—and the availability of properly measured data—specifically, $S_{\dot{\phi}}(\omega)$ and $(\text{p.f.})_\phi$—characterizing the instantaneous frequency perturbations of the oscillation. It should be clear that this interpretation is only a user guideline and definitely not the criterion for the choice of "adequate" time interval in tests aimed at furnishing $S_{\dot{\phi}}(\omega)$ and $(\text{p.f.})_\phi$. The experimental determination of the instability characteristics of oscillators and the associated interpretation of "short term" will be discussed later on in this paper.

For specific illustrations, consider first the situation in which

$$S_{\dot{\phi}}(\omega) = a^2 = \text{constant for all } \omega. \tag{14}$$

Then if

$$B_{lp} \gg 1/t_d \tag{15}$$

the integral in (13) is closely approximated by $a^2 2\pi/t_d$. Solution for t_d yields

$$t_d = \left(\frac{2\pi f_0 \rho_\theta}{a(\text{p.f.})_\phi}\right)^2 \stackrel{\Delta}{=} t_{d_p}. \tag{16}$$

On the other hand, if

$$B_{lp} \leq (1/t_d)/10 \tag{17}$$

then the integral in (13) becomes closely approximated by $4\pi a^2 B_{lp}$, and the corresponding t_{d_p} becomes

$$t_{d_p} = \frac{2\pi f_0 \rho_\theta}{(\text{p.f.})_\phi a\sqrt{2B_{lp}}}\cdot \tag{18}$$

Consider next the situation in which

$$S_{\dot{\phi}}(\omega) = b^2\omega^2, \qquad \text{for all } \omega. \tag{19}$$

Substitution in (13) yields

$$\frac{(\text{p.f.})_\phi b}{2\pi f_0}\sqrt{\frac{1}{2\pi}\int_{-\infty}^{\infty} |H_{lp}(j\omega)|^2(1-\cos\omega t_d)d\omega} = \rho_\theta. \tag{20}$$

The quantity under the square-root sign equals

$$\tfrac{1}{2}\alpha[1-e^{-\alpha t_d}] \quad \text{for } H_{lp}(j\omega) = \frac{\alpha}{\alpha + j\omega} \tag{21}$$

$$\sqrt{4\pi\alpha}[1-e^{-\alpha t_d^2}] \text{ for } |H_{lp}(j\omega)|^2 = e^{-\omega^2/4\alpha} \tag{22}$$

$$2B_{lp}\left[1 - \frac{\sin(2\pi B_{lp}t_d)}{(2\pi B_{lp}t_d)}\right] \quad \text{for } |H_{lp}(j\omega)| = 1, \ |\omega| \leq 2\pi B_{lp}; \quad = 0, \ |\omega| > 2\pi B_{lp}. \tag{23}$$

From (21) and (22) we conclude that when the smoothing filter is a simple RC or a Gaussian low-pass filter, the instability error associated with the spectral density specified by (19) will always rise monotonically with t_d. With the rectangular filter, (23) shows that the error rises monotonically with t_d for $0 \leq t_d < (4.5/2\pi B_{lp})$, but undulates thereafter with decreasing amplitude about the value $(\text{p.f.})_\phi b\sqrt{2B_{lp}}/(2\pi f_0)$. It is interesting to observe that in each case the instability error approaches a bounded value as $t_d \to \infty$. For each filter, an expression for t_{d_p} may be obtained by substitution from (21), (22), and (23) into (13).

Two final conclusions may be drawn from inspection of the left-hand member of (13). Values of $S_{\dot{\phi}}(\omega)$ at and near $\omega = 0$ will influence the value of the mean-square of the filtered error, unless $H_{lp}(j\omega)$ filters that region out. This means that in applications where comparisons are made of phase samples t_d apart in time, oscillator frequency instability phenomena that vary so slowly as to be almost constant within the interval t_d ("long-term" effects) must be considered if the observation filter, $H_{lp}(j\omega)$, passes the spectrum at and near $\omega = 0$. Moreover, phenomena that vary rapidly within t_d ("short-term" effects) may be filtered out by $H_{lp}(j\omega)$ or reduced by the $(\sin x/x)^2$ function and hence may not be of importance. These conclusions contradict many statements made by other writers about the essential effect of oscillator instability in such applications.

B. Effect of Instability on Precision Range-Rate Measurement

The radial velocity, or range rate, of a moving target may be determined from a measurement of the Doppler frequency shift of a tone returned by the target, relative to the frequency of a current segment of the continuing output of the source of the original transmitted tone. Frequency comparisons are also employed in coherent Doppler and CW systems, in timing and in synchronization operations. Just as the mean-square spectral density of $\dot{\phi}(t)$ of (5) determines the mean-square instability error in measuring increments of $\theta(t)$, the mean-square spectral density of $\ddot{\phi}(t)$ may be expected to determine the mean-square error in the measurement of changes in $\dot{\theta}(t)$. That this is indeed the case is readily shown by starting with the fact that the desired information is now conveyed by

$$\begin{aligned}\Delta\dot{\theta}(t, t_d) &= \dot{\theta}(t+t_d) - \dot{\theta}(t)\\ &= \ddot{\theta}(t) \otimes p_{\text{rect}, t_d}(t + t_d/2).\end{aligned} \tag{24}$$

Thus, the error in the observed Doppler shift caused by frequency instability in the transmitted tone is given by

$$\Delta f_{\text{Dop}} = \frac{1}{4\pi}\ddot{\phi}(t) \otimes p_{\text{rect}, t_d}(t + t_d/2) \tag{25}$$

with an rms value given by

$$\sigma_{fd} = \frac{t_d}{4\pi}\sqrt{\frac{1}{2\pi}\int_{-\infty}^{\infty}\omega^2 S_{\dot{\phi}}(\omega)\left(\frac{\sin \omega t_d/2}{\omega t_d/2}\right)^2 \mid H_{lp}(j\omega)\mid^2 d\omega}$$

$$= \frac{1}{2\pi}\sqrt{\frac{1}{2\pi}\int_{-\infty}^{\infty} S_{\dot{\phi}}(\omega)\mid H_{lp}(j\omega)\mid^2 \sin^2(\omega t_d/2)d\omega} \quad (26)$$

in which we have introduced the effect of a postdetection low-pass filter aimed at reducing the effects of extraneous additive noise.

Again we observe that

1) the rms instability error in the Doppler shift measurement is determined by the mean-square spectral density $S_{\dot{\phi}}(\omega)$ of the *instantaneous frequency perturbations* of the unstable transmitted tone, and
2) a filtering effect, $\sin^2(\omega t_d/2)$, weights the spectral density in a manner that favors alternate intervals of duration $2\pi/t_d$ rad/s, one of which is defined by $|\omega| \leq \pi/t_d$.

The term "short-term" may also be given a user quantitative interpretation relative to Doppler measurements. Here we seek the value of t_d for which the effects of instability become noticeable in a measurement of Doppler shift. The equation to be solved for this value of t_d is

$$(\text{frequency shift cumulated in } t_d \text{ sec}) = \rho_{\dot{\theta}} \quad (27)$$

in which $\rho_{\dot{\theta}}$ defines the instrumental precision in measuring frequency changes. As before, we may assume a peak factor $(\text{p.f.})_{\dot{\theta}}$ for the frequency fluctuations caused by instability, and write $(\text{p.f.})_{\dot{\theta}}\sigma_{f_d}$ for the left-hand member of (27).

C. Applications Affected by Oscillator RF Spectrum

As we have mentioned previously, there are applications in which the spectral composition of an oscillator signal is of primary importance. Doppler resolution radar is a good example. This system is used to resolve several targets moving with different radial velocities on the basis of the different Doppler shifts of the associated returned replicas of the illuminating signal. A bank of filters is used to resolve the spectrum into "velocity regions," and the output of an individual filter will present only those targets whose radial velocities cause the returned signals to fall into the passband of the filter. The system works perfectly if the RF spectrum of the oscillator is a pure line. Amplitude, phase and frequency instabilities of the illuminating signal broaden the "line width" and spectral occupancy of this signal and thereby limit the ultimate velocity resolution of the system. It is quite clear that, to determine the effect of oscillator instability in such an application, we should work directly with the RF spectrum of the oscillator signal.

Spreading of the oscillator spectrum by unwanted amplitude and phase fluctuations also adds to the spectral density of background signals and noise that interfere with a desired return. This degrades the possibility of discriminating a desired target return from other unavoidable obstacle and sidelobe returns (or clutter) and from relatively strong additive background noise.

It should be clear that, in applications that are affected by the spectral splatter of the oscillation, the frequency or phase instabilities of the oscillation may or may not account for the significant sidebands and diffused line width of the oscillation. Only after it has been established in a clear manner that amplitude fluctuation effects on the spectrum are entirely negligible is it safe to attribute the oscillation spectral characteristics exclusively to the frequency or phase fluctuations.

D. Specification of Oscillator Instability Characteristics

For applications of the type discussed in Sections II A and B the most direct determinant of the effects of oscillator instability is $S_{\dot{\phi}}(\omega)$. Since $\dot{\phi}(t) = d\phi/dt$,

$$S_{\dot{\phi}}(\omega) = \omega^2 S_{\phi}(\omega). \quad (28)$$

Hence, either spectral density, $S_{\dot{\phi}}(\omega)$ or $S_{\phi}(\omega)$, would suffice as a specification.

One particularly interesting form of presentation of $S_{\dot{\phi}}(\omega)$ is as a polynomial approximation such as

$$S_{\dot{\phi}}(\omega) \approx \alpha_0 + \alpha_1\omega + \alpha_{-1}\omega^{-1} + \alpha_2\omega^2 + \alpha_{-2}\omega^{-2} + \cdots \quad (29)$$

or as a piecewise representation using selected terms from this polynomial over specified frequency intervals. The values of the various coefficients plus the RF bandwidth of the oscillator signal (limited by filtering or amplifier bandwidths) could serve as characteristic parameters of $S_{\dot{\phi}}(\omega)$. This approach is suggested naturally by the theoretical models to be discussed in Section III. Integrals of $S_{\dot{\phi}}(\omega)$ that are suitable for specified applications may be sufficient in some cases.

Of the two related functions, $S_{\dot{\phi}}(\omega)$ and $S_{\phi}(\omega)$, the one preferred as a general oscillator specification for a particular source is the one that is easier to measure economically and with low error. This question is often clearly resolvable on the basis of which of the two is expected to vary more slowly or uniformly with ω. For example, in unlocked oscillators, $\phi(t)$ is characterized by a random-walk behavior, and hence $S_{\phi}(\omega)$ will exhibit a nonintegrable singularity at the origin. In other cases, the principal effect of the instability on the oscillation is to result in a tantalizingly unending $1/\omega$ behavior of $S_{\dot{\phi}}(\omega)$ near $\omega = 0$.

For some applications, the properties of the RF spectrum of the oscillator signal must be specified. This is most directly and simply achieved by direct measurements on the RF spectrum itself. Unfortunately, current practice is heavily oriented toward indirect approaches involving the measurement of quantities of questionable significance—such as rms frequency deviation—based exclusively on the phase or frequency fluctuations. The difficulty with these indirect approaches arises from two fundamental considerations:

1) the possibility that the amplitude fluctuations cannot be neglected, and
2) the fact that the RF spectrum per se is not uniquely determined by any one number (or even a few numbers) obtained by averaging some power of the phase or frequency fluctuations.

One might suppose that inasmuch as the emphasis really is often on the RF spectral density *levels* rather than *shape*, it may be possible to determine upper bounds on the density level in terms of some average quantity derived from the frequency fluctuations. This approach is quickly discouraged by the fact that the effect of components in any part of the RF spectrum upon the frequency fluctuations depends not only on the spectral level of those components, but also upon their frequency separations from the undisturbed position of the oscillation frequency. Thus, if a given value of mean-square frequency deviation is to be used in bounding the level of RF spectral density, then it will be found that no upper bound can be associated with it. In fact, concentrating the energy in one disturbing side-component for bounding purposes would result in an upper bound that varies as $1/(\omega-\omega_0)^2$ around the long-time average oscillation frequency, ω_0, going to ∞ as $|\omega-\omega_0|\rightarrow 0$.

III. Models and Characteristics of Unstable Oscillations

Having discussed the manner in which oscillator instabilities introduce errors in applications, and having identified the intrinsic characteristics of the instabilities that determine the errors, we now turn to the problem of mathematical modeling of unstable oscillations and the characterization of instabilities caused by various mechanisms.

There are two approaches to the modeling of oscillations with frequency instabilities. These approaches may be called the *causal approach* and the *black-box approach.*

In the causal approach, a functional structure of the oscillation source is postulated, and various causes of instability are identified and combined with an assumed ideal oscillation signal. The model is thus centered on postulated occurrences *within* a box, and the consequences of these occurrences on the properties of the output signal are then pursued. This approach is essential for an understanding of the reasons for the instability, especially in the search for developing ultrastable sources.

In the black-box approach, the model is centered on the resultant disturbed oscillation at the output terminals of the box, the point where the oscillation is available to the user. Thus, a number of output signal models are postulated, their properties explored, and the actual source output is identified piecewise (for different frequency regions) with one (or perhaps more) of these models.

A. Causal Approach

In the causal approach to the modeling of unstable oscillations, two types of disturbances that give rise to frequency instability are distinguished; namely, *additive noise* and *multiplicative noise.* In loop-controlled oscillators, the reference signal input may also contain a nonnegligible noise component added to the reference signal.

Additive noise models take account of noise added to the oscillator signal either in the oscillator loop or in the buffer amplifier following the oscillator, as illustrated in Fig. 3. The spectrum of this type of noise occupies a frequency region that often extends widely around the oscillator frequency. The noise introduced into (or originating in) the oscillator loop will be called internal additive noise; the noise added in the amplifier following the oscillator will be called external additive noise.

In addition to the additive noise near the oscillator frequency, multiplicative low-frequency processes are also present which effectively frequency modulate the oscillator. Examples of such processes are current and voltage fluctuations (flicker noise), mechanical vibrations, temperature variations, etc. The deviation ratio of the frequency modulation caused by these processes is usually large. Consequently, the shape of the resulting RF spectrum is essentially identical with the shape of the probability density function of the resultant baseband noise process.

1) Additive-Noise Model: External Noise: For this model, it is assumed that the oscillator delivers a perfectly stable signal, to which relatively white Gaussian noise is added. If the oscillator signal is expressed as

$$e_{\text{osc}}(t) = A \cos \omega_0 t \tag{30}$$

the additive noise may be expressed in the form

$$n_e(t) = n_c(t) \cos \omega_0 t + n_q(t) \sin \omega_0 t \tag{31}$$

where $n_c(t)$ and $n_q(t)$ are white Gaussian processes. The sum of the oscillator signal and the additive noise is

$$e_{\text{out}}(t) = [A + n_c(t)] \cos \omega_0 t + n_q(t) \sin \omega_0 t. \tag{32}$$

Since we are dealing with very stable oscillators, it can be assumed that the rms value of the noise is much smaller than the rms value of the oscillator signal. Hence, only $n_q(t) \sin \omega_0 t$ influences the phase of $e_{\text{out}}(t)$ significantly and

$$e_{\text{out}}(t) \approx A[1 + n_c(t)/A] \cos \{\omega_0 t + [n_q(t)/A]\}. \tag{33}$$

Thus, a relatively weak externally added noise component results in envelope fluctuations, $n_c(t)/A$, and phase fluctuations, $\phi_e(t)$ given by

$$\phi_e(t) = n_q(t)/A. \tag{34}$$

If, over the frequency range of interest, the mean-square one-side spectral density of the added noise is N_0 volts^2/c/s, then the two-side spectral density of $n_q(t)$ will also be N_0 volts^2/c/s, and the phase perturbations of $e_{out}(t)$ will have a spectral density N_0/A^2 rad^2/c/s. The added noise will of course be modified by the RF filtering in the buffer amplifier. This filtering restricts the bandwidth of the phase perturbations so that their spectrum is confined to a frequency region extending up to one-half of the amplifier RF bandwidth. Thus, the mean-square spectral density $S_\phi(\omega)$ of the output signal phase fluctuations will be as illustrated in Fig. 4, where B_a c/s denotes one-half of the amplifier bandwidth.

Since the instantaneous frequency fluctuations are given by the time derivative of the phase fluctuations, the mean-square spectral density $S_{\dot\phi}(\omega)$ of the frequency fluctuations is given by

$$S_{\dot\phi}(\omega) = \omega^2 S_\phi(\omega). \tag{35}$$

In the present case

$$S_{\dot\phi}(\omega) = \omega^2 N_0/A^2, \qquad \text{for } |\omega| < 2\pi B_a. \tag{36}$$

If the RF amplifier frequency response beyond $\pm 2\pi B_a$ falls off faster than $1/\omega^2$, the mean-square spectral density of the frequency fluctuations will be as illustrated in Fig. 4.

It is clear that this type of noise in unstable oscillator outputs can be characterized by the following two gross parameters:

a) the relative noise density N_0/A^2,
b) the buffer amplifier bandwidth $2B_a$.

2) *Additive-Noise Model: Internal Noise:* In this model the additive noise enters into the oscillator loop as shown in Fig. 5. The noise $n_i(t)$ is assumed white in a frequency region around the oscillator frequency. The additive noise $n_i(t)$ can be expressed in the form

$$n_i(t) = V_n(t)\cos[\omega_0 t + \theta_n(t)] \tag{37}$$

where $V_n(t)$ and $\theta_n(t)$ are low-pass processes. The sum of the output of the limiter and the noise is

$$\begin{aligned} e_2(t) &= e_1(t) + n_i(t) \\ &= \{A + V_n(t)\cos[\phi(t) - \theta_n(t)]\}\cos[\omega_0 t + \phi(t)] \\ &\quad + \{V_n(t)\sin[\phi(t) - \theta_n(t)]\}\sin[\omega_0 t + \phi(t)]. \end{aligned} \tag{38}$$

With $V_n(t) \ll A$ almost all of the time, the instantaneous phase of $e_2(t)$ is approximately $[\omega_0 t + \phi(t) + \eta(t)]$, in which

$$\eta(t) \approx \frac{V_n(t)}{A}\sin[\phi(t) - \theta_n(t)]. \tag{39}$$

Thus the effect of the additive noise is to introduce a fluctuating phase shift into the loop. Recognizing this, we can replace the adder in Fig. 5 by an equivalent phase modulator as in Fig. 6.

When the phase shift $\eta(t)$ is equal to zero, the oscillator will oscillate at the frequency for which the phase shift through the resonator is zero (i.e., at its resonance frequency). However, when $\eta(t) \neq 0$, the introduction of the phase shift will change the oscillator frequency so that the phase shift in the resonator is equal to $-\eta(t)$ and the total phase shift around the loop is zero. Thus the variation of $\eta(t)$ will cause fluctuations of the oscillator frequency. In fact, Fig. 6 represents the block diagram of a familiar frequency-modulated oscillator. In this case, however, the control signal

$$V_n(t)\sin[\phi(t) - \theta_n(t)]$$

is dependent on the phase $\phi(t)$ of the oscillator signal.

The frequency-modulation sensitivity is determined by the phase slope (vs. frequency) of the resonator transfer function. Since the resulting frequency fluctuations will be small, only the part of the phase characteristic near the resonant frequency is of interest. In that region the phase characteristic can be assumed linear. For a resonator represented by a single tuned circuit with a half-power bandwidth of B_R c/s, the phase shift $\phi(t)$ caused by a small frequency variation around the resonant frequency is approximately

$$\phi(t) \approx \dot\phi(t)/\pi B_R. \tag{40}$$

Thus the phase condition for oscillations becomes

$$\eta(t) - (1/\pi B_R)\dot\phi(t) = 0 \tag{41}$$

whence

$$\dot\phi(t) = \pi B_R[V_n(t)/A]\sin[\phi(t) - \theta_n(t)]. \tag{42}$$

This is a nonlinear differential equation with random input that seems very hard to solve. However, some observations about the mean-square spectral density of $\dot\phi(t)$ may be made. If, for example, $\phi(t)$ was absent from the right-hand side of (42), then the mean-square spectral density of $\dot\phi(t)$ would be equal to the spectral density of $\pi B_R[V_n(t)/A]\sin\theta_n(t)$; i.e., it would be white with a mean-square spectral density equal to $(\pi B_R/A)^2 N_0$. The presence of $\phi(t)$ in the right-hand side will spread out the spectrum with a downward slope away from $\omega = 0$. Aside from the tendency for peaking at $\omega = 0$ [because of the spectral convolutions associated with the nonlinearities inherent in (42)], the general level of the spectral density of $\dot\phi(t)$ will be lower than $(\pi B_R/A)^2 N_0$. Since it seems reasonable to assume that this spectral density is approximately uniform over the frequency range of interest, we can express it as $\alpha(\pi B_R/A)^2 N_0$, where α is a positive constant smaller than unity.

As in the case of the external noise, it is clear that the amplifier following the oscillator will limit the spectrum of $\dot\phi(t)$ essentially to the frequencies up to B_a c/s. The appearance of the mean-square spectral density of the phase and frequency fluctuations due to an internal noise source is shown in Fig. 7. In the region where $S_{\dot\phi}(f)$ is uniform, the spectrum of the phase is given by

$$S_\phi(f) = S_{\dot\phi}(f)/4\pi^2 f^2.$$

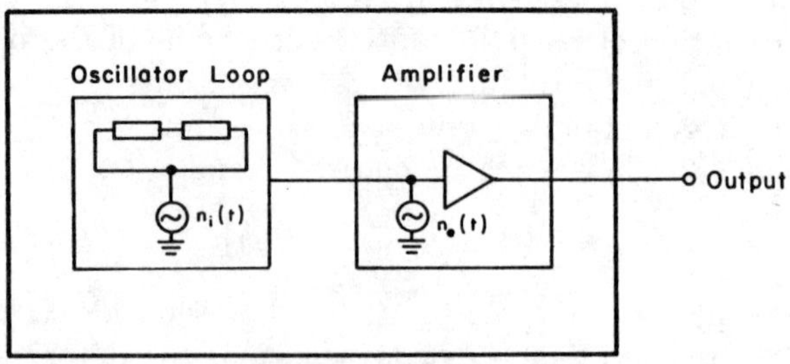

Fig. 3. Block illustration of an oscillator with a buffer amplifier, showing additive noise sources. $n_i(t)$ = internal noise source and $n_e(t)$ = external noise source.

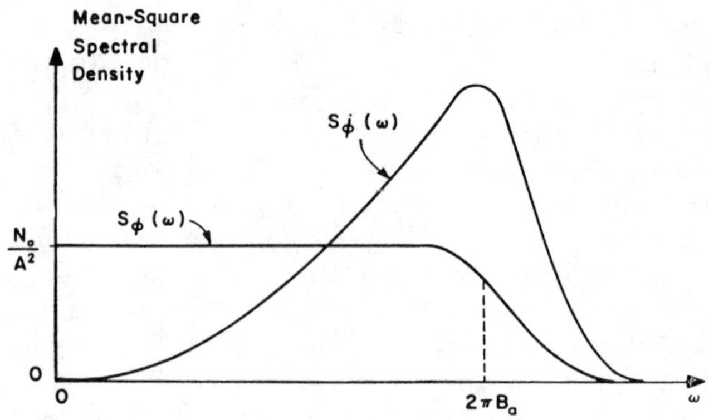

Fig. 4. Example of mean-square spectral density of phase fluctuations due to additive external noise.

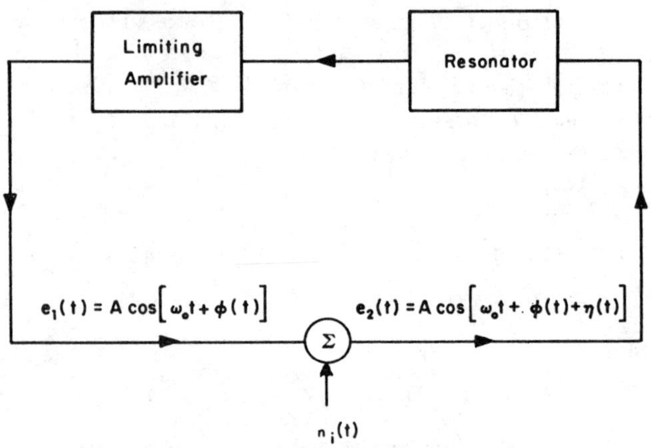

Fig. 5. Block diagram of oscillator loop with internal additive noise.

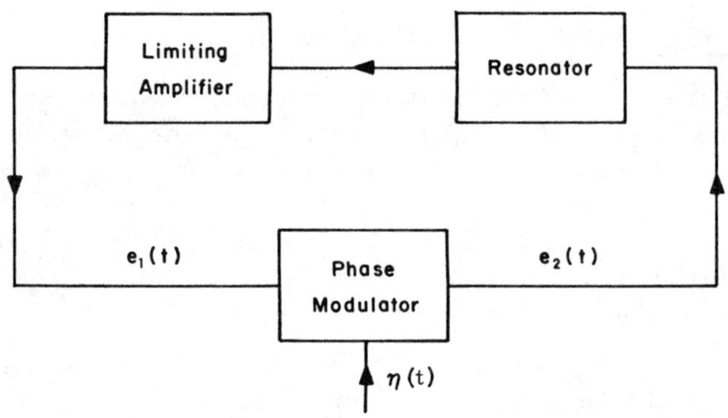

Fig. 6. Equivalent diagram of the oscillator in Fig. 5.

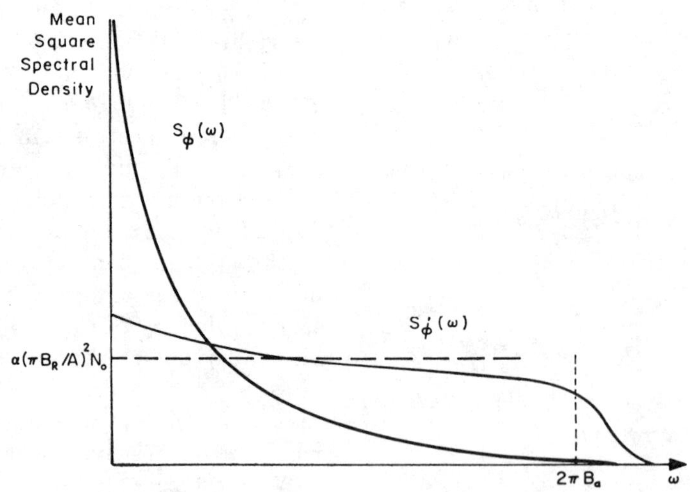

Fig. 7. Mean-square spectral densities of phase and frequency fluctuations due to internal noise source.

The foregoing type of oscillator instability can then be characterized by the following parameters:

a) The mean-square spectral density $\alpha(\pi B_R/A)^2 N_0$ of the oscillator frequency fluctuations.

b) The buffer amplifier bandwidth $2B_a$.

3) Multiplicative Noise Model: The frequency of an oscillator is always somewhat susceptible to variations in circuit parameters. An ultrastable oscillator is, of course, designed so that the sensitivity to the parameter variations is very small. Considerable effort is also directed to keeping the parameters as constant as possible. Examples of phenomena that can cause parameters variations in an oscillator are temperature variations, voltage supply variations, current variations, magnetic field variations, mechanical vibrations, etc. These low-frequency processes essentially frequency modulate the oscillator. Some of them (as for example voltage supply variations due to hum or periodic vibrations) are periodic functions and will thus yield a discrete mean-square spectral density for the oscillator frequency or phase fluctuations. The other processes usually have continuous spectral densities that decrease monotonically with frequency.

The most common of the foregoing processes are the low-frequency current fluctuations in the electronic tubes and transistors used in crystal oscillators. This type of fluctuation is called flicker noise. It has a spectral density of the form $S(\omega)=\beta/\omega$, where β is a constant. Since the energy of the flicker noise must be finite, this spectral density must ultimately become flat as $\omega \rightarrow 0$. Thus, the spectrum of the frequency fluctuations resulting from flicker noise can be characterized by the gross parameter β.

4) Loop-Controlled (APC and AFC) Oscillators [1]: There are two types of loop-controlled oscillators that are of wide interest, namely, automatic phase-controlled reference oscillators (APC) and automatic frequency-controlled reference oscillators (AFC). The performance of these systems in the presence of a strong reference signal is well known: the instantaneous frequency noise caused by relatively white input additive Gaussian noise is also Gaussian with a mean-square spectral density that rises as the square of frequency and is ultimately bounded by the filtering effect of the driving filter in cascade with the low-pass equivalent closed-loop system function.

It is of interest here to consider the situation in which the input reference signal, if any, is weak compared with the noise present.

a) *APC oscillator:* Consider the functional diagram of an APC loop shown in Fig. 8. Let

$$e_{\text{in}}(t) = [E_s + n_{c,\text{if}}(t)] \cos [\omega_{\text{if}} t + \psi(t)] - n_{q,\text{if}}(t) \sin [\omega_{\text{if}} t + \psi(t)] \tag{43}$$

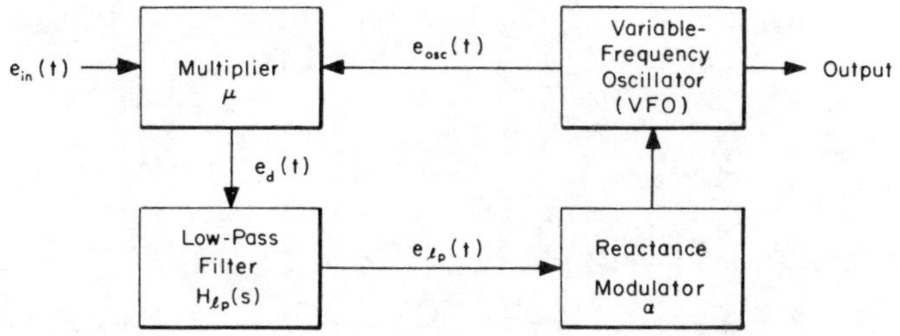

Fig. 8. Block diagram of APC oscillator.

in which

E_s = constant amplitude of input reference signal

$\psi(t)$ = instantaneous phase fluctuations of input reference signal,

and $n_{c,\text{if}}(t)$ and $n_{q,\text{if}}(t)$ are instantaneous amplitudes of cophasal and quadrature components of the input noise. Also let the oscillation be represented by

$$e_{\text{osc}}(t) = E_{\text{osc}} \cos [\omega_{\text{osc}} t + \phi_{\text{osc}}(t)]. \tag{44}$$

Under the assumption of stable feedback operation, and in the absence of spurious by-products in the output of the multiplier, we have, with $\omega_{\text{if}} = \omega_{\text{osc}}$,

$$\frac{1}{\mu\alpha} \dot{\phi}_{\text{osc}}(t) = \left\{ \left[1 + \frac{1}{E_s} n_{c,\text{if}}(t) \right] \sin [\psi(t) - \phi_{\text{osc}}(t)] + \frac{1}{E_s} n_{q,\text{if}}(t) \cos [\psi(t) - \phi_{\text{osc}}(t)] \right\} \otimes h_{lp}(t) \tag{45}$$

where $\otimes$ denotes convolution and $h_{lp}(t)$ is the impulse response of the low-pass filter. The existence of a feedback steady-state condition in which (45) gives an adeadequate description of loop performance requires that

$$| \psi(t) - \phi_{\text{osc}}(t) | \le \pi/2. \tag{46}$$

If a perfect band-pass limiter operates on the sum of signal and noise before it is applied to the APC loop, (45) is replaced by

$$\frac{1}{\mu\alpha} \dot{\phi}_{\text{osc}}(t) = h_{lp}(t) \otimes \sin [\psi(t) + \theta_{n,\text{if}}(t) - \phi_{\text{osc}}(t)] \tag{47}$$

where

$$\theta_{n,\text{if}}(t) = \tan^{-1} \frac{n_{q,\text{if}}(t)}{E_s + n_{c,\text{if}}(t)} \cdot \tag{48}$$

It is generally convenient to set

$$\phi_{\text{osc}}(t) = \phi_{\text{osc},s}(t) + \phi_{\text{osc},n,d}(t) \tag{49}$$

in which $\phi_{\text{osc},n,d}(t)$ combines all the distortion and noise products in the phase of the oscillation and $\phi_{\text{osc},n,d}(t)$ represents a pure signal term given by $\psi(t) \otimes h_{\text{eq}}(t)$, $h_{\text{eq}}(t)$ being the impulse response of the low-pass-linear equivalent strong-signal model of the closed-loop system.

When the input signal component is much weaker than the noise, (45) can be reduced to

$$\frac{1}{\mu\alpha}\dot{\phi}_{osc,n,d}(t) \approx h_{lp}(t) \otimes \left\{\frac{1}{E_s} V_{n,if}(t) \sin\left[\theta_{n,if}(t) - \phi_{osc,n,d}(t)\right]\right\} \quad (50)$$

where

$$\theta_{n,if}(t) \approx \tan^{-1} n_{q,if}(t)/n_{c,if}(t) \quad (51)$$

and

$$V_{n,if}(t) \approx \sqrt{n^2_{c,if}(t) + n^2_{q,if}(t)}. \quad (52)$$

Formally, (50) differs from (42) for the effect of an internal noise source only by the low-pass filtering effect $h_{lp}(t)$.

The introduction of a perfect band-pass limiter to operate upon the sum of signal and noise before the APC loop transforms (50) to

$$\frac{1}{\mu\alpha}\dot{\phi}_{osc,n,d}(t) \approx h_{lp}(t) \otimes \sin\left[\theta_{n,if}(t) - \phi_{osc,n,d}(t)\right]. \quad (53)$$

In order to understand the significance of (50) and (53), we first note that in this discussion we are concerned primarily with the situations in which either

$$\phi_{osc,n,d}(t) \text{ fluctuates much more slowly than } \theta_{n,if}(t) \quad (54)$$

or

$$\left|\phi_{osc,n,d}(t)\right| \ll \left|\theta_{n,if}(t)\right| \leq \pi/2. \quad (55)$$

These conditions correspond to an APC loop whose low-pass transmittance to the phase of the oscillation attenuates strongly all frequencies that are not much smaller than the bandwidth of the IF system preceding the loop.

Thus, under either of conditions (54) and (55), (50) and (53) can be approximated by

$$\frac{1}{\mu\alpha}\dot{\phi}_{osc,n,d}(t) \approx h_{lp}(t) \otimes \left[\frac{1}{E_s} x_{q,if}(t)\right] \quad (56)$$

and

$$\frac{1}{\mu\alpha}\dot{\phi}_{osc,n,d}(t) \approx h_{lp}(t) \otimes \sin\theta_{n,if}(t). \quad (57)$$

Integration yields

$$\phi_{osc,n,d}(t) \approx \frac{\mu\alpha}{E_s}\int^t h_{lp}(u) \otimes x_{q,if}(u)du \quad (58)$$

in the absence of prelimiting, and

$$\phi_{osc,n,d}(t) \approx \mu\alpha\int^t h_{lp}(u) \otimes \sin\theta_{n,if}(u)du \quad (59)$$

when the input signal plus noise is prelimited. Equations (58) and (59) show that irrespective of prelimiting or the lack of it, the APC loop treats the noise presented to it in an essentially linear manner with an equivalent linear filter system function given by

$$H_{ol}(s) = \mu\alpha H_{lp}(s)/s \quad (60)$$

as long as either of conditions (54) and (55) is satisfied. Note that $H_{ol}(s)$ is the "open-loop" transfer function of the APC loop.

Some remarks about the physical significance of the foregoing results are in order. First, note that $H_{lp}(s)/s$ has a pole at the origin as long as $H_{lp}(s)$ is not allowed to have a zero there. The effect of this pole is to cause $|H_{ol}(j\omega)|^2$ to weight a nonzero noise density around $\omega=0$ in a hyperbolic manner. Thus, even though $|H_{lp}(j\omega)/\omega|$ cuts off more rapidly than $|H_{lp}(j\omega)|$ at the higher frequencies, the behavior near $\omega=0$ does not allow the integral that yields the mean-squared value of the output noise to converge. The physical meaning of this is that, although the mean value of $\phi_{osc,n,d}(t)$ will be zero, the unboundedness of the mean square indicates unbounded excursions that require a re-examination of condition (55). Physically, the oscillation beats continuously with the input noise. Heavy filtering of the higher-frequency noise beats by $H_{lp}(s)/s$ causes $\phi_{osc,n,d}(t)$ to be rather slow compared with $\phi_{n,if}(t)$, thus allowing the noise modulation of the oscillator frequency to have an essentially Gaussian character and permitting the approximation of (50) by (56) and (53) by (57). The oscillator phase fluctuations wander out of step with the input noise fluctuations causing the loop to treat the noise essentially in an open-loop manner. If the input signal component is much weaker than the noise so that its contribution at the low-pass filter output is dominated by the noise, the loop will be unable to distinguish the signal from the equivalent of a weak noise component. No closed-loop control can therefore be established by the signal, and the oscillator frequency is modulated mainly by the Gaussian noise as filtered by the low-pass filter $h_{lp}(t)$. The effect of this modulation of the oscillator frequency is known to cause the oscillator phase to stray in random walk-like manner. But, even though the mean-square error in identifying the oscillator phase will diverge (directly, for random walk) with the length of the observation interval [because of the cumulative action of the integration in (58) and (59)], the mean-square error in identifying the oscillator frequency will go to zero (inversely) with increasing length of the observation interval. Comparison of (59) and (58) also shows that prelimiting increases the noise level in the output.

b) AFC oscillator: The functional diagram of an oscillator with an automatic frequency control loop is shown in Fig. 9. We assume that the closed-loop system can lock properly to the desired signal in the absence of noise and let $e_{in}(t)$ and $e_{osc}(t)$ be described by (43) and (44), respectively. The AFC loop is assumed to be driven with no prior amplitude limiting of signal plus noise. Under these conditions, we further assume

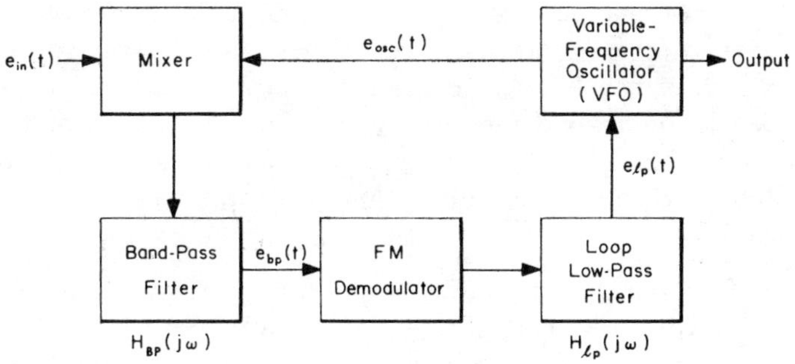

Fig. 9. Block diagram of AFC oscillator.

that the mixer delivers in the nominal passband of the loop band-pass filter only the beat components

$$[E_s + c_{c,\text{if}}(t)] \cos [\omega_{bp}t + \psi(t) - \phi_{\text{osc}}(t)] - x_{q,\text{if}}(t) \sin [\omega_{bp}t + \psi(t) - \phi_{\text{osc}}(t)]$$

in which $\omega_{bp} \equiv \omega_{\text{if}} - \omega_{\text{osc}}$ = nominal center frequency of the band-pass filter within the loop. If an amplitude-insensitive linear FM demodulator and a linear feedback from the discriminator output to the instantaneous frequency of the VFO are assumed, we can show that

$$\phi_{\text{osc}}(t) = k_d\beta_{fb}\left[\tan^{-1}\frac{A_q(t) \otimes h_{LP}(t)}{A_c(t) \otimes h_{LP}(t)}\right] \otimes h_{lp}(t) \tag{61}$$

where

$$A_c(t) \equiv \left[1 + \frac{1}{E_s}x_{c,\text{if}}(t)\right] \cos [\psi(t) - \phi_{\text{osc}}(t)] - \frac{1}{E_s}x_{q,\text{if}}(t) \sin [\psi(t) - \phi_{\text{osc}}(t)]$$

$$A_q(t) \equiv \left[1 + \frac{1}{E_s}x_{c,\text{if}}(t)\right] \sin [\psi(t) - \phi_{\text{osc}}(t)] + \frac{1}{E_s}x_{q,\text{if}}(t) \cos [\psi(t) - \phi_{\text{osc}}(t)]$$

and $h_{LP}(t)$ is the impulse response of the low-pass analog of the band-pass filter.

Now let the input reference signal be much weaker than the noise. Two cases are considered. In the first, we assume that $|\psi(t) - \phi_{\text{osc}}(t)| \ll 1$. We then have from (61)

$$\phi_{\text{osc}}(t) \approx k_d\beta_{fb}h_{lp}(t) \otimes \tan^{-1}\frac{\{n_{q,\text{if}}(t) + n_{c,\text{if}}(t)[\psi(t) - \phi_{\text{osc}}(t)]\} \otimes h_{LP}(t)}{\{n_{c,\text{if}}(t) - n_{q,\text{if}}(t)[\psi(t) - \phi_{\text{osc}}(t)]\} \otimes h_{LP}(t)}. \tag{62}$$

If we introduce some additional assumptions, this expression can be simplified still further. For example, if the loop low-pass filter bandwidth is relatively "sharply" restricted to a value below one-half of the band-pass filter bandwidth and if $k_d\beta_{fb}$ is not high enough to broaden the bandwidth of $\phi_{\text{osc}}(t)$ above one-half of the band-pass filter bandwidth, then

$$\begin{aligned}\phi_{\text{osc}}(t) &\approx k_d\beta_{fb}h_{lp}(t) \otimes \tan^{-1}\left\{\psi(t) - \phi_{\text{osc}}(t) + \frac{n_{q,\text{if}}(t) \otimes h_{LP}(t)}{n_{c,\text{if}}(t) \otimes h_{LP}(t)}\right\} \\ &\approx k_d\beta_{fb}h_{lp}(t) \otimes [\psi(t) - \phi_{\text{osc}}(t)] + k_d\beta_{fb}h_{lp}(t) \otimes \phi_{n,LP}(t)\end{aligned}$$

where

$$\phi_{n,LP}(t) = \tan^{-1}\frac{n_{q,\text{if}}(t) \otimes h_{LP}(t)}{n_{c,\text{if}}(t) \otimes h_{LP}(t)}.$$

Rearrangement of terms yields (with $u_0(t)$ = unit impulse)

$$[u_0(t) + k_d\beta_{fb}h_{lp}(t)] \otimes \phi_{\text{osc}}(t) \approx k_d\beta_{fb}h_{lp}(t) \otimes [\psi(t) + \phi_{n,LP}(t)]$$

whence

$$\phi_{\text{osc}}(t) = h_{\text{eq}'}(t) \otimes [\psi(t) + \phi_{n,LP}(t)] \tag{63}$$

where $h_{\text{eq}'}(t)$ is the impulse response of a linear filter with transmittance

$$H_{\text{eq}'}(s) = \frac{k_d\beta_{fb}H_{lp}(s)}{1 + k_d\beta_{fb}H_{lp}(s)}. \tag{64}$$

This system function differs from the more complete closed-loop system function associated with strong-signal models of AFC systems and frequency-compressive feedback demodulators only in that the transmission effects of the loop band-pass filter upon the frequency fluctuations of the input reference signal have been neglected [1].

If instead of requiring $|\psi(t) - \phi_{\text{osc}}(t)| \ll 1$, we express $\phi_{\text{osc}}(t)$ as a sum [see (49)] of a pure signal-modulation component $\psi_{\text{osc},s}(t)$, and noise and distortion component $\phi_{\text{osc},n,d}(t)$ and require only that

$$|\psi(t) - \phi_{\text{osc},s}(t)|_{\max} \ll 1$$

then (61), with the noise assumed much stronger than the signal, leads to

$$\phi_{\text{osc},s}(t) \approx \psi(t) \otimes h_{\text{eq}'}(t)$$

and

$$\phi_{\text{osc},n,d}(t) \approx k_d\beta_{fb}h_{lp}(t) \otimes \tan^{-1}\frac{h_{LP}(t) \otimes \sin [\phi_{n,\text{if}}(t) - \phi_{\text{osc},n,d}(t)]}{h_{LP}(t) \otimes \cos [\phi_{n,\text{if}}(t) - \phi_{\text{osc},n,d}(t)]} \tag{65}$$

where

$$\phi_{n,\text{if}}(t) \equiv \tan^{-1}\frac{n_{q,\text{if}}(t)}{n_{c,\text{if}}(t)}.$$

In each case, the phase fluctuations of the weak input phase reference are transmitted, filtered by $h_{\text{eq}'}(t)$, to the controlled oscillator phase without compression by the noise. However, in practice the FM noise impulses and the dips of the noise envelope below the drive threshold of the FM demodulator within the loop will prevent this FM demodulator from operating in the "amplitude-insensitive linear" manner assumed in the analysis. Moreover, the various factors that contribute to improper AFC tracking of signal plus noise will all be in evidence below the AFC noise threshold. Consequently, the comparatively severe disruptive noise effects may well completely mask the frequency fluctuations contributed by the input reference signal.

B. Black-Box Approach

In the black-box approach to the modeling of unstable oscillations, the source output is represented in various frequency ranges by signal models whose properties fit the properties of the source output over the specified portions of the spectrum. This approach is useful because it provides an alternate route to the characterization of source outputs with emphasis on the user needs. It also adds further insight into the mechanisms of instability in the signal structure, provides representations that are particularly suitable for analysis in subsequent parts of a system, bypasses the difficulties of dealing with nonlinear differential equations by assuming applicable signal representations or forms, and is naturally suited to the analysis of problems of spectral purity.

A common way to explain how an oscillation comes about in a properly designed loop without visible "provocation" is to rationalize that unavoidable random current or voltage fluctuations in various parts of the loop provide an excitation in a feedback loop with sharp selectivity (determined principally by the phase shift of a high-Q resonator) about the frequency of in-phase feedback. The regeneration around the loop builds up the feeble provocation to a strong signal whose amplitude is limited ultimately by automatic nonlinear gain control action or by actual saturation. This, together with the fact that the regenerative loop selectivity is of nonzero (albeit small) width about the frequency of in-phase feedback, suggests that the resulting signal may actually consist of a band of Gaussian-like noise with or without a distinct sinewave at the center of the band. In the immediate vicinity of the oscillation frequency, the oscillation signal may therefore be modeled by *a band of amplitude-limited Gaussian noise*.

Other models for representing the signal over various frequency intervals are:

1) sinewave plus Gaussian noise,
2) constant-amplitude carrier frequency-modulated by narrow-band Gaussian noise,
3) sinewave plus one or more discrete-frequency components, and
4) combinations of the foregoing.

Applicability of these various models depends upon the origin of the signal. For example, if the signal is a processed form of an originally extremely stable source, then frequency processing adds noise with a bandwidth extending over at least ten times the original source width of band. Therefore, sinewave plus Gaussian noise would be a good model. If, however, the width of the source band is not too narrow, relative to bandwidths of later processing circuits, then the appropriate model may be

1) a band of noise, if passive filters only are used for tone isolation;
2) a constant-amplitude signal frequency modulated by noise if a phase locked loop is used for "tone isolation."

We now discuss some of the foregoing models to bring out their characteristic properties.

1) *Amplitude-Limited Narrow-Band Gaussian Noise:* If the center frequency is denoted ω_0 rad/s, band-pass amplitude-limited narrow-band Gaussian noise may be expressed as

$$e_l(t) = A_l \cos [\omega_0 t + \eta(t)] \tag{66}$$

where $\eta(t)$ is the instantaneous phase of a Gaussian noise process and hence has a uniform distribution over the range $|\eta| \leq \pi$. The mean-square spectral density, $S_{\dot{\eta}}(\omega)$, of $\dot{\eta}(t)$ has been derived in a number of publications [2]–[5] for rectangular as well as for Gaussian shapes of the spectrum before limiting. Illustrative curves are sketched in Fig. 10(a) and (b) ($\rho = 0$). It is interesting to observe the $1/f$ shape of $S_{\dot{\eta}}(\omega)$ over a narrow frequency range surrounding $f=0$ and extending principally over a few times the bandwidth of the prelimited process (defined by the resonator in the oscillator circuit).

The peak factor, $(\text{p.f.})_\eta$, of $\eta(t)$ is readily shown to be $\sqrt{3}$ whether η is considered to range over $|\eta| \leq \pi$ or over $0 \leq \eta \leq 2\pi$.

The RF spectrum of this type of signal has also been studied [2], [6] extensively and is illustrated in Fig. 11.

2) *Sinewave Plus Gaussian Noise:* The properties of the resultant of sinewave plus Gaussian noise are well known [3], [7]. Of principal interest here are the following facts. Under conditions of very high ratio of the mean-square value of the sinewave to the mean-square value of the Gaussian noise,

1) the statistics of the phase and frequency fluctuations of the resultant signal are practically Gaussian, and
2) the mean-square spectral density of the phase fluctuations practically differs only by a constant multiplier from the sum of the positive-frequency half and the negative-frequency half of the spectral density of the input noise, after both halves have been shifted down to $\omega = 0$.

If the amplitude-limited resultant of the sinewave plus the Gaussian component is expressed as in (66) and

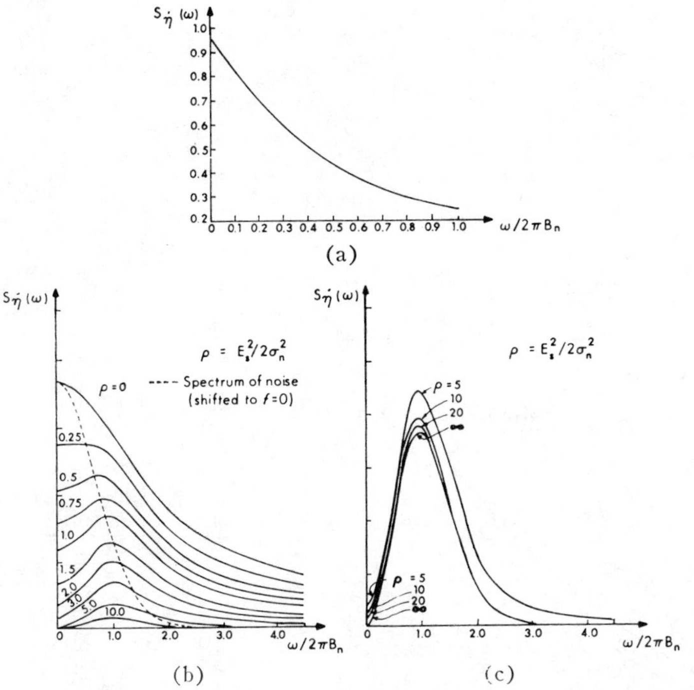

Fig. 10. (a) Mean-square spectral density of instantaneous frequency of narrow-band Gaussian noise with a rectangular input spectrum, (adapted from [2]). (b) Mean-square spectral density of the frequency fluctuations, $\dot{\eta}(t)$, of the resultant of sinewave plus Gaussian noise for different relative sinewave powers (adapted from [5]). (c) The same as (a) but for strong sinewave (note change in scale). B_n is equal to half of the noise bandwidth before limiting, (adapted from [5]).

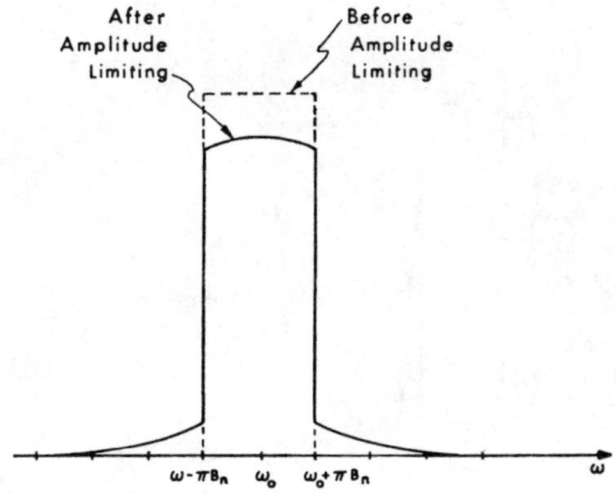

Fig. 11. Shape of RF spectral density of band-pass limited Gaussian noise for rectangular prelimiting noise spectrum (adapted from [6]).

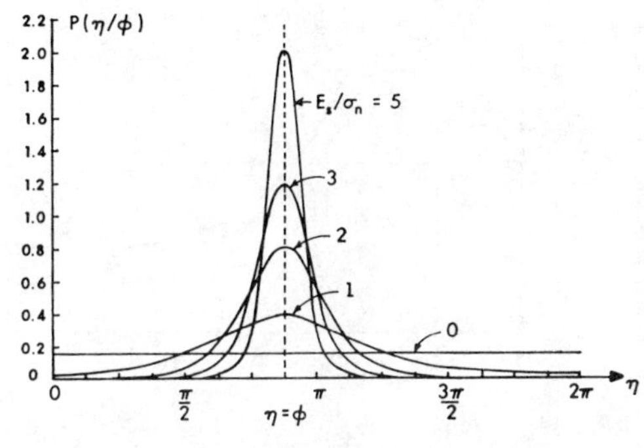

Fig. 12. Probability density distribution of the total phase η, conditioned on the phase ϕ of the sinewave, for various strengths of the sinewave (adapted from [5]).

$e_l(t)$ is used to represent the output of an oscillator, the "initial" phase ϕ of the sinewave must be assumed to be uniformly distributed over $-\pi \leq \phi \leq \pi$, since all values of ϕ are equally likely. Then the unconditional probability density function $p(\eta)$ of the total phase will therefor also be uniform. However, the conditional probability density $P(\eta/\phi)$ of the total phase $\eta(t)$, given the phase ϕ of the sinewave, is shown in Fig. 12 for different relative strengths of the sinewave.

The mean-square spectral density $S_{\dot{\eta}}(\omega)$ of $\dot{\eta}(t)$ is shown in Figs. 10(b) and 10(c) for the case in which the noise spectral density before limiting has a Gaussian shape. The shape of the input noise spectrum about the center frequency ω_0 is superimposed as a dashed curve on the low-frequency curves in Fig. 10(b).

It is interesting to note from Fig. 10(b) that, outside of the frequency range defined for each solid curve by the intersection with the dashed curve, the solid curve under consideration shows a $1/\omega$ dropoff with ω, whereas inside of this range the solid curve either flattens out or actually bends down as $\omega \to 0$. The mean-square spectral density will always have a nonzero value for $\omega = 0$ because of the impulsive component present in the instantaneous frequency of the resultant. (Only in the case of an infinitely strong sinewave is the spectral density strictly zero for zero frequency.) In view of the extremely narrow postdetection processing bandwidths normally encountered in applications of ultrastable oscillations, no distinction between the impulsive component and the smooth component of the instantaneous frequency need be made.

For a nonzero spectral density for $\dot{\eta}(t)$ at $\omega = 0$, the mean-square spectral density of $\eta(t)$ is infinite at $\omega = 0$. This follows from the fact that the spectrum of the phase is obtained by dividing the frequency modulation spectrum by ω^2. The infinite spectral density for $\omega = 0$ results in an infinite rms value of the phase in any frequency region that includes $\omega = 0$. These properties are characteristic of a so-called *random-walk* process.

C. Combined Effect of Various Mechanisms

The various mechanisms of oscillator frequency instability modeled in the preceding sections may all be expected to be present in varying degrees in practical oscillators. The net effect of the combined mechanisms on the spectral density of the resulting instantaneous frequency fluctuations is illustrated in Fig. 13.

Of the various parts of the overall curve in Fig. 13, the part contained within $0 \leq \omega \leq 2\pi B_{osc}$ merits further explanation. By definition, B_{osc} is a measure of the "width" of the oscillation band and is determined by the shape of the nominally "infinite-Q" selectivity curve

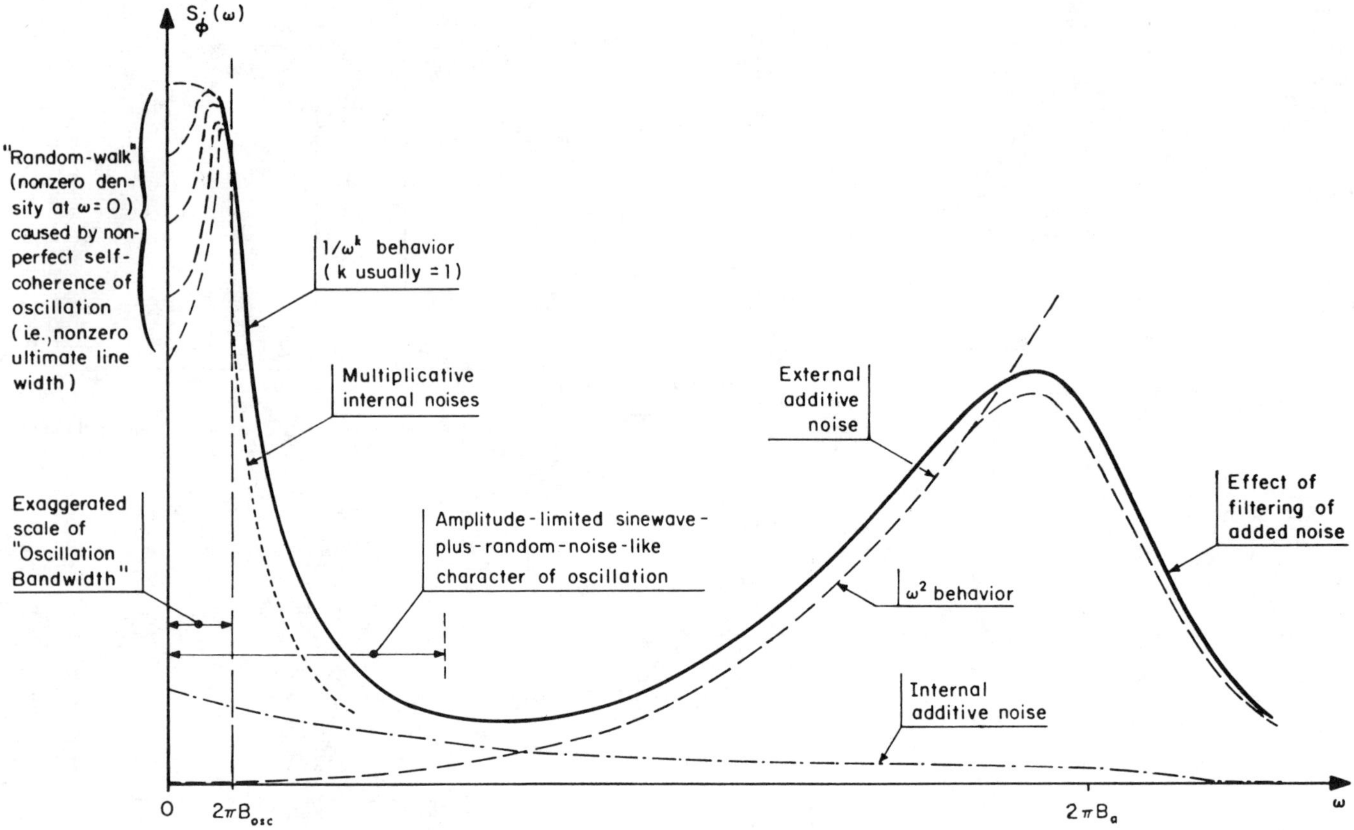

Fig. 13. Spectral density (solid curve) of instantaneous frequency fluctuations resulting from combination of various instability mechanisms. (Arbitrarily chosen relative levels for illustration only.)

for feedback around the oscillating loop. In an ultrastable oscillator, B_{osc} will be an extremely small fraction of a c/s, perhaps corresponding to periods of hours, days, or longer. The scale of B_{osc} in Fig. 13 is greatly exaggerated in order to illustrate a *guess* about what the heretofore unmeasured part of the spectral density $S_{\dot\phi}(\omega)$ might look like. The suggested extensions of the curve of $S_{\dot\phi}(\omega)$ toward $\omega=0$ are based on a model of the oscillation as a sinewave plus a band of Gaussian noise arising from the various irregularities of the electronic phenomena associated with the loop components and shaped spectrally by the selectivity of the feedback under conditions of sustained oscillation. Two significant consequences of such a physically motivated model are:

1) The ultimate "line width" of the oscillation, as gauged by B_{osc}, is nonzero. This causes $S_{\dot\phi}(\omega)$ to have a *nonzero* value at $\omega=0$, thus accounting for the inherent random-walk phenomenon in unlocked oscillators. $S_{\dot\phi}(0)=0$ (and hence the random-walk effect is absent) only if the oscillation is perfectly self-coherent, i.e., its ultimate line width, B_{osc}, is zero.

2) The extremely slow fluctuations of the extremely narrow band of noise representing the ultimate oscillation signal are practically indistinguishable from the slow multiplicative noise phenomena (flicker noise, etc.). If one were to filter out the flicker and related noise effects within the loop by some high-pass filtering action, hyperbolic behavior of the spectral density of the frequency fluctuations immediately beyond $\omega=2\pi B_{osc}$ would persist.

IV. Measurement of Oscillator Instabilities

Any theoretical characterization of oscillator instabilities is of little practical value if some effective means cannot be devised to measure the pertinent parameters or properties brought out by the characterization. In this section, we examine the measurement problem for the purpose of developing satisfactory means for measuring the principal characteristics of unstable oscillations. The importance of the RF spectrum of the entire oscillator signal and of the mean-square spectral density of only the frequency fluctuations of the oscillator signal for different applications has been discussed in previous sections. Some general aspects of RF spectrum measurement are presented in this section. However, in view of the fact that frequency instability is the subject of paramount interest in this paper, major attention is directed to the measurement of the spectral densities $S_\phi(\omega)$ and $S_{\dot\phi}(\omega)$ of the instantaneous phase and frequency fluctuations. We shall identify the features that should be exhibited by techniques for measuring these spectral densities and then outline the form and discuss the particular technique that is most suitable. We shall compare this technique to that most commonly advocated and employed by others. Finally, we shall point out the general limitations of all approaches to the measurement problem that exist at the present time.

A. Representation of Oscillator Signal

In general, a perturbed oscillation signal may be expressed in the form

$$e(t) = A(t) \cos [2\pi \bar{f} t + \phi(t)] \tag{67}$$

where $A(t)$ is the amplitude of the signal, $\bar{f}$ is the constant *long-time average* frequency, and $\phi(t)$ embodies the instantaneous phase fluctuations with zero *long-time* mean. It is important to observe the emphasis made here on *long-time averages*. In Section II, we treated a typical "one-shot" measurement in specified applications and emphasized that the error due to instability would, strictly speaking, result only from changes in the continuing oscillator signal that would cumulate during the time separation between the two compared samples of the oscillation. For that reason, we employed a representation of the oscillation phase in which $\phi(t) = 0$ at the time of the first sample; namely

$$2\pi f_0 t + \phi(t)$$

where f_0 = instantaneous frequency at the time of the first sample. The difference between the two representations is philosophically important, although it may turn out to be of no practical significance in a number of applications. Thus, if for the time being the $\phi(t)$ associated with f_0 is denoted $\phi_0(t)$ and that with $\bar{f}$ is denoted $\phi_-(t)$, then over a very long time interval (assuming the instability process to be at least wide-sense stationary)

$$\bar{f} - f_0 = \overline{\dot{\phi}_0(t)} \tag{68}$$

sine $\overline{\dot{\phi}_-(t)} = 0$ and $\overline{\bar{f}} = \bar{f}$, where the overbar denotes "*long-time* average." This suggests that f_0 may not be convenient to work with for general-purpose measurement of the characteristics of oscillator frequency instability, because it is too tied down to the timing of the first sample in a specific one-spot application, and it may be expected to vary randomly from "one-shot" to "one-shot" in a sequence of application measurements. Consequently, for the purpose of general-purpose measurements of oscillator frequency instability characteristics, the representation

$$2\pi \bar{f} + \phi(t)$$

is preferable. The effect of the use of such general-purpose data in the computations of a specific application that entails a sequence of individual one-shot comparisons of time-spaced samples of oscillator phase or frequency is a problem in the well-developed field of statistical sampling and therefore will not be considered here.

B. RF Spectrum Measurement

Direct spectral analysis is the most obvious way of measuring the RF spectrum of an oscillator signal. This technique has instrumentation problems which are treated extensively in the literature. In order to avoid these difficulties, it is frequently supposed that the RF spectrum can be determined in a straightforward manner from the spectrum of the instantaneous frequency fluctuations, $S_{\dot{\phi}}(\omega)$. The supporting argument is based on the assumption that the phase fluctuations caused by the frequency instabilities are sufficiently small so that only the first-order RF sidebands will be significant. We shall now show that the preceding assumption about the effects of the instabilities is untenable, and hence that the RF spectrum is *not* related in the usually assumed manner to $S_{\dot{\phi}}(\omega)$.

Starting with the representation in (67), we assume that the AM effects are negligible. Trigonometric expansion yields

$$e(t) = A[\cos \phi(t) \cos 2\pi \bar{f} t - \sin \phi(t) \sin 2\pi \bar{f} t]. \tag{69}$$

If it can be assumed that

$$\overline{\phi^2(t)} \ll 1 \tag{70}$$

then we can write

$$e(t) \approx A[\cos 2\pi \bar{f} t - \phi(t) \sin 2\pi \bar{f} t] \tag{71}$$

which represents the so-called low-modulation-index approximation to $e(t)$. This shows that the spectrum of $e(t)$ would consist of a spectral line at $\omega = 2\pi \bar{f}$, plus the spectrum of $\phi(t)$ translated to this carrier frequency.

It is well known, however, that $\phi(t)$ always exhibits the characteristics of a random walk, so that

$$\lim_{\omega \to 0} S_\phi(\omega) \to \infty \tag{72}$$

and the rms value of $\phi(t)$ is unbounded. Thus, condition (70) never holds, and the phase fluctuations of the carrier cannot be assumed to be sufficiently small so that approximation (71) is valid. Severe spectral spreading results from the highly nonlinear modulation characteristics indicated in (69). Effects of AM, which we have thus far neglected, may also be present. Hence, a simple, unique relationship between the spectrum of the frequency (or phase) fluctuations of an oscillator signal and the RF spectrum of the complete signal cannot be established.

C. Fundamental Properties of Satisfactory Measurement Systems for $S_\phi(\omega)$ and $S_{\dot{\phi}}(\omega)$

The first requirement for measuring $S_\phi(\omega)$ and $S_{\dot{\phi}}(\omega)$ is to extract $\phi(t)$ and $\dot{\phi}(t)$ for direct examination, free of the effects of $2\pi \bar{f} t$ and $A(t)$. Other desired characteristics of the measurement technique are sensitivity and accuracy, sensitivity being the more important because it bears on the observability of the fluctuations sought. Finally, provisions for monitoring the measurement errors are desirable.

We can evolve the basic structure of the measurement system by considering how to realize the preceding properties. By forming the product of two oscillator signals we can obtain a signal $e_d(t)$ such that

$$e_d(t) = k \cos [2\pi(\bar{f}_1 - \bar{f}_2)t + \phi_1(t) - \phi_2(t)] \tag{73}$$

where the subscripts refer to the individual oscillators, at the sacrifice of including the instability of two oscillator signals in each measurement. Providing that the statistical independence of all oscillator signals is assured, two-at-a-time measurements upon three members of a given class of oscillators are sufficient to determine the deviation spectra of each. In the limit as $\bar{f}_1 - \bar{f}_2$ approaches zero, the product operation becomes synchronous detection. If the inputs to the detector are in quadrature, (73) becomes

$$e_d(t) = k \sin [\phi_1(t) - \phi_2(t)] \approx k\Delta\phi(t) \tag{74}$$

where

$$\Delta\phi(t) = \phi_1(t) - \phi_2(t) \tag{75}$$

providing $\Delta\phi(t)$ is sufficiently small. The resulting signal can be processed to yield the power spectrum of $\Delta\phi(t)$, $S_{\Delta\phi}(\omega)$, by several means. Differentiation of $e_d(t)$ results in a signal that can be processed for $S_{\Delta\dot{\phi}}(\omega)$. Because of the expected shapes of instability spectra for a highly stable oscillator, $S_{\Delta\dot{\phi}}(\omega)$ should be measured for spectral regions of relatively small ω, and $S_{\Delta\phi}(\omega)$ should be measured for relatively large ω.

The assurance of synchronous detection of quadrature signals establishes a self-error evaluation property for the system. If *one* signal is introduced at both inputs (except that one input is shifted 90° in phase) to the detector, the power spectrum of the resulting output signal should be uniformly zero. A residual spectrum produced by the measurement circuitry will result, however, from the noise and distortion that are not completely cross-correlated at the inputs to the detector. Evaluation of this residual spectrum establishes the measurement error as well as the effective measurement sensitivity of the system. Inasmuch as the introduction of one signal as an input to both signal channels of the system can be accomplished at any point before the detector, the residual spectrum contributed by predetection gain circuits such as amplifiers and multipliers, as well as frequency synthesizers that convert atomic resonance frequencies to standard outputs, can be measured in this way. Thus, we can provide additional gain, and perhaps sensitivity, by incorporating multiplication techniques before the detector.

Since both oscillator signals must be independent, the inputs to the product device are not guaranteed to be coherent or in quadrature. However, a system that ensures these necessary relationships without affecting the independence of the signals has been conceived and analyzed by ADCOM, Inc.

The limitations of this type of system are the state of the art in circuitry required for its implementation and the general problem of spectral analysis of real-time data.

It is our opinion that this spectrum measuring technique does not suffer in comparison to any others. In some cases, it displays significant advantages. We will compare it with the period counting technique as an example of other methods for measuring the frequency instabilities of oscillators.

D. Period Counting Techniques

We will consider the common technique of period measurement with a counter, although our arguments will pertain to the whole class of measurements that incorporate a moving-average calculation.

A standard period counting test setup is shown in Fig. 14.

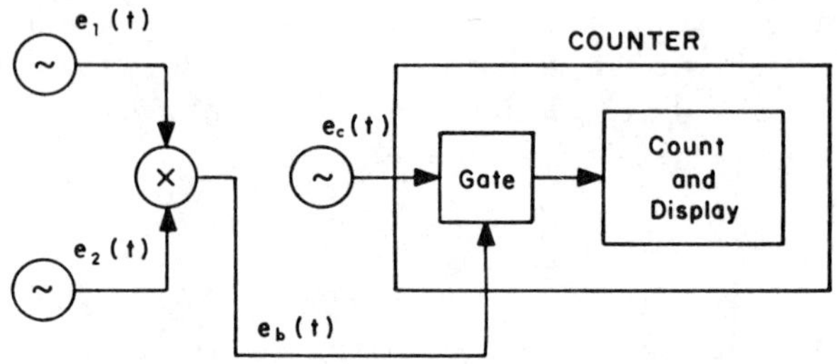

Fig. 14. A standard period counting test setup.

Two independent oscillator signals, $e_1(t)$ and $e_2(t)$, are mixed to produce a beat signal, $e_b(t)$, whose average frequency is much lower than that of the input signals. In this way, much of the average value contamination will be removed. $e_b(t)$ is introduced as a triggering signal to a gate in the counter. At one zero-crossing of $e_b(t)$, the gate opens to pass the high-frequency-counter oscillator signal, $e_c(t)$, which is "accumulated," or counted in the counter register. After an integral number of periods of $e_b(t)$, the gate is closed and the total count read. As the period of $e_c(t)$ will be a simple decimal fraction of a second, say 0.1 μs, the counter will display the total number of 0.1 μs that equal the integral number of $e_b(t)$ periods. We define the length of time the gate is open as τ. Usually several consecutive measurements of length τ are made, and the resulting data reduced to yield the average value and the mean-square deviation of the accumulated periods. The rms value of the deviation divided by the average of the period accumulated in a time τ is called $I(\tau)$, the frequency instability in a time τ of an oscillator signal. The assumption here is that the normalized rms deviation of frequency is essentially equal to the normalized rms deviation of period, which is valid for highly stable oscillator signals.

There is existing literature describing the sensitivity of measurement of this technique in terms of the imperfections of counters [8]. Here we are interested in comparing this technique with the spectral measurement we advocate.

In the first place, the two techniques measure different quantities, so we shall start by relating these quantities. Let us assume a perfect counter with such a high internal oscillator frequency that we can consider the period accumulation to be a continuous time process. In addition, let us assume that the display has so many digits that we can resolve the deviation in period by measuring one oscillator signal directly, instead of the beat between two oscillators. Let us represent the oscillator signal as in (67). Then we may define the instantaneous frequency of the oscillator signal, $f(t)$, as

$$f(t) = \bar{f} + \delta f(t) \tag{76}$$

where $\bar{f}$ is the average value and $\delta f(t)$ the deviation from it. We can define the instantaneous period of the signal as $T(t)$, such that

$$T(t) = \bar{T} + \delta T(t) \tag{77}$$

and by definition,

$$T(t) = \frac{1}{f(t)} \cdot \tag{78}$$

We can show that

$$N(\tau, t) = \int_t^{t+\tau} T(\mu) d\mu \tag{79}$$

where $N(\tau, t)$ is the accumulation in the display for a given count of length τ. Following the same procedure as in Section II, we can show that, as a consequence of (79),

$$I(\tau) = \frac{1}{2\pi \bar{f}} \sqrt{\frac{1}{2\pi} \int_{-\infty}^{+\infty} S_{\dot\phi}(\omega) \frac{\sin^2(\omega\tau/2)}{(\omega\tau/2)^2} d\omega}. \tag{80}$$

Having established the relationship between the two measured quantities, we are in a position to make some comparisons.

1) *$S_{\dot\phi}(\omega)$ is a more fundamental parameter than $I(\tau)$.* Given $S_{\dot\phi}(\omega)$, (80) shows that $I(\tau)$ can easily be calculated from it. Reversing the process would indeed be difficult. Only after we have specified a form of the spectrum with a small number of parameters, such as

$$S_{\dot\phi}(\omega) = \frac{a}{\omega} + b + c\omega^2, \tag{81}$$

can we obtain spectral information from $I(\tau)$. Then by solving (80) we could obtain $I(\tau)$ in terms of complicated functions of a, b, c, and ω. The three chosen parameters could be evaluated through a high-order least error curve fit to $I(\tau)$. Even so, this is not a direct calculation of $S_{\dot\phi}(\omega)$ from $I(\tau)$, inasmuch as $S_{\dot\phi}(\omega)$ has been forced to conform to a preconceived notion whose validity may be questioned. For evaluation of a variety of applications from one basic test, we must compute $S_{\dot\phi}(\omega)$.

2) *$I(\tau)$ may well be ambiguous* in that there may be an additional degree of freedom inherent in $I(\tau)$ which is not a property of the oscillator.

To illustrate this second point, let us consider a simplified spectrum of $S_{\dot\phi}(\omega) = c\omega^2$. Then (80) reduces to the square-root of the area under the curve shown in Fig. 15.

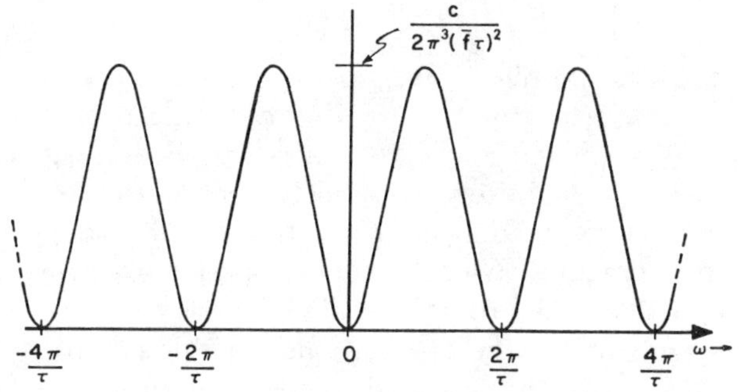

Fig. 15. The instability integral associated with a parabolic spectral density for the frequency fluctuations.

The area under the curve is infinite. But in a real situation, there will be some filtering in a measurement system that will band-limit $S_{\dot\phi}(\omega)$. Let us assume that this filtering is approximately rectangular with a rather wide bandwidth, such that

$$\omega_a \gg \frac{2\pi}{\tau}$$

where ω_a is the cutoff radian frequency of the filter. In this case, the curve shown in Fig. 15 will go through many cycles as ω approaches ω_a. The area will then be approximately the average value of the curve multiplied by $2\omega_a$. Thus

$$I(\tau) \approx \frac{1}{2\pi \bar{f} \tau} \sqrt{\frac{2c\omega_a}{\pi}} \cdot \tag{82}$$

This result shows a familiar $1/\tau$ shape, but the presence of ω_a makes (82) unspecific about c. Thus any description of $I(\tau)$ must contain the effective filtering characteristic as additional information in the data. In general, we cannot expect $I(\tau)$ to remain unchanged if the bandwidth of the measuring system is changed. Thus a direct $I(\tau)$ measurement may not provide a specification of the oscillator alone and is generally unacceptable as a general-purpose specification of the oscillator frequency instability.

Others have shown that the moving average or "rectangular time window" which describes the period counting technique is an inefficient and relatively unstable operation for determining the power spectrum of a process [9].

3) *The period counting technique is not self-checking.* In Section IV-A we showed that the determination of the residual spectrum of the test equipment requires the introduction of two identical, completely cross-correlated input signals to the equipment. It is obvious that this is useless in the

case of a period measuring system such as illustrated in Fig. 15, because the system requires the presence of a beat frequency output from the mixer. (A modified period-counting technique that avoids this difficulty has been described by Cutler [10].)

In some applications, operations that are indistinguishable from the basic period counting operation are actually carried out. For these applications, period counting is definitely a useful testing method provided that it is designed as an accurate imitation of the eventual operations. More generally, period-counting tests may be designed to imitate a wide variety of *spot measurements* in specific applications. In this regard, period counting can be exploited as a useful simulation tool. But outside of these rather restricted circumstances, data on oscillator instability obtained by the period-counting technique are practically useless.

E. General Limitations of Spectral Measurements of Phase and Frequency Fluctuations

As we have stated previously, the general limitations of the measurement of oscillator frequency instability are those of extracting spectral information from real-time signals. Thus there are upper and lower limits upon the magnitude of ω for which spectral information can be obtained. Since the analysis of oscillator frequency instability effects in coherent ranging and related systems involves integration of $S_\phi(\omega)$ around $\omega=0$, the lower limit is far more important. This limit is caused by the extreme averaging time required to resolve the spectra close to the origin and the complicated nonstationary statistics of oscillator drift phenomena. Some authors have attempted to establish a lower limit on the magnitude of ω related to the total time that an oscillator can be operated without recalibration [11], [12]. The assumption is made that this lower limit bounds a region of the spectrum centered about the origin over which integration for analysis will not be needed, as the oscillator will not be operated for intervals longer than the time corresponding to this bound. The extension toward $\omega=0$ of resolution in the spectrum of frequency instability is worthy of further effort.

Acknowledgment

The authors are indebted to J. G. Gregory of the Marshall Space Flight Center for his interest in and many helpful discussions of this work; to A. R. Chi of Goddard Space Flight Center for his encouragement and invitation of this paper for presentation at the Symposium on the Definition and Measurement of Short-Term Frequency Stability, Goddard Space Flight Center, Greenbelt, Md., November 1964; and to A. F. Ghais of ADCOM, Inc. for helpful discussions and criticisms.

References

[1] Baghdady, E. J., Theoretical comparison of exponent demodulation by phase-lock and frequency-compressive feedback techniques, *1964 IEEE Internatl. Conv. Rec.*, pt 6, pp 402–421.
[2] Wang, M. C., *Threshold Signals*, J. L. Lawson and G. E. Uhlenbeck, Ed., vol 18, Rad. Lab. Series. New York: McGraw-Hill, 1950.
[3] Rice, S. O., Statistical properties of a sinewave plus random noise, *Bell Syst. Tech. J.*, vol 27, Jan 1948, pp 109–157.
[4] Stumpers, F. L. H. M., Theory of frequency-modulation noise, *Proc. IRE*, vol 36, Sep 1948, pp 1081–1092.
[5] Middleton, D., *Statistical Communication Theory*. New York: McGraw-Hill, 1960.
[6] Price, R., A note on the envelope and phase-modulated components of narrow-band Gaussian noise, *IRE Trans. on Information Theory*, vol 1, Sep 1955, pp 9–13.
[7] North, D. O., An analysis of the factors which determine signal/noise discrimination in pulsed-carrier systems, Tech. Rept. PTR-6C, RCA Labs., Princeton, N. J., Jun 25, 1943; republished in *Proc. IEEE*, vol 51, Jul 1963, pp 1016–1027.
[8] Tauzman, H. D., Short term stability measurements, *1961 IRE Iternatl. Conv. Rec.*, pt 9, pp 191–206.
[9] Blackman, R. B., and J. W. Tukey, *The Measurement of Power Spectra*. New York: Dover, 1959.
[10] Cutler, L. S., Some aspects of the theory and measurement of frequency fluctuations in frequency standards, presented at the Symp. on the Definition and Measurement of Short-Term Frequency Stability, Goddard Space Flight Center, Greenbelt, Md., Nov 1964.
[11] Atkinson, W. R., L. Fey, and J. Newman, Spectrum analysis of extremely low frequency variations of quartz oscillators, *Proc. IEEE (Correspondence)*, vol 51, Jan 1963, p. 379.
[12] Malakhov, A. N., Spectra of flickernoise, *Radiotekh. i Elektron.*, vol 4, Jan 1959, pp 54–62.

Characterization of Frequency Stability

JAMES A. BARNES, SENIOR MEMBER, IEEE, ANDREW R. CHI, SENIOR MEMBER, IEEE, LEONARD S. CUTLER, MEMBER, IEEE, DANIEL J. HEALEY, MEMBER, IEEE, DAVID B. LEESON, SENIOR MEMBER, IEEE, THOMAS E. McGUNIGAL, MEMBER, IEEE, JAMES A. MULLEN, JR., SENIOR MEMBER, IEEE, WARREN L. SMITH, SENIOR MEMBER, IEEE, RICHARD L. SYDNOR, MEMBER, IEEE, ROBERT F. C. VESSOT, AND GERNOT M. R. WINKLER, MEMBER, IEEE

Abstract—Consider a signal generator whose instantaneous output voltage $V(t)$ may be written as

$$V(t) = [V_0 + \epsilon(t)] \sin [2\pi\nu_0 t + \varphi(t)]$$

where V_0 and ν_0 are the nominal amplitude and frequency, respectively, of the output. Provided that $\epsilon(t)$ and $\dot{\varphi}(t) = (d\varphi/(dt)$ are sufficiently small for all time t, one may define the fractional instantaneous frequency deviation from nominal by the relation

$$y(t) \equiv \frac{\dot{\varphi}(t)}{2\pi\nu_0}.$$

A proposed definition for the measure of frequency stability is the spectral density $S_y(f)$ of the function $y(t)$ where the spectrum is considered to be one sided on a per hertz basis.

An alternative definition for the measure of stability is the infinite time average of the sample variance of two adjacent averages of $y(t)$; that is, if

$$\bar{y}_k = \frac{1}{\tau} \int_{t_k}^{t_k+\tau} y(t)\, dt$$

where τ is the averaging period, $t_{k+1} = t_k + T$, $k = 0, 1, 2 \cdots$, t_0 is arbitrary, and T is the time interval between the beginnings of two successive measurements of average frequency; then the second measure of stability is

$$\sigma_y^2(\tau) \equiv \left\langle \frac{(\bar{y}_{k+1} - \bar{y}_k)^2}{2} \right\rangle$$

where $\langle\ \rangle$ denotes infinite time average and where $T = \tau$.

In practice, data records are of finite length and the infinite time averages implied in the definitions are normally not available; thus estimates for the two measures must be used. Estimates of $S_y(f)$ would be obtained from suitable averages either in the time domain or the frequency domain. An obvious estimate for $\sigma_y^2(\tau)$ is

$$\sigma_y^2(\tau) \approx \frac{1}{m} \sum_{k=1}^{m} \frac{(\bar{y}_{k+1} - \bar{y}_k)^2}{2}.$$

Parameters of the measuring system and estimating procedure are of critical importance in the specification of frequency stability. In practice, one should experimentally establish confidence limits for an estimate of frequency stability by repeated trials.

Manuscript received December 1, 1970. The authors of this paper are members of the Subcommittee on Frequency Stability of the Technical Committee on Frequency and Time of the IEEE Group on Instrumentation and Measurement.

J. A. Barnes is with the Time and Frequency Division, Institute for Basic Standards, NBS, Boulder, Colo. 80302.
A. R. Chi is with the Head Timing Systems Section, NASA Goddard Space Flight Center, Greenbelt, Md. 20771.
L. S. Cutler is with the Physical Research Laboratory, Hewlett-Packard Company, Palo Alto, Calif. 94304.
D. J. Healey is with the Aerospace Division, Westinghouse Electric Corporation, Baltimore, Md. 21203.
D. B. Leeson is with California Microwave, Inc., Sunnyvale, Calif. 94086.
T. E. McGunigal is with the RF and Quantum Techniques Section, NASA Goddard Space Flight Center, Greenbelt, Md. 21715.
J. A. Mullen is with the Research Division, Raytheon Company, Waltham, Mass. 02154.
W. L. Smith is with Bell Telephone Laboratories, Allentown, Pa. 18108.
R. L. Sydnor is with the Telecommunications Department, Jet Propulsion Laboratory, Pasadena, Calif. 91103.
G. M. R. Winkler is with the Time Service Division, U. S. Naval Observatory, Washington, D. C. 20390.
R. F. C. Vessot is with the Smithsonian Astrophysical Observatory, Cambridge, Mass. 02138.

Glossary of Symbols

$B_1(N, r, \mu)$, $B_2(r, \mu)$	Bias function for variances based on finite samples of a process with a power-law spectral density. (See [13].)
C_α	A real constant defined by (70).
c_0, c_1	Real constants.
$c(t)$	A real, deterministic function of time.
$D_x^2(\tau)$	Expected value of the squared second difference of $x(t)$ with lag time τ. See (80).
$f \equiv \omega/2\pi$	Fourier frequency variable.
f_h	High-frequency cutoff of an idealized infinitely sharp cutoff low-pass filter.
f_l	Low-frequency cutoff of an idealized infinitely sharp cutoff, high-pass filter.
$g(t)$	A real function of time.
h_α	Positive real coefficient of f^α in a power series expansion of the spectral density of the function $y(t)$.
i, j, k, m, n	Integers, often a dummy index of summation.
M	Positive integer giving the number of cycles averaged.
N	Positive integer giving the number of data points used in obtaining a sample variance.
$n(t)$	A nondeterministic function of time.
$R_y(\tau)$	Autocovariance function of $y(t)$. See (58).
r	Positive real number defined by $r \equiv T/\tau$.

Reprinted from *IEEE Trans. Instrum. Meas.*, vol. IM-20, pp. 105–120, May 1971.

S	An intermediate term used in deriving (23). The definition of S is given by (64).
$S_g(f)$	One-sided (power) spectral density on a per hertz basis of the pure real function $g(t)$. The dimensions of $S_g(f)$ are the dimensions of $g^2(t)/f$.
$S_y(f)$	A definition for the measure of frequency stability. One-sided (power) spectral density of $y(t)$ on a per hertz basis. The dimensions of $S_y(f)$ are Hz^{-1}.
T	Time interval between the beginnings of two successive measurements of average frequency.
t	Time variable.
t_0	An arbitrary fixed instant of time.
t_k	The time coordinate of the beginning of the kth measurement of average frequency. By definition, $t_{k+1} = t_k + T$, $k = 0, 1, 2 \cdots$.
u	Dummy variable of integration; $u \equiv \pi f \tau$.
$V(t)$	Instantaneous output voltage of signal generator. See (2).
V_0	Nominal peak amplitude of signal generator output. See (2).
$V_r(t)$	Instantaneous voltage of reference signal. See (40).
V_{0r}	Peak amplitude of reference signal. See (40).
$v(t)$	Voltage output of ideal product detector.
$v'(t)$	Low-pass filtered output of product detector.
$x(t)$	Real function of time related to the phase of the signal $V(t)$ by $x(t) \equiv [\varphi(t)]/(2\pi\nu_0)$.
$\hat{x}(t)$	A predicted value for $x(t)$.
$y(t)$	Fractional frequency offset of $V(t)$ from the nominal frequency. See (7).
$\bar{y}_k$	Average fractional frequency offset during the kth measurement interval. See (9).
$\langle \bar{y} \rangle_N$	The sample average of N successive values of $\bar{y}_k$. See (76).
$z_n(t)$	Nondeterministic (noise) function with (power) spectral density given by (25).
α	Exponent of f for a power-law spectral density.
γ	Positive real constant.
$\delta_k(r-1)$	The Kronecker δ function defined by $\delta_k(r-1) \equiv \begin{cases} 1, & \text{if } r = 1 \\ 0, & \text{otherwise.} \end{cases}$
$\epsilon(t)$	Amplitude fluctuations of signal. See (2).
μ	Exponent of τ. See (29).
$\nu(t)$	Instantaneous frequency of $V(t)$. Defined by $$\nu(t) \equiv \frac{1}{2\pi}\frac{d}{dt}\Phi(t).$$
ν_0	Nominal (constant) frequency of $V(t)$.
$\kappa(t)$	The Fourier transform of $n(t)$.
$\sigma_y^2(N, T, \tau)$	Sample variance of N averages of $y(t)$, each of duration τ, and spaced every T units of time. See (10).
$\langle \sigma_y^2(N, T, \tau) \rangle$	Average value of the sample variance $\sigma_y^2(N, T, \tau)$.
$\sigma_y^2(\tau)$	A second choice of the definition for the measure of frequency stability. Defined by $\sigma_y^2(\tau) \equiv \langle \sigma_y^2(N = 2, T = \tau, \tau) \rangle$.
$\sigma_x^2(\tau)$	Time stability measure defined by $\sigma_x^2(\tau) \equiv \tau^2 \sigma_y^2(\tau)$.
τ	Duration of averaging period of $y(t)$ to obtain $\bar{y}_k$. See (9).
$\Phi(t)$	Instantaneous phase of $V(t)$. Defined by $\Phi(t) \equiv 2\pi\nu_0 t + \varphi(t)$.
$\varphi(t)$	Instantaneous phase fluctuations about the ideal phase $2\pi\nu_0 t$. See (2).
$\psi_x^2(T, \tau)$	Mean-square time error for Doppler radar. See (82).
$\omega \equiv 2\pi f$	Angular Fourier frequency variable.

I. Introduction

THE measurement of frequency and fluctuations in frequency has received such great attention for so many years that it is surprising that the concept of frequency stability does not have a universally accepted definition. At least part of the reason has been that some uses are most readily described in the frequency domain and other uses in the time domain, as well as in combinations of the two. This situation is further complicated by the fact that only recently have noise models been presented that both adequately describe performance and allow a translation between the time and frequency domains. Indeed, only recently has it been recognized that there can be a wide discrepancy between commonly used time domain measures themselves. Following the NASA–IEEE Symposium on Short-Term Stability in 1964 and the Special Issue on Frequency Stability in the PROCEEDINGS OF THE IEEE, February 1966, it now seems reasonable to propose a definition of frequency stability. The present paper is presented as technical background for an eventual IEEE standard definition.

This paper attempts to present (as concisely as practical) adequate, self-consistent definitions of frequency stability. Since more than one definition of frequency stability is presented, an important part of this paper (perhaps the most important part) deals with translations among the suggested definitions of frequency stability. The applicability of these definitions to the more common noise models is demonstrated.

Consistent with an attempt to be concise, the references cited have been selected on the basis of being of most value to the reader rather than on the basis of being exhaustive. An exhaustive reference list covering the subject of frequency stability would itself be a voluminous publication.

Almost any signal generator is influenced to some extent by its environment. Thus observed frequency instabilities may be traced, for example, to changes in ambient temperature, supply voltages, magnetic field, barometric pressure, humidity, physical vibration, or even output loading, to mention the more obvious. While these environmental influences may be extremely important for many applications, the definition of frequency stability presented here is independent of these causal factors. In effect, we cannot hope to present an exhaustive list of environmental factors and a prescription for handling each even though, in some cases, these environmental factors may be by far the most important. Given a particular signal generator in a particular environment, one can obtain its frequency stability with the measures presented below, but one should not then expect an accurate prediction of frequency stability in a new environment.

It is natural to expect any definition of stability to involve various statistical considerations such as stationarity, ergodicity, average, variance, spectral density, etc. There often exist fundamental difficulties in rigorous attempts to bring these concepts into the laboratory. It is worth considering, specifically, the concept of stationarity since it is a concept at the root of many statistical discussions.

A random process is mathematically defined as stationary if every translation of the time coordinate maps the ensemble onto itself. As a necessary condition, if one looks at the ensemble at one instant of time t, the distribution in values within the ensemble is exactly the same as at any other instant of time t'. This is not to imply that the elements of the ensemble are constant in time, but, as one element changes value with time, other elements of the ensemble assume the previous values. Looking at it in another way, by observing the ensemble at some instant of time, one can deduce no information as to when the particular instant was chosen. This same sort of invariance of the *joint* distribution holds for any set of times $t_1, t_2, \cdots, t_n$ and its translation $t_1 + \tau, t_2 + \tau, \cdots, t_n + \tau$.

It is apparent that any ensemble that has a finite past as well as a finite future cannot be stationary, and this neatly excludes the real world and anything of practical interest. The concept of stationarity does violence to concepts of causality since we implicitly feel that current performance (i.e., the applicability of stationary statistics) cannot be logically dependent upon future events (i.e., if the process is terminated some time in the distant future). Also, the verification of stationarity would involve hypothetical measurements that are *not* experimentally feasible, and therefore the concept of stationarity is not directly relevant to experimentation.

Actually the utility of statistics is in the formation of idealized models that *reasonably* describe significant observables of real systems. One may, for example, consider a hypothetical ensemble of noises with certain properties (such as stationarity) as a model for a particular real device. If a model is to be acceptable, it should have at least two properties: first, the model should be tractable; that is, one should be able to easily arrive at estimates for the elements of the models; and second, the model should be consistent with *observables* derived from the real device that it is simulating.

Notice that one does not need to know that the device was selected from a stationary ensemble, but only that the observables derived from the device are *consistent* with, say, elements of a hypothetically stationary ensemble. Notice also that the actual model used may depend upon how clever the experimenter–theorist is in generating models.

It is worth noting, however, that while some texts on statistics give "tests for stationarity," these tests are almost always inadequate. Typically, these tests determine only if there is a substantial fraction of the noise power in Fourier frequencies whose periods are of the same order as the data length or longer. While this may be very important, it is *not* logically essential to the concept of stationarity. If a nonstationary model actually becomes common, it will almost surely be because it is useful or convenient and not because the process is "actually nonstationary." Indeed, the phrase "actually nonstationary" appears to have no meaning in an operational sense. In short, stationarity (or nonstationarity) is a property of models, *not* a property of data [1].

Fortunately, many statistical models exist that adequately describe most present-day signal generators; many of these models are considered below. It is obvious that one cannot guarantee that all signal generators are adequately described by these models, but the authors do feel they are adequate for the description of most signal generators presently encountered.

II. Statement of the Problem

To be useful, a measure of frequency stability must allow one to predict performance of signal generators used in a wide variety of situations as well as allow one to make meaningful relative comparisons among signal generators. One must be able to predict performance in devices that may most easily be described either in the time domain, or in the frequency domain, or in a combination of the two. This prediction of performance may involve actual distribution functions, and thus second moment measures (such as power spectra and variances) are not totally adequate.

Two common types of equipment used to evaluate the performance of a frequency source are (analog) spectrum analyzers (frequency domain) and digital electronic counters (time domain). On occasion the digital counter data are converted to power spectra by computers. One must realize that any piece of equipment simultaneously has certain aspects most easily described in the time domain and other aspects most easily described in the frequency domain. For example, an electronic counter has a high-frequency limitation, an experimental spectra are determined with finite time averages.

Research has established that ordinary oscillators demonstrate noise, which appears to be a superposition of causally generated signals and random nondeterministic noises. The random noises include thermal noise, shot noise, noises of undetermined origin (such as flicker noise), and integrals of these noises.

One might well expect that for the more general cases one would need to use a nonstationary model (not stationary even in the wide sense, i.e., the covariance sense). Nonstationarity would, however, introduce significant difficulties in the passage between the frequency and time domains. It is interesting to note that, so far, experimenters have seldom found a nonstationary (covariance) model useful in describing actual oscillators.

In what follows, an attempt has been made to separate general statements that hold for any noise or perturbation from the statements that apply only to specific models. It is important that these distinctions be kept in mind.

III. Background and Definitions

To discuss the concept of frequency stability immediately implies that frequency can change with time and thus one is not considering Fourier frequencies (at least at this point). The conventional definition of instantaneous (angular) frequency is the time rate of change of phase; that is

$$2\pi\nu(t) \equiv \frac{d\Phi(t)}{dt} \equiv \dot{\Phi}(t) \qquad (1)$$

where $\Phi(t)$ is the instantaneous phase of the oscillator. This paper uses the convention that time-dependent frequencies of oscillators are denoted by $\nu(t)$ (cycle frequency, hertz), and Fourier frequencies are denoted by ω (angular frequency) or f (cycle frequency, hertz) where $\omega \equiv 2\pi f$. In order for (1) to have meaning, the phase $\Phi(t)$ must be a well-defined function. This restriction immediately eliminates some "nonsinusoidal" signals such as a pure random uncorrelated ("white") noise. For most real signal generators, the concept of phase is reasonably amenable to an operational definition and this restriction is not serious.

Of great importance to this paper is the concept of spectral density, $S_g(f)$. The notation $S_g(f)$ is to represent the one-sided spectral density of the (pure real) function $g(t)$ on a per hertz basis; that is, the total "power" or mean-square value of $g(t)$ is given by

$$\int_0^\infty S_g(f)\, df.$$

Since the spectral density is such an important concept to what follows, it is worthwhile to present some important references on spectrum estimation. There are many references on the estimation of spectra from data records, but worthy of special note are [2]–[5].

IV. Definition of Measures of Frequency Stability (Second-Moment Type)

A. General

Consider a signal generator whose instantaneous output voltage $V(t)$ may be written as

$$V(t) = [V_0 + \epsilon(t)] \sin [2\pi\nu_0 t + \varphi(t)] \qquad (2)$$

where V_0 and ν_0 are the nominal amplitude and frequency, respectively, of the output and it is assumed that

$$\left|\frac{\epsilon(t)}{V_0}\right| \ll 1 \qquad (3)$$

and

$$\left|\frac{\dot{\varphi}(t)}{2\pi\nu_0}\right| \ll 1 \qquad (4)$$

for substantially all time t. Making use of (1) and (2) one sees that

$$\Phi(t) = 2\pi\nu_0 t + \varphi(t) \qquad (5)$$

and

$$\nu(t) = \nu_0 + \frac{1}{2\pi}\dot{\varphi}(t). \qquad (6)$$

Equations (3) and (4) are essential in order that $\varphi(t)$ may be defined conveniently and unambiguously (see measurement section).

Since (4) must be valid even to speak of an instantaneous frequency, there is no real need to distinguish stability measures from instability measures. That is, any fractional frequency stability measure will be far from unity, and the chance of confusion is slight. It is true that in a very strict sense people usually measure instability and speak of stability. Because the chances of confusion are so slight, the authors have chosen to continue in the custom of measuring "instability" and speaking of stability (a number always much less than unity).

Of significant interest to many people is the radio frequency (RF) spectral density $S_V(f)$. This is of direct concern in spectroscopy and radar. However, this is *not* a good primary measure of frequency stability for two reasons. First, fluctuations in the amplitude $\epsilon(t)$ contribute directly to $S_V(f)$; and second, for many cases when

$\epsilon(t)$ is insignificant, the RF spectrum $S_V(f)$ is not uniquely related to the frequency fluctuations [6].

B. General: First Definition of the Measure of Frequency Stability—Frequency Domain

By definition, let

$$y(t) \equiv \frac{\dot{\varphi}(t)}{2\pi\nu_0}, \tag{7}$$

where $\varphi(t)$ and ν_0 are as in (2). Thus $y(t)$ is the instantaneous fractional frequency deviation from the nominal frequency ν_0. A proposed definition of frequency stability is the spectral density $S_y(f)$ of the instantaneous fractional frequency fluctuations $y(t)$. The function $S_y(f)$ has the dimensions of Hz^{-1}.

One can show [7] that if $S_\varphi(f)$ is the spectral density of the phase fluctuations, then

$$\begin{aligned} S_y(f) &= \left(\frac{1}{2\pi\nu_0}\right)^2 S_{\dot{\varphi}}(f) \\ &= \left(\frac{1}{\nu_0}\right)^2 f^2 S_\varphi(f). \end{aligned} \tag{8}$$

Thus a knowledge of the spectral density of the phase fluctuations $S_\varphi(f)$ allows a knowledge of the spectral density of the frequency fluctuations $S_y(f)$, the first definition of frequency stability. Of course, $S_y(f)$ cannot be *perfectly* measured—this is the case for any physical quantity; useful estimates of $S_y(f)$ are, however, easily obtainable.

C. General: Second Definition of the Measure of Frequency Stability—Time Domain

The second definition is based on the sample variance of the fractional frequency fluctuations. In order to present this measure of frequency stability, define $\bar{y}_k$ by the relation

$$\bar{y}_k \equiv \frac{1}{\tau}\int_{t_k}^{t_k+\tau} y(t)\,dt = \frac{\varphi(t_k+\tau)-\varphi(t_k)}{2\pi\nu_0\tau}, \tag{9}$$

where $t_{k+1} = t_k + T$, $k = 0, 1, 2, \cdots$, T is the repetition interval for measurements of duration τ, and t_0 is arbitrary. Conventional frequency counters measure the number of cycles in a period τ; that is, they measure $\nu_0\tau(1+\bar{y}_k)$. When τ is 1 s they count the number of $\nu_0(1+\bar{y}_k)$. The second measure of frequency stability, then, is defined in analogy to the sample variance by the relation

$$\langle\sigma_y^2(N, T, \tau)\rangle \equiv \left\langle \frac{1}{N-1}\sum_{n=1}^{N}\left(\bar{y}_n - \frac{1}{N}\sum_{k=1}^{N}\bar{y}_k\right)^2\right\rangle, \tag{10}$$

where $\langle g\rangle$ denotes the infinite time average of g. This measure of frequency stability is dimensionless.

In many situations it would be wrong to assume that (10) converges to a meaningful limit as $N \rightarrow \infty$. First, of course, one cannot practically let N approach infinity and, second, it is known that some actual noise processes contain substantial fractions of the total noise power in the Fourier frequency range below one cycle per year. In order to improve comparability of data, it is important to specify particular N and T. For the preferred definition we recommend choosing $N = 2$ and $T = \tau$ (i.e., no dead time between measurements). Writing $\langle\sigma_y^2(N=2, T=\tau, \tau)\rangle$ as $\sigma_y^2(\tau)$, the Allan variance [8], the proposed measure of frequency stability in the time domain may be written as

$$\sigma_y^2(\tau) = \left\langle \frac{(\bar{y}_{k+1}-\bar{y}_k)^2}{2}\right\rangle \tag{11}$$

for $T = \tau$.

Of course, the experimental estimate of $\sigma_y^2(\tau)$ must be obtained from finite samples of data, and one can never obtain perfect confidence in the estimate; the true time average is not realizable in a real situation. One estimates $\sigma_y^2(\tau)$ from a finite number (say, m) of values of $\sigma_y^2(2, \tau, \tau)$ and averages to obtain an estimate of $\sigma_y^2(\tau)$. Appendix I shows that the ensemble average of $\sigma_y^2(2, \tau, \tau)$ is convergent (i.e., as $m \rightarrow \infty$) even for noise processes that do not have convergent $\langle\sigma_y^2(N, \tau, \tau)\rangle$ as $N \rightarrow \infty$. Therefore, $\sigma_y^2(\tau)$ has greater utility as an idealization than does $\langle\sigma_y^2(\infty, \tau, \tau)\rangle$ even though both involve assumptions of infinite averages. In effect, increasing N causes $\sigma_y^2(N, T, \tau)$ to become more sensitive to the low-frequency components of $S_y(f)$. In practice, one must distinguish between an experimental estimate of a quantity (say, of $\sigma_y^2(\tau)$) and its idealized value. It is reasonable to believe that extensions to the concept of statistical ("quality") control [9] may prove useful here. One should, of course, specify the actual number m of independent samples used for an estimate of $\sigma_y^2(\tau)$.

In summary, therefore, $S_y(f)$ is the proposed measure of (instantaneous) frequency stability in the (Fourier) frequency domain and $\sigma_y^2(\tau)$ is the proposed measure of frequency stability in the time domain.

D. Distributions

It is natural that people first become involved with second moment measures of statistical quantities and only later with actual distributions. This is certainly true with frequency stability. While one can specify the argument of a distribution function to be, say $(\bar{y}_{k+1} - \bar{y}_k)$, it makes sense to postpone such a specfication until a real use has materialized for a particular distribution function. This paper does not attempt to specify a preferred distribution function for frequency fluctuations.

E. Treatment of Systematic Variations

1) General: The definition of frequency stability $\sigma_y^2(\tau)$ in the time domain is useful for many situations. However, some oscillators, for example, exhibit an aging or almost linear drift of frequency with time. For some applications, this trend may be calculated and should be removed [8] before estimating $\sigma_y^2(\tau)$.

In general, a systematic trend is perfectly deterministic (i.e., predictable) while the noise is nondeterministic. Consider a function $g(t)$, which may be written in the form

$$g(t) = c(t) + n(t) \tag{12}$$

where $c(t)$ is some deterministic function of time and $n(t)$, the noise, is a nondeterministic function of time. We will define $c(t)$ to be the *systematic trend* to the function $g(t)$. A problem of significance here is to determine when and in what sense $c(t)$ is measurable.

2) Specific Case—Linear Drift: As an example, if we consider a typical quartz crystal oscillator whose fractional frequency deviation is $y(t)$, we may let

$$g(t) = \frac{d}{dt}\, y(t). \tag{13}$$

With these conditions, $c(t)$ is the drift rate of the oscillator (e.g., 10^{-10}/day) and $n(t)$ is related to the frequency "noise" of the oscillator by a time derivative. One sees that the time average of $g(t)$ becomes

$$\frac{1}{T}\int_{t_0}^{t_0+T} g(t)\,dt = c_1 + \frac{1}{T}\int_{t_0}^{t_0+T} n(t)\,dt \tag{14}$$

where $c(t) = c_1$ is assumed to be the constant drift rate of the oscillator. In order for c_1 to be an observable, it is natural to expect the average of the noise term to vanish, that is, converge to zero.

It is instructive to assume [8], [10] that in addition to a linear drift, the oscillator is perturbed by a flicker noise, i.e.,

$$S_y(f) = \begin{cases} h_{-1}f^{-1}, & 0 < f \le f_h \\ 0, & f > f_h \end{cases} \tag{15}$$

where h_{-1} is a constant (see Section V-A-2) and thus,

$$S_n(f) = \begin{cases} (2\pi)^2 h_{-1} f, & 0 \le f \le f_h \\ 0, & f > f_h \end{cases} \tag{16}$$

for the oscillator we are considering. With these assumptions, it is seen that

$$\lim_{T\to\infty} \frac{1}{T}\int_{t_0}^{t_0+T} n(t)\,dt = \kappa(0) = 0 \tag{17}$$

and that

$$\lim_{T\to\infty}\left\{\text{variance}\left[\frac{1}{T}\int_{t_k}^{t_k+T} n(t)\,dt\right]\right\} = 0 \tag{18}$$

where $\kappa(f)$ is the fourier transform of $n(t)$. Since $S_n(0) = 0$, $\kappa(0)$ must also vanish both in probability and in mean square. Thus, not only does $n(t)$ average to zero, but one may obtain arbitrarily good confidence on the result by longer averages.

Having shown that one can reliably estimate the drift rate c_1 of this (common) oscillator, it is instructive to attempt to fit a straight line to the frequency aging. That is, let

$$g(t) = y(t) \tag{19}$$

and thus

$$g(t) = c_0 + c_1(t - t_0) + n'(t) \tag{20}$$

where c_0 is the frequency intercept at $t = t_0$ and c_1 is the drift rate previously determined. A problem arises here because

$$S_{n'}(f) = S_y(f) \tag{21}$$

and

$$\lim_{T\to\infty}\left\{\text{variance}\left[\frac{1}{T}\int_{t_k}^{t_k+T} n'(t)\,dt\right]\right\} = \infty \tag{22}$$

for the noise model we have assumed. This follows from the fact that the (infinite N) variance of a flicker noise process is infinite [7], [8], [10]. Thus, c_0 cannot be measured with any realistic precision, at least, in an absolute sense.

We may interpret these results as follows. After experimenting with the oscillator for a period of time one can fit an empirical equation to $y(t)$ of the form

$$y(t) = c_0 + tc_1 + n'(t),$$

where $n'(t)$ is nondeterministic. At some later time it is possible to reevaluate the coefficients c_0 and c_1. According to what has been said, the drift rate c_1 should be reproducible to within the confidence estimates of the experiment regardless of when it is reevaluated. For c_0, however, this is not true. In fact, the more one attempts to evaluate c_0, the larger the fluctuations are in the result.

Depending on the spectral density of the noise term, it may be possible to predict future measurements of c_0 and to place realistic confidence limits on the prediction [11]. For the case considered here, however, these confidence limits tend to infinity when the prediction interval is increased. Thus, in a certain sense, c_0 is "measurable" but it is not in statistical control (to use the language of the quality control engineer [9]).

V. Translations Among Frequency Stability Measures

A. Frequency Domain to Time Domain

1) General: It is of value to define $r = T/\tau$; that is, r is the ratio of the time interval between successive measurements to the duration of the averaging period. Cutler has shown (see Appendix I) that

$$\langle\sigma_y^2(N, T, \tau)\rangle = \frac{N}{(N-1)}\int_0^\infty df\, S_y(f)\,\frac{[\sin^2(\pi f\tau)]}{(\pi f\tau)^2}\left\{1 - \frac{\sin^2(\pi r f N\tau)}{N^2\sin^2(\pi r f\tau)}\right\}. \tag{23}$$

Equation (23) in principle allows one to calculate the time-domain stability $\langle\sigma_y^2(N, T, \tau)\rangle$ from the frequency-domain stability $S_y(f)$.

2) Specific Model: A model that has been found useful [8], [10]–[13] consists of a set of five independent noise processes $z_n(t)$, $n = -2, -1, 0, 1, 2$, such that

$$y(t) = z_{-2}(t) + z_{-1}(t) + z_0(t) + z_1(t) + z_2(t) \tag{24}$$

and the spectral density of z_n is given by

$$S_{z_n}(f) = \begin{cases} h_n f^n, & 0 \le f \le f_h \\ 0, & f > f_h, \; n = -2, -1, 0, 1, 2, \end{cases} \tag{25}$$

where the h_n are constants. Thus, $S_y(f)$ becomes

$$S_y(f) = h_{-2}f^{-2} + h_{-1}f^{-1} + h_0 + h_1 f + h_2 f^2, \tag{26}$$

for $0 \le f \le f_h$ and $S_y(f)$ is assumed to be negligible beyond this range. In effect, each z_n contributes to both $S_y(f)$ and $\langle \sigma_y^2(N, T, \tau)\rangle$ independently of the other z_n. The contributions of the z_n to $\langle \sigma_y^2(N, T, \tau)\rangle$ are tabulated in Appendix II.

Any electronic device has a finite bandwidth and this certainly applies to frequency-measuring equipment also. For fractional frequency fluctuations $y(t)$ whose spectral density varies as

$$S_y(f) \sim f^\alpha, \qquad \alpha \ge -1 \tag{27}$$

for the higher Fourier components, one sees (from Appendix I) that $\langle \sigma_y^2(N, T, \tau)\rangle$ may depend on the exact shape of the frequency cutoff. This is true because a substantial fraction of the noise "power" may be in these higher Fourier components. As a simplifying assumption, this paper assumes a sharp cutoff in noise "power" at the frequency f_h for the noise models. It is apparent from the tables of Appendix II that the time domain measure of frequency stability may depend on f_h in a very important way, and, in some practical cases, the actual shape of the frequency cutoff may be very important [7]. On the other hand, there are many practical measurements where the value of f_h has little or no effect. Good practice, however, dictates that the system noise bandwidth f_h should be specified with any results.

In actual practice, the model of (24)–(26) seems to fit almost all real frequency sources. Typically, only two or three of the h-coefficients are actually significant for a real device and the others can be neglected. Because of its applicability, this model is used in much of what follows. Since the z_n are assumed to be independent noises, it is normally sufficient to compute the effects for a general z_n and recognize that the superposition can be accomplished by simple additions for their contributions to $S_y(f)$ or $\langle \sigma_y^2(N, T, \tau)\rangle$.

B. Time Domain to Frequency Domain

1) General: For general $\langle \sigma_y^2(N, T, \tau)\rangle$ no simple prescription is available for translation into the frequency domain. For this reason, one might prefer $S_y(f)$ as a general measure of frequency stability. This is especially true for theoretical work.

2) Specific Model: Equations (24)–(26) form a realistic model that fits the random nondeterministic noises found on most signal generators. Obviously, if this is a good model, then the tables in Appendix II may be used (in reverse) to translate into the frequency domain.

Allan [8] and Vessot [12] showed that if

$$S_y(f) = \begin{cases} h_\alpha f^\alpha, & 0 \le f \le f_h \\ 0, & f > f_h \end{cases} \tag{28}$$

where α is a constant, then

$$\langle \sigma_y^2(N, T, \tau)\rangle \sim |\tau|^\mu, \qquad 2\pi\tau f_h \gg 1 \tag{29}$$

for N and $r = T/\tau$ held constant. The constant μ is related to α by the mapping shown[1] in Fig. 1. If (28) and (29) hold over a reasonable range for a signal generator, then (28) can be substituted into (23) and evaluated to determine the constant h_α from measurements of $\langle \sigma_y^2(N, T, \tau)\rangle$. It should be noted that the model of (28) and (29) may be easily extended to a superposition of similar noises as in (26).

C. Translations Among the Time-Domain Measures

1) General: Since $\langle \sigma_y^2(N, T, \tau)\rangle$ is a function of N, T, and τ (for some types of noise f_h is also important), it is very desirable to be able to translate among different sets of N, T, and τ (f_h held constant). This is, however, not possible in general.

2) Specific Model: It is useful to restrict consideration to a case described by (28) and (29). Superpositions of independent noises with different power-law types of spectral densities (i.e., different α) can also be treated by this technique, e.g., (26). One may define two "bias functions," B_1 and B_2 by the relations [13]

$$B_1(N, r, \mu) \equiv \frac{\langle \sigma_y^2(N, T, \tau)\rangle}{\langle \sigma_y^2(2, T, \tau)\rangle} \tag{30}$$

and

$$B_2(r, \mu) \equiv \frac{\langle \sigma_y^2(2, T, \tau)\rangle}{\langle \sigma_y^2(2, \tau, \tau)\rangle} \tag{31}$$

where $r \equiv T/\tau$ and μ is related to α by the mapping of Fig. 1. In words, B_1 is the ratio of the average variance for N samples to the average variance for two samples (everything else held constant), while B_2 is the ratio of the average variance with dead time between measurements ($r \ne 1$) to that of no dead time ($r = 1$ and with $N = 2$ and τ held constant). These functions are tabulated in [13]. Figs. 2 and 3 show a computer plot of $B_1(N, r = 1, \mu)$ and $B_2(r, \mu)$.

Suppose one has an experimental estimate of $\langle \sigma_y^2(N_1, T_1, \tau_1)\rangle$ and its spectral type is known, i.e., (28) and (29) form a good model and μ is known. Suppose also that one wishes to know the variance at some other set of measurement parameters N_2, T_2, τ_2. An unbiased estimate of $\langle \sigma_y^2(N_2, T_2, \tau_2)\rangle$ may be calculated by

[1] It should be noted that in Allan [8], the exponent α corresponds to the spectrum of *phase* fluctuations while variances are taken over average *frequency* fluctuations. In the present paper, α is identical to the exponent $\alpha + 2$ in [8].

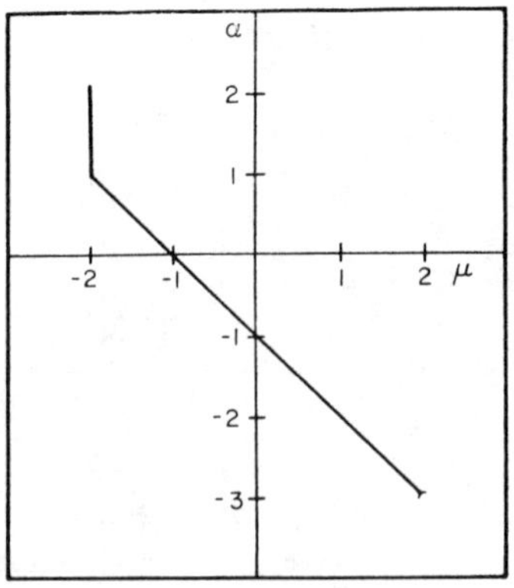

Fig. 1. $\mu - \alpha$ mapping.

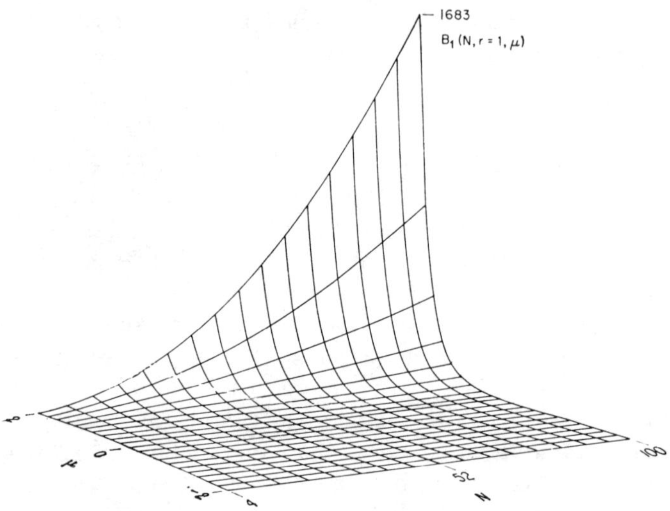

Fig. 2. Function $B_1(N, r = 1, \mu)$.

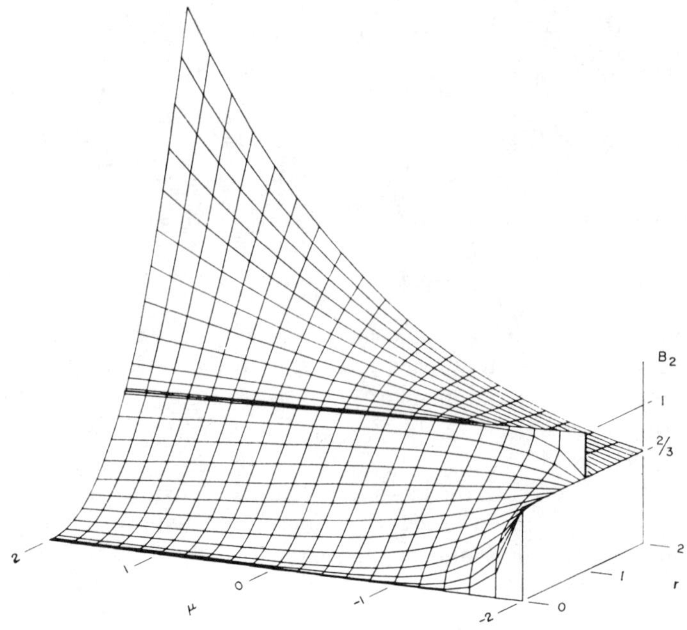

Fig. 3. Bias function $B_2(r, \mu)$.

$$\langle\sigma_y^2(N_2, T_2, \tau_2)\rangle = \left(\frac{\tau_2}{\tau_1}\right)^{\mu} \cdot \left[\frac{B_1(N_2, r_2, \mu)B_2(r_2, \mu)}{B_1(N_1, r_1, \mu)B_2(r_1, \mu)}\right]\langle\sigma_y^2(N_1, T_1, \tau_1)\rangle, \tag{32}$$

where $r_1 = T_1/\tau_1$ and $r_2 = T_2/\tau_2$.

3) General: While it is true that the concept of the bias functions B_1 and B_2 could be extended to other processes besides those with the power-law types of spectral densities, this generalization has not been done. Indeed, spectra of the form given in (28) [or superpositions of such spectra as in (26)] seem to be the most common types of nondeterministic noises encountered in signal generators and associated equipment. For other types of fluctuations (such as causally generated perturbations), translations must be handled on an individual basis.

VI. Applications of Stability Measures

Obviously, if one of the stability measures is exactly the important parameter in the use of a signal generator, the stability measure's application is trivial. Some nontrivial applications arise when one is interested in a different parameter, such as in the use of an oscillator in Doppler radar measurements or in clocks.

A. Doppler Radar

1) General: From its transmitted signal, a Doppler radar receives from a moving target a frequency-shifted return signal in the presence of other large signals. These large signals can include clutter (ground return) and transmitter leakage into the receiver (spillover). Instabilities of radar signals result in noise energy on the clutter return, on spillover, and on local oscillators in the equipment.

The limitations of subclutter visibility (SCV) rejections due to the radar signals themselves are related to the RF power spectral density $S_V(f)$. The quantity typically referred to is the carrier-to-noise ratio and can be mathematically approximated by the quantity

$$\frac{S_V(f)}{\int_0^\infty S_V(f')df'}.$$

The effects of coherence of target return and other radar parameters are amply considered in the literature [14]–[17].

2) Special Case: Because FM effects generally predominate over AM effects, this carrier-to-noise ratio is approximately given by [6]

$$\frac{S_V(f)}{\int_0^\infty S_V(f')\,df'} \approx \tfrac{1}{2}S_\varphi(|f - \nu_0|), \tag{33}$$

for many signal sources provided $|f - \nu_0|$ is sufficiently greater than zero. (The factor of $\frac{1}{2}$ arises from the fact that $S_\varphi(f)$ is a one-sided spectrum.) Thus, if $f - \nu_0$ is a frequency separation from the carrier, the carrier-to-noise ratio at that point is approximately

$$\tfrac{1}{2}S_\varphi(|f-\nu_0|) = \frac{1}{2}\left(\frac{\nu_0}{f-\nu_0}\right)^2 S_y(|f-\nu_0|). \tag{34}$$

B. Clock Errors

1) General: A clock is a device that counts the cycles of a periodic phenomenon. Thus, the reading error $x(t)$ of a clock run from the signal given by (2) is

$$x(t) = \frac{\varphi(t)}{2\pi\nu_0} \tag{35}$$

and the dimensions of $x(t)$ are seconds.

If this clock is a secondary standard, then one could have available some past history of $x(t)$, the time error relative to the standard clock. It often occurs that one is interested in predicting the clock error $x(t)$ for some future date, say $t_0 + \tau$, where t_0 is the present date. Obviously, this is a problem in pure prediction and can be handled by conventional methods [3].

2) Special Case: Although one could handle the prediction of clock errors by the rigorous methods of prediction theory, it is more common to use simpler prediction methods [10], [11]. In particular, one often predicts a clock error for the future by adding to the present error a correction that is derived from the current rate of gain (or loss) of time. That is, the predicted error $\hat{x}(t_0+\tau)$ is related to the past history of $x(t)$ by

$$\hat{x}(t_0+\tau) = x(t_0) + \tau\left[\frac{x(t_0) - x(t_0 - T)}{T}\right]. \tag{36}$$

It is typical to let $T = \tau$.

Thus, the mean-square error of prediction for $T = \tau$ becomes

$$\begin{aligned}&\langle[x(t_0+\tau) - \hat{x}(t_0+\tau)]^2\rangle \\ &\quad = \langle[x(t_0+\tau) - 2x(t_0) + x(t_0-\tau)]^2\rangle,\end{aligned} \tag{37}$$

which, with the aid of (11), can be written in the form

$$\langle[x(t_0+\tau) - \hat{x}(t_0+\tau)]^2\rangle = 2\tau^2\sigma_y^2(\tau). \tag{38}$$

One can define a *time* stability measure $\sigma_x^2(\tau)$ by

$$\sigma_x^2(\tau) \equiv \tau^2\sigma_y^2(\tau). \tag{39}$$

Clearly, however, the actual errors of prediction of clock readings are dependent on the prediction algorithm used and the utility of such a definition as $\sigma_x^2(\tau)$ is not great. Caution should be used in employing this definition.

VII. Measurement Techniques for Frequency Stability

A. Heterodyne Techniques (General)

It is possible for oscillators to be very stable and values of $\sigma_y(\tau)$ can be as small as 10^{-14} in some state-of-the-art equipment. Thus, one often needs measuring techniques capable of resolving very small fluctuations in $y(t)$. One of the most common techniques is a heterodyne or beat frequency technique. In this method, the signal from the oscillator to be tested is mixed with a reference signal of almost the same frequency as the test oscillator in order that one is left with a lower average frequency for analysis without reducing the frequency (or phase) fluctuations themselves. Following Vessot *et al.* [18], consider an ideal reference oscillator whose output signal is

$$V_r(t) = V_{0r}\sin 2\pi\nu_0 t \tag{40}$$

and a second oscillator whose output voltage $V(t)$ is given by (2): $V(t) = [V_0 + \epsilon(t)]\sin[2\pi\nu_0 t + \varphi(t)]$. Let these two signals be mixed in a product detector; that is, the output of the product detector $v(t)$ is equal to the product $\gamma V(t) \times V_r(t)$, where γ is a constant (see Fig. 4).

Let $v(t)$, in turn, be processed by a sharp low-pass filter with cutoff frequency f_h' such that

$$0 < f_h < f_h' < \nu_0. \tag{41}$$

One may write

$$\begin{aligned}&\gamma V(t)\cdot V_r(t) \\ &\quad = \gamma V_{0r}(V_0 + \epsilon)[\sin 2\pi\nu_0 t][\sin(2\pi\nu_0 t + \varphi)] \\ &\quad = v(t) = \gamma\frac{(V_{0r}V_0)}{2}\left(1 + \frac{\epsilon}{V_0}\right)[\cos\varphi - \cos(4\pi\nu_0 t + \varphi)].\end{aligned} \tag{42}$$

Assume that $\cos[\varphi(t)]$ has essentially no power in Fourier frequencies f in the region $f \geq f_h'$. The effect of the low-pass filter then is to remove the second term on the extreme right of (42); that is

$$v'(t) = \gamma\frac{V_{0r}V_0}{2}\left(1 + \frac{\epsilon}{V_0}\right)\cos\varphi(t). \tag{43}$$

This separation of terms by the filter is correct only if $|[\dot{\varphi}(t)/2\pi\nu_0]| \ll 1$ for all t (4).

The following two cases are of interest.

Case I: The relative phase of the oscillators is adjusted so that $|\varphi(t)| \ll 1$ (in-phase condition) during the period of measurement. Under these conditions

$$v'(t) \approx \frac{\gamma}{2}V_{0r}V_0 + \frac{\gamma}{2}V_{0r}\epsilon(t), \tag{44}$$

since $\cos\varphi(t) \approx 1$. That is to say one detects the amplitude noise $\epsilon(t)$ of the signal.

Case II: The relative phase of the oscillators is adjusted to be in approximate quadrature; that is

$$\varphi'(t) = \varphi(t) + \frac{\pi}{2} \tag{45}$$

where $|\varphi'(t)| \ll 1$. Under these conditions,

$$\cos\varphi(t) = \sin\varphi'(t) \approx \varphi'(t) \tag{46}$$

and

$$v'(t) = \frac{\gamma}{2}V_{0r}V_0\varphi'(t) + \frac{\gamma}{2}V_{0r}\varphi'(t)\epsilon(t). \tag{47}$$

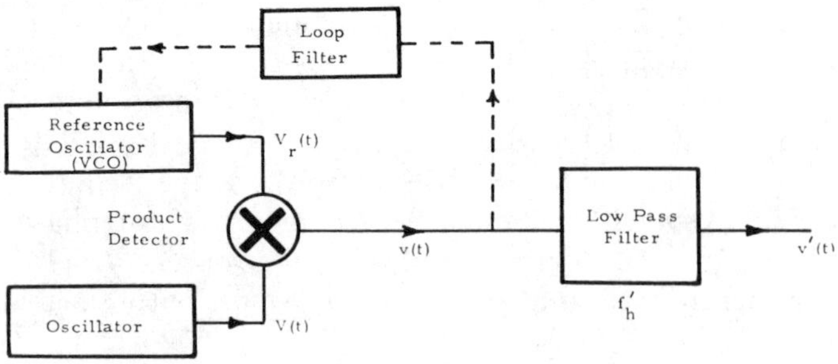

Fig. 4. Heterodyne scheme.

If it is true that $|[\epsilon(t)/V_0]| \ll 1$ for all t (3), then (47) becomes

$$v'(t) \approx \frac{\gamma}{2} V_{0r} V_0 \varphi'(t); \tag{48}$$

that is, $v'(t)$ is proportional to the phase fluctuations. Thus, in order to observe $\varphi'(t)$ by this method, (3) and (4) must be valid. For different average phase values, mixtures of amplitude and phase noise are observed.

In order to maintain the two signals in quadrature for long observational periods, the reference oscillator can be a voltage-controlled oscillator (VCO) and one may feed back the phase error voltage as defined in (48) to control the frequency of the VCO [19]. In this condition of the phase-locked oscillator, the voltage $v'(t)$ is the analog of the *phase* fluctuations for Fourier frequencies above the loop cutoff frequency of the locked loop. For Fourier frequencies below the loop cutoff frequency of the loop, $v'(t)$ is the analog of *frequency* fluctuations. In practice, one should measure the complete servo-loop response.

B. Period Measurement

Assume one has an oscillator whose voltage output may be represented by (2). If $|[\epsilon(t)/V_0]| \ll 1$ for all t and the total phase

$$\Phi(t) = 2\pi\nu_0 t + \varphi(t) \tag{5}$$

is a monotonic function of time (that is, $|[\dot{\varphi}(t)/2\pi\nu_0]| \leq 1$), then the time t between successive positive going zero crossings of $V(t)$ is related to the average frequency during the interval τ. Specifically

$$\frac{1}{\tau} = \nu_0(1 + \bar{y}_n). \tag{49}$$

If one lets τ be the time between a positive going zero crossing of $V(t)$ and the Mth successive positive going zero crossing, then

$$\frac{M}{\tau} = \nu_0(1 + \bar{y}_n). \tag{50}$$

If the variations $\Delta\tau$ of the period are small compared to the average period τ_0, Cutler and Searle [7] have shown that one may make a reasonable approximation to $\langle\sigma_y^2(N, T, \tau_0)\rangle$ using period measurements.

C. Period Measurement With Heterodyning

Suppose that $\varphi(t)$ is a monotonic function of time. The output of the filter of Section VII-A (43) becomes

$$v'(t) \approx \gamma \frac{V_{0r} V_0}{2} \cos \varphi(t) \tag{51}$$

if $|[\epsilon(t)/V_0]| \ll 1$. Then one may measure the period τ of two successive positive zero crossings of $v'(t)$. Thus

$$\frac{1}{\tau} = \nu_0 \, |\bar{y}_n| \tag{52}$$

and for the Mth positive crossover

$$\frac{M}{\tau} = \nu_0 \, |\bar{y}_n|. \tag{53}$$

The magnitude bars appear because $\cos \varphi(t)$ is an even function of $\varphi(t)$. It is impossible to determine by this method alone whether φ is increasing with time or decreasing with time. Since $\bar{y}_n$ may be very small ($\sim 10^{-11}$ or 10^{-12} for very good oscillators), τ may be quite long and thus measurable with a good relative precision.

If the phase $\varphi(t)$ is not monotonic, the true $\bar{y}_n$ may be near zero but one could still have many zeros of $\cos \varphi(t)$ and thus (52) and (53) would not be valid.

D. Frequency Counters

Assume the phase (either Φ or φ) is a montonic function of time. If one counts the number M of positive going zero crossings in a period of time τ, then the average frequency of the signal is M/τ. If we assume that the signal is $V(t)$ as defined in (2), then

$$\frac{M}{\tau} = \nu_0(1 + \bar{y}_n). \tag{54}$$

If we assume that the signal is $v'(t)$ as defined in (48), then

$$\frac{M}{\tau} = \nu_0 \, |\bar{y}_n|. \tag{55}$$

Again, one measures only positive frequencies.

E. Frequency Discriminators

A frequency discriminator is a device that converts frequency fluctuations into an analog voltage by means of a dispersive element. For example, by slightly detuning a resonant circuit from the signal $V(t)$ the frequency fluctuations $(1/2\pi)]\dot{\varphi}(t)$ are converted to amplitude fluctuations of the output signal. Provided the input amplitude fluctuations $[\epsilon(t)]/V_0$ are insignificant, the output amplitude fluctuations can be a good measure of the frequency fluctuations. Obviously, more sophisticated frequency discriminators exist (e.g., the cesium beam).

From the analog voltage one may use analog spectrum analyzers to determine $S_y(f)$, the frequency stability. By converting to digital data, other analyses are possible on a computer.

F. Common Hazards

1) Errors Caused by Signal-Processing Equipment: The intent of most frequency stability measurements is to evaluate the source and not the measuring equipment. Thus, one must know the performance of the measuring system. Of obvious importance are such aspects of the measuring equipment as noise level, dynamic range, resolution (dead time), and frequency range.

It has been pointed out that the noise bandwidth f_h is very essential for the mathematical convergence of certain expressions. Insofar as one wants to measure the signal *source*, one must know that the measuring system is not limiting the frequency response. At the very least, one must recognize that the frequency limit of the measuring system may be a very important, implicit parameter for either $\sigma_y^2(t)$ or $S_y(f)$. Indeed, one must account for any deviations of the measuring system form ideality such as a "nonflat" frequency response of the spectrum analyzer itself.

Almost any electronic circuit that processes a signal will, to some extent, convert amplitude fluctuations at the input terminals into phase fluctuations at the output. Thus, AM noise at the input will cause a time-varying phase (or FM noise) at the output. This can impose important constraints on limiters and automatic gain control (AGC) circuits when good frequency stability is needed. Similarly, this imposes constraints on equipment used for frequency stability measurements.

2) Analog Spectrum Analyzers (Frequency Domain): Typical analog spectrum analyzers are very similar in design to radio receivers of the superheterodyne type, and thus certain design features are quite similar. For example, image rejection (related to predetection bandwidth) is very important. Similarly, the actual shape of the analyzer's frequency window is important since this affects spectral resolution. As with receivers, dynamic range can be critical for the analysis of weak signals in the presence of substantial power in relatively narrow bandwidths (e.g., 60 Hz).

The slewing rate of the analyzer must be consistent with the analyzer's frequency window and the post-detection bandwidth. If one has a frequency window of 1 Hz, one cannot reliably estimate the intensity of a bright line unless the slewing rate is much slower than 1 Hz/s. Additional post-detection filtering will further reduce the maximum usable slewing rate.

3) Spectral Density Estimation from Time Domain Data: It is beyond the scope of this paper to present a comprehensive list of hazards for spectral density estimation; one should consult the literature [2]–[5]. There are a few points, however, which are worthy of special notice: a) data aliasing (similar to predetection bandwidth problems); b) spectral resolution; and c) confidence of the estimate.

4) Variances of Frequency Fluctuations $\sigma_y^2(\tau)$: It is not uncommon to have discrete frequency modulation of a source such as that associated with the power supply frequencies. The existence of discrete frequencies in $S_y(f)$ can cause $\sigma_y^2(\tau)$ to be a very rapidly changing function of τ. An interesting situation results when τ is an exact multiple of the period of the modulation frequency (e.g., one makes $\tau = 1$ s and there exists 60-Hz frequency modulation on the signal). In this situation, $\sigma_y^2(\tau = 1 \text{ s})$ can be very optimistic relative to values with slightly different values of τ.

One also must be concerned with the convergence properties of $\sigma_y^2(\tau)$ since not all noise processes will have finite limits to the estimates of $\sigma_y^2(\tau)$ (see Appendix I). One must be as critically aware of any "dead time" in the measurement process as of the system bandwidth.

5) Signal Source and Loading: In measuring frequency stability one should specify the exact location in the circuit from which the signal is obtained and the nature of the load used. It is obvious that the transfer characteristics of the device being specified will depend on the load and that the measured frequency stability might be affected. If the load itself is not constant during the measurements, one expects large effects on frequency stability.

6) Confidence of the Estimate: As with any measurement in science, one wants to know the confidence to assign to numerical results. Thus, when one measures $S_y(f)$ or $\sigma_y^2(\tau)$, it is important to know the accuracies of these estimates.

a) The Allan Variance: It is apparent that a single sample variance $\sigma_y^2(4, \tau, \tau)$ does not have good confidence, but, by averaging many independent samples, one can improve the accuracy of the estimate greatly. There is a key point in this statement, "independent samples." For this argument to be true, it is important that one sample variance be independent of the next. Since $\sigma_y^2(2, \tau, \tau)$ is related to the first difference of the frequency (11), it is sufficient that the noise perturbing $y(t)$ have "independent increments," i.e., that $y(t)$ be a random walk. In other words, it is sufficient that $S_y(f) \sim f^{-2}$ for low frequencies. One can show that for noise processes that are more divergent at low frequencies than f^{-2}, it is difficult (or impossible) to gain good confidence on estimates of $\sigma_y^2(\tau)$. For noise processes that are less divergent than f^{-2}, no problem exists.

It is worth noting that if we were interested in $\sigma_y^2(N = \infty, \tau, \tau)$, then the limit noise would become $S_y(f) \sim f^0$ instead of f^{-2} as it is for $\sigma_y^2(2, \tau, \tau)$. Since most real signal generators possess low-frequency divergent noises, $\langle \sigma_y^2(2, \tau, \tau) \rangle$ is more useful than $\sigma_y^2(N = \infty, \tau, \tau)$.

Although the sample variances $\sigma_y^2(2, \tau, \tau)$ will not be normally distributed, the variance of the average of m

independent (nonoverlapping) samples of $\sigma_y^2(2, \tau, \tau)$ (i.e., the variance of the Allan variance) will decrease as $1/m$ provided the conditions on low-frequency divergence are met. For sufficiently large m, the distribution of the m sample averages of $\sigma_y^2(2, \tau, \tau)$ will tend toward normal (central limit theorem). It is thus possible to estimate confidence intervals based on the normal distribution.

As always, one may be interested in τ values approaching the limits of available data. Clearly, when one is interested in τ values of the order of a year, one is severely limited in the size of m, the number of samples of $\sigma_y^2(2, \tau, \tau)$. Unfortunately, there seems to be no substitute for many samples and one extends τ at the expense of confidence in the results. "Truth in packaging" dictates that the sample size m be stated with the results.

b) Spectral Density: As before, one is referred to the literature for discussions of spectrum estimation [2]–[5]. It is worth pointing out, however, that for $S_y(f)$ there are basically two different types of averaging that can be employed: sample averaging of independent estimates of $S_y(f)$, and frequency averaging where the resolution bandwidth is made much greater than the reciprocal data length.

VIII. Conclusions

A good measure of frequency stability is the spectral density $S_y(f)$ of fractional frequency fluctuations $y(t)$. An alternative is the expected variance of N sample averages of $y(t)$ taken over a duration τ. With the beginning of successive sample periods spaced every T units of time, the variance is denoted by $\sigma_y^2(N, T, \tau)$. The stability measure, then, is the expected value of many measurements of $\sigma_y^2(N, T, \tau)$ with $N = 2$ and $T = \tau$; that is, $\sigma_y^2(\tau)$. For all real experiments one has a finite bandwidth. In general, the time domain measure of frequency stability $\sigma_y^2(\tau)$ is dependent on the noise bandwidth of the system. Thus, there are four important parameters to the time domain measure of frequency stability.

- N Number of sample averages ($N = 2$ for preferred measure).
- T Repetition time for successive sample averages ($T = \tau$ for preferred measure).
- τ Duration of each sample average.
- f_h System noise bandwidth.

Translations among the various stability measures for common noise types are possible, but there are significant reasons for choosing $N = 2$ and $T = \tau$ for the preferred measure of frequency stability in the time domain. This measure, the Allan variance, ($N = 2$) has been referenced by [12], [20]–[22] and more.

Although $S_y(f)$ appears to be a function of the single variable f, actual experimental estimation procedures for the spectral density involve a great many parameters. Indeed, its experimental estimation can be at least as involved as the estimation of $\sigma_y^2(\tau)$.

Appendix I

We want to derive (23) in the text. Starting from (10) we have

$$\begin{aligned}\langle\sigma_y^2(N, T, \tau)\rangle &= \left\langle \frac{1}{N-1}\sum_{n=1}^{N}\left(\bar{y}_n - \frac{1}{N}\sum_{k=1}^{N}\bar{y}_k\right)^2\right\rangle \\ &= \frac{1}{N-1}\left\{\sum_{n=1}^{N}\langle\bar{y}_n^2\rangle - \frac{1}{N}\sum_{i=1}^{N}\sum_{j=1}^{N}\langle\bar{y}_i\bar{y}_j\rangle\right\} \\ &= \frac{1}{(N-1)\tau^2}\left\{\sum_{n=1}^{N}\int_{t_n}^{t_n+\tau}dt''\int_{t_n}^{t_n+\tau}dt'\,\langle y(t')y(t'')\rangle\right. \\ &\quad \left. - \frac{1}{N}\sum_{i=1}^{N}\sum_{j=1}^{N}\int_{t_i}^{t_i+\tau}dt''\int_{t_j}^{t_j+\tau}dt'\,\langle y(t')y(t'')\rangle\right\} \end{aligned} \tag{56}$$

where (9) has been used. Now

$$\langle y(t')y(t'')\rangle = R_y(t' - t'') \tag{57}$$

where $R_y(\tau)$ is the autocorrelation function of $y(t)$ and is the Fourier transform of $S_y(f)$, the power spectral density of $y(t)$. Equation (57) is true provided that $y(t)$ is stationary (at least in the wide or covariance sense), and that the average exists. If we assume the power spectral density of $y(t)$, $S_y(f)$ has low and high frequency cutoffs f_l and f_h (if necessary) so that

$$\int_0^\infty S_y(f)\,df$$

exists, then if y is a random variable, the average does exist and we may safely assume stationarity.

In practice, the high-frequency cutoff f_h is always present either in the device being measured or in the measuring equipment itself. When the high-frequency cutoff is necessary for convergence of integrals of $S_y(f)$ (or is too low in frequency), the stability measure will depend on f_h. The latter case can occur when the measuring equipment is too narrow-band. In fact, a useful type of spectral analysis may be done by varying f_h purposefully [18].

The low-frequency cutoff f_l may be taken to be much smaller than the reciprocal of the longest time of interest. The results of calculations as well as measurements will be meaningful if they are independent of f_l as f_l approaches zero. The range of exponents in power law spectral densities for which this is true will be discussed and are given in Fig. 1.

To continue, the derivation requires the Fourier transform relationships between the autocorrelation function and the power spectral density

$$\begin{aligned} S_y(f) &= 4\int_0^\infty R_y(\tau)\cos 2\pi f\tau\,d\tau \\ R_y(\tau) &= \int_0^\infty S_y(f)\cos 2\pi f\tau\,df. \end{aligned} \tag{58}$$

Using (58) and (57) in (56) gives

$$\langle\sigma_y^2(N, T, \tau)\rangle = \frac{1}{(N-1)\tau^2}\left\{\sum_{n=1}^{N}\int_0^\infty df\, S_y(f)\int_{t_n}^{t_n+\tau} dt''\right.$$

$$\cdot\int_{t_n}^{t_n+\tau} dt' \cos 2\pi f(t'-t'') - \frac{1}{N}\sum_{i=1}^{N}\sum_{j=1}^{N}\int_0^\infty df\, S_y(f)$$

$$\left.\cdot\int_{t_i}^{t_i+\tau} dt''\int_{t_j}^{t_j+\tau} dt' \cos 2\pi f(t'-t'')\right\}$$

$$= \frac{1}{(N-1)\tau^2}\left\{\sum_{n=1}^{N}\left[\int_0^\infty df\, S_y(f)\frac{\sin^2 \pi f\tau}{(\pi f)^2}\right]\right.$$

$$-\frac{1}{N}\sum_{i=1}^{N}\sum_{j=1}^{N}\left[\int_0^\infty df\, \frac{S_y(f)}{(2\pi f)^2}\,(2\cos 2\pi fT(j-i)\right.$$

$$\left.\left.-\cos 2\pi f[T(j-i)+\tau] - \cos 2\pi f[T(j-i)-\tau])\right]\right\}. \tag{59}$$

(The interchanges in order of integration are permissible here since the integrals are uniformly convergent with the given restrictions on $S_y(f)$.) The first summation in the curly brackets is independent of the summation index n and thus gives just

$$N\int_0^\infty df\, S_y(f)\frac{\sin^2 \pi f\tau}{(\pi f)^2}. \tag{60}$$

The kernel in the second term in the curly brackets may be further simplified

$$2\cos 2\pi fT(j-i) - \cos 2\pi f(T(j-i)+\tau)$$
$$- \cos 2\pi f(T(j-i)-\tau) = 4\sin^2 \pi f\tau \cos 2\pi fT(j-i). \tag{61}$$

The second term is then

$$-\frac{1}{N}\left(\int_0^\infty df\, \frac{S_y(f)}{(\pi f)^2}\sin^2 \pi f\tau \sum_{i=1}^{N}\sum_{j=1}^{N}\cos 2\pi fT(j-i)\right). \tag{62}$$

(The interchange of summation and integration is justified.) We must now do the double sum. Let

$$j - i = k$$
$$2\pi fT = x. \tag{63}$$

Changing summation indices from i and j to i and k gives for the sum

$$S \equiv \sum_{i=1}^{N}\sum_{j=1}^{N}\cos x(j-i) = \sum_{i=1}^{N}\sum_{k=1-i}^{N-i}\cos kx. \tag{64}$$

The region of summation over the discrete variables i and k is shown in Fig. 5 for $N = 4$.

The summand is independent of i so that one may interchange the order of summation and sum over i first. The summand is even in k and the contributions for $k < 0$ are equal to those for $k > 0$, and so we may pull out the term for $k = 0$ separately and write

$$S = 2\left(\sum_{k=1}^{N-1}\cos kx \sum_{i=1}^{N-k} 1\right) + \sum_{i+1}^{N} 1$$

$$= 2\left(\sum_{k=1}^{N-1}(N-k)\cos kx\right) + N. \tag{65}$$

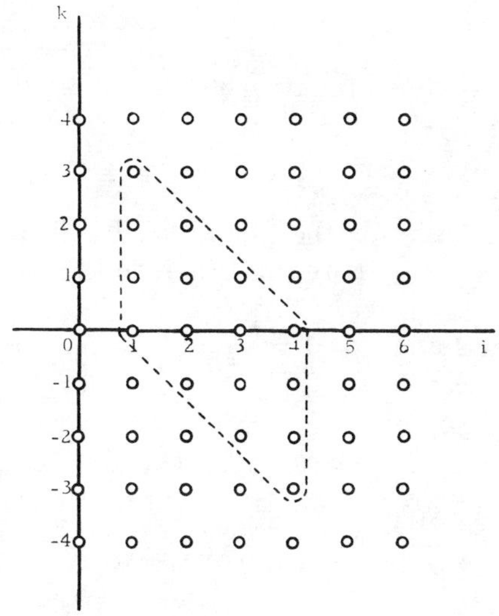

Fig. 5. Region of summation for i and k for $N = 4$.

This may be written as

$$S = N + 2\,\mathrm{Re}\left[N - \frac{1}{i}\frac{d}{dx}\right]\sum_{k=1}^{N-1} e^{ikx} \tag{66}$$

where $\mathrm{Re}[U]$ means the real part of U and d/dx is the differential operator. The series is a simple geometric series and may be summed easily, giving

$$S = N + 2\,\mathrm{Re}\left\{\left[N - \frac{1}{i}\frac{d}{dx}\right]\frac{e^{ix} - e^{iNx}}{1 - e^{ix}}\right\}$$

$$= N + 2\,\mathrm{Re}\left\{\frac{1 - e^{iNx} - N(1 - e^{ix})}{4\sin^2 x/2}\right\}$$

$$= \frac{\sin^2 Nx/2}{\sin^2 x/2}. \tag{67}$$

Combining everything we get, after some rearrangement,

$$\langle\sigma_y^2(N, T, \tau)\rangle$$

$$= \frac{N}{N-1}\int_0^\infty df\, S_y(f)\frac{\sin^2 \pi f\tau}{(\pi f\tau)^2}\left[1 - \frac{\sin^2 \pi r f N\tau}{N^2 \sin^2 \pi r f\tau}\right] \tag{68}$$

where $r = T/\tau$. This is the result given in (23).

We can determine a number of things very easily from this equation. First let us change variables. Let $\pi f\tau = u$, then

$$\langle\sigma_y^2(N, T, \tau)\rangle$$

$$= \frac{N}{(N-1)\pi\tau}\int_0^\infty du\, S_y\!\left(\frac{u}{\pi\tau}\right)\frac{\sin^2 u}{u^2}\left\{1 - \frac{\sin^2 Nru}{N^2\sin^2 ru}\right\}. \tag{69}$$

The kernel behaves like u^2 as $u \to 0$ and like u^{-2} as $u \to \infty$. Therefore $\langle\sigma_y^2(N, T, \tau)\rangle$ is convergent for power law spectral densities, $S_y(f) = h_\alpha f^\alpha$, without any low- or high-frequency cutoffs for $-3 < \alpha < 1$. Using (69) for power law spectral densities we find

$$\langle\sigma_y^2(N, T, \tau)\rangle = \tau^{-\alpha-1}h_\alpha C_\alpha, \qquad -3 < \alpha < 1$$
$$= \tau^{\mu}h_\alpha C_\alpha, \qquad \mu \equiv -\alpha - 1$$

and

$$C_\alpha \equiv \frac{N}{(N-1)\pi^{\alpha+1}} \int_0^\infty du\, u^\alpha \frac{\sin^2 u}{u^2} \left\{1 - \frac{\sin^2 Nru}{N^2 \sin^2 ru}\right\}. \tag{70}$$

This is the basis for the plot in Fig. 1 in the text of μ versus α. For $\alpha \geq 1$ we must include the high-frequency cutoff f_h.

For $N = 2$ and $r = 1$ the results are particularly simple. We have

$$\langle \sigma_y^2(2, \tau, \tau)\rangle = \tau^{-\alpha-1} h_\alpha \frac{2}{\pi^{\alpha+1}} \int_0^\infty du\, u^{\alpha-2} \sin^4 u \tag{71}$$

for power law spectral densities. For $N = 2$ and general r we get

$$\langle \sigma_y^2(2, T, \tau)\rangle$$

$$= \frac{1}{2\pi\tau} \int_0^\infty du\, S_y\left(\frac{u}{\pi\tau}\right)$$

$$\cdot \frac{1 - \cos 2u - \cos 2ru + \dfrac{\cos 2u(r+1)}{2} + \dfrac{\cos 2u(r-1)}{2}}{u^2}$$

$$= \frac{2}{\pi\tau} \int_0^\infty du\, S_y\left(\frac{u}{\pi\tau}\right) \frac{\sin^2 u \sin^2 ru}{u^2}. \tag{72}$$

The first form in (72) is particularly simple and is also useful for $r = 1$ in place of (71).

Let us discuss the case for $\alpha \geq 1$ in a little more detail. As mentioned above we must include the high-frequency cutoff f_h for convergence. The general behavior can be seen most easily from (68). After placing the factor τ^{-2} outside the integral and combining the factor f^{-2} with $S_y(f)$ we find that the remaining part of the kernel consists of some constants and some oscillatory terms. If $2\pi f_h \tau \gg 1$ it is apparent that the rapidly oscillating terms contribute very little to the integral. Most of the contribution comes from the integral over the constant term causing the major portion of the τ dependence to be the τ^{-2} factor outside the integral. This is the reason for the vertical slope at $\mu = -2$ in the μ versus α plot in Fig. 1 in the text.

One other point deserves some mention. The constant term of the kernel discussed in the preceding paragraph is different for $r = 1$ from the value for $r \neq 1$. This is readily seen from (72) for $N = 2$; for $r = 1$ the constant term is 3/2 while for $r \neq 1$ it is 1. This is the reason for $\delta_k(r-1)$, which appears in some of the results of Appendix II. In practice, $\delta_k(r-1)$ does not have zero width but is smeared out over a width of approximately $(2\pi f_h \tau)^{-1}$. If there must be dead time $r \neq 1$, it is wise to choose $(r-1) \gg (2\pi f_h\tau)^{-1}$ or $(r-1) \ll (2\pi f_h \tau)^{-1}$ but with $2\pi f_h \tau \gg 1$. In the latter case, one may assume $r \approx 1$.

Appendix II

Let $y(t)$ be a sample function of a random noise process with a spectral density $S_y(f)$. The function $y(t)$ is assumed to be pure real and $S_y(f)$ is a one-sided spectral density relative to a cycle frequency (i.e., the dimensions of $S_y(f)$ are that of y^2 per hertz). (For additional information see Appendix I, [7], [8], [18].)

Let $x(t)$ be defined by the equation

$$\dot{x}(t) \equiv \frac{dx}{dt} \equiv y(t). \tag{73}$$

Define the following. t_0 is arbitrary instant of time and

$$t_{n+1} \equiv t_n + T, \qquad n = 0, 1, 2, \cdots, \tag{74}$$

$$\bar{y}_n \equiv \frac{1}{\tau} \int_{t_n}^{t_n+\tau} y(t)\, dt = \frac{x(t_n + \tau) - x(t_n)}{\tau} \tag{75}$$

$$\langle \bar{y} \rangle_N \equiv \frac{1}{N} \sum_{n=1}^N \bar{y}_n, \tag{76}$$

and let f_h be a high-frequency cutoff (infinitely sharp) with $2\pi f_h \tau \gg 1$.

Definition:

$$\langle \sigma_y^2(N, T, \tau)\rangle \equiv \left\langle \frac{1}{N-1} \sum_{n=1}^N (\bar{y}_n - \langle \bar{y}\rangle_N)^2 \right\rangle. \tag{77}$$

Special Case:

$$\langle \sigma_y^2(2, T, \tau)\rangle = \left\langle \frac{(\bar{y}_2 - \bar{y}_1)^2}{2} \right\rangle. \tag{78}$$

Special Case:

$$\sigma_y^2(\tau) \equiv \langle \sigma_y^2(2, \tau, \tau)\rangle$$

$$= \left\langle \frac{[x(t_0 + 2\tau) - 2x(t_0 + \tau) + x(t_0)]^2}{2\tau^2} \right\rangle. \tag{79}$$

Definition:

$$D_x^2(\tau) \equiv \langle [x(t_0 + 2\tau) - 2x(t_0 + \tau) + x(t_0)]^2\rangle. \tag{80}$$

Consequence of Definitions:

$$D_x^2(\tau) = 2\tau^2 \sigma_y^2(\tau) \equiv 2\sigma_x^2(\tau). \tag{81}$$

Definition:

$$\psi_x^2(T, \tau) \equiv \langle [x(t_0 + T + \tau) - x(t_0 + T) - x(t_0 + \tau) + x(t_0)]^2\rangle. \tag{82}$$

Consequence of Definitions:

$$\psi_x^2(T, \tau) = 2\tau^2 \langle \sigma_y^2(2, T, \tau)\rangle. \tag{83}$$

Special Case:

$$\psi_x^2(\tau, \tau) = D_x^2(\tau). \tag{84}$$

Random Walk y

$$S_y(f) = \frac{h_{-2}}{f^2} \qquad \left(S_x(f) = \frac{h_{-2}}{(2\pi)^2 f^4}\right)$$

$$r \equiv \frac{T}{\tau}, \qquad 0 \leq f \leq f_h .$$

Quantity	*Relation*	
$\langle\sigma_y^2(N, T, \tau)\rangle$	$h_{-2}\cdot\frac{(2\pi)^2\,\lvert\tau\rvert}{12}\,[r(N+1)-1], \quad r \geq 1$	(85)
$\langle\sigma_y^2(N, \tau, \tau)\rangle$	$h_{-2}\cdot\frac{(2\pi)^2\,\lvert\tau\rvert}{12}\cdot N, \quad r = 1$	(86)
$\sigma_y^2(\tau)$	$h_{-2}\cdot\frac{(2\pi)^2\,\lvert\tau\rvert}{6}, \quad N = 2, r = 1$	(87)
$D_x^2(\tau) = 2\sigma_x^2(\tau)$	$h_{-2}\cdot\frac{2(2\pi)^2\,\lvert\tau\rvert^3}{6}$	(88)
$\psi_x^2(T, \tau)$	$h_{-2}\cdot\frac{(2\pi)^2\,\lvert\tau\rvert^3}{6}\,(3r-1), \quad r \geq 1$	
	$h_{-2}\cdot\frac{(2\pi)^2\,\lvert T\rvert^3}{6}\left(\frac{3}{r}-1\right), \quad r \leq 1\,.$	(89)

Flicker y

$$S_y(f) = \frac{h_{-1}}{f} \qquad \left(S_x(f) = \frac{h_{-1}}{(2\pi)^2 f^3}\right)$$

$$r = T/\tau, \qquad 0 \leq f \leq f_h\,.$$

Quantity	*Relation*	
$\langle\sigma_y^2(N, T, \tau)\rangle$	$h_{-1}\cdot\frac{1}{N(N-1)}\sum_{n=1}^{N}(N-n)$	
	$\cdot[-2(nr)^2\ln(nr) + (nr+1)^2\ln(nr+1)$	
	$+ (nr-1)^2\ln\lvert nr-1\rvert]$	(90)
$\langle\sigma_y^2(N, \tau, \tau)\rangle$	$h_{-1}\cdot\frac{N\ln N}{N-1}, \quad (r = 1)$	(91)
$\sigma_y^2(\tau)$	$h_{-1}\cdot 2\ln 2, \; (N = 2, r = 1)$	(92)
$D_x^2(\tau) = 2\sigma_x^2(\tau)$	$h_{-1}\cdot 4\tau^2\ln 2$	(93)
$\psi_x^2(T, \tau)$	$h_{-1}\cdot\tau^2[-2r^2\ln r + (r+1)^2\ln(r+1)$	
	$+ (r-1)^2\ln\lvert r-1\rvert]$	(94)
	$\sim h_{-1}\cdot 2\tau^2(2+\ln r), \quad r \gg 1$	
	$\sim h_{-1}\cdot 2T^2(2-\ln r), \quad r \ll 1\,.$	(95)

White y (Random Walk x)

$$S_y(f) = h_0 \qquad S_x(f) = \frac{h_0}{(2\pi)^2 f^2}$$

$$r = T/\tau, \qquad 0 \leq f \leq f_h\,.$$

Quantity	*Relation*	
$\langle\sigma_y^2(N, T, \tau)\rangle$	$\frac{h_0}{2}\cdot\lvert\tau\rvert^{-1}, \quad r \geq 1$	
	$h_0\cdot\frac{1}{6}r(N+1)\,\lvert\tau\rvert^{-1}, \quad Nr \leq 1$	(96)
$\langle\sigma_y^2(N, \tau, \tau)\rangle$	$\frac{h_0}{2}\cdot\lvert\tau\rvert^{-1}, \quad r = 1$	(97)
$\sigma_y^2(\tau)$	$\frac{h_0}{2}\cdot\lvert\tau\rvert^{-1}, \quad N = 2, r = 1$	(98)
$D_x^2(\tau) = 2\sigma_x^2(\tau)$	$h_0\cdot\lvert\tau\rvert$	(99)
$\psi_x^2(T, \tau)$	$h_0\cdot\lvert\tau\rvert, \quad r \geq 1$	
	$h_0\cdot T, \quad r \leq 1\,.$	(100)

Flicker x

$$S_y(f) = h_1\,\lvert f\rvert\left(S_x(f) = \frac{h_1}{(2\pi)^2 f}\right)$$

$$r = T/\tau, \qquad 2\pi f_h\tau \gg 1, \qquad 2\pi f_h T \gg 1, \qquad 0 \leq f \leq f_h\,.$$

Quantity	*Relation*	
$\langle\sigma_y^2(N, T, \tau)\rangle$	$h_1\cdot\frac{2}{(2\pi\tau)^2}\Big\{0.577 + \ln(2\pi f_h\tau)$	
	$+ \frac{1}{N(N-1)}\sum_{n=1}^{N-1}(N-n)$	
	$\cdot\ln\left[\frac{n^2r^2}{n^2r^2-1}\right]\Big\}, \quad r \gg 1$	(101)
$\langle\sigma_h^2(N, \tau, \tau)\rangle$	$h_1\cdot\frac{2(N+1)}{N\tau^2(2\pi)^2}\left[0.577 + \ln(2\pi f_h\tau) - \frac{\ln N}{N^2-1}\right],$	
	$r = 1$	(102)
$\sigma_y^2(\tau)$	$h_1\cdot\frac{1}{\tau^2(2\pi)^2}\left\{3[0.577 + \ln(2\pi f_h\tau)] - \ln 2\right\},$	
	$N = 2, r = 1$	(103)
$D_x^2(\tau) = 2\sigma_x^2(\tau)$	$\frac{h_1}{(2\pi)^2}\cdot 2\left\{3[0.577 + \ln(2\pi f_h\tau)] - \ln 2\right\}$	(104)
$\psi_x^2(T, \tau)$	$h_1\cdot\frac{4}{(2\pi)^2}\,[0.577 + \ln(2\pi f_h\tau)], \quad r \gg 1$	
	$h_1\cdot\frac{2}{(2\pi)^2}\left\{3[0.577 + \ln(2\pi f_h\tau)] - \ln 2\right\},$	
	$r = 1$	(105)
	$h_1\cdot\frac{4}{(2\pi)^2}\,[0.577 + \ln(2\pi f_h T)], \quad r \ll 1\,.$	

White x

$$S_y(f) = h_2 f^2 \qquad \left(S_x(f) = \frac{h_2}{(2\pi)^2}\right)$$

$$r = T/\tau;\; \delta_k(r-1) = \begin{cases}1, & \text{if } r = 1\\ 0, & \text{otherwise}\end{cases}$$

$$2\pi f_h\tau \gg 1, \qquad 0 \leq f \leq f_h\,.$$

Quantity	*Relation*	
$\langle\sigma_y^2(N, T, \tau)\rangle$	$h_2\cdot\frac{N + \delta_k(r-1)}{N(2\pi)^2}\cdot\frac{2f_h}{\tau^2}$	(106)
$\langle\sigma_y^2(N, \tau, \tau)\rangle$	$h_2\cdot\frac{N+1}{N(2\pi)^2}\cdot\frac{2f_h}{\tau^2}, \quad r = 1$	(107)
$\sigma_y^2(\tau)$	$h_2\cdot\frac{3f_h}{(2\pi)^2\tau^2}, \quad N = 2, r = 1$	(108)
$D_x^2(\tau) = 2\sigma_x^2(\tau)$	$h_2\cdot\frac{6f_h}{(2\pi)^2}$	(109)
$\psi_x^2(T, \tau)$	$h_2\cdot[2 + \delta_k(r-1)]\,\frac{2f_h}{(2\pi)^2}$	(110)

ACKNOWLEDGMENT

The authors are particularly indebted to D. W. Allan, Dr. D. Halford, Dr. S. Jarvis, and Dr. J. J. Filliben of the National Bureau of Standards. The authors are also indebted to Mrs. Carol Wright for preparing the many revised copies of this paper.

REFERENCES

[1] E. T. Jaynes, "Information theory and statistical mechanics," *Phys. Rev.*, vol. 108, sec. 15, Oct. 1957, pp. 171–190.
[2] C. Bingham, M. D. Godfrey, and J. W. Tukey, "Modern techniques of power spectrum estimation," *IEEE Trans. Audio Electrocoust.*, vol. AU-15, June 1967, pp. 56–66.
[3] R. B. Blackman, *Linear Data Smoothing and Prediction in Theory and Practice*. Reading, Mass.: Addison-Wesley, 1965.
[4] R. B. Blackman and J. W. Tukey, *The Measurement of Power Spectra*. New York: Dover, 1958.
[5] E. O. Brigham and R. E. Morrow, "The fast Fourier transform," *IEEE Spectrum*, vol. 4, Dec. 1967, pp. 63–70.
[6] E. J. Baghdady, R. D. Lincoln, and B. D. Nelin, "Short-term frequency stability: theory, measurement, and status," *Proc. IEEE–NASA Symp. on Short-Term Frequency Stability* (NASA SP-80), Nov. 1964, pp. 65–87; also in *Proc. IEEE*, vol. 53, July 1965, pp. 704–722 and *Proc. IEEE* (Corresp.), vol 53, Dec. 1965, pp. 2110–2111.
[7] L. Cutler and C. Searle, "Some aspects of the theory and measurement of frequency fluctuations in frequency standards," *Proc. IEEE*, vol. 54, Feb. 1966, pp. 136–154.
[8] D. W. Allan, "Statistics of atomic frequency standards," *Proc. IEEE*, vol. 54, Feb. 1966, pp. 221–230.
[9] N. A. Shewart, *Economic Control of Quality of Manufactured Product*. Princeton, N. J.: Van Nostrand, 1931, p. 146.
[10] J. A. Barnes, "Atomic timekeeping and the statistics of precision signal generators," *Proc. IEEE*, vol. 54, Feb. 1966, pp. 207–220.
[11] J. A. Barnes and D. W. Allan, "An approach to the prediction of coordinated universal time," *Frequency*, vol. 5, Nov.–Dec. 1967, pp. 15–20.
[12] R. F. C. Vessot *et al.*, "An intercomparison of hydrogen and cesium frequency standards," *IEEE Trans. Instrum. Meas.*, vol. IM-15, Dec. 1966, pp. 165–176.
[13] J. A. Barnes, "Tables of bias functions, B_1 and B_2, for variances based on finite samples of processes with power law spectral densities," NBS, Washington, D. C., Tech. Note 375, Jan. 1969.
[14] D. B. Leeson and G. F. Johnson, "Short-term stability for a Doppler radar: requirements, measurements, and techniques," *Proc. IEEE*, vol. 54, Feb. 1966, pp. 244–248.
[15] W. K. Saunders in *Radar Handbook*, M. I. Skolnik, Ed. New York: McGraw Hill, 1970, ch. 16.
[16] R. S. Raven, "Requirements on master oscillators for coherent radar," *Proc. IEEE*, vol. 54, Feb. 1966, pp. 237–243.
[17] D. B. Leeson, "A simple model of feedback oscillator noise spectrum," *Proc. IEEE* (Lett.), vol. 54, Feb. 1966, pp. 329–330.
[18] R. F. C. Vessot, L. Mueller, and J. Vanier, "The specification of oscillator characteristics from measurements made in the frequency domain," *Proc. IEEE*, vol. 54, Feb. 1966, pp. 199–207.
[19] Floyd M. Gardner, *Phaselock Techniques*. New York: Wiley, 1966.
[20] C. Menoud, J. Racine, and P. Kartaschoff, "Atomic hydrogen maser work at L.S.R.H., Neuchatel, Switzerland," *Proc. 21st Ann. Symp. Frequency Control*, Apr. 1967, pp. 543–567.
[21] A. G. Mungall, D. Morris, H. Daams, and R. Bailey, "Atomic hydrogen maser development at the National Research Council of Canada," *Metrologia*, vol. 4, July 1968, pp. 87–94.
[22] R. F. C. Vessot, "Atomic hydrogen masers, an introduction and progress report," *Hewlett-Packard J.*, vol. 20, Oct. 1968, pp. 15–20.

Theory of Oscillator Instability Based Upon Structure Functions

WILLIAM C. LINDSEY, FELLOW, IEEE, AND CHAK MING CHIE, STUDENT MEMBER, IEEE

Abstract—Structure functions are apparently unfamiliar to most engineers and the unifying role they play in oscillator instability theory has largely gone unrecognized. This paper introduces and places into perspective the role which *Kolmogorov structure functions* have in theory. It is demonstrated that the rms fractional frequency deviation (phase accumulation) introduced by Cutler and Searle is related to the first phase structure function; the two-sample Allan variance is related to the second phase structure function. In addition, it is shown how the two-sample Allan variance is related to the rms fractional frequency deviation under suitable conditions. The L-sample Allan variance is also identified in terms of the first phase structure function; it is shown to be an asymptotically unbiased estimator of the rms fractional frequency deviation squared if the latter exists. The utility of higher order structure functions of frequency and phase in the theory of instability is also demonstrated; in particular, how the frequency drift and "flicker"-type noise convergence problems can be overcome.

I. Introduction

IN CONTEMPORARY telecommunication engineering, it is becoming of greater interest to study and account for the deleterious effects which oscillator instability produces on high-capacity system performance; especially those operating in the S, X, and K bands. The same is true in advanced radar and navigation systems as well as in predicting the requirements and specifications for future frequency and timing technology. In doing so, it is often increasingly necessary to take into account the random fluctuations and instability in the parameters which characterize the oscillation, viz., frequency and phase.

In the last decade, a large number of papers and reports pertaining to frequency instability have been published. The state of affairs prior to 1966 are largely surveyed and reviewed in [1]. The main results presented therein stem from the important work of Cutler and Searle [2], Allan [3], Barnes [4], Barnes and Allan [5], Baghdady *et al.* [6], Sydnor *et al.* [7], and Tausworthe and Gray [8], [9]. In addition, other contributions can be found in [13]–[39] dealing with measurements, system applications, and modeling. We will elaborate on these later.

Characterization of oscillator frequency and phase instability is developed herein in terms of the Nth-order structure functions of frequency and phase. The interconnections and similarity of this new approach with previous characterizations and performance criteria is given; the generality and unifying role of this new approach is also presented. The following are primary reasons for introducing structure functions.

(1) Structure functions of the random processes involved are fundamental to definitions involving oscillator instability and serve to alleviate certain enigmas associated with previous characterizations.

(2) The random processes which affect oscillator instability have large amounts of energy in the low-frequency region, i.e., their spectrum may have a power singularity of the form $1/|f|^\nu$ at the origin.[1] Use of structure functions can avoid the mathematical difficulties due to these singularities.

(3) The random processes which affect oscillator instability have mean values which are functions of time. Applications of structure functions can alleviate the problem of detrending measured data due to the oscillator drift and the associated error.

(4) A unified theory can be developed in terms of random process theory without recourse to special and *ad hoc* definitions.

For the most part, these facts have been overlooked in previous mathematical characterizations.

Mathematical transformations are derived which relate the structure functions of frequency and phase instability to the power spectral density (PSD) of frequency instability. It is shown that the PSD of the frequency instabilities together with the highest order of frequency drift uniquely determine the appropriate structure function of phase and frequency. As such, it is fundamental to the theory as well as to measurement techniques.

A new algorithm is formulated to determine the drift order and the PSD of frequency instability from standard frequency-counted measurements. As an additional bonus, it provides a means to test, within the resolving power of the measuring device, whether the frequency noise process has a lower cut-off frequency or if it is a truly "flicker" type process.

To demonstrate our results, a "flicker"-type noise example is given to show how the frequency drift and the convergence problems are overcome. This serves to point out the advantages of our approach and justify the necessary departure from the familiar second moment theory of random processes. In special cases of our theory, the true and measured rms fractional frequency deviation introduced by Cutler and Searle, and the two-sample Allan variance as determined from frequency-counted measurements, are discussed in terms of the conditions under which these quantities are meaningful. The L-sample Allan variance is also identified in terms of the first phase structure function and it is shown to be an asymptotically unbiased estimator of the rms fractional frequency deviation squared, if it exists. The bias is explicitly identified and it is shown under what conditions the L-sample Allan variance will converge. In the Appendices, cross-reference formulas between these "standards" and the structure functions are derived whenever it is legitimate. They would be useful in certain special applications. Also, we comment on the bias of these quantities in connection with the general

Manuscript received January 3, 1975; revised June 30, 1976. This work was performed for LinCom Corporation, Pasadena, CA, under Contract N000173-75-C0481 sponsored by the Satellite Communication Branch of the Naval Research Laboratory, Washington, DC. It was also supported in part by the Office of Naval Research under Contract N00014-75-C-0519.

The authors are with the Department of Electrical Engineering, University of Southern California, Los Angeles, CA 90007.

[1] We shall term it "flicker"-type noise.

Reprinted from *Proc. IEEE*, vol. 64, pp. 1652–1666, Dec. 1976.

theory, and show when and how the Allan variance and rms fractional frequency deviation are related.

Broad categories of frequency and time standards users which depend upon frequency instability include radar and navigation systems, radio astronomy, microwave spectroscopy, precision cartography, coherent and noncoherent communication systems, clock manufacturers, geodetics and measuring device characteristics of TWT and klystron amplifiers, electronic instrumentation and measuring systems, etc. In addition, measurements of the Earth's wobble can be studied in terms of the theory given herein.

A. Representation of the Oscillator Motion

The oscillators considered here belong to the category of the sine-wave generators, i.e., oscillators whose output is "close" to being sinusoidal. The oscillator output $s[t, \Phi(t)] = \sqrt{2}\, A(t) \sin \Phi(t)$ differs from the output $\sqrt{2}\, A_0 \sin \omega_0 t$ (A_0 and ω_0 constant) of an ideal oscillator because of the slowly varying but constantly accumulating random fluctuations of the *amplitude function* $A(t) = A_0 + \delta A(t)$, the *phase function* $\Phi(t)$, and radian *frequency function* $\dot{\Phi}(t)$. For a good oscillator, these random fluctuations are small for several periods $T_0 \triangleq 2\pi/\omega_0$, but with the passage of time the amplitude and phase change completely and the oscillation $s[t, \Phi(t)]$ goes into a "new" one. The random amplitude and phase fluctuations which arise are due to many factors: instability of circuit parameters; internal shot and flicker noise; power supply instabilities; ambient conditions; channel effects; etc.

For purposes of what follows we model the oscillator output by the "periodic" motion

$$s[t, \Phi(t)] = \sqrt{2}\, A(t) \sin \Phi(t) \tag{1.1}$$

where

$$A(t) \triangleq A_0 + \delta A(t) \tag{1.2}$$

and $\delta A(t)$ serves to characterize the *amplitude scatter* about the fixed amplitude A_0 and $(\Phi(t) - \Phi(0))$ serves to characterize the total *phase accumulated* during the time interval $[0, t]$. The instantaneous radian *frequency process* $\dot{\Phi}(t)$[2] of the oscillator is assumed to be of the form

$$\dot{\Phi}(t) = \underbrace{\omega_0}_{\substack{\text{mean}\\\text{frequency}}} + \underbrace{\sum_{k=1}^{N-1} \frac{\Omega_k}{k!} t^k}_{\substack{\text{long-term}\\\text{frequency}\\\text{drift}}} + \underbrace{\dot{\psi}(t)}_{\substack{\text{short-term}\\\text{frequency}\\\text{instabilities}}} \tag{1.3}$$

where ω_0 is assumed to be the constant mean frequency, the Ω_k's describe a set of (time-independent) random variables modelling the kth-order frequency drifts, and $\dot{\psi}(t)$ is a stationary,[3] zero mean random process used to characterize the short-term oscillator instability. Notice that as t increases $\Omega_{N-1} t^{N-1}/(N-1)!$ is the dominant term in the polynomial representation of the long-term frequency drift in (1.3). Hence it is plausible that the variance of Ω_{N-1} is related to the "long-term" oscillator instability.

In what follows, we develop a theory for assessing the phase and frequency stability of the oscillation defined by (1.1), (1.2), and (1.3) when the amplitude scatter $\delta A(t)$ can be considered negligible. The presentation is given in terms of structure functions of phase and frequency which have apparently gone unrecognized by engineers as being the appropriate mathematical development to embed the theory.

[2] The "overdot" denotes differentiation with respect to time.

[3] To include "flicker"-type noise and Brownian motion in this discussion, we only require a stationary process $\dot{\psi}(t)$ to have an autocorrelation function that depends only on time-difference. If $\dot{\psi}(t)$ has a finite variance as well, we say it is a *second-order* stationary process.

B. Structure Function of the Nth Increment of the Phase Process

Consider a general *phase process* $\Phi(t)$. If one wishes to characterize its first two moments from its statistical sample average or to talk about its power spectral density with any meaning, one must assume that Wiener's second moment theory and ergodicity hold [11], [15], [16]. This requires that the process $\Phi(t)$ be stationary in some sense. We note that the most general class of random processes that can provide us with such an opportunity is the class of stochastic processes with stationary Nth-order increments [15], [16]. For those unfamiliar with the notion of structure functions of random processes with stationary Nth-order increments, a short discussion is given in Appendix I. In particular, a wide-sense stationary random process is a (wide-sense) stationary zeroth-order increment process, and a process with stationary Nth increments is also stationary for higher increments.

Let us now consider a radian frequency process $\dot{\Phi}(t)$ with stationary $(N-1)$th-order increments, which we assume takes the form given in (1.3).[4] We will show later on in this section that a higher order structure function of $\dot{\psi}(t)$ can be bounded even though $\dot{\psi}(t)$ may have infinite variance. Integrating (1.3) from 0 to t, we have the oscillator *phase process*

$$\Phi(t) = \omega_0 t + \underbrace{\sum_{k=2}^{N} \frac{\Omega_{k-1}}{k!} t^k}_{\substack{\text{long-term}\\\text{phase drift}}} + \underbrace{[\psi(t) - \psi(0)]}_{\substack{\text{short-term}\\\text{phase instability}}} + \Phi(0) \tag{1.4}$$

for $N \geq 2$.

First of all, we note that the first phase increment $\Delta\psi(t) \triangleq \Delta\psi(t; \tau) \triangleq \psi(t+\tau) - \psi(t)$ in (1.4) is stationary[5] [16, p. 552]. Hence, the Nth increment of $\psi(t)$ is stationary and the Nth structure function of phase instability is given by[6] (see equation (I.10) in Appendix I)

$$D_\psi^{(N)}(\tau) \triangleq E\{[\Delta^N \psi(t)]^2\} = E\{[\Delta^{N-1}(\Delta\psi(t))]^2\}$$

$$= \sum_{k=-(N-1)}^{N-1} (-1)^k \binom{2(N-1)}{N-1+k} R_{\Delta\psi}(k\tau) \tag{1.5}$$

where $R_{\Delta\psi}(k\tau) \triangleq E[\Delta\psi(t; \tau)\Delta\psi(t+k\tau; \tau)]$ is the autocorrelation function[7] of the first increment $\Delta\psi(t; \tau)$. A simple calculation shows that if $S_{\dot{\psi}}(\omega)$ is the two-sided PSD of the

[4] Actually the following development still holds (with appropriate minor modifications) even if the short term instability represented by $\dot{\psi}(t)$ in (1.3) is not stationary. The only requirement is that $\dot{\psi}(t)$ be the output of some mth-order cascaded integrators driven by a stationary process $u(t)$, i.e., $(d^m/dt^m)\, \dot{\psi}(t) = u(t)$. In that case, the process $\dot{\psi}(t)$ does not possess a power spectral density in the usual sense and our discussions have to be based solely upon structure functions and the PSD of the stationary process $u(t)$.

[5] If the process $\psi(t)$ is stationary, then the results that follow still hold as special cases.

[6] Here $E\{\cdot\}$ denotes the (mathematical) expectation operator.

[7] If $\Delta\psi(t)$ is not a second-order process, $R_{\Delta\psi}(k\tau)$ must be interpreted as a generalized function [40]. In any case, it is a strict forward (but tedious) mathematical exercise to check that (1.7) and (1.8) remain valid in the usual sense provided that the integrals involved are finite.

stationary frequency process $\dot{\psi}(t)$, then (see equation (II.5) in Appendix II)

$$R_{\Delta\psi}(k\tau) = 4\int_{-\infty}^{\infty} e^{-jk\omega\tau}\frac{S_{\dot{\psi}}(\omega)}{\omega^2}\sin^2\left(\frac{\omega\tau}{2}\right)\frac{d\omega}{2\pi}. \quad (1.6)$$

Using (1.6) in (1.5), and making some straightforward manipulations (see (I.7) in Appendix I), one finally gets

$$D_\Phi^{(N)}(\tau) = \tau^{2N}E[\Omega_{N-1}^2] + D_\psi^{(N)}(\tau)$$

$$= \tau^{2N}E[\Omega_{N-1}^2] + 2^{2N}\int_{-\infty}^{\infty}\sin^{2N}\left(\frac{\omega\tau}{2}\right)\frac{S_{\dot{\psi}}(\omega)}{\omega^2}\frac{d\omega}{2\pi}. \quad (1.7)$$

The first term on the right-hand side of (1.7) is related to the long term phase stability in (1.4), while the second one is attributable to the short term phase fluctuations in (1.4). For $M > N$, (1.7) reduces to

$$D_\Phi^{(M)}(\tau) = D_\psi^{(M)}(\tau) = 2^{2M}\int_{-\infty}^{\infty}\sin^{2M}\left(\frac{\omega\tau}{2}\right)\frac{S_{\dot{\psi}}(\omega)}{\omega^2}\frac{d\omega}{2\pi}. \quad (1.8)$$

From (1.8) we note that the structure functions of order $M > N$ are determined solely from a knowledge of the PSD $S_{\dot{\psi}}(\omega)$. Furthermore, (1.7) reduces to a well known result for $M = 1$ [2, eq. (17a)]. In the next section we relate (1.7) to the Allan variance and the rms fractional frequency deviation introduced by Cutler and Searle [1], [2].

Let us comment briefly on $S_{\dot{\psi}}(\omega)$ in terms of the convergence of (1.8). Due to the inherent cutoff frequency ω_{co} in any physical system, $S_{\dot{\psi}}(\omega) \cong 0$ for $|\omega| \geq \omega_{co}$. If var $(\dot{\psi}(t)) = \int_{-\infty}^{\infty} S_{\dot{\psi}}(\omega)\, d\omega/2\pi$ is infinite, it can only be due to the singularity effect near the origin, i.e. $S_{\dot{\psi}}(\omega)$ behaves like $|\omega|^{-\nu}$, $\nu \geq 1$, in the small frequencies region.

Let us now examine (1.8) in more detail. For a fixed τ, the term $\sin^{2M}(\omega\tau/2)/\omega^2 \sim$ constant $\cdot\, \omega^{2M-2}$ for $|\omega|$ small. So, as long as $(2M - 1) > \nu$, the expression for $D_\Phi^{(M)}(\tau)$ is finite. We will have the occasion to demonstrate this point relating to convergence more vividly when we deal with "flicker"-type noise processes in Section I-D.

1) Oscillator with Linear Frequency Drift: As an example of our present results, consider an oscillator which has a linear frequency drift component, that is, $N = 2$ in (1.3). Now $D_\Phi^{(1)}(\tau)$ is time-dependent. However, from (1.7) and (1.8) we can write

$$D_\Phi^{(2)}(\tau) = \tau^4 E[\Omega_1^2] + \frac{8}{\pi}\int_{-\infty}^{\infty}\sin^4\left(\frac{\omega\tau}{2}\right)\frac{S_{\dot{\psi}}(\omega)}{\omega^2}d\omega \quad (1.9)$$

and

$$D_\Phi^{(3)}(\tau) = \frac{32}{\pi}\int_{-\infty}^{\infty}\sin^6\left(\frac{\omega\tau}{2}\right)\frac{S_{\dot{\psi}}(\omega)}{\omega^2}d\omega. \quad (1.10)$$

For $M \geq N = 3$, then (1.8) holds.

C. Structure Function of the Nth Increment of the Frequency Process

Although the frequency process $\dot{\Phi}(t)$ has not explicitly been used in the past to characterize oscillator stability, we demonstrate its utility and the role it plays in terms of its structure functions. Consider now the frequency process $\dot{\Phi}(t)$ defined in (1.3) with stationary $(N-1)$th-order frequency increments. Since[8]

$$R_{\dot{\psi}}(\lambda) = \int_{-\infty}^{\infty} e^{j\omega\lambda} S_{\dot{\psi}}(\omega)\frac{d\omega}{2\pi} \quad (1.11)$$

then using (II.6) in Appendix II, we can show that the $(N-1)$th-structure function for frequency instability satisfies

$$D_{\dot{\Phi}}^{(N-1)}(\tau) = \tau^{2(N-1)}E[\Omega_{N-1}^2] + D_{\dot{\psi}}^{(N-1)}(\tau)$$

$$= \tau^{2(N-1)}E[\Omega_{N-1}^2] + 2^{2(N-1)} \cdot \int_{-\infty}^{\infty}\sin^{2(N-1)}\left(\frac{\omega\tau}{2}\right)S_{\dot{\psi}}(\omega)\frac{d\omega}{2\pi}. \quad (1.12)$$

The first term on the right-hand side of (1.12) is related to the long-term frequency stability of (1.3), while the second term is attributable to the short-term frequency instability in (1.3). For $M \geq N$, it follows that

$$D_{\dot{\Phi}}^{(M)}(\tau) = D_{\dot{\psi}}^{(M)}(\tau) = 2^{2M}\int_{-\infty}^{\infty}\sin^{2M}\left(\frac{\omega\tau}{2}\right)S_{\dot{\psi}}(\omega)\frac{d\omega}{2\pi}. \quad (1.13)$$

Again we note that the PSD $S_{\dot{\psi}}(\omega)$ plays a key role in defining the *M*th order frequency structure function.

Now suppose $S_{\dot{\psi}}(\omega)$ behaves like $|\omega|^{-\nu}$ with $\nu \geq 1$ as $|\omega| \to 0$; $D_{\dot{\Phi}}^{(N)}(\tau)$ is bounded if $\nu < 2N + 1$. If $\nu \geq 2N + 1$, we can always take high enough increments, say M, such that $2M > \nu - 1$ to obtain a bounded structure function. Hence, besides combatting the problems associated with the long-term frequency drift, the structure function approach is useful in treating "flicker"-type noise.[9] Most importantly, if $D_{\dot{\Phi}}^{(M)}(\tau) < \infty$ for $M \geq N$, the *sample structure function*, say

$$\hat{D}_{\dot{\Phi}}^{(M)}(\tau) \triangleq \frac{1}{L}\sum_{i=1}^{L}[\Delta^M\dot{\Phi}(t_i)]^2 \quad (1.14)$$

where L is the number of samples considered, *can converge* to $D_{\dot{\Phi}}^{(M)}(\tau)$.

We would like to point out that for a frequency process with $(N-1)$th-order drifts, the $(N-1)$th increment of the *frequency* process is stationary. Yet, the *M*th increment of the *phase* process is not unless $M \geq N$. Also, comparing (1.13) with (1.8), it is sometimes convenient to *define formally* the PSD of the phase process:

$$S_\psi(\omega) \triangleq \frac{S_{\dot{\psi}}(\omega)}{\omega^2} \quad (1.15)$$

and use this in (1.8) so that it conforms to (1.13). However, we should bear in mind that in our model $\psi(t)$ is not stationary; hence, $\psi(t)$ *does not* possess a PSD in the usual sense. On the other hand, if (1.15) holds in the usual sense, i.e., if $\psi(t)$

[8] If $\dot{\psi}(t)$ is not of second-order, $R_{\dot{\psi}}(\lambda)$ must be interpreted as a generalized function. In any case, (1.12) and (1.13) remain valid in the usual sense provided that the integrals involved are finite.

[9] Later we give an explicit example which demonstrates how the convergence problem is overcome.

is stationary, then the theory and equations remain valid as a special case.

1) Oscillator with Linear Frequency Drift: Letting $N = 2$ in (1.3) yields a model for an oscillator frequency process with linear drift. Then from (1.12), the first structure function for the frequency process becomes

$$D_{\dot{\Phi}}^{(1)}(\tau) = \tau^2 E[\Omega_1^2] + \frac{2}{\pi}\int_{-\infty}^{\infty} \sin^2\left(\frac{\omega\tau}{2}\right) S_{\dot{\psi}}(\omega)\, d\omega \tag{1.16}$$

and for $M \geqslant N = 2$, (1.13) relates the structure function of the frequency process to the PSD of the short-term frequency instabilities. Recall from Section I-B(1) that $D_{\Phi}^{(1)}(\tau)$ is time-dependent for $N = 2$ in (1.3); we therefore conclude that the structure functions $D_{\psi}^{(1)}(\tau)$, $D_{\dot{\psi}}^{(1)}(\tau)$, $D_{\dot{\Phi}}^{(1)}(\tau)$, $D_{\dot{\Phi}}^{(2)}(\tau)$, $D_{\Phi}^{(2)}(\tau)$, and $D_{\Phi}^{(3)}(\tau)$ serve to characterize the time-independent stability of an oscillator with linear frequency drift provided that they are finite.

D. Structure Functions for "Flicker"-Type Noise

We now give an example concerning the convergence problem for a "flicker"-type noise process. It serves to illustrate the versatility of the structure function approach compared to other approaches [1]-[9] when higher order frequency drifts and "flicker"-type noise are present.

Consider the frequency process with second-order drift

$$\dot{\Phi}(t) = \omega_0 + \Omega_1 t + \Omega_2 \frac{t^2}{2} + \dot{\psi}(t) \tag{1.17}$$

and assume that $S_{\dot{\psi}}(\omega) = h_{-3}|\omega|^{-3}$. Now $D_{\dot{\Phi}}^{(1)}(\tau)$, $D_{\Phi}^{(1)}(\tau)$, and $D_{\Phi}^{(2)}(\tau)$ are time-dependent; however, the second frequency structure function for the frequency process is found from (1.12) to be

$$D_{\dot{\Phi}}^{(2)}(\tau) = \tau^4 E[\Omega_2^2] + 4h_{-3}\frac{\ln 2}{\pi}\tau^2. \tag{1.18}$$

The third frequency structure function for the frequency process is found from (1.13) to be

$$D_{\dot{\Phi}}^{(3)}(\tau) = \frac{3}{\pi} h_{-3}(8 \ln 2 - 3 \ln 3)\tau^2 \tag{1.19}$$

while the third phase structure function of the corresponding phase process is found from (1.7) to be

$$D_{\Phi}^{(3)}(\tau) = \tau^6 E[\Omega_2^2] + \frac{1}{4\pi} h_{-3}(27 \ln 3 - 32 \ln 2)\tau^4. \tag{1.20}$$

It is interesting to note that all standard formulas given in the literature [1]-[6] will not yield meaningful results unless the frequency acceleration term in (1.17) is neglected and a low cutoff frequency for $S_{\dot{\psi}}(\omega) = h_{-3}|\omega|^{-3}$ is introduced.

We shall show in Section II that the two-sample Allan variance [1], [3] is proportional to the second phase structure function $D_{\Phi}^{(2)}(\tau)$ of the phase noise process. Therefore, when (1.17) holds, it does not make sense to use the Allan variance because:

1) the second phase increment of (1.17) is *not* stationary so the Allan variance is a function of t as well as τ;
2) $D_{\Phi}^{(2)}(\tau)$ is unbounded due to the $|\omega|^{-3}$ behavior so one cannot make any meaningful short or long-term second phase increment measurements.[10]

Thus it would appear that $D_{\dot{\Phi}}^{(2)}(\tau)$, $D_{\dot{\Phi}}^{(3)}(\tau)$, $D_{\Phi}^{(3)}(\tau)$, and $D_{\Phi}^{(4)}(\tau)$ are the meaningful τ-domain quantities to deal with for oscillators whose frequency process is modeled by (1.17).

II. RMS Frequency Deviation and the Allan Variance in Terms of Structure Functions

In this section, we give new interpretations to the recommended definitions of stability [1] in terms of the theory based upon structure functions. In particular, we discuss the conditions to be imposed on the phase or frequency process in order to attach meaning to the rms fractional frequency deviation and the two-sample and L-sample Allan variance [1], [3]. We generalize the results of this section in Section III.

A. RMS Fractional Frequency Deviation

If the radian frequency process $\dot{\Phi}(t)$ can be modeled as a constant ω_0 plus a stationary frequency component $\dot{\psi}(t)$ whose PSD is well behaved near $\omega = 0$, i.e., $S_{\dot{\psi}}(\omega) \sim |\omega|^{-\nu}$, $\nu < 1$, then

$$D_{\psi}^{(1)}(\tau) \triangleq E\{[\psi(t+\tau) - \psi(t)]^2\} < \infty \tag{2.1}$$

is the expected value of the square of the phase $\psi(t)$ accumulated in τ seconds. This can be interpreted as the variance of the total phase accumulated $(\Phi(t+\tau) - \Phi(t))$ from its mean value $\omega_0\tau$; and when normalized by τ^2, it has been suggested as the square of the rms frequency deviation [2]. As we shall presently see, the variance, say $\sigma_{\dot{\psi}}^2(\tau)$, of frequency counted data is closely related to the first phase structure function $D_{\psi}^{(1)}(\tau)$. In particular, the (*true*) *rms fractional frequency deviation* defined by Cutler-Searle [2] is, in our notation,

$$\frac{\delta f(\tau)}{f_0} \triangleq \sqrt{\frac{E\{[\psi(t+\tau) - \psi(t)]^2\}}{(\omega_0\tau)^2}} = \sqrt{\frac{D_{\psi}^{(1)}(\tau)}{(\omega_0\tau)^2}} \tag{2.2}$$

which follows from (2.1). Furthermore, the statistical average of the *measured rms fractional frequency deviation* $\widehat{\delta f}(\tau)/f_0$ as found using frequency counted data, is easily shown to satisfy (see Appendix IV),

$$E\left\{\left[\frac{\widehat{\delta f}(\tau)}{f_0}\right]^2\right\} = \frac{D_{\psi}^{(1)}(\tau) - D_{\psi}^{(1)}(L\tau)/L^2}{(\omega_0\tau)^2} = \left[\frac{\delta f(\tau)}{f_0}\right]^2 - \left[\frac{\delta f(L\tau)}{f_0}\right]^2 \tag{2.3}$$

where L is the number of samples. From (2.3) it is obvious that the measurement bias is given by

$$\left[\frac{\delta f(L\tau)}{f_0}\right]^2 = \frac{D_{\psi}^{(1)}(L\tau)}{L^2(\omega_0\tau)^2} \tag{2.4}$$

from which we observe that

$$\lim_{L\to\infty} E\left\{\left[\frac{\widehat{\delta f}(\tau)}{f_0}\right]^2\right\} = \left[\frac{\delta f(\tau)}{f_0}\right]^2 \tag{2.5}$$

[10] We are not in a position to comment on the validity of the assumed mathematical model; this question of an oscillator model seems to remain open.

if $D_{\psi}^{(1)}(\tau) < \infty$ for all τ by our assumptions. Hence, the estimate $\widehat{\delta f}(\tau)/f_0$ is an asymptotically unbiased estimator of $\delta f(\tau)/f_0$. Further comments on the various relationships between the rms fractional frequency deviation, the structure functions $D_{\psi}^{(1)}(\tau)$ and $D_{\Phi}^{(1)}(\tau)$, and the PSD $S_{\dot{\psi}}(\omega)$ can be found in Appendix IV.

B. The Two-Sample Allan Variance

If the frequency process $\dot{\Phi}(t)$ is modeled with no higher than a linear frequency drift term and the PSD $S_{\dot{\psi}}(\omega)$ of the "flicker"-type noise behaves like $|\omega|^{-\nu}$ for $\nu < 3$ near $\omega = 0$, then

$$\frac{D_{\Phi}^{(2)}(\tau)}{(\omega_0\tau)^2} \triangleq \frac{E\{[\Delta^2\Phi(t)]^2\}}{(\omega_0\tau)^2} = E\left\{\left[\underbrace{\frac{\Phi(t+2\tau)-\Phi(t+\tau)}{\omega_0\tau}}_{\text{term no. 1}} - \underbrace{\frac{\Phi(t+\tau)-\Phi(t)}{\omega_0\tau}}_{\text{term no. 2}}\right]^2\right\} < \infty \tag{2.6}$$

is the expected squared fractional deviation of the average frequency at time $t+\tau$ (term no. 1) from its initial average frequency at time t (term no. 2). For small τ, this can be interpreted as the variance of the fractional frequency deviation accumulated in τ seconds from its nominal value ($\omega_0 + \Omega_1\tau$) due to the linear frequency drift component Ω_1.

The two-sample Allan variance $E\{\sigma_A^2(2,\tau,\tau)\}$ defined in [1] is,[11] in our notation

$$E\{\sigma_A^2(2,\tau,\tau)\} \triangleq \frac{1}{2}E\left\{\left[\frac{\Phi(t+2\tau)-\Phi(t+\tau)}{\omega_0\tau} - \frac{\Phi(t+\tau)-\Phi(t)}{\omega_0\tau}\right]^2\right\}. \tag{2.7}$$

Using (2.6) in (2.7), we have

$$E\{\sigma_A^2(2,\tau,\tau)\} = \frac{D_{\Phi}^{(2)}(\tau)}{2(\omega_0\tau)^2}. \tag{2.8}$$

Except for a factor of $\frac{1}{2}$ and the normalization constant $(\omega_0\tau)^2$, the two-sample Allan variance is exactly the second phase structure function. Notice that, except for the factor of $\frac{1}{2}$, (2.8) conforms to (2.2). Thus we have given the interpretation of the Allan variance in terms of a well defined function in random process theory.

In addition, from

$$\frac{1}{\tau^2}\frac{D_{\Phi}^{(2)}(\tau)}{\tau^2} = E\left\{\left[\frac{[\Phi(t+2\tau)-\Phi(t+\tau)]/\tau - [\Phi(t+\tau)-\Phi(t)]/\tau}{\tau}\right]^2\right\} \tag{2.9}$$

and (2.6), we note that the two-sample Allan variance is a measure of the (ensemble) average of the squared acceleration associated with the phase process $\Phi(t)$.

It is shown in Appendix V that if the assumptions regarding the frequency process $\dot{\Phi}(t)$ in Section II-A hold, then the two-sample Allan variance and the rms fractional frequency deviation form a one-to-one correspondence via

$$E\{\sigma_A^2(2,\tau,\tau)\} = 2\left\{\left[\frac{\delta f(\tau)}{f_0}\right]^2 - \left[\frac{\delta f(2\tau)}{f_0}\right]^2\right\}. \tag{2.10}$$

The equation can be inverted to yield

$$\left[\frac{\delta f(\tau)}{f_0}\right]^2 = \frac{1}{2}\sum_{k=0}^{\infty} E\{\sigma_A^2(2, 2^k\tau, 2^k\tau)\} = \frac{1}{2}E\{\sigma_A^2(2,\tau,\tau)\} + \frac{1}{2}\sum_{k=1}^{\infty}\frac{D_{\Phi}^{(2)}(2^k\tau)}{(\omega_0 2^k\tau)^2}. \tag{2.11}$$

The authors believe that this fact has not been recognized. Further comments on the two-sample Allan variance in terms of the present theory based upon structure functions can be found in Appendix V.

C. The L-Sample Allan Variance

The L-sample Allan variance $\hat{\sigma}_A^2(L,\tau,\tau)$ [3] is the normalized statistical variance of L phase increment samples given by

$$\hat{\sigma}_A^2(L,\tau,\tau) = \frac{1}{(L-1)}\left\{\sum_{k=0}^{L-1}\left[\frac{\Delta\psi(t+k\tau;\tau)}{\omega_0\tau}\right]^2 - \frac{1}{L}\sum_{k=0}^{L-1}\left[\frac{\Delta\psi(t+k\tau;\tau)}{\omega_0\tau}\right]^2\right\}. \tag{2.12}$$

Because of its time-dependence, it cannot accommodate frequency drifts, and if "flicker"-type noise is present, $S_{\dot{\psi}}(\omega)$ must behave like $|\omega|^{-\nu}$, $\nu < 3$. In addition, we show in Appendix V that the L-sample Allan variance is not related to the two-sample Allan variance in a simple way, but rather it is a biased estimator of the rms fractional frequency deviation squared provided the latter is well defined. Explicitly, the mean of the L-sample Allan variance is

$$E\{\hat{\sigma}_A^2(L,\tau,\tau)\} = \frac{L}{(L-1)}\left[\left(\frac{\delta f(\tau)}{f_0}\right)^2 - \left(\frac{\delta f(L\tau)}{f_0}\right)^2\right] \tag{2.13}$$

where $\delta f(\tau)/f_0$ is the rms fractional frequency deviation. The estimator bias is given by

$$\frac{1}{(L-1)}\left[\frac{\delta f(\tau)}{f_0}\right]^2 - \frac{L}{(L-1)}\left[\frac{\delta f(L\tau)}{f_0}\right]^2 = \frac{1}{(\omega_0\tau)^2}\cdot\left[\frac{1}{(L-1)}D_{\psi}^{(1)}(\tau) - \frac{1}{L(L-1)}D_{\psi}^{(1)}(L\tau)\right]. \tag{2.14}$$

In the limit as $L \to \infty$, the bias is zero if $D_{\psi}^{(1)}(\tau) < \infty$ for all τ. In other words, $\hat{\sigma}_A^2(L,\tau,\tau)$ is an asymptotically unbiased estimator of $(\delta f(\tau)/f_0)^2$. In that sense, the term "L-sample Allan variance" is misleading since it is not an estimate for the "Allan variance," which appears to be used exclusively for the two-sample Allan variance. Interconnections between the L-sample Allan variance, the corresponding structure functions, and the PSD $S_{\dot{\psi}}(\omega)$ are developed in Appendix V.

III. Oscillator Frequency and Phase Instabilities Characterization in Terms of Structure Functions

In the general case, the nonstationary frequency process $\dot{\Phi}(t)$ of (1.3) is assumed to have higher order drift components. In addition, assume the stationary frequency process $\dot{\psi}(t)$ may have a PSD of the form

$$S_{\dot{\psi}}(\omega) = \sum_{i=1}^{P} h_i|\omega|^{-\nu_i} \tag{3.1}$$

[11] Actually, it is defined on a process $\tilde{\Phi}(t) = \Phi(t) - \omega_0 t$. However, since $\Delta^2\tilde{\Phi}(t) = \Delta^2\Phi(t)$, the definition given here is equivalent.

where some of the ν_i's are positive. If one of the ν_i's is greater than or equal to one, $R_{\dot{\psi}}(0)$ is clearly infinite so the variance of this stationary frequency component is of no practical value. The quantities of interest for characterizing the τ-domain stability defined by (1.1), (1.3), and (1.4) are the PSD $S_{\dot{\psi}}(\omega)$ and the order of the highest frequency drift, $(N - 1)$. For once they are known, we can determine a host of important relationships involving structure functions from the previously developed theory.

In addition, we outline a meaningful testing algorithm for obtaining $S_{\dot{\psi}}(\omega)$ and $(N - 1)$ based upon observing the behavior of the statistical samples of the $(M - 1)$th frequency structure function and/or the Mth phase structure function. We stress the word meaningful because within the scope of this theory, if M is large enough, $D_{\Phi}^{(M)}(\tau)$ and $D_{\dot{\Phi}}^{(M-1)}(\tau)$ are well defined, time-independent and well behaved quantities which can be approximated by their sample counterparts. What is more attractive is that frequency counted data used in determining the rms fractional frequency deviation or the Allan variance are sufficient for our test.

1) Measurement Considerations: Standard measurement techniques that give phase and frequency increment data over the measurement time τ with no dead time between samples are well documented [1]–[6]. These types of experimental data are sufficient for our purpose. From these L pairs of increments $\Delta\dot{\Phi}(t)$ and/or $\Delta\Phi(t)$ for $t = t_0, t_0 + \tau, \cdots, t_0 + (L - 1)\tau$, numerical computation is all that is needed to generate recursively the nth order sample structure functions:

$$\hat{D}_{\dot{\Phi}}^{(n)}(\tau) \triangleq \frac{1}{L - n + 1} \sum_{i=0}^{L-n} [\Delta^{n-1}(\Delta\dot{\Phi}(t = t_0 + i\tau; \tau))]^2 \tag{3.2}$$

and

$$\hat{D}_{\Phi}^{(n)}(\tau) \triangleq \frac{1}{L - n + 1} \sum_{i=0}^{L-n} [\Delta^{n-1}(\Delta\Phi(t = t_0 + i\tau; \tau))]^2 \tag{3.3}$$

for any $n \leq L$. Obviously, we can also generate $\hat{D}_{\dot{\Phi}}^{(n)}(l\tau)$ and $\hat{D}_{\Phi}^{(n)}(l\tau)$, for all l of interest.

2) An Algorithm to Test for the Highest-Order of Frequency Drift and for the Presence of "Flicker"-Type Noise: Once we get the estimates for $D_{\Phi}^{(n)}(\tau)$ and/or $D_{\dot{\Phi}}^{(n)}(\tau)$, using test data in (3.2) and (3.3), we can proceed to test for the highest frequency drift rate, $(N - 1)$ and the exponent, ν of the most dominant divergent component of the "flicker"-type noise spectrum in (3.1) as ω approaches zero. The following algorithm based on testing $\hat{D}_{\Phi}^{(n)}(\tau)$ serves to illustrate the principle behind such test procedures involving structure functions. The idea behind it is relatively simple;

1) $D_{\Phi}^{(n)}(\tau)$ is not bounded if $(2n - 1) \leq \nu$;
2) $D_{\Phi}^{(m)}(\tau)$ is time-dependent if $m < N$.

A. Algorithm T

*T*1: [Initialize]. Set $i = 1$.

*T*2: [Test for stationarity]. If $\hat{D}_{\Phi}^{(i)}(\tau)$ is not time-independent, increase i by 1 and go back to *T*2. Otherwise, set order of frequency drift to $(i - 1)$. Go to *T*3.

*T*3: [Test for boundedness]. If $\hat{D}_{\Phi}^{(i)}(\tau)$ is not bounded, increase i by 1. Go back to *T*3. Otherwise, set $M = i$ and go to *T*4.

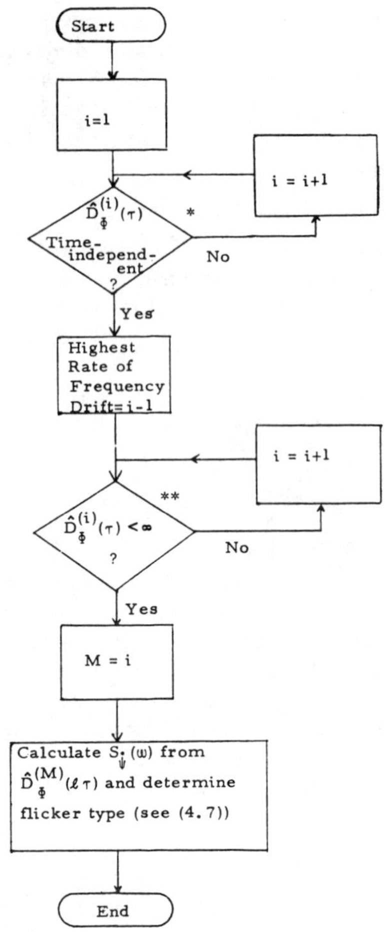

Fig. 1. Flow chart for algorithm *T*. *Determined by comparing $\hat{D}_{\Phi}^{(i)}(\tau)$ from two separate measurements. **Use some threshold test.

*T*4: [End]. Calculate[12] $S_{\dot{\psi}}(\omega)$ from $\hat{D}_{\Phi}^{(M)}(l\tau)$, $l = 1, 2, \ldots, l_{\max}$. Determine "flicker" type.

Fig. 1 represents a flow chart for this algorithm.

After M is determined, the Mth sample phase structure function or the $(M - 1)$th sample frequency structure function can also be used as a measure of instability. If τ is small, from the definition of the nth derivative of a function $f(t)$, notice that $\Delta^n f(t; \tau)$ is related via

$$\frac{d^n f(t)}{dt^n} \triangleq \lim_{\tau \to 0} \frac{\Delta^n f(t; \tau)}{\tau^n}. \tag{3.4}$$

Thus $D_{\Phi}^{(M)}(\tau)$ is really a measure of the average of the square of the Mth rate of change of the phase process $\Phi(t)$. Alternatively, it can be interpreted as the average of the square of the $(M - 1)$th rate of change of the phase process $\Phi(t)$ accumulated over τ seconds. In special cases (see Section II), we have already shown that $D_{\Phi}^{(M)}(\tau)$ can be associated with certain standard recommended definitions for oscillator instability. Similarly, $D_{\dot{\Phi}}^{(M)}(\tau)$ is a measure of the square of the Mth average rate of change or the $(M - 1)$th average rate of change accumulated over τ seconds, of the frequency process $\dot{\Phi}(t)$.

Assuming (1.3), (1.4), and (3.1) to be the appropriate models for the frequency process $\dot{\Phi}(t)$, the phase process $\Phi(t)$, and the "flicker"-type noise PSD, respectively, a more

[12] See Section IV.

TABLE I

INTERPRETATION OF QUANTITIES ASSOCIATED WITH THE FREQUENCY AND PHASE PROCESS

Quantity	Interpretation	Statistical Consideration	Meaning of $\sqrt{D_{\psi}^{(\,)}(\tau)}$, $\sqrt{D_{\dot\psi}^{(\,)}(\tau)}$
Frequency $\dot\Phi(t)$	Instantaneous oscillator frequency.	Stationary only for $N = 1$. $R_{\dot\Phi}(\tau) = E[\omega_0^2] + R_{\dot\psi}(\tau)$ defined for $\nu < 1$.	(Does not apply.)
First phase increment $\Delta\Phi(t) = \Phi(t+\tau) - \Phi(t)$	Oscillator phase accumulated in τ seconds. $\Delta\Phi(t)/\tau$ is the approximate short term average oscillator frequency.	Stationary only for $N = 1$. $D_{\Phi}^{(1)}(\tau) = \tau^2 E[\omega_0^2] + D_{\psi}^{(1)}(\tau)$ defined for $\nu < 1$; equivalent to rms frequency deviation.	Average phase accumulated in τ seconds; characterizes stability of phase.
First frequency increment $\Delta\dot\Phi(t) = \dot\Phi(t+\tau) - \dot\Phi(t)$	Oscillator frequency accumulated in τ seconds. $\Delta\dot\Phi(t)/\tau$ is the approximate short term average oscillator frequency drift (phase acceleration).	Stationary only for $N \leqslant 2$. $D_{\dot\Phi}^{(1)}(\tau) = \tau^2 E[\Omega_1^2] + D_{\dot\psi}^{(1)}(\tau)$ defined for $\nu < 3$.	Average frequency accumulated in τ seconds; characterizes stability of frequency.
Second phase increment $\Delta^2\Phi(t) = \Phi(t+2\tau) - 2\Phi(t+\tau) + \Phi(t)$	$\Delta^2\Phi(t)/\tau$ is the approximate short term average oscillator frequency accumulated in τ seconds. $\Delta^2\Phi(t)/\tau^2$ is the approximate short term phase acceleration (frequency drift).	Stationary only for $N \leqslant 2$. $D_{\Phi}^{(2)}(\tau) = \tau^4 E[\Omega_1^2] + D_{\psi}^{(2)}(\tau)$ defined for $\nu < 3$; equivalent to two-sample Allan variance.	
Second frequency increment $\Delta^2\dot\Phi(t) = \dot\Phi(t+2\tau) - 2\dot\Phi(t+\tau) + \dot\Phi(t)$	$\Delta^2\dot\Phi(t)/\tau$ is the approximate short term average oscillator frequency drift accumulated in τ seconds. $\Delta^2\dot\Phi(t)/\tau^2$ is the approximate short term oscillator frequency acceleration (phase "jerk").	Stationary only for $N \leqslant 3$. $D_{\dot\Phi}^{(2)}(\tau) = \tau^4 E[\Omega_2^2] + D_{\dot\psi}^{(2)}(\tau)$ defined for $\nu < 5$.	Average phase acceleration accumulated in τ seconds; characterizes stability of frequency drift.
Third phase increment $\Delta^3\Phi(t) = \Phi(t+3\tau) - 3\Phi(t+2\tau) + 3\Phi(t+\tau) - \Phi(t)$	$\Delta^3\Phi(t)/\tau^2$ is the approximate short term oscillator frequency drift (phase acceleration) accumulated in τ seconds. $\Delta^3\Phi(t)/\tau^3$ is the approximate short term oscillator frequency acceleration (phase "jerk").	Stationary only for $N \leqslant 3$. $D_{\Phi}^{(3)}(\tau) = \tau^6 E[\Omega_2^2] + D_{\psi}^{(3)}(\tau)$ defined for $\nu < 5$; interpreted by Barnes as variance of clock error.[a]	

[a]Barnes [4] interpreted $(\Phi(t+2\tau) - \Phi(t+\tau))/\tau$ as the mean frequency used for a clock. So the total phase accumulated indicated by the clock over the time interval $(t, t+3\tau)$ is $3\tau(\Phi(t+2\tau) - \Phi(t+\tau))/\tau = 3(\Phi(t+2\tau) - \Phi(t+\tau))$. However, the actual phase accumulated is $(\Phi(t+3\tau) - \Phi(t))$. Hence, the difference (clock error) is $\Phi(t+3\tau) - \Phi(t) - 3(\Phi(t+2\tau) - \Phi(t+\tau)) = \Delta^3\Phi(t)$.

general interpretation of the Mth increment and associated structure functions of the phase and frequency process based on (3.4) is given in Table I for the first few values of M.

IV. τ-DOMAIN TO FREQUENCY-DOMAIN TRANSFORMATIONS

Most of the results in previous sections have been expressed in terms of the PSD $S_{\dot\psi}(\omega)$ of the stationary frequency process $\dot\psi(t)$. The next step is to develop the machinery needed to enable one to write down $S_{\dot\psi}(\omega)$ (or equivalently $R_{\dot\psi}(\tau)$ when it is finite) from the Mth-order phase structure function $D_{\Phi}^{(M)}(\tau)$, i.e., convert τ-domain stability to the frequency-domain. Once $S_{\dot\psi}(\omega)$ is determined, we shall see that the rms fractional frequency deviation (phase accumulation) as defined by Cutler and Searle [2] can be found provided the potential singularity at the origin in $S_{\dot\psi}(\omega)$ causes no convergence problems.

We now explore the τ-domain to frequency-domain relationships between $R_{\dot\psi}(\tau)$, $S_{\dot\psi}(\omega)$, $D_{\Phi}^{(M)}(\tau)$, and $D_{\dot\Phi}^{(M)}(\tau)$, and later summarize the results. In the following discussion, we assume that the order of increments M is chosen large enough such that:

(a) both $D_{\Phi}^{(M)}(\tau)$ and $D_{\dot\Phi}^{(M)}(\tau)$ are finite, so there is no convergence problem associated with possible "flicker"-type spectrum for $S_{\dot\psi}(\omega)$;

(b) $M > N$, where $(N-1)$ is the highest component of frequency drift rates, so the random increment processes $\Delta^M\Phi(t)$ and $\Delta^M\dot\Phi(t)$ are stationary, and (1.8) and (1.13) apply.

A. Relationships in the τ-Domain

We state mainly the results here as their derivations are quite detailed. The interested reader can consult Appendix II.[13] Then,

$$D_{\Phi}^{(M)}(\tau) = 2 \cdot \sum_{k=1}^{M} (-1)^{k+1} \binom{2M}{M-k} \int_0^{k\tau} (k\tau - \lambda) R_{\dot\psi}(\lambda)\, d\lambda \quad (4.1)$$

and $D_{\Phi}^{(M)}(\tau)$ is the solution of the differential equation

$$\frac{\partial^2}{\partial\tau^2} D_{\Phi}^{(M)}(\tau) = 2M(2M-1) D_{\Phi}^{(M-1)}(\tau) - M^2 D_{\Phi}^{(M)}(\tau) \quad (4.2)$$

with initial conditions $D_{\Phi}^{(M)}(0) = 0$ and

$$\frac{\partial}{\partial\tau} D_{\Phi}^{(M)}(\tau)\Big|_{\tau=0} = 0.$$

[13] As before, if $R_{\dot\psi}(\tau)$, $D_{\psi}^{(1)}(\tau)$, or $D_{\dot\psi}^{(1)}(\tau)$ are unbounded, they have to be interpreted as generalized functions.

In addition,

$$D_{\Phi}^{(M)}(\tau) = \sum_{k=1}^{M} (-1)^{k+1} \binom{2M}{M-k} D_{\psi}^{(1)}(k\tau) \tag{4.3}$$

and

$$D_{\dot{\Phi}}^{(M)}(\tau) = \sum_{k=1}^{M} (-1)^{k+1} \binom{2M}{M-k} D_{\dot{\psi}}^{(1)}(k\tau). \tag{4.4}$$

For completeness, we modify (I.10) in Appendix I for the present purpose and note that

$$D_{\dot{\Phi}}^{(M)}(\tau) = \sum_{j=-M}^{M} (-1)^{j} \binom{2M}{M+j} R_{\dot{\psi}}(j\tau). \tag{4.5}$$

If convergence is not a problem in certain types of oscillators so that $D_{\psi}^{(1)}(\tau)$, $D_{\dot{\psi}}^{(1)}(\tau)$, or $R_{\dot{\psi}}(\tau)$ exist, and higher order structure functions are used only for combatting frequency drifts, these three quantities all serve to characterize the frequency instability of the oscillator (see (4.3)–(4.5)). In particular, (4.3) and (2.2) interconnect the rms fractional frequency deviation with $D_{\Phi}^{(M)}(\tau)$ provided $D_{\psi}^{(1)}(\tau)$ is well defined. For $M \leqslant 5$, a technique is developed in Appendix III to invert (4.3)–(4.5). With this technique, $D_{\psi}^{(1)}(\tau)$, $D_{\dot{\psi}}^{(1)}(\tau)$, and $R_{\dot{\psi}}(\tau)$ can be obtained from $D_{\Phi}^{(M)}(\tau)$ and $D_{\dot{\Phi}}^{(M)}(\tau)$, which are readily approximated by measurements discussed in Section III-1.

B. Inverse Transform Relationships Between $D_{\Phi}^{(M)}(\tau)$, $D_{\dot{\Phi}}^{(M)}(\tau)$, and $S_{\dot{\psi}}(\omega)$ or $R_{\dot{\psi}}(\tau)$

Consider the relationships given in (1.8) and (1.13). Treating them as integral equations, the general solution can be given in terms of Mellin transforms. If $\mathfrak{M}_x(s)$ denotes the Mellin transform of $x(t)$, then

$$S_{\dot{\psi}}(\omega) = \frac{\pi}{2^{2M}} \mathfrak{M}^{-1} \left[\frac{\mathfrak{M}_g(1-s)}{\mathfrak{M}_k(1-s)} \right] \tag{4.6}$$

where $\mathfrak{M}^{-1}[\ ]$ denotes the inverse Mellin transform and

$$g(\tau) = D_{\dot{\Phi}}^{(M)}(\tau)$$
$$k(\omega\tau) = \sin^{2M}(\omega\tau/2).$$

Alternatively,

$$S_{\dot{\psi}}(\omega) = \frac{\omega^2 \pi}{2^{2M}} \mathfrak{M}^{-1} \left[\frac{\mathfrak{M}_h(1-s)}{\mathfrak{M}_k(1-s)} \right] \tag{4.7}$$

where $h(\tau) = D_{\Phi}^{(M)}(\tau)$. Taking the inverse Mellin transform in (4.6) and (4.7) is difficult in general and an alternate approach based upon the solution to a functional equation similar to (4.5) is presented in Appendix III.

Fig. 2 summarizes the calculus relating the frequency spectrum $S_{\dot{\psi}}(\omega)$, the frequency correlation function $R_{\dot{\psi}}(\tau)$, the Mth phase structure function $D_{\Phi}^{(M)}(\tau)$, and the Mth frequency structure function $D_{\dot{\Phi}}^{(M)}(\tau)$. In particular, these four quantities are equivalent in the sense that if one is known, the rest are uniquely determined. As already noted, the PSD $S_{\dot{\psi}}(\omega)$ is the key element in the theory as it relates directly to all other functions of interest.

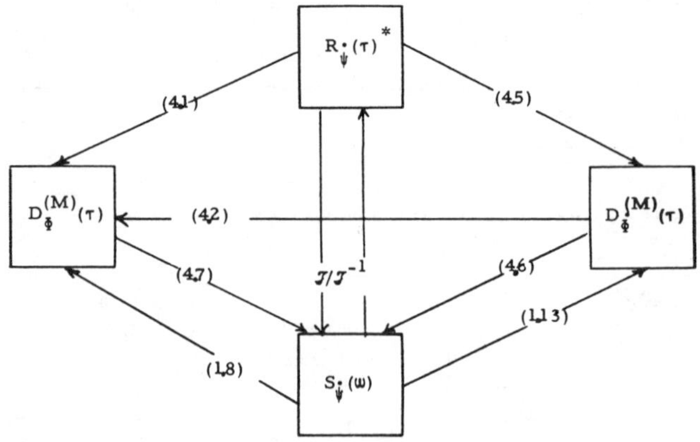

Fig. 2. Transform theory of oscillator instability in terms of structure functions, the PSD $S_{\dot{\psi}}(\omega)$ and correlation function $R_{\dot{\psi}}(\tau)$. *See footnote 13.

APPENDIX I

STRUCTURE FUNCTIONS

For those unfamiliar with the notion of the structure function of a random process with stationary Nth-order increments, a short discussion on the basic definitions and certain properties that are pertinent to the development of our theory is given here. The mathematically inclined reader can consult the works of Doob [16], Yaglom [11], and Gikhman and Skorokhod [15].

A. Definition of the Nth *Increment of* $x(t)$

Let $\Delta x(t;\tau)$ be the first increment of a function $x(t)$ defined by

$$\Delta x(t;\tau) \triangleq x(t+\tau) - x(t). \tag{I.1}$$

Here τ is a fixed parameter, and is subsequently suppressed for convenience in the notation, i.e., $\Delta x(t) \triangleq \Delta x(t;\tau)$. The Nth increment, $N > 1$, of $x(t)$ is defined recursively by

$$\Delta^N x(t) \triangleq \Delta^{N-1}(\Delta x(t)) \tag{I.2}$$

and, explicitly,

$$\Delta^N x(t) = \sum_{k=0}^{N} (-1)^k \binom{N}{k} x(t + (N-k)\tau). \tag{I.3}$$

It is easy to verify that the Mth increment of a Nth-order polynomial is zero for $M > N$. We will make use of this fact in what follows.

B. A Random Process with Stationary Nth *Increments*

Consider the random process

$$x(t) = \sum_{k=0}^{N} \frac{A_k}{k!} t^k + x_s(t) \tag{I.4}$$

where $x_s(t)$ is a stationary process and the A_k's constitute a set of random variables. The Nth increment of $x(t)$ is given by

$$\Delta^N x(t) = A_N \tau^N + \Delta^N x_s(t) \tag{I.5}$$

while for $M > N$

$$\Delta^M x(t) = \Delta^M x_s(t). \tag{I.6}$$

Therefore, the expectation

$$E\{[\Delta^N x(t)]^2\} = \tau^{2N} E[A_N^2] + E\{[\Delta^N x_s(t)]^2\} \tag{I.7}$$

if A_N and $x_s(t)$ are uncorrelated, and either one has zero mean. Also, for $M > N$,

$$E\{[\Delta^M x(t)]^2\} = E\{[\Delta^M x_s(t)]^2\}. \tag{I.8}$$

Since $x_s(t)$ is stationary, it can be verified easily that $\Delta^k x_s(t)$ is also stationary for every $k > 0$. Hence the Mth increment $\Delta^M x(t)$ of the random process $x(t)$ is stationary for $M \geqslant N$. Thus the problem of finding $E\{[\Delta^N x(t)]^2\}$ reduces to one of finding $E\{[\Delta^N x_s(t)]^2\}$. In Appendix II-A, it is shown that

$$E\{[\Delta^N x_s(t)]^2\} = (-1)^N \Delta^{2N} R_{x_s}(\lambda)|_{\lambda=-N\tau} \tag{I.9}$$

$$= \sum_{j=-N}^{N} (-1)^j \binom{2N}{N+j} R_{x_s}(j\tau) \tag{I.10}$$

where $R_{x_s}(\lambda) \triangleq E[x_s(t)x_s(t+\lambda)]$. Thus

$$E\{[\Delta^N x(t)]^2\} = \tau^{2N} E[A_N^2] + (-1)^N \Delta^{2N} R_{x_s}(\lambda)|_{\lambda=-N\tau} \tag{I.11}$$

which is time-independent.

C. Structure Functions

In general, a random process $x(t)$ is called *a process with stationary* Nth *increments* if the expectations

$$E\{\Delta^N x(t;\tau)\} = c(\tau) \tag{I.12}$$

$$E\{\Delta^N x(t;\tau)\Delta^N x(t+s;\tau)\} = D_x^{(N)}(s;\tau) \tag{I.13}$$

exist for all real s, τ and do not depend on t. The function $D_x^{(N)}(s;\tau)$ is the *structure function* of the Nth increment. Briefly, it means that $\Delta^N x(t)$ has a time-independent mean and its autocorrelation depends only on the time difference; hence it is wide-sense stationary. In the main text, we are interested only in $D_x^{(N)}(\tau) \triangleq D_x^{(N)}(s=0;\tau)$ which is referred to as *the* Nth *structure function* of the process $x(t)$.

Appendix II

The equations that are too lengthy to derive in the main text are developed here. The headings indicate the equations. We also mention the general trigonometric expansion

$$2^{2n} \sin^{2n}\left(\frac{A}{2}\right) = \sum_{k=-n}^{n} (-1)^k \binom{2n}{n+k} \cos kA \tag{II.1}$$

which is used on many occasions.

A. Equation (I.9)

We want to derive (I.9) in Appendix I. Starting from the definition (I.3),

$$E\{[\Delta^N x_s(t)]^2\} = E\left\{\left[\sum_{i=0}^{N} (-1)^i \binom{N}{i} x_s(t+(N-i)\tau)\right] \cdot \left[\sum_{j=0}^{N} (-1)^j \binom{N}{j} x_s(t+(N-j)\tau)\right]\right\}$$

$$= \sum_{i,j=0}^{N} (-1)^{i+j} \binom{N}{i}\binom{N}{j} E\{x_s(t+(N-i)\tau) \cdot x_s(t+(N-j)\tau)\}. \tag{II.2}$$

Since $(-1)^{i+j} = (-1)^{i-j}$, after manipulating with the order of summation,

$$E\{[\Delta^N x_s(t)]^2\} = \sum_{i=0}^{N} \binom{N}{i}^2 R_{x_s}(0) - \sum_{i=0}^{N-1} \binom{N}{i}\binom{N}{i+1} \cdot [R_{x_s}(\tau) + R_{x_s}(-\tau)] + \cdots + (-1)^N \cdot [R_{x_s}(N\tau) + R_{x_s}(-N\tau)]. \tag{II.3}$$

However,

$$\sum_{i=0}^{N-k} \binom{N}{i}\binom{N}{i+k} = \binom{2N}{N-k}, \qquad \text{for } k = 0, 1, \cdots, N.$$

So, we can simplify further to get (I.9):

$$E[\Delta^N x_s(t)]^2 = \sum_{j=-N}^{N} (-1)^j \binom{2N}{N+j} R_{x_s}(j\tau) = (-1)^N \Delta^{2N} R_{x_s}(\lambda)|_{\lambda=-N\tau}. \tag{II.4}$$

B. Equation (1.6)

Let $S_{\dot\psi}(\omega)$ be the PSD of the $\dot\psi(t)$ process, i.e., $S_{\dot\psi}(\omega) = \int_{-\infty}^{\infty} R_{\dot\psi}(\tau)e^{-j\omega\tau}\, d\tau$. Then,

$$R_{\Delta\psi}(k\tau) \triangleq E[\Delta\psi(t)\Delta\psi(t+k\tau)]$$

$$= \int_t^{t+\tau} dx \int_{t+k\tau}^{t+k\tau+\tau} dy R_{\dot\psi}(y-x)$$

$$= \int_{-\infty}^{\infty} S_{\dot\psi}(\omega)\left[\int_t^{t+\tau} e^{+j\omega x}\, dx \int_{t+k\tau}^{t+k\tau+\tau} e^{-j\omega y}\, dy\right] \cdot \frac{d\omega}{2\pi} = 4\int_{-\infty}^{\infty} e^{-jk\omega\tau} \frac{S_{\dot\psi}(\omega)}{\omega^2} \sin^2\left(\frac{\omega\tau}{2}\right) \frac{d\omega}{2\pi} \tag{II.5}$$

which is (1.6).

C. Equations (1.7) and (1.13)

Since

$$R_{\dot\psi}(k\tau) + R_{\dot\psi}(-k\tau) = \int_{-\infty}^{\infty} 2\cos(k\omega\tau) S_{\dot\psi}(\omega) \frac{d\omega}{2\pi}$$

from (II.4),

$$D_{\psi}^{(M)}(\tau) = \int_{-\infty}^{\infty} \left[\binom{2M}{M} - 2\binom{2M}{M-1}\cos\omega\tau + \cdots + 2(-1)^M \cdot \cos M\omega\tau\right] S_{\dot\psi}(\omega) \frac{d\omega}{2\pi} = \int_{-\infty}^{\infty} 2^{2M} \sin^{2M}\left(\frac{\omega\tau}{2}\right) S_{\dot\psi}(\omega) \frac{d\omega}{2\pi} \tag{II.6}$$

which leads to (1.13). Similarly, from (II.5),

$$R_{\Delta\psi}(k\tau) + R_{\Delta\psi}(-k\tau) = 4\int_{-\infty}^{\infty} 2\cos k\omega\tau \frac{S_{\dot\psi}(\omega)}{\omega^2} \cdot \sin^2\left(\frac{\omega\tau}{2}\right) \frac{d\omega}{2\pi}. \tag{II.7}$$

So

$$D_{\psi}^{(M)}(\tau) = E\{\Delta^{M-1}(\Delta\psi(t))^2\}$$

$$= \binom{2M-2}{M-1} R_{\Delta\psi}(0) - \sum_{k=1}^{M-1} \binom{2M-2}{M-1-k} (-1)^k \cdot [R_{\Delta\psi}(k\tau) + R_{\Delta\psi}(-k\tau)]$$

$$= 4\int_{-\infty}^{\infty} 2^{2(M-1)} \sin^{2(M-1)}\left(\frac{\omega\tau}{2}\right) \frac{S_{\dot\psi}(\omega)}{\omega^2} \sin^2\left(\frac{\omega\tau}{2}\right) \frac{d\omega}{2\pi} \tag{II.8}$$

which gives (1.7).

D. Equation (4.1)

Consider the definite integral

$$I(\lambda) = \int_{-\infty}^{\infty} \cos \omega\lambda \, \frac{S_{\dot{\psi}}(\omega)}{\omega^2} \frac{d\omega}{2\pi} \tag{II.9}$$

which is assumed to be bounded. It can be shown that

$$I(\lambda) = I(0) - \int_0^{\lambda} dy \int_0^{y} dx R_{\dot{\psi}}(x)$$

$$= I(0) - \int_0^{\lambda} (\lambda - x) R_{\dot{\psi}}(x)\, dx. \tag{II.10}$$

From the representation (1.8),

$$D_{\psi}^{(M)}(\tau) = 2^{2M} \int_{-\infty}^{\infty} \sin^{2M}\left(\frac{\omega\tau}{2}\right) \frac{S_{\dot{\psi}}(\omega)}{\omega^2} \frac{d\omega}{2\pi}$$

$$= \int_{-\infty}^{\infty} \left[\sum_{k=-M}^{M} (-1)^k \binom{2M}{M+k} \cos k\omega\tau\right] \frac{S_{\dot{\psi}}(\omega)}{\omega^2} \frac{d\omega}{2\pi}$$

$$= -\sum_{k=-M}^{M} (-1)^k \binom{2M}{M+k} \int_0^{k\tau} (k\tau - x) R_{\dot{\psi}}(x)\, dx \tag{II.11}$$

which gives (4.1). Alternatively, $D_{\psi}^{(M)}(\tau)$ can be expressed as $(-1)^{M+1}\Delta^{2M}[\int_0^{\lambda} (\lambda - x) R_{\dot{\psi}}(x)\, dx]_{\lambda=-M\tau}$ which is suitable for numerical evaluation.

E. Equation (4.2)

From (1.8),

$$\frac{\partial}{\partial\lambda} D_{\psi}^{(M)}(\lambda) = 2^{2M} \int_{-\infty}^{\infty} M\omega \sin^{2M-1}\left(\frac{\omega}{2}\lambda\right) \cos\left(\frac{\omega}{2}\lambda\right) \frac{S_{\dot{\psi}}(\omega)}{\omega^2} \frac{d\omega}{2\pi} \tag{II.12}$$

and

$$\frac{\partial^2}{\partial\lambda^2} D_{\psi}^{(M)}(\lambda) = 2^{2M} \int_{-\infty}^{\infty} \frac{\omega^2}{2} M\left[(2M-1)\sin^{2M-2}\left(\frac{\omega\lambda}{2}\right) - 2M \sin^{2M}\left(\frac{\omega\lambda}{2}\right)\right] \frac{S_{\dot{\psi}}(\omega)}{\omega^2} \frac{d\omega}{2\pi}$$

$$= 2M(2M-1)2^{2(M-1)} \int_{-\infty}^{\infty} \sin^{2(M-1)}\left(\frac{\omega\lambda}{2}\right) \cdot S_{\dot{\psi}}(\omega) \frac{d\omega}{2\pi} - M^2 2^{2M} \int_{-\infty}^{\infty} \sin^{2M}\left(\frac{\omega\lambda}{2}\right) \cdot S_{\dot{\psi}}(\omega) \frac{d\omega}{2\pi}$$

$$= 2M(2M-1) D_{\psi}^{(M-1)}(\lambda) - M^2 D_{\psi}^{(M)}(\lambda). \tag{II.13}$$

It is obvious that

$$D_{\psi}^{(M)}(0) = 0 \quad \text{and} \quad \frac{\partial}{\partial\lambda} D_{\psi}^{(M)}(\lambda)|_{\lambda=0} = 0.$$

F. Equations (4.3) and (4.4)

To derive (4.3) and (4.4), we need the following identity,

$$2^{2(n-1)} \sin^{2n}\alpha = \binom{2n}{n-1} \sin^2\alpha - \binom{2n}{n-2} \sin^2 2\alpha + \cdots + (-1)^{n-1} \sin^2 n\alpha. \tag{II.14}$$

This can be obtained by expanding $\sin^{2n}\alpha$ via De Moivre's formula. Using the binomial theorem, we have

$$\sin^{2n}\alpha = \left(\frac{e^{j\alpha} - e^{-j\alpha}}{2j}\right)^{2n} - \left(\frac{1-1}{2j}\right)^{2n}$$

$$= \frac{1}{2^{2n}} (-1)^n \sum_{k=0}^{n-1} \binom{2n}{k} (-1)^k (e^{j2(n-k)\alpha} + e^{-j2(n-k)\alpha} - 2)$$

$$= \frac{1}{2^{2(n-1)}} \sum_{k=0}^{n-1} \binom{2n}{k} (-1)^{n-1+k} \sin^2 (n-k)\alpha \tag{II.15}$$

which gives (II.14). Finally, using (II.14) in (1.8) and (1.13), we obtain

$$D_{\Phi}^{(M)}(\tau) = 2^{2M} \int_{-\infty}^{\infty} \sin^{2M}\left(\frac{\omega\tau}{2}\right) \frac{S_{\dot{\psi}}(\omega)}{\omega^2} \frac{d\omega}{2\pi}$$

$$= \sum_{k=1}^{M} (-1)^{k-1} \binom{2M}{M-k} 2^2 \int_{-\infty}^{\infty} \sin^2\left(\frac{k\omega\tau}{2}\right) \cdot \frac{S_{\dot{\psi}}(\omega)}{\omega^2} \frac{d\omega}{2\pi} \tag{II.16}$$

and

$$D_{\Phi}^{(M)}(\tau) = 2^{2M} \int_{-\infty}^{\infty} \sin^{2M}\left(\frac{\omega\tau}{2}\right) S_{\dot{\psi}}(\omega) \frac{d\omega}{2\pi}$$

$$= \sum_{k=1}^{M} (-1)^{k-1} \binom{2M}{M-k} 2^2 \int_{-\infty}^{\infty} \sin^2\left(\frac{k\omega\tau}{2}\right) \cdot S_{\dot{\psi}}(\omega) \frac{d\omega}{2\pi} \tag{II.17}$$

from which (4.3) and (4.4) follow immediately.

Appendix III

Solution of a Functional Equation and Its Applications

In this Appendix, we derive the solution of a functional equation which is applicable for inverting (4.3)-(4.5) for $M \leqslant 5$. A similar method can also be used for (1.8).

Theorem: Consider a functional equation of the form

$$\sum_{ijk \in I_0} a_{ijk} f(2^i \cdot 3^j \cdot 5^k t) = g(t) \tag{III.1}$$

where a_{ijk}'s are constants, a_{000} is normalized to 1, and $I_0 = \{000, 100, 200, 010, 001\}$ is the index set that the subscripts

ijk can take on. Explicitly, we have

$$f(t) + a_{100}f(2t) + a_{200}f(4t) + a_{010}f(3t) + a_{001}f(5t) = g(t). \tag{III.2}$$

If $|g(t)| < g_{\max}$ for all t, and $\Sigma_{ijk\in I} |a_{ijk}| = \alpha < 1$ where $I \triangleq I_0 \backslash \{000\}$, then

$$f(t) = \sum_{a,b,c\geqslant 0} D(a,b,c)g(2^a \cdot 3^b \cdot 5^c \cdot t) \tag{III.3}$$

where a, b, c are nonnegative integer indices and $D(a, b, c)$ is determined recursively by

$$D(a,b,c) = -\sum_{ijk\in I} a_{ijk}D(a-i, b-j, c-k) \tag{III.4}$$

with initial conditions

$$D(0,0,0) = 1$$

and

$$D(a,b,c) = 0 \text{ for any negative } a, b, \text{ or } c. \tag{III.5}$$

Furthermore, if we approximate (III.3) by the truncated sum, i.e.,

$$f(t) = \sum_{a+b+c\geqslant 0}^{n} D(a,b,c) \cdot g(2^a \cdot 3^b \cdot 5^c t) + e(n) \tag{III.6}$$

then,

$$|e(n)| \leqslant K \cdot (n+1)^2 \alpha^{-(n+1)} \longrightarrow 0 \text{ as } n \longrightarrow \infty \tag{III.7}$$

where K is some bounded constant.

Proof: To simplify the notation, let

$$f(x,y,z) \triangleq f(2^x \cdot 3^y \cdot 5^z \cdot t)$$

and

$$g(x,y,z) \triangleq g(2^x \cdot 3^y \cdot 5^z \cdot t) \tag{III.8}$$

where x, y, z are dummy integer variables. First of all, we show that $f(t)$ is bounded. By iterating the recursion (III.4), one can show that

$$|D(a,b,c)| \leqslant \left\{ \sum_{ijk\in I} |a_{ijk}| \right\}^n = \alpha^n. \tag{III.9}$$

If we truncate the infinite sum in (III.3), we get

$$f(0,0,0) = \sum_{a+b+c\geqslant 0}^{n} D(a,b,c)g(a,b,c) + e(n). \tag{III.10}$$

The first term is always finite. By the Schwarz's inequality

$$|e(n)| \leqslant g_{\max} \sum_{a+b+c=n+1}^{\infty} \alpha^{a+b+c}. \tag{III.11}$$

It is a straightforward exercise to show that the number of different choices of (a, b, c) for $a + b + c = m$ is $(m + 1)(m + 2)/2$ for $a \geqslant 0$, $b \geqslant 0$, and $c \geqslant 0$. So

$$|e(n)| \leqslant g_{\max} \sum_{k=n+1} \frac{(k+1)(k+2)}{2} \alpha^k$$
$$\leqslant K \cdot (n+1)^2 \alpha^{-(n+1)} \tag{III.12}$$

which has a zero limit as n becomes infinite for $\alpha < 1$.

Now we verify that $f(t)$ as given by (III.3) satisfies (III.1). From (III.3), if $i, j, k \geqslant 0$,

$$f(i,j,k) = \sum_{a,b,c\geqslant 0} D(a,b,c) \cdot g(a+i, b+j, c+k)$$
$$= \sum_{a,b,c\geqslant 0}^{\infty} D(a-i, b-j, c-k)g(a,b,c) \tag{III.13}$$

where the last step is obtained by manipulating a, b, c and using (III.5). Finally, using (III.13),

$$\sum_{ijk\in I_0} a_{ijk}f(i,j,k) = \sum_{a,b,c\geqslant 0} a_{000}D(a,b,c)g(a,b,c) + \sum_{a,b,c\geqslant 0} \cdot g(a,b,c)\left(\sum_{ijk\in I} a_{ijk}D(a-i,b-j,c-k)\right)$$
$$= \sum_{a,b,c\geqslant 0} D(a,b,c)g(a,b,c) - \sum_{a+b+c>0} \cdot D(a,b,c)g(a,b,c) \tag{III.14}$$
$$= g(0,0,0).$$

Q.E.D.

Corollary: $D(a, b, c)$ is given by

$$D(a,b,c) = \sum_{k=0}^{I_a} (-1)^{a+b+c+k} \binom{a+b+c-k}{a-2k} \binom{b+c+k}{k} \binom{b+c}{b} \cdot A(a-2k, k, b, c) \tag{III.15}$$

for $a \geqslant 0$, $b \geqslant 0$, and $c \geqslant 0$. And

$$A(a-2k, k, b, c) \triangleq a_{100}^{a-2k} \cdot a_{200}^{k} \cdot a_{010}^{b} \cdot a_{001}^{c}$$

with the convention

$$a_{ijk}^d = \begin{cases} 0, & d > 0 \\ 1, & d = 0 \end{cases} \quad \text{for } a_{ijk} = 0$$

and

$$I_a = \begin{cases} a/2, & \text{for } a \text{ even} \\ (a-1)/2, & \text{for } a \text{ odd} \end{cases}.$$

Proof: It is easy to check that $D(a, b, c)$ as given by (III.15) is consistent with (III.5). One can then verify that $D(a, b, c)$ satisfies (III.4). First of all, $D(a, b, c)$ in (III.15) is used in the right hand side of (III.4). By absorbing the a_{ijk}'s, $A(a - 2k, k, b, c)$ is obtained as a common part of each individual term. Finally, when the terms are added in the right sequence, the binomial coefficient products collapse into the desired form. Q.E.D.

The expression for $D(a, b, c)$ simplifies for special cases. If $a_{001} = 0$, we have $D(a, b, c) = 0$ for $c > 0$ so (III.15) reduces to

$$D(a,b) = \sum_{k=0}^{I_a} (-1)^{a+b+k} \binom{a+b-k}{a-2k} \binom{b+k}{k} a_{100}^{a-2k} a_{200}^{k} a_{010}^{b}. \tag{III.16}$$

Similarly, if $a_{001} = a_{200} = 0$, we have

$$D(a,b) = (-1)^{a+b} \binom{a+b}{a} a_{100}^{a} \cdot a_{200}^{b}. \tag{III.17}$$

Finally, if $a_{001} = a_{200} = a_{010} = 0$,

$$D(a) = (-1)^a \cdot a_{100}^a. \qquad \text{(III.18)}$$

Our result can be extended to handle equations of the form

$$f(t) + \sum_{k=2}^{n} a_k f(kt) = g(t) \qquad \text{(III.19)}$$

for $n > 5$. However, for the following applications, the condition $\sum_{k=2}^{n} |a_k| < 1$ is not satisfied for $n > 5$, and the theorem is given in the present form.

Applications

Let us apply (III.15)–(III.18) to invert (4.5)[14], assuming that $R_{\dot\psi}(t) = R_{\dot\psi}(-t)$ bounded for all t. We shall give a solution of $R_{\dot\psi}(\tau)$ in terms of $D_{\dot\Phi}^{(M)}(\tau)$ for the cases $M = 2, 3, 4, 5$:

(i) $M = 2$. We have

$$D_{\dot\Phi}^{(2)}(\tau) = 6R_{\dot\psi}(0) - 8R_{\dot\psi}(\tau) + 2R_{\dot\psi}(2\tau)$$

or

$$R_{\dot\psi}(\tau) - \tfrac{1}{4}R_{\dot\psi}(2\tau) = \tfrac{1}{8}[6R_{\dot\psi}(0) - D_{\dot\Phi}^{(2)}(\tau)].$$

Using (III.18) and $D_{\dot\psi}^{(M)}(0) = 0$,

$$R_{\dot\psi}(\tau) = R_{\dot\psi}(0) - \tfrac{1}{8} \sum_{a=0}^{\infty} (\tfrac{1}{4})^a D_{\dot\Phi}^{(2)}(2^a \tau). \qquad \text{(III.20)}$$

(ii) $M = 3$. We have

$$R_{\dot\psi}(\tau) = R_{\dot\psi}(0) - \tfrac{1}{30} \sum_{a,b \geq 0}^{\infty} (-1)^b \binom{a+b}{a} (\tfrac{2}{5})^a (\tfrac{1}{15})^b \cdot D_{\dot\Phi}^{(3)}(2^a \cdot 3^b \cdot \tau). \qquad \text{(III.21)}$$

(iii) $M = 4$. We have

$$R_{\dot\psi}(\tau) = R_{\dot\psi}(0) - \tfrac{1}{112} \sum_{a,b \geq 0}^{\infty} H(a,b) D_{\dot\Phi}^{(4)}(2^a \cdot 3^b \cdot \tau) \qquad \text{(III.22)}$$

where

$$H(a,b) = \sum_{k=0}^{I_a} (-1)^b \binom{b+k}{k} \binom{a+b-k}{a-2k} (\tfrac{1}{2})^{a-2k} (\tfrac{1}{7})^k (\tfrac{1}{56})^b.$$

(iv) $M = 5$. We have

$$R_{\dot\psi}(\tau) = R_{\dot\psi}(0) - \tfrac{1}{420} \sum_{a,b,c \geq 0}^{\infty} H(a,b,c) D_{\dot\Phi}^{(5)}(2^a \cdot 3^b \cdot 5^c \cdot \tau) \qquad \text{(III.23)}$$

where

$$H(a,b,c) = \sum_{k=0}^{I_a} (-1)^{b+c} \binom{a+b+c-k}{a-2k} \binom{b+c+k}{k} \binom{b+c}{b} \cdot (\tfrac{4}{7})^{a-2k} (\tfrac{1}{21})^k (\tfrac{3}{14})^b (\tfrac{1}{210})^c.$$

A similar procedure can be used to solve (1.8), which can be written as

$$D_{\dot\Phi}^{(M)}(\tau) = \int_{-\infty}^{\infty} \sum_{k=-M}^{M} (-1)^k \binom{2M}{M+k} \cos(k\omega\tau) \frac{S_{\dot\psi}(\omega)}{\omega^2} \frac{d\omega}{2\pi}.$$

[14] (4.3) and (4.4) are similar and will be omitted.

If

$$\int_{-\infty}^{\infty} \frac{S_{\dot\psi}(\omega)}{\omega^2} \frac{d\omega}{2\pi} < \infty,$$

we let

$$V(k\tau) = \int_{-\infty}^{\infty} \cos(k\omega\tau) \frac{S_{\dot\psi}(\omega)}{\omega^2} \frac{d\omega}{2\pi}$$

and the method still applies to

$$\sum_{k=-M}^{M} (-1)^k \binom{2M}{M+k} V(k\tau) = D_{\Phi}^{(M)}(\tau) \qquad \text{(III.24)}$$

for obtaining $V(\tau)$ in terms of $D_{\Phi}^{(M)}(\tau)$. Finally, $S_{\dot\psi}(\omega) = \omega^2 \mathfrak{F}[V(\tau)]$ where $\mathfrak{F}$ is the Fourier transform operator.

Appendix IV

In this appendix, various mathematical relationships between the PSD $S_{\dot\psi}(\omega)$, the structure functions $D_{\psi}^{(1)}(\tau)$ and $D_{\Phi}^{(1)}(\tau)$, the correlation function $R_{\dot\psi}(\tau)$, the rms fractional frequency deviation $[\delta f(\tau)]/f_0$ and the measured (L-sample) rms fractional frequency deviation $[\delta \hat{f}(\tau)]/f_0$ will be derived under the assumptions discussed in Section II-A.

The time average of the frequency deviation over τ seconds [2, (4a)] is, in our notations

$$\frac{\Delta\psi(t)}{\tau} \triangleq \frac{1}{\tau} \int_t^{t+\tau} \dot\psi(\lambda)\, d\lambda. \qquad \text{(IV.1)}$$

Its variance $\sigma_{\dot\psi}^2(\tau)$ is, by the definition of $D_{\psi}^{(1)}(\tau)$,

$$\sigma_{\dot\psi}^2(\tau) \triangleq E\left\{\left[\frac{\Delta\psi(t)}{\tau}\right]^2\right\} = \frac{1}{\tau^2} D_{\psi}^{(1)}(\tau) \qquad \text{(IV.2)}$$

$$= \frac{1}{\tau^2} D_{\Phi}^{(1)}(\tau) - \omega_0^2 \qquad \text{(IV.3)}$$

since $D_{\Phi}^{(1)}(\tau) = \tau^2 \omega_0^2 + D_{\psi}^{(1)}(\tau)$. Using (4.1) with $M = 1$, we have

$$\sigma_{\dot\psi}^2(\tau) = \frac{2}{\tau} \int_0^{\tau} \left(1 - \frac{\lambda}{\tau}\right) R_{\dot\psi}(\lambda)\, d\lambda \qquad \text{(IV.4)}$$

which is (16b) in [2]. Using (1.7) for $N = 1$, we have

$$\sigma_{\dot\psi}^2(\tau) = \int_{-\infty}^{\infty} \left(\frac{\sin(\omega\tau/2)}{\omega\tau/2}\right)^2 S_{\dot\psi}(\omega) \frac{d\omega}{2\pi} \qquad \text{(IV.5)}$$

which is [2, eq. (17b)]. From (II.12) with $M = 1$, we have

$$\frac{\partial D_{\psi}^{(1)}(\tau)}{\partial\tau} = 2\int_{-\infty}^{\infty} \sin(\omega\tau) \frac{S_{\dot\psi}(\omega)}{\omega} \frac{d\omega}{2\pi} \qquad \text{(IV.6)}$$

and thus

$$\frac{\partial^2 D_{\psi}^{(1)}(\tau)}{\partial\tau^2} = 2\int_{-\infty}^{\infty} \cos(\omega\tau) S_{\dot\psi}(\omega) \frac{d\omega}{2\pi}. \qquad \text{(IV.7)}$$

Applying Fourier inversion formulas to (IV.6) and (IV.7), we have

$$S_{\dot{\psi}}(\omega) = \omega \int_0^\infty \frac{\partial}{\partial\tau}(\tau^2 \sigma_{\dot{\psi}}^2(\tau)) \sin \omega\tau \, d\tau$$

$$= \int_0^\infty \frac{\partial^2}{\partial\tau^2}(\tau^2 \sigma_{\dot{\psi}}(\tau)) \cos \omega\tau \, d\tau. \quad \text{(IV.8)}$$

The sample variance $\hat{\sigma}_{\dot{\psi}}^2(\tau)$ based on observing L samples, which determines the measured rms fractional frequency deviation

$$\frac{\delta\hat{f}(\tau)}{f_0} \triangleq \frac{\sqrt{\hat{\sigma}_{\dot{\psi}}^2(\tau)}}{\omega_0} \quad \text{(IV.9)}$$

is defined in terms of the frequency-counted measurements (compare with [2, eq. (74)])

$$\hat{\sigma}_{\dot{\psi}}^2(\tau) \triangleq \frac{1}{L}\sum_{k=0}^{L-1}\left(\frac{\Delta\psi(t+k\tau;\tau)}{\tau}\right)^2 - \left(\frac{1}{L}\sum_{k=0}^{L-1}\frac{\Delta\psi(t+k\tau;\tau)}{\tau}\right)^2 \quad \text{(IV.10)}$$

and represents a random variable by which $\sigma_{\dot{\psi}}^2(\tau)$ can be estimated. The (ensemble) mean $E[\hat{\sigma}_{\dot{\psi}}^2(\tau)]$ of the sample variance can be related to the PSD $S_{\dot{\psi}}(\omega)$ through [2, eq. (77)]

$$E[\hat{\sigma}_{\dot{\psi}}^2(\tau)] = \int_{-\infty}^{\infty}\left[\left(\frac{\sin(\omega\tau/2)}{\omega\tau/2}\right)^2 - \left(\frac{\sin(\omega L\tau/2)}{L\omega\tau/2}\right)^2\right] S_{\dot{\psi}}(\omega)\frac{d\omega}{2\pi}. \quad \text{(IV.11)}$$

By comparing (IV.11) with (1.8) we find that the (ensemble) mean of the sample variance is related to the structure function of the phase instability through

$$E[\hat{\sigma}_{\dot{\psi}}^2(\tau)] = \frac{1}{\tau^2}\left[D_{\psi}^{(1)}(\tau) - \frac{1}{L^2}D_{\psi}^{(1)}(L\tau)\right]$$

$$= \sigma_{\dot{\psi}}^2(\tau) - \sigma_{\dot{\psi}}^2(L\tau). \quad \text{(IV.12)}$$

The measurement bias is given by

$$\sigma_{\dot{\psi}}^2(\tau) - E[\hat{\sigma}_{\dot{\psi}}^2(\tau)] = \frac{D_{\psi}^{(1)}(L\tau)}{(L\tau)^2} \quad \text{(IV.13)}$$

from which we observe that

$$\lim_{L\to\infty} E[\hat{\sigma}_{\dot{\psi}}^2(\tau)] = \sigma_{\dot{\psi}}^2(\tau)$$

if $D_{\psi}^{(1)}(\tau) < \infty$ for all τ, i.e., the estimate is asymptotically unbiased.

If we use the measurement $\hat{\sigma}_{\dot{\psi}}^2(\tau)$ as an estimate $\hat{D}_{\psi}^{(1)}(\tau)$ of $D_{\psi}^{(1)}(\tau)$, then we have (see (IV.2))

$$\hat{D}_{\psi}^{(1)}(\tau) = \tau^2 \hat{\sigma}_{\dot{\psi}}^2(\tau) \quad \text{(IV.14)}$$

so that the PSD $S_{\dot{\psi}}(\omega)$ is estimated from (IV.8) via

$$\hat{S}_{\dot{\psi}}(\omega) = \int_0^\infty \left(\frac{\partial^2 \hat{D}_{\psi}^{(1)}(\tau)}{\partial\tau^2}\right)\cos\omega\tau \, d\tau$$

$$= \omega\int_0^\infty\left(\frac{\partial\hat{D}_{\psi}(\tau)}{\partial\tau}\right)\sin\omega\tau\,d\tau. \quad \text{(IV.15)}$$

Finally, the variance of the frequency instability estimate is

$$\hat{R}_{\dot{\psi}}(\tau) = \int_{-\infty}^{\infty}\hat{S}_{\dot{\psi}}(\omega)\cos\omega\tau\frac{d\omega}{2\pi}. \quad \text{(IV.16)}$$

Appendix V

Under the assumptions discussed in Section II-B, we establish various mathematical relationships between the following quantities: the PSD $S_{\dot{\psi}}(\omega)$; the first and second structure functions of the process $\Phi(t)$ and $\psi(t)$; the two-sample Allan variance $E\{\sigma_A^2(2,\tau,\tau)\}$ and its measurement estimate $\hat{\sigma}_A^2(2,\tau,\tau)$; and the L-sample Allan variance $\hat{\sigma}_A^2(L,\tau,\tau)$ and its (ensemble) mean $E\{\hat{\sigma}_A^2(L,\tau,\tau)\}$. In addition, the rms fractional frequency deviation and the Allan variance are shown to be related to one another under certain conditions.

Using (1.9) in (2.8), we have

$$E\{\sigma_A^2(2,\tau,\tau)\} = \frac{1}{2(\omega_0\tau)^2} \cdot \left[\tau^4 E[\Omega_1^2] + 2^4\int_{-\infty}^{\infty}\sin^4\left(\frac{\omega\tau}{2}\right)\frac{S_{\dot{\psi}}(\omega)}{\omega^2}\frac{d\omega}{2\pi}\right] \quad \text{(V.1)}$$

which relates the two-sample Allan variance $E\{\sigma_A^2(2,\tau,\tau)\}$ and the PSD $S_{\dot{\psi}}(\omega)$. If $E[\Omega_1^2]$ can be determined, then (V.1) can be inverted to yield $S_{\dot{\psi}}(\omega)$ using the Mellin transform or the technique discussed in Appendix III. Furthermore, the two-sample Allan variance can be related to the rms fractional frequency deviation provided that the latter is well defined, i.e., $E[\Omega_1^2] = 0$ and $\nu < 1$ in the PSD $S_{\dot{\psi}}(\omega)$ as in Section II-A. For, with $M = 2$ in (4.3), we have[15]

$$D_{\Phi}^{(2)}(\tau) = 4D_{\psi}^{(1)}(\tau) - D_{\psi}^{(1)}(2\tau) \quad \text{(V.2)}$$

from which we deduce, see (2.2) and (2.8),

$$E\{\sigma_A^2(2,\tau,\tau)\} = 2\left[\left(\frac{\delta f(\tau)}{f_0}\right)^2 - \left(\frac{\delta f(2\tau)}{f_0}\right)^2\right] \quad \text{(V.3)}$$

where $\delta f(\tau)/f_0$ is the rms fractional frequency deviation defined by Cutler and Searle [2]. The relationship (V.2) can be inverted using (III.18) to yield

$$\left(\frac{\delta f(\tau)}{f_0}\right)^2 = \frac{1}{2}\sum_{k=0}^{\infty}E\{\sigma_A^2(2,2^k\tau,2^k\tau)\} \quad \text{(V.4)}$$

which express the rms fractional frequency deviation in terms of the two-sample Allan variance. The measured two-sample

[15] (V.2) is still valid even if $1 \leq \nu < 3$. Although $D_{\psi}^{(1)}(\tau)$ is not well defined, the difference $4D_{\psi}^{(1)}(\tau) - D_{\psi}^{(1)}(2\tau)$ is. For, if t is small:

$$4D_{\psi}^{(1)}(\tau) - D_{\psi}^{(1)}(2\tau)$$

$$= 4\int_{-\infty}^{\infty}4\sin^2\left(\frac{\omega\tau}{2}\right)\frac{S_{\dot{\psi}}(\omega)}{\omega^2}\frac{d\omega}{2\pi} - 4\int_{-\infty}^{\infty}\sin^2(\omega\tau)\frac{S_{\dot{\psi}}(\omega)}{\omega^2}\frac{d\omega}{2\pi}$$

$$= 4\int_{-t}^{t}\left[4\sin^2\left(\frac{\omega\tau}{2}\right) - \sin^2(\omega\tau)\right]\frac{S_{\dot{\psi}}(\omega)}{\omega^2}\frac{d\omega}{2\pi} + \text{bounded term}$$

$$\cong \int_{-t}^{t}(\omega\tau)^4\frac{S_{\dot{\psi}}(\omega)}{\omega^2}\frac{d\omega}{2\pi} + \text{bounded term} < \infty.$$

Allan variance is given by

$$\hat{\sigma}_A^2(2, \tau, \tau) = \frac{1}{2m(\omega_0 \tau)^2} \left\{ \sum_{k=0}^{m-1} [\Delta^2 \Phi(t + k\tau; \tau)]^2 \right\}. \quad \text{(V.5)}$$

Its (ensemble) mean is clearly $E\{\sigma_A^2(2, \tau, \tau)\}$ and so is an unbiased estimate of the two-sample Allan variance.

The L-sample Allan variance is defined by Allan [3, eq. (11)] (when normalized by ω_0^2) as

$$\omega_0^2 \hat{\sigma}_A^2(L, \tau, \tau) = \frac{1}{(L-1)} \left\{ \sum_{k=0}^{L-1} \left[\frac{\Delta\psi(t + k\tau; \tau)}{\tau} \right]^2 - \frac{1}{L} \left[\sum_{k=0}^{L-1} \frac{\Delta\psi(t + k\tau; \tau)}{\tau} \right]^2 \right\} \quad \text{(V.6)}$$

over the L data samples.[16] The interested reader can verify that if $\Omega_1 \neq 0$, then $\omega_0^2 \hat{\sigma}_A^2(L, \tau, \tau)$ is not time-independent. Except for a factor of $L/(L-1)$, the normalized L-sample Allan variance is seen to be exactly the sample variance $\hat{\sigma}_{\dot{\psi}}^2(\tau)$ of frequency counted measurements when one compares (V.6) with (IV.10). For $L = 2$, this is the extra factor of two found when comparing the two-sample Allan variance with the second phase structure function $D_\Phi^{(2)}(\tau)$. For L large, $L/(L-1)$ is approximately one, so the L-sample Allan variance is indistinguishable from the sample variance $\hat{\sigma}_{\dot{\psi}}^2(\tau)$.

The (ensemble) mean of $\hat{\sigma}_A(L, \tau, \tau)$ has been worked out [1, eq. (68), $r = 1$]; namely, in our notation:

$$\begin{aligned} \omega_0^2 E\{\hat{\sigma}_A^2(L, \tau, \tau)\} &= \frac{L}{(L-1)} \int_{-\infty}^{\infty} S_{\dot{\psi}}(\omega) \frac{\sin^2(\omega\tau/2)}{(\omega\tau/2)^2} \cdot \left[1 - \frac{\sin^2(\omega L\tau/2)}{L^2 \sin^2(\omega\tau/2)} \right] \frac{d\omega}{2\pi} \\ &= \frac{L}{(L-1)} \frac{1}{\tau^2} 2^2 \int_{-\infty}^{\infty} \left[\left(\sin^2\left(\frac{\omega\tau}{2}\right) - \frac{\sin^2(\omega L\tau/2)}{L^2} \right) \frac{S_{\dot{\psi}}(\omega)}{\omega^2} \right] \frac{d\omega}{2\pi}. \end{aligned} \quad \text{(V.7)}$$

In terms of structure functions:

$$(\omega_0\tau)^2 E\{\hat{\sigma}_A^2(L, \tau, \tau)\} = \frac{L}{(L-1)} \left[D_\psi^{(1)}(\tau) - \frac{1}{L^2} D_\psi^{(1)}(L\tau) \right] \quad \text{(V.8)}$$

provided that $D_\psi^{(1)}(\tau)$ is time-independent and finite for all τ. In that case, if the number of samples L increases, in the limit:

$$\lim_{L\to\infty} (\omega_0\tau)^2 E\{\hat{\sigma}_A^2(L, \tau, \tau)\} = D_\psi^{(1)}(\tau). \quad \text{(V.9)}$$

Hence, $\hat{\sigma}_A^2(L, \tau, \tau)$ is an asymptotically unbiased estimator of $D_\psi^{(1)}(\tau)/(\omega_0\tau)^2 = [\delta f(\tau)/f_0]^2$ which is the rms fractional frequency deviation squared. In view of our comment earlier concerning the sample variance of frequency counted measurements $\hat{\sigma}_{\dot{\psi}}^2(\tau)$ and the L-sample Allan variance, this result is not at all surprising. Although the estimate for the two-sample Allan variance defined in (V.5) is an unbiased estimator, the L-sample Allan variance in (V.6) is misleading in the sense that it is unrelated to the two-sample Allan variance. However, it is related to the rms fractional frequency deviation as we have shown here.

[16] Unlike the two-sample Allan variance which is defined as the (probabilistic) expectation of a function of two random samples, see (2.7), the L-sample Allan variance is really a "sample" variance (see [3, p. 223, footnote 2]).

Acknowledgment

The authors wish to express their thanks to Dr. J. A. Barnes and D. W. Allan of the National Bureau of Standards for their comments, discussions, and earlier criticisms of this work. We are also indebted to W. E. Leavitt and P. M. Hooten, Satellite Communications Branch of the Naval Research Laboratory, for their support, encouragement, and for participating in various stimulating discussions throughout the course of this work. Finally, the authors are grateful for the helpful mathematical suggestions made by Prof. L. Welch of the University of Southern California in connection with solving a functional equation.

References

[1] J. A. Barnes *et al.*, "Characterization of frequency stability," *IEEE Trans. Instrum. Meas.*, vol. IM-20, pp. 105–120, May 1971.

[2] L. S. Cutler and C. L. Searle, "Some aspects of the theory and measurement of frequency fluctuations in frequency standards," *Proc. IEEE*, vol. 54, pp. 136–154, Feb. 1966.

[3] D. W. Allan, "Statistics of atomic frequency standards," *Proc. IEEE*, vol. 54, pp. 221–230, Feb. 1966.

[4] J. A. Barnes, "Atomic timekeeping and the statistics of precision signal generators," *Proc. IEEE*, vol. 54, pp. 207–220, Feb. 1966.

[5] J. A. Barnes and D. W. Allan, "A statistical model of flicker noise," *Proc. IEEE*, vol. 54, pp. 176–178, Feb. 1966.

[6] E. J. Baghdady, R. N. Lincoln, and B. D. Nelin, "Short-term frequency stability: Characterization, theory, and measurement," *Proc. IEEE*, vol. 53, pp. 704–722, July 1965.

[7] R. Sydnor, J. J. Caldwell, and B. E. Rose, "Frequency stability requirements for space communications and tracking systems," *Proc. IEEE*, vol. 54, pp. 231–236, Feb. 1966.

[8] R. M. Gray and R. C. Tausworthe, "Frequency-counted measurements and phase-locking to noisy oscillators," *IEEE Trans. Commun. Technol.*, vol. COM-19, pp. 21–30, Feb. 1971.

[9] R. C. Tausworthe, "Convergence of oscillator spectral estimators for counted-frequency measurements," *IEEE Trans. Commun. Technol.*, vol. COM-20, pp. 214–217, Apr. 1972.

[10] A. N. Kolmogorov, "Dissipation of energy in locally isotropic turbulence," *Dokl. Nauk. SSR*, vol. 32, pp. 16–18, 1941. Reprinted (in English) in *Turbulence: Classic Papers on Statistical Theory*, S. K. Frielander and L. Topper, Eds., New York: Interscience, 1951, pp. 159–161.

[11] A. M. Yaglom, *Theory of Stationary Random Functions.* Englewood Cliffs, NJ: Prentice-Hall, 1962.

[12] V. I. Tatarski, *Wave Propagation in a Turbulent Medium.* New York: McGraw-Hill, 1961.

[13] W. C. Lindsey, *Synchronization Systems in Communication and Control.* Englewood Cliffs, NJ: Prentice-Hall, 1972.

[14] W. C. Lindsey and M. K. Simon, *Telecommunication Systems Engineering.* Englewood Cliffs, NJ: Prentice-Hall, 1973.

[15] I. I. Gikhman and A. V. Skorokhod, *Introduction to the Theory of Random Processes.* Philadelphia, PA: W. B. Saunders, 1969, ch. 1.

[16] J. L. Doob, *Stochastic Processes.* New York: Wiley, 1953.

[17] W. C. Lindsey and J. Lewis, "Modeling, characterization and measurement of oscillator frequency instability," Naval Research Laboratory, Washington, DC, Code 5436, Washington, DC, June 1974. Also see *LinCom Technical Report* 08-7507-1, by W. C. Lindsey, Aug. 1975.

[18] W. C. Lindsey and C. Chie, "Transform theory of oscillator instability," USC Technical Report, University of Southern California, Los Angeles, CA, Jan. 1975.

[19] E. C. Titchmarsh, *Theory of Fourier Integrals.* Amen House, London, England: Oxford Univ. Press, 1962.

[20] T. E. Firle and H. Winston, "Noise measurements in semiconductors at very low frequencies," *J. Appl. Phys.*, vol. 26, p. 716, June 1955.

[21] M. W. E. Strandberg, "Noise spectrum of phase-locked oscillators," *Proc. IRE*, vol. 48, pp. 1168–1169, June 1960.

[22] W. A. Edson, "Noise in oscillators," *Proc. IRE*, vol. 48, pp. 1464–1466, Aug. 1960.

Characterization of Phase and Frequency Instabilities in Precision Frequency Sources: Fifteen Years of Progress

JACQUES RUTMAN, MEMBER, IEEE

Abstract—Precision frequency sources such as quartz oscillators, masers, and passive atomic frequency standards are affected by phase and frequency instabilities including both random and deterministic components. It is of prime importance to have a comprehensive characterization of these instabilities in order to be able to assess the potential utility of each source. For that purpose, many parameters have been proposed especially for dealing with random fluctuations. Some of them have been recommended by the IEEE Subcommittee on Frequency Stability and later by Study Group 7 on "Standard Frequencies and Time Signals" of the International Radio Consultative Committee (CCIR). Others are not so widely used but show interesting capabilities. This paper aims at giving a broad review of parameters proposed for phase and frequency instability characterization, including both classical widely used concepts and more recent less familiar approaches. Transfer functions that link frequency-domain and time-domain parameters are emphasized because they provide improved understanding of the properties of a given time-domain parameter or facilitate introducing of new parameters. As far as new approaches are concerned, an attempt has been made to demonstrate clearly their respective advantages. To this end, some developments that did not appear in the original references are presented here, e.g., the modified three sample variance $\Sigma_y^2(\tau)$, the expressions of $\langle \delta \bar{y}_T^2 \rangle$, the interpretation of structure functions of phase and its relations with $\Sigma_y^2(\tau)$ and the Hadamard variance. The effects of polynomial phase and frequency drifts on various parameters have also been pointed out in parallel with those of random processes modeled by power-law spectral densities.

I. Introduction

PHASE AND FREQUENCY instability characterization has become of great concern to many engineers working in various fields since an increasing number of systems rely upon high-quality time and frequency sources such as quartz-crystal oscillators, frequency synthesizers, atomic frequency standards and clocks; also, the advent of frequency stabilized lasers has provided us with frequency standards in the optical range.

The following nonexhaustive list of systems illustrates the wide range of users [1], [2]:

1) Doppler radar systems with a narrow bandwidth receiver tuned to detect the shifted frequency return need high-performance transmitter oscillators and receiver local oscillators since any instability limits range resolution and sensitivity.

2) Oscillators are used in missiles and spacecrafts for a variety of purposes including guidance, tracking, and communications; frequency instabilities are harmful in all cases because they degrade system performance.

3) In range measurements where ranging signals are phase compared relative to a reference signal, instabilities in any of the oscillators involved introduce an uncertainty in the range estimate.

4) In communication systems, interference is reduced and performance is improved by better frequency control of the carrier frequencies. In digital communications, emphasis is put on the timing capability of the network clocks: an adequate performance measure, the maximum time interval error, is related to clocks phase and frequency instabilities. For long-term frequency fluctuations, a fractional value of 1 part in 10^{11} or better is recommended by the International Telegraph and Telephone Consultative Committee (CCITT) for the interconnection of several synchronous networks on the international level.

5) Of course, one should not forget the field of time and frequency metrology where sophisticated laboratory-type time and frequency standards are designed, constructed and operated, e.g., long cesium-beam and hydrogen devices. In fact, with few exceptions, most of the widely used stability measures have been developed by scientists working in the field.

A practical problem is that several groups of people with different backgrounds have had to find a common language for oscillator specifications.

The question was how to develop a useful and comprehensive characterization of phase and frequency instabilities that can be understood and applied by everyone. More specifically, in each field, everyone wanted to know how instabilities affect system performance and how the possible instability measures can be used to assess system performance. Of course, no single answer can be given and much work has been carried out during the last fifteen years to provide some useful answers.

Frequency stability was already recognized as an important problem at the beginning of the sixties: the special IEEE–NASA Symposium on Short-Term Frequency Stability held at Goddard Space Flight Center on November 23–24, 1964 [1] appeared as the first opportunity for cross-fertilization of ideas. Following this Symposium, an IEEE Subcommittee on Frequency Stability of the Technical Committee on Frequency and Time was created with the ultimate aim of providing a set of recommendations for standards and definitions on both short-term and long-term stability. A Special Issue of the PROCEEDINGS OF THE IEEE was devoted to frequency stability in February 1966 [2] to promote further exchange of information, the Subcommittee serving as Editorial Committee.

Manuscript received April 15, 1978; revised May 25, 1978. The submission of this paper was encouraged after submission of an advanced proposal. This paper is an expanded version of "Oscillator specifications: A review of classical and new ideas," which appeared in *Proceedings of the 31st Annual Frequency Control Symposium*, June 1–3, 1977 [47].

The author is with the Laboratoire Primaire du Temps et des Fréquences, Bureau National de Métrologie, Observatoire de Paris, Paris 75014, France.

Reprinted from *Proc. IEEE*, vol. 66, pp. 1048–1075, Sept. 1978.

Several basic papers dealing explicitly with frequency stability characterization in both frequency and time domains and including the translations between them, were then published [3]–[7]. Of particular interest, the use of sample variances for time-domain characterization was developed in [7].

In May 1971, the Subcommittee issued a paper[1] [8] presented as technical background for an eventual IEEE standard definition (not yet adopted): two definitions of frequency stability were given together with translation relationships that play an important theoretical and practical role. Both have gained wide acceptance among manufacturers and users of precision frequency sources.

In the present paper, following a description of the mathematical models and basic definitions given in Section II, the parameters proposed as frequency stability measures will be studied in Section III (Fourier frequency domain) and Section IV (time domain).

In Section V, emphasis will be put on the role of the transfer functions which allow one to calculate the time-domain parameters (the variances) from the knowledge of the frequency-domain parameters (the spectral densities).

Up to now, we have mentioned only the basic concepts in wide use today namely the spectral densities of phase and frequency fluctuations and the two sample variance of averaged (fractional) frequency fluctuations. But in the meantime, other researchers have proposed new concepts leading to new time domain measures which are believed to exhibit some specific advantages relatively to the two-sample variance. These approaches will be presented in Sections VI–VIII with special emphasis on the relevant transfer functions relating these new parameters to the spectral densities. The links with more conventional concepts will be also outlined. It must be recognized that most of these new parameters are not widely used, possibly because their properties have not been pointed out clearly enough: our intent is not to recommend strongly the use of any new parameter but rather to demonstrate as clearly as possible their potential interest and usefulness. Readers are encouraged to try them whenever possible to test their utility.

From the experimental point of view, many test sets have been designed to measure frequency instability in both frequency and time domains. Although detailed technical descriptions are beyond the scope of this paper, some typical features of these systems will be recalled in Section IX.

To summarize, this paper aims at giving a broad review coverage of the published material on phase and frequency instability characterization, including both classical concepts and less familiar approaches. Since many points of view have been developed with great detail in the literature, only the main features can be covered here. An attempt has been made to present the material in a self-contained form that can be understood by nonspecialists. For more details, the reader is referred to the references[2] cited which have not been selected on the basis of being exhaustive but rather to point out the significant contributions.

II. Mathematical Models and Basic Definitions

Before dealing with mathematical models, let us define rather loosely that by frequency instability, we mean any unwanted frequency departure from a nominal value ν_0. In other words, frequency stability is the degree to which a source produces a constant frequency over a specified time interval. In practice, one often speaks of stability whereas applications are limited by instabilities; also, the measurement results which are much smaller than unity, e.g., 10^{-11} and so on, are indeed values of fractional (or relative or normalized) frequency instability.

The problem is hence the characterization of the unwanted frequency departures which are time-dependent because of the various physical mechanisms to be presented in this section.

A mathematical model is required for the oscillator quasi-sinusoidal output signal since the mere concept of frequency instability immediately implies that the signal is no longer a pure sinewave. Before establishing this model, it is useful to emphasize first the dichotomy between deterministic and random variations of the oscillator output frequency.

A. Deterministic Versus Random Frequency Variations.

Due to several physical mechanisms, the output frequency of any real source (even of the best quality) is continuously changing with time. Typical changes are as follows.

1) Systematic variations, also known as drifts or trends: they may be due to the aging of the resonator material (e.g., in quartz oscillators), but are also found in atomic frequency standards (e.g., some commercial cesium units exhibit a frequency drift of a few parts in 10^{13} per year). These extremely slow changes are often referred to as "long-term instability" and expressed in terms of parts in 10^x of frequency change per hour, day, month, or year, according to the device or the application. No statistical treatment is needed for the evaluation of these deterministic processes.

2) Deterministic periodic variations due to unwanted frequency modulation (FM) by periodic signals, such as the power-supply frequency and its harmonics. In some cases, quasi periodic frequency changes may arise from induced FM that may be traced to temperature, vibrations, pressure, etc. Of course, these environmental factors may induce more irregular frequency fluctuations.

3) Random fluctuations due to noise sources such as thermal, shot, and flicker noises encountered in electronic components. The related frequency fluctuations are often referred to as "short-term" instability since they become more and more significant when shorter time intervals are considered. Due to their random nature, statistical treatment is needed for their characterization.

4) It has been suggested that the frequency of an oscillator may suddently take on a new and permanent average value. These frequency steps have limited documentation and it is not clear whether they are additional to the other parts of the model or just an unsuspected visual aspect of the data [10], [12]. Since no significant errors appear when these steps are ignored, they will not be introduced in the following models.

It is worth noting that expressions such as long-term or short-term have no absolute meaning; no objective limit can be given, valid for any oscillators or any applications. It is prefered to state explicitly the durations involved. The statistical parameters presented in the next sections are measures of instability due to random noises. However, in relation with the preceding discussion, some of their limitations must be pointed out.

1) Any frequency generator is influenced to some extent by its environment: since we cannot hope to give a unique prescription for handling every possible case, the proposed

[1] Also published as NBS Technical Note 394, October 1970.

[2] The original notations have sometimes been changed to get a uniform terminology throughout this paper.

definitions of frequency instability will be independent of environmental factors. With proper design, most of these effects are secondary in importance in a laboratory environment, but in some applications (e.g., airborne) they may be by far the most important. Of course, high-quality oscillators are designed to minimize environmental sensitivities (e.g., magnetic shields in atomic frequency standards). In short, one should not expect an accurate prediction of frequency stability in a new environment when measurements have been made in another one [8].

2) A failure to separate systematic trends (drifts) from random fluctuations will, in most cases, affect any statistical measure in a very misleading way because of the dependency of statistical parameters on the number of samples. Whenever possible, systematics should be removed before statistical treatments. Linear frequency drifts can be rather easily substracted but the elimination of trends is sometimes a tricky problem: then, it is wise to take off only the most obvious overall trend [9]. In any case, deterministic elements are not discarded: they are just recognized, evaluated, substracted before statistical analysis and reintroduced in the final statement of oscillators specifications [10].

3) Measures of random frequency fluctuations may have a time dependence due to aging, e.g., frequency fluctuations in cesium clocks may increase after several years of continuous operation [9]. "Short term" measurements are hence valid only for a limited period.

4) As will be shown in Section IV, statistical measures are also affected by periodic frequency modulations.

Since we have an idea about the expected frequency departures, we are now in position to develop a model for the oscillator quasi-sinusoidal signal.

B. Model for the Oscillator Signal–Basic Definitions

The output signal of an ideal (noiseless, nondrifting) oscillator could be modeled as a pure sinewave:

$$V(t) = V_0 \sin 2\pi\nu_0 t \tag{2.1}$$

where V_0 and ν_0 are the nominal amplitude and frequency, respectively, $(\omega_0 = 2\pi\nu_0)$. For real oscillators, existing departures from V_0 and ν_0 have to be included in the model. The following general expression can be used:

$$V(t) = [V_0 + \epsilon(t)] \sin [2\pi\nu_0 t + \Phi(t)] \tag{2.2}$$

$\epsilon(t)$ is a random process[3] denoting amplitude fluctuations around V_0, also known as amplitude noise (see Section II-D). Before dealing with $\Phi(t)$, let us emphasize that any frequency variation immediately implies a related phase variation: more precisely, the instantaneous frequency is the time rate of change of phase divided by 2π. Therefore, $\Phi(t)$ denotes the phase modulation that may be traced to frequency departures introduced in Section II-A. A specific example reads as

$$\Phi(t) = D_1 t^2 + \Delta\Phi \sin 2\pi f_m t + \varphi(t) \tag{2.3}$$

where phase drift is modeled by a second-order polynomial (related to a linear frequency drift), periodic phase modulation is just a sinewave modulation and $\varphi(t)$ denotes a random process[3] modeling the so-called phase noise associated with random frequency fluctuations.

In this paper, we are mainly interested in random instabilities. However, the other terms will be dealt with in Sections V and VIII since they may greatly affect the statistical measures.

Assuming negligible amplitude noise, the following simplified model may be used to study the random phase and frequency fluctuations [8]:

$$V(t) = V_0 \sin [2\pi\nu_0 t + \varphi(t)]. \tag{2.4}$$

By definition, the signal instantaneous frequency is

$$\nu(t) = \frac{1}{2\pi}\frac{d}{dt}(2\pi\nu_0 t + \varphi(t)) = \nu_0 + \frac{1}{2\pi}\frac{d\varphi(t)}{dt} \tag{2.5}$$

which may be rewritten as

$$\nu(t) = \nu_0 + \Delta\nu(t) \tag{2.6}$$

with

$$\Delta\nu(t) = \frac{1}{2\pi}\frac{d\varphi(t)}{dt} \tag{2.7}$$

where $\Delta\nu(t)$ is a random process modeling frequency noise. Of course, for high-quality oscillators:

$$|\Delta\nu(t)| << \nu_0 \tag{2.8}$$

for substantially all time t.

A useful parameter is the instantaneous fractional (or normalized) frequency deviation $y(t)$ defined as:

$$y(t) = \frac{\Delta\nu(t)}{\nu_0}. \tag{2.9}$$

The interest of such a dimensionless quantity is that it remains unchanged under frequency multiplications and divisions commonly encountered in systems (assuming noiseless multipliers and dividers). Also, it allows easier comparisons among sources having different nominal frequencies. It is sometimes of interest to introduce the parameter [8], [68]

$$x(t) = \frac{\varphi(t)}{2\pi\nu_0} \tag{2.10}$$

which is pulse expressed in units of time (sometimes referred to as phase time). Also, $x(t)$ is the instantaneous time error of a clock run from the oscillator having an instantaneous frequency $\nu(t)$. The following relation holds:

$$y(t) = \frac{dx(t)}{dt}. \tag{2.11}$$

C. Mathematical Difficulties (Model Pathologies)

The concepts of phase noise and frequency noise have just been introduced and modeled by random processes $\varphi(t)$ and $\Delta\nu(t)$, respectively. The theory of random processes is well developed[4] and should be applied with as much mathematical rigor as possible. Related problems are discussed below but first, let us emphasize that models are used to represent the

[3] In many books on statistics, a real random process is denoted by: $\{z_k(t)\}$, $-\infty < t < +\infty$. It is an ensemble of real-valued functions which can be characterized through its probability structure. Each particular function $z_k(t)$ where k is fixed and t is variable is called a sample function: it may be thought of as the observed result of a single experiment. For the special class of ergodic random process, it is possible to derive statistical information about the entire random process from appropriate analysis of a single arbitrary sample function. For brevity, we will not include the symbol $\{\ \}$ and the index k in the text. For an ergodic process, the index k is not needed since any one sample function is representative of all other sample functions.

[4] Many textbooks exist on the subject and should be consulted for more details.

physical world which is so complex that many details are ignored in the model: otherwise, the latter would become intractable. On the other hand, properties that have no direct meaningful counterparts in the real world have to be included in the model to make it tractable (stationarity of random processes is a well-known example). These problems have been discussed by Slepian [11] and by Barnes [10], [12] who gave a discussion of model pathologies for stability measurements.

1) Stationarity of Phase Noise: Stationarity of a random process is a precisely defined mathematical property[4] meaning that statistical measures are time-independent. Clearly, lifetime limitations of physical experiments show that stationarity can only be a property of the model. The question is then whether and how one can use stationary models. Since we are looking for models that reasonably describe significant observables of real systems [8], and since stationary models are easy to use, stationarity must not be rejected from the lifetime argument.

For oscillator noise modeling, it is very convenient to assume the stationarity of $\varphi(t)$ since many theoretical results are valid only in this case, especially those related to correlation functions and spectral densities [5].

However, one should not misuse this assumption when dealing with phase and frequency fluctuations related by (2.7) which may be rewritten as [13]

$$\varphi(t) = \varphi_0 + \int_0^t 2\pi \Delta\nu(\theta)\, d\theta. \tag{2.12}$$

From this equation, and assuming that frequency noise is modeled by a stationary process $\Delta\nu(t)$ as can be justified by physical considerations, the random process $\varphi(t)$ is not generally stationary. In this case, one should not even use the correlation function $R_\varphi(\tau)$ of phase noise, a quantity that has been widely used in papers on frequency stability. As a particular case, white noise internal to the oscillator loop yields a so-called phase diffusion process analog to the mathematically ideal Brownian motion [12], [13] at least when one forgets bandlimiting effects existing in real oscillators.

To summarize, it is convenient to assume the stationarity of $\varphi(t)$ but one must check that this is not in conflict either with other parts of the model (e.g., properties of $\Delta\nu(t)$) or with some physical arguments.

2) Existence of the Instantaneous Frequency: A stationary random process can quite well have no derivative (it is differentiable in the mean-square sense if its correlation function has derivatives of order up to two). When this happens to phase noise, it means that the instantaneous frequency is not defined from the mathematical point of view. This may occur when discrete phase jumps are included in the model [13] but also with the ideal Brownian motion phase diffusion process [14]. In this case, one cannot even use the concept of instantaneous frequency defined by (2.5).

However, one has to realize that this is a model pathology, not a device pathology [12]. Bandwidth limiting that exists in real circuits yields fluctuations that can be modeled by processes where all orders of differentiation can be assumed to exist.

In conclusion, dealing with the instantaneous frequency, a key concept for frequency stability characterization, is valid only if it is properly defined mathematically.

3) Nonstationary Models: The limitations of the stationarity assumption have been pointed out, and one might well expect that one would need nonstationary models for some problems. However, such models have seldom been found useful in describing oscillators frequency instability, except maybe those with stationary increments which belong to a particular class of nonstationary processes [15], [16]. In fact, some kind of nonstationarity has to be specified to get a tractable model.

Very often, flicker noise and other low-frequency (LF) divergent noises (defined in Section III-E) are associated with nonstationary properties but here again, experimental data and mathematical models should not be confused; a given set of data can often be modeled by either a stationary or a nonstationary model [12]. In terms of observation, neither model is more correct nor incorrect than the other: one usually prefers stationary models for simplicity (e.g., a stationary model has been developed for flicker frequency noise [17]).

Today, nonstationary random processes belong to a less familiar branch of statistics although much work has already been done in this field (see for example [18]). One has to be aware about these developments since they may prove useful for oscillator specification.

D. Amplitude Noise

In most treatments of frequency stability, the amplitude noise $\epsilon(t)$ appearing in (2.2) is neglected since:

1) it does not contribute directly to frequency instability although AM to PM conversion can occur in nonlinear devices;
2) most high-quality oscillators have some kind of amplitude stabilization: the degradation of spectral purity (defined in Section III-F) due to $\epsilon(t)$ usually is smaller than the one due to $\varphi(t)$. However, calculation shows that they have equal contributions when an external additive noise process [3], [19] is predominant;
3) limiter stages are used in many systems as an interface with a frequency standard, thus removing most AM noise.

To conclude, one has to worry about AM noise when the signal is processed by nonlinear devices or when interest is focused on signal spectral purity since AM noise degrades directly this performance (see (3.13)).

III. Characterization of Frequency Stability in the Fourier Frequency Domain

In this section, we are dealing with the random fluctuations of phase, frequency, and fractional frequency defined respectively in (2.4), (2.7), and (2.9). Since they are random quantities, statistical parameters are needed for their description, some of which are classically used in stationary random process theory, e.g., correlation functions and spectral densities.

Analysis in the Fourier frequency domain is of great importance both for theoretical purposes (in terms of included information) and for application purposes (in terms of power spreading in the frequency domain, a primary specification for many engineers). For these reasons, the spectral density concept has been widely used for oscillator stability characterization [3], [5], [8].

From here on, the word "frequency" will be used with two different meanings not to be confused: the time-dependent instantaneous frequency $\nu(t)$ of an oscillator, and the time-independent Fourier frequency, denoted by f, that will appear in spectral densities.

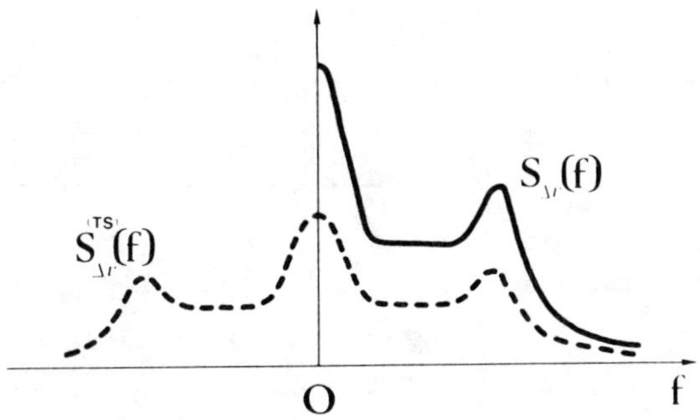

Fig. 1. Two-sided and one-sided spectral densities of frequency fluctuations.

A. Spectral Density of (Fractional) Frequency Noise

Assuming that the frequency noise is modeled by a stationary process $\Delta\nu(t)$, the spectral density $S_{\Delta\nu}(f)$ is defined as the Fourier transform of the correlation function [5]

$$R_{\Delta\nu}(\tau) = \langle \Delta\nu(t)\, \Delta\nu(t - \tau) \rangle \tag{3.1}$$

where $\langle\ \rangle$ denotes the average of the quantity inside the brackets. More precisely, this approach yields the two-sided (TS) spectral density defined for $-\infty < f < +\infty$; it is a real non-negative and an even function of f (Fig. 1):

$$S^{(TS)}_{\Delta\nu}(f) = \int_{-\infty}^{+\infty} R_{\Delta\nu}(\tau) \exp(-i2\pi f\tau)\, d\tau \tag{3.2}$$

For theoretical considerations, TS spectral densities often simplify the calculations. For experimental purposes, the one-sided spectral density $S_{\Delta\nu}(f)$ defined as follows is used:

$$S_{\Delta\nu}(f) = 2S^{(TS)}_{\Delta\nu}(f), \qquad \text{for } 0 \leqslant f < \infty, \qquad \text{otherwise zero.} \tag{3.3}$$

In practice, this quantity may be measured for example by filtering, squaring, and averaging operations on sample records of $\Delta\nu(t)$ or by digital techniques involving the fast Fourier transform (FFT). Of course, as for any statistical parameter, only useful estimates may be obtained experimentally [20].

The following equation links the mean-square value of $\Delta\nu(t)$ to the area under $S_{\Delta\nu}(f)$:

$$\langle \Delta\nu^2(t) \rangle = \int_0^{\infty} S_{\Delta\nu}(f)\, df. \tag{3.4}$$

The dimensions of $S_{\Delta\nu}(f)$ are Hz^2/Hz, i.e., Hz.

In the same way, the spectral density $S_y(f)$ of $y(t)$ may be introduced: it has the dimensions of 1/Hz.

The IEEE Subcommittee has proposed to use $S_y(f)$ as a definition for the measure of frequency stability in the Fourier frequency domain: oscillators at different nominal frequencies may thus be meaningfully compared by plotting their $S_y(f)$ on the same figure.

It must be recognized that correlation functions and spectral densities carry exactly the same information about the random process: however, the spectral density format is very often more desirable for engineering purposes.

B. Spectral Density of Phase Noise

Despite the problems that may arise concerning the stationarity of phase noise, it is very convenient to assume that $\varphi(t)$ is a stationary process: its one-sided spectral density $S_\varphi(f)$ is then defined from $R_\varphi(\tau)$ and has the dimensions of $radian^2/Hz$.

The concept of $S_\varphi(f)$ is widely used for several reasons:

1) it is measured with the well-known phase detector technique implemented in many laboratories (see Section IX);
2) it is very simply related to $S_y(f)$, the recommended measure (3.7);
3) under certain assumptions, it provides an estimate of signal spectral purity whose direct measurement is often difficult for high-quality sources (see Section III-F).

C. Translations among Frequency-Domain Measures

From (2.9):

$$S_y(f) = \frac{1}{\nu_0^2}\, S_{\Delta\nu}(f). \tag{3.5}$$

Since $2\pi\Delta\nu(t)$ is the time derivative of $\varphi(t)$ (2.7), the following relationship holds in the frequency domain:

$$S_{\Delta\nu}(f) = f^2 S_\varphi(f). \tag{3.6}$$

Combining (3.5) and (3.6) yields

$$S_y(f) = \left(\frac{f}{\nu_0}\right)^2 S_\varphi(f). \tag{3.7}$$

Thus these three spectral densities carry essentially the same information, each of them being best suited for a given class of applications.

Also, the spectral density $S_x(f)$ of time error $x(t)$ defined in (2.10) may be introduced and:

$$S_y(f) = 4\pi^2 f^2 S_x(f). \tag{3.8}$$

Last, a general remark is that the modulating processes $\varphi(t)$ and $\Delta\nu(t)$ are slow relatively to the carrier $\sin 2\pi\nu_0 t$: therefore, their spectral densities take significant values only for f much smaller than ν_0.

D. Estimation of Spectral Densities

Spectral densities are theoretical concepts involving infinite duration processes, infinite frequency range and true averages. In practice, only finite-duration processes are available and spectrum analyzers have nonzero-bandwidth frequency windows, limited dynamic ranges, slewing rates (for swept models), lower and upper Fourier frequency limits, etc.

As a consequence, experimental knowledge of spectral densities suffers from several kinds of limitations.

1) The Fourier frequency range is limited: a lower limit of about $f = 1$ Hz is reached by several LF analog spectrum analyzers and about 10^{-3} Hz may be reached with digital techniques using the FFT. An upper limit of tens of kilohertz is typical for LF analyzers.

2) Statistical errors result from the use of a finite sample of observations [20]: error bars should be specified with any experimental result. Other errors are due to data acquisition and processing techniques: systematic errors greater than 10 dB can occur, a few dB appearing as a typical value. An accuracy of 0.2 dB can be obtained only by exercising considerable caution including the study of the analyzer's filter shape effects [21].

3) Some analyzers yield only the spectral densities for discrete values of f: narrow lines due to periodic modulation may be overlooked with such instruments.

Nevertheless, the spectral density concept plays a key role in stability characterization since it provides an unambigous identification of the noise processes encountered in real oscillators as shown in the next paragraph.

E. The Power-Law Spectral Density Model

From spectral density measurements made in many laboratories on various sources including quartz-crystal oscillators, masers, passive atomic standards, and other microwave oscillators, it appears that experimental results may quite well be modeled by power law curves. For $S_y(f)$, the following model has been found useful [8]:

$$S_y(f) = h_\alpha f^\alpha. \tag{3.9}$$

The exponent α typically takes the integral values $-2, -1, 0, +1, +2$ and is a characteristic of the kind of noise. The constant h_α is a measure of the noise level. Nonintegral values of α may also be considered.

For a given oscillator, $S_y(f)$ is the sum of two or three such terms, the others being negligible.

1) Classification of Power Laws: The integral power laws may be designated using the classical terminology of "white noise" for a noise whose spectral density is a constant (independent of f), "flicker noise" when it varies as f^{-1} and "random walk" when it varies as f^{-2}. Table I summarizes these laws.

TABLE I
THE POWER-LAW MODEL FOR SPECTRAL DENSITIES

$S_y(f)$	$S_\varphi(f)$	Designation
$h_{-2} f^{-2}$	$\nu_0^2 h_{-2} f^{-4}$	Random walk frequency noise
$h_{-1} f^{-1}$	$\nu_0^2 h_{-1} f^{-3}$	Flicker frequency noise
h_0	$\nu_0^2 h_0 f^{-2}$	White frequency noise
$h_1 f$	$\nu_0^2 h_1 f^{-1}$	Flicker phase noise
$h_2 f^2$	$\nu_0^2 h_2$	White phase noise

Note that the quantity of interest (frequency versus phase) must be explicitly stated since both might exist simultaneously. Experimental data are usually plotted on log-log scales where power laws appear as straight lines: the slopes are then easily recognized (if measurement accuracy is not too bad) and hence the kinds of noise present in the oscillator.

2) Physical Origins: White and flicker electrical noise sources are ever present in electronic components. In oscillators, depending on their location in the circuits, their additive contribution to the signal yields either phase modulation (for external noises) or frequency modulation (for internal noise) by the original noise processes [3]-[5]. Moreover, the so-called "multiplicative" noises directly modulate the phase or the frequency according to the physical mechanisms considered.

More specific statements can be made [22].

a) Random walk frequency noise usually relates to the oscillator environment (temperature, vibration, shocks, etc.).

b) Flicker frequency noise sources are not yet fully understood but are thought to be related to the resonator in quartz oscillators and to electronics and environment in atomic frequency standards [23].

c) White frequency noise arises from additive white noise sources internal to the oscillator loop, such as thermal noise [3]. It is also found in passive atomic standards where white noise sources directly modulate the locked oscillator output frequency via the control element [24].

d) Flicker phase noise is usually added by noisy electronics, e.g., by output amplifiers or frequency multipliers, and can be reduced by radio frequency (RF) negative feedback and component selection [25].

e) White phase noise is usually due to additive white noise sources external to the oscillator loop [3]. Bandpass filtering the oscillator output signal is then useful.

3) Power-Law Pathologies [12]: Power laws are useful models for data obtained over a limited Fourier frequency interval. Extrapolation over the whole frequency domain from $f = 0$ to $f = \infty$, i.e., beyond the range of applicability given by observations, yields mathematical difficulties such as infinite power and divergence of some time-domain stability measures. But this results from the use of an unrealistic model ignoring the constraints of the real world such as finite bandwidths and finite duration which prevent both high-frequency (HF) and LF divergence from being observable. Also, the instantaneous frequency generated by any real source is bounded and thus, models of $S_y(f)$ with infinite area are physically unrealistic.

In practice, the power-law model is often used as follows [8]:

$$S_y(f) = \begin{cases} \sum_{\alpha=-2}^{+2} h_\alpha f^\alpha, & 0 \leqslant f \leqslant f_h \\ 0, & f > f_h \end{cases} \tag{3.10}$$

where a sharp upper cutoff frequency f_h is introduced. Sometimes, the actual shape of the cutoff is of importance [6] and must then be specified. As will be seen in the next section, time domain stability measures sometimes depend on f_h which must then be given with any numerical result, although no recommendation has been made for its value (1 to 10 kHz is typical).

A lower cutoff frequency is usually not included in the power-law model since useful time-domain measures have been defined such that convergence arises for $S_y(f)$ given by (3.10). Moreover, no experimental results indicate clearly the value that should be included: measurements have just shown that flicker-type spectral densities extend down to very-low Fourier frequencies, e.g., smaller than one cycle per month as observed on quartz oscillators [26]. No flattening has yet been observed.

F. Oscillator Spectral Purity

When dealing with the frequency domain for oscillator specification, it is of value to introduce the concept of spectral purity, i.e., a frequency-domain estimate of the quality of the oscillator quasi-sinusoidal output signal $V(t)$.

To this end, the spectral density $S_v(f)$ of the complete signal $V(t)$ is used and often referred to as RF spectrum[5] since it

[5] Strictly speaking, ν_0 is not restricted to the RF range and may be in the microwave, infrared, or even visible according to the source under test.

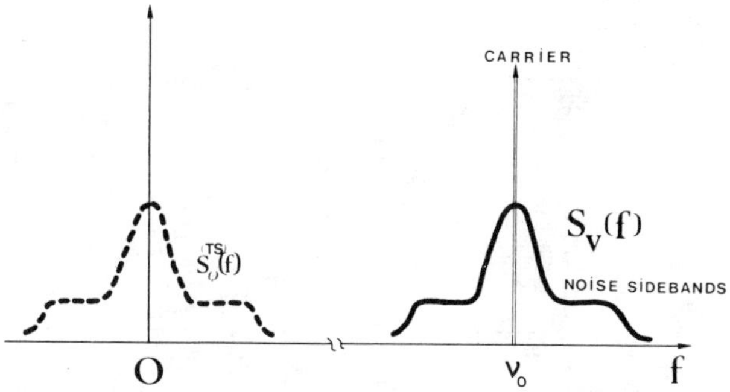

Fig. 2. Signal spectral purity: The RF spectrum.

takes significant values only around ν_0. Knowledge of it is of prime importance in several applications such as radar, spectroscopy, frequency synthesis and communications.

Nevertheless, $S_v(f)$ is not considered as being a good primary measure of frequency stability since amplitude fluctuations $\epsilon(t)$ also contribute to $S_v(f)$ [8]. Moreover, it is not simply related with other measures of frequency stability in the most general cases.

It is only when AM noise is negligible and the root-mean-square (rms) value of $\varphi(t)$ is much smaller than one radian that a simple approximate relationship may be derived [3], [5]:

$$S_v(f) \simeq \frac{V_0^2}{2} \{\delta(f - \nu_0) + S_\varphi^{(TS)}(f - \nu_0)\}. \quad (3.11)$$

As pictured on Fig. 2, the continuous noise sidebands around the discrete carrier centered at ν_0 have then the shape of the spectral density of phase noise. This so-called low-modulation-index approximation may be used for high-quality frequency standards for which the above assumptions are reasonably fulfilled. Sometimes, the complete formula is needed, e.g., to forecast spectral purity of a signal after very high ratio frequency multiplication [27] as realized for laser-frequency measurements.

Neglecting AM noise, the complete relationship between phase noise spectral density and RF spectrum[6] reads as [28]

$$S_v(f) = \frac{V_0^2}{2} e^{-\langle\varphi^2\rangle} \Big\{\delta(f - \nu_0) + S_\varphi^{(TS)}(f - \nu_0) + \sum_{n=2}^{\infty} \frac{1}{n!} [S_\varphi^{(TS)}(f) \overset{n-1}{\otimes} S_\varphi^{(TS)}(f)]_{\nu_0}\Big\}. \quad (3.12)$$

In this equation, $\langle\varphi^2\rangle$ denotes the mean-square value of $\varphi(t)$ which is assumed to be a stationary Gaussian random process. $[S_\varphi^{(TS)}(f) \overset{n-1}{\otimes} S_\varphi^{(TS)}(f)]_{\nu_0}$ denotes $n - 1$ convolutions of $S_\varphi(f)$ with itself, followed by a translation around the carrier frequency ν_0. Due to the infinite sum, the noise sidebands no longer have the shape of $S_\varphi(f)$.

More precisely, an increase of $\langle\varphi^2\rangle$ yields:

1) a decrease of carrier power (due to the exponential term);
2) a change of the noise sidebands shape (due to the increasing contribution of the convolutions);
3) a decrease of carrier-to-noise power ratio since the total area under $S_v(f)$ remains constant ($V_0^2/2$).

[6] RF spectra of signals randomly modulated in amplitude, phase and frequency have been studied by several authors with other backgrounds than frequency standards specification, see for example Middleton [29].

If necessary, the contribution of AM noise may be taken into account as shown in the following equation:

$$S_v(f) \simeq \frac{V_0^2}{2} \{\delta(f - \nu_0) + S_\varphi^{(TS)}(f - \nu_0) + S_A^{(TS)}(f - \nu_0)\} \quad (3.13)$$

where $S_A^{(TS)}(f)$ is the two-sided spectral density of the fractional amplitude fluctuations $A(t) = \epsilon(t)/V_0$.

Direct measurements of high-quality oscillator's spectral purity is very difficult because of the ultralow sideband levels encountered and the limited capabilities of even the best RF spectrum analyzers. The measurement of $S_\varphi(f)$ is often used as a means to estimate $S_v(f)$ through (3.11).

IV. Characterization of Frequency Stability in the Time Domain

Since instabilities in oscillators are time variations of the quantities of interest (phase, frequency, fractional frequency), they may be characterized by a measure of the variations that occur over a specified time interval τ: this is the basis of the so-called time-domain characterization (time-interval or averaging-time domain would be more appropriate). Since random phenomena are involved, the relevant measures are again given in terms of statistical parameters. These values are usually plotted versus τ which may vary from say milliseconds to days, months, or years.

Some important applications of frequency standards directly rely on the time domain behavior of the output signal, e.g., timekeeping [30]. In this case, a time-domain instability measure is really what is needed, rather than a frequency-domain one. The techniques described below are not restricted to power-law spectral densities and may be used with other noise types.

A major difficulty will arise from the fact that many parameters may indeed be considered as time-domain measures: one gives here a presentation showing their respective properties and relationships.

A. Basic Measurements

The oscillator instantaneous frequency $\nu(t)$ defined in (2.5) is not an observable since any frequency-measurement technique does involve a finite time interval over which the measurement is performed; for example, a digital frequency counter counts the number of cycles n_k of the input signal during the time interval τ begining at t_k provided by its time base driven by a reference oscillator.

Therefore, the average value of $\nu(t)$ over a time interval τ begining at t_k provides a more useful quantity directly related to an experimental result:

$$\langle\nu(t)\rangle_{t_k,\tau} = \nu_0 + \frac{1}{\tau}\int_{t_k}^{t_k+\tau} \Delta\nu(\theta)\, d\theta = \frac{n_k}{\tau}. \quad (4.1)$$

The normalized quantity $\bar{y}_k$ defined as follows is widely used [8]:

$$\bar{y}_k = \frac{1}{\tau}\int_{t_k}^{t_k+\tau} y(\theta)\, d\theta \quad (4.2)$$

and then

$$\langle\nu(t)\rangle_{t_k,\tau} = \nu_0(1 + \bar{y}_k) = \frac{n_k}{\tau}. \quad (4.3)$$

From (2.7) and (2.9), it can be demonstrated that

$$\bar{y}_k = \frac{\varphi(t_k + \tau) - \varphi(t_k)}{2\pi \nu_0 \tau} \tag{4.4}$$

where the numerator denotes the phase error accumulated from t_k to $t_k + \tau$, also known as the first phase increment or first difference of the phase [30], [31].

Since $\bar{y}_k$ is easily related to experimental results given by counting techniques, it will be used in the following to define the time-domain parameters. More precisely, one individual measurement of duration τ provides one sample $\bar{y}_k$; repeated measurements of $\bar{y}_k$ are necessary for a statistical treatment yielding a meaningful measure of instability over τ.

B. The True Variance [3]–[6]

Due to the random fluctuations of $y(t)$ in real oscillators, repeated measurements of $\bar{y}_k$ give numerical values that are randomly scattered i.e., they are samples of a random variable: for each value of τ, a statistical characteristic of the dispersion of the $\bar{y}_k$ provides a time-domain measure of instability over τ.

To this end, the variance σ^2 (or the standard-deviation σ) is widely used in statistics. Assuming that $y(t)$ and hence the $\bar{y}_k$ have a zero mean, the variance is equal to the mean square value of $\bar{y}_k$:

$$\sigma^2[\bar{y}_k] = \langle \bar{y}_k^2 \rangle. \tag{4.5}$$

The bracket $\langle\ \rangle$ denotes either a statistical average calculated over an infinite number of samples at a given instant t_k, or an infinite time average made over one sample of $y(t)$ (ergodicity of $y(t)$ is assumed). Because of the infinite number of samples or infinite duration implied in its definition, this variance is an idealization often referred to as the true variance. It will be denoted as $I^2(\tau)$ to indicate that it is a measure of instability over a time interval τ.

For stationary frequency fluctuations, the general shape of $I(\tau)$ reported on Fig. 3 shows the following limits:

$$\lim_{\tau \to 0} I(\tau) = \sqrt{\langle y^2(t) \rangle} \tag{4.6}$$

$$\lim_{\tau \to \infty} I(\tau) = 0. \tag{4.7}$$

The first limit corresponds to ideal instantaneous frequency measurements ($\tau = 0$) and hence is equal to the rms value (i.e., the standard deviation) of $y(t)$. The upper limit means that stationary fluctuations would be completely averaged out for $\tau = \infty$ and hence the dispersion of the results would be zero. Of course, none of these limits are observables. The decrease of $I(\tau)$ as τ increases indicates better averaging of stationary random fluctuations.

Properties but also limitations[7] of the usefulness of $I(\tau)$ will be studied in Section V through its relationship to $S_y(f)$.

C. The Sample Variance [7], [8]

The true variance $I^2(\tau)$ is a theoretical idealization since it involves an infinite number of data: practical estimates of it must be based upon a finite number of samples $\bar{y}_k$. The so-called sample variance is defined from an ensemble of N samples $\bar{y}_k$ with $k = 1, 2, 3, \cdots, N$ and $t_{k+1} = t_k + T$. The corresponding measurement cycle is shown on Fig. 4, where T is the repetition interval for individual measurements of duration τ ($T = \tau +$ dead time between measurements) and t_1 is arbitrary.

[7]Especially, $I^2(\tau)$ does not converge for some kinds of noise such as flicker frequency noise.

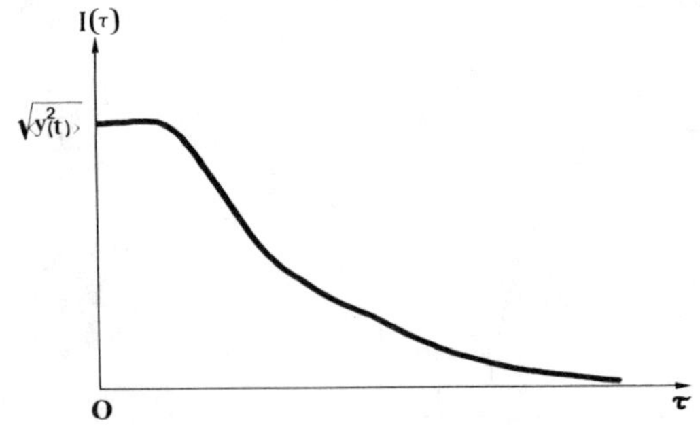

Fig. 3. Instability in the time domain: Square-root of the true variance for stationary frequency noise.

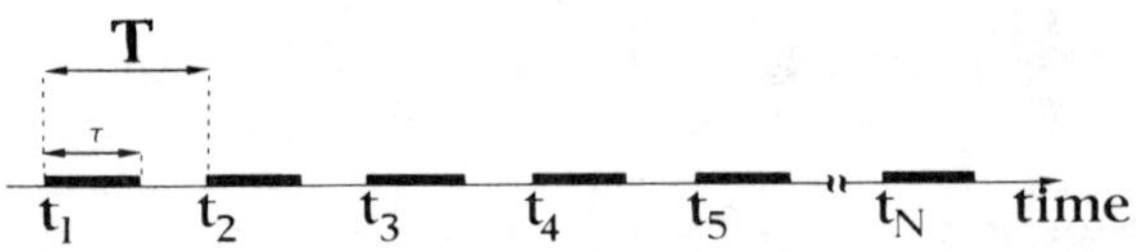

Fig. 4. The measurement cycle in the time domain.

Several related definitions may be given for the sample variance [32].

1) First Definition: Following the general definition of the variance of a random variable x, namely $\sigma^2(x) = \langle (x - \langle x \rangle)^2 \rangle$, a logical definition of the sample variance reads as

$$\sigma_{(1)}^2(N, T, \tau) = \frac{1}{N} \sum_{i=1}^{N} \left(\bar{y}_i - \frac{1}{N} \sum_{j=1}^{N} \bar{y}_j \right)^2. \tag{4.8}$$

This quantity is itself a random variable, N being the sample size. Several factors may be used to characterize the "goodness" of an estimator [20], bias being one of them: an estimator is unbiased if its mean value equals the true parameter, namely if

$$\langle \sigma_{(1)}^2(N, T, \tau) \rangle = I^2(\tau). \tag{4.9}$$

This equality should hold for any value of N, T, and τ and any kind of noise. Thus the relation between $\langle \sigma_{(1)}^2(N, T, \tau) \rangle$ and $I^2(\tau)$ has to be studied carefully: let us consider the specific case of great importance where the samples are adjacent ($T = \tau$); then, the following relation holds [7], [32]:

$$\langle \sigma_{(1)}^2(N, \tau, \tau) \rangle = I^2(\tau) - I^2(N\tau). \tag{4.10}$$

For stationary frequency noise (Fig. 5):

$$\lim_{N \to \infty} \langle \sigma_{(1)}^2(N, \tau, \tau) \rangle = I^2(\tau). \tag{4.11}$$

Hence, for finite N, the above estimator is biased for any kind of stationary frequency noise and the bias may be calculated from the theoretical knowledge of $I^2(N\tau)$. The other definitions of sample variance given below are such that the bias is indeed zero for a given definition and a corresponding kind of noise.

2) Second Definition and White Frequency Noise: For white frequency noise (see Table I), equation (5.6) yields $I^2(\tau) = h_0/2\tau$ and hence,

$$\langle \sigma_{(1)}^2(N, \tau, \tau) \rangle = \left(1 - \frac{1}{N}\right) I^2(\tau). \tag{4.12}$$

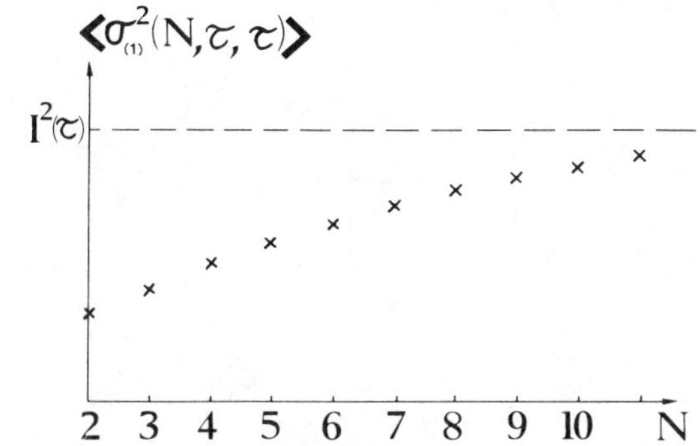

Fig. 5. The sample variance versus N for stationary frequency noise (first definition).

In this case, the following definition provides an unbiased estimator:

$$\sigma^2_{(2)}(N, T, \tau) = \frac{1}{N-1} \sum_{i=1}^{N} \left(\bar{y}_i - \frac{1}{N} \sum_{j=1}^{N} \bar{y}_j \right)^2. \quad (4.13)$$

For $T = \tau$ *and* white frequency noise

$$\langle \sigma^2_{(2)}(N, \tau, \tau) \rangle = I^2(\tau). \quad (4.14)$$

The factor $1 - 1/N$ appearing in (4.12) is found also for other stationary noises when T is much greater than the correlation time of $y(t)$, i.e., when the samples $\bar{y}_k$ are independent: this is a classical result in statistics [20].

3) Third Definition and White Phase Noise [32]: For white phase noise including a cutoff frequency f_h, equation (5.6) yields $I^2(\tau) = h_2 f_h / 2\pi^2 \tau^2$ for $2\pi f_h \tau >> 1$; then,

$$\langle \sigma^2_{(1)}(N, \tau, \tau) \rangle = \left(1 - \frac{1}{N^2} \right) I^2(\tau). \quad (4.15)$$

In this case, the following sample variance is unbiased for any $N \geqslant 2$ and $T = \tau$:

$$\sigma^2_{(3)}(N, T, \tau) = \frac{N}{N^2 - 1} \sum_{i=1}^{N} \left(\bar{y}_i - \frac{1}{N} \sum_{j=1}^{N} \bar{y}_j \right)^2. \quad (4.16)$$

To summarize, we are faced with several definitions of the sample variance, many values of $N \geqslant 2$, and a wide range of possible values of $T \geqslant \tau$: also, the value of f_h is sometimes of prime importance (e.g., for white and flicker phase noises).

To achieve the goal of recommending a unique time-domain measure, some choices must be made.

D. The Zero-Dead-Time Two-Sample Variance

To allow meaningful comparisons of measurements in the time domain, one sample variance must be chosen with specified values of N, T, and f_h(τ being the independent variable).

Following the work of Allan [7], the IEEE Subcommittee on Frequency Stability has proposed the following time-domain measure [8]:

1) the average value of $\sigma^2_{(2)}(N, T, \tau)$ given by (4.13) is used, namely $\langle \sigma^2_{(2)}(N, T, \tau) \rangle$ often denoted as $\langle \sigma^2_y(N, T, \tau) \rangle$;
2) the sample size is taken as $N = 2$;
3) adjacent samples (zero-dead-time) are assumed, and hence $T = \tau$.

The resulting measure reads as

$$\langle \sigma^2_y(2, \tau, \tau) \rangle = \left\langle \sum_{i=1}^{2} \left(\bar{y}_i - \frac{1}{2} \sum_{j=1}^{2} \bar{y}_j \right)^2 \right\rangle. \quad (4.17)$$

The short notation $\sigma^2_y(\tau)$ is widely used for that quantity

$$\sigma^2_y(\tau) = \tfrac{1}{2} \langle (\bar{y}_2 - \bar{y}_1)^2 \rangle. \quad (4.18)$$

It is also known as the Allan variance or the pair variance.

Several comments are of value.

1) Since infinite duration is implied in the average denoted as $\langle \ \rangle$, $\sigma^2_y(\tau)$ is a theoretical measure in the same sense as $I^2(\tau)$: however, as shown in Section V, it has a greater practical utility since it exists for the five power laws encountered in real oscillators (Table I). Also, simple experimental estimators may be devised for $\sigma^2_y(\tau)$ (see Section IV-E).

2) From the discussion on the various sample variances, it appears that $\sigma^2_y(\tau)$ is biased except for white frequency noise, but one has to remember that other noises are found as well in oscillators; from (4.10) and (4.17), one gets:

$$\sigma^2_y(\tau) = 2[I^2(\tau) - I^2(2\tau)]. \quad (4.19)$$

For white frequency noise and white phase noise, the following relations are valid, respectively,

$$\sigma^2_y(\tau) = I^2(\tau) \qquad \text{(no bias)} \quad (4.20)$$

and

$$\sigma^2_y(\tau) = \tfrac{3}{2} I^2(\tau) \qquad 2\pi f_h \tau >> 1 \qquad \text{(bias)}. \quad (4.21)$$

3) The choice of $N = 2$ in the preferred definition yields an easy measurement process since only pairs of samples are involved: this choise is really the key feature in the definition of $\sigma^2_y(\tau)$.

4) Although no recommendation was made for the value of f_h, it has to be specified with any experimental results.

E. Estimates of the Two-Sample Variance

Only estimates of $\sigma^2_y(\tau)$ can be obtained from a finite number of samples $\bar{y}_k$ and an inherent statistical uncertainty exists. With m values of $\bar{y}_k$, a possible estimator reads as [8]

$$\hat{\sigma}^2_y(\tau, m) = \frac{1}{2(m-1)} \sum_{i=1}^{m-1} (\bar{y}_{i+1} - \bar{y}_i)^2. \quad (4.22)$$

This quantity is itself a random variable whose variance (i.e., the variance of the variance) may be used to calculate the error bars on the plot of $\sigma_y(\tau)$ versus τ. For Gaussian noises, Lesage and Audoin [33] have shown that the error bars (confidence intervals) for power laws are given by

$$E_\alpha \simeq \sigma_y(\tau) K_\alpha m^{-1/2}, \qquad \text{for } m > 10 \quad (4.23)$$

with $K_2 = K_1 = 0.99$, $K_0 = 0.87$, $K_{-1} = 0.77$, $K_{-2} = 0.75$.

Also, other estimators of $\sigma^2_y(\tau)$ are possible and have been considered [34].

For long-term stability (τ of the order of months or even a year) one is severely limited in the size of m. In any case, m should be stated with any results.

F. Translations Among the Time-Domain Measures

For the five integral power laws shown on Table I, as well as intermediate values of α, two "bias functions" have been introduced by Barnes [35] to translate from $\langle \sigma^2_y(N_1, T_1, \tau_1) \rangle$ to $\langle \sigma^2_y(N_2, T_2, \tau_2) \rangle$ (the second definition given by (4.13) is used):

$$B_1 = \frac{\langle \sigma^2_y(N, T, \tau) \rangle}{\langle \sigma^2_y(2, T, \tau) \rangle} \quad (4.24)$$

and

$$B_2 = \frac{\langle \sigma_y^2(2, T, \tau) \rangle}{\langle \sigma_y^2(2, \tau, \tau) \rangle}. \tag{4.25}$$

They have been computed and tabulated in [35].

Note that B_2 allows one to calculate the recommended measure $\sigma_y^2(\tau)$ from a two-sample variance measured with dead-time, provided the spectral density slope is known:

$$\sigma_y^2(\tau) = \frac{\langle \sigma_y^2(2, T, \tau) \rangle}{B_2}. \tag{4.26}$$

This is of practical interest since counting techniques usually have nonzero dead-time between successive measurements.

V. Translations Among Frequency-Domain and Time-Domain Measures: The Transfer Function Concept

To cover adequately the needs of oscillator users, two domains have been considered, namely the Fourier frequency and the (averaging) time domains. Moreover, several parameters have been introduced in each case. Translations among frequency domain measures and then among time domain measures were dealt with, respectively, in Sections III-C and IV-F.

It is now of prime importance to study the translations between the two domains for several reasons:

1) these relations provide a unified picture of frequency stability characterization;
2) the power laws which have been identified in the frequency domain may be translated into specific laws of $I(\tau)$ or $\sigma_y(\tau)$ through these relations;
3) stability measurements are sometimes made in only one domain due to equipment availability or capability: relations between parameters can then give an estimate of the performance in the other domain.

As will be demonstrated, a so-called "transfer function" appears in the general relationships between frequency domain and time-domain parameters.

This concept will be used in the following, either to study the properties of classical time-domain measures, or even to introduce new time-domain measures offering new measurement capabilities (Section VI).

A. *True Variance Versus Spectral Density*

First, the true variance $I^2(\tau)$ will be expressed in terms of $S_y(f)$ as

$$I^2(\tau) = \int_0^{\infty} S_y(f) \, |H_I(f)|^2 \, df \tag{5.1}$$

where $H_I(f)$ is a transfer function.

1) Demonstration: Equation (4.5) may be rewritten as

$$I^2(\tau) = \left\langle \left(\frac{1}{\tau} \int_{t_k}^{t_k+\tau} y(\theta) \, d\theta \right)^2 \right\rangle \tag{5.2}$$

or again

$$I^2(\tau) = \left\langle \left(\int_{-\infty}^{\infty} y(t) \, h_I(t_k - t) \, dt \right)^2 \right\rangle \tag{5.3}$$

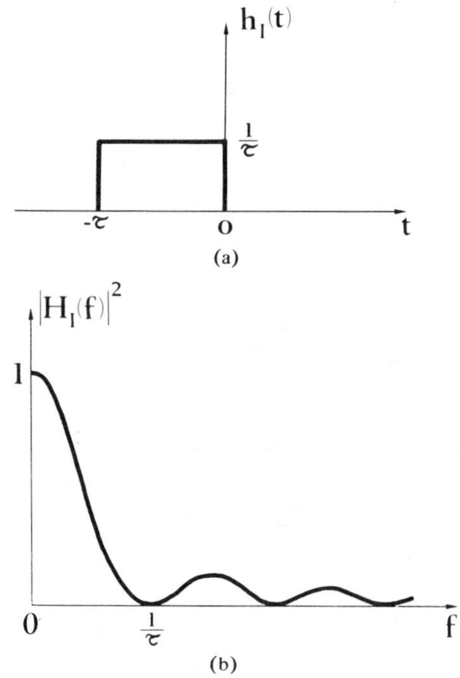

Fig. 6. (a) Measurement sequence of the true variance. (b) Transfer function of the true variance.

with

$$h_I(t) = \begin{cases} 0, & t < -\tau \\ \dfrac{1}{\tau}, & -\tau \leqslant t \leqslant 0 \\ 0, & t > 0. \end{cases} \tag{5.4}$$

Equation (5.3) includes a convolution integral where $y(t)$ is convolved with a function $h_I(t)$ which resembles the basic measurement sequence of one sample $\bar{y}_k$ (Fig. 6(a)). This integral may represent the output signal of a hypothetical filter with impulse response $h_I(t)$ receiving an input signal $y(t)$: it is well known that the spectral density of the output signal is then given by the product $S_y(f) \, |H_I(f)|^2$ where the filter transfer function $H_I(f)$ is the Fourier transform of $h_I(t)$ and hence (Fig. 6(b)):

$$|H_I(f)|^2 = \left(\frac{\sin \pi \tau f}{\pi \tau f} \right)^2. \tag{5.5}$$

The true variance which is the mean-square value of this pseudo-output signal (5.3) is then equal to the area under its spectral density, i.e.,

$$I^2(\tau) = \int_0^{\infty} S_y(f) \left(\frac{\sin \pi \tau f}{\pi \tau f} \right)^2 df. \tag{5.6}$$

Thus the true variance may be calculated from $S_y(f)$ through an integral including the transfer function $(\sin \pi \tau f / \pi \tau f)^2$ [3], [5], [6] and this is not restricted to power law spectral densities.

The above relationship cannot be reversed in a closed form in the most general cases, i.e., $S_y(f)$ cannot be expressed in terms of $I^2(\tau)$; this is an example of the key role played by spectral densities for random process characterization.

2) Applications: Some limitations of the usefulness of $I^2(\tau)$ appear from the fact that the transfer function is equal to one

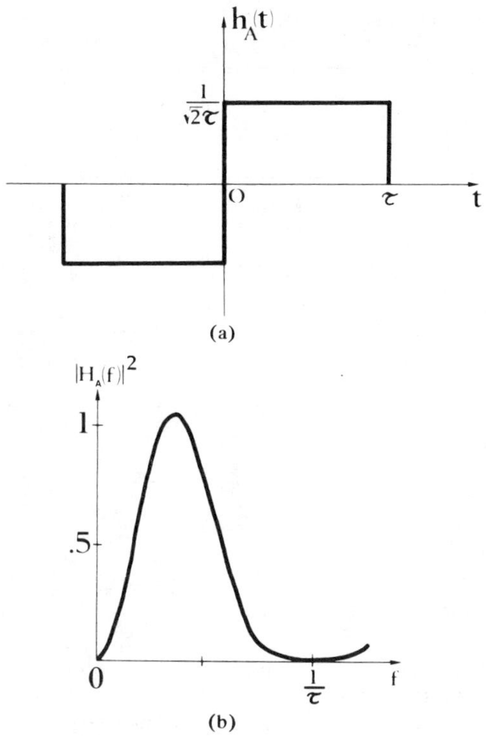

Fig. 7. (a) Measurement sequence of the two-sample variance. (b) Transfer function of the two-sample variance.

for $\pi\tau f \ll 1$: the true variance is therefore very sensitive to the low Fourier frequency components of $S_y(f)$, i.e., to the very slow fluctuations of the oscillator output frequency $\nu(t)$.

More specifically, a major problem arises for flicker frequency noise since the model $S_y(f) = h_{-1}f^{-1}$ yields $I^2(\tau) = \infty$. Thus the true variance is not suited to study real oscillators since most of them are perturbed by this noise [23]. Of course, this is a model pathology but the practical utility of $I^2(\tau)$ is also severely limited since actual measurements of $I^2(\tau)$ depend on the total number of samples and increase indefinitely with it when flicker frequency noise exists [5], [7].

For this reason, the true variance has never been widely used as a practical stability measure, although it may be useful for theoretical work with well-behaved spectral density models. As an example, the RF spectrum $S_v(f)$ may be deduced from $I^2(\tau)$ when $\varphi(t)$ and $y(t)$ are Gaussian processes with $y(t)$ stationary [28]:

$$S_v^{(TS)}(f) = \frac{V_0^2}{2}\int_{-\infty}^{\infty} \exp\left(-\frac{\omega_0^2}{2}\tau^2 I^2(\tau)\right) \cdot \cos \omega_0\tau \exp(-i2\pi f\tau)\, d\tau. \quad (5.7)$$

The laws found for white frequency noise and white phase noise are, respectively,

$$I^2(\tau) = \frac{h_0}{2\tau} \quad (5.8)$$

and

$$I^2(\tau) = \frac{h_2 f_h}{2\pi^2\tau^2}\left[1 - \frac{\sin 2\pi f_h\tau}{2\pi f_h\tau}\right]. \quad (5.9)$$

In the last case, f_h is necessary for convergence and $I(\tau) \sim \tau^{-1}$ for $2\pi f_h\tau \gg 1$.

For flicker phase noise, a HF cutoff is also needed for convergence and $I(\tau) \sim \tau^{-1}$ for $2\pi f_h\tau \gg 1$ [6]. Thus, both white and flicker phase noises provide the same slope on an $I(\tau)$ curve, and thus cannot be unambiguously recognized from such a plot.

B. Sample Variances Versus Spectral Density

Let us consider the second definition of the sample variance given by (4.13) since it has been recommended as a basis for time domain measures. The relation between its average value and $S_y(f)$ may be demonstrated very simply for adjacent samples[8] [37]; combining the definitions (4.8) and (4.13) with the law expressed by equation (4.10):

$$\langle \sigma_{(2)}^2(N, \tau, \tau)\rangle = \frac{N}{N-1}[I^2(\tau) - I^2(N\tau)]. \quad (5.10)$$

The substitution of (5.6) into this relation gives:

$$\langle \sigma_y^2(N, \tau, \tau)\rangle = \frac{N}{N-1} \cdot \int_0^{\infty} S_y(f)\left(\frac{\sin \pi\tau f}{\pi\tau f}\right)^2 \left[1 - \left(\frac{\sin N\pi\tau f}{N \sin \pi\tau f}\right)^2\right] df. \quad (5.11)$$

The key point is the transfer function which behaves now as $[N(N+1)/3](\pi\tau f)^2$ for $N\pi\tau f \ll 1$.

Convergence of the integral at the lower limit is thus ensured even for flicker frequency noise and random walk frequency noise: the sample variance is thus a useful tool for time domain characterization. Greater values of N increase the sensitivity of the sample variance to low Fourier frequency components: in fact, the true variance is approached as $N \to \infty$.

C. Two-Sampled Variance Versus Spectral Density

Since $\sigma_y^2(\tau)$ is by far the most widely used time-domain stability measure, let us consider its relationship with $S_y(f)$ and apply it to the typical kinds of frequency variations encountered in real oscillators.

Making $N = 2$ in (5.11) gives directly

$$\sigma_y^2(\tau) = \int_0^{\infty} S_y(f)\, \frac{2 \sin^4 \pi\tau f}{(\pi\tau f)^2}\, df. \quad (5.12)$$

The transfer function is such that the integral converges at the lower limit for the five power laws reported in Table I. Here again, the relation cannot be inverted for the most general cases.

As shown on Fig. 7, the transfer function $|H_A(f)|^2$ appearing in (5.12) is the |Fourier Transform|^2 of a step-wise time function $h_A(t)$ that resembles the measurement sequence of the quantity $(\bar{y}_2 - \bar{y}_1)/\sqrt{2}$ involved in the definition of $\sigma_y^2(\tau)$ (4.18).

This general property will be used later to develop new time-domain parameters.

[8] A complete derivation was given by Cutler for nonadjacent samples [8]:

$$\langle \sigma_y^2(N, T, \tau)\rangle = \frac{N}{N-1}\int_0^{\infty} S_y(f)\left(\frac{\sin \pi\tau f}{\pi\tau f}\right)^2\left[1 - \left(\frac{\sin N\pi Tf}{N \sin \pi Tf}\right)^2\right] df$$

1) Application to the Power-Law Model: The substitution of the widely used power-law spectral density model into (5.12) yields the expressions shown in Table II.

From these results, it appears that:

1) An upper cutoff frequency f_h is necessary for white and flicker phase noises since a substantial fraction of the noise "power" lies in the higher Fourier frequencies when $S_y(f)$ varies as f^2 or f. It has to be specified and the reported expressions are valid only for $2\pi f_h \tau >> 1$.

2) $\sigma_y^2(\tau)$ obeys to power laws with a slight modification for flicker phase noise because of the logarithm. Thus a log–log plot of $\sigma_y(\tau)$ contains segments of straight lines whose slopes may be easily identified.

3) The results may be used to translate from time domain into the frequency domain for the power law model (3.9), although the general relationship (5.12) is not reversible. Since both white and flicker phase noises give very similar slopes, there is some ambiguity in noise identification when time-domain measurements give $\sigma_y(\tau) \sim \tau^{-1}$ (varying f_h purposely may help to distinguish between white and flicker phase noises [6]).

4) For flicker frequency noise, $\sigma_y(\tau)$ is independent of τ and the corresponding flat part on the plot is often referred to as the flicker floor [23].

For a given oscillator, $\sigma_y^2(\tau)$ is the sum of two or three terms, e.g., cesium beam standards are often well-modeled by

$$\sigma_y^2(\tau) = \frac{h_0}{2\tau} + 2 \ln 2 h_{-1} \tag{5.13}$$

and the values of h_0 and h_{-1} may be deduced from measurements of $\sigma_y(\tau)$ over a sufficient range of τ.

5) $\sigma_y^2(\tau)$ can be a useful measure even when the model of (3.9) is not valid.

2) Application to Sinusoidal FM: Even the best sources are frequency modulated by unwanted sinusoidal signals. Although the above stability measures were developed to deal with random processes, sinusoidal instabilities do have an influence on them.

a) Fourier frequency domain: Assuming that

$$y(t) = \frac{\Delta\nu_0}{\nu_0} \sin 2\pi f_m t \tag{5.14}$$

the spectral density $S_y(f)$ contains a discrete line at the modulating frequency f_m:

$$S_y(f) = \frac{1}{2}\left(\frac{\Delta\nu_0}{\nu_0}\right)^2 \delta(f - f_m). \tag{5.15}$$

Looking for discrete lines in spectral densities provides a convenient means to identify periodic variations. Their presence usually does not interfer with the identification of the slopes due to random noise.

b) Time domain: The substitution of (5.15) into (5.12) gives[9]

$$\sigma_y^2(\tau) = \left(\frac{\Delta\nu_0}{\nu_0}\right)^2 \frac{\sin^4 \pi f_m \tau}{(\pi f_m \tau)^2} \tag{5.16}$$

Thus the effect of sinusoidal FM is zero when τ equals the modulation period $T_m = f_m^{-1}$ or one of its harmonics since the modulating signal is completely averaged out.

The worst case occurs when τ is near $T_m/2$ or one of its harmonics [19], [38].

As a practical consequence, when caution has not been exercized about the relation between the experimental values of τ and the expected value of T_m, there may be some scatter of the data because of the oscillating behavior of (5.16) (of course, this scatter is added to the law(s) due to the random noise(s) present).

[9] Direct calculation through (4.18) and (5.14) gives the same result.

TABLE II
TWO-SAMPLE VARIANCE $\sigma_y^2(\tau)$ FOR THE POWER-LAW SPECTRAL DENSITY MODEL (ADAPTED FROM [8]; SEE ALSO [36])

$S_y(f)$	$\sigma_y^2(\tau)$	Slope of $\sigma_y^2(\tau)$ VS. τ on log- log plot
$h_2 f^2$	$\frac{3 h_2 f_h}{4 \pi^2 \tau^2}$	-2
$h_1 f$	$\frac{h_1}{4 \pi^2 \tau^2} \left\{ 1.038 + 3 \ln 2 \pi f_h \tau \right\}$	≈ -2
h_0	$\frac{h_0}{2\tau}$	-1
$h_{-1} f^{-1}$	$2 \ln 2 \; h_{-1}$	0
$h_{-2} f^{-2}$	$\frac{2 \pi^2 h_{-2} \tau}{3}$	$+1$

3) Application to Linear Frequency Drift: Equation (5.12) is not useful for evaluating $\sigma_y^2(\tau)$ when linear frequency drift exists (i.e., $y(t) = d_1(t)$) since no tractable model seems to exist for $S_y(f)$ in this case.

Direct calculation in the time domain from (4.18) gives:

$$\sigma_y(\tau) = \frac{d_1}{\sqrt{2}} \tau. \tag{5.17}$$

Thus, linear frequency drift yields a τ^{+1} law for the square root of the two-sample variance. Such a law is observed for great values of τ when linear frequency drift has not been removed before statistical treatment. (Notice that no power law spectral density yields the same law; from Table II, it should be $S_y(f) = h_{-3} f^{-3}$ but $\sigma_y^2(\tau)$ diverges in this case due to its transfer function).

Higher order polynomial drifts will be treated in Section VIII in relation with structure functions.

VI. TRANSFER FUNCTIONS AS A TOOL TO DEFINE NEW PARAMETERS

Up to now, we have dealt only with the basics of frequency stability characterization.

1) Two domains have been considered and parameters have been defined independently in both domains.

2) Relationships have been established among these parameters that include, for each time-domain parameter, a specific transfer function which is the Fourier transform of the measurement sequence. Let us emphasize that the shapes of the transfer functions were not chosen *a priori*, but result from the definition of the time-domain parameters.

3) From the experimental point of view, spectrum analyzers are commonly associated with frequency-domain measurements and digital electronic counters with time-domain measurements.

In this section, the transfer function concept will be used as a tool to define new parameters giving new possibilities to existing measurement test sets.

A. The Hadamard Variance

The Hadamard variance has been developed by Baugh [39] for making high-resolution spectral analysis of $y(t)$ from measurements of $\bar{y}_k$, i.e., the frequency-domain parameter $S_y(f)$ is estimated from time-domain data provided by digital counters.

First, the limitations of the two-sample variance $\sigma_y^2(\tau)$ for such a spectral analysis must be emphasized and understood.

1) Limitations of $\sigma_y^2(\tau)$ for Spectral Analysis: Equation (5.12) shows that $\sigma_y^2(\tau)$ is not well suited for high-resolution spectral analysis of $y(t)$ since the main lobe of the transfer function (Fig. 7(b)) is wide. As a consequence, for a given value τ_c of the averaging time, the measured estimate $\hat{\sigma}_y^2(\tau_c)$ will not provide a precise estimate of $S_y(f_c)$ where $f_c \simeq 0.37\, \tau_c^{-1}$ is the main lobe center frequency for $\tau = \tau_c$.

For power-law spectral densities, the fact that both white phase noise and flicker phase noise yield nearly the same slope $\sigma_y(\tau) \sim \tau^{-1}$ results from this lack of selectivity.

However, let us remember that $\sigma_y(\tau)$ was introduced and recommended as a practical meaningful time-domain measure and not as a powerful spectral analysis tool.

2) Definition of the Hadamard Variance: Using the link between measurement sequences and transfer functions, the Hadamard variance is defined by a measurement sequence such that the corresponding transfer function contains a narrow lobe well-suited for spectral analysis [39]:

$$\langle \sigma_H^2(N, T, \tau) \rangle = \langle (\bar{y}_1 - \bar{y}_2 + \bar{y}_3 - \cdots - \bar{y}_N)^2 \rangle. \qquad (6.1)$$

The Hadamard variance[10] is thus calculated from groups of N samples $\bar{y}_k (k = 1 \text{ to } N)$ and is related to $S_y(f)$ by:

$$\langle \sigma_H^2(N, T, \tau) \rangle = \int_0^\infty S_y(f) |H_H(f)|^2 \, df \qquad (6.2)$$

where the transfer function $H_H(f)$ is the Fourier transform of the measurement sequence $h_H(t)$ (shown on Fig. 8(a) for $N = 6$ and $T = \tau$).

Sauvage [40] has shown that:

$$|H_H(f)|^2 = \left(\frac{\sin \pi \tau f}{\pi \tau f} \right)^2 \left(\frac{\sin N\pi T f}{\cos \pi T f} \right)^2. \qquad (6.3)$$

The transfer function exhibits a narrow main lobe centered at the Fourier frequency $f_1 = (2T)^{-1}$; its bandwidth decreases with increasing N as demonstrated from the calculation of an equivalent bandwidth [40].

Before emphasizing the practical utility of $\langle \sigma_H^2(N, T, \tau) \rangle$, its limitations must be recognized and accounted for.

3) Limitations of the Hadamard Variance: Here again, the limitations are a consequence of the transfer function shape. Although it has a narrow main lobe, it contains also other unwanted features:

[10] T and τ have the same meaning as for the sample variance. $\sigma_y^2(\tau)$ is one half times $\langle \sigma_H^2(N, \tau, \tau) \rangle$ with $N = 2$.

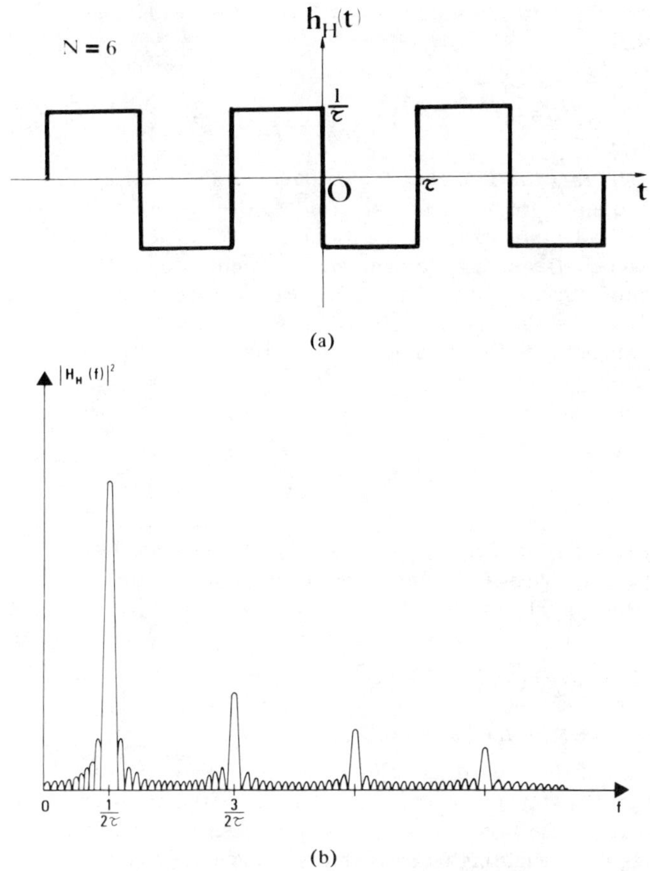

Fig. 8. (a) Measurement sequence of the Hadamard variance. (b) Transfer function of the Hadamard variance.

a) For adjacent samples ($T = \tau$), spurious responses appear at odd harmonics of the main lobe center frequency f_1. When measurements of $\langle \sigma_H^2(N, \tau, \tau) \rangle$ are made, they all contribute to the measured value and thus may lead to important errors in the estimate of $S_y(f_1)$. For white phase noise where $S_y(f) = h_2 f^2$, each harmonic lobe yields the same contribution to the measured Hadamard variance as the main lobe (which ideally should be alone);

b) As shown on Fig. 8(b), the transfer function has also large sidelobes around the main lobe centered at f_1.

4) Improvements of the Hadamard Variance: Limitation (a) can be partly eliminated with a dead time between successive samples of $\bar{y}_k$, i.e., with $T > \tau$. The optimum dead time $\tau_D = \tau/2$ eliminates the 3rd, 9th, 15th, etc., harmonic responses; in this case $T = 3\tau/2$.

More complicated measurement sequences have also been proposed to reduce the harmonic content wherein a pseudo-sinusoidal weighting of the samples $\bar{y}_k$ suppresses the 3rd, 5th, 7th, and 9th harmonics [31], [40]. Analog filtering techniques have also been used to filter out the spurious harmonic responses [41].

Limitation b) may be partly or even completely eliminated by multiplying each $\bar{y}_k$ within the set of N samples by an appropriate weighting factor. The unwanted sidelobes are completely suppressed by weighting the samples with binomial coefficients (BC) [39].

$$\langle \sigma_{H\,BC}^2(N, T, \tau) \rangle = \left\langle \left(\sum_{k=1}^{N} (-1)^{k-1} \binom{N-1}{k-1} \bar{y}_k \right)^2 \right\rangle \qquad (6.4)$$

and the corresponding transfer function reads then as [31] [40]:

$$|H_{H_{BC}}(f)|^2 = 2^{2(N-1)} \left(\frac{\sin \pi\tau f}{\pi\tau f}\right)^2 \sin^{2(N-1)} \pi T f. \quad (6.5)$$

5) Practical Utility of the Hadamard Variance: Provided adequate care has been exercized to overcome its limitations, the Hadamard variance is a useful tool for experiments since it extends frequency domain measurements down to very low Fourier frequencies of the order of 10^{-3} Hz just by increasing T and τ in the time domain, whereas analog spectrum analyzers are usually limited to about 1 to 10 Hz by hardware problems.

For the optimum dead time $\tau_D = \tau/2$, a working approximation of equation (6.2) giving an estimate of $S_y(f)$ reads as:

$$\hat{S}_y(f_1) \simeq \frac{\tau}{N}\left\langle \sigma_H^2\left(N, \frac{3\tau}{2}, \tau\right)\right\rangle \quad (6.6)$$

where $T = 3\tau/2$ and $f_1 = 1/2T = 1/3\tau$. Through this equation, spectral density estimates are made from measurements done with digital counters in the time domain. Of course, the Hadamard variance is itself estimated from a finite number of groups of N samples $\bar{y}_k$.

B. The Transfer Function Approach

In all the preceding approaches to time domain, any parameter, say $\sigma^2(\tau)$, was defined by its measurement sequence $h(t)$ (a step-wise function involving quantities such as N, T, τ, and possibly weighting factors) and related to $S_y(f)$ by

$$\sigma^2(\tau) = \int_0^\infty S_y(f)|H(f)|^2 \, df \quad (6.7)$$

where $H(f)$ is the Fourier transform of $h(t)$. The parameter was measured with a counter programmed to follow the chosen measurement sequence.

In the so-called "transfer function approach" developed by Rutman [42], the opposite point of view has been adopted: the parameter $\sigma^2(\tau)$ appearing in (6.7) is defined by the transfer function $H(f)$ whose shape is chosen *a priori*, even if no corresponding measurement sequence exists in the time domain (i.e., even if the inverse Fourier transform of $H(f)$ is not a step-wise function). One can thus consider a broader class of transfer functions since they need not be the Fourier transform of step-wise functions: however, counting techniques can no longer be used to measure the parameter defined in this way since the $\bar{y}_k$'s do not explicitly appear in this approach. The relevant measurement technique will be described.

To illustrate the usefulness of this approach, let us study two variances defined by their transfer functions, the "high-pass variance" and the "bandpass variance."

1) The High-Pass Variance: First, it is convenient to rewrite (6.7) as

$$\sigma^2(\tau) = \frac{8}{\omega_0^2\tau^2}\int_0^\infty S_\varphi(f)|H_\varphi(f)|^2 \, df \quad (6.8)$$

since (5.12) can itself be rewritten as

$$\sigma_y^2(\tau) = \frac{8}{\omega_0^2\tau^2}\int_0^\infty S_\varphi(f)\sin^4 \pi\tau f \, df. \quad (6.9)$$

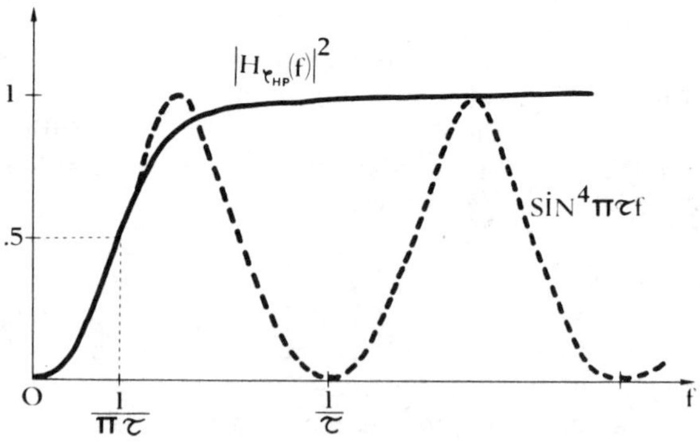

Fig. 9. Transfer function of the high-pass variance.

TABLE III
HIGH-PASS VARIANCE FOR THE POWER-LAW SPECTRAL DENSITY MODEL [42]

$S_y(f)$	$\sigma^2_{HP}(\tau)$	Slope vs. τ	$\frac{\sigma_{HP}(\tau)}{\sigma_y(\tau)}$
$h_2 f^2$	$\frac{2h_2 f_h}{\pi^2\tau^2}$	-2	1.63
$h_1 f$	$\frac{h_1}{2\pi^2\tau^2}\left[\ln\left(1+(\pi\tau f_h)^4\right)\right]$	≈ -2	1.36 to 1.63 ($\pi\tau f_h$ = 10 to ∞)
h_0	$\frac{h_0}{\sqrt{2}\,\tau}$	-1	1.19
$h_{-1} f^{-1}$	$\frac{\pi h_{-1}}{2}$	0	1.06
$h_{-2} f^{-2}$	$\frac{\pi^2 h_{-2}\tau}{\sqrt{2}}$	$+1$	1.03

* Convergence for higher negative slopes can be obtained by taking a higher-order high-pass Butterworth filter. Second-order was chosen here for greater similarity with $\sin^4 \pi\tau f$ for $\pi\tau f << 1$.

The subscript φ in $|H_\varphi(f)|^2$ indicates that $|H_\varphi(f)|^2$ multiplies the spectral density of phase noise. The new variances will be named according to the shape of their $|H_\varphi(f)|^2$.

The high-pass variance $\sigma^2_{HP}(\tau)$ is defined by (6.8) with a second-order high-pass Butterworth filter:

$$|H_{\varphi HP}(f)|^2 = \frac{f^4}{f_c^4 + f^4} \quad (6.10)$$

where the cutoff frequency f_c is such that $|H_{\varphi HP}(f_c)|^2 = 0.5$. The relation between f_c and the value of τ in $\sigma^2_{HP}(\tau)$ is chosen as $f_c = (\pi\tau)^{-1}$ by definition[11]; the above $|H_{\varphi HP}(f)|^2$ and $\sin^4 \pi\tau f$ appearing in equation (6.9) are illustrated on Fig. 9: they are very close for $\pi\tau f \lesssim 1$, but $|H_{\varphi HP}(f)|^2$ is not periodic.

The expressions of $\sigma^2_{HP}(\tau)$ for the power-law spectral density model are reported in Table III and show that $\sigma_y^2(\tau)$ and $\sigma^2_{HP}(\tau)$ have the same general behavior: same laws versus τ, same dependence or independence versus f_h, same range of convergence for power laws (due to the choice of a second-order filter for $\sigma^2_{HP}(\tau)$), and same order of magnitude as shown by the ratio $\sigma_{HP}(\tau)/\sigma_y(\tau)$ which may be interpreted as a bias coefficient when $\sigma_{HP}(\tau)$ is used as an estimate of $\sigma_y(\tau)$.

[11] Notice that $\sin^4 \pi\tau f_c = 0.5$ for $f_c = (\pi\tau)^{-1}$.

From this similarity, one gets a better understanding on why $\sigma_y^2(\tau)$ is not well suited for high-resolution spectral analysis: despite the bandpass shape of each lobe, the $\sin^4 \pi\tau f$ transfer function has indeed a global high-pass behavior which is poorly selective (in particular, white phase noise and flicker phase noise cannot be completely resolved by this filter).

Moreover, the high-pass variance demonstrates clearly that instability (due to random noise) over a time interval τ is equally due to all Fourier frequency components of $S_\varphi(f)$ lying above $f_c = (\pi\tau)^{-1}$, e.g., above 318 Hz for $\tau = 1$ ms, (and under f_h). This result has been used to estimate $\sigma_y(\tau)$ from records of the RF spectrum assuming the low modulation index is valid (3.11) [28]; a working formula for white phase noise reads as

$$\sigma_y^2(\tau) \simeq \frac{3 f_h\ 10^{-|\beta_0|/10}}{2\pi^2\ \nu_0^2 \tau^2} \tag{6.11}$$

where β_0 denotes the RF spectrum level under the carrier expressed in dB/Hz. Formulas are given in [28] for other power-law noises.

2) The Bandpass Variance: Having recognized the lack of selectivity of $\sigma_y(\tau)$ and $\sigma_{\mathrm{HP}}(\tau)$, one may use the transfer function approach to introduce a more selective variance through (6.8): the bandpass variance $\sigma_{\mathrm{BP}}^2(\tau)$ defined by its second-order bandpass Butterworth transfer function with a center frequency $f_0 = (2\tau)^{-1}$ and a constant Q factor (by analogy with the first lobe of $\sin^4 \pi\tau f$ centered at $f = (2\tau)^{-1}$ and for which $Q \simeq 1.37$). Fig. 10 shows the new transfer function.

The expressions of $\sigma_{\mathrm{BP}}^2(\tau)$ for the power-law model are reported in Table IV and show significant departures from those of $\sigma_y^2(\tau)$ or $\sigma_{\mathrm{HP}}^2(\tau)$.

The higher selectivity of $|H_{\varphi\,\mathrm{BP}}(f)|^2$ has the following consequences.

a) The high-frequency cutoff f_h is no longer necessary for convergence in the cases of white and flicker phase noises and therefore does not appear anymore in the relevant expressions of $\sigma_{\mathrm{BP}}^2(\tau)$.

b) The slope of $\sigma_{\mathrm{BP}}^2(\tau)$ takes a different value for each kind of noise including white phase and flicker phase noises for which it is equal to -3 and -2, respectively, (instead of -2 and ≈ -2 for $\sigma_y^2(\tau)$). For white phase noise, it is worth noting that $\sigma_{\mathrm{BP}}(\tau)$ does not provide an estimate of $\sigma_y(\tau)$, even with a constant bias, since the slopes are different.

For the three last kinds of noise, $\sigma_{\mathrm{BP}}(\tau)$ nevertheless provides an estimate of $\sigma_y(\tau)$ with bias factors near unity as shown in the last column.

To summarize, $\sigma_{\mathrm{BP}}(\tau)$ should not be considered as an estimator of $\sigma_y(\tau)$ but rather as a useful tool in describing the statistical properties of oscillators since it reveals the presence of white and/or flicker phase noises on a time domain plot [43].

These properties of $\sigma_{\mathrm{BP}}(\tau)$ are not surprising if we realize that the bandpass variance is indeed nothing but a constant percentage bandwidth (constant Q) spectral analysis of phase noise, the result of which is plotted in the time domain through (6.8).

Also, the bandpass variance may be viewed as an ideal Hadamard variance since the goal of any improvements on the latter is to get a transfer function with a single narrow lobe without any spurious responses: thus there is a conceptual link between the two approaches relying on the transfer function concept (of course, the measurement techniques are different as shown below).

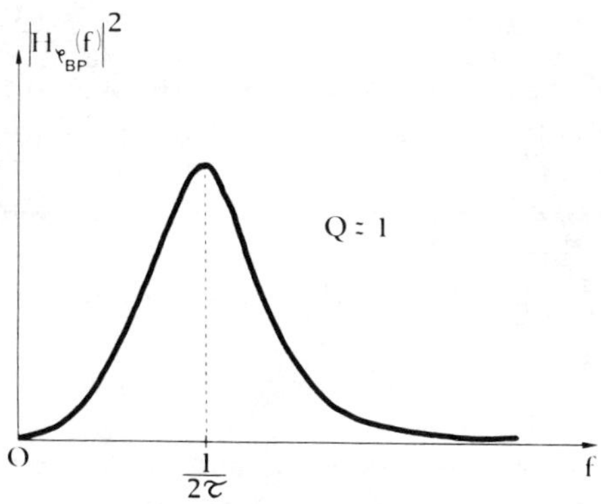

Fig. 10. Transfer function of the bandpass variance.

TABLE IV
BANDPASS VARIANCE FOR THE POWER-LAW SPECTRAL DENSITY MODEL [42]
($Q = 1$)

$S_y(f)$	$\sigma_{\mathrm{BP}}^2(\tau)$	Slope vs. τ	$\dfrac{\sigma_{\mathrm{BP}}(\tau)}{\sigma_y(\tau)}$
$h_2 f^2$	$\dfrac{h_2}{2\sqrt{2}\pi\tau^3}$	-3	$\dfrac{1.22}{\sqrt{\tau f_h}}$
$h_1 f$	$\dfrac{h_1}{\pi\sqrt{2}\tau^2}$	-2	$\dfrac{3}{\sqrt{1.038 + 3\ln 2\pi f_h \tau}}$
h_0	$\dfrac{\sqrt{2} h_0}{\pi\tau}$	-1	0.95
$h_{-1} f^{-1}$	$\dfrac{5 h_{-1}}{\pi\sqrt{2}}$	0	0.90
$h_{-2} f^{-2}$	$\dfrac{8\sqrt{2} h_{-2}\tau}{\pi}$	$+1$	0.74

3) Practical Utility of the Transfer Function Approach: Beyond its conceptual interest, the practical utility of this approach is in the additional possibilities it gives to the well-known phase detector measurement technique (see Section IX).

From (6.8), the parameter $\sigma(\tau)$ may be viewed as the rms value of the phase fluctuations $\varphi(t)$ filtered by a suitable transfer function $H_\varphi(f)$ are multiplied by the constant $\sqrt{8}\,(\omega_0\tau)^{-1}$. Thus this approach leads to a data acquisition technique relying on a suitable filtering of the demodulated phase noise and therefore the phase detector technique may be used since it provides an analog output voltage proportional to the oscillator phase fluctuations [22], [44], [45].

An experimental test set designed for that purpose has been developed [43] and includes the following.

a) The well-known phase detector system with a loose phase-lock loop: the output voltage is proportional to the oscillator phase noise for Fourier frequencies greater than about 1 Hz when the loop time constant is about one second. Such a system is often used for frequency-domain measurements.

b) A filter whose transfer function $H_\varphi(f)$ may be adjusted to several shapes, especially high-pass and bandpass.

c) A true rms voltmeter as the final measurement apparatus.

According to the filter response which has been selected, it is possible to measure the following.

a) $S_\varphi(f)$ with a narrow bandpass filter, and this is indeed the classical application of this test set.

b) $\sigma_{HP}(\tau)$ with a high-pass filter having a cutoff $f_c = (\pi\tau)^{-1}$; the upper cutoff frequency f_h is usually provided by another low-pass filter;

c) $\sigma_{BP}(\tau)$ with a constant Q bandpass filter having a center frequency $f_0 = (2\tau)^{-1}$.

In the last two cases, $\sigma(\tau)$ is simply given by

$$\sigma(\tau) = \frac{\sqrt{8}}{k\omega_0\tau}\, v_{rms} \tag{6.12}$$

where k is a calibration constant and v_{rms} is the measured rms value of the filtered voltage analog to $\varphi(t)$.

Thus a unique experimental test set provides stability measures both in frequency and time domains, without any digital counter and associated statistical treatment of data in the latter case. Following the same basic idea, bandpass filtering of $y(t)$ at the output of a frequency discriminator has also been used to provide an estimate of $\sigma_y(\tau)$ [46].

VII. Other Suggested Time-Domain Measures

In this section, we describe two more time-domain approaches relying on specific statistical treatments of the $\bar{y}_k$ usually provided by counting techniques. In both cases, the new parameter is believed to be more "efficient" than $\sigma_y(\tau)$ in some respect to be discussed; the transfer function appears again as a convenient means to interpret their respective properties.

The links with the Hadamard variance and the sample variance will be given when appropriate.

The expressions of the new parameters will be presented for the power-law spectral density model allowing thus easier comparisons with the recommended time-domain measure $\sigma_y(\tau)$.

A. Finite-Time Frequency Control

This method introduced by Boileau and Picinbono [13] is based upon a combination of the $\bar{y}_k$ that may be interpreted as a modified sample variance.[12] To illustrate this approach, a specific modified sample variance $\Sigma_y^2(\tau)$ has been considered in [47], the expressions of which are given below in Table V for the power-law spectral density model.

1) The Modified Sample Variance: From N discrete values of $\bar{y}_k$, the following quantity is defined as a modified sample variance:

$$\Sigma_y^2(N, T, \tau) = \left(\bar{y}_{(N+1)/2} - \frac{1}{N}\sum_{i=1}^{N} \bar{y}_i\right)^2 \tag{7.1}$$

where N is odd and $\bar{y}_{(N+1)/2}$ denotes the central sample within a set of N samples.

T and τ have the same meaning as in Fig. 4. Of course, $\Sigma_y^2(N, T, \tau)$ is a random variable and its infinite time average is related to $S_y(f)$ by the following integral [13]:

$$\left\langle \Sigma_y^2(N, T, \tau) \right\rangle = \int_0^\infty S_y(f) \left(\frac{\sin \pi\tau f}{\pi\tau f}\right)^2 \left(1 - \frac{\sin N\pi T f}{N \sin \pi T f}\right)^2 df \tag{7.2}$$

TABLE V

Modified Three-Adjacent-Sample Variance $\Sigma_y^2(\tau)$ for the Power-Law Spectral Density Model ($\pi\tau f_h >> 1$)

$S_y(f)$	$\Sigma_y^2(\tau)$	Slope vs. τ	$\frac{\Sigma_y(\tau)}{\sigma_y(\tau)}$
$h_2 f^2$	$\frac{5 h_2 f_h}{9\pi^2\tau^2}$	-2	0.86
$h_1 f$	$\frac{5 h_1}{9\pi^2\tau^2}\left[.964 + \ln \pi\tau f_h\right]$	≈ -2	0.85 to 0.86 ($\pi\tau f_h = 10$ to ∞)
h_0	$\frac{h_0}{3\tau}$	-1	0.82
$h_{-1} f^{-1}$	$\frac{8 \ln 2 - 3 \ln 3}{3} h_{-1}$	0	0.74
$h_{-2} f^{-2}$	$\frac{2}{9}\pi^2 h_{-2}\tau$	$+1$	0.58
$h_{-3} f^{-3}$	$\frac{27 \ln 3 - 32 \ln 2}{9}\pi^2 h_{-3}\tau^2$	$+2$	–
$h_{-4} f^{-4}$	$\frac{44}{90}\pi^4 h_{-4}\tau^3$	$+3$	–

The new transfer function behaves now as $(N^2 - 1)^2 (\pi T f)^4/36$ for $N\pi T f << 1$, i.e., for the lowest Fourier frequencies. Due to this behavior, this time-domain parameter is convergent for the power-law model $S_y(f) = h_\alpha f^\alpha$ even for $\alpha = -3$ and $\alpha = -4$ (whereas the classical sample variance was limited to $\alpha = -2$). It is interesting to note that the law $S_y(f) = h_{-3} f^{-3}$ has been referred to as a possible model for a high-level quartz-crystal oscillator at Fourier frequencies around 1 Hz [21].

Thus the finite-time frequency-control method allows one to deal with spectral densities having greater negative slopes (sometimes interpreted as "stronger nonstationarity") than the classical sample variance approach.

2) The Modified Three-Adjacent-Sample Variance: For the same reasons as it appeared useful to recommend the use of one particular sample variance $\sigma_y^2(\tau)$, comparability of data will be improved by specifying the values of N and T in $\langle \Sigma_y^2(N, T, \tau)\rangle$, with τ remaining the independent variable [47].

Let us consider the modified sample variance with $T = \tau$ (adjacent samples as in $\sigma_y^2(\tau)$) and $N = 3$ (the smallest possible value of N for this measurement sequence where N is odd).[13] Equations (7.1) and (7.2) then read as:

$$\Sigma_y^2(3, \tau, \tau) = \frac{1}{9}(2\bar{y}_2 - \bar{y}_1 - \bar{y}_3)^2 \tag{7.3}$$

$$\Sigma_y^2(\tau) \equiv \left\langle \Sigma_y^2(3, \tau, \tau)\right\rangle = \int_0^\infty S_y(f)\, \frac{16}{9}\, \frac{\sin^6 \pi\tau f}{(\pi\tau f)^2}\, df. \tag{7.4}$$

The transfer function is the Fourier transform of the time-domain measurement sequence as shown on Fig. 11.

[12] Continuous expressions were also developed in [13] and reviewed in [47]. Only the discrete realization based on discrete values of $\bar{y}_k$ is presented here.

[13] A so-called curvature variance has been introduced by Kramer [60], [61], which is equal to 1.5 times $\Sigma_y^2(\tau)$. The second difference of $\bar{y}_k$ or the third difference of φ_k was also discussed by Barnes in [30] and [69].

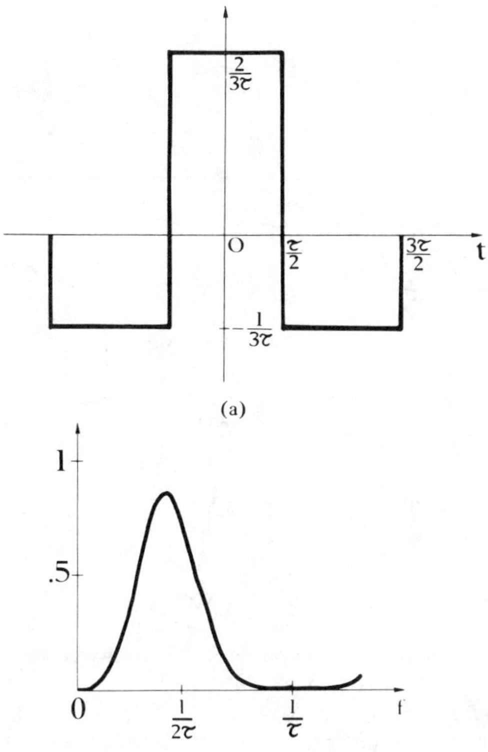

Fig. 11. (a) Measurement sequence of the modified three sample variance. (b) Transfer function of the modified three sample variance.

Table V gives the expressions of $\Sigma_y^2(\tau)$ for the power-law model, including the cases $\alpha = -3$ and $\alpha = -4$ for which $\sigma_y(\tau) = \infty$. For the five other encountered laws, $\Sigma_y(\tau)$ is an estimator of $\sigma_y(\tau)$ with constant bias coefficients reported in the right column: both follow the same laws vs τ and f_h and have the same order of magnitude. Especially, $\Sigma_y(\tau)$ is not better suited than $\sigma_y(\tau)$ to distinguish between white phase noise and flicker phase noise that yield nearly the same slope.

3) Relations with Previous Approaches: First, it must be recognized that $\Sigma_y^2(3, \tau, \tau)$ is quite different from $\sigma_y^2(3, \tau, \tau)$ derived from (4.13) with $N = 3$ and $T = \tau$:

$$\sigma_y^2(3, \tau, \tau) = \frac{1}{3}\left(\bar{y}_1^2 + \bar{y}_2^2 + \bar{y}_3^2 - \bar{y}_1\bar{y}_2 - \bar{y}_1\bar{y}_3 - \bar{y}_2\bar{y}_3\right). \tag{7.5}$$

However, the Hadamard variance weighted by binomial coefficients (6.4) with $N = 3$ and $T = \tau$ reads as

$$\langle \sigma_{H\,\mathrm{BC}}^2(3, \tau, \tau)\rangle = \langle(\bar{y}_1 - 2\bar{y}_2 + \bar{y}_3)^2\rangle \tag{7.6}$$

and hence

$$\Sigma_y^2(\tau) = \frac{1}{9}\langle \sigma_{H\,\mathrm{BC}}^2(3, \tau, \tau)\rangle. \tag{7.7}$$

It is easy to verify that making $N = 3$ and $T = \tau$ in (6.5) and then dividing by 9 yields the transfer function appearing in the above integral (7.4).

This link with the weighted Hadamard variance will be interpreted in the next section from the study of structure functions of phase.

As far as the transfer function approach is concerned, expression (7.4) may be rewritten in terms of the spectral density of phase noise:

$$\Sigma_y^2(\tau) = \frac{64}{9\omega_0^2\tau^2}\int_0^\infty S_\varphi(f)\sin^6 \pi\tau f\, df. \tag{7.8}$$

Thus $\Sigma_y^2(\tau)$ can be estimated by high-pass filtering the demodulated phase noise, but a third-order Butterworth filter is needed to yield the same convergence properties as the $\sin^6 \pi\tau f$ transfer function.

To summarize, the new parameter $\Sigma_y(\tau)$ whose experimental estimation by counting techniques is not very complicated, has the same behavior as $\sigma_y(\tau)$ for power-law spectral densities, except that it converges for two more negative slopes (one of which has been mentioned in [21]); therefore, its use may be recommended whenever a law such as $S_y(f) = h_{-3}f^{-3}$ (or $h_{-4}f^{-4}$) is expected from some theoretical and/or experimental consideration; then, variations of $\Sigma_y(\tau) \sim \tau$ (or $\tau^{3/2}$) may be observed for the greater values of τ. Note that these laws cannot be observed for $\sigma_y(\tau)$ since $+\frac{1}{2}$ is the highest positive slope that arises with power-law models; this point will be discussed in greater detail in Section VIII in connection with the effects of polynomial drifts.

B. Frequency Instability Over a Time Interval T

In all the preceding time domain approaches, the parameter of interest ($I(\tau)$, $\sigma_y(\tau)$, $\sigma_{\mathrm{HP}}(\tau)$, $\sigma_{\mathrm{BP}}(\tau)$, or $\Sigma_y(\tau)$) was plotted versus the averaging time interval τ which may vary from say 10^{-3} s. to thousands of seconds or even days or months. The decreasing parts of the curves show then the averaging of stationary random noises and, beyond the flicker floor, the increasing parts are due to nonstationary effects (either random or deterministic).

Another point of view has been proposed by De Prins and Cornelissen [48] who studied the frequency fluctuations over a time interval T for a fixed value of the averaging time interval τ: the parameter is thus plotted versus T (which has in this paragraph the same meaning as in Fig. 4).

Let us consider first the case $\tau = 0$ for simplicity and then the case $\tau \neq 0$ which is physically realistic.

1) Zero Averaging Time ($\tau = 0$): An ideal sampling of $y(t)$ at $t_k (k = 1, 2, 3, \cdots)$ such that $t_{k+1} - t_k = T$ yields the instantaneous values $y(t_k)$ denoted as y_k. This idealized measurement process corresponds to instantaneous frequency measurements ($\tau = 0$).

The difference $\delta y_T = y_2 - y_1$ is a measure of the instantaneous fractional frequency instability over the time interval T from t_1 to t_2. This is of course a random quantity whose mean-square value may be used as a statistical meaningful measure of frequency instability over T:

$$\langle \delta y_T^2\rangle = \langle (y_2 - y_1)^2\rangle. \tag{7.9}$$

Since only one sample of $y(t)$ is usually available to the experimenter, a suitable measurement process has been recommended [49] to avoid artefacts due to correlation between data.

This parameter is simply related to $R_y(\tau)$ and $S_y(f)$ by

$$\langle \delta y_T^2\rangle = 2[R_y(0) - R_y(T)] \tag{7.10}$$

$$\langle \delta y_T^2\rangle = \int_0^\infty S_y(f)\, 4\sin^2 \pi T f\, df. \tag{7.11}$$

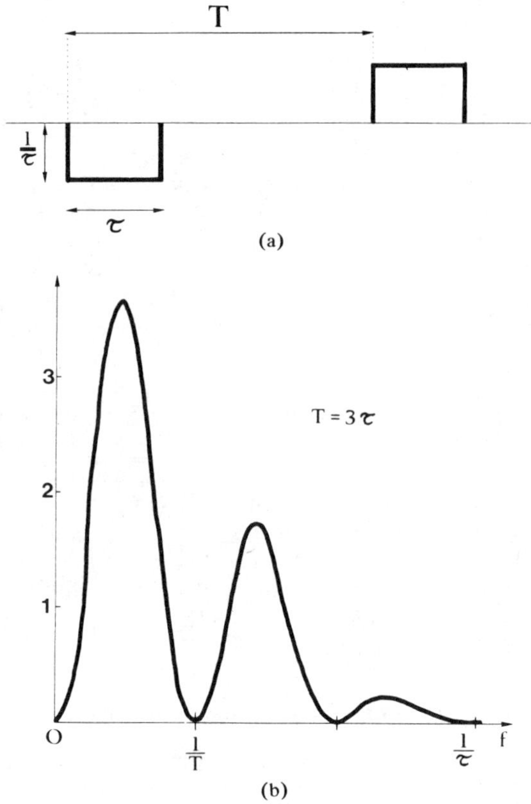

Fig. 12. (a) Measurement sequence of $\langle \delta \bar{y}_T^2 \rangle$. (b) Transfer function of $\langle \delta \bar{y}_T^2 \rangle$ for $T = 3\tau$.

Assuming that $y(t)$ is a zero-mean stationary process with finite variance $\langle y^2(t) \rangle = R_y(0)$, the following limit holds (see (7.10)):

$$\lim_{T \to \infty} \langle \delta y_T^2 \rangle = 2 \langle y^2(t) \rangle. \tag{7.12}$$

Assuming that $S_y(f)$ has an upper frequency cutoff f_h, it is easy to show from (7.11) that:

$$\langle \delta y_T^2 \rangle \simeq 4\pi^2 \beta_y T^2, \qquad \text{for } \pi T f_h << 1 \tag{7.13}$$

where β_y is a constant equal to[14]

$$\beta_y = \int_0^{f_h} f^2 S_y(f)\, df. \tag{7.14}$$

Thus the evolution of $\langle \delta y_T^2 \rangle$ with T is quite different from the laws of any other time-domain measure versus the averaging time interval τ and this may be traced to the shape of the periodic transfer function appearing in (7.11). This evolution may be said to be "cumulative" in the sense that all the possible values of $y(t)$ are scanned and contribute to $\langle \delta y_T^2 \rangle$ as T increases (instead of being averaged out with increasing τ).

2) Nonzero Averaging Time ($\tau \neq 0$): Since experimental test sets provide averaged samples $\bar{y}_k$ (4.2), it is of practical interest to generalize the definition of $\langle \delta y_T^2 \rangle$ as follows:

$$\langle \delta \bar{y}_T^2 \rangle = \langle (\bar{y}_2 - \bar{y}_1)^2 \rangle. \tag{7.15}$$

The measurement sequence and transfer function are shown on

[14] For a power law $S_y(f) = h_\alpha f^\alpha$, the constant equals $\beta_{y,\alpha} = h_\alpha f_h^{\alpha+3} / (\alpha + 3)$.

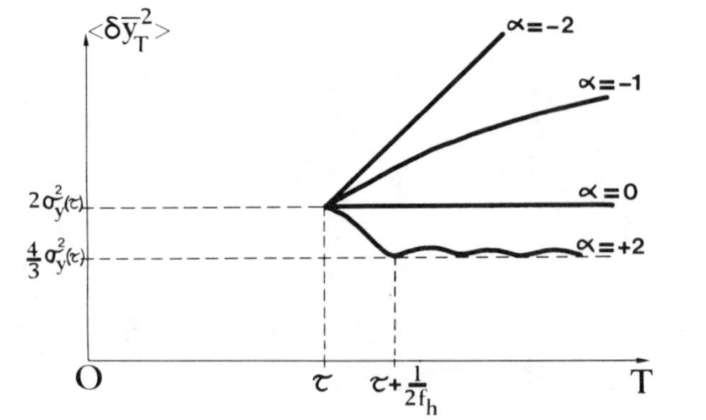

Fig. 13. The time-domain parameter $\langle \delta \bar{y}_T^2 \rangle$ versus T for the power law model.

Fig. 12, and the relationship with $S_y(f)$ now reads as

$$\langle \delta \bar{y}_T^2 \rangle = \int_0^\infty S_y(f)\, 4 \left(\frac{\sin \pi \tau f}{\pi \tau f} \right)^2 \sin^2 \pi T f\, df. \tag{7.16}$$

As shown below, this parameter is simply related to one particular sample variance.

3) Relation Between $\langle \delta \bar{y}_T^2 \rangle$ and the Sample Variance: From the following definition of the sample variance

$$\sigma_y^2(N, T, \tau) = \frac{1}{N-1} \sum_{i=1}^{N} \left(\bar{y}_i - \frac{1}{N} \sum_{j=1}^{N} \bar{y}_j \right)^2 \tag{7.17}$$

it is easy to demonstrate that

$$\langle \delta \bar{y}_T^2 \rangle = 2 \langle \sigma_y^2(2, T, \tau) \rangle. \tag{7.18}$$

In other words, $\langle \delta \bar{y}_T^2 \rangle$ is just twice the two-sample variance with nonadjacent samples ($T > \tau$) which will be here studied versus T for a fixed value of τ. Thus (7.16) is a particular case of the general formula given by Cutler (equation (23) with $N = 2$ in [8]; see also footnote 8).

Except for white phase noise for which a more complete formula is given here, the expressions of $\langle \delta \bar{y}_T^2 \rangle$ reported in Table VI are thus deduced from $\langle \sigma_y^2(2, T, \tau) \rangle$ given in [8], Appendix II.

For $T = \tau$, these expressions are equal to $2\sigma_y^2(\tau)$ as expected from (7.18), except the one for flicker phase noise valid only for $T >> \tau$. For increasing values of T, the evolution of $\langle \delta \bar{y}_T^2 \rangle$ is pictured on Fig. 13: it is linear for random walk frequency noise, logarithmic for flicker frequency noise, independent of T for white frequency noise and it approaches the limit value of (4/3). $\sigma_y^2(\tau)$ for white and flicker phase noises (this evolution also appears on [35], Fig. 2 where the bias function B_2 (4.25) is plotted for $0 < T \leqslant 2\tau$).

These expressions have been used by Cornelissen [49] to predict the long-term behavior of frequency standards.

For real oscillators wherein flicker frequency noise is ever present, a logarithmic law of $\langle \delta \bar{y}_T^2 \rangle$ versus T may be expected for the greater values of T (when not masked by random walk of frequency or by deterministic drifts). In particular, linear frequency drift modeled by $y(t) = d_1\, t$ yields:

$$\langle \delta \bar{y}_T^2 \rangle = d_1^2 T^2 \tag{7.19}$$

To summarize, despite its simple mathematical relationship with the two-nonadjacent sample variance, the parameter $\langle \delta \bar{y}_T^2 \rangle$ has been introduced on a quite different conceptual

TABLE VI
THE TIME-DOMAIN PARAMETER $<\delta y_T^{-2}>$ FOR THE POWER-LAW SPECTRAL DENSITY MODEL

$S_y(f)$	$<\delta \bar{y}_T^2>$ for $T>\tau$	
$h_2 f^2$	$\frac{h_2 f_h}{\pi^2 \tau^2}\left(1+\frac{1}{2}\cdot\frac{\sin 2\pi (T-\tau) f_h}{2\pi (T-\tau) f_h}\right)$	for $\pi \tau f_h >> 1$
$h_1 f$	$\frac{h_1}{\pi^2 \tau^2}\left(0.577+\ln 2\pi f_h \tau+\frac{1}{2}\ln\frac{(T/\tau)^2}{(T/\tau)^2-1}\right)$	for $\pi \tau f_h >> 1$ and $T >> \tau$
h_0	$\frac{h_0}{\tau}$	
$h_{-1} f^{-1}$	$h_{-1}\left[-2\left(\frac{T}{\tau}\right)^2 \ln\frac{T}{\tau}+\left(\frac{T}{\tau}+1\right)^2 \ln\left(\frac{T}{\tau}+1\right)+\left(\frac{T}{\tau}-1\right)^2 \ln\left(\frac{T}{\tau}-1\right)\right]$	$\approx 2 h_{-1} \ln\frac{T}{\tau}$ for $T >> \tau$
$h_{-2} f^{-2}$	$2\pi^2 h_{-2}\left(T-\frac{\tau}{3}\right)$	

basis where emphasis is put on the time evolution of the fluctuations under study. Therefore, it may appear more useful than $\sigma_y(\tau)$ for certain kinds of prediction problems [49], and exhibits unfamiliar laws reported on Table VI.

VIII. THE STRUCTURE FUNCTION APPROACH

In the preceding sections, the discrete samples $\bar{y}_k$ provided by a suitable measurement process have been used to define the following quantities that serve several purposes:

1) the true variance given by (4.5), as a theoretical measure of time-domain frequency instability;
2) the sample variances given by (4.8), (4.13), or (4.16), as practical measures of time-domain instability; $\sigma_y^2(\tau)$ given by (4.18) is just a particular case;
3) the Hadamard variance (6.1) or (6.4), as a time-domain measure leading to spectral density estimates;
4) the modified sample variance (7.1), as a time-domain measure which converges even for $S_y(f) \sim f^{-3}$ and f^{-4}, with $\Sigma_y^2(\tau)$ given by (7.4) being a particular case;
5) $\langle \delta \bar{y}_T^2 \rangle$ given by (7.11) as a different means to present instability data in the time domain.

In the original references, these parameters have been introduced to serve one of the above mentioned purposes but no unified presentation had been looked for.

The structure function approach developed by Lindsey and Chie [50], [51] plays such a unifying role in the sense that the most important time-domain parameters appear as particular cases of one general concept, namely the so-called structure function.

This theory is presented below with special emphasis on its relationships with previous approaches and on its application to deterministic polynomial drifts.

A. Definition of Structure Functions

For $M \geqslant 1$, the Mth increment of a random process $g(t)$ is defined by

$$\Delta^{(M)} g(t;\tau) = \sum_{k=0}^{M} (-1)^k \binom{M}{k} g(t+(M-k)\tau). \qquad (8.1)$$

The process $g(t)$ has a stationary Mth increment [15], [16], if the following averages exist for all real T and τ and do not depend on the instant t:

$$\langle \Delta^{(M)} g(t;\tau) \rangle = \alpha(\tau) \qquad (8.2)$$

$$\langle \Delta^{(M)} g(t;\tau) \cdot \Delta^{(M)} g(t+T;\tau) \rangle = D_g^{(M)}(T;\tau). \qquad (8.3)$$

In other words, the Mth increment has a time independent mean and its autocorrelation depends only on the time difference (wide sense stationarity).

By definition, $D_g^{(M)}(T;\tau)$ is the structure function of the Mth increment. In the following, the Mth structure function of the random process $g(t)$ will be defined as $D_g^{(M)}(T=0;\tau)$ and denoted as $D_g^{(M)}(\tau)$:

$$D_g^{(M)}(\tau) = \langle (\Delta^{(M)} g(t;\tau))^2 \rangle. \qquad (8.4)$$

B. Application to Oscillators Instabilities: Phase Instability Versus Frequency Instability

When dealing with instabilities in quasi-sinusoidal oscillators, the structure function concept may be applied to the phase fluctuations $\varphi(t)$ and to the (fractional) frequency fluctuations $y(t)$, yielding two sets of parameters [50]. Distinct definitions may thus be given for phase instability and frequency instability, whereas these two concepts have often been confused in the literature. To illustrate this point, let us consider the first ($M=1$) structure functions of $\varphi(t)$ and of $y(t)$.

1) Phase Instability: Phase instability over a time interval τ may be defined as the ratio of the rms value of the phase noise accumulated during τ seconds to the phase accumulated by a noiseless oscillator ($\omega_0 \tau$). This quantity is simply related to the first structure function of phase noise:

$$\frac{\sqrt{\langle [\varphi(t+\tau)-\varphi(t)]^2 \rangle}}{\omega_0 \tau} = \frac{1}{\omega_0 \tau}\sqrt{D_\varphi^{(1)}(\tau)}. \qquad (8.5)$$

But (4.4) and (4.5) show that the above quantity is indeed nothing but the true standard deviation $I(\tau)$ usually interpreted as a theoretical measure of frequency instability (the limitations of which have been pointed out).

2) Frequency Instability: Fractional frequency instability over a time interval T may be defined as the rms value of the fractional frequency fluctuations $y(t)$ accumulated during T

TABLE VII
STRUCTURE FUNCTIONS OF PHASE: RELATIONS WITH OTHER TIME-DOMAIN PARAMETERS

M	$\Delta^{(M)}\varphi(t;\tau)$	$D^{(M)}_{\varphi}(\tau)$	Relations with other measures
1	$\omega_0\tau\bar{y}_1$	$\omega_0^2\tau^2\langle\bar{y}_1^2\rangle$	$I(\tau)=\frac{1}{\omega_0\tau}\sqrt{D^{(1)}_{\varphi}(\tau)}$
2	$\omega_0\tau(\bar{y}_2-\bar{y}_1)$	$\omega_0^2\tau^2\langle(\bar{y}_2-\bar{y}_1)^2\rangle$	$\sigma_y(\tau)=\frac{1}{\sqrt{2}\,\omega_0\tau}\sqrt{D^{(2)}_{\varphi}(\tau)}$
3	$\omega_0\tau(\bar{y}_1-2\bar{y}_2+\bar{y}_3)$	$\omega_0^2\tau^2\langle(\bar{y}_1-2\bar{y}_2+\bar{y}_3)^2\rangle$	$\Sigma_y(\tau)=\frac{1}{3\,\omega_0\tau}\sqrt{D^{(3)}_{\varphi}(\tau)}$
M	$\omega_0\tau\sum_{i=1}^{M}(-1)^i\binom{M-1}{i-1}\bar{y}_i$	$\omega_0^2\tau^2\left\langle\left(\sum_{i=1}^{M}(-1)^i\binom{M-1}{i-1}\bar{y}_i\right)^2\right\rangle$	$\langle\sigma^2_{H_{BC}}(M,\tau,\tau)\rangle^{1/2}=\frac{1}{\omega_0\tau}\sqrt{D^{(M)}_{\varphi}(\tau)}$

seconds; this is just the first structure function of $y(t)$ [52]:

$$\sqrt{\langle[y(t+T)-y(t)]^2\rangle}=\sqrt{D^{(1)}_y(T)}. \tag{8.6}$$

Moreover, it is equal to $\sqrt{\langle\delta\bar{y}_T^2\rangle}$ defined in (7.9); that is why we kept the symbol T instead of τ in (8.6).

The parameter $\langle\delta\bar{y}_T^2\rangle$ defined by (7.11) is related to phase structure functions as follows [47]:

$$\langle\delta\bar{y}_T^2\rangle=\frac{2}{\omega_0^2\tau^2}\,[D^{(1)}_{\varphi}(\tau)-D^{(1)}_{\varphi}(T;\tau)]. \tag{8.7}$$

When dealing with $\bar{y}_k$, structure functions of phase appear because of the link between $\bar{y}_k$ and the first increment of $\varphi(t)$ expressed by (4.4). Structure functions of phase are hence of prime importance for oscillators instability characterization.

C. Structure Function of Phase

First, it must be remembered that taking the Mth increments of the phase had been proposed by Barnes [30] as a powerful means of classifying the statistical fluctuations in oscillators: Structure functions of phase are just another way of rewriting these quantities.

1) Mth Increment of Phase Versus $\bar{y}_k$: Since the $\bar{y}_k$ are meaningful data, it is of interest to rewrite $\Delta^{(M)}\varphi(t;\tau)$ in terms of the M values of $\bar{y}_k$ involved. By definition, $\Delta^{(M)}\varphi(t;\tau)$ is calculated from the instantaneous values of $\varphi(t)$ at the instants $t, t+\tau, t+2\tau, \cdots, t+M\tau$. The ith value $\varphi(t+(i-1)\tau)$, where $i=1$ to $M+1$, is weighted by $(-1)^i\binom{M}{i-1}$ when M is odd and by $(-1)^{i+1}\binom{M}{i-1}$ when M is even; the term $\varphi(t+M\tau)$ has always a positive sign from the definition of $\Delta^{(M)}\varphi(t;\tau)$.

By using relation (4.4), $\Delta^{(M)}\varphi(t;\tau)$ may then be rewritten in terms of the M values of $\bar{y}_k$ (with $k=1$ to M) involved in its definition [47]:

$$\Delta^{(M)}\varphi(t;\tau)=\omega_0\tau\sum_{k=1}^{M}(-1)^{k+1}\binom{M-1}{k-1}\bar{y}_k,\qquad M\text{ odd} \tag{8.8}$$

$$\Delta^{(M)}\varphi(t;\tau)=\omega_0\tau\sum_{k=1}^{M}(-1)^{k}\binom{M-1}{k-1}\bar{y}_k,\qquad M\text{ even.} \tag{8.9}$$

These expressions are used next to relate $D^{(M)}_{\varphi}(\tau)$ to some of the time-domain measures.

2) Interpretation of $D^{(M)}_{\varphi}(\tau)$ in Terms of Time-domain Measures: From (8.8) and (8.9), the expressions of $D^{(M)}_{\varphi}(\tau)$ calculated for $M=1, 2, 3$, etc., are reported in Table VII and show that some of the previous time domain measures may be re-interpreted in terms of structure functions of phase.

For $M=1$ and $M=2$, the structure function is related to the true variance $I^2(\tau)$ and to $\sigma_y^2(\tau)$, respectively, as pointed out by Lindsey and Chie [50].

For $M=3$ and for M even ($\geqslant 4$), it is related respectively to $\Sigma_y^2(\tau)$ and to the Hadamard variance with adjacent samples weighted by binomial coefficients as pointed out in [47].

The unifying role of the structure function concept is then clear from the right column of Table VII.

3) Relation Between $D^{(M)}_{\varphi}(\tau)$ and $S_y(f)$: The relation between structure functions of phase and $S_y(f)$ may be easily deduced from (6.5) with $T=\tau$ and from the relation between $D^{(M)}_{\varphi}(\tau)$ and $\langle\sigma^2_{H\,BC}(M,\tau,\tau)\rangle$ reported in Table VII:

$$D^{(M)}_{\varphi}(\tau)=2^{2(M-1)}(\omega_0\tau)^2\int_0^{\infty}S_y(f)\,\frac{\sin^{2M}\pi\tau f}{(\pi\tau f)^2}\,df. \tag{8.10}$$

The corresponding inverse transform relationship can be given in terms of Mellin transforms as shown in [50]; however, the formula involved are rather complex to use.

Because of the exponent $2M$ appearing in the transfer function, the desirability of taking structure functions with higher orders becomes obvious from the point of view of convergence with power-law spectral densities having steep negative slopes: this general relationship in connection with the expressions reported in Table VII explains why $\Sigma_y(\tau)$ is "more convergent" than $\sigma_y(\tau)$ which is itself "more convergent" than $I(\tau)$.

When $S_{\varphi}(f)$ may be defined unambiguously, equation (8.10) may be rewritten as

$$D^{(M)}_{\varphi}(\tau)=2^{2M}\int_0^{\infty}S_{\varphi}(f)\sin^{2M}\pi\tau f\,df. \tag{8.11}$$

4) Application of $D^{(M)}_{\varphi}(\tau)$ to Polynomial Drifts: Up to now, we have been concerned mainly with the characterization of

TABLE VIII
TIME-DOMAIN PARAMETERS FOR POLYNOMIAL PHASE DRIFT

	ORDER OF THE PHASE DRIFT POLYNOMIAL (P)			
ORDER OF THE PHASE INCREMENT (M)	2 (Linear frequency drift)	3 (quadratic frequency drift)	4 (cubic frequency drift)	P
1	time dependent	time dependent	time dependent	time dependent
2	$\sigma_y(\tau) = \frac{d_1}{\sqrt{2}}\tau$	time dependent	time dependent	time dependent
3	0	$\Sigma_y(\tau) = \frac{2}{3} d_2 \tau^2$	time dependent	time dependent
4	0	0	$\sigma_{HBC} = 6 d_3 \tau^3$	time dependent
M = P	0	0	0	$\sigma_{HBC} = (P-1)!\, d_{P-1}\, \tau^{P-1}$

random fluctuations in oscillators: however, nonrandom variations (drifts) of the output frequency do exist in most devices and can be sometimes modeled by deterministic polynomial functions [9]. In this context, structure functions are also of interest since they provide a means to obtain insight into the highest order drift coefficient, i.e., into the long-term instability of the source [50].

Let us assume that the oscillator frequency drift is modeled by a $(P-1)$th degree polynomial $(P \geqslant 2)$:

$$y(t) = \sum_{i=1}^{P-1} d_i t^i. \tag{8.12}$$

Integrating from 0 to t yields an Pth degree polynomial for the phase drift:

$$\varphi(t) = \sum_{i=1}^{P-1} D_i t^{i+1} \tag{8.13}$$

where the coefficients D_i are related to the d_i by

$$D_i = \frac{\omega_0 d_i}{i+1} \tag{8.14}$$

A frequency measurement beginning at t_k and lasting τ seconds yields the average value (ignoring random fluctuations):

$$\bar{y}_k = \sum_{i=1}^{P-1} \frac{d_i}{(i+1)\tau}\,[(t_k+\tau)^{i+1} - t_k^{i+1}]. \tag{8.15}$$

Substituting this expression into (8.8) or (8.9) allows one to study the properties of the successive phase increments in the presence of polynomial drifts:

a) the phase increments of order M smaller than P are time dependent and hence are of little use as measures of the drift constant coefficients;
b) the phase increments of order M greater than P are all equal to zero and hence are meaningless;
c) only the Pth phase increment $(M = P)$ as time-independent and nonzero:

$$\Delta^{(P)}\varphi(t;\tau) = (P-1)!\,\omega_0 d_{P-1}\tau^P. \tag{8.16}$$

It is directly related to the highest order drift coefficient d_{P-1}, and provides thus a measure of long-term instability since $d_{P-1}t^{P-1}$ becomes the dominant term in (8.12) as t increases [50]. Of course, real oscillators are equally perturbed by random noises and the higher phase increments will not be zero but rather will provide a measure of the noise through relation (8.10).

For a better comparison among drifts and random instabilities, it is of interest to calculate parameters such as $\sigma_y(\tau)$, $\Sigma_y(\tau)$ or others for polynomial drifts; for that, expression (8.16) may be substituted into the quantities appearing in Table VII. The results are shown in Table VIII: for a given value $P-1$ of the degree of $y(t)$, only one parameter is of interest and varies as τ^{P-1}: e.g., $\sigma_y(\tau)$ varies as τ for linear frequency drift and $\Sigma_y(\tau)$ as τ^2 for quadratic frequency drift.

The above results have been used to make an algorithm to test for the highest order of frequency drift (and also for the presence of power-law noises) based upon observing the behavior of samples of structure functions [50].

D. Structure Function of Frequency

From the mathematical point of view, it is also possible to deal with the successive increments of $y(t)$ defined as [52]:

$$\Delta^{(M)}y(t;\tau) = \sum_{k=0}^{M} (-1)^k \binom{M}{k} y(t+(M-k)\tau). \tag{8.17}$$

However, this quantity involves the instantaneous values of $y(t)$ which are not observables and hence its practical utility is limited. Nevertheless, calculations show that $D_y^{(M)}(\tau)$ has general properties very similar to those of $D_\varphi^{(M)}(\tau)$ which is related to observables through equations (8.8) and (8.9).

1) Random Instabilities: In this case, $D_y^{(M)}(\tau)$ is related to $S_y(f)$ by

$$D_y^{(M)}(\tau) = 2^{2M} \int_0^\infty S_y(f) \sin^{2M} \pi\tau f \, df \qquad (8.18)$$

which is analog to (8.11) for phase noise. Again, higher values of M allow one to deal with more divergent power laws.

2) Polynomial Drifts: From the model given by equation (8.12), it can be demonstrated that:

a) the frequency increments of orders M smaller than $P - 1$ are time-dependent and hence not very useful;

b) the frequency increments of orders M greater than $P - 1$ are zero, and hence meaningless;

c) only the $(P-1)$th frequency increment is both nonzero and time-independent and reads as:

$$\Delta^{(P-1)}y(t;\tau) = (P-1)!\, d_{P-1} \tau^{P-1}. \qquad (8.19)$$

It is directly related to the highest order drift coefficient in $y(t)$, and is equal to $\sigma_{H_{BC}}$ appearing in Table VIII.

As particular cases, for $P = 2$ (linear frequency drift):

$$\Delta^{(1)}y(t;\tau) = \sqrt{2}\, \sigma_y(\tau) \qquad (8.20)$$

and for $P = 3$ (quadratic frequency drift):

$$\Delta^{(2)}y(t;\tau) = 3 \textstyle\sum_y (\tau). \qquad (8.21)$$

Thus structure functions of frequency applied to polynomial drifts, provide essentially the same information as structure functions of phase.

E. Long-Term Instability of Oscillators

For great values of τ (say, days, months, etc), significant frequency instabilities may arise from both very slow random fluctuations and very slow deterministic frequency drifts; they are modeled, respectively, by power-law spectral densities with negative slopes and by polynomials. Structure functions allow one to deal with both kinds of processes as shown above (see (8.10) and (8.16)): it is then of interest to compare the laws obtained for the relevant parameters.

A problem is that the Mth-order structure function of phase allows one to deal with several spectral densities $S_y(f) = h_\alpha f^\alpha$, namely down to $\alpha > -(2M - 1)$, whereas it yields a significant value only for the Mth-order polynomial phase drift $\varphi(t) = D_{M-1} t^M$. As a consequence, with a unique statistical parameter (i.e., derived from a fixed order structure function) it is not possible to establish a full correspondence between power laws and polynomial drifts in the sense that one sample of each would yield a significant law for the parameter of interest.

The results presented in Table IX show the laws of $\sigma_y(\tau)$, $\Sigma_y(\tau)$ and $\sigma_{H_{BC}}(N, \tau, \tau)$ for power laws with integral negative slopes and polynomial drifts: generally speaking, these parameters follow a positive-slope power law versus τ, with increasing slope when either the negative slope of $S_y(f)$ or the order of the polynomial drift increases.

More specifically, for random processes, the maximum positive slope versus τ of the standard-deviation derived from the Mth-order structure function of phase is $M - 3/2$ for integral values of α (e.g., $\sigma_y(\tau)$ derived from the second order phase structure function has a maximum positive slope of 1/2). For nonintegral values, the actual maximum is $< M - 1$. On the other hand, for a drift $y(t) = d_{P-1} t^{P-1}$, only one standard-deviation is of interest and follows the law τ^{P-1}.

TABLE IX
LONG-TERM INSTABILITY: SLOPES OF TIME-DOMAIN PARAMETERS FOR RANDOM NOISES AND POLYNOMIAL DRIFTS

$S_y(f)$	$\sigma_y(\tau)$	$\Sigma_y(\tau)$
$h_{-1} f^{-1}$	τ^0	τ^0
$h_{-2} f^{-2}$	$\tau^{1/2}$	$\tau^{1/2}$
$h_{-3} f^{-3}$	∞	τ
$h_{-4} f^{-4}$	∞	$\tau^{3/2}$
$h_{-5} f^{-5}$	∞	∞
max slope for integral values of α $M - \frac{3}{2}$	1/2	3/2

$y(t) = d_{P-1} t^{P-1}$	Parameters of interest Slope : P−1
$d_1 t$	$\sigma_y(\tau) \sim \tau$
$d_2 t^2$	$\Sigma_y(\tau) \sim \tau^2$
$d_3 t^3$	$\sigma_{H_{BC}} \sim \tau^3$
$d_4 t^4$	$\sigma_{H_{BC}} \sim \tau^4$

Thus it is impossible to have a specific power law (α fixed) and a specific drift order yielding the same slope versus τ for the same time-domain parameter (of course, different parameters may follow the same law, e.g., $\Sigma_y(\tau) \sim \tau$ for $S_y(f) = h_{-3} f^{-3}$ and $\sigma_y(\tau) \sim \tau$ for $y(t) = d_1 t$).

As a consequence, a careful study of the laws of $\sigma_y(\tau)$, $\Sigma_y(\tau)$ and $\sigma_{H_{BC}}(N, \tau, \tau)$ versus τ should allow one to determine unambiguously whether the frequency standard is disturbed by such or such random noise and/or by such or such polynomial drift; these conclusions are summarized in Table X. Note that these comments are true only for integral values of α in the power law spectra. If α can take on fractional powers, say $\alpha = -2.8$, then $\sigma_y(\tau) \sim \tau^{+0.9}$ and it is convergent. As α gets close to -3 but $\alpha > -3$, $\sigma_y(\tau) \to \tau^{+1}$ and remains convergent.

Of course, visual inspection of an historical graph of $\bar{y}_k$ versus time may be very helpful.

Also, for finite data sets, a spectra of $h_{-3} f^{-3}$ or $h_{-5} f^{-5}$ may well give false diagnosis even though $\sigma_y(\tau)$ and $\Sigma_y(\tau)$ are not convergent for these noises.

TABLE X
IDENTIFICATION OF RANDOM AND DETERMINISTIC POLYNOMIAL VARIATIONS FORM THE LAWS OF $\sigma_y(\tau)$ AND $\Sigma_y(\tau)$

The law of $\sigma_y(\tau)$	indicates the presence of
$\tau^{1/2}$	random walk frequency noise
τ	linear frequency drift

The law of $\Sigma_y(\tau)$	indicates the presence of
$\tau^{1/2}$	random walk frequency noise
τ	$S_y(f) = h_{-3} f^{-3}$
$\tau^{3/2}$	$S_y(f) = h_{-4} f^{-4}$
τ^2	quadratic frequency drift

F. Summary

The structure function approach introduced by Lindsey and Chie appeared very fruitful in terms of:

a) theoretical unification of the various parameters proposed as time domain measures;
b) new possibilities for dealing with power-law spectral densities of random noises and with polynomial drifts.

In this paper, the unification process has been developed by showing the relations between $D_\varphi^{(M)}(\tau)$ and previous parameters for $M \geqslant 3$ (see Table VII).

Also, the use of structure functions for identifying power law type random noises when polynomial drifts are present has been outlined (Tables IX and X), but great care must be exercized since false diagnosis may arise.[15]

IX. Frequency Instability Measurement

Even the most sophisticated statistical parameters are of no practical value for oscillator specification if they cannot be measured experimentally, i.e., estimated from finite samples of data. To this end, in parallel with the theoretical developments reviewed in this paper, much work has been devoted in the laboratories to the implementation of measurement test-sets capable of resolving the very small fluctuations of precision frequency sources. These systems rely heavily on the development of modern components and instrumentation such as low noise mixers/amplifiers and digital counters or spectrum analyzers.

It is not our intent to give here technical descriptions of these systems that have been described with great details in many references (see for example: [1], [2], [19], [21], [22], [38], [41], [44], [45], [54], [55], [57]). The goal is rather to emphasize some of their features. Readers are referred to original references for circuit descriptions.

A. The Need for a Reference Oscillator

To measure the frequency of one oscillator, at least two oscillators are indeed needed, e.g., one of them providing the time base of the digital counter (this is of course a particular case of the need for a reference in any physical measurement).

As far as the reference oscillator has a better stability than the oscillator under test, measured instabilities are due mainly to the latter. But when dealing with state-of-the-art sources, both oscillators are often of the same type and have nearly the same quality; the assumption is then made that they have uncorrelated random frequency fluctuations with identical statistical properties: the measured standard deviations must then be divided by $\sqrt{2}$ to get the contribution of one oscillator. A method has been proposed for estimating frequency instability of an individual oscillator, using data obtained by comparing it with at least two other oscillators [53], assuming their FM noises are uncorrelated; the method works even if the "reference" oscillators are less stable than the oscillator under test, although confidence in the results deteriorates rapidly if the test oscillator is much better than the references.

B. Frequency Versus Time-Domain Measurements

The dichotomy that appeared in the theory is of course found again in measurement apparatus: classically, digital counters are used for time-domain measurements and low-frequency spectrum analyzers for frequency-domain measurements.

However, theoretical developments leading to parameters such as the Hadamard variance or the high-pass variance have attenuated this dichotomy, since spectral densities may be measured by counting techniques [31], [39], [40] and variances may be measured by filtering techniques [43], [46].

Generally speaking, frequency-domain and time-domain measurement systems have several common parts even when

[15] During the review process of this paper, reference [70] has been published where emphasis is put on the identification of power-law type spectral densities. Note that the parameter $D_x^{(3)}(\tau)$ appearing in their Table I is related to $\Sigma_y^2(\tau)$ in our Table V by $\Sigma_y^2(\tau) = D_x^{(3)}(\tau)/9\tau^2$.

the output measurement apparatus (e.g., a counter or a spectrum analyzer) is not the same.

C. The Need for Demodulation

Since we want to measure the phase or frequency fluctuations of the oscillator quasi-sinusoidal output signal, techniques for demodulating these fluctuations are required before measurement and statistical analysis.

A PLL wherein a reference oscillator is locked to the oscillator under test, the two signals being then in phase quadrature at the phase detector inputs, is often used for that purpose: the PLL time constant is chosen as to provide a phase detector output analog voltage proportional either to the phase or to the frequency fluctuations over the Fourier frequency range of interest [22], [44], [45], [59]. According to the settings and the measurement apparatus used, the PLL system may be used either for frequency or for time domain measurements.

Sometimes, dispersive elements are used as frequency discriminators to convert frequency noise into an analog voltage [54].

D. The Need for High-Performance Systems

The state-of-the-art of time and frequency standards has advanced to such a level that a single measurement apparatus is unable to resolve directly their very low fluctuations by lack of resolution, sensitivity or dynamic range.

Techniques are therefore needed to enhance the fluctuations before measurement: for example, frequency multipliers followed by heterodyning, or frequency error multipliers have been used for that purpose. For the same reason, a very low noise amplifier is included between the phase detector in the PLL technique and the measurement apparatus.

The system noise must also be kept at a lower level than the oscillator noise level to be measured; in this respect, the advent of Schottky barrier diode mixers (used as phase detectors) has been a significant breakthrough in implementing low-noise measurement test-sets. It is recommended to measure system noise before any measurement on oscillators as shown in [55].

E. Automated Measurements

Following a general trend, automated measurement systems have been developed for both frequency and time domains [38], [54], [56], [58]. Automated systems are really useful in this field since great numbers of data have to be processed, various statistical treatments are made yielding different kinds of parameters and graphical plots of results are often needed. For these reasons, a calculator program may include frequency selection, bandwidth settings, measurement sequence, statistical treatments, results plotting and even system calibration.

F. A Hierarchy of Measurement Systems

While developing a high-resolution time difference measurement system, Allan and Daams [59] have set up the following hierarchy of measurement systems according to the quantity that they can measure:

a) time;
b) time fluctuations;
c) frequency;
d) frequency fluctuations.

For a system capable of measuring say time fluctuations, frequency and frequency fluctuations can be deduced from the results, but time cannot; and so on for the other systems in this hierarchy. The systems of status a) with state-of-the-art performance provide the greatest flexibility in data processing, i.e., systems where time differences can be measured with adequate precision. However, a system of status (d) is quite satisfactory when frequency instability is the primary concern: this is the case of the tight PLL system where the time constant is of the order of a few milliseconds: the output voltage is then proportional to the frequency (and not to the phase) fluctuations for averaging times longer than the loop time constant. This system is useful for time-domain measurements with τ of the order of one second and longer.

X. Conclusions

We have reviewed several proposed phase and frequency stability measures including both well-known widely used parameters and new concepts. For the latter, we have tried to point out clearly their specific advantages through the study of their transfer functions. In connection with structure functions, the effects of polynomial drifts of great importance for long term have been outlined and compared with those of LF divergent random noises.

In all the approaches, spectral densities play a key role in the sense that the other measures may be deduced from them whereas general inversion of the formulas is usually very difficult if not impossible.

In February 1978, Study Group 7 on "Standard frequencies and time signals" of the International Radio Consultative Committee (CCIR) has adopted a new document (to be included in [62]) recommending the use of spectral densities and, for time domain, the use of the two-sample standard deviation $\sigma_y(\tau)$. This new recommendation is certainly of great importance at the international level for nonspecialist engineers faced with the problem of stability characterization.

Nevertheless, it is felt that experimenters should be encouraged to try some of the new parameters since they may prove useful, especially for research work where better insight into the oscillator fundamental properties is looked for.

At this point, it must be emphasized that other analysis techniques not covered in this paper have been developed with possible applicability to frequency sources specification. Especially, the fitting of Auto-Regressive Integrated Moving Average (ARIMA) models to the data provides a powerful means of computer simulation and future values optimal prediction [63]. Barnes has discussed their applicability to time and frequency data, including the relation between the model parameters and $S_y(f)$ [12]. With more details, ARIMA models have been used by Percival [64] for predictions of fractional frequency of commercial cesium beam standards with lead times from 1 to 64 days, and by Hübner [65] for prediction in the realization of time scales.

Of special interest in connection with frequency stability characterization, Percival has proposed to use prediction errors as a measure of frequency instability, i.e., frequency instability is associated with a measure of frequency unpredictability which automatically incorporates the systematic terms.

Other new approaches have been recently proposed such as the finite-time variance defined as the mean of a general quadratic form of the measurements [66], and the suggested applicability of ambiguity functions for stability characterization when the oscillator is included in a global system whose performance are looked for [67].

At present, the existing models for random gaussian noises are well documented and provide a good background for actual measurements interpretation. However, future researches will certainly include the development of more sophisticated models that may possibly improve our understanding of some oscillator fundamental properties, e.g., of flicker noises and possible sporadic elements. These new models will probably call for new stability measures.

Thus although it appears to be mature, the field is still widely open for research. As in the past, no doubt that the development of complex systems, such as digital communications [68], with stringent requirements as regards frequency sources and clocks specification will provide a strong motivation.

Glossary of Symbols

Many symbols are used throughout this paper. The most important ones are defined below. For clarity, they are presented under several general headings.

Symbols	Definitions	See Equation
	Signal	
$V(t)$	Instantaneous output voltage of a (quasi) sinusoidal signal generator.	(2.1); (2.2); (2.4)
V_0	Nominal amplitude of $V(t)$.	(2.1); (2.2); (2.4)
ν_0	Nominal frequency of $V(t)$ (angular frequency $\omega_0 = 2\pi\nu_0$).	(2.1); (2.2); (2.4)
	Fluctuations, Modulations, and Drifts	
$\epsilon(t)$	Random instantaneous fluctuations of amplitude (amplitude noise).	(2.2)
$\varphi(t)$	Random instantaneous fluctuations of phase about $2\pi\nu_0 t$ (phase noise).	(2.3); (2.4)
$\Delta\nu(t)$	Random instantaneous fluctuations of frequency about ν_0 (frequency noise).	(2.6); (2.7)
$\nu(t)$	Instantaneous frequency of $V(t)$.	(2.6)
$x(t)$	Random instantaneous time error.	(2.10)
$y(t)$	Fractional fluctuations of frequency.	(2.9)
$A(t)$	Fractional fluctuations of amplitude.	(3.13)
f_m	Modulating frequency for a sinusoidal FM or φM.	(2.3); (5.14)
d_i	Polynomial drift coefficient in fractional frequency (ith order).	(5.17); (8.12)
D_i	Polynomial drift coefficient in phase ($i + 1$)th order.	(2.3); (8.13)
	Frequency Domain	
$S_g^{(TS)}(f)$	Two-sided spectral density on a per hertz basis of the real function $g(t)$.	(3.2)
$S_g(f)$	One-sided spectral density of $g(t)$. Applied in the paper to: $\varphi(t)$, $\Delta\nu(t)$, $x(t)$, $y(t)$, $A(t)$, $V(t)$.	(3.3)
f	Fourier frequency ($\omega = 2\pi f$).	
α	Exponent of f for a power-law spectral density (usually: $\alpha = 2, 1, 0, -1, -2$).	(3.9)
h_α	Positive real coefficient of f^α in a power-law spectral density.	(3.9)
f_h	HF cutoff of an ideal sharp cutoff low-pass filter.	(3.10)
	Measurement Sequence	
τ	Averaging time interval for frequency measurements.	(4.1)
t_k	The instant of the beginning of the kth measurement of frequency ($k = 1, 2, 3, \cdots, N$).	(4.1)
T	Time interval between the beginnings of two successive measurements of frequency ($T = t_{k+1} - t_k$).	Fig. 4
N	Number of data for calculating a sample variance, a modified sample variance or a Hadamard variance (positive integer).	Fig. 4
m	For $N = 2$, $m + 1$ data are used to estimate $\sigma_y^2(\tau)$.	(4.22)
$h(t)$	A real function of time that resembles the measurement sequence of a time-domain parameter. Applied to several variances in the text.	(5.3)
	Time Domain	
y_k	Instantaneous value of $y(t)$ at $t = t_k$.	(7.9)
$\bar{y}_k$	Fractional frequency fluctuations averaged from t_k to $t_k + \tau$.	(4.2)
n_k	A counting result: number of input signal cycles between t_k and $t_k + \tau$.	(4.1)
$I^2(\tau)$	The true variance of $\bar{y}_k$.	(4.5)
$\sigma_y^2(N, T, \tau)$	Sample variance of N samples $\bar{y}_k$ ($k = 1$ to N). Three definitions are given in the text. This symbol applies to the second one.	(4.13)

$\sigma_y^2(\tau)$	The above for $N = 2$ and $T = \tau$. A recommended measure of instability in the time domain.	(4.17); (4.18)
$\hat{\sigma}_y^2(\tau, m)$	An estimate of $\sigma_y^2(\tau)$ obtained from $m + 1$ measurements of $\bar{y}_k$.	(4.22)
E_α	Confidence interval for the above estimate and a power law spectral density $h_\alpha f^\alpha$.	(4.23)
$\langle \sigma_H^2(N, T, \tau) \rangle$	The Hadamard variance with N samples $\bar{y}_k (k = 1 \text{ to } N)$.	(6.1)
$\langle \sigma_{H\,\mathrm{BC}}^2(N, T, \tau) \rangle$	The Hadamard variance where the samples are weighted by binomial coefficients.	(6.4)
$\sigma_{HP}^2(\tau)$	The high-pass variance (measured by filtering).	(6.8); (6.10)
$\sigma_{BP}^2(\tau)$	The bandpass variance (measured by filtering).	(6.8); Table IV
$\Sigma_y^2(N, T, \tau)$	A modified sample variance of N samples $\bar{y}_k (k = 1 \text{ to } N)$.	(7.1)
$\Sigma_y^2(\tau)$	The above for $N = 3$ and $T = \tau$.	(7.4)
$\langle \delta y_T^2 \rangle$	The mean square of the fractional frequency increment $\delta y_T = y(t_{k+1}) - y(t_k)$ where $T = t_{k+1} - t_k$.	(7.9)
$\langle \delta \bar{y}_T^2 \rangle$	The mean square of the difference between two successive non adjacent $\bar{y}_k (T > \tau)$.	(7.15)
$R_g(\tau)$	Correlation function of the real function $g(t)$.	(3.1)
	Transfer Functions	
$H(f)$	The transfer function associated with a variance and appearing in its relationship with $S_y(f)$. It is the Fourier Transform of $h(t)$. Applied to several variances in the text.	(3.1); (6.7)
$H_\varphi(f)$	The transfer function appearing in the relationship between a variance and $S_\varphi(f)$.	(6.8); (6.9); (6.10)
	Structure Functions	
M	The order of an increment or of a structure function (positive integer).	
$\Delta^{(M)} g(t; \tau)$	The Mth increment of the real function $g(t)$.	(8.1)
$D_g^{(M)}(T; \tau)$	The structure function of the Mth increment.	(8.3)
$D_g^{(M)}(\tau)$	The Mth structure function of $g(t)$ defined as $D_g^{(M)}(T = 0; \tau)$.	(8.4)
P	The order of a polynomial drift.	(8.12)

Acknowledgment

The author would like to express his thanks to numerous people with whom he had very fruitful discussions during the last few years, in particular to D. W. Allan, Dr. J. A. Barnes, Prof. J. de Prins and Prof. B. Picinbono. Also, the constructive discussions with Dr. A. R. Chi, Dr. P. Kartaschoff, and Dr. G. M. R. Winkler during the last two meetings of Study Group 7 of the CCIR were greatly appreciated. Lastly, special thanks are due to Prof. J. Uebersfeld for his constant interest and encouragement.

References

[1] "IEEE-NASA Symposium on Short-Term Frequency Stability," Goddard Space Flight Center, Greenbelt, MD, Nov. 23-24, 1964 (Document NASA SP-80).

[2] Special Issue on Frequency Stability, *Proc. IEEE*, vol. 54, pp. 101-338, Feb. 1966.

[3] E. J. Baghdady, R. N. Lincoln, and B. D. Nelin, "Short term frequency stability: Characterization, theory and measurement," in [1], pp. 65-88.

[4] L. S. Cutler, "Some aspects of the theory and measurement of frequency fluctuations in frequency standards," in [1], pp. 89-100.

[5] L. S. Cutler and C. L. Searle, "Some aspects of the theory and measurement of frequency fluctuations in frequency standards," in [2], pp. 136-154.

[6] R. Vessot, L. Mueller, and J. Vanier, "The specification of oscillator characteristics from measurements made in the frequency domain," in [2], pp. 199-207.

[7] D. W. Allan, "Statistics of atomic frequency standards," in [2], pp. 221-230.

[8] J. A. Barnes, A. R. Chi, L. S. Cutler, D. J. Healey, D. B. Leeson, T. E. McGunigal, J. A. Mullen, W. L. Smith, R. L. Sydnor, R. Vessot and G. M. R. Winkler, "Characterization of frequency stability," *IEEE Trans. Instrum. Meas.*, vol. IM-20, pp. 105-120, May 1971.

[9] G. M. R. Winkler, "A brief review of frequency stability measures," in *Proc. 8th Annu. PTTI Applications and Planning Meeting*, (U.S. Naval Research Laboratory, Washington DC), Nov. 30-Dec. 2, 1976, pp. 489-527.

[10] J. A. Barnes, "A review of methods of analyzing frequency stability," in *Proc. 9th Annu. PTTI Applications and Planning Meeting*, Nov. 29-Dec. 1, 1977.

[11] D. Slepian, "On bandwidth," *Proc. IEEE*, vol. 64, pp. 292-300, Mar. 1976.

[12] J. A. Barnes, "Models for the interpretation of frequency stability measurements," NBS Tech. Note 683, Aug. 1976.

[13] E. Boileau and B. Picinbono, "Statistical study of phase fluctuations and oscillator stability," *IEEE Trans. Instrum. Meas.*, vol. IM-25, no. 1, pp. 66-75, March 1976.

[14] J. L. Doob, "The Brownian movement and stochastic equations," *Ann. Math.*, vol. 43, no. 2, Apr. 1942.

[15] A. M. Yaglom, "Correlation theory of processes with random stationary Nth increments," *Ann. Math. Soc. Trans.*, vol. 8, pp. 87-141, 1958.

[16] B. Picinbono, "Processus à accroissements stationnaires," *Ann. des Télécom.*, Tome 30, no. 7-8, pp. 211-212, July-Aug. 1975.

[17] J. Rutman and J. Uebersfeld, "A model for flicker frequency noise of oscillators," *Proc. IEEE*, vol. 60, pp. 233-235, Feb. 1972.

[18] "Annales des Télécommunications," Special issue on analysis of non-stationary signals, Tome 30, no. 7-8, pp. 203-308, July-Aug. 1975.

[19] F. Walls, "New horizons and old pitfalls in frequency and time metrology," in *Proc. 2nd Frequency Standards and Metrology Symp.*, Copper Mountain, CO, July 5-7, 1976, pp. 489-517.

[20] J. S. Bendat and A. G. Piersol, *Random Data: Analysis and Measurement Procedures*. New York: Wiley-Interscience, 1971.

[21] F. L. Walls and S. R. Stein, "Accurate measurements of spectral density of phase noise in devices," in *Proc. 31st Annu. Frequency Control Symp.*, (Atlantic-City, NJ), June 1-3, 1977, pp. 335-343.

[22] D. A. Howe, "Frequency domain stability measurements: A tutorial introduction," NBS Technical Note 679, Mar. 1976.

[23] D. W. Allan, "A review of flicker noise frequency instabilities in precision frequency standards," in *Proc. Symp. 1/f Fluctuations* (Sasakawa Hall, Tokyo, Japan), July 11-13, 1977.

[24] R. F. Lacey, A. L. Helgesson, and J. H. Holloway, "Short term stability of passive atomic frequency standards," in [2], pp. 170-176.
[25] D. Halford, A. E. Wainwright, and J. A. Barnes, "Flicker noise of phase in RF amplifiers and frequency multipliers: Characterization, cause and cure" (summary), in *Proc. 22nd Annu. Frequency Control Symp.*, (Atlantic-City, NJ), Apr. 1968, pp. 340-341.
[26] W. K. Attkinson, L. Fey, and J. Newman, "Spectrum analysis of extremely low-frequency variations of quartz oscillators," *Proc. IEEE*, vol. 51, p. 379, Feb. 1963.
[27] F. Clerc, "Pureté spectrale et stabilité des harmoniques de rang élevé d'un oscillateur métrologique," Thèse de Doctorat de 3e Cycle, Université Paris VI, Sept. 30, 1977.
[28] J. Rutman, "Relations between spectral purity and frequency stability," in *Proc. 28th Annu. Frequency Control Symp.*, (Atlantic-City, NJ), May 29-31, 1974, pp. 160-165.
[29] D. Middleton, *Introduction to Statistical Communication Theory.* New York: McGraw-Hill, 1960.
[30] J. A Barnes, "Atomic time-keeping and the statistics of precision signal generators," in [2], pp. 207-220.
[31] P. Lesage and C. Audoin, "Instabilité de fréquence des oscillateurs: mesure dans le domaine fréquence par une méthode temporelle," *L'Onde Electrique*, vol. 55, no. 2, pp. 82-89, Feb. 1975.
[32] J. Rutman, "Instabilité de fréquence des oscillateurs," *l'Onde Electrique*, vol. 52, no. 11, pp. 480-487, Dec. 1972.
[33] P. Lesage and C. Audoin, "Characterization of frequency stability: uncertainty due to the finite number of measurements," *IEEE Trans. Instrum. Meas.*, vol. IM-22, pp. 157-161, June 1973. (See also the corrections published in Mar. 1974 and Sept. 1976).
[34] K. Yoshimura, "Characterization of frequency stability: Uncertainty due to the autocorrelation function of the frequency fluctuations," *IEEE Trans. Instrum. Meas.*, vol. IM-27, no. 1, pp. 1-7, Mar. 1978.
[35] J. A. Barnes, "Tables of bias functions, B_1 and B_2, for variances based on finite samples of processes with power law spectral densities," NBS Tech. Note 375, Jan. 1969.
[36] A. R. Chi, "The mechanics of translation of frequency stability measures between frequency and time domain measurements," in *Proc. 9th Annu. PTTI Applications and Planning Meeting*, Nov. 29-Dec. 1, 1977.
[37] J. Rutman, "Comment on: Characterization of frequency stability," *IEEE Trans. Instrum. Meas.*, vol. IM-21, p. 85, Feb. 1972.
[38] M. C. Fisher, "Frequency stability measurement procedures," in *Proc. 8th Annu. PTTI Applications and Planning Meeting*, (U.S. Naval Research Laboratory, Washington DC), Nov. 30-Dec. 2, 1976, pp. 575-617.
[39] R. A. Baugh, "Frequency modulation analysis with the Hadamard variance," in *Proc. 25th Annu. Frequency Control Symp.*, (Atlantic-City, NJ), Apr. 1971, pp. 222-225.
[40] G. Sauvage and J. Rutman, "Analyse spectrale du bruit de fréquence des oscillateurs par la variance de Hadamard," *Ann. Télécom.*, tome 28, no. 7-8, pp. 301-314, July-Aug. 1973.
[41] J. Groslambert, "A new spectrum analyser using both analog and digital filtering via Hadamard variance," *IEEE Trans. Instrum. Meas.*, vol. IM-25, pp. 480-483, Dec. 1976.
[42] J. Rutman, "Characterization of frequency stability: a transfer function approach and its application to measurement via filtering of phase noise," *IEEE Trans. Instrum. Meas.*, vol. IM-23, pp. 40-48, Mar. 1974.
[43] J. Rutman and G. Sauvage, "Measurement of frequency stability in time and frequency domains via filtering of phase noise," *IEEE Trans. Instrum. Meas.*, vol. IM-23, pp. 515-518, Dec. 1974.
[44] D. W. Allan, "The measurement of frequency and frequency stability of precision oscillators," NBS Tech. Note 669, May 1975.
[45] D. G. Meyer, "A test-set for the accurate measurement of phase noise on high quality signal sources," *IEEE Trans. Instrum. Meas.*, vol. IM-19, pp. 215-227, Nov. 1970.
[46] R. G. Wiley, "A direct time domain measure of frequency stability: the modified Allan variance," *IEEE Trans. Instrum. Meas.*, vol. IM-26, pp. 38-41, Mar. 1977.
[47] J. Rutman, "Oscillator specifications: A review of classical and new ideas," in *Proc. 31st Annu. Frequency Control Symp.*, (Atlantic-City, NJ), June 1-2-3, 1977, pp. 291-301.
[48] J. De Prins and G. Cornelissen, "Power spectrum, frequency stability and flicker noise," in *Proc. Frequency Standards and Metrology Seminar* (Univ. Laval, Québec, Canada), Aug. 30-31-Sept. 1, 1971, pp. 368-387.
[49] G. Cornelissen, "Analyse de signaux et application aux problèmes de définition de la stabilité de fréquence," Thèse de Doctorat ès Sciences Physiques, Univ. Libre de Bruxelles, Belgique, Année académique 1975-1976.
[50] W. C. Lindsey and C. M. Chie, "Theory of oscillator instability based upon structure functions," *Proc. IEEE*, vol. 64, pp. 1652-1666, Dec. 1976.
[51] W. C. Lindsey, C. M. Chie, and W. E. Leavitt, "Interpretation and application of oscillator instability measures using structure functions," in *Proc. 8th Annu. PTTI Applications and Planning Meeting*, (U.S. Naval Res. Lab., Washington, DC), Nov. 30-Dec. 2, 1976, pp. 535-560.
[52] W. C. Lindsey and C. M. Chie, "Structure function of oscillator fractional phase and frequency fluctuations," *IEEE Trans. Instrum. Meas.*, vol. IM-26, pp. 420-421, Dec. 1977.
[53] J. E. Gray and D. W. Allan, "A method for estimating the frequency stability of an individual oscillator," in *Proc. 28th Annu. Frequency Control Symp.*, (Atlantic-City, NJ), May 29-31, 1974, pp. 243-246.
[54] A. L. Lance, W. D. Seal, F. G. Mendoza and N. W. Hudson, "Automating phase noise measurements in the frequency domain," in *Proc. 31st Annu. Frequency Control Symp.*, (Atlantic-City, NJ), June 1-3, 1977, pp. 347-358.
[55] F. L. Walls, S. R. Stein, J. E. Gray, and D. J. Glaze, "Design considerations in state-of-the-art signal processing and phase noise measurement system," in *Proc. 30th Annu. Frequency Control Symp.*, (Atlantic City, NJ), June 1976, pp. 269-274.
[56] J. Groslambert, M. Olivier, and J. Uebersfeld, "Automatic plotting systems for frequency stability curves and discrete spectral density of frequency standards," *IEEE Trans. Instrum. Meas.*, vol. IM-21, no. 4, pp. 405-409, Nov. 1972.
[57] J. Groslambert, G. Marianneau, and M. Olivier, "Méthodes expérimentales de caractérisation des oscillateurs," *L'Onde Electrique*, vol. 55, no. 2, pp. 74-81, Feb. 1975.
[58] D. Babitch and J. Oliverio, "Phase noise of various oscillators at very low Fourier frequencies," in *Proc. 28th Annu. Frequency Control Symp.*, (Atlantic City, NJ), May 29-31 1974, pp. 150-159.
[59] D. W. Allan and H. Daams, "Picosecond time difference measurement system," in *Proc. 29th Annu. Frequency Control Symp.*, (Atlantic City, NJ), May 28-30, 1975, pp. 404-411.
[60] G. Kramer, "Digitale Bestimmung von Schwankungs-spektren," *PTB, Jahresbericht* 1975, p. 157.
[61] ——, "Charakterisierung von Schwankungs-erscheinungen," *PTB-Bericht*, PTB-Me-13, Feb. 1977.
[62] C.C.I.R., XIVth Plenary Assembly, Kyoto 1978, Volume VII "Standard Frequencies and Time Signals," published by the International Telecommunication Union, Geneva, Switzerland.
[63] G. E. P. Box and G. M. Jenkins, *Time Series Analysis: Forecasting and Control.* San Francisco, CA: Holden-Day, 1976 (revised edition).
[64] D. B. Percival, "Prediction error analysis of atomic frequency standards," in *Proc. 31st Annu. Frequency Control Symp.*, (Atlantic City, NJ), USA, June 1-3, 1977, pp. 319-326.
[65] U. Hübner, "Models and predictions for the realization of time scales," in *Proc. 31st Annu. Frequency Control Symp.*, (Atlantic City, NJ), June 1-3, 1977, pp. 327-334.
[66] E. Boileau and Y. Lecourtier, "Description statistique de la stabilité des oscillateurs," in *Proc. Int. Symp. Measurements in Telecommunications* (organized by URSI, Lannion, France), Oct. 3-7, 1977, pp. 231-236.
[67] R. Moreau, "Comparaison des oscillateurs ultrastables. Intérêt relatif des mesures temporelles et fréquentielles," in *Proc. Int. Symp. Measurements in Telecommunications* (organized by URSI, Lannion, France), Oct. 3-7, 1977, pp. 219-225.
[68] P. Kartaschoff, "Frequency control and timing requirements for communications," in *Proc. 31st Annu. Freq. Control Symp.*, (Atlantic City, NJ), June 1-3, 1977, pp. 478-483.
[69] J. A. Barnes and D. W. Allan, "An approach to the prediction of coordinated universal time," *Frequency*, vol. 5, pp. 15-20, Nov.-Dec. 1967.
[70] W. C. Lindsey and C. M. Chie, "Identification of power-law type oscillator phase noise spectra from measurements," *IEEE Trans. Instrum. Meas.*, vol. IM-27, pp. 46-53, Mar. 1978.

Other References of General Interest

Books

P. Kartaschoff, *Frequency and Time.* London, England: Academic Press, 1978.
B. E. Blair, Ed., *Time and Frequency: Theory and Fundamentals* (NBS Monograph no. 140) Washington DC: U.S. Government Printing Office.
G. M. R. Winkler, *Timekeeping and Its Applications, Advances in Electronics and Electron Physics*, vol. 44. New York: Academic Press, 1977.
R. B. Blackman and J. W. Tukey, *The Measurement of Power Spectra.* New York: Dover Publ., 1958.
F. M. Gardner, *Phaselock Techniques.* New York: Wiley, 1966.
A. Papoulis, *Probability, Random Variables and Stochastic Processes.* New York: McGraw-Hill, 1965.

E. Oran Brighman, *The Fast Fourier Transform.* Englewood Cliffs, NJ: Prentice-Hall, 1974.

G. A. Campbell and R. M. Foster, *Fourier Integrals for Practical Applications.* Princeton NJ: Van Nostrand, 1948.

W. B. Davenport and W. L. Root, *Random Signals and Noise.* New York: McGraw-Hill, 1958.

Reviews

Proc. IEEE, Special Issue on Time and Frequency: Generation, Dissemination, Applications, vol. 60, no. 5, pp. 473-648, May 1972.

IEEE Trans. Instrum. Meas., Special Issue on Conference on Precision Electromagnetic Measurements, Every December of even-numbered years.

L'Onde Electrique, Special Issue on ultra stable oscillators and their applications, vol. 52, no. 11, pp. 479-536, Dec. 1972 and vol. 53, no. 2, pp. 39-78, Feb. 1973.

L'Onde Electrique, Special Issue on Frequency Stability and Spectral Purity Measurement Techniques, vol. 55, no. 2, pp. 65-126, Feb. 1975.

Symposium

Proceedings of the Annual Frequency Control Symposium:

Year	No	Available at	Document No
1956	10	NTIS[a]	AD 298 322
7	11	"	AD 298 323
8	12	"	AD 298 324
9	13	"	AD 298 325
1960	14	"	AD 246 500
1	15	"	AD 265 455
2	16	"	PB 162 343
3	17	"	AD 423 381
4	18	"	AD 450 341
5	19	"	AD 471 229
6	20	"	AD 800 523
7	21	"	AD 659 792
8	22	"	AD 844 911
9	23	"	AD 746 209
1970	24	EIA[b]	
1	25	NTIS	AD 746 211
2	26	"	AD 771 043
3	27	"	AD 771 042
4	28	EIA	
5	29	"	
6	30	"	
7	31	"	

[a]NTIS, National Technical Information Service
Sills Building
5285 Port Royal Road
Springfield, VA 22161.

Proceedings of the Annual Precise Time and Time Interval Applications and Planning Meeting.[c]

[b]EIA, Publications Committee
Annual Frequency Control Symposium
Electronic Industries Association
2001 Eye Street, N.W.
Washington DC 20006.

[c]Technical Information and Administrative Support Division, Code 250, Goddard Space Flight Center, Greenbelt, MD 20771.

PROPERTIES OF SIGNAL SOURCES AND MEASUREMENT METHODS

D. A. Howe, D. W. Allan, and J. A. Barnes

Time and Frequency Division
National Bureau of Standards
Boulder, Colorado 80303

X. PITFALLS IN DIGITIZING THE DATA

The advent and prolific use of digital computers has changed the manner in which processing of analog signals takes place if a computer is used. This section addresses the most common problems in such analyses.

10.1 Discrete-Continuous Processes

Digital processing implies that data must be presented to a computer or other processor as an array of numbers whether in a batch or in a time series. If the data are not already in this form (it usually is not when considering frequency stability measurements), then it is necessary to transform to this format by digitizing. Usually, the signal available for analysis is a voltage which varies with frequency or phase difference between two oscillators.

10.2 Digitizing the Data

Digitizing the data is the process of converting a continuous waveform into discrete numbers. The process is completed in real time using an analog-to-digital converter (ADC). Three considerations in the ADC are of importance here:

1. Conversion time
2. Resolution (quantization uncertainty)
3. Linearity

An ADC "looks at" an incoming waveform at equispaced intervals of time T. Ideally, the output of the ADC is the waveform (denoted by y(t)) multiplied by a series of infinitely narrow sampling intervals of unit height as in figure 10.1. We have at t = T

$$y_1(t) = y(t)\delta(t-T) = y(T)\delta(t-T) \qquad (10.1)$$

where $\delta(t-T)$ is a delta function. If y(t) is continuous at t = nT and n = 0, ±1, ±2,..., then

$$y_\ell(t) = \sum_{n=-\infty}^{\infty} y(nT)\,\delta\,(t-nT) \qquad (10.2)$$

ℓ = integer

The delta function respresentation of a sampled waveform eq (10.2) is useful when a subsequent continuous integration is performed using it.[6]

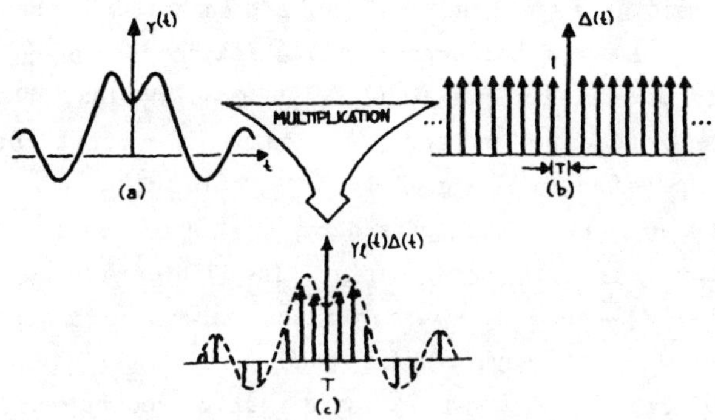

FIGURE 10.1

In ADC's, the input signal is sampled during an aperture time and held for conversion to a digital number, usually binary. Sampling and processing takes time which is specified as the conversion time. This is the total time required for a complete measurement at one sample to achieve a given level of accuracy. If $y_\ell(t)$ is the ideal discrete-time representation of continuous process y(t), then the ADC output denoted by $y_\ell'(t)$ is:

$$y_\ell' = y_{\ell-d} \pm \varepsilon\left(\frac{dy}{dt}\right) \qquad (10.3)$$

where "d" is the conversion time and $\varepsilon \cdot \frac{dy}{dt}$ the accuracy tolerance at "d" as a function of rate-of-change in y(t). In general a trade-off exists between d and ε. For example, for a commonly available, high-quality 10-bit ADC, a conversion time of d = 10 μs yields a maximum error of 3%. Whereas given a 30 μs conversion time, we can obtain 0.1% maximum error.

The error due to conversion time "d" is many times negligible since processing in digital filters and spectrum analysis takes place after the converter. Conversion time delay can be of critical concern, however, where real-time processing at speeds of the order of "d" become important

Reprinted with permission from *Proc. 35th Annu. Symp. Frequency Contr.*, Philadelphia, PA, May 27–29, 1981, pp. A26–A35.

such as in digital servo loops where corrections are needed for fast changing errors.

A portion of the conversion-time error is a function of the rate of change $\frac{dy}{dt}$ of the process if the sample-and-hold portion of the ADC relies on the charging of a capacitor during an aperture time. This is true because the charge cycle will have a finite time-constant and because of aperture time uncertainty. For example, if the time-constant is 0.1 ns (given by say a 0.1 Ω source resistance charging a 0.001 μfd capacitor), then a 0.1% nominal error will exist for slope $\frac{\Delta y}{\Delta t}$ = 1V/μs due to charging. With good design, this error can be reduced. The sampling circuit (before charge) is usually the dominant source of error and logic gate-delay jitter creates an aperture time uncertainty. The jitter typically is between 2-5 ns which means an applied signal slewing at, say, 1 V/μs produces an uncertainty of 2-5 mV. Since $\varepsilon \cdot \frac{dy}{dt}$ is directly proportional to signal slewing rate, it can be anticipated that high-level, high-frequency components of y(t) will have the greatest error in conversion. For typical ADC's, less than 0.1% error can be achieved by holding $\frac{\Delta y}{\Delta t}$ to less than 0.2 V/μs.

The continuous process y(t) is partitioned into 2^n discrete ranges for n-bit conversion. All analog values within a given range are represented by the same digital code, usually assigned to the nominal midrange value. There is, therefore, an inherent quantization uncertainty of ± ½ least-significant bit (LSB), in addition to other conversion errors. For example, a 10-bit ADC has a total of 1024 discrete ranges with a lowest order bit then representing about 0.1% of full scale and quantization uncertainty of ± 0.05%.

We define the dynamic range of a digital system as the ratio between the maximum allowable value of the process (prior to any overflow condition) and the minimum discernable value. The dynamic range when digitizing the data is set by the quantizing uncertainty, or resolution, and is the ratio of 2^n to ½ LSB. (If additive noise makes coding ambiguous to the ½ LSB level, then the dynamic range is the ratio of 2^n to the noise uncertainty, but this is usually not the case.) For example, the dynamic range of a 10-bit system is 2^n = 1024 to ½, or 2048 to 1. Expressed in dB's, this is

$$20 \log 2048 = 66.2 \text{ dB}$$

if referring to a voltage-to-code converter.

The converter linearity specifies the degree to which the voltage-to-code transfer approximates a straight line. The nonlinearity is the deviation from a straight line drawn between the end points (all zeros to all ones code). It is usually not acceptable to have nonlinearity greater than ½ LSB which means that the sum of the positive errors or the sum of the negative errors of the individual bits must not exceed ½ LSB (or ± ½ LSB). The linearity specification used in this context includes all effects such as temperature errors under expected operating temperature extremes and power supply sensitivity errors under expected operating supply variations.

10.3 Aliasing

Figure 10.1 illustrates equispaced sampling of continuous process y(t). It is important to have a sufficient number of samples/second to properly describe information in the high frequencies. On the other hand, sampling at too high a rate may unnecessarily increase the processing labor. As we reduce the rate, we see that sample values could represent low or high frequencies in y(t). This property is called aliasing and constitutes a source of error similiar to "imaging" which occurs in analog frequency mixing schemes (i.e., in the multiplication of two different signals).

If the time between samples (k) is T seconds, then the sampling rate is $\frac{1}{T}$ samples per second. Then useful data in y(t) will be from 0 to $\frac{1}{2T}$ Hz and frequencies higher than $\frac{1}{2T}$ Hz will be folded into the lower range from 0 to $\frac{1}{2T}$ Hz and confused with data in this lower range. The cutoff frequency is then given by

$$f_s = \frac{1}{2T} \tag{10.4}$$

and is sometimes called the "Nyquist frequency."

We can use the convolution theorem to simply illustrate the existence of aliases. This theorem states that multiplication in the time domain corresponds to convolution in the frequency domain, and the time domain and frequency domain representations are Fourier transform pairs.[7] The Fourier transform of y(t) in figure 10.1(a) is denoted by Y(f); thus:

$$Y(f) = \int_{-\infty}^{\infty} y(t)e^{-j2\pi ft}dt \qquad (10.5)$$

and

$$y(t) = \frac{1}{2\pi}\int_{-\infty}^{\infty} Y(f)e^{j2\pi ft}df \qquad (10.6)$$

The function Y(f) is depicted in figure 10.2(a). The Fourier transform of $\Delta(t)$ is shown in figure 10.2(b) and is given by $\Delta(f)$ where applying the discrete transform yields:

$$\Delta(f) = \frac{1}{T}\sum_{n=-\infty}^{\infty} \delta(f - \frac{n}{T}), \qquad (10.7)$$

recalling that

$$\Delta(t) = \sum_{n=-\infty}^{\infty} \delta(t - nT), \qquad (10.8)$$

from eq (10.2).

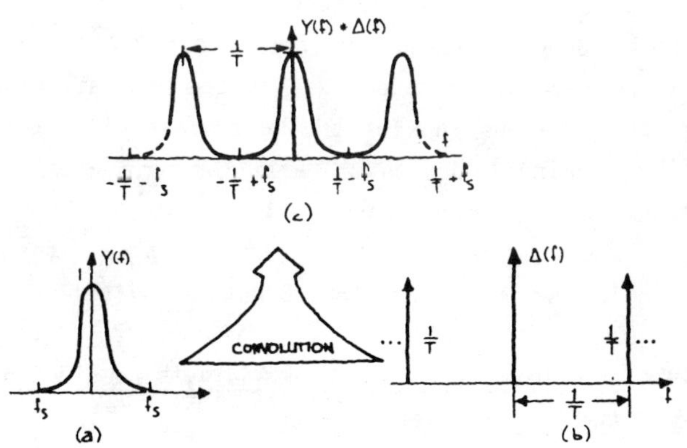

FIGURE 10.2

Y(f) convolved with Δf is denoted by $Y(f)* \Delta(f)$ and is shown in figure 10.2(c). We see that the transform Y(f) is repeated with origins at $f = \frac{n}{T}$. Conversely, high frequency data with information around $f = \frac{n}{T}$ will fold into the data around the origin between $-f_s$ and $+f_s$. In the computation of power spectra, we encounter errors as shown in figure (10.3).

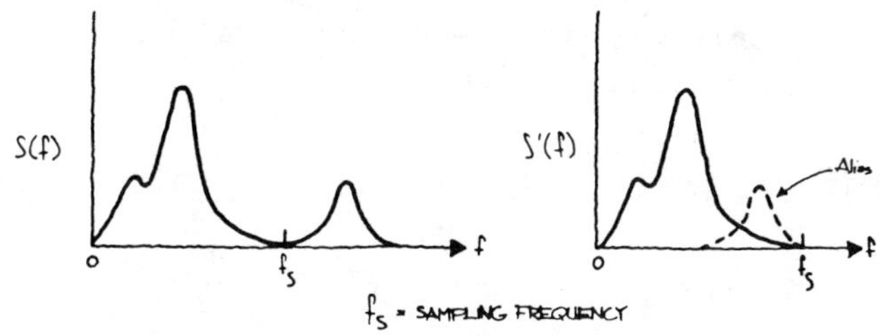

FIGURE 10.3
Aliased power spectra due to folding. (a) True Spectra, (b) Aliased Spectra.

Two pioneers in information theory, Harold Nyquist and Claude Shannon, developed design criteria for discrete-continuous processing systems. Given a specified accuracy, we can convey time-domain process y(t) through a finite bandwidth whose upper limit f_N is the highest significant spectral component of y(t). For discrete-continuous process $y_k(t)$, ideally the input signal spectrum should not extend beyond f_s, or

$$f_N \leq f_s \qquad (10.9)$$

where f_s is given by eq (10.4). Equation (10.9) is refered to as the "Shannon limit."

In practice, there is never a case in which there is absolutely no signal or noise component above f_N. Filters are used before the ADC in order to suppress components above f_N which fold into the lower bandwidth of interest. This so-called anti-aliasing filter usually must be quite sophisticated in order to have low ripple in the passband, constant phase delay in the passband, and steep rolloff characteristics. In examining the rolloff requirements of the anti-aliasing filter, we can apply a fundamental filter property that the output spectrum is equal to the input spectrum multiplied by the square of the frequency response function; that is,

$$S(f)\,[H(f)]^2 = S(f)_{out} \qquad (10.10)$$

The filter response must be flat to f_N and attenuate aliased noise components at $\frac{n}{T} \pm f = 2nf_s \pm f$. In digitizing the data, the observed spectra will be the sum of the baseband spectrum (to f_N) and all spectra which are folded into the baseband spectrum observed

$$S(f) = S_o(f) + S_{-1}\,(2f_s - f) + S_{+1}\,(2f_s + f)$$

$$+ S_{-2}\,(4f_s - f)\ldots. + S_i\,(2\,i\,(f_s + \frac{1}{2i}f))$$

$$(10.11)$$

$$= S_o(f) + \sum_{i=-M}^{M} S_i\,(2\,i\,(f_s + \frac{f}{2i}))$$

where M is an appropriate finite limit.

For a given rejection at an upper frequency, clearly the cutoff frequency f_c for the anti-aliasing filter should be as low as possible to relax the rolloff requirements. Recall that an nth order low-pass filter has frequency response function

$$H(f) = \frac{1}{1 + j\left(\frac{f}{f_c}\right)^n} \qquad (10.12)$$

and output spectrum

$$S(f)_{out} = \frac{S(f)}{1 + \left(\frac{f}{f_c}\right)^{2n}} \qquad (10.13)$$

and after sampling, we have (applying eq (10.11))

$$S(f)_{observed} = \frac{S_o(f)}{1 + \left(\frac{f}{f_c}\right)^{2n}} + \sum_{i=-M}^{M} \frac{S_i(2\,i\,(f_s + \frac{1}{2i}f))}{1 + \left[\frac{2\,i\,(f_s + \frac{1}{2i}f)}{f_c}\right]^{2n}}$$

$$(10.14)$$

If f_c is chosen to be higher than f_N, then the first term (baseband spectrum) is negligibly affected by the filter, which is our hope. It is the second term (the sum of the folded in spectra) which causes an error.

As an example of the rolloff requirement, consider the measurement of noise process n(t) at f = 400 Hz in a 1 Hz bandwidth on a digital spectrum analyzer. Suppose n(t) is white; that is,

$$S_n(f) = k_o \qquad (10.15)$$

$$k_o = \text{constant}$$

Suppose further that we wish to only measure the noise from 10 Hz to 1 kHz; thus f_N = 1 kHz. Let us assume a sampling frequency of $f_s = 2f_N$ or 2 kHz. If we impose a 1 dB error limit in $S_{observed}$ and have 60 dB of dynamic range, then we can tolerate an error limit of 10^{-6} due to aliasing effects in this measurement, and the second term in eq (10.14) must be reduced to this level. We can choose f_c = 1.5kHz and obtain

$$S(f)_{observed} = k_o + \sum_{i=-M}^{M} \frac{k_o}{1+\left[\frac{2\,i\,(f_s + \frac{1}{2i}f)}{f_c}\right]^{2n}} \qquad (10.16)$$

The term in the series which contributes most is at i = -1, the nearest fold-in. The denominator must be 10^6 or more to realize the allowable error limit and at $n \geq 8$ this condition is met. The next most contributing term is i = +1 at which the error is $< 10^{-7}$ for n = 8, a negligible contribution. The error increases as f increases for a fixed n because the nearest fold-in (i = -1) is coming down in frequency (note fig. 10.2(c)) and power there is filtered less by the anti-aliasing filter. Let us look at the worst case (f = 1kHz) to determine a design criteria for this example. At f = 1 kHz, we must have $n \geq 10$.

Thus the requirement in this example is for a 10-pole low-pass filter (60 dB/octave rolloff).

10.4 Some History of Spectrum Analysis Leading to the Fast Fourier Transform

Newton in his Principia (1687) documented the first mathematical treatment of wave motion although the concept of harmonics in nature was pointed out by Pythagoras, Kepler, and Galileo. However, it was the work of Joseph Fourier in 1807 which showed that almost any function of a real variable could be represented as the sum of sines and cosines. The theory was rigorously treated in a document in 1822.

In using Fourier's technique, the periodic nature of a process or signal is analyzed. Fourier analysis assumes we can apply fixed amplitudes, frequencies, and phases to the signal.

In the early 1900's two relatively independent developments took place: (1) radio electronics and electric power hardware were fast growing technologies; and (2) statistical analysis of events or processes which were not periodic became increasingly understood. The radio engineer explored signal and noise properties of a voltage or current into a load by means of the spectrum analyzer and measurement of the power spectrum. On the other hand, statisticians explored deterministic and stochastic properties of a process by means of the variance and self-correlation properties of the process at different times. Wiener (1930) showed that the variance spectrum (i.e., the breakdown of the variance with Fourier frequency) was the Fourier transform of the autocorrelation function of the process. He also theorized that the variance spectrum was the same as the power spectrum normalized to unit area. Tukey (1949) advocated the use of the variance spectrum in the statistical treatment of all processes because (1) it is more easily interpreted than correlation-type functions; and (2) it fortuitously is readily measureable by the radio engineer.

The 1950's saw rigorous application of statistics to communication theory. Parallel to this was the rapid advancement of digital computer hardware. Blackman and Tukey (1959) and Welch (1961) elaborated on other useful methods of deriving an estimate for the variance spectrum by taking the ensemble time-average sampled, discrete line spectra. The approach assumes the random process is ergodic. Some digital approaches estimate the variance spectrum using Wiener's theorem if correlation-type functions are useful in the analysis, but in general the time-averaged, sample spectrum is the approach taken since its implementation is direct and straightforward. Most always, ergodicity can be assumed.[8,9]

The variance of process y(t) is related to the total power spectrum by

$$\sigma^2[y(t)] = \int_{-\infty}^{\infty} S_y(f)\, df. \tag{10.17}$$

Since

$$\sigma^2[y(t)] = \lim_{T\to\infty} \frac{1}{2T} \int_{-T}^{T} y^2(t)\, dt \tag{10.18}$$

we see that if y(t) is a voltage or current into a 1-ohm load, then the mean power of y(t) is the integral of $S_y(f)$ with respect to frequency over the entire range of frequencies $(-\infty,\infty)$. $S_y(f)$ is, therefore, the power spectrum of process y(t). The power spectrum curve shows how the variance is distributed with frequency and should be expressed in units of watts per unit of frequency, or volts squared per unit of frequency when the load is not considered.

Direct estimation of power spectra has been carried out for many years through the use of analog instruments. These have variously been referred to as sweep spectrum analyzers, harmonic analyzers, filter banks, and wave analyzers. These devices make use of the fact that the spectrum of the output of a linear system (analog filter) is the spectrum of the input multiplied by the square of the system's frequency response function (real part of the transfer characteristic). Note eq (10.10). If y(t) has spectrum $S_y(f)$ feeding a filter with frequency response function H(f), then its output is

$$S(f)_{\text{filtered}} = [H(f)]^2\, S_y(f) \tag{10.19}$$

If H(f) is rectangular in shape with width Δf, then we can measure the contribution to the total power spectrum due to $S_y(f \pm \frac{\Delta f}{2})$.

The development of the fast Fourier transform (FFT) in 1965 made digital methods of spectrum estimation increasingly attractive. Today the choice between digital or analog methods depends more on the objectives of the analysis rather than on technical limitations. However, many aspects of digital spectrum analysis are not well known by the casual user in the laboratory while the analog analysis methods and their limitations are understood to a greater extent.

Digital spectrum analysis is realized using the discrete Fourier transform (DFT), a modified version of the continuous transform depicted in eqs (10.5) and (10.6). By sampling the input waveform y(t) at discrete intervals of time $t_\ell = \Delta t$ representing the sampled waveform by eq (10.2) and integrating eq (10.5) yields

$$Y(f) = \sum_{\ell=-\infty}^{\infty} y(\ell t)e^{-j2\pi f\ell T} \qquad (10.20)$$

Equation (10.20) is a Fourier series expansion. Because f(t) is specified as being bandlimited, the Fourier transform as calculated by eq (10.20) is as accurate as eq (10.5); however, it cannot extend beyond the Nyquist frequency, eq (10.4).

In practice we cannot compute the Fourier transform to an infinite extent, and we are restricted to some observation time T consisting of $n\Delta t$ intervals. This produces a spectrum which is not continuous in f but rather is computed with resolution Δf where

$$\Delta f = \frac{1}{n\Delta t} = \frac{1}{T} \qquad (10.21)$$

With this change, we get the discrete finite transform

$$Y(m\Delta f) = \sum_{n=0}^{N-1} y_\ell(t)e^{-j2\pi m\Delta fnt} \qquad (10.22)$$

The DFT computes a sampled Fourier series, and eq (10.22) assumes that the function y(t) repeats itself with period T. $Y(m\Delta f)$ is called the "line spectrum." A comparison of the DFT with the continuous Fourier transform is shown later in part 10.7.

The fast Fourier transform (FFT) is an algorithm which efficiently computes the line spectrum by reducing the number of adds and multiplies involved in eq (10.22). If we choose $T/\Delta t$ to equal a rational power of 2, then a symmetric matrix can be derived through which $y_\ell(t)$ passes and quickly yields $Y(m\Delta f)$. An N-point transformation by the direct method requires a processing time proportional to N^2 whereas the FFT requires a time proportional to $N \log_2 N$. The approximate ratio of FFT to direct computing time is given by

$$\frac{N \log_2 N}{N^2} = \frac{\log_2 N}{N} = \frac{\gamma}{N} \qquad (10.23)$$

where $N = 2^\gamma$. For example, if $N = 2^{10}$, the FFT requires less than 1/100 of the normal processing time.

We must calculate both the magnitude and phase of a frequency in the line spectrum, i.e., the real and imaginary part at the given frequency. N points in the time domain allow N/2 complex quantities in the frequency domain.

The power spectrum of y(t) is computed by squaring the real and imaginary components, adding the two together and dividing by the total time T. We have

$$S_y(m\Delta f) = \frac{R[Y(m\Delta f)]^2 + I[Y(m\Delta f)]^2}{T} \qquad (10.24)$$

This quantity is the sampled power spectrum and again assumes periodicity in process y(t) with total period T.[10]

10.5 Leakage

Sampled digital spectrum analysis always involves transforming a finite block of data. Continuous process y(t) is "looked at" for T time through a data window which can functionally be described by

$$y'(t) = w(t)\cdot y(t) \qquad (10.25)$$

where w(t) is the time domain window. The time-discrete counterpart to eq (10.25) is

$$y'_\ell(t) = w_\ell(t)\cdot y_\ell(t) \qquad (10.26)$$

and $w_\ell(t)$ is now the sampled version of w(t) derived similarly to eq (10.2). Equation (10.26) is equivalent to convolution in the frequency domain, or

$$Y'(m\Delta f) = W(m\Delta f)*Y(m\Delta f) \qquad (10.27)$$

$Y'(m\Delta f)$ is called the "modified" line spectrum due to convolution of the original line spectrum with the Fourier transform of the time-domain window function.

Suppose the window function is rectangular, and

$$\begin{aligned} w_\ell(t) &= 1, \qquad \frac{-T}{2} \le t \le \frac{T}{2} \\ &= 0, \qquad t > \left|\frac{T}{2}\right| \end{aligned} \qquad (10.28)$$

This window is shown in figure 10.4(a). The Fourier transform of this window is

$$W(m\Delta f) = T\,\frac{\sin \pi\, m\Delta f\, NT}{\pi m\Delta f NT} \qquad (10.29)$$

and is shown in figure 10.4(b). If y(t) is a sine wave, we convolve the spectrum of the sinusoid, a delta function, with $W(m\Delta f)$.

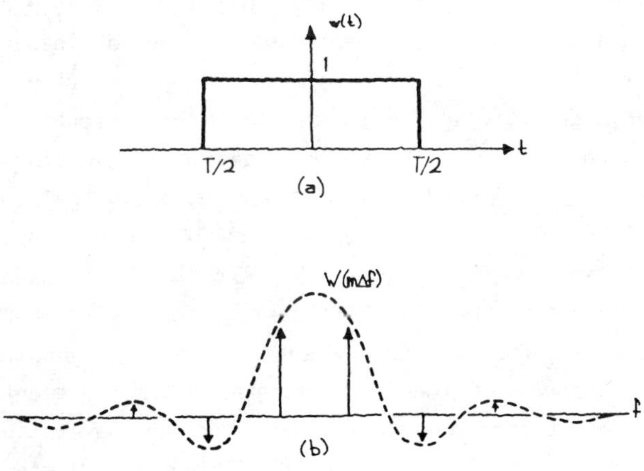

FIGURE 10.4

The transform process (eq 10.22) treats the sample signal as if it were periodically extended. Discontinuities usually occur at the ends of the window function in the extended version of the sampled waveform as in figure 10.5(c). Sample spectra thus represent a periodically extended sampled waveform, complete with discontinuites at its ends, rather than the original waveform.

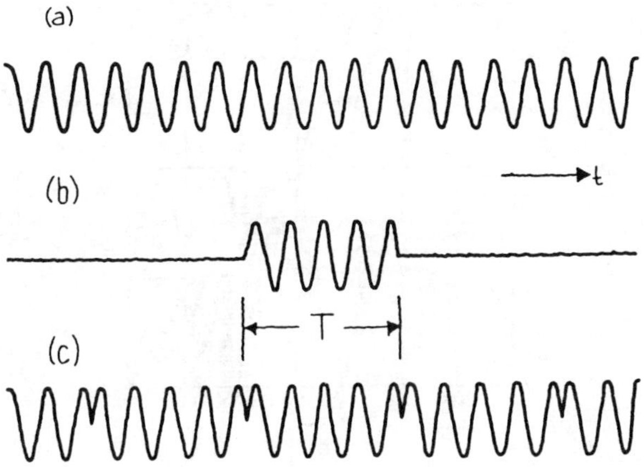

FIGURE 10.5

Spurious components appear near the sinusoid spectrum and this is referred to as "leakage." Leakage results from discontinuites in the periodically extended sample waveform.

Leakage cannot be eliminated entirely, but one can choose an appropriate window function w(t) in order to minimize its effect. This is usually done at the expense of resolution in the frequency domain. An optimum window for most cases is the Hanning window given by:

$$w(t) = [\tfrac{1}{2} - \tfrac{1}{2}\cos(\tfrac{2\pi t}{T})]^a \qquad (10.30)$$

for $0 \le t \le T$ and "a" designates the number of times the window is implemented. Figure 10.6(a) shows the window function and 10.6(b) shows the Hanning line shape in the frequency domain for various numbers of "Hanns." Note that this window

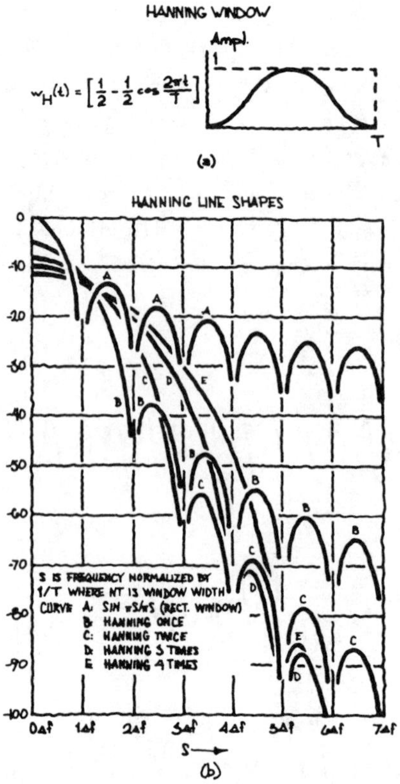

FIGURE 10.6

eliminates discontinuities due to the ends of sample length T.

Each time the Hanning window is applied, the sidelobes in the transform are attenuated by 12 dB/octave, and the main lobe is widened by 2Δf. The amplitude uncertainty of an arbitrary sine wave input is reduced as we increase the number of Hanns; however, we trade off resolution in frequency.

The effective noise bandwidth indicates the departure of the filter response away from a true rectangularly shaped filtered response (frequency domain). Table 10-I lists equivalent noise bandwidth corrections for up to three applications of the Hanning window.[11]

Number of Hanns	Equivalent Noise Bandwidth
1	1.5 Δf
2	1.92 Δf
3	2.31 Δf

TABLE 10.I

10.6 Picket-Fence Effect

The effect of leakage discussed in the previous section gives rise to a sidelobe type response that can be tailored according to the time-window function through which the analyzed signal passes as a block to be transformed to the frequency domain. Using the Hanning window diminishes the amplitudes of the sidelobes, however, it increases the effective bandwidth of the passband around the center frequency. This is because the effective time-domain window length is shorter than a perfect rectangular window. Directly related to the leakage (or sidelobe) effect is one called the "picket-fence" effect. This is because the sidelobes themselves resemble a frequency response which has geometry much like a picket fence.

The existence of both sidelobe leakage and the resultant picket-fence effect are an artifact of the way in which the FFT analysis is performed. Frequency-domain analysis using analog filters involves a continuous signal in and a continuous signal out. On the other hand, FFT analysis involves a continuous signal in, but the transform to the frequency domain is performed on blocks of data. In order to get discrete frequency information from a block, the assumption is made that the block represents one period of a periodic signal. The picket-fence effect is a direct consequence of this assumption. For example, consider a sinewave signal which is transformed from a time-varying voltage to a frequency-domain representation through an FFT. The block of data to be transformed will be length, T, in time. Let's say that the block, T, represents only 4½ cycles of the input sinewave as in figure 10.5. Artificial sidebands will be created in the transform to the frequency domain, whose frequency spacing equals $\frac{1}{T}$, or the reciprocal of the block length. This represents a worst-case condition for sidelobe generation and creates a large number of spurious discrete frequency components as shown in figure 10.7(b).

If, on the other hand, one changes the block time, T, so that the representation is an integral number of cycles of the input sinewave, then the transform will not contain sidelobe leakage compo-

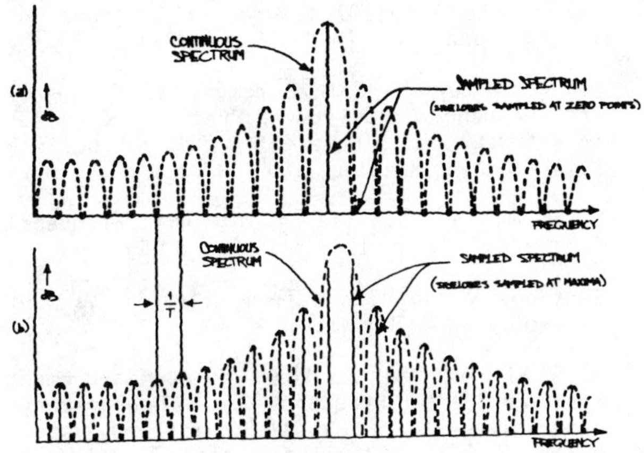

FIGURE 10.7

nents and the artificial sideband frequency components disappear. In practice, when looking at complex input signals, the block time, T, is not synchronous with any component of the transformed part of the signal. As a result, discrete frequency components in the frequency domain have associated with them sidebands which come and go depending on the phase of the time window, T, relative to the sine components of the incoming signal. The effect is much like looking through a picket fence at the sidebands.[12]

An analogy to the sidelobe leakage and picket-fence effect is to record the incoming time-varying signal on a tape loop, which has a length of time, T. The loop of tape then repeats itself with a period of T. This repeating signal is then coupled to a scanning or filter-type spectrum analyzer. The phase discontinuity between the end of one passage of the loop and the beginning of the loop on itself again represents a phase-modulation component, which gives rise to artificial sidebands in the spectrum analysis. A word of caution — this is not what actually happens in a FFT analyzer (i.e., there is no recirculating memory). However, the Fourier transform treats the incoming block as if this were happening.

10.7 Time Domain-Frequency Domain Transforms

A. Integral transform

Figure 10.8 shows the well-known integral transform, which transforms a continuous time-domain signal extending over all time into a

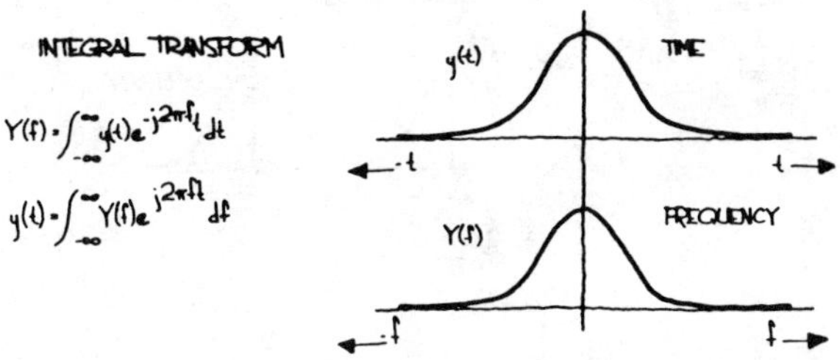

FIGURE 10.8

continuous frequency-domain signal extending over all frequency. This is the ideal transform. In practice, however, one deals with finite times and bandwidths. The integral transform then, at best, is an estimate of the transform and is so for only short, well-behaved signals. That is, the signal goes to zero at infinite time and at infinitely high frequency.

B. Fourier series

The Fourier-series transform assumes periodicity in the time-domain signal for all time. Only one period of the signal (for time T) is required for this kind of transform. The Fourier series treats the incoming signal as periodic with period, T, and continuous. The transformed spectrum is then discrete with infinite harmonic components with frequency spacing of $\frac{1}{T}$. This is shown in figure 10.9.

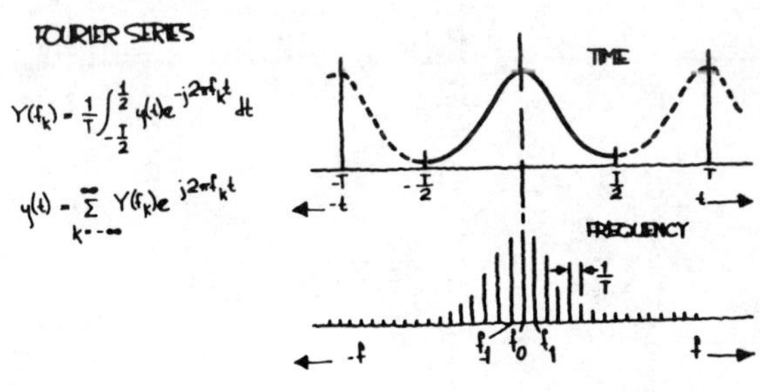

FIGURE 10.9

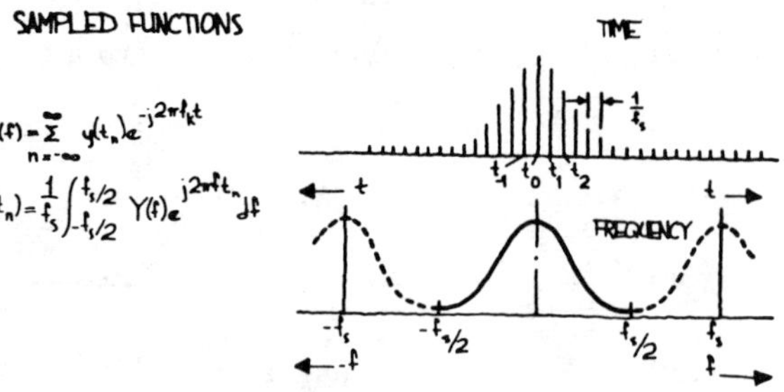

FIGURE 10.10

Figure 10.10 shows the transform from a sampled time-domain signal to the frequency domain. Note that in the frequency domain, the result is repetitive in frequency. This effect, commonly called aliasing, is discussed earlier in section 10.3. Figure 10.9 and figure 10.10 show the symmetry between the time- and frequency-domain transforms of discrete lines.

C. Discrete fourier transform

Finally, we have the sampled, periodic (assumed) time-domain signal which is transformed to a discrete and repetitive (aliased) frequency-domain representation. This is shown in figure 10.11

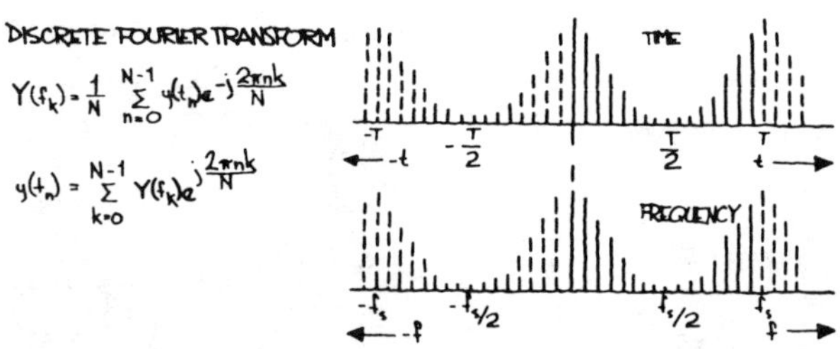

FIGURE 10.11

References

1. J. A. Barnes, Andrew R. Chi, Leonard S. Cutler, Daniel J. Healey, David B. Leeson, Thomas E. McGunigal, James A. Mullen, Jr., Warren L. Smith, Richard L. Sydnor, Robert F. C. Vessot, and Gernot M. R. Winkler, "Characterization of Frequency Stability," NBS Tech. Note 394 (1970); Proc. IEEE Trans. on Instrum. and Meas. IM-20, 105 (1971).

2. D. W. Allan, "The Measurement of Frequency and Frequency Stability of Precision Oscillators," NBS Tech. Note 669 (1975); Proc. 6th PTTI Planning Meeting, 109 (1975).

3. D. W. Allan, "Statistics of Atomic Frequency Standards," Proc. IEEE 54, 221 (1966).

4. Sections V and VI are from notes to be published by J. A. Barnes.

5. S. R. Stein, "An NBS Phase Noise Measurement System Built for Frequency Domain Measurements Associated with the Global Positioning System, NBSIR 76-846 (1976). See also F. L. Walls and S. R. Stein, "Servo Technique in Oscillators and Measurement Systems," NBS Tech. Note 692 (1976).

6. C. D. McGillen and G. R. Cooper, Continous and Discrete Signal and System Analysis, Holt, Rinehart, and Winston, Inc. (1974).

7. E. O. Brigham, The Fast Fourier Transform, Prentice-Hall, Inc., New Jersey (1974).

8. Private communication with Norris Nahman, June 1981.

9. G. M. Jenkins and D. G. Watts, Spectral Analysis and Its Application, Holden-Day, San Francisco, 1968.

10. P. F. Panter, Modulation, Noise, and Spectral Analysis, McGraw-Hill, New York, 1965, pp. 137-141.

11. D. Babitch and J. Oliverio, "Phase Noise of Various Oscillators at Very Low Frequencies, Proc. 28th Annual Symp. on Frequency Control, May 1974. U. S. Army Electronics Command, Ft. Monmouth, NJ.

12. N. Thrane, "The Discrete Fourier Transform and FFT Analyzers," Bruel and Kjaer Technical Review, No. 1 (1979).

13. Much of the material presented here is contained in NBS Tech. Note 679, "Frequency Domain Stability Measurements: A Tutorial Introduction," D. A. Howe (1976).

Characterization of Frequency Stability: Uncertainty due to the Finite Number of Measurements

PATRICK LESAGE AND CLAUDE AUDOIN

Abstract—We consider the characterization of the frequency stability of signal generators by means of the Allan variance of their fractional frequency fluctuations. As the number m of measurements is always finite, one can only get an estimate of this variance, which is a random function of m. We calculate the variance of the Allan variance for commonly encountered spectral densities of frequency fluctuations and deduce the relative uncertainty in the evaluation of frequency stability, as a function of m. The theory is experimentally checked for white noise of phase and flicker noise of frequency. The main hypothesis of the calculations, i.e., the Gaussian character of frequency fluctuations is verified. The m dependence of the uncertainty (one sigma) in the evaluation of frequency stability agrees with the predicted value.

I. Introduction

THE frequency stability of oscillators is currently characterized, in the time domain, by the Allan variance [1] of the frequency fluctuations. The definition, properties, and examples of application of the Allan variance are summarized in [2].

Manuscript received January 29, 1973. This work was supported in part by the Centre National de la Recherche Scientifique, Paris, and in part by the Direction des Recherches et Moyens D'Essais.

P. Lesage is with the Laboratoire de l'Horloge Atomique, Université Paris-Sud, Orsay, France.

C. Audoin is with the Laboratoire de l'Horloge Atomique, Université Paris-Sud, Orsay, France, and also with the Centre National de la Recherche Scientifique, Paris, France.

The Allan variance of fractional frequency fluctuations is defined by

$$\sigma_y^2(\tau) = \langle \sigma_y^2(2, \tau, \tau) \rangle = \left\langle \frac{(\bar{y}_{k+1} - \bar{y}_k)^2}{2} \right\rangle \tag{1}$$

where we use the same notation as in [2]. The quantity $\bar{y}_k$ is the mean value of the fractional frequency fluctuation over the time interval τ.

The definition of the variance involves the calculation of a mathematical mean over an infinite number of samples, while, of course, experiment gives a finite number of measurements. One can therefore only calculate an estimate of the variance; the confidence of which improves as the number of measurements is increased.

Since, in practice, one is concerned with the square root of the experimentally estimated Allan variance when one draws the graph of frequency stability [3] $\sigma_y(2, \tau, \tau) = f(\tau)$, we calculate in this paper the uncertainty of this quantity as a function of the number of measurements.

We consider a model of frequency fluctuations which consists in a set of five independent noise processes, with the following

Reprinted from *IEEE Trans. Instrum. Meas.*, vol. IM-22, pp. 157–161, June 1973.

power-law type of spectral density:

$$S_y'(f) = h_n f^n, \qquad n = -2, -1, 0, 1, 2, \; f > 0. \tag{2}$$

We take into account the finite bandwidth of the signal source, or of the circuits, which might be used to filter the signal before its processing by the frequency fluctuation measuring system. We assume that the transfer function $H(f)$ of this low-pass filter has a pole,[1] that is

$$H(f) = \frac{1}{1 + jf/f_c}. \tag{3}$$

The spectral density of the analyzed fractional frequency fluctuations is then

$$S_y(f) = h_n f^n \frac{1}{(1 + f/f_c)^2} = h_n \left(\frac{\omega}{2\pi}\right)^n \frac{1}{(1 + \omega/\omega_c)^2} \tag{4}$$

where $f_c = \omega_c/2\pi$ is the 3-dB bandwidth of the filter.

In Section V, we show that the theoretical results agree with the statistical analysis of experimental data. A calculation of the variance of an N-sample variance has been published [4] elsewhere. But, for $N = 2$, this work deals with the quantity $\bar{y}_{2k+1} - \bar{y}_{2k}$ instead of $\bar{y}_{k+1} - \bar{y}_k$ which appears in the definition 1 of the Allan variance.

II. Calculation of the Variance of the Estimated Allan Variance

As in [2], we assume that the amplitude fluctuations of the signal are negligible and that its frequency fluctuations are sufficiently small [2]. We let

$$x(t) = \frac{\varphi(t)}{2\pi\nu_0} \tag{5}$$

where $x(t)$ is the instantaneous phase fluctuations of the signal, the mean frequency of which is ν_0. The instantaneous fractional frequency deviation $y(t)$ is then

$$y(t) = \mathring{x}(t) \tag{6}$$

and the mean value of $y(t)$ over the time interval $(t_k, t_k + \tau)$ is equal to

$$\bar{y}_k = \frac{1}{\tau} \int_{t_k}^{t_k+\tau} y(t)\, dt = \frac{1}{\tau} [x(t_k + \tau) - x(t_k)]. \tag{7}$$

The quantity $\bar{y}_k$ is directly accessible to experiment by frequency counting or period measurement [2], [3]. For example, if τ_k is the measured period of the signal, we have

$$\bar{y}_k = 1 - \frac{\tau_k}{\tau}, \qquad \tau = \langle \tau_k \rangle. \tag{8}$$

[1] Although this point is not very important, this transfer function seems to be more realistic, in many cases, than the sharp high-frequency cutoff transfer function of [2]. Nevertheless, one must take care that other band-limiting devices do not contribute significantly to $H(f)$. The value of the Allan variance (Table I) is then modified for $n = 2$ only, by the ratio of the 3-dB bandwidth to the equivalent noise bandwidth.

The true Allan variance of the fractional frequency fluctuations can be written as

$$\sigma_y^2(\tau) = \tfrac{1}{2} \langle \alpha_k^2 \rangle \tag{9}$$

where α_k is the first difference of frequency fluctuation measurements

$$\alpha_k(\tau) = \bar{y}_{k+1} - \bar{y}_k. \tag{10}$$

In fact, however, only a finite number m of measurements of $\bar{y}_k$ has been performed, and the estimated Allan variance, which is a function of m, is calculated by the following formula:

$$\sigma_y^2(\tau, m) = \frac{1}{2(m-1)} \sum_{i=1}^{m-1} \alpha_i^2(\tau). \tag{11}$$

The true Allan variance is therefore the asymptotic value of $\sigma_y^2(\tau, m)$ (in the mean-square sense)

$$\lim_{m\to\infty} [\sigma_y^2(\tau, m)] = \sigma_y^2(\tau). \tag{12}$$

When m is finite, the estimated Allan variance $\sigma_y^2(\tau, m)$ is a random variable. We define its variance accordingly to the common understanding of a variance by

$$\sigma^2[\sigma^2(\tau, m)] = \langle [\sigma_y^2(\tau, m) - \langle \sigma_y^2(\tau, m) \rangle]^2 \rangle. \tag{13}$$

With the expression (11) of $\sigma_y^2(\tau, m)$, we get

$$\sigma^2[\sigma_y^2(\tau, m] = \left[\frac{1}{2(m-1)}\right]^2 \langle A(\tau) \rangle \tag{14}$$

where

$$\langle A(\tau) \rangle = \sum_{i=1}^{m-1} \sum_{j=1}^{m-1} [\langle \alpha_i^2(\tau) \alpha_j^2(\tau) \rangle - \langle \alpha_i^2(\tau) \rangle \langle \alpha_j^2(\tau) \rangle].$$

For further calculation, we need to assume that the quantities $\alpha_k(\tau)$ are normally distributed. It is reasonable to make the hypothesis that $y(t)$ and therefore $\bar{y}_k$ is a Gaussian process because the frequency fluctuations are the result of a very large number of independent perturbations. Furthermore, we have experimentally checked that $\bar{y}_k$ is a Gaussian process in the cases where $n = 2$ and -1 (Section V). The Gaussian character being retained by linear filtering, (10) shows that $\alpha_k(\tau)$ is also normally distributed, with a zero-mean value. We can then transform the fourth-order momentum $\langle \alpha_i^2(\tau) \alpha_j^2(\tau) \rangle$ in a sum of second-order momenta, and we obtain

$$\langle A(\tau) \rangle = 2 \sum_{i=1}^{m-1} \sum_{j=1}^{m-1} [\langle \alpha_i(\tau) \alpha_j(\tau) \rangle^2]. \tag{15}$$

III. Relation Between the Variance of the Allan Variance and the Spectral Density of Frequency Fluctuations

We consider, at first, noise processes leading to a stationary phase, i.e., $n = 1$ and 2 [5] in (2) or (4). The autocorrelation function of $x(t)$ is then defined

$$R_x(\theta) = \langle x(t) \cdot x(t - \theta) \rangle \tag{16}$$

and $\langle A(\tau)\rangle$ can be written as

$$\langle A(\tau)\rangle = 2 \sum_{i=1}^{m-1} \sum_{j=1}^{m-1} \{6R_x[(i-j)\tau] - 4R_x[(i-j-1)\tau] - 4R_x[(i-j+1)\tau] + R_x[(i-j-2)\tau] + R_x[(i-j+2)\tau]\}^2. \tag{17}$$

The autocorrelation function of $R_x(\theta)$ is connected to the one-sided power spectral density of fractional frequency fluctuations by

$$R_x(\theta) = \frac{1}{2\pi} \int_0^{\infty} \frac{1}{\omega^2} S_y(\omega) \cos \omega\theta \, d\omega. \tag{18}$$

Introducing (17) into (14), we obtain the useful expression of the variance of the estimated Allan variance

$$[2\pi(m-1)\tau^2]^2 \sigma^2 [\sigma_y^2(\tau, m)]$$

$$= \frac{1}{2}(m-1) \left\{ \int_0^{\infty} \frac{S_y(\omega)}{\omega^2} (6 - 8\cos\omega\tau + 2\cos 2\omega\tau)\, d\omega \right\}^2$$

$$+ \sum_{k=1}^{m-2} (m-k-1) \left\{ \int_0^{\infty} \frac{S_y(\omega)}{\omega^2} [6\cos k\omega\tau - 4\cos(k-1)\omega\tau - 4\cos(k+1)\omega\tau + \cos(k-2)\omega\tau + \cos(k+2)\omega\tau]\, d\omega \right\}^2 \tag{19}$$

with $k = i - j$.

For $n = 0, -1$, and -2, the phase is not stationary; one must consider $y(t)$ [6] which is normally distributed. It is possible to recalculate [7] the expression of $\langle A(\tau)\rangle$ in this case, working with the autocorrelation function of $y(t)$. As this calculation is lengthy, and straightforward, it will not be reported here. The important result is that (19) is also valid for a nonstationary phase. It is interesting to notice that the integrals involved in (19) converge (for $\omega = 0$ and $\omega \to \infty$) for the spectral densities of frequency fluctuation considered in (2). The variance of the estimated Allan variance is therefore determined without any approximation.

IV. Relative Uncertainty on the Characterization of Frequency Stability

A. *Relative Uncertainty on Estimated Allan Variance*

We introduce Δ the relative deviation of $\sigma_y^2(\tau, m)$, relative to $\sigma_y^2(\tau)$

$$\Delta = \frac{\sigma_y^2(\tau, m) - \sigma_y^2(\tau)}{\sigma_y^2(\tau)}. \tag{20}$$

The quantity Δ is a random function of m. The standard deviation $\sigma(\Delta)$ defines the relative uncertainty in the determination of $\sigma_y^2(\tau)$ due to the finite number of measurements.

Let us link $\sigma(\Delta)$ to the variance of the Allan variance. From (20), we obtain

$$\sigma_y^2(\tau, m) = (1 + \Delta)\sigma_y^2(\tau). \tag{21}$$

TABLE I
Summary of Theoretical Results

Spectral density of fractional freq. fluctuations	Allan variance	variance of the estimated Allan variance, as a function of the number of measurements	relative uncertainty on the characterization of frequency stability
$S_y(f)$	$\sigma_y^2(\tau)$	$\sigma^2\left[\sigma_y^2(\tau, m)\right]$	$\sigma\left[\Delta(m)\right]$
$h_2\left(\frac{\omega}{2\pi}\right)^2$	$\frac{3h_2\omega_c}{16\pi^2\tau^2}$ $\omega_c\tau \gg 1$	$\left[\frac{3h_2\omega_c}{16\pi^2\tau^2}\right]^2 \times \frac{35m-53}{9(m-1)^2}$ $\omega_c\tau \gg 1$	$\frac{1}{3(m-1)}\sqrt{35m-53}$ $\omega_c\tau \gg 1$
$h_1\left(\frac{\omega}{2\pi}\right)$	$\frac{3h_1 \operatorname{Log}\omega_c\tau}{4\pi^2\tau^2}$ $\omega_c\tau \gg 1$	$\left[\frac{3h_1 \operatorname{Log}\omega_c\tau}{4\pi^2\tau^2}\right]^2 \times \frac{35m-53}{9(m-1)^2}$ $\omega_c\tau \gg 1$	$\frac{1}{3(m-1)}\sqrt{35m-53}$ $\omega_c\tau \gg 1$
h_0	$\frac{h_0}{2\tau}$ $\omega_c\tau \gg 1$	$\left[\frac{h_0}{2\tau}\right]^2 \times \frac{3m-4}{(m-1)^2}$ $\omega_c\tau \gg 1$	$\frac{1}{(m-1)}\sqrt{3m-4}$ $\omega_c\tau \gg 1$
$h_{-1}\left(\frac{\omega}{2\pi}\right)^{-1}$	$2h_{-1}\operatorname{Log} 2$	$\left[2h_{-1}\operatorname{Log} 2\right]^2 \times \frac{5{,}7m-31}{(m-1)^2}$	$\frac{1}{(m-1)}\sqrt{5{,}7m-31}$
$h_{-2}\left(\frac{\omega}{2\pi}\right)^{-2}$	$\frac{2\pi^2\tau h_{-2}}{3}$	$\left[\frac{2\pi^2\tau h_{-2}}{3}\right]^2 \times \frac{9m-10}{4(m-1)^2}$	$\frac{1}{2(m-1)}\sqrt{9m-10}$

The signal is filtered by a low pass filter, the transfer function of which is $H(f) = 1/(1 + j\frac{f}{f_c})$.

The variance of which is

$$\sigma^2[\sigma_y^2(\tau, m)] = [\sigma_y^2(\tau)]^2 \sigma^2(1 + \Delta). \tag{22}$$

We then obtain the following relation between $\sigma(\Delta)$ and the variance of the estimated Allan variance

$$\sigma(\Delta) = \frac{1}{\sigma_y^2(\tau)} \{\sigma^2[\sigma_y^2(\tau, m)]\}^{1/2}. \tag{23}$$

The values of $\sigma(\Delta)$ are shown in Table I for the considered spectral densities of frequency noise. They can be used to calculate the relative uncertainty on the estimated Allan variance, even in the case of a small number of measurements, for example, $m = 2$.

B. *Relative Uncertainty on Square Root of Estimated Allan Variance.*

In practice, when one draws the graph of frequency stability of a frequency source, one is concerned with the square root of the experimentally estimated Allan variance, i.e., with the standard deviation $\sigma_y(\tau, m)$. Therefore, we also introduce δ the relative deviation of $\sigma_y(\tau, m)$, relative to $\sigma_y(\tau)$

$$\delta = \frac{\sigma_y(\tau, m) - \sigma_y(\tau)}{\sigma_y(\tau)}. \tag{24}$$

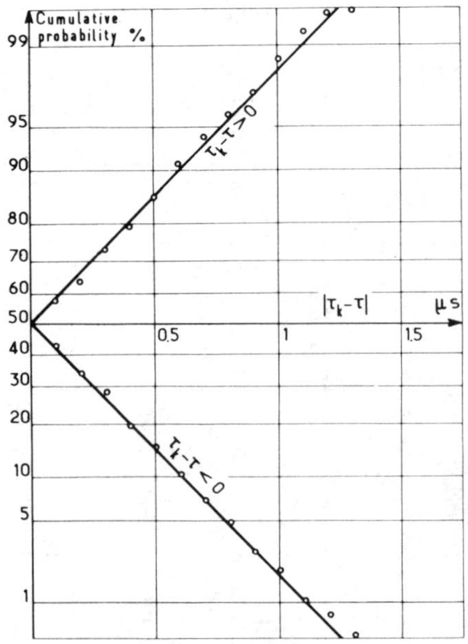

Fig. 1. Distribution of counting times for white noise of phase ($n = 2$; $\tau = 1$ s), in Galtonian coordinates. The circles represent the cumulative probability corresponding to $|\tau_k - \tau|$, with $\tau = \langle \tau_k \rangle$. Solid line corresponds to the normal distribution of the same width.

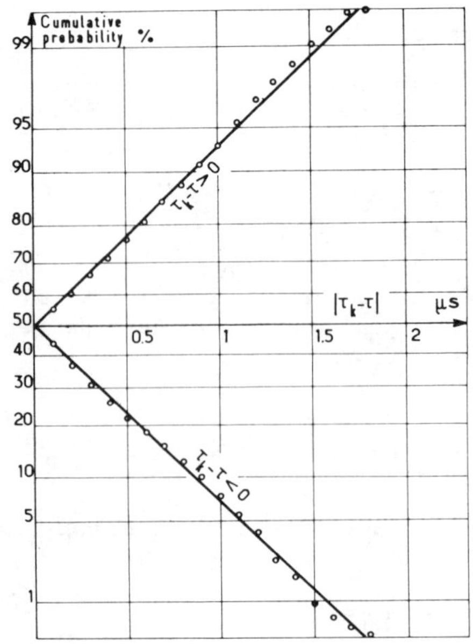

Fig. 2. Distribution of counting times for flicker noise of frequency ($n = -1$; $\tau = 8$ s) in Galtonian coordinates. Circles represent cumulative probability corresponding to $|\tau_k - \tau|$, with $\tau = \langle \tau_k \rangle$. Solid line corresponds to normal distribution of same width.

From this equation we obtain

$$\sigma^2 [\sigma_y^2(\tau, m)] = [\sigma_y^2(\tau)]^2 \sigma^2 [(1 + \delta)^2]. \tag{25}$$

We assume, later on, that the condition $\delta << 1$ is fulfilled, i.e., that the number of measurements is sufficiently large to get a good confidence on the estimated Allan variance. We then have

$$\sigma^2 [(1 + \delta)^2] = \langle [(1 + \delta)^2 - \langle (1 + \delta)^2 \rangle]^2 \rangle.$$

$$\simeq 4\sigma^2(\delta) \tag{26}$$

We therefore obtain the following relation between $\sigma(\delta)$ and the variance of the Allan variance:

$$\sigma(\delta) \simeq \frac{1}{2\sigma_y^2(\tau)} \{\sigma^2 [\sigma_y^2(\tau, m)]\}^{1/2}, \qquad \delta << 1 \tag{27}$$

and

$$\sigma(\delta) \simeq \tfrac{1}{2}\, \sigma(\Delta), \qquad \delta << 1. \tag{28}$$

From Table I and (28) it is easy to show that we can write, within an accuracy better than 10 percent, for m larger than 10,

$$\sigma(\delta) \simeq K_n m^{-1/2}. \tag{29}$$

The functional dependence $\sigma(\delta) = f(m)$ has already been stated in [2]. Our results allow to specify the numerical value of K_n, for the different spectral densities considered. We have

$$K_2 = K_1 = 0.99$$

$$K_0 = 0.87$$

$$K_{-1} = 1.19$$

$$K_{-2} = 0.75. \tag{30}$$

The different values of K_n can be experimentally distinguished as it will be shown in the next section, for $n = 2$ and $n = -1$.

V. Experimental Results

A high-quality quartz-crystal oscillator is a convenient frequency source to check the theoretical results. One can test two different noise processes ($n = 2$ white noise of phase and $n = -1$ flicker noise of frequency) for given values of the time interval τ of the order of 1 s. The reference oscillator is a hydrogen maser showing a frequency stability $\sigma_y(2, \tau, \tau, f_c = 4 \text{ Hz})$ of $(1/\tau) \times 4 \times 10^{-13}$, for $\tau < 100$ s, 5×10^{-15}, for $100 \text{ s} < \tau < 1000$ s, and with daily frequency variations of the order of 10^{-14}.

The maser signal is mixed in a triple heterodyne receiver with signals derived from the quartz crystal under test by frequency synthesis. A beat note at 5.75 kHz is obtained at the output of a selective amplifier, with a bandwidth of 40 Hz. The time duration for a given number of periods of this signal is measured. Counting results are punched on a tape for computer analysis.

In the preceding arrangement, the measured quartz-crystal fractional frequency stability is equal to $(1.2 \times 10^{-12})/\tau$, for $\tau < 3$ s ($n = 2$ white noise of phase)[2] and to 4×10^{-13} for $\tau > 3$ s ($n = -1$ flicker noise of frequency). The Gaussian character of the frequency fluctuations were first tested for working time intervals of $\tau = 1$ s and $\tau = 8$ s.

In that experiment it is necessary that systematic frequency drifts, such as those related to crystal aging and room temperature variations be negligible. For this purpose, the quartz crystal under test is phase locked [8] to a second hydrogen

[2]The τ^{-1} dependence can be determined by the white noise of phase $n = 2$ or the flicker noise of phase $n = 1$. As $K_2 = K_4$, we did not try to distinguish between these two possibilities.

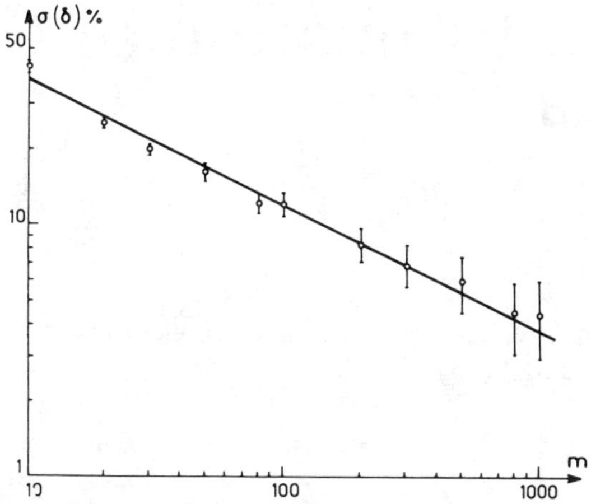

Fig. 3. Experimental dependence of $\sigma[\delta(m)]$ on number m of measurements for white noise of phase ($n = 2$; $\tau = 1$ s). Circles are experimental points. Error bar characterizes confidence in experimental value of $\sigma[\delta(m)]$. Solid line is adjusted by method of least squares. Its slope is -0.46, and experimental value of K_2 equals 0.97.

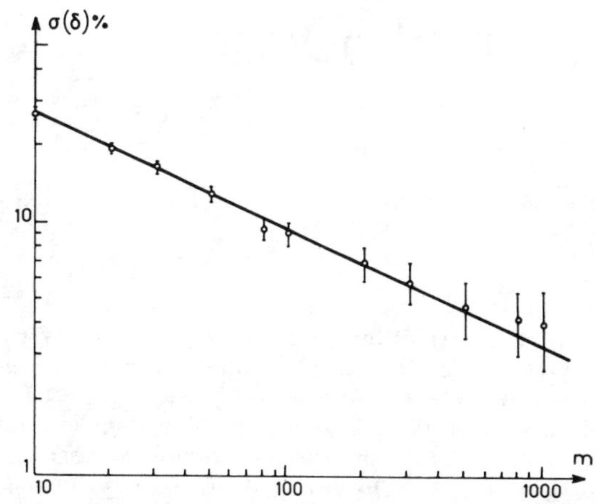

Fig. 4. Experimental dependence of $\sigma[\delta(m)]$ on number m of measurements, for flicker noise of frequency ($n = -1$; $\tau = 8$ s). Circles are experimental points. Error bar characterizes confidence (one sigma) in experimental value of $\sigma[\delta(m)]$. Solid line is adjusted by method of least squares. Its slope is -0.50, and experimental value of K_{-1} equals 1.07.

maser, similar to the first one. The loop is of second order, adjusted to critical damping with a time constant of 30 s.

For $\tau < 30$ s, the spectral density of the quartz-crystal frequency fluctuations is therefore not appreciably modified by phase locking, while systematic drifts are removed.

A total number of 8000 measurements has been performed, for $\tau = 1$ s ($n = 2$) and $\tau = 8$ s ($n = -1$). Figs. 1 and 2 show that the counting times τ_k are normally distributed. As one goes from τ_k to $\bar{y}_k$ by linear filtering (8), the quantity $\bar{y}_k$ is also normally distributed. Furthermore, one can demonstrate that $\bar{y}_k$ being normally distributed implies that $y(t)$ possesses the same property.

Afterward, for given τ, the measurements have been separated into 8 000/m sets of m measurements, with $10 < m < 1\,000$. The estimated Allan variance has been computed for each set of given m, and the relative uncertainted $\sigma(\delta)$ has been determined from the 8000/m values of the estimated Allan variance. The comparison between the theoretical and the experimental values of $\sigma(\delta)$ is given on Figs. 3 and 4. They show that $\sigma(\delta)$ varies as $m^{-1/2}$ ($m > 10$) and that the values of the coefficients K_2 and K_{-1} are in very good agreement with the theory. The experimental values are $K_2 = 0.97$ and $K_{-1} = 1.07$.

VI. Conclusion

We have calculated the confidence of the estimated Allan variance, as a function of the available number of measurements. The theoretical results have been verified by statistical analysis of the frequency fluctuations of a high-quality frequency source in the case of the white noise of phase and in the case of the flicker noise of frequency. We have experimentally proved that the frequency fluctuations of the source is normally distributed, for these noise processes. It is recommended to specify the error bars (one sigma) by the use of (25) and (26) in the $\sigma_y(2, \tau, \tau)$ graph of frequency stability, according to the number of available measurements.

Acknowledgment

The authors wish to thank Professor B. Picinbono for helpful discussion and Drs. J. A. Barnes and D. W. Allan for their valuable comments on the manuscript.

References

[1] D. W. Allan, "Statistics of atomic frequency standards," *Proc. IEEE*, vol. 54 pp. 221–230, Feb. 1966.
[2] J. A. Barnes, A. R. Chi, L. S. Cutler, D. J. Healey, D. B. Leeson, T. E. McGunigal, J. A. Mullen, W. L. Smith, R. L. Sydnor, R. F. C. Vessot, and G. M. R. Winkler, "Characterization of frequency stability," *IEEE Trans. Instrum. Meas.*, vol. IM-20, p. 105–120, May 1971.
[3] L. S. Cutler and C. L. Searle, "Some aspects of the theory and measurement of frequency fluctuations in frequency standards," *Proc. IEEE*, vol. 54, pp. 136–154, Feb. 1966.
[4] R. C. Tausworthe, "Convergence of oscillator spectral estimators for counted-frequency measurements," *IEEE Trans. Commun.* (Concise Paper), vol. COM-20, pp. 214–217, Apr. 1972.
[5] B. Picinbono, "Processus de diffusion et stationnarité–Application à la représentation du bruit des oscillateurs," *C.R. Acad. Sci.*, vol. 271, p. 661, 1970.
[6] J. Rutman, "Bruit dans les oscillateurs–Application à la métrologie des fréquences et à la spectroscopie hertzienne," thèse, Paris, Mar. 1972.
[7] P. Lesage, Thèse de 3e cycle–Orsay–Requirements to be fulfilled in 1973.
[8] F. M. Gardner, *Phaselock Technique*. New York: Wiley, 1966.

Effect of Dead-Time on the Estimation of the Two-Sample Variance

PAUL LESAGE AND CLAUDE AUDOIN

Abstract—**Expressions for the averaged two-sample variance as well as expressions for the uncertainty on its estimation are given, in cases where dead time is present between measurements. The effect of this dead time on the precision of the time domain measure of frequency stability, in the presence of a limited number of data, is specified. The possibility to distinguish between white and flicker noises is pointed out.**

I. Introduction

THIS WORK completes the results published in a previous article [1], where the uncertainty on the estimation of the two-sample variance [2], [3] (Allan variance) was calculated under the assumption that i) no dead time is present between measurements and ii) the counting time τ is large compared to the inverse $1/2\pi f_c$ of the angular spectrum-width of the analyzed signal. However, it appears in many experimental circumstances that the two above conditions are not fulfilled. For instance, if the counting time τ equals the period of the processed signal, it is then common that the dead time equals τ. On the other hand, the condition $2\pi f_c \tau \gg 1$ does not hold if τ is smaller than the time constant of a servo phase or frequency loop included in the considered frequency generator, such as in atomic frequency standards.

In this article we specify the uncertainty on the estimation of the two-sample variance when the two aforementioned conditions are removed. We will show that, for a given number m of available measurements, the precision in the estimation of the two-sample variance depends on the dead time and the value of $2\pi f_c \tau$.

The effect of the correlation between measurements involved in the characterization of frequency stability has been considered in a recent work [4], but with a peculiar definition of the dead time.

II. Averaged Two-Sample Variance and Estimated Average of the Two-Sample Variance

The averaged two-sample variance (ATSV) of the fractional frequency fluctuations is defined as [3]

$$\langle \sigma_y^2(2, T, \tau) \rangle = \tfrac{1}{2}\langle (\bar{y}_{k+1} - \bar{y}_k)^2 \rangle \tag{1}$$

where $\bar{y}_k$ represents the average value of the instantaneous fractional frequency fluctuations $y(t)$ over the time interval τ, according to [3]. We have

$$\bar{y}_k = \frac{1}{\tau}\int_{t_k}^{t_k+\tau} y(t)\, dt, \qquad t_k = t_0 + kT \tag{2}$$

where T is the time interval between the beginnings of two successive measurements and k is a positive integer. We set

$$r = \frac{T}{\tau}, \qquad r \geq 1 \tag{3}$$

As in practice, only m counting results are available, one calculates the *estimated* average of the two-sample variance (EATSV) as follows:

$$\hat{\sigma}_y^2(2, T, \tau, m) = \frac{1}{2(m-1)} \sum_{k=1}^{m-1} (\bar{y}_{k+1} - \bar{y}_k)^2. \tag{4}$$

Relation (4) shows that the EATSV is a random function of m.

It is clear, that the expected value of the EATSV equals the ATSV. Therefore, the finite number of measurements does not introduce bias in the estimation of the two-sample variance.

III. Variance of the Estimated Two-Sample Variance

In order to characterize the uncertainty in the estimation, the m dependence of the variance of the EATSV is considered. We set:

$$\sigma^2[\hat{\sigma}_y^2(2, T, \tau, m)] = \langle [\hat{\sigma}_y^2(2, T, \tau, m) - \langle \sigma_y^2(2, T, \tau)\rangle]^2 \rangle. \tag{5}$$

Equation (5) can also be written as:

$$\sigma^2[\hat{\sigma}_y^2(2, T, \tau, m)] = \frac{1}{m-1}\left[\frac{\Gamma_0}{4} + \frac{1}{2(m-1)} \sum_{k=1}^{m-2} (m-1-k)\Gamma_k\right] \tag{6}$$

with

$$\Gamma_k = \langle \beta_i \beta_{i-k} \rangle \tag{7}$$

and

$$\beta_i = (\bar{y}_{i+1} - \bar{y}_i)^2 - 2\langle \sigma_y^2(2, T, \tau)\rangle \tag{8}$$

Γ_k, which does not depend on m, represents the auto correlation coefficient of β_i and β_{i-k}.

Under the assumption that Γ_k decreases with k, equation (6) shows that a $1/(m-1)$ dependence occurs but asympto-

Manuscript received November 28, 1977; revised October 6, 1978.

The authors are with the Laboratoire de l'Horloge Atomique, Equipe de Recherche du CNRS, Batiment 221, Université Paris Sud, 91405 Orsay, France.

Reprinted from *IEEE Trans. Instrum. Meas.*, vol. IM-28, pp. 6–10, Mar. 1979.

tically for large enough values of $(m-1)$. We shall derive the exact dependence in the following.

We assume that quantities $\bar{y}_{i+1} - \bar{y}_i$ are normally distributed. This hypothesis has been checked for three different noise processes: white phase, white frequency and flicker frequency [1], [5].

Relation (6), therefore gives

$$(m-1)^2\sigma^2[\hat{\sigma}_y^2(2, T, \tau, m)]$$
$$= 8(m-1)\left\{\int_0^\infty S_y(f)\, \frac{\sin^2 \pi f T \sin^2 \pi f \tau}{(\pi f \tau)^2}\, df\right\}^2$$
$$+ 16 \sum_{k=1}^{m-2} (m-1-k)$$
$$\cdot \left\{\int_0^\infty S_y(f) \cos 2k\pi f T \sin^2 \pi f T\, \frac{\sin^2 \pi f \tau}{(\pi f \tau)^2}\, df\right\}^2 \quad (9)$$

where $S_y(f)$ denotes the one-sided power spectral density of the low-pass filtered fractional frequency fluctuations, f_c being the 3-dB bandwidth of the filter. We have

$$S_y(f) = \frac{h_\alpha f^\alpha}{1 + (f/f_c)^2} \quad (10)$$

where α takes the values $-2, -1, 0, 1, 2$ in the conventional model considered for noise processes in frequency generators.

IV. Relative Uncertainty on the Characterization of Frequency Stability in the Time Domain

We consider Δ, the fractional deviation of $\hat{\sigma}_y^2(2, T, \tau, m)$ relative to $\langle\sigma_y^2(2, T, \tau)\rangle$ defined as follows:

$$\Delta = \frac{\hat{\sigma}_y^2(2, T, \tau, m) - \langle\sigma_y^2(2, T, \tau)\rangle}{\langle\sigma_y^2(2, T, \tau)\rangle}. \quad (11)$$

The quantity Δ being a random function of m, the standard deviation $\sigma(\Delta)$ defines the relative uncertainty in the determination of $\langle\sigma_y^2(2, T, \tau)\rangle$ due to the finite number of measurements. We have:

$$\sigma(\Delta) = \frac{1}{\langle\sigma_y^2(2, T, \tau)\rangle} \{\sigma^2[\hat{\sigma}_y^2(2, T, \tau, m)]\}^{1/2}. \quad (12)$$

For each considered value of α, expressions of $\sigma(\Delta)$ have been calculated within 4 commonly encountered practical conditions

a) $2\pi f_c \tau \gg 1$ and $2\pi f_c(T-\tau) \gg 1$
b) $2\pi f_c \tau \ll 1$ and $2\pi f_c(mT-\tau) \ll 1$
c) $2\pi f_c \tau \ll 1$ and $2\pi f_c(T-\tau) \gg 1$
d) $2\pi f_c \tau \gg 1$ and $2\pi f_c(T-\tau) \ll 1$.

Expressions of ATSV are given in Table I [6]. Expressions of $\sigma(\Delta)$, which are valid for $m \geq 2$ are summarized in Tables II and III.

As a matter of fact, we have expressed the ATSV and the quantity $\sigma(\Delta)$ for all possible values of $2\pi f_c \tau$ and $2\pi f_c(T-\tau)$. However, if the conditions specified in Tables II and III are not fulfilled, expressions are lengthy and they will not be given here.

Tables II shows that for m large enough $\sigma(\Delta)$ varies as $1/m$, except within the above conditions b) and for $\alpha = -1$ and -2. In these specific conditions $\sigma(\Delta)$ does not depend on m and equals $\sqrt{2}$. A different proof of this peculiar result is given in the Appendix. This simply means that the characterization of frequency stability cannot be performed within conditions b), for $\alpha = -1$ and -2. Under conditions $2\pi f_c \tau \gg 1$ and $2\pi f_c(T-\tau) \ll 1$ expressions of $\sigma(\Delta)$ agree with previously published results [1].

In practice, the time-domain frequency stability of a frequency source is characterized by the square root of the experimentally determined EATSV i.e., by $[\hat{\sigma}_y^2(2, T, \tau, m)]^{1/2}$. We therefore consider δ defined as:

$$\delta = \frac{[\hat{\sigma}_y^2(2, T, \tau, m)]^{1/2} - \langle\sigma_y^2(2, T, \tau)\rangle^{1/2}}{\langle\sigma_y^2(2, T, \tau)\rangle^{1/2}} \quad (13)$$

$\sigma(\delta)$ specifies the precision of the frequency stability measure, as a function of the number of available data.

If, furthermore, the condition $2\pi f_c \tau \gg 1$ is fulfilled, one obtains the following simple dependence of $\sigma(\delta)$ with m, without any condition on T:

$$\sigma(\delta) \simeq K_\alpha m^{-1/2}. \quad (14)$$

The values of K_α depend on the noise process considered[1] as well as the values of $2\pi f_c \tau$ and r. This is depicted on Figs. 1 to 3 for different values of $2\pi f_c \tau$. These figures show that K_α depends on the dead time between measurements as it was experimentally observed in the special case of white phase noise [5]. This is specially pronounced in the vicinity of $r = 1$ for $\alpha = 2, 1, 0$. Furthermore, variations of K_α with r depend on the value of α, and for $r > 1$, K'_e differs from K_1 (while $K_2 = K_1$ for $r = 1$). This gives a means to distinguish between the white and the flicker phase noises ($\alpha = 2$ and 1, respectively) without changing the bandwidth of the measurement set-up, whereas this is not practically possible by looking at the τ-dependence of the two-sample variance with constant f_c.

V. Conclusion

We have considered the variance of the EATSV, in presence of a dead time between measurements, as a function of the number of available data, for the commonly encountered frequency noise processes.

For conditions which have been specified, we have also derived the expressions for the two-sample variance and the relative uncertainty in the estimation of the time-domain measure of frequency stability with a finite number of measurements.

It appears possible to discern white and flicker phase noises by measurements with dead time.

Appendix

For $\alpha = -1$ or -2, the spectral density $S_y(f)$ of $y(t)$ varies very rapidly, as f^{-3} or f^{-4}, respectively, for $f > f_c$.

[1] As the expressions of K_α are lengthy, they will not be reported here, but can be obtained from the authors upon request.

TABLE I

conditions of validity \ α	2	1	0	− 1	− 2
$2\pi f_c\tau >> 1$ $2\pi f_c(T-\tau) >> 1$	$\frac{h_2 f_c}{4\pi\tau^2}$	$\frac{h_1 \ln(2\pi f_c\tau)}{2\pi^2\tau^2}$	$\frac{h_o}{2\tau}$	$h_{-1}\ln(T/\tau)$ $T >> \tau$	$\pi^2 h_{-2} T$
$2\pi f_c\tau << 1$ $2\pi f_c(mT-\tau) << 1$	$\frac{h_2 f_c^2}{2\tau}$	$2h_1 f_c^2 \ln(2)$ for $T = \tau$ $h_1 f_c^2 \ln(T/\tau)$ for $T >> \tau$	$\frac{2}{3}\pi^2 h_o f_c^2 \tau$ for $T = \tau$ $\pi^2 h_o f_c^2 T$ for $T >> \tau$	$2\pi^2 h_{-1} f_c^2 T^2$	$\pi^3 h_{-2} f_c T^2$
$2\pi f_c\tau << 1$ $2\pi f_c(T-\tau) >> 1$	$\frac{h_2 f_c^2}{2\tau}$	$h_1 f_c^2 \ln(2\pi f_c\tau)$	$\frac{\pi}{2} h_o f_c$	$h_{-1}\ln(2\pi f_c T)$	$\pi^2 h_{-2} T$
$2\pi f_c\tau >> 1$ $2\pi f_c(T-\tau) << 1$	$\frac{3h_2 f_c}{8\pi\tau^2}$	$\frac{3h_1 \ln(2\pi f_c\tau)}{4\pi^2\tau^2}$	$\frac{h_o}{2\tau}$	$2h_{-1}\ln(2)$	$\frac{2\pi^2 h_{-2}\tau}{3}$

Expressions of the averaged two sample variance of fractional frequency fluctuations. The low-pass filtered spectral density is given by:

$$S_y(f) = \frac{h_\alpha f^\alpha}{1 + (f/f_c)^2}$$

When both conditions $2\pi f_c(mT - \tau) \ll 1$ and $T \gg \tau$ are considered, they must be fulfilled simultaneously.

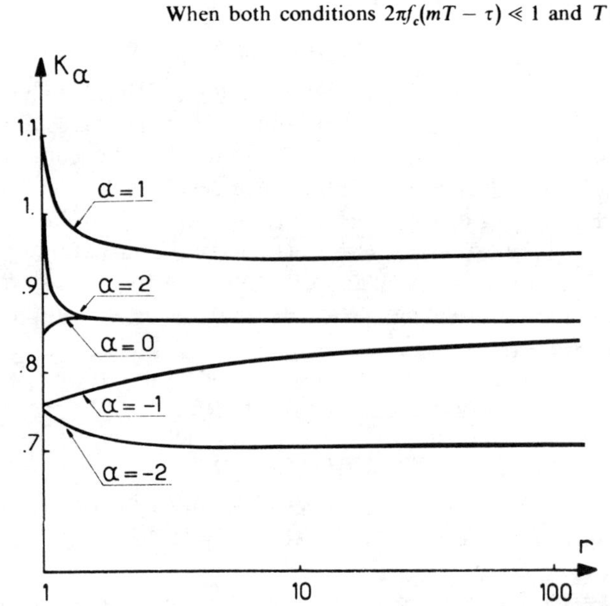

Fig. 1. r-dependence of coefficients K_α under condition $2\pi f_c\tau = 10$.

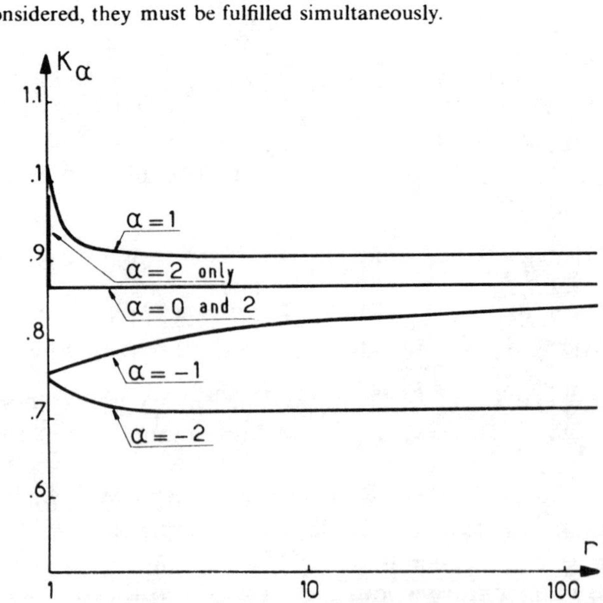

Fig. 2. r-dependence of coefficients K_α under condition $2\pi f_c\tau = 100$.

TABLE II

conditions of validity \ α	2	1	0	- 1	- 2
$2\pi f_c \tau \gg 1$ $2\pi f_c (T-\tau) \gg 1$	$\frac{\sqrt{3m-4}}{m-1}$	$\frac{\sqrt{3m-4}}{m-1}$	$\frac{\sqrt{3m-4}}{m-1}$	$\frac{\sqrt{3m-4}}{m-1}$ $T \gg \tau$	$\frac{\sqrt{2}}{\sqrt{m-1}}$
$2\pi f_c \tau \ll 1$ $2\pi f_c (mT - \tau) \ll 1$	$\frac{\sqrt{3m-4}}{m-1}$	$\simeq \frac{\sqrt{2.3m-2.6}}{m-1}$ for $m > 5$*, $T=\tau$ $\frac{\sqrt{3m-4}}{m-1}$ for $T \gg \tau$	$\frac{\sqrt{9m-10}}{2(m-1)}$ for $T = \tau$ $\frac{\sqrt{2}}{\sqrt{m-1}}$ for $T \gg \tau$	$\sqrt{2}$	$\sqrt{2}$
$2\pi f_c \tau \ll 1$ $2\pi f_c (T-\tau) \gg 1$	$\frac{\sqrt{3m-4}}{m-1}$	$\frac{\sqrt{3m-4}}{m-1}$	$\frac{\sqrt{3m-4}}{m-1}$	$\frac{\sqrt{3m-4}}{m-1}$	$\frac{\sqrt{2}}{\sqrt{m-1}}$
$2\pi f_c \tau \gg 1$ $2\pi f_c (T-\tau) \ll 1$	$\frac{\sqrt{35m-53}}{3(m-1)}$	$\frac{\sqrt{35m-53}}{3(m-1)}$	$\frac{\sqrt{3m-4}}{m-1}$	$\simeq \frac{\sqrt{2.3m-2.5}}{m-1}$ $m > 5$*	$\frac{\sqrt{9m-10}}{2(m-1)}$

Expressions of the relative uncertainty $\sigma(\Delta)$ on the estimated two-sample variance of fractional frequency fluctuations. The low pass filtered spectral density is given by:

$$S_y(f) = \frac{h_\alpha f^\alpha}{1 + (f/f_c)^2}$$

When both conditions $2\pi f_c(mT - \tau) \ll 1$ and $T \gg \tau$ are considered, they must be fulfilled simultaneously.
* For $m < 5$, values of $\sigma(\Delta)$ are shown in Table III.

TABLE III

m	2	3	4	5
$\sigma(\Delta)$	1.41	1.02	0.85	0.74

Values of $\sigma(\Delta)$ for small m and for the two following conditions:

i) $\alpha = 1$, $2\pi f_c \tau \ll 1$, $2\pi f_c(mT - \tau) \ll 1$, $T = \tau$

ii) $\alpha = -1$, $2\pi f_c \tau \gg 1$, $2\pi f_c(T - \tau) \ll 1$

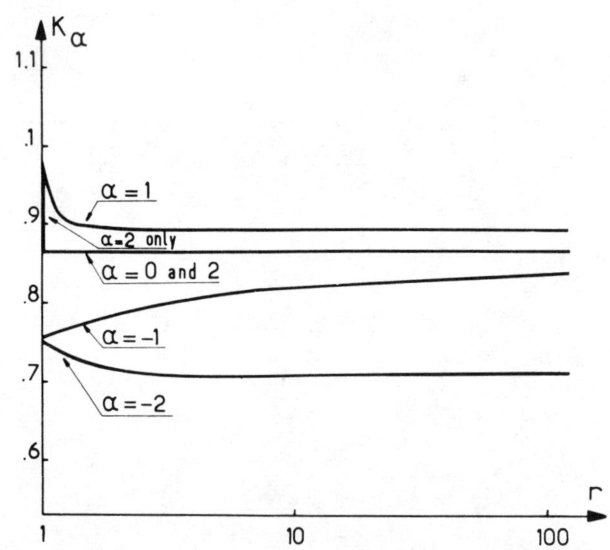

Fig. 3. r-dependence of coefficients K_α under condition $2\pi f_c \tau = 1000$.

The spectral density $S_a(f)$ of the quantities $a_i = \bar{y}_{i+1} - \bar{y}_i$ is given by [5]:

$$S_a(f) = 2 \sin^2 \pi f T \frac{\sin^2 \pi f \tau}{(\pi f \tau)^2} S_y(f). \tag{15}$$

It is then easy to see that the power spectrum of $S_a(f)$ has negligible values for $f \gtrsim f_c$. The Fourier transform $R_a(\theta)$ of $S_a(f)$ is therefore, almost a constant over the time interval

$$-(2\pi f_c)^{-1} \lesssim \theta \lesssim (2\pi f_c)^{-1}.$$

On the other hand, starting from (7) and (8), and assuming that the a_i's are normally distributed, one can show that we have

$$\Gamma_k = 2[\langle a_i a_{i-k} \rangle]^2 \tag{16}$$

where $\langle a_i a_{i-k} \rangle$ is the value of $R_a(kT)$. k and T are as defined in the text in equation (2).

The condition b) ($2\pi f_c \tau \ll 1$ and $2\pi f_c(mT - \tau) \ll 1$) implies that we have

$$kT \ll \frac{k}{2\pi m f_c}. \tag{17}$$

As k is at most equal to m, we see that the useful values of kT are included within the time interval where $R_a(\theta)$ can be considered as a constant. We conclude that all the values of Γ_k, for $0 < k \leq m - 2$, are such as:

$$\Gamma_k \simeq \Gamma_0 \simeq 8[\langle \sigma_y^2(2, T, \tau) \rangle]^2. \tag{18}$$

The value of Γ_k being assumed a constant, and equal to Γ_0, we obtain, from equation (6):

$$\sigma^2[\hat{\sigma}_y^2(2, T, \tau, m)] \simeq 2[\langle \sigma_y^2(2, T, \tau) \rangle]^2. \tag{19}$$

It then follows

$$\sigma(\Delta) \simeq \sqrt{2}. \tag{20}$$

It is worth noticing that the application of the Schwarz inequality to the random variable Γ_k gives, without assuming any condition on $2\pi f_c \tau$ and $2\pi f_c(T - \tau)$, the following result:

$$\sigma(\Delta) \leq \sqrt{2}. \tag{21}$$

References

[1] P. Lesage and C. Audoin, "Characterization of frequency stability: Uncertainty due to the finite number of measurements," *IEEE Trans. Instrum. Meas.*, vol. IM-22, pp. 157–161, June 1973; vol. IM-24, p. 86, Mar. 1975; aux. vol. IM-25, p. 270, Sept. 1976.

[2] D. W. Allan, "Statistics of atomic frequency standards," *Proc. IEEE*, vol. 54, pp. 221–230, Feb. 1966.

[3] J. A. Barnes, A. R. Chi, L. S. Cutler, D. J. Healey, D. B. Leason, T. E. McGunigal, J. A. Mullen, W. L. Smith, R. L. Sydnor, R. F. C. Vessot, and G. M. R. Winkler, "Characterization of frequency stability," *IEEE Trans. Instrum. Meas.*, vol. IM-20, pp. 105–120, May 1971.

[4] K. Yoshimura, "Characterization of frequency stability: Uncertainty due to the auto-correlation of the frequency fluctuations," *IEEE Trans. Instrum. Meas.*, vol. IM-27, pp. 1–7, Mar. 1978.

[5] P. Lesage and C. Audoin, "Estimation of the two sample variance with a limited number of data," in *Proc. 31st Annu. Symp. Frequency Control* (Atlantic City, NJ), pp. 311–318, June 1977.

[6] P. Lesage, "Caractérisation de la stabilité de fréquence d'oscillateurs dans le domaine temps et dans le domaine fréquence," Thèse de Doctorat de 3e cycle, Orsay 1973.

Characterization of Frequency Stability: Uncertainty Due to the Autocorrelation of the Frequency Fluctuations

KAZUYUKI YOSHIMURA

Abstract—We considered the general sampling form for the estimate of the Allan variance which is the proposed measure of frequency stability in the time domain, and we defined a variable proportional to the difference between the average fractional frequency fluctuations over the time interval τ to derive the autocorrelation coefficient of the process to which the variable belongs. Calculations of the variance of the estimated Allan variance proved that it may be convergent to its true value with infinite sample number for considered spectral densities of frequency noise. We also applied the results to estimations of frequency measurements to know the influence of the autocorrelation of the process considered.

In order to obtain some direct estimates of the confidence of the estimate, distributions of the estimate were plotted by means of computer simulations, and were compared with the chi-square distribution. Those results suggested that for white- and flicker-phase noises (and white-frequency noise) we have to take into account the autocorrelation of the process, while for flicker- or random-walk-frequency noise we may regard the process as a nearly independent (and Gaussian) one.

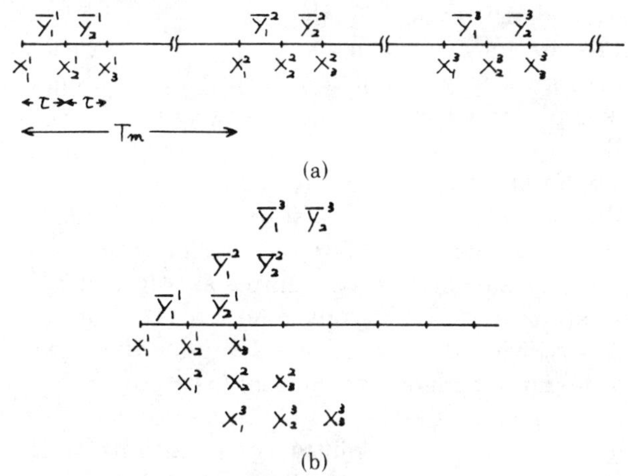

Fig. 1. (a) General sampling form for the estimated Allan variance. (b) Special case of (a) with $T_m = \tau$.

I. Introduction

THE ALLAN variance, the proposed measure of frequency stability in the time domain, is defined by [1]

$$\sigma_y{}^2(\tau) = \langle (\bar{y}_k - \bar{y}_{k+1})^2 \rangle / 2 \tag{1}$$

where $\langle\ \rangle$ denotes infinite time average and the quantity $\bar{y}_k$ is the mean value of the fractional frequency fluctuation $y(t)$ over the time interval τ during kth measurement interval. In practice, data records are of finite length, and the infinite time average implied in the definition is not normally available; an estimate for the measure must be used. The estimate given by [1] is

$$\sigma_y{}^2(\tau) \simeq \frac{1}{M} \sum_{k=1}^{M} (y_k - y_{k+1})^2 / 2 \tag{2}$$

where M is a finite number of measurements of $\bar{y}_k$. We see that (2) is quite similar in form to (1).

Lesage and Audoin [2] calculated the variance of the estimate in (2) with a straightforward manipulation for commonly encountered spectral densities of frequency fluctuations, and deduced the relative uncertainty in the evaluation of frequency stability, as a function of M.

Manuscript received June 28, 1976.
The author is with the Frequency Standard Division, Radio Research Laboratories, Ministry of Posts and Telecommunications, Koganei-shi, Tokyo 184, Japan.

Here we have to notice that the average contained in (2) is a kind of ensemble average, compared with the time average in (1). More strictly speaking, a long record from a single member of an ensemble, measured in the time domain, can be cut into a sequence of strips so that each strip may give, for example, each term in the summation of (2). Then this sequency of samples drawn from a single record can be taken as an ensemble, and according to the ergotic hypothesis this new ensemble is statistically equivalent to the original ensemble [3]. In other words, the average of the new ensemble may be nearly equivalent to the infinite time average as shown in (1). Thus it is clear that the terms in the summation of (2) are expected to be mutually uncorrelated, and moreover (2) cannot be the only estimate form for the Allan variance given in (1).

We considered the general sampling form for the estimate of the Allan variance as shown in Fig. 1(a) and defined a variable z_i as

$$z_i \equiv (\bar{y}_1{}^i - \bar{y}_2{}^i)/\sqrt{2}\,. \tag{3}$$

Accordingly, the estimate is expressed in general form as

$$\sigma_y{}^2(\tau) \simeq \sum_{i=1}^{M} z_i{}^2/M. \tag{4}$$

We see that (2) is given as a special case of (4)—a case of $r_m = T_m/\tau = 1$ in Fig. 1(a) (see Fig. 1(b)). In general, frequency fluctuations or their measured values may be

Reprinted from *IEEE Trans. Instrum. Meas.*, vol. IM-27, pp. 1–7, Mar. 1978.

autocorrelated, and in case of (2) any one of $\bar{y}_2, \bar{y}_3, \cdots$ is commonly used in two adjacent terms in the summation. Therefore z_i or z_i^2 cannot be uncorrelated of each other.

We defined the correlation coefficient of the process $z_1, z_2 \cdots$ for use in the evaluation of the confidence of the estimate in (4) and for general evaluations of variances associated with the process, and expressed the coefficient using the stability measure $\langle \sigma_y^2(N,\tau,\tau)\rangle$ in the time domain [1] to calculate it as a function of r_m for considered spectral densities of frequency noises.

The results showed that with $r_m = 1$ and 2, the coefficients for white-phase noise had the largest value, and for flicker-phase noise the next largest, and so on. Random-walk frequency-noise process always had the zero coefficients for any r_m.

The use of the autocorrelation coefficients calculated made it possible to evaluate the variance of the estimate in (4), and the variances of sample averages of z_i and $\bar{y}_i$. The estimate proved its convergence to the Allan variance for any r_m with $M \to \infty$, on the assumption that the process had the multidimensional normal distribution.

For some direct estimates of the confidence of the estimate given in (4), we plotted distributions of Σz_i^2 by means of computer simulations for the frequency noises with commonly encountered spectral densities, and examined their deviations from the chi-square distributions with M degrees of freedom.

From those results we came to the conclusion that for white- and flicker-phase noises (and for white-frequency noise) we have to consider the autocorrelation of the process to which the variable z_i belongs, at least with $r_m = 1$ and 2, but for flicker- and random-walk-frequency noises we may neglect such autocorrelation, so that the process can be regarded as a nearly independent (and Gaussian) one, which denotes that the Allan variance is a good measure for those divergent frequency noises.

II. Correlation Coefficients

Fig. 1(a) shows the general sampling form for the Allan variance and Fig. 1(b) is a special case of $r_m = T_m/\tau = 1$ in Fig. 1(a), where X^i indicates time. From Fig. 1(a),

$$\bar{y}_1^i = (X_1^i - X_2^i)$$
$$\bar{y}_2^i = (X_2^i - X_3^i). \tag{5}$$

The autocorrelation coefficient of the process $z_1, z_2, \cdots$, defined by (3), is written as

$$\rho_{ij} = \langle z_i z_j\rangle / \langle z_i^2\rangle \tag{6}$$

where $\langle\ \rangle$ denotes infinite time average. We will assume ergotic stationarity in our discussions.

By definition,

$$\langle z_i^2\rangle = \sigma_y^2(\tau) \tag{7}$$

and, on the assumption of stationarity,

$$\rho_{ij} = \rho_{ji} = \rho_k(k = |i-j|). \tag{8}$$

Substitution of (3) into (6) gives

$$\begin{aligned}\langle z_i z_j\rangle &= \langle(\bar{y}_1^i - \bar{y}_2^i)(\bar{y}_1^j - \bar{y}_2^j)\rangle/2 \\ &= [\langle \bar{y}_1^i\bar{y}_1^j\rangle - \langle \bar{y}_1^i\bar{y}_2^j\rangle - \langle \bar{y}_2^i\bar{y}_1^j\rangle + \langle \bar{y}_2^i\bar{y}_2^j\rangle]/2 \end{aligned}\tag{9}$$

The terms on the right-hand side of (9) can be expressed using the time variable X^i given in (5); for example,

$$\begin{aligned}\tau^2\langle \bar{y}_1^i\bar{y}_1^j\rangle = \langle X_1^iX_1^j\rangle &- \langle X_1^iX_2^j\rangle \\ &- \langle X_2^iX_1^j\rangle + \langle X_2^iX_2^j\rangle\end{aligned} \tag{10}$$

Since each term on the right hand side of (10) indicates the autocorrelation function of the time process, we may write them using the notation R_X. For example, we obtain, from Fig. 1,

$$R_X(kr_m\tau) = \langle X_1^iX_1^j\rangle = \langle X_2^iX_2^j\rangle \tag{11}$$

where $k = |i - j|$.

If we let

$$U_X(Z) \equiv 2[R_X(0) - R_X(Z)] \tag{12}$$

then $U_X(n\tau)$ can be related to the stability measure $\langle \sigma_y^2(n,\tau,\tau)\rangle$ [4]:

$$U_X(n\tau) = -n(n-1)\tau^2\langle\sigma_y^2(n,\tau)\rangle + n^2U_X(\tau) \tag{13}$$

where $\langle\sigma_y^2(n,\tau,\tau)\rangle$ is abbreviated to $\langle\sigma_y^2(n,\tau)\rangle$ for simplicity and n is not necessarily an integer [5].

Substitution of the relations given in (11)–(13) to (10) yields

$$\begin{aligned}\langle \bar{y}_1^i\bar{y}_1^j\rangle &= kr_m(kr_m - 1)\langle\sigma_y^2(kr_m,\tau)\rangle \\ &- (kr_m + 1)kr_m\langle\sigma_y^2(kr_m + 1,\tau)\rangle/2 \\ &- (kr_m - 1)(kr_m - 2)\langle\sigma_y^2(kr_m - 1,\tau)\rangle/2 \\ &+ U_X(\tau)/\tau^2.\end{aligned} \tag{14}$$

In (14) $U_X(\tau)/\tau^2$ agrees with $\langle\sigma_y^2(\infty,\tau)\rangle$ when it exists. $\langle\bar{y}_1^i\bar{y}_1^j\rangle$ denotes the autocorrelation function of the fractional-frequency fluctuations.

The other terms on the right hand side of (9) are expressed similarly to (14), and the results will be

$$\begin{aligned}\langle z_iz_j\rangle &= 1.5a(a-1)\langle\sigma_y^2(a,\tau)\rangle \\ &- (a+1)a\langle\sigma_y^2(a+1,\tau)\rangle \\ &- (a-1)(a-2)\langle\sigma_y^2(a-1,\tau)\rangle + 0.25(a \\ &+ 2)(a+1)\langle\sigma_y^2(a+2,\tau)\rangle \\ &+ 0.25(a-2)(a-3)\langle\sigma_y^2(a-2,\tau)\rangle\end{aligned} \tag{15}$$

where $a \equiv kr_m$ and all the terms $U_X(\tau)/\tau^2$ canceled out.

Barnes *et al.* [1] tabulate the stability measure $\langle\sigma_y^2(n,\tau)\rangle$ for the following power-law type of spectral density:

$$S_y(f) = \begin{cases} h_\alpha f^\alpha, & 0 \le f \le f_h \\ 0, & f > f_h, \quad \alpha = -2,-1,0,1,2 \end{cases} \tag{16}$$

where $S_y(f)$ is the spectral density of $y(t)$ and h_α is a constant. f_h shows high-frequency cutoff of an idealized infinitely sharp cutoff low-pass filter. Thus, we can calculate ρ_{ij} in (6) by substituting $\langle\sigma_y^2(n,\tau)\rangle$, which is a function of h_α, n, τ, and f_h, into (15). (See Table I.)

TABLE I

(a)—Autocorrelation coefficient ρ_k.

a \ noise	f^0 PM	f^{-1} PM	f^0 FM	f^{-1} FM	f^{-2} FM
$a \neq 0,1,2$	0	$\frac{1}{c_2}[2\ln(a^2-1)-3\ln(a)-\frac{1}{2}\ln(a^2-4)]$	0	$\frac{1}{4\ln 2}[3a^2\ln(a)-2(a+1)^2\ln(a+1)-2(a-1)^2\ln\lvert a-1\rvert+\frac{1}{2}(a+2)^2\ln(a+2)+\frac{1}{2}(a-2)^2\ln\lvert a-2\rvert]$	0
$a \to 1$	$-\frac{2}{3}$	$\frac{-1}{c_2}[2c_1+\frac{1}{2}\ln 3-2\ln 2]$	$-\frac{1}{2}$		
$a \to 2$	$\frac{1}{6}$	$\frac{1}{4c_2}[2c_1-16\ln 2+8\ln 3]$			

(b)—Numerical value of ρ_k for flicker-phase and flicker-frequency noises. $E-d \equiv \times 10^{-d}$. $c_1 = 3/2 + \ln(2\pi f_h \tau)$, $c_2 = 9/2 + 3\ln(2\pi f_h\tau) - \ln 2$, $2\pi f_h \tau \gg 1$. $a \equiv kr_m$.

noise	f^{-1}FM	f^{-1}PM		
a \ $2\pi f_h\tau$		10	10^2	10^3
1	-0.217	-0.632	-0.645	-0.651
2	-0.132	0.124	0.141	0.148
3	-0.455E-1	0.544E-2	0.331E-2	0.238E-2
4	-0.241E-1	0.138E-2	0.838E-3	0.602E-3
5	-0.150E-1	0.516E-3	0.314E-3	0.226E-3
10	-0.364E-2	0.290S-4	0.176E-4	0.127E-4
100	-0.385E-4	0.282E-8	0.172E-8	0.123E-8

For phase noises ($\alpha = 1,2$), the coefficient ρ_{ij} are larger with $a = kr_m = 1,2$ than with any other values of a, the reason for which is that a time variable X^i used for a sample variance of $(\bar{y}_k - \bar{y}_{k+1})^2/2$ may be again used for the next sample (see Fig. 1(b) for $a = 1$), which makes the correlation between X^i's maximum. See Appendix A for the calculations of ρ_{ij} in those cases. We see from Table I that ρ_{ij} becomes rather small with $a \geq 3$ even for flicker-phase noise ($\alpha = 1$).

For white-frequency noise ($\alpha = 0$), $\rho_{ij} = 0$, except for the case where $a = 1$ in which a certain $\bar{y}^i$ may be commonly used in two adjacent sample variances (see Fig. 1(b)). While for random-walk-frequency noise ($\alpha = -2$), the coefficient always becomes zero. It may depend on the following reason: z_i is given by the difference between two average fractional frequency fluctuations as shown in (3), so the power spectral density of z_i,$S_z(f)$, will be nearly proportional to $f^2 \cdot S_y(f)$. It follows that for random-walk frequency noise the process of z_i ($i = 1,2,\cdots$) may become equivalent to a white noise process in which $\rho_{ij} = 0$ except for the case where $i = j$.

Similarly, $S_z(f) \propto f$ for flicker-frequency noise ($\alpha = -1$), which indicates that the process of $z_1, z_2, \cdots$ is a convergent one. On the other hand, $S_z(f) \propto f^3$ for flicker-phase noise, which denotes an increasingly rapid variation of the process.

Comparisons among the autocorrelation coefficients with $a = 1$ and 2 indicate that ρ_{ij} for white-phase noise had the largest value and for flicker-phase noise the next largest, and so on. These are related to the fact that in two adjacent sample variances there are more number of time variables of X^i's which are commonly used for phase noises than $\bar{y}^i$'s for frequency noises. In other words, the fact that z_i is given by the difference between two $\bar{y}^i$'s may result in rather large correlation for phase-noise processes.

III. Variance of Estimated Allan Variance

The variance of the estimate of the Allan variance which is given in (4) will be written as

$$\sigma^2 = \left\langle \left[\sum_{i=1}^{M} z_i{}^2/M - \sigma_y{}^2(\tau) \right]^2 \right\rangle$$

$$= \langle z_i{}^4 \rangle / M + \sum_{i \neq j} \langle z_i{}^2 z_j{}^2 \rangle / M^2 - [\sigma_y{}^2(\tau)]^2. \quad (17)$$

Let us assume that two variables z_i and z_j have the two-dimensional normal distribution. Then we obtain the following relations [6]:[1]

$$\langle z_i{}^4 \rangle = 3[\sigma_y{}^2(\tau)]^2$$

$$\langle z_i{}^2 z_j{}^2 \rangle = [\sigma_y{}^2(\tau)]^2 (1 + 2\rho_{ij}{}^2). \quad (18)$$

Consequently, (17) becomes

$$\sigma^2 = (2/M)[\sigma_y{}^2(\tau)]^2 Fct \quad (19)$$

[1]If variables $z_1, z_2, \cdots, z_{2k}$, where k is an integer, have the multidimensional normal distribution, then the ensemble average of $z_1 z_2 \cdots z_{2k}$ may be calculated using the following relation:

$$\overline{z_1 z_2 \cdots z_{2k}} = \Sigma\, \overline{z_1 z_2}\ \overline{z_3 z_4} \cdots \overline{z_{2k}}$$

where the bar denotes the ensemble average and Σ implies to sum up every combination of the moment of any two variables multiplied by that of the residuals.

TABLE II

(a)—Factor for the variance of the estimated Allan variance; $Fct = 1 + (2/M)\,\Sigma_{k=1}{}^{M-1}(M-k)\rho_k{}^2$.

noise / r_m	f^0 PM	f^{-1} PM	f^0 FM	f^{-1} FM	f^{-2} FM
1	$\frac{35}{18} - \frac{1}{M}$	see (b)	$\frac{3}{2} - \frac{1}{2M}$	see (b)	1
2	$\frac{19}{18} - \frac{1}{18M}$	≈1	1	≈1	1
≥3	1	≈1	1	≈1	1

(b)—Numerical value of Fct for flicker-phase and frequency noises.

noise	f^{-1} FM			f^{-1} PM ($2\pi f_h\tau = 10^2$)		
M \ r_m	1	2	3	1	2	3
2	1.047	1.018	1.002	1.417	1.020	1.000
5	1.098	1.029	1.004	1.690	1.032	1.000
10	1.117	1.033	1.004	1.781	1.036	1.000
100	1.134	1.036	1.004	1.864	1.039	1.000

where

$$Fct \equiv 1 + (2/M) \sum_{k=1}^{M-1} (M-k)\rho_k{}^2 \qquad (20)$$

with the relation of

$$\sum_{i \neq j} \rho_{ij}{}^2 = 2 \sum_{i<j} \rho^2{}_{k=j-i} = 2 \sum_{k=1}^{M-1} (M-k)\rho_k{}^2 \qquad (21)$$

Note that $Fct = 1$ with $\rho_k = 0$ for all k. In Table II, Fct is tabulated as a parameter of r_m for the considered spectral densities of frequency noises. Although Table II(b) indicates only the case of $2\pi f_h\tau = 10^2$ for flicker-phase noise, it is true that Fct does not much depend on $f_h\tau$. It is readily seen from Table II that Fct is convergent to a constant as $M \to \infty$ in each case. So the variance in (17) approaches zero as $M \to \infty$, which results in the estimated Allan variance to converge to the true value (Chebyshev's theorem).

Lesage and Audoin [2] and Tausworthe [7] calculated σ^2 in (17) with $r_m = 1$ and 2, respectively, by means of straightforward manipulations, but their results do not necessarily seem to agree with those in Table II.

IV. Distributions of the Estimated Allan Variance by Means of Computer Simulations

In general the estimation of confidence intervals needs the knowledge of the distribution of the process. If the M-sample averages of $z_i{}^2$ (4) have the normal distribution (probably by central limit theorem for sufficiently large M [1]) the confidence interval may be easily estimated on the normal distribution using the variance in (19). But it seems to be more natural to consider that the estimate (4) has the chi-square distribution even when the variables z_i's are mutually independent and have normal distribution. In that case, the variance of the estimated Allan variance, calculated in Section III, may not be very useful for the estimation of confidence interval.

Here we will show the distribution of (4) by means of computer simulations, where z_i's are not necessarily independent of each other. White-noise process used for the simulations are made to have the normal distribution, which seems to agree to actual cases that we have obtained

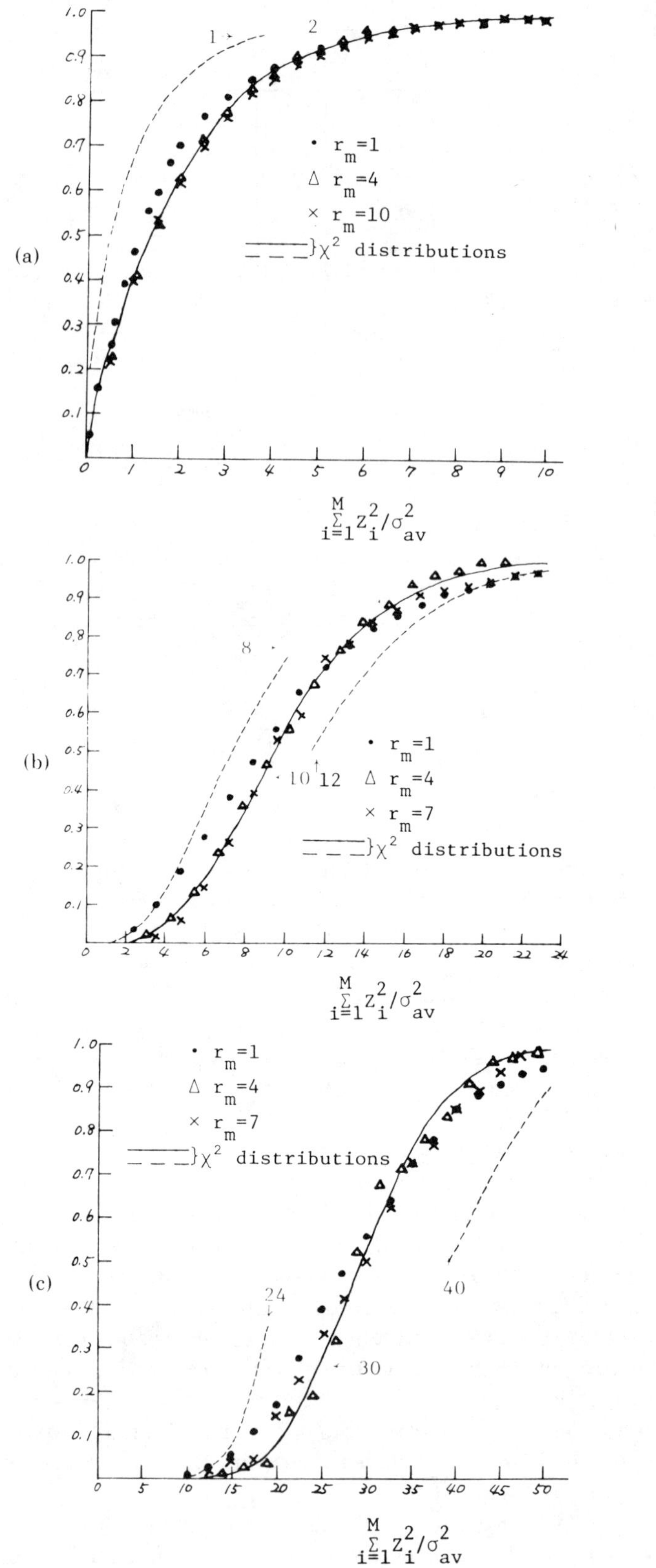

Fig. 2. Distributions of $\Sigma z_i{}^2/\sigma_{av}{}^2$ for a white phase noise by means of computer simulations with (a) $M = 2$, (b) $M = 10$ and (c) $M = 30$. The vertical axes denote probability distribution functions and $\sigma_{av}{}^2$ is supposed to be close to the Allan variance. The numbers around the curves express degrees of freedom for χ^2 distributions.

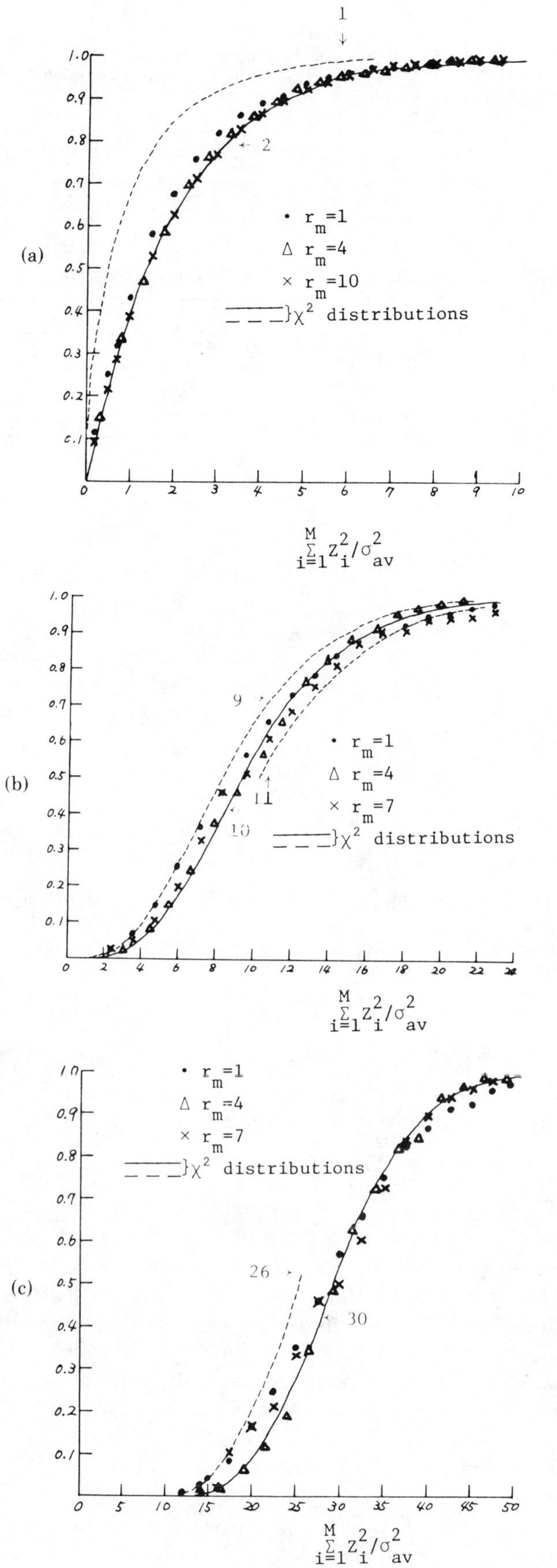

Fig. 3. Distributions of $\Sigma z_i{}^2/\sigma_{av}{}^2$ for a white-frequency noise by means of computer simulations with (a) $M = 2$, (b) $M = 10$, and (c) $M = 30$. The vertical axes denote probability distribution functions and $\sigma_{av}{}^2$ is supposed to be close to the Allan variance. The numbers around the curves express degrees of freedom for χ^2 distribution.

so far. Flicker-noise process used is generated by proper filtering of the white-noise process [8], and it seems to show a deviation from the normal distribution because of its characteristic of divergence occurring with time passage. There are some reports, however, in which actual data on measurements of flicker-frequency noises indicated nearly normal distributions [2], [9].

We gave examples of distributions of $\Sigma z_i{}^2$ by computer simulations in Fig. 2 for white-phase noise and in Fig. 3 for white-frequency noise as a parameter of r_m, and omitted to show those for flicker-phase or frequency noise. In each of the figures, the vertical axis denotes the probability distribution function and the variance of $\sigma_{av}{}^2$, a normalized factor, may be regarded as the true variance $\sigma_y{}^2(\tau)$, and the variance $\sigma_{av}{}^2$ is estimated by the following: First we calculate (4) with $M = 10$ to 100 and with a fixed r_m for one process, and then we average those for 10 independent processes. Let us denote it $\sigma_{r_m}{}^2$. Then $\sigma_{av}{}^2$ is given by

$$\sigma_{av}{}^2 = \frac{1}{5}\sum_{r_m} \sigma_{r_m}{}^2, \qquad r_m = 10,20,30,50,100$$

$\sigma_{r_m}{}^2$ for each r_m coincides with $\sigma_{av}{}^2$ in the error less than 0.5 percent.

In Figs. 2 and 3 larger M and r_m make smaller the number of elements which constitute the distributions, so the curves of the distribution function come to show lack of smoothness. The curves with $M = 30$ seem to be close to normal distribution (central limit theorem).

We see from Figs. 2 and 3 that for white-frequency noise the distributions are closer to chi-square distribution of M degrees of freedom which are also shown in the diagrams for comparisons than those for white-phase noise. The distributions for flicker-phase noise which are omitted here can be said to be in between those in Fig. 2 and Fig. 3. The distributions for flicker-frequency noise are such that they are closer to chi-square distributions of M degrees of freedom than for white-frequency noise. It is easily seen, on the assumption of normal distribution of white-frequency noise and from the discussions mentioned in Section II on random-walk-frequency noise process, that for random walk frequency noise $\Sigma z_i{}^2$ will have chi-square distribution of M degrees of freedom.

The distributions at least with $r_m = 1$ have a tendency to coincide with chi-square distribution of less than M degrees of freedom around the bottom of the curves and of more than M degrees of freedom around the top of the curves. Consequently, confidence limit of the estimated Allan variance may become broader in these cases than in chi-square distribution of M degrees of freedom.

V. Estimations of Frequency Values Measured

A. *Estimation of Frequency Difference*

Consider that we observe step variations of output frequency by changing parameters of an oscillator in a periodic manner and sample them in the form shown in Fig. 1 to obtain a series of difference frequency data $\Delta y_1, \Delta y_2,$

$\cdots$. Let us assume that there exists the true average of the difference frequencies, ΔY_0;

$$\Delta Y_0 = \langle \Delta y_i \rangle \tag{22}$$

And the variance of $\Delta y_i, \sigma_{\Delta y}{}^2$, which is supposed to be independent of i, is given by

$$\sigma_{\Delta y}{}^2 = \langle (\Delta y_i - \Delta Y_0)^2 \rangle = \langle \Delta y_i{}^2 \rangle - \Delta Y_0{}^2 \tag{23}$$

(23) is connected with the Allan variance defined by (1):

$$\sigma_{\Delta y}{}^2 = 2\sigma_y{}^2(\tau). \tag{24}$$

It follows that even for flicker or random-walk frequency noise the variance $\sigma_{\Delta y}{}^2$ is finite, so ΔY_0 can be observable.

Let ΔY_M be the M-sample average of Δy_i;

$$\Delta Y_M = \sum_{i=1}^{M} \Delta y_i / M. \tag{25}$$

Then the variance of $\Delta Y_M, \sigma_{\Delta Y_M}{}^2$, is written as

$$\sigma_{\Delta Y_M}{}^2 = \langle (\Delta Y_M - \Delta Y_0)^2 \rangle \tag{26}$$

we define a correlation coefficient as

$$\rho_{ij} = \rho_{k=|i-j|} = \langle (\Delta y_i - \Delta Y_0)(\Delta y_j - \Delta Y_0) \rangle / \sigma_{\Delta y}{}^2. \tag{27}$$

Substitution of (22), (23), and (25) into (26) gives

$$\sigma_{\Delta Y_M}{}^2 = \sigma_{\Delta y}{}^2 Fct'/M \tag{28}$$

where, using a similar relation to (21),

$$Fct' \equiv 1 + (2/M) \sum_{k=1}^{M-1} (M-k)\rho_k. \tag{29}$$

Since $Fct' = 1$ with $\rho_k = 0$ for all k, (28) takes the familiar form for independent variables.

Notice that the right hand side of (25) will be $(\bar{y}_1{}^1 - \bar{y}_2{}^M)/M$ for $r_m = 1$ (see Fig. 1(b)), implying that all the intermediate terms of $\bar{y}^i$ disappear. Hence it is necessary to take the sum of the Δy_i's properly by changing their signs for actual applications. Thus Fct' should be written as

$$Fct' \equiv 1 + (2/M) \sum_{k=1}^{M-1} (-1)^b (M-k)\rho_k \tag{30}$$

where b is an odd number for r_m, and k odd, and otherwise an even number. Equation (30) can be calculated by using the autocorrelation coefficients evaluated in Section II. See Table III. With M infinite, $\sigma_{\Delta Y_M}{}^2 \to 0$ since Fct' is convergent to a finite value. It follows that $\Delta Y_M \to \Delta Y_0$ by Chebyshev's theorem.

We may estimate the confidence interval of the average difference frequency using the variance in (28).

B. Estimation of Average Frequency

The sample average of N measurements of $\bar{y}_i$ is given by

$$Y_N = \sum_{i=1}^{N} \bar{y}_i / N. \tag{31}$$

TABLE III

(a)—Factor for the variance of the average difference frequency; $Fct' = 1 + (2/M)\,\Sigma_{k=1}{}^{M-1}\,(-1)^b (M-k)\rho_k$, where b is an odd number for r_m and k odd and otherwise an even.

noise / r_m	f PM	f^{-1} PM	f FM	f^{-1} FM	f^{-2} FM
1	$\frac{8}{3} - \frac{2}{M}$	see (b)	$2 - \frac{1}{M}$	see (b)	1
2	$\frac{4}{3} - \frac{1}{3M}$		1		1
≥3	1	≈1	1	≈1	1

(b)—Numerical value of Fct for flicker-phase and frequency noises.

noise	f^{-1} FM			f^{-1} PM ($2\pi f_h \tau = 10$)		
M \ r_m	1	2	3	1	2	3
2	1.217	0.868	1.046	1.645	1.141	0.997
5	1.215	0.749	1.063	2.199	1.226	0.995
10	1.223	0.694	1.070	2.383	1.255	0.994
100	1.229	0.627	1.076	2.548	1.280	0.994

TABLE IV

Correlation coefficient ρ_k', factor Fct'' and variance $\sigma_{Y_N}{}^2$ of average frequency for white-phase noise.

kr_m	ρ_k'	Fct''	$\sigma_{YN}{}^2$
1	-1/2	1/N	$\sigma_{\bar{y}}{}^2/N^2$
>1	0	1	$\sigma_{\bar{y}}{}^2/N$

Then the variance of $Y_N, \sigma_{Y_N}{}^2$, will be expressed in the same way as (28):

$$\sigma_{Y_N}{}^2 = Fct'' \sigma_{\bar{y}}{}^2 / N \tag{32}$$

where $\sigma_{\bar{y}}{}^2$, the variance of $\bar{y}_i$, equals the stability measure $\langle \sigma_y{}^2(\infty, \tau) \rangle$ in (13), and

$$Fct'' \equiv 1 + 2 \sum_{k=1}^{N-1} (N-k)\rho_k'/N \tag{33}$$

which is similar to (29) in form, but ρ_k' is the correlation coefficient of $\bar{y}_i$ in this case to be given by

$$\rho_{ij}' = \rho_k'(k = |i-j|) = \langle (\bar{y}_i - Y_0)(\bar{y}_j - Y_0) \rangle / \sigma_{\bar{y}}{}^2 \tag{34}$$

where $Y_0 \equiv \langle Y_N \rangle$ is the true value of Y_N, assuming that it exists.

For white-frequency noise, $Fct'' = 1$ because $\rho_k' = 0$ for all k and $\sigma_{\bar{y}}{}^2$ agrees with $\sigma_y{}^2(\tau)$ [1]. For white-phase noise, ρ_k' and Fct'' can be evaluated by using (14). The results are shown in Table IV with $\sigma_{Y_N}{}^2$ in (32). Notice that $\sigma_{\bar{y}}{}^2 = 2\sigma_y{}^2(\tau)/3$ for white-phase noise [1]. It is supposed that the result for flicker-phase noise will be similar to those for white-phase noise. Consequently, M-sample average of frequency measured may converge into its true value with $M \to \infty$.

While for flicker- or random-walk-frequency noise $\sigma_{\bar{y}}{}^2 (= \langle \sigma_y{}^2(N \to \infty, \tau) \rangle)$ will be infinite. Only a finite number for N holds the stability measure in time domain finite and makes it significant. Hence the variance of N-sample average Y_N for a mixture of white-frequency noise and flicker- or random-walk-frequency noise may be expressed as

$$\sigma_{Y_N}{}^2 = \sigma_y{}^2(\tau)_{f^0FM}/N + \langle \sigma_y{}^2(N,T,\tau)\rangle_{f^\alpha(\alpha<0)FM} \quad (35)$$

We see from (35) that there exist optimum N and τ to make $\sigma_{Y_N}{}^2$ minimum.

VI. Conclusion

We have calculated the autocorrelation coefficients between the variables z_i's which constitute the estimated Allan variance and applied it to the estimations of frequency values measured. The calculation of the variance of the estimate has proved its convergence into the true value (the Allan variance) with M infinite for considered spectral densities of frequency noises. We have plotted the distributions of $\Sigma z_i{}^2$, the estimated Allan variance multiplied by M, by means of computer simulations.

Those results have led to the conclusion that for phase noises (and white-frequency noise) we have to consider the autocorrelation of the process especially with $r_m = 1$ in the case of which the estimated Allan variance is given by (2). While for flicker- or random-walk-frequency noise we may deal with the process as nearly independent (and Gaussian) one, which reveals the superiority of the Allan variance, the stability measure in time domain.

Appendix A

Calculations of ρ_k with $a = 1$ and 2 for Phase Noises

Note that each term in the right-hand side of (15) takes the form of $n(n-1)\langle\sigma_y{}^2(n,\tau)\rangle$, and the evaluations of (15) with $a = kr_m = 1$ or 2 are reduced to those of $n(n-1)\langle\sigma_y{}^2(n,\tau)\rangle$ with $n = 1$ and 2.

From (13), we obtain

$$n(n-1)\langle\sigma_y{}^2(n,\tau)\rangle = \frac{1}{\tau^2}[n^2U_X(\tau) - U_X(n\tau)]. \quad (A1)$$

It follows that $n(n-1)\langle\sigma_y{}^2(n,\tau)\rangle \to 0$ with $n \to 1$ for any noises if $U_X(\tau)$ exists.

With $n \to 0$, (A1) will become

$$n(n-1)\langle\sigma_y{}^2(n,\tau)\rangle \to -U_X(n\tau)/\tau^2 \equiv -n^2U_X(\tau')/\tau'^2 = -n^2\langle\sigma_y{}^2(\infty,\tau')\rangle, \qquad \tau' \equiv n\tau \quad (A2)$$

assuming U_X is finite. Since $\langle\sigma_y{}^2(\infty,\tau')\rangle$ is constant on τ' for bandlimited white-phase noise [10], we obtain

$$n(n-1)\langle\sigma_y{}^2(n\to 0,\tau)\rangle \to 0. \quad (A3)$$

While for flicker-phase noise $U_X(\tau')$ can be written as [12]

$$U_X(\tau') \propto [2\beta^2\ln\beta - (1+\beta)^2\ln(1+\beta) - (1-\beta)^2\ln|1-\beta|] \quad (A4)$$

where $\beta = \omega_h\tau'$. The right-hand side of (A4) approaches $2\beta^2(\beta - 2)$ with $\tau' \to 0$, so we again obtain (A3) for flicker-phase noise.

After all, the use of the above results for (15) yields the correlation coefficients with $a = 1$ and 2 as shown in Table I for flicker-phase noise, and for white-phase noise, we obtain with $a = 1$,

$$\langle z_iz_j\rangle \simeq -2\langle\sigma_y{}^2(2,\tau)\rangle + \frac{3}{2}\langle\sigma_y{}^2(3,\tau)\rangle \quad (A5)$$

which gives, from (6) and (7),

$$\rho_k(kr_m \to 1) = -2/3 \quad (A6)$$

and with $a = 2$,

$$\langle z_iz_j\rangle = 3\langle\sigma_y{}^2(2,\tau)\rangle - 6\langle\sigma_y{}^2(3,\tau)\rangle + 3\langle\sigma_y{}^2(4,\tau)\rangle \quad (A7)$$

which gives

$$\rho_k(kr_m \to 2) = 1/6. \quad (A8)$$

Acknowledgment

The author wishes to thank Dr. H. Akaike and Y. Yasuda for their helpful discussions. The author also wishes to acknowledge the significant contribution of Dr. J. A. Barnes in the Time and Frequency Division of NBS, Boulder, CO, in making manuscript criticisms.

References

[1] J. A. Barnes *et al.*, "Characterization of frequency stability," *IEEE Trans. Instrum. Meas.*, vol. IM-20, p. 105, May 1971.
[2] P. Lesage and C. Audoin, "Characterization of frequency stability: uncertainty due to the finite number of measurements," *IEEE Trans. Instrum. Meas.*, vol. IM-22, p. 157, June 1973.
[3] J. S. Bendat, *Principles and Applications of Random Noise Theory.* New York: Wiley, 1958.
[4] D. W. Allan, "Statistics of atomic frequency standards," *Proc. IEEE*, vol. 54, p. 221, Feb. 1966.
[5] K. Yoshimura, "The generation of an accurate and uniform time scale with calibrations and prediction," NBS Tech. Note 626, Nov. 1972.
[6] K. Miyawaki, *Mathematical Analysis of Random Noise.* Tokyo: Asakura, Inc., 1970 (in Japanese).
[7] R. C. Tausworthe, "Convergence of oscillator spectral estimators for counted-frequency measurements," *IEEE Trans. Commun.* (Concise Paper), vol. COM-20, p. 214, Apr. 1972.
[8] J. A. Barnes and S. Jarvis, "Efficient numerical and analog modeling of flicker noise processes," NBS Tech. Note 604, June 1971.
[9] D. W. Allan *et al.*, "Some performance data for the standards in the National Bureau of Standards atomic time scale system," in *Proc. 27th Annu. Symp. Frequency Control,* Fort Monmouth, N.J., June 12–14, 1973.
[10] L. S. Cutler and C. L. Searl, "Some aspects of the theory and measurements of frequency fluctuations in frequency standards," *Proc. IEEE*, vol. 54, p. 136, Feb. 1966.

Part IV
Measurement of Frequency Stability

ANALOG measurement of frequency stability, i.e., the direct measurement of the phase noise PSD $S_\varphi(f)$, is easily improvised in the laboratory with a modest cost. The advantage is the excellent sensitivity and low background noise; some difficulty may be caused by the calibration procedure. Another drawback is the low frequency limit which is in the neighborhood of about 10 Hz.

An actual arrangement is discussed by Meyer in the first of the reprint papers in this part. Additional information about laboratory equipment may be found, e.g., in NBS publications [IV-1]–[IV-3], particularly in the latter one in which diagrams of actual circuits as used in NBS are described.

The extensive application of digital techniques in the mid-70's made possible the automation of frequency stability measurements on one hand and extension of phase-noise PSD measurements as close to the carrier as 100 or 20 mHz (the major obstacle being the long averaging time) on the other hand. The former problem, i.e., merely the calibration and measurement steps being controlled by a calculation, is dealt with by Lance *et al.* in the second reprint paper and in a concise paper by M. McNamee [IV-4] which refers to application of a desktop computer.

Extension of PSD measurements below the 10 Hz limit is based on the indirect evaluation of the time domain measurements; consider the conversion table, Table III-1, between two-sample (or Allan) variance and PSD $S_y(f)$ or $S_\varphi(f)$. The introduction of the computing counter systems [IV-5] made application of filtering properties of the Hadamard transform easy. An actual system with the calculated background noise (due to quantization) is discussed in the third reprint paper, by Peregrino. The difficulty with the background quantization noise can be overcome by frequency multiplication [IV-6], by phase error multiplication [IV-7], or by dc amplification of the phase detector output with the following voltage-to-frequency conversion as proposed by Grosslambert in the fourth reprint paper. Let us mention that there exists another possibility for getting below the 10 Hz limit, namely with the assistance of digital filters [IV-8] replacing the analog analyzer.

Further perfection of the time domain measurements proposed in the mid-70's by Allan deal with instances where the nominal frequencies of the measured and reference generators are effectively the same (typical examples are atomic frequency standards); see the fifth reprint paper. Allan used an auxiliary transfer oscillator (synthesizer), generated two beat frequencies, and measured the time interval between the nearly coincident zero crossings (to eliminate noise of the auxiliary oscillator). The advantages and disadvantages with respect to other frequency stability measurement systems are summarized in Tables IV-1 and IV-2 taken from [III-10]. Actual circuits of this "dual mixer time difference measurement system" as used in NBS have been revealed in [IV-9] and a discussion of its possible extension to an automated phase measurement system comprised of more than two oscillators is revealed in [IV-10].

The complexity of digital time domain frequency stability measurements instigated the search for analog methods (see Part III, paper 5, [III-9], and [IV-11]). In the sixth reprint paper, by Wiley, a particularly simple filter, the transfer function of which closely resembles that computed for the two-sample variance, makes it possible to estimate the time domain stability quite easily even for short "averaging" times. Another system using more complicated active filters is discussed by Youlin and Juebang in the seventh reprint paper where the actual circuits with components values are given.

In the mid-70's large scale integration was perfected such that spectral analyzers based on fast Fourier transform (FFT) became a reality. Babitch and Oliverio were the first to compare the close to carrier phase noise measurements made with the assistance of a commercial FFT processor, converted to the time domain with data found directly via the computing counter system. As the sensitivity of the present-day FFT analyzers is not sufficient for such precise measurements, they took the recourse of enhancing the phase noise by frequency multiplication; see the eighth reprint paper. Soon afterwards, the introduction of microprocessors into the measurement technique made possible the construction of commercial FFT analyzers, and as a consequence, paved the way even for a nonspecialist to perform complicated frequency stability measurements [IV-12] and [IV-13].

The last reprint paper in this part, by Lindsey and Chie, deals with some practical applications of structure functions in frequency stability measurements, namely with the mutual conversion between the time domain and frequency domain in the use of fractional exponents in piecewise linearized characteristics and with removal of long-term drifts. Conversion tables between higher order structure functions D_x^3 and D_x^4 (i.e., between the

TABLE IV-1[1]

COMPARISON OF FREQUENCY STABILITY MEASUREMENT SYSTEMS

Hierarchy status	Example of measurement method	Data deducible from the measurement			
		Time	δTime	Frequency	δFrequency
1	Dual mixer time diff. or Time interval counter	$x(t) = \frac{\Delta t}{\tau \nu}$	$\delta\, x(t,\tau) = x\,(t+\tau) - x(t)$	$y(t,\tau) = \frac{\delta\, x(t,\tau)}{\tau}$	$\delta\, y(t,\tau) = y(t+\tau,\tau) - y(t,\tau)$
2	Loose phase-locked reference oscillator	can't measure	$\delta\, x(t,\tau) = \frac{\delta\, \phi(t,\tau)}{2\pi\nu}$	''	''
3	Heterodyne or beat frequency	can't measure	can't measure	$\lvert y(t,\tau) \rvert = \frac{\nu_{beat}}{\nu_0}$	''
4	Tight phase-locked reference oscillator	can't measure	can't measure	can't measure	$\delta\, y(t,\tau) = c \cdot \delta V$

[1]Tables IV-1 and IV-2 are reprinted from [III-10, p. A-8].

TABLE IV-2

ADVANTAGES AND DISADVANTAGES OF DIFFERENT FREQUENCY STABILITY MEASUREMENT SYSTEMS

Measurement Method	Time accuracy	Time stability	Frequency accuracy (1 day)	Frequency stability	Advantages	Disadvantages	Approximate Cost (k$)
Dual mixer time difference	~100 ps	0.1 ps	$\sim 10^{-16}$	$\sim 1.10^{-13}\ x\ \tau^{-1}$	No dead time; many choices of sample times (1 ms to as large as desired); oscillators may be at zero beat or different; measurement bandwidth easily changed; measurement noise typically below oscillators noise for tau=1 s and longer; measures times, time stability, frequency, and frequency stability.	More complex than other methods, and hence is more susceptible to extraneous signal pickup, e.g., ground loops; the time difference is modulo the beat period, e.g., 200 ns @ 5 MHz.	~5 to 30
Time interval counter	~100 ps	10 ps	$\sim 10^{-15}$	$\sim 1.10^{-11}\ x\ \tau^{-1}$	Wide bandwidth input allows a variety of signals; simple to use; cycle ambiguity almost never a problem; measures time, time stability, frequency, and frequency stability.	Measurement noise typically in excess of oscillator instabilities for tau values up to and of the order of several thousand seconds; hence, typically not useful for short term stability analysis.	1 to 7
Loose phase-lock reference[2] oscillator	—	~0.1 ps	depends on calibration of varicap curve	$\sim 1.10^{-11}\ x\ \tau^{-1}$	Useful for short term time stability analysis as well as spectrum analysis; excellent phase sensitivity—typically better than most oscillators.	Calibration depends on appropriate reference oscillator varicap curve characteristics.	<1
Heterodyne or beat frequency	—	—	$\sim 10^{-18}$	$\sim 1.10^{-13}\ x\ \tau^{-1}$	Measurement noise can typically be made less than oscillator instabilities for tau = 1 s and longer.	Minimum tau determined by period of beat frequency—typically not adjustable; can not compare oscillators near zero beat; cannot tell which oscillator is high or low in frequency; dead time often associated with these measurements.	0.2 to 1
Tight phase-locked reference[2] oscillator	—	—	—	$\sim 1.10^{-13}\ x\ \tau^{-1}$	Measurement noise typically less than oscillator instabilities for tau=1 s and longer; good measurement system bandwidth control; dead time can be made small or negligible.	Need voltage controlled reference oscillator; frequency sensitivity is a function of varicap tuning curve, hence not conducive to measuring absolute frequency differences.	<1

[2]Assumes reference oscillator is available to the metrologist.

binomial weighted "Hadamard variances" with $N = 3$ and $N = 4$) and the "h" noise coefficients are given.

Up to now we have not mentioned the possibility of measuring the frequency stability with only one oscillator. The method is generally used in microwave ranges and is discussed in detail in papers reprinted in the next part. Nevertheless, it can also be used in lower ranges, and one system for the measurement of crystal oscillators has been suggested and realized by Parzen and Hou [IV-14].

References

[IV-1] J. H. Shoaf *et al.*, "Frequency stability specification and measurement: High frequency and microwave signals," NBS Tech. Note 632, Jan. 1973.

[IV-2] B. F. Blair, Ed., *Time and Frequency: Theory and Fundamentals,* NBS Monograph 140, May 1974.

[IV-3] S. R. Stein, "An NBS phase noise measurement system built for frequency domain measurements associated with the global positioning system," NBS Rep. NBSIR 76-846, Aug. 1976.

[IV-4] M. McNamee, "Automate for improved phase-noise measurement," *Microwaves*, pp. 80–85, May 1979.

[IV-5] D. Martin, "Frequency stability measurements by computing counter system," *Hewlett-Packard J.* pp. 9–14, Nov. 1971.

[IV-6] M. B. Bloch and A. Vulcan, "Low noise measurement techniques," in *Proc. 28th Annu. Frequency Contr. Symp.,* Atlantic City, NJ, May 1974, pp. 184–189.

[IV-7] V. F. Kroupa, *Frequency Synthesis, Theory, Design and Applications.* London: Griffin, 1973; New York: Wiley, 1973.

[IV-8] V. F. Kroupa *et al.*, "Digital filters used for frequency stability measurements," in *Proc. Int. Conf. Digital Signal Processing,* Florence, Italy, Sept. 1981, pp. 831–838.

[IV-9] D. W. Allan, "Report on NBS dual mixer time difference system (DMTD) built for time-domain measurements associated with phase 1 of GPS," NBS Rep. NBSIR 75-827, Jan. 1976.

[IV-10] S. Stein *et al.*, "Performance of an automated high accuracy measurement system," in *Proc. 36th Annu. Frequency Contr. Symp.,* 1982, pp. 314–320; also in *Conf. Precision Electromag. Meas. CPEM-82 Dig.,* pp. N4–N6.

[IV-11] J. Rutman and G. Sauvage, "Measurement of frequency stability in time and frequency domains via filtering of phase noise," *IEEE Trans. Instrum. Meas.,* vol. IM-23, pp. 515–518, Dec. 1974.

[IV-12] N. A. Pendergrass and J. S. Farnbach, "A high-resolution, low-frequency spectrum analyzer," *Hewlett-Packard J.,* pp. 2–13, Sept. 1978.

[IV-13] J. Gibbs and R. Temple, "Frequency domain yields its data to phase-locked synthesizer," *Electronics,* pp. 107–113, Apr. 27, 1978.

[IV-14] B. Parzen and J. P.-W. Hou, "A technique for a self phase noise measuring system for signal sources," in *Proc. 32nd Annu. Frequency Contr. Symp.,* Atlantic City, NJ, May–June 1978, pp. 432–438.

A Test Set for the Accurate Measurement of Phase Noise on High-Quality Signal Sources

DONALD G. MEYER, MEMBER, IEEE

Abstract—The test set described here is capable of measuring the spectral density of phase noise on carrier frequencies from 1 to 500 MHz, for offset frequencies from 20 Hz to 50 kHz. Measurements to 50 MHz are described. The test set has a residual single-sideband phase-noise-power-to-signal-power ratio of −142 dB/Hz at 20 Hz offset from the carrier, which decreases to a floor of −172 dB/Hz at offset frequencies greater than 5 kHz. The estimated calibration accuracy achievable is ±0.8 dB, exclusive of random reading errors due to the Gaussian distribution of the phase fluctuations being observed. The estimated 1 σ repeatability of a measurement is 0.7 dB (70 percent of the observations on a given test will fall within ±0.7 dB of the average value). This test set is capable of characterizing the phase-noise performance of existing atomic frequency standards, crystal oscillators, frequency synthesizers, and other high-quality sources more accurately than has previously been possible. The increased accuracy has been achieved by a system design that minimizes readout fluctuations, allows for the accurate measurement of correction factors used to reduce systematic errors, and minimizes the possibility of operator error and bias.

I. Introduction

THE PHASE-NOISE modulation sidebands on modern high-quality signal sources from 1 to 500 MHz lie mostly in the region between −80 dB/Hz and −160 dB/Hz with offset frequencies in the range of 20 Hz–50 kHz. Special measurement techniques involving some method of carrier suppression must be used in order to perform spectral analysis of such signals, since present day RF power spectrum analyzers do not have sufficient dynamic range or resolution for this type of measurement. Since we basically wish to examine the phase fluctuations present on high-quality sources, an excellent technique is to drive a low-noise phase detector with two sources having the same output frequency, and then perform an audio frequency analysis of the demodulated signal at the output of the phase detector [1]. Due to the large separation between the signal and modulation frequencies, the carrier and sum frequencies in the phase-detector output are easily removed by low-pass filtering. The dynamic range problem is virtually eliminated by this technique, and the sensitivity of the measurement is limited only by the residual noise of the phase detector, provided reasonable care is used in the selection of the audio analyzer, the low-noise amplifier, and the coaxial phase shifters.

Previous measurements made by this technique have been limited in accuracy due to large random fluctuations of the observed noise and inaccuracies involved in calibration techniques. The design of the test set described in this paper has resulted in achieved accuracies of approxi-

Manuscript received June 4, 1970.
The author is with the John Fluke Manufacturing Company, Seattle, Wash. 98133.

Reprinted from *IEEE Trans. Instrum. Meas.*, vol. IM-19, pp. 215–227, Nov. 1970.

mately ±1 dB over a carrier-frequency range of 1–50 MHz. The same techniques can be used easily to 500 MHz. In order to achieve this performance, the following basic design concepts were used:

1) design for a high ratio of predetection to post-detection bandwidth to reduce the readout fluctuations, and to increase the statistical confidence of the measurement;
2) design a system that is stable and gives reproducible results, so that meaningful correction factors can be measured and applied to increase measurement accuracy;
3) design a system that can be used with a minimum possibility of operator error and bias.

In the body of the paper, the definitions of the spectral density of phase noise and the signal-power-to-total-phase-noise-sideband-power ratio are discussed. The theory of operation of the test set is described in more detail, and the equations that relate the test-set parameters to the final results desired are derived. Several types of tests, which can be made with this technique, are described. Special methods of calibration to reduce the systematic errors due to each part of the system are shown. The application of the test set for general laboratory measurements is discussed, and an example of the application of the test set to the production testing of a high-performance frequency synthesizer is given.

II. Measures of Phase-Noise Performance

This test set was designed to measure signal sources whose primary output contaminant is phase noise, which is indeed the case with most high-quality sources. Throughout the remainder of this paper the following important assumptions about the nature of the signals to be measured will be made: the power in the AM sidebands is much less than the power in the PM sidebands for all offset frequencies of interest, and rms phase deviation of the signal is much less than 1 radian. Such a signal $V(t)$ may be expressed mathematically as is shown in (1), where $\phi(t)$ represents the random fluctuation of the phase of the signal from the nominal value $\omega_0 t$ of the carrier phase. A further assumption, which will be made, is that the time average of $\phi(t)$ is zero.

$$V(t) = A_0 \sin [\omega_0 t + \phi(t)]. \tag{1}$$

One of the simplest measures of the quality of a source is the RF power spectrum of the source. In the case of a signal that is composed primarily of a carrier and phase-noise sidebands of low intensity, the RF power spectrum is a direct measure of the phase-noise performance. We can describe the quality of such a signal quantitatively in terms of the ratio of the *single-sideband* phase-noise power in a 1-Hz bandwidth to the signal power, as a function of the offset frequency f, and referred to a specified carrier frequency. In this paper we use the symbol $\mathcal{L}(f)$ for this quantity. Our usage of $\mathcal{L}(f)$ is similar to that of Halford [2], [3], except for the choice of the specified carrier frequency. In this paper, we use $\mathcal{L}(f)$ to refer to the carrier frequency at the output of a device under test, whereas Halford defines $\mathcal{L}(f)$ to refer to the carrier frequency at the input of a device under test. Since the modulation spectrum is symmetrical, it is only necessary to specify $\mathcal{L}(f)$ for positive modulation frequencies in order to completely characterize the signal. Note that $\mathcal{L}(f)$ is defined in terms of a power ratio per hertz bandwidth. If we express the power ratio in decibels, it is correct to speak of $\mathcal{L}(f)$ in units of decibels per hertz.

Another widely used measure of phase-noise performance is the *two-sided* spectral density $S\phi(\omega)$ of the random phase fluctuations of $\phi(t)$ [4], [5]. Since the dimension of $\phi(t)$ is radians, the units of $S\phi(\omega)$ are radians squared per hertz. If we express this spectral density in decibels per hertz as a function of f (where $\omega \equiv 2\pi f$), it can be shown that for the case of a signal having only phase noise that $S\phi(f)$ expressed in radians squared per hertz is numerically equal to $\mathcal{L}(f)$ expressed in hertz^{-1}. The conceptual difference between these two definitions lies in the fact that $S\phi(f)$ is the "mathematical" spectral density of $\phi(t)$, and hence there is no physical significance to negative values of f in terms of modulation sidebands because there is no "carrier" present. In contrast, $\mathcal{L}(f)$ is defined specifically in terms of spectral sideband power above and below the carrier frequency, and does have "physical" significance for positive and negative values of f.

An important formula that relates $S\phi(\omega)$ to the mean-squared value of $\phi(t)$ can be derived by means of the Weiner–Khintchine theorem [6] that states that the autocorrelation function $R\phi(\tau)$ and the spectral density $S\phi(\omega)$ of a random variable are Fourier transforms of each other. Using this theorem we may write:

$$S\phi(\omega) = \int_{-\infty}^{\infty} e^{-i\omega\tau} R\phi(\tau)\, d\tau \tag{2}$$

$$R\phi(\tau) = \frac{1}{2\pi} \int_{-\infty}^{\infty} e^{i\omega\tau} S\phi(\omega)\, d\omega. \tag{3}$$

For the case of $\tau = 0$, (3) reduces to:

$$R\phi(0) = \frac{1}{2\pi} \int_{-\infty}^{\infty} S\phi(\omega)\, d\omega. \tag{4}$$

The autocorrelation function $R\phi(\tau)$ is defined as:

$$R\phi(\tau) \equiv \lim_{T\to\infty} \frac{1}{T} \int_{-T/2}^{T/2} \phi(t)\phi(t+\tau)\, dt. \tag{5}$$

For the case of $\tau = 0$, (5) reduces to:[1]

$$R\phi(0) = \lim_{T\to\infty} \frac{1}{T} \int_{-T/2}^{T/2} \phi^2(t)\, dt = \langle \phi^2(t) \rangle. \tag{6}$$

[1] Angular brackets ⟨ ⟩ are used in this paper to denote the infinite time average (or expected value) of the enclosed quantity.

Combining (4) and (6) we obtain:

$$\langle \phi^2(t) \rangle = \frac{1}{2\pi} \int_{-\infty}^{\infty} S\phi(\omega)\, d\omega. \tag{7}$$

Utilizing the symmetry of the limits and integrand, and making the change of variable $\omega \equiv 2\pi f$ (where f is the offset frequency from the carrier frequency)

$$\langle \phi^2(t) \rangle = 2 \int_{0}^{\infty} S\phi(f)\, df. \tag{8}$$

This value of $\langle \phi^2(t) \rangle$ is the value that would be obtained from an average over infinite time, since the spectral density $S\phi(f)$ is defined as the infinite time averaged value [6]. In practice we do not experimentally evaluate the integral (8) for extremely low or high values of f due to practical averaging time and bandwidth limitations. A more practical measurement is obtained by integrating over a finite frequency range f_L–f_H. Denoting the mean-squared value of the phase fluctuations due only to the region of $S\phi(f)$ where $f_L < |f| < f_H$ as $\langle \phi^2(t, f_L, f_H) \rangle$, we define:

$$\langle \phi^2(t, f_L, f_H) \rangle \equiv 2 \int_{f_L}^{f_H} S\phi(f)\, df. \tag{9}$$

Denoting the reciprocal of this quantity as $S/N(f_L, f_H)$ we can further define:

$$S/N(f_L, f_H) \equiv \frac{1}{\langle \phi^2(t, f_L, f_H) \rangle}. \tag{10}$$

It can be shown that this quantity represents the ratio of the signal power to the total upper-plus-lower sideband power contained in the modulation frequency range $f_L < |f| < f_H$. This quantity is commonly used as a measure of performance of high-quality sources, and sometimes is stated as the signal to phase-noise-power ratio in a bandwidth of $2f_H$, excluding a bandwidth of $2f_L$ centered on the carrier. An estimate of $\langle \phi^2(t, f_L, f_H) \rangle$ can be made directly with the test set, and this can be converted easily to $S/N(f_L, f_H)$ by means of (10).

In a similar manner we can denote the mean-squared value of the phase fluctuations due only to the region of $S\phi(f)$ where $(f - F_N/2) < |f| < (f + F_N/2)$ as $\langle \phi^2 [t, f - (F_N/2), f + (F_N/2)] \rangle$, and define

$$\left\langle \phi^2\left(t, f - \frac{F_N}{2}, f + \frac{F_N}{2}\right)\right\rangle \equiv 2 \int_{f - F_N/2}^{f + F_N/2} S\phi(f)\, df, \tag{11}$$

where F_N is the noise-equivalent bandwidth of the analyzer.

An estimate of the value of $\langle \phi^2[t, f - (F_N/2), f + (F_N/2)] \rangle$ can be made directly with the test set. In order to be able to estimate $S\phi(f)$ from the measured data, it is necessary to develop an explicit relation between these two quantities. If we assume that $S\phi(f)$ is constant over the narrow analyzer bandwidth F_N, we can evaluate the integral in (11).

$$\left\langle \phi^2\left(t, f - \frac{F_N}{2}, f + \frac{F_N}{2}\right)\right\rangle = 2F_N S\phi(f). \tag{12}$$

Solving for $S\phi(f)$:

$$S\phi(f) = \frac{\left\langle \phi^2\left(t, f - \frac{F_N}{2}, f + \frac{F_N}{2}\right)\right\rangle}{2F_N}. \tag{13}$$

III. Measurement of $S/N\ (f_L, f_H)$

There are two basic modes of operation of the test set, both of which can be explained with the help of the block diagram shown in Fig. 1. In order to operate the test set, the two input signals to be examined should be applied to inputs A and B at levels of $\approx$1 V rms. With the level-meter switch set to the ON position the two inputs are applied to the 50-ohm input impedance voltage-doubler detectors. The levels into the detectors are then set to exactly +10 dBm by adjusting the two 0–10 dB variable attenuators until the level meters read precisely full scale. The level-meter switch is then set to the OFF position, which applies the two precisely adjusted +10-dBm signals to the high-level mixer. Next the phase-meter switch is set to the ON position, which connects the dc component of the high-level mixer output to the phase meter. The phase relationship between the two input signals is now adjusted so that the phase meter reads zero (the method for doing this varies according to the test setup, as will be described later). The phase-meter switch is now returned to the OFF position, which applies the output of the high-level mixer to the input of the low-noise amplifier.

The total random phase fluctuation $\phi(t)$ between the two input signals is now being transduced as a voltage at the output of the phase detector. This voltage is then amplified by the gain $A_1 = 50$ dB of the low-noise audio amplifier. The output of the low-noise amplifier can then be switched through two possible signal paths before being measured by the rms voltmeter. If we wish to measure $S/N(f_L, f_H)$, the filter–analyzer switches are placed in the filter position. The value of f_H to be used for the measurement can then be set by means of the selectable bandwidth low-pass filter to be 100 Hz, 1 kHz, 15 kHz, or 50 kHz. These values of f_H are the noise-equivalent bandwidths of the single-pole RC low-pass filters used ($f_H = (\pi/2)f_3$, where f_3 is the 3 dB point of the RC filter used). The frequency f_L is the noise-equivalent high-pass frequency due to the ac coupling used in the low-noise amplifier and the rms voltmeter and is $\approx$0.5 Hz. We can now derive the formula that relates the voltage measured by the rms voltmeter to $S/N(f_L, f_H)$ as defined by (10).

With the test-set controls positioned as described above, let us denote the time-varying voltage at the input of the rms voltmeter as $v_x(t, f_L, f_H)$. This voltage is related to the phase fluctuations at the output of the high-level mixer by the phase slope $K\phi$ (volts per radian)

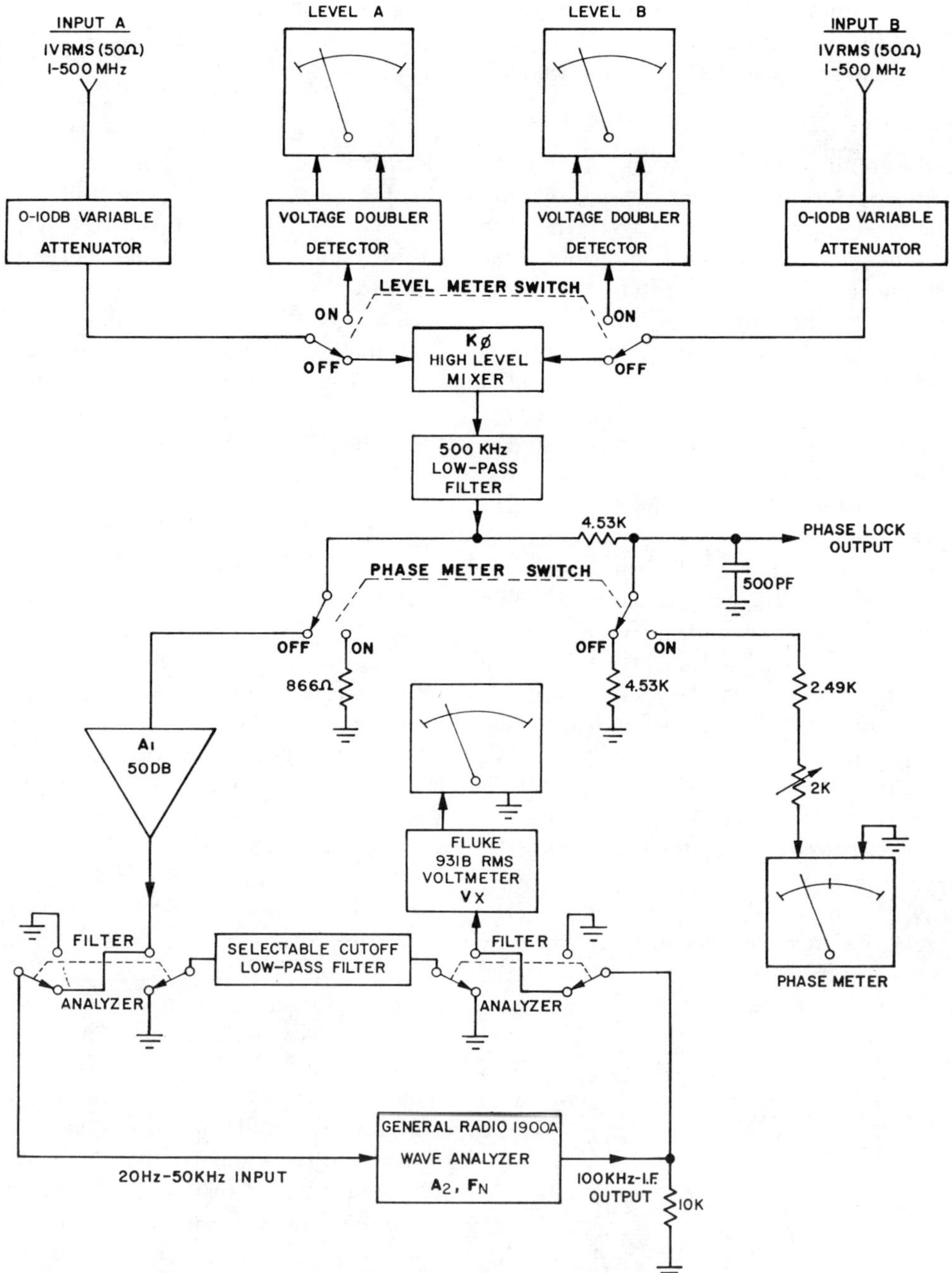

Fig. 1. Block diagram of phase-noise test set.

of the mixer, and the gain A_1 of the low-noise amplifier as is shown below:

$$v_x(t, f_L, f_H) = K\phi A_1 \phi(t, f_L, f_H), \tag{14}$$

where $\phi(t, f_L, f_H)$ represents the fluctuations of phase between the two mixer inputs due to spectral components between f_L and f_H. Squaring both sides, averaging, and solving for the reciprocal of $\langle \phi^2(t, f_L, f_H) \rangle$ we obtain the following:

$$\frac{1}{\langle \phi^2(t, f_L, f_H) \rangle} = \frac{(K\phi A_1)^2}{\langle v_x^2(t, f_L, f_H) \rangle}. \tag{15}$$

By use of the definition given by (10), this can be written:

$$S/N(f_L, f_H) = \frac{(K\phi A_1)^2}{\langle v_x^2(t, f_L, f_H) \rangle}. \tag{16}$$

Then expressing this ratio in decibels as a function of $\overline{v_x(t, f_L, f_H)}$:[2]

$$S/N(f_L, f_H) = 20 \log_{10} \left[\frac{K\phi A_1}{\overline{v_x(t, f_L, f_H)}} \right] \text{ dB}. \tag{17}$$

[2] A horizontal line over a variable is used to indicate the rms value of that variable.

This expression now allows us to calculate the value of $S/N(f_L, f_H)$ from the measured voltage $v_x(t, f_L, f_H)$ and the known parameters $K\phi$ and A_1 of the test set.

IV. Measurement of $\mathcal{L}(f)$

Assuming that the input levels have been set and the phase meter zeroed as described in the previous section, the test set can be used to measure $\mathcal{L}(f)$ by setting the filter–analyzer switches to the analyzer position. This switches the selectable cutoff low-pass filter out of the signal path, and applies the output of the low-noise amplifier to the input of the wave analyzer. Also, the 100-kHz IF output of the wave analyzer is connected to the rms voltmeter. The effect of the wave analyzer is to translate the voltage components at the analyzer input in a noise-equivalent bandwidth of F_N centered at the frequency f to which the analyzer is tuned, to the 100-kHz IF output frequency. Let us denote the time-varying voltage at the input of the rms voltmeter as $v_x[t, f - (F_N/2), f + (F_N/2)]$, and the corresponding phase fluctuations between the mixer input signals as $\phi[t, f - (F_N/2), f + (F_N/2)]$. The exact expression relating these two quantities is not quite as simple to write as was (14), due to the frequency translation involved in passing through the wave analyzer. We will avoid this difficulty by simply defining the transfer voltage gain A_2 of the wave analyzer as the ratio of the rms value of the IF output voltage at 100 kHz, to the rms value of the input voltage in a noise-equivalent bandwidth of F_N centered at f (where f is the frequency to which the analyzer is tuned). We may now write the relationship between the mean-squared values of these variables directly:

$$\left\langle v_x^2\left(t, f - \frac{F_N}{2}, f + \frac{F_N}{2}\right)\right\rangle = (K\phi A_1 A_2)^1 \left\langle \phi^2\left(t, f - \frac{F_N}{2}, f + \frac{F_N}{2}\right)\right\rangle. \tag{18}$$

Solving (18) for $\langle \phi^2[t, f - (F_N/2), f + (F_N/2)]\rangle$:

$$\left\langle \phi^2\left(t, f - \frac{F_N}{2}, f + \frac{F_N}{2}\right)\right\rangle = \frac{\left\langle v_x^2\left(t, f - \frac{F_N}{2}, f + \frac{F_N}{2}\right)\right\rangle}{(K\phi A_1 A_2)^2} \tag{19}$$

and substituting this value into (13) we obtain:

$$S\phi(f) = \frac{\left\langle v_x^2\left(t, f - \frac{F_N}{2}, f + \frac{F_N}{2}\right)\right\rangle}{2F_N(K\phi A_1 A_2)^2}. \tag{20}$$

As it was pointed out in Section II, $S\phi(f)$ in radians squared per hertz is numerically equal to $\mathcal{L}(f)$ in hertz^{-1} due to the restrictions we have placed on the nature of the signal. Therefore $\mathcal{L}(f)$ is equal to the expression on the right-hand side of (20). Expressing $\mathcal{L}(f)$ in decibels per hertz as a function of the rms value of $v_x[t, f - (F_N/2), f + (F_N/2)]$:

$$\mathcal{L}(f) = 20 \log_{10} \cdot \left[\frac{v_x[t, f - (F_N/2), f + (F_N/2)]}{K\phi A_1 A_2 \sqrt{2F_N}}\right] \text{ dB/Hz}. \tag{21}$$

This expression now allows us to calculate the value of $\mathcal{L}(f)$ in decibels per hertz from the measured voltage $v_x[t, f - (F_N/2), f + (F_N/2)]$ and the known parameters $K\phi$, A_1, A_2, and F_N of the test set. It should be noted at this point that the test set can also be used to measure discrete spurious signals that represent phase modulation, such as line-frequency modulation components. Equation (21) can be used to do this by simply leaving out the bandwidth F_N, since no bandwidth correction is needed in this case. If the spurious signal is well above the noise, the value of the single-sideband spurious to signal ratio can be accurately measured in this manner.

V. Types of Device or System Measurements That Can be Made With the Test Set

There are three basic types of measurements possible with this test set. It is also possible to measure the residual noise of the test set itself. The block diagrams for these four types of tests are shown in Fig. 2. The general mode of operation of the test set is identical in all of these tests. The input levels are always set to +10 dBm by use of the level meters, and the phase angle between the carrier signals must always be set for a zero indication on the phase meter. The filter–analyzer switches are set either to the filter or analyzer positions, depending on whether one wishes to measure $S/N(f_L, f_H)$ or $\mathcal{L}(f)$.

An important point, which must be understood in order to interpret the measured data from a given test correctly, is the question of whether the measured data are the result of only one source of phase fluctuations, or two sources. For instance, let us consider the measurements shown in tests Fig. 2(a) and (d). In Fig. 2(a), only one device is under test, and hence we can consider the entire voltage indicated by the test set as being due to the device under test. In Fig. 2(d), we are interested in obtaining the measurement limitation of the test set, and hence we can again attribute the entire voltage indicated as being due to the test set. In Fig. 2(b) and (c), there are two devices under test that might be contributing nearly equal amounts of noise, so that in general, we must consider the system output voltage to be a result of the phase fluctuations of both devices under test. The next point we must then consider is how to determine what portion of the indicated voltage should be attributed to each device under test.

It is not possible to tell from just one test using two devices of fairly similar performance [as indicated in Fig. 2(b) or (c)] what the performance of each individual device is. In order to determine accurately individual unit performance, at least three units of similar performance are necessary. Suppose we have three units under test, which we will denote as units A, B, and C. We

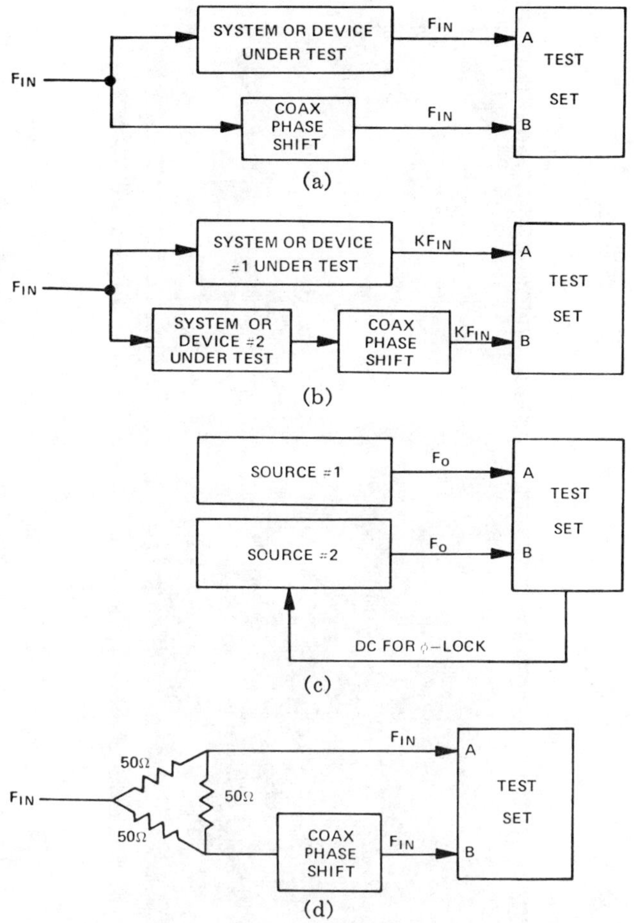

Fig. 2. Some measurements possible with the test set. (a) Residual noise of systems with $F_{out} = F_{in}$. (b) Residual noise of systems with $F_{out} = KF_{in}$. (c) Measurement of two independent sources at F_0. (d) Residual noise of test set.

must run three tests in order to be able to characterize any particular unit. First we use A and B as the devices under test and denote the voltage indicated by the test set as V_{AB}. Then we measure V_{BC} due to units B and C under test, and V_{CA} due to units C and A. We can now use the fact that the random phase fluctuations due to each individual source are statistically independent variables to write the following equations, where V_A, V_B, and V_C are the rms voltages that would be indicated by the test set due only to unit A, B, or C, respectively:

$$(V_{AB})^2 = V_A^2 + V_B^2 \tag{22}$$

$$(V_{BC})^2 = V_B^2 + V_C^2 \tag{23}$$

$$(V_{CA})^2 = V_C^2 + V_A^2. \tag{24}$$

We can now solve this set of equations to obtain the rms voltages V_A, V_B, and V_C in terms of the measured rms voltages V_{AB}, V_{BC}, and V_{CA}:

$$V_A = \sqrt{\frac{(V_{CA})^2 + (V_{AB})^2 - (V_{BC})^2}{2}} \tag{25}$$

$$V_B = \sqrt{\frac{(V_{BC})^2 + (V_{AB})^2 - (V_{CA})^2}{2}} \tag{26}$$

$$V_C = \sqrt{\frac{(V_{CA})^2 + (V_{BC})^2 - (V_{AB})^2}{2}}. \tag{27}$$

We can now use these equations to obtain the value of voltage that would be indicated by the test set due only to any one of these individual sources. Once such a value is determined, (17) or (21) can be used in the normal manner to calculate $S/N(f_L, f_H)$ or $\mathcal{L}(f)$ from the data as is appropriate. When using (25)–(27), it is important that the performance of the three sources under test be very nearly equal, due to the finite statistical fluctuations of the voltage estimates being made. It is particularly important to avoid the situation where $V_A \approx V_B$ and V_C is $\ll V_A$ or V_B. In this case the value under the radical sign of (27) fluctuates plus and minus around zero due to the statistical variations of V_A and V_B. Under these conditions the value calculated for V_C has no meaning, but V_A and V_B can still represent good estimates if the value of V_C is disregarded. Either can be determined by a single measurement against V_C.

Another very important point to notice with regard to the tests shown in Fig. 2(a), (b), and (d) is that the phase noise present on the input signal F_{IN} tends to be cancelled out due to the symmetry of the test set. This cancellation often allows measurement of very low values of $S/N(f_L, f_H)$ or $\mathcal{L}(f)$ even in the presence of noisy signals used for F_{IN}. In the test shown in Fig. 2(a), a source with very low noise must be used if the device under test has a narrow bandwidth. In the test of Fig. 2(b), a low noise source may be needed if the devices under test are not similar in nature. The term "residual noise" is used for these tests since the noise measured is the residual noise after removing the spectrum of the input signal F_{IN}. Further, the tests shown in Fig. 2(a), and (b) are intended for systems or devices that maintain a time-invariant phase relationship between input and output terminals. An example of the test shown in Fig. 2(a) is the evaluation of residual phase noise introduced by a standard-frequency distribution amplifier with 5-MHz input and output frequencies. An example of the use of the test shown in Fig. 2(b) is the evaluation of the residual noise present on two frequency synthesizers with a common 5-MHz frequency standard input, and output frequencies of 49.9 MHz. The phase-meter output is set to zero in the tests shown in Fig. 2(a), (b), and (d) by simply inserting the proper length of coaxial cable in one signal path.

The test shown in Fig. 2(c) indicates the method used to measure the noise on independent sources. Since the two signals of frequency F_0 are not phase related, the dc output of the high-level mixer is held at approximately zero volts dc by use of a narrow phase-lock loop. The loop is established by using the phase-lock output shown in Fig. 1 as a fine frequency control for one of the sources under test. In practice we have shown that the measured value of $S/N(f_L, f_H)$ is not affected as long as the loop bandwidth is <0.4 Hz, when high-quality sources are being tested. This means that for use with this test set, the modulation sensitivity of the fine frequency adjust should be <1 Hz/V. Many other types of tests may be made by using combinations of the general principles described here, as long as one is very careful in the interpretation of the test results.

VI. General Physical Description of the Test Set

The entire test set is self-contained in a standard 19-inch (≈48 cm) equipment rack (see Fig. 3), complete with two line-isolation transformers to power equipment under test. The top panel in the rack is the main test panel to which the input signals are connected. The rms voltmeter is mounted directly under the main test panel. The large panel meter mounted next to the rms voltmeter is driven from the chart recorder output of the voltmeter and is used to read out all measurements made with the test set. The wave analyzer and a chart recorder are located just below the rms voltmeter. The cables shown are the only interconnections required to operate the system. In general, all that is necessary to make a measurement is to connect the input cables from the devices under test, plug in the ac power cords (if used), adjust the input levels and phase, and set the other controls for the desired measurement. The only calibration that must be made at other than the periodic maintenance intervals is the setting of the transfer gain of the wave analyzer, which is described later.

VII. Design Factors Affecting the Test-Set Parameter $K\phi$

The portions of the system that affect the phase slope $K\phi$ are the high-level mixer, the voltage-doubler detectors and level meters, the 10-dB variable attenuators, the level-meter switch, and the input impedance to the 500-kHz low-pass filter. The high-level mixer (see Fig. 4) is double balanced, and uses hot-carrier diodes (also known as schottky-barrier or metal-semiconductor diodes). One basic effect utilized in this part of the system is that if the mixer input signals are supplied at levels of exactly +10-dBm available power from 50-ohm sources, the phase slope $K\phi$ of the mixer in volts per radian is extremely linear over the range of ±0.44 volt dc output. The other basic effect used is that if a good 50-ohm system supplies the +10-dBm available power to the mixer, and distortion on the input signals is less than ≈30 dB, the phase slope $K\phi$ at a given carrier frequency is extremely reproducible with time and in different test setups. The primary factor here is reproducibility, since the phase slope over the output range of ±0.44 volt dc is always linear within $\approx\frac{1}{2}$ percent with these drive conditions and can be measured with an accuracy of ≈0.1 percent.

In order to ensure reproducibility, good solid hardware must be used that depends on basic effects not likely to change with time. Weinschel model 905 variable attenuators are used to provide very smooth and stable adjustment, along with very low SWR up to 1 GHz.

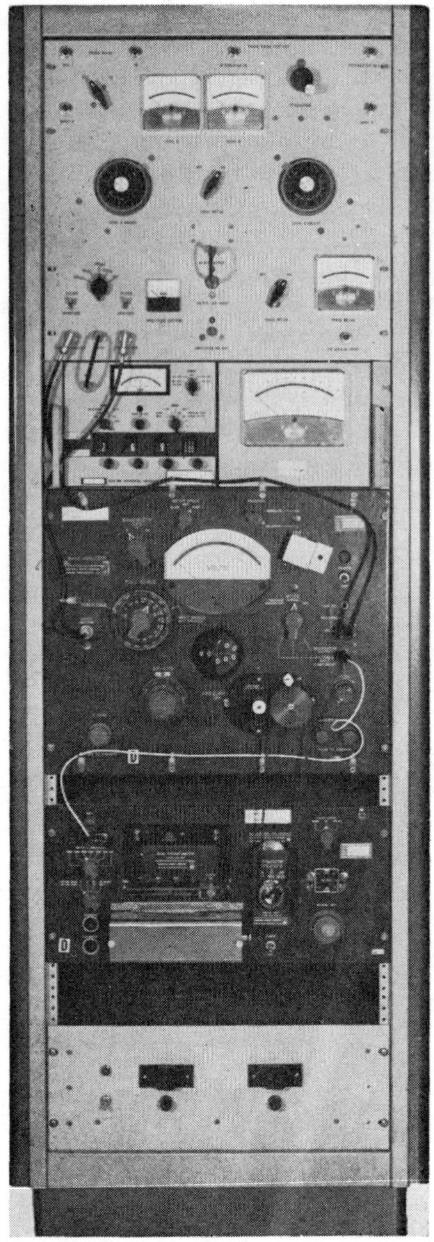

Fig. 3. The phase-noise test set.

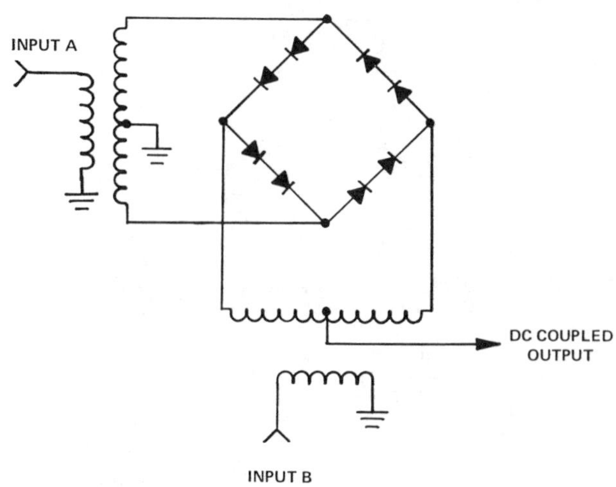

Fig. 4. Schematic of high-level mixer.

The voltage-doubler detectors (see Fig. 5) use hot-carrier diodes, and are flat within ±0.3 dB from 1 to 500 MHz. Input SWR is less than 1.1 up to 500 MHz. The dc output across the diodes is ≈1.25 volts with +10 dBm inputs. The loading of the detectors is kept low by the use of 50-μA movements in the level meters. Since the level meters utilize taut-band movements with mirror-backed 3-inch scales, and the input voltage is always adjusted for an exact full-scale reading, level-set errors of ½ percent are very easily discernible. The level-meter switch is a Texscan TS-100A coaxial switch that has a good solid switching action and SWR <1.1 at 1 GHz. The RF load impedance seen by the sum frequency coming out of the mixer is invariant with time for a given carrier frequency, since there are no significant parameter drifts in the LC components of the low-pass filter. The low-frequency loading at the output of the low-pass filter is ≈866 ohms due to the low-noise amplifier input impedance. An 866-ohm resistor is substituted for the low-noise amplifier loading when the phase slope is being measured to avoid overloading the amplifier and to maintain a constant value of load impedance.

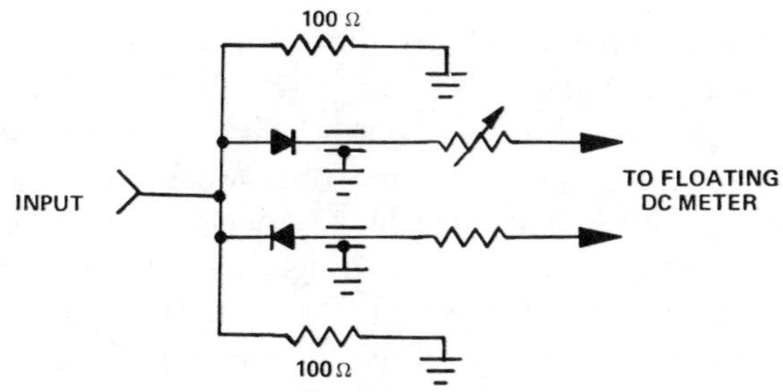

Fig. 5. Voltage-doubler detector schematic.

VIII. Method of Measuring $K\phi$

The basic method used to measure $K\phi$ is shown in Fig. 6. Two very low hysteresis rate-independent current-mode Schmitt trigger circuits are set so that their upper and lower trigger points are precisely symmetrical about +0.44 and −0.44 volt dc, respectively. In order to measure $K\phi$, two +10-dBm input signals at carrier frequency f_0, but with a very small offset Δf, are applied to the high-level mixer (we use two frequency synthesizers offset by exactly 1 Hz). With the phase-meter switch set to ON, the voltage across the 866-ohm resistor shown in Fig. 1 is then applied to the inputs of the Schmitt triggers, which have input impedance high enough to avoid significant loading. All the circuitry required to generate A, $\bar{A}$, B, $\bar{B}$ and finally the $T+$ and $T-$ waveforms are built into a small test jig, along with means for accurate calibration of the trigger points. As can be seen from Fig. 6, the $T+$ and $T-$ waveforms are related to the positive and negative slopes of the phase detectors that we will denote as $K\phi(+)$ and $K\phi(-)$, respectively. The times $T+$ and $T-$ indicated in the figure are measured by use of a digital time-interval meter, by starting on the positive edge and stopping on the negative edge. The period T_0 can be measured by starting and stopping on only the positive edges of either $T+$ or $T-$. The values of $K\phi(+)$ and $K\phi(-)$ can now be calculated very simply by using the following expressions:

$$K\phi(+) = \frac{0.88T_0}{2\pi T+} \quad (28)$$

$$K\phi(-) = \frac{-0.88T_0}{2\pi T-}. \quad (29)$$

In practice, the values of $K\phi(+)$ and $K\phi(-)$ we have measured have never differed by more than ≈0.1 percent at a given carrier frequency. Using this technique, the values of $K\phi$ we measure vary by +1.7 and −0.9 dB from a nominal value of 0.8 V/rad over the frequency range of 1–50 MHz. The same technique can be used up to 500 MHz, due to the wide bandwidth of the high-level mixer and the other RF components.

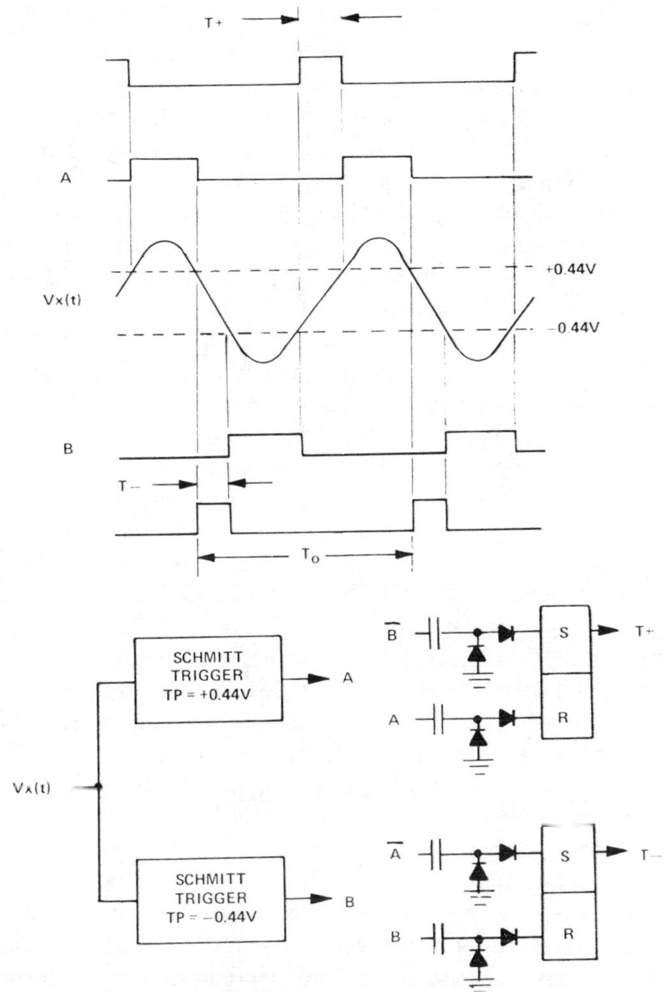

Fig. 6. Method used to measure phase slope $K\phi$ of high-level mixer output voltage.

IX. Design Factors Affecting the Parameter A_1

Let us define the voltage gain A_1 as the overall voltage gain from the input of the 500 kHz low-pass filter, to the output of the low-noise amplifier. Due to the fact that the output impedance of the high-level mixer and the input impedance of the low-noise amplifier both vary with frequency, it is very difficult to design the system so that the frequency response of A_1 is flat for all values of offset and carrier frequency. A further complication is placed on the design by the requirement that the carrier feedthrough and sum frequencies from the high-level mixer must be attenuated greatly before reaching the low-noise amplifier input over the entire carrier frequency range of 1–500 MHz. This is necessary in order to keep the low-noise amplifier input stage from being overdriven by the RF signals, which would cause the effective low-frequency gain of the amplifier to change. Other restrictions are also imposed by the requirements for low-noise performance and gain stability with time.

A computer simulation of the entire low-noise amplifier and low-pass filter combination was used to achieve the best practical performance compromise possible with the 50–90-ohm output impedance variation exhibited by the high-level mixer, and the various other design requirements. A schematic of the low-pass filter, low-noise amplifier combination is shown in Fig. 7. Note that the transistors are operated with zero collector-to-base voltage to minimize flicker noise, and that negative feedback has been used for gain stabilization. The amplifier has ≈ 1.8 nV/$\sqrt{\text{Hz}}$ noise, referred to the input, when driven from a source whose resistance is 50 ohms, over the range of 20 Hz–50 kHz. It is battery powered and mounted inside a section of iron pipe to avoid 60-Hz line-frequency pickup. The gain A_1 can be easily set within ± 1 percent and remains very stable over the operating life (≈ 700 hours) of the mercury batteries. A push-button battery-power switch and a meter to indicate when the battery voltage is within acceptable limits are mounted on the main test panel.

As in the case of all the test-set parameters, the most important factor is reproducibility. The overall amplifier and filter design as just described results in a system that maintains a stable gain and frequency response with time. Once the frequency response has been measured, the same correction factors can be used even after making minor gain adjustments and changing the batteries. However, the response with offset frequency does vary depending on the carrier frequency with which the mixer is being driven. Appropriate correction factors must be measured and used to reduce this source of error for the most accurate measurements.

X. Calibration and Measurement of A_1

The method that is used to set the midband voltage gain of A_1 to exactly 50 dB is shown in Fig. 8. In this type of calibration the 50-dB attenuator serves as a reference standard, since the voltage gain of the amplifier is basically adjusted to match the insertion loss of the attenuator. The value of 97.4 mV rms to which the output is adjusted is used to compensate for the loading across the 50-ohm resistor, which is caused by the ≈ 866-ohm input impedance of the amplifier. If this correction were

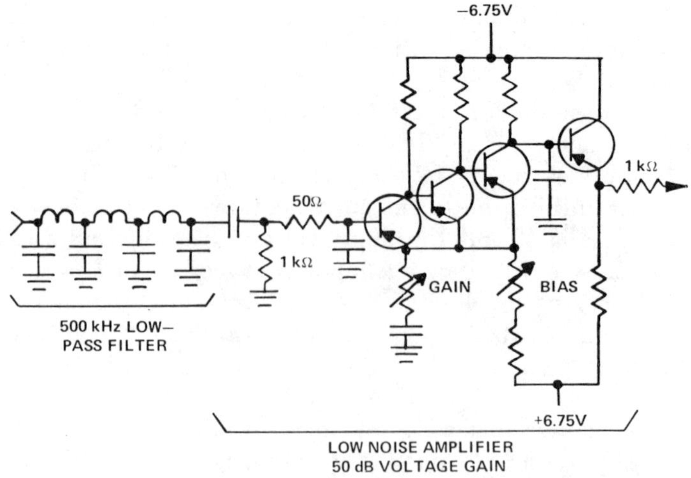

Fig. 7. Low-pass filter, low-noise amplifier schematic.

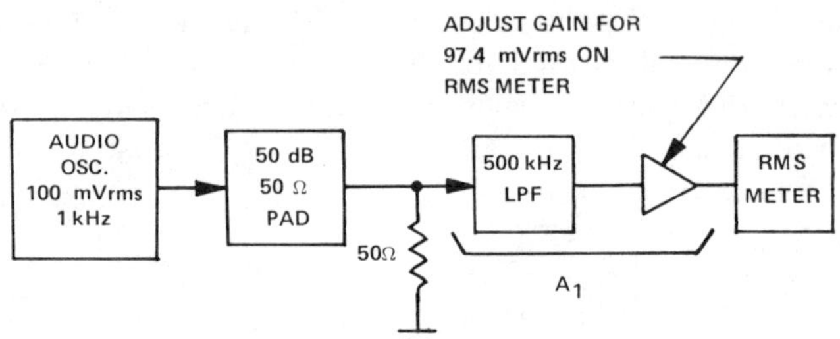

Fig. 8. Method of setting low-noise amplifier gain to 50 dB.

not used, the gain would be set high by about 2.6 percent.

In order to establish the frequency response of A_1 accurately, extreme care must be used to maintain the system in its exact operating condition while the measurement is being made. This can be accomplished by means of the technique illustrated in Fig. 9. Since all three of the synthesizers shown use a common 5-MHz frequency standard, the phase angle between the outputs of synthesizers 1 and 2 is fixed when both are set to synthesize f_0. With the 5-MHz frequency standard signal removed from synthesizer 3, the amplitude and phase of inputs A and B to the test set are set in the normal manner. The phase can be conveniently adjusted by switching the search oscillator of synthesizer 1 or 2 momentarily into a low-order digit until the desired phase is achieved and then switching it back out. The filter–analyzer switches are set to the filter position, and the selectable low-pass filter (see Fig. 10) is switched to the OUT position.

Now the output of synthesizer 3 is turned on by reconnecting it to the 5-MHz frequency standard. Note that the loading on input B, due to synthesizer 3, is constant throughout this procedure, and is minimized by the 1.5-kΩ series resistor shown in Fig. 9.

The effect of the $f_0 + f$ signal is to introduce a single sideband spurious signal at input B of the high-level mixer. The PM portion of this spurious signal is demodulated by the high-level mixer, and appears at the mixer output at frequency f. The magnitude of this voltage is large enough to be well out of the noise, but small enough to maintain linear operation of the mixer to the spurious

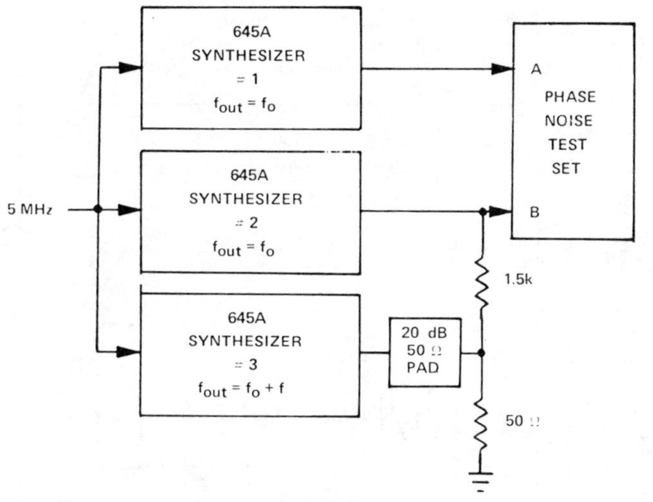

Fig. 9. Technique used to measure frequency response of A_1.

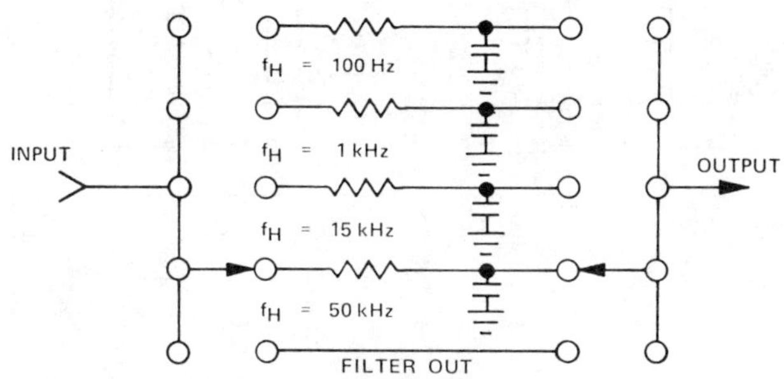

Fig. 10. Selectable cutoff low-pass filter schematic.

PM signal. With f set at 1 kHz, the voltage indicated on the rms voltmeter is recorded and used as the zero-decibel reference value, since the gain A_1 has been accurately calibrated to be 50 dB at 1 kHz. The ratio of this reading to the readings with other values of f now can be used to calculate the appropriate decibel correction factors. The procedure must be repeated for different values of f_0 and other appropriate calibration factors calculated as necessary.

This correction ranges from -0.1 to -0.4 dB at f = 20 kHz, and -0.8 to -1.6 dB for f = 50 kHz, for carrier frequencies from 1–50 MHz. Note that the response of the selectable cutoff low-pass filter can be measured in an identical manner by simply setting it for a particular value of f_H, and checking the frequency response as before. This is not necessary, however, since the RC low-pass characteristics are precisely controlled by use of 1 percent tolerance components.

XI. Calibration and Measurement of A_2 and F_N

With the rest of the system calibrated as has been described, the controls of the wave analyzer can be set in such a way as to obtain a dynamic range sufficient to measure values of $\mathcal{L}(f)$ from -100 dB/Hz down to the residual noise floor of the test set, without further analyzer adjustments. The dynamic range stated is attained by setting the analyzers' input attenuator to 0 dB, IF attenuator to 50 dB, and the noise-equivalent IF bandwidth to the 3-Hz position. If we adjust the value of A_2 (and measure F_N) with the controls in this setting, virtually all of the errors due to the analyzer can be calibrated out.

For convenience, we calibrate the transfer gain A_2 so that if $K\phi$ = 0.8 V/rad and A_1 = 50 dB, a measured IF output voltage of 5 V rms corresponds precisely to an input spectral density of $\mathcal{L}(f) = -100$ dB/Hz. This allows a 10-dB safety factor of dynamic range (the IF output is linear for up to 15 V rms outputs) to prevent compression of the peaks of the Gaussian noise-modulated 100 kHz IF output signal. In order to make this calibration, we must know the value of F_N accurately. F_N can be measured by the use of the test setup shown in Fig. 11. The wave analyzer controls are set as has been indicated, and it is tuned to f = 100 Hz. Next, the IF output voltage is measured for input frequencies of 94–106 Hz in 1-Hz steps, and a computer is used to calculate F_N from this data, assuming the highest value noted as the zero-decibel reference value.

With the assumed values of $K\phi$ and A_1 as was indicated, the output of the amplifier for F_N = 3 Hz can be calculated by use of (21) to be 6.204 mV rms, if $\mathcal{L}(f) = -100$ dB/Hz. Now if we turn the analyzer input attenuator to the CAL position, an internal 60-Hz calibration signal of 9.487 mV rms is applied to the analyzer input. If we now set the input attenuator to CAL and very carefully peak the wave analyzer tuning at 60 Hz, the analyzers' relative-gain control can be used to set the IF output voltage to the value V_{CAL}, which will result in the desired calibration. V_{CAL} can be calculated by use of (30) below:

$$V_{CAL} = 7.646\sqrt{(F_N/3)} \quad \text{V rms.} \tag{30}$$

Using the same test setup as shown in Fig. 11, we easily can measure the frequency response of the wave analyzer directly. The analyzer should be calibrated and the controls set as indicated in the last three paragraphs. We simply set the synthesizer to 60 Hz and record the IF output voltage when the analyzer is carefully peaked up on the input signal, noting this as the zero-decibel reference level. Now we can record the IF output voltage in a similar manner for other values of f and calculate the appropriate decibel correction factors. At 20 Hz the correction for A_2 is ≈ -0.1 dB, at 20 kHz ≈ 0.3 dB, at 50 kHz ≈ 0.6 dB, and it is negligible from 50 Hz to 15 kHz.

XII. Design of the Test-Set Ground System

It is very difficult to avoid problems due to ac power ground loops with a system of this type. This is because of the very low level ac measurements that are being made and the fact that the power-line frequency and its harmonics lie within the band of interest. The problem is further complicated when one or both of the instruments under test require ac power. Several difficult and unusual problems arise when we try to connect several individually powered ac instruments into a ground system. None of the instruments was designed with internal isolation

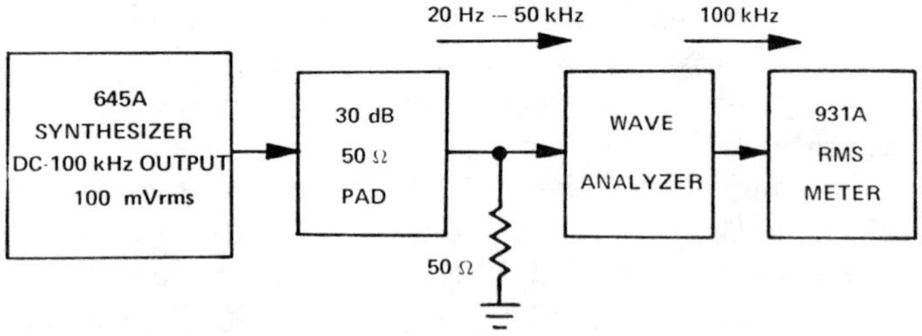

Fig. 11. Technique used to measure frequency response of A_2.

transformers, since they are not necessary in the normal applications for which the instruments are intended. This means we are forced to use external isolation transformers to help reduce the ground loops, which makes the overall system fairly complex to analyze.

The ground system shown in Fig. 12 was designed specifically to allow this type of measurement to be made simply, without much knowledge about ground-loop problems on the part of the operator. The first thing of importance to notice is that each of the individual instrument chassis and the main test-set panel are isolated from the equipment-rack cabinet except for the connections shown on Fig. 12. The second thing to notice is that each ac powered piece of equipment has only one dc path to ground, which is the ground shield of the coaxial cable by which the input or output of that piece of equipment is connected to the main test panel. The third thing to notice is that the local ground, the test-set cabinet ground, and the bus line from all of the center shield windings are all wired to a single "system ground" point located on the test-set main panel. Two pertinent observations can now be made about the ground currents in the system. First, currents flowing in the three wires connected to the system ground point can not cause ground potential differences between any of the test instruments or the main test panel, and hence can not create ground loops in the measurement system. Second, if no currents flow in the coax-cable ground shields there will be essentially no ground loops in the entire system. The conclusion, which then follows, is that the remainder of the design should be aimed at minimizing the coax-cable ground-shield currents.

The system involves on the order of 25 different stray capacities and about 8 or 10 current loops that can cause trouble. We will only attempt here to discuss briefly some of the most difficult ground-loop problems, which are greatly alleviated by the techniques shown. First, the primary-to-secondary capacity of the isolation transformer is reduced by connecting all the center shields to the system ground point. Second, notice that one side of the primary transformer in the instruments being powered is often connected through a line filter to ground (denoted by C_f in Fig. 12). This capacity is often fairly large, so that one side of the isolation transformer secondary is essentially at full ac line potential with respect to the chassis. The capacity between the isolation transformer secondary and shield 3 is about 500 pF, which means that if shield 3 were connected to the system ground point, a large circulating current would flow through the coax-cable ground shield. This is avoided by connecting shield 3 of each transformer back to the chassis ground of the instrument the transformer is powering. Last, note that capacity from any of the ac power cords (used to connect the instruments to the isolation transformers to the cabinet of the test set) causes a problem similar to the case just discussed. This is minimized by using power cords that have as the third wire ground a coaxial shield that is connected back to the chassis of the instrument being powered. One other problem that is not shown in Fig. 12 is that the isolation transformers themselves must be placed well away from the test equipment to avoid magnetic coupling.

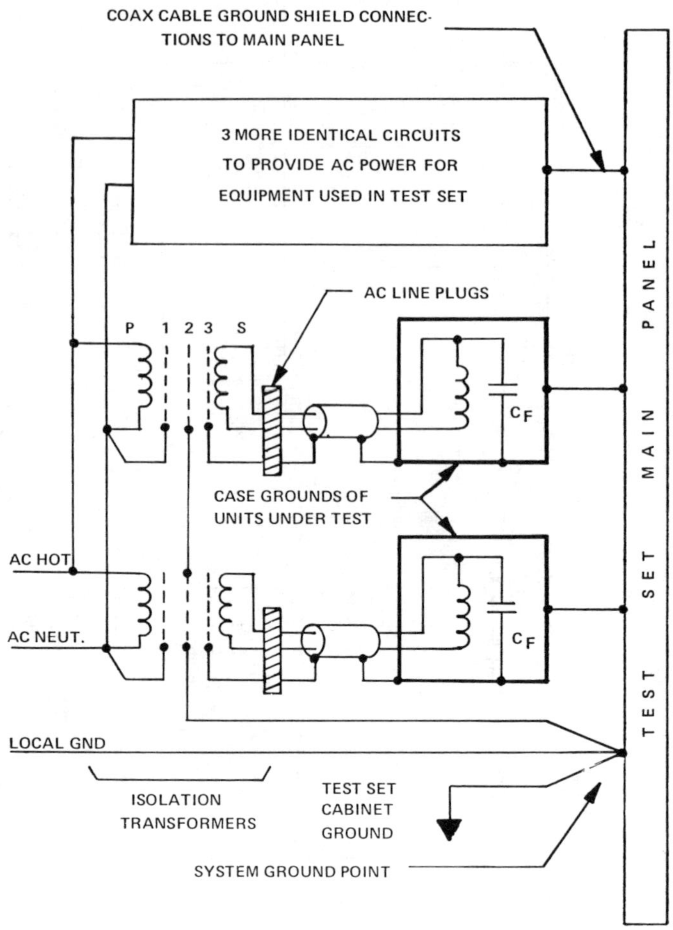

Fig. 12. Phase-noise test-set ground system.

XIII. Overall Performance of Test Set

The overall residual $\mathcal{L}(f)$ due to the test set, as measured by the technique shown in Fig. 2(d) is shown in Fig. 13. The solid line gives the total system residual noise, and the dashed line shows the portion of the noise that is due only to the low-noise amplifier and the spectrum analyzer. The main source of noise is the high-level mixer. Experiments have shown that the noise is definitely multiplicative in nature. This type of noise is due to the hot-carrier diodes in the mixer. Its basic cause remains to be proven. It definitely varies among different types of

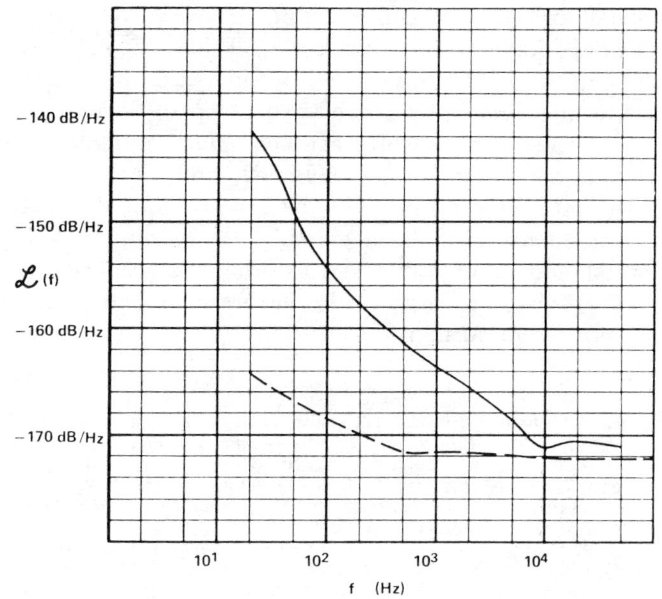

Fig. 13. Residual noise performance of test set. Solid line—total system noise floor. Dashed line—amplifier and analyzer only (R_s = 50 ohms).

hot-carrier diodes and would be far worse if p-n junction diodes were used in place of the hot-carrier diodes. The residual line-frequency components due to ground loops are equivalent to observing coherent sidebands on the order of 120 dB down from the carrier, even when two instruments under test using as much as 100 watts each are being powered from the isolation transformers in the test set.

The statistical confidence of the estimate of the rms voltage (which is made by observing the output of the voltmeter as indicated on the large panel meter) can be greatly improved by switching in a special 14 second averaging time constant that has been built into the rms voltmeter. The method we use to estimate the magnitude of the fluctuating voltage is to set up the measurement and switch in the 14-second time constant, then wait 50 seconds for the transient to settle out, and estimate the average reading during a 20-second observation time. A timer is used to determine these times roughly, and thereby maintain a more constant operator bias function. The overall performance of the system is good enough that we have been able to reproduce measurements taken of the residual noise due to two synthesizers within ± 1 dB, where the data were taken about four months apart.

XIV. Use of the Test Set for General Laboratory Measurements

In order to make the use of the test set more convenient for general laboratory applications, we have generated several different charts (by means of a set of computer programs) that allow rapid and accurate interpretation of the measured data. Three of the charts were made assuming the nominal parameter values that were indicated in the text, and the fourth provides total correction factors to increase the measurement accuracy. The first chart covers the range of $\mathcal{L}(f)$ from -90 to -170 dB/Hz in 0.2 dB steps, and gives the corresponding value of $v_x[t, f - (F_N/2), f + (F_N/2)]$ for each value of $\mathcal{L}(f)$. A second chart does the same thing for $S/N(f_L, f_H)$ and $v_x(t, f_L, f_H)$ for the range of 50–120 dB. The third chart gives the single-sideband coherent spurious modulation-power-to-signal-power ratio as a function of the rms voltage observed on the test set. The correction chart gives the total decibel correction factors for $S/N(f_L, f_H)$ for carrier frequencies from 1–50 MHz in 1 MHz steps, and also gives total decibel correction factors for $\mathcal{L}(f)$ at 11 offset frequencies (in a 2, 5, 10 sequence from 20 Hz to 50 kHz) at each of the same carrier frequencies. With these charts, any reading taken on the test set can be reduced to $\mathcal{L}(f)$ or $S/N(f_L, f_H)$ in a few minutes.

XV. Application of the Test Set to Production Synthesizer Testing

This test set was originally designed for the purpose of making rapid and accurate measurements of phase noise on the Fluke model 645A frequency synthesizer in a production environment. The end goal in this application is to be able to verify that the phase-noise performance of each synthesizer is better than the published specification. In order to do this, the first step was to select three synthesizers having superior and nearly equal performance. These were then characterized by measurements and computer programs based on the techniques developed in this paper. One unit was selected to be maintained as a "noise-standard" instrument. The performance of any production unit can then be determined by only one comparison measurement with the noise-standard unit, since the performance of the noise standard is accurately known.

A particularly simple way of making these tests was developed. A computer program was written that uses the test-set calibration data, the instrument performance specification, and the measured performance data of the noise standard to generate a typed test-report form on which the observed voltages from the actual testing are recorded. The observed voltages from the actual testing are then compared with the maximum allowable voltages, which are printed directly on the test report form. If the observed values are less than the maximum allowable readings, then the unit under test meets the specifications. The accuracy of this determination is essentially the full accuracy of the test set, since the performance of the noise-standard unit is quite a bit better than the published specification. Each instrument is checked for residual noise at 15 different carrier frequencies and at four offset frequencies for each carrier frequency. The test frequencies are arranged in such a way that the residual noise of every signal processing module in the synthesizer is thoroughly tested, so that noise performance can be verified over the entire range of 1–50 MHz. A typical plot of the measured performance of one of these synthesizers is shown in Fig. 14 as an example of an actual application of the test set [7].

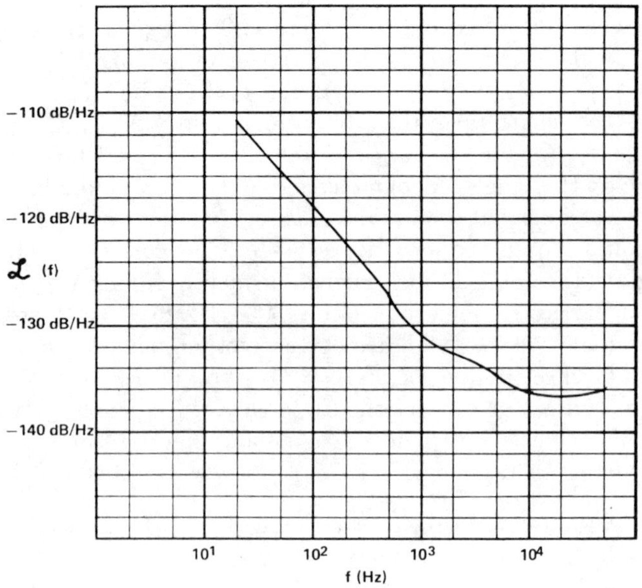

Fig. 14. Residual noise of 645A synthesizer at 49.9 MHz.

XVI. Conclusions

An accurate method for the measurement of phase noise has been explained and the theoretical relationships between the electrical parameters of the equipment involved and the measures of phase noise we desire to estimate have been developed. Practical calibration techniques that can be used to increase the accuracy of the measurement have been developed. Several types of performance tests which can be made on different types of equipment have been explained, and some examples of the actual performance of the test technique for actual measurements have been given. The techniques derived here should be of direct use to other researchers in this field who wish to make more accurate measurements of phase-noise performance in various types of equipments.

References

[1] V. Van Duzer, "Short term stability measurements," *Proc. 1964 IEEE–NASA Symp. on Definition and Measurement of Short-Term Frequency Stability*. Washington, D. C.: U. S. Government Printing Office, NASA SP-80, 1965, pp. 269–272.

[2] D. Halford, *NBS Seminar on Frequency and Time Stability*, lecture notes, 1969 (unpublished).

[3] D. J. Glaze, "Improvements in atomic cesium beam frequency standards at the National Bureau of Standards," *IEEE Trans. Instrumentation and Measurement*, vol. IM-19, pp. 156–160, August 1970.

[4] L. S. Cutler and C. L. Searle, "Some aspects of the theory and measurement of frequency fluctuations in frequency standards," *Proc. IEEE*, vol. 54, pp. 136–154, February 1966.

[5] D. W. Allan, "Statistics of atomic frequency standards," *Proc. IEEE*, vol. 54, pp. 221–230, February 1966.

[6] M. Schwartz, *Information Transmission, Modulation, and Noise*. New York: McGraw-Hill, 1959, pp. 429–432.

[7] D. G. Meyer, "An ultra low noise direct frequency synthesizer," *Proc. 24th Ann. Frequency Control Symp.* (unpublished).

AUTOMATING PHASE NOISE MEASUREMENTS IN THE FREQUENCY DOMAIN

Algie L. Lance, Wendell D. Seal,
Frank G. Mendoza and Neal W. Hudson

TRW Defense and Space Systems Group
Equipment Management Center, Metrology
One Space Park
Redondo Beach, California 90278

Summary

Automation of phase noise measurements has been developed with satisfactory results using two types of measurement systems. Measurements are performed in the frequency domain using a spectrum analyzer which provides a frequency window following the phase or frequency detector. One system measures the combined phase noise characteristics of two sources. The two source signals are applied in quadrature to a phase sensitive detector (double balanced mixer) and the voltage fluctuations analog to the phase fluctuations are measured at the detector output. One measurement system is designed to measure the phase noise characteristics of a single oscillator. The single-oscillator measurement system is designed using the delay line as an FM discriminator. Voltage fluctuations analog to frequency fluctuations are measured at the detector output.

The calibration and measurement steps are controlled by a calculator program. The calibration sequence requires several manual operations. The software program controls frequency selection, bandwidth settings, settling time, amplitude ranging, measurements, calculations, graphics, and data plotting.

A quasi-continuous plot of phase noise is obtained by performing measurements at Fourier frequencies separated by the IF bandwidth of the spectrum analyzer used during the measurement. Plots of other defined parameters can be obtained and plotted as desired. Data plots will be used to show the effects of phase lock loops, inadequate isolation of two sources when using the two-oscillator technique, and the limiting frequency range of measurements performed using the delay line as an FM discriminator.

High quality sources can be measured without multiplication to enhance the phase noise prior to downconverting and measuring at baseband frequencies.

Measurements can be extended through the millimeter frequency band. The integrated phase noise can be calculated for any selected range of Fourier frequencies. This capability is being added to the present software program.

Special reference -- Automated Phase Noise Measurements, The Microwave Journal, Vol. 20, No. 6, June, 1977.

Introduction

In this presentation the Greek letter nu (ν) represents frequency for carrier-related measures. Modulation-related frequencies are designated (f). If the carrier is considered as dc, the frequencies measured with respect to the carrier are referred to as baseband, offset from the carrier, modulation, noise, or Fourier frequencies.

The term frequency stability encompasses the concepts of random noise, intended and incidental modulation, and any other fluctuations of the output frequency of a device. In general, frequency stability is the degree to which an oscillating source produces the same value of frequency throughout a specified period of time.

Long-term stability is usually expressed in terms of parts per million per hour, day, week, month or year. This stability represents phenomenon due to the aging process of circuit elements and of the material used in the frequency determining element. Short-term stability relates to frequency changes of less than a few seconds duration about the nominal frequency. The measurement of frequency stability can be accomplished in both the time domain and frequency domain. In the frequency domain, measurements are performed using a spectrum analyzer which provides a frequency window following the detector. In the time domain, measurements are performed with a gated counter which provides a time window following the detector.

Phase noise is the term most widely used to describe the characteristic randomness of frequency. The term spectral purity refers to the signal power to phase noise sideband power ratio. Spurious signals are outputs in the spectrum of a source that are neither part of the carrier nor its harmonics, and they may be discrete frequencies or bands of frequencies. The spectral content of the noise modulating signals is finite in bandwidth and does not have the same bandwidth and same amplitude at all frequencies as one finds with white noise.

Practical oscillators demonstrate noise which appears to be a combination of casually generated signals and random, non-deterministic noises. The random noises include thermal noise, shot noise, noises of undetermined origin (such as flicker noise), and integrals of those noises. The end result is time dependent phase and amplitude fluctuations.

General Theory and Definitions

The instantaneous output voltage V(t) of a signal generator or oscillator may be written as [1]

$$V(t) = [V_o + \varepsilon(t)] \sin [2\pi\nu_o t + \phi(t)] \quad (1)$$

where V_o and ν_o are the nominal amplitude and frequency respectively and $\varepsilon(t)$ and $\phi(t)$ are the instantaneous amplitude and phase fluctuations of the signal. The definition of instantaneous fractional frequency deviation y(t) is based on the provision that $\varepsilon(t)$ and $d\phi/dt$ are sufficiently small for all time t. $\phi(t)$ is a varying real function of time, representing angle modulation, i.e., frequency and phase fluctuation.

Frequency fluctuations ($\delta\nu$) are related to phase fluctuations ($\delta\phi$) by

$$\delta\nu \equiv \delta\Omega/2\pi = 1/2\pi \; d(\delta\phi)/dt \quad (2)$$

i.e., angular frequency deviation $\delta\Omega$ is equal to the rate of change of phase deviation (the first time derivative of the instantaneous phase deviation).

y is defined as the fractional frequency fluctuation or normalized frequency deviation. It is the value of $\delta\nu$ normalized to the average (nominal) signal frequency ν_o.

$$y \equiv \delta\nu/\nu_o \quad \text{[dimensionless]} \quad (3)$$

Reprinted with permission from *Proc. 31st Annu. Symp. Frequency Contr.*, Atlantic City, NJ, June 1-3, 1977, pp. 347-358.

y(t) is the instantaneous fractional frequency deviation from the nominal frequency ν_o.

A representation of fluctuations in the frequency domain is a graph called spectral density. Spectral density is the distribution of variance versus frequency.

The spectral density $S_y(f)$ of the instantaneous fractional frequency fluctuations y(t) is defined as a measure of frequency stability[1].

$S_y(f)$ is the one-sided spectral density of fractional frequency fluctuations on a "per hertz" basis; the dimensionality is Hz^{-1}.

$$S_y(f) = S_{\delta\nu}(f)/\nu_o^2 \qquad [Hz^{-1}] \qquad (4)$$

$S_{\delta\nu}(f)$, in Hz^2/Hz, is the one-sided spectral density of frequency fluctuations $\delta\nu$. It is calculated as $(\delta\nu_{rms})^2$/(Bandwidth used in the measurement of $\delta\nu_{rms}$).

The spectral density of phase fluctuations is a frequency domain measure of phase fluctuations defined as follows:

$S_{\delta\phi}(f)$, in rad^2/Hz, is the one-sided spectral distribution of the phase fluctuations on a "per hertz" basis. It is calculated as

$$S_{\delta\phi}(f) = \frac{(\delta\phi_{rms})^2}{\text{Bandwidth used to meas. } \delta\phi_{rms}} \qquad [rad^2/Hz] \qquad (5)$$

The phase and fractional frequency fluctuation spectral densities are related by:

$$S_{\delta\phi}(f) = (\nu_o^2/f^2)\, S_y(f) \qquad [rad^2/Hz] \qquad (6)$$

$S_{\delta\Omega}(f)$, in $(rad/s)^2/Hz$, is the spectral density of angular frequency fluctuations $\delta\Omega$. The defined spectral densities have the following interconnecting relationships.

$$S_{\delta\nu}(f) = \nu_o^2\, S_y(f) = (1/2\pi)^2\, S_{\delta\Omega}(f)$$

$$= f^2\, S_{\delta\phi}(f) \qquad [Hz^2/Hz] \qquad (7)$$

and

$$S_y(f) = S_{\delta\nu}(f)/\nu_o^2 = (1/2\pi)^2(1/\nu_o)^2\, S_{\delta\Omega}(f)$$

$$= \frac{f^2}{\nu_o^2}\, S_{\delta\phi}(f) \qquad [Hz^{-1}] \qquad (8)$$

A useful measure of frequency stability relates the sideband power associated with phase fluctuations to the carrier power level. The defined measurand is called Script $\mathcal{L}(f)$. $\mathcal{L}(f)$ is defined as the ratio of the power in one sideband, referred to the input carrier frequency, on a per hertz of bandwidth spectral density basis, to the total signal power, at Fourier frequency f from the carrier, per one device[2]. It is a normalized frequency domain measure of phase fluctuation sideband power.

$$\mathcal{L}(f) \equiv \frac{\text{Power Density (one phase modulation sideband)}}{\text{Carrier Power}} \qquad [Hz^{-1}] \qquad (9)$$

For the condition that the phase fluctuations occuring at rates f and faster are small compared to one radian, a good approximation is[2]

$$S_{\delta\phi}(f)\Big|_{\text{one device}} = (2\ rad^2)\,\mathcal{L}(f) \quad [rad^2/Hz] \qquad (10)$$

If the small angle condition is not met, Bessel function algebra must be used to relate $S_{\delta\phi}(f)$ to $\mathcal{L}(f)$.

Script $\mathcal{L}(f)$ often is expressed in decibels relative to the carrier per hertz (dBc/Hz) which is calculated as

$$\mathcal{L}(f) = 10 \log\left[\frac{S_{\delta\phi}(f)\Big|_{\text{one unit}}}{2\ rad^2}\right] \qquad [dBc/Hz] \qquad (11)$$

It is very important to note that the theory, definitions and equations given above relate to the noise of a single device.

Noise Processes

The spectral density plot of a typical oscillator's output is usually a combination of different noise processes. It is very useful and meaningful to categorize these processes since the first job in evaluating a spectral density plot is to determine which type of noise exists for the particular range of Fourier frequencies. The two basic categories of noise are the discrete frequency noise and the power-law noise process.

Discrete frequency noise is a type of noise in which there is a dominant observable probability i.e., deterministic in that they can usually be related to the mean frequency, power line frequency, vibration frequencies or AC magnetic fields, or Fourier components of the nominal frequency. These frequencies can have their own spectral density plots which can be defined as noise on noise.

Power-law noise process - Types of noise which produce a certain slope on the one sided spectral density plot. They are characterized by their dependence on frequency. The spectral density plot of a typical oscillator output is usually a combination of the different power-law processes. In general, we can classify the power-law processes into five categories. These five processes are illustrated in Figure 1 which can be referred to with respect to the following description of each process[2].

Random Walk FM - (random walk of frequency) - The plot goes down as $1/f^4$. This noise is usually very close to the carrier and is difficult to measure. It is usually related to the oscillator's physical environment (mechanical shock, vibration, temperature, or other environmental effects).

Flicker FM - (flicker of frequency) - The plot goes down as $1/f^3$. This noise is typically related to the physical resonance mechanism of the active oscillator or the design or choice of parts used for the electronics or power supply, or even environmental properties. The time domain frequency stability over extended periods is constant. In high quality oscillators this noise may be masked by white FM ($1/f^2$) or flicker phase modulation ϕM (1/f). It may be masked by drift in low quality oscillators.

White FM - (white of frequency) - Random Walk of Phase - Plot goes down as $1/f^2$. A common type of noise found in passive-resonator frequency standards. Cesium and Rubidium frequency standards have white FM noise characteristics since the oscillator (usually quartz) is locked to the resonance feature of these devices. This noise gets better as a function of time until it (usually) becomes flicker FM ($1/f^3$) noise.

Flicker ϕM - (flicker (modulation) of phase) - The plot goes down as 1/f. This noise may relate to the physical resonance mechanism in an oscillator. It is common in the highest quality oscillators. This noise can be introduced by noisy electronics, amplifiers necessary to bring the signal amplitude up to a usable level, and frequency multipliers. This noise can be reduced by careful design and hand-selecting all components.

White ϕM - (white of phase) - White phase noise plot is flat f^0. Broadband phase noise is generally produced in the same way as flicker ϕM (1/f). Late stages of amplification are usually responsible. This noise can be kept low by careful selection of components and narrow band filtering at the output.

The interconnecting relationships of modulation index, rms frequency deviation, peak frequency deviation, and spectral density of phase fluctuations are shown in Figure 1. These relationships are derived in Appendix A.

Basic Two-Oscillator Technique

A functional block diagram of the measurement system employing two oscillators is shown in Figure 2. NBS has performed phase noise measurements using this basic type system since 1967.

Assume that the reference oscillator is perfect (no phase noise) and that it can be adjusted in frequency. Also, assume that both oscillators are extremely stable so that phase quadrature can be maintained without the use of a phase-lock loop.

The double balanced mixer acts as a phase-sensitive detector so that when two signals are identical in frequency and nominally are in phase quadrature, the mixer output is a small fluctuating voltage δv centered on approximately zero volts. This small fluctuating voltage represents the phase modulation PM sideband components of the signal.

If the two oscillator signals applied to the mixer of Figure 2 are slightly out of zero beat, a slow sinusoidal voltage with a peak-to-peak voltage of V_{ptp} can be measured at the mixer output. Under the conditions for which Equation (10) is valid, the ralationship between mean-square fluctuations of phase $\delta\phi$ and voltage δv interpreted in a spectral density fashion has been shown to be[1]

$$S_{\delta\phi}(f) = S_{\delta v}(f)/2(V_{rms})^2 \qquad [rad^2/Hz] \quad (12)$$

Here, $S_{\delta v}(f)$, in volts squared per hertz, is the spectral density of the voltage fluctuations at the mixer output. Since the spectrum analyzer measures rms voltage in a bandwidth B, the analyzer reading is in units of volts per $\sqrt{\text{Bandwidth}}$. Therefore,

$$S_{\delta v}(f) = [\delta v_{rms}/\sqrt{B}]^2 = \frac{(\delta v_{rms})^2}{B} \qquad [V^2/Hz] \quad (13)$$

where B is the noise power bandwidth used in the measurement.

Since it was assumed that the reference oscillator did not contribute any phase noise, the voltage fluctuations $(\delta v)_{rms}$ represent the oscillator under test and the spectral density of the phase fluctuations in terms of the voltage measurements performed with the spectrum analyzer is

$$S_{\delta\phi}(f) = \frac{1}{2}\,\frac{(\delta v_{rms})^2}{B(V_{rms})^2} \qquad [rad^2/Hz] \quad (14)$$

Equation 12 is sometimes expressed as[3]

$$S_{\delta\phi}(f) = S_{\delta v}(f)/K^2 \qquad [rad^2/Hz] \quad (15)$$

where K is the calibration factor in volts per radian.

For sinusoidal beat signals, the peak voltage of the signal equals the slope of the zero crossings in volts per radian. Therefore, $(V_{peak})^2 = 2(V_{rms})^2$, which is the same as the denominator in Equation 12.

$S_{\delta\phi}(f)$ can be expressed in terms of decibels relative to one square radian per hertz by calculating 10 log $S_{\delta\phi}(f)$ of the previous equation.

$S_{\delta\phi}(f)$ in decibels relative to 1 rad^2/Hz =

$$[20 \log[(\delta v)_{rms}] - 20 \log(V_{rms}) - 10 \log(B) - 3\text{ dB}] \quad [dB(rad^2/Hz] \quad (16)$$

A correction of +2.5 dB to the $(\delta v)_{rms}$ term is required for the Tracking Spectrum Analyzer used in these measurement systems. See HP Application Note 207.[3]

$\mathscr{L}(f)$ can be expressed in terms of decibels relative to the carrier power per hertz (dBc/Hz).

$\mathscr{L}(f)$ in decibels relative to 1 Hz^{-1} =

$$[20 \log[(\delta v)_{rms}] - 20 \log(V_{rms}) - 10 \log(B) - 6\text{ dB}] \quad [dBc/Hz] \quad (17)$$

The -6 dB correction occurs because the operation of the mixer, when it is driven at quadrature, is such that the amplitudes of the two phase sidebands add linearly in the output of the mixer. This results in four times as much power in the output as would be present if only one of the phase sidebands were allowed to contribute to the output of the mixer.

Spectrum analyzer corrections must also be added to this equation.

Two Noisy Oscillators

The measurement system of Figure 2 yields the output noise from both oscillators. If the reference oscillator is superior in performance as assumed in the previous discussions, then one obtains a direct measure of the noise characteristics of the oscillator under test.

If the reference and test oscillators are the same type, a useful approximation is to assume that the measured noise power is twice that which is associated with one noisy oscillator. This approximation is in error by no more than 3 dB for the noisier oscillator even if one oscillator is the major source of noise. The equation for the spectral density of measured phase fluctuations is

$$S_{\delta\phi}(f)\Big|_{\#1} + S_{\delta\phi}(f)\Big|_{\#2} = \frac{S_{\delta v}(f)\Big|_{\text{two devices}}}{2(V_{rms})^2}$$

$$\doteq 2\,S_{\delta\phi}(f)\Big|_{\text{one device}} \qquad [rad^2/Hz] \quad (18)$$

The measured value is therefore divided by two in order to obtain the value for the single oscillator. A determination of the noise of each oscillator can be made if one has three oscillators that can be measured in all pair combinations.

Spectral Density of Amplitude Fluctuations

The spectral density ($S_{\delta\epsilon}(f)$) of the amplitude fluctuations of a signal follows the same general derivation previously given for the spectral density of phase fluctuations[1].

When two signals are slightly different in frequency, a slow, almost sinusoidal beat is produced at the mixer output and the peak-to-peak voltage swing is defined as (V_{ptp}).

If the two signals are now tuned to colinear phase (0 or 180° phase angle difference), the mixer output is a fluctuating voltage centered on $V_{ptp}/2$ volts. There is no requirement that the output fluctuations be small compared to $V_{ptp}/2$. In order to obtain linearity in the measurements of AM, and to make the measurement sensitive to the test oscillator only, the reference signal into the balanced mixer should be at least 10 dB greater than the test oscillator signal input to the mixers.

The normalized frequency domain measure of fractional amplitude fluctuation sidebands of a signal is called Script $\mathcal{M}(f)$. It is defined to be the ratio of the spectral density of one amplitude modulation sideband to the total signal power, at Fourier frequency difference f from the signal's average frequency ν_o, for a single specified signal or device. The dimensionality is per hertz. Because here f is a frequency difference, the range of f is from minus ν_o to plus infinity. Script $\mathcal{L}(f)$ and Script $\mathcal{M}(f)$ are similar functions; the former is a measure of phase modulation (PM) sidebands, the latter is a corresponding measure of amplitude modulation (AM) sidebands. We introduce the symbol Script $\mathcal{M}(f)$ in order to have useful terminology for the important concept of normalized AM sideband power.

For the types of signals under consideration, by definition the two amplitude fluctuation sidebands (lower sideband and upper sideband, at -f and +f from ν_o, respectively) of a signal are coherent with each other. They are of equal intensity also. The operation of the mixer when it is driven at colinear phase is such that the amplitudes of the two AM sidebands add linearly in the output of the mixer, resulting in four times as much power in the output as would be present if only one of the AM sidebands were allowed to contribute to the output of the mixer. It has been shown that $\mathcal{M}(f)$ is related to the spectral density of the amplitude fluctuations $S_{\delta\epsilon}(f)$ by[2]

$$\mathcal{M}(f) = (1/2V_o^2) \; S_{\delta\epsilon}(f) \tag{19}$$

where V_o is the average (nominal) amplitude also of the signal.

$$\mathcal{M}(f) = 2(\delta v \text{ (one unit)}/V_{ptp})^2 = S_{\delta v}(f)/4\,V_{rms}^2 \tag{20}$$

Automated Phase Noise Measurement System

The TRW Metrology automated phase noise measurement system is shown in Figure 3. It is program controlled by the Hewlett-Packard 9830 Programmable Calculator. Each step of the calibration and measurement sequence is included in the program. The software program controls frequency selection, bandwidth settings, settling time, amplitude ranging, measurements, calculations, graphics, and data plotting. The system will be described as it is used to obtain a direct plot of Script $\mathcal{L}(f)$. The direct measurement of $\mathcal{L}(f)$ is represented by the following equation.

$\mathcal{L}(f)$ in decibels relative to 1 Hz^{-1} =

(Noise Power Level)/(Carrier Power Level) in dB

$$-6 \text{ dB} + 2.5 \text{ dB} - 10\log(B) - 3 \text{ dB} \quad [\text{dBc/Hz}] \tag{21}$$

The noise power is measured relative to the carrier power level and the remaining terms of the equation represent corrections that must be applied due to the type measurement and the characteristics of the measurement equipment as follows.

- o The basic measurement of noise sidebands with the signals in phase quadrature requires the -6 dB correction as explained after Equation (17).
- o The nonlinearity of the spectrum analyzer logarithmic IF amplifier results in compression of the noise peaks which, when average detected, require the +2.5 dB correction for the HP 3571A Tracking Spectrum Analyzer.
- o The bandwidth correction is required because the spectrum analyzer measurements of random noise are a function of the particular bandwidth used in the measurement.
- o The -3 dB correction is required since this is a measurement of $\mathcal{L}(f)$ using two oscillators, assuming that the oscillators are of a similar type and that the noise contribution is the same for each oscillator. If one oscillator is sufficiently superior to the other, this correction is not used.

The Calibration and Measurement Sequence

The mixer is driven with the maximum power that will result in a 50 ohm output impedance of the mixer. We used equal power inputs to the mixer when maximum noise floor was required. However, most sources were measured with a reference power of about 12 mw and a signal power level between 1 and 3 mw.

1. Measure the noise power bandwidth for each IF bandwidth setting on the Tracking Spectrum Analyzer.
2. Obtain a carrier power reference level (referenced to the output of the mixer).
3. Adjust for phase quadrature of the two signals applied to the mixer.
4. Noise power is measured at the selected Fourier frequencies, the calculations are performed, and the data is plotted (or stored) using calculator and program control (fully automated).
5. Measure and plot the system noise floor characteristics if desired.

Measurement of Noise Power Bandwidth - The precise IF noise power bandwidth of the Tracking Spectrum Analyzer must be known when performing phase noise measurements. The basic automated measurement is as follows:

a) A tee junction is inserted at the HP 3330B synthesizer 1-MHz output and the 1-MHz signal is applied to the input of the HP 3571A Tracking Spectrum Analyzer.

b) The calculator controls the synthesizer for the desired incremental changes in frequency. The power output is recorded for each frequency setting over the range indicated in Figure 4. The range of measurements is illustrated for equal dB values on each side of the 1-MHz center frequency. One should choose points greater than 40 dB below the carrier. We use 100 increments in frequency. Our experience has indicated that the 40 dB level and the 100 increments in frequency are not the absolute minimum permissible values.

c) The above recorded data is plotted for each IF bandwidth as illustrated in Figure 4. The noise power bandwidth is calculated as,

$$\text{Noise bandwidth} = \frac{(P_1 + P_2 + P_3 + ---P_{100})\Delta f}{\text{Peak Power Reading}} \quad [\text{Hz}] \quad (22)$$

where Δf is the frequency increment (Hz) and the peak power is the maximum measured point obtained during the measurements. All power values are in watts.

Setting Carrier Power Reference Level - Essentially, the carrier reference power level is established at the output of the low-pass filter in Figure 3 as follows:

a) The precision IF step attenuator is set to a high value to prevent overloading the spectrum analyzer (assume 50 dB as our example).

b) Approximately equal power is applied to the inputs of the mixer as required to obtain the 50 ohm output impedance of the mixer in our 50 ohm system. Special techniques are required to obtain the correct impedance when lower impedance is involved.

c) If the frequency of one of the oscillators can be adjusted, adjust its frequency for an IF output frequency in the range of 10 to 20 kHz.

   If neither oscillator is adjustable, replace the oscillator under test with one that can be adjusted as required and that can be set to the identical power level of the oscillator under test.

d) The resulting IF power level is measured by the spectrum analyzer and the measured value is corrected for the attenuator setting which was assumed to be 50 dB. The correction is necessary since this attenuator will be set to its Zero dB indication during the measurements of noise power. Assuming a spectrum analyzer reading of -40 dBm, the carrier power reference level is calculated as,

$$\text{Carrier reference power level} = 50\text{ dB} - 40\text{ dBm} = 10\text{ dBm} \quad (23)$$

Phase Quadrature Adjustment - After the carrier power reference has been established, the oscillator under test and the reference oscillator are tuned to the same frequency and the original reference levels that were used during calibration are re-established. The quadrature adjustment depends upon the type of system used. Three possibilities are illustrated in Figure 3.

a) If the oscillators are very stable, have high resolution tuning, and are not phase-locked, the frequency of one oscillator is adjusted for zero dc voltage output of the mixer as indicated by the sensitive oscilloscope.

b) If the common reference frequency is used, as illustrated in Figure 3, then it is necessary to include a phase shifter in the line between one of the oscillators and the mixer (preferably between the attenuator and mixer). The phase shifter is adjusted to obtain zero dc output of the mixer.

c) If one oscillator is phase-locked using a phase-lock loop as shown dotted in on Figure 3, the frequency of the unit under test is adjusted for zero dc output of the mixer as indicated on the oscilloscope.

The phase-lock loop is an electronic servomechaniam that acts to null out any phase error between the two inputs to the mixer (phase detector). The time constant of the loop can be adjusted as needed by varying amplifier gain and RC filtering within the loop.

A loose phase-lock loop is characterized by:

- The correction voltage varies as phase (in the short term) and phase variations are therefore observed directly.
- The bandwidth of the servo response is small compared to the Fourier frequency to be measured.
- The response time is very slow.

The tight phase-lock loop is characterized by:

- The correction voltage of the servo loop varies as frequency.
- The bandwidth of the servo response is relatively large.
- The response time is much smaller than the smallest time interval (τ) at which measurements are performed.

NOTE: Throughout the measurement process one should check and maintain phase quadrature.

Measurements, Calculations, and Data Plots

The measurement sequence is automated except for the case where manual adjustments are required to maintain phase quadrature of the signals.

Automated measurements are executed after phase quadrature is obtained and the Precision IF Step Attenuator is set to its Zero dB indication.

As an example, our previous assumptions resulted in a carrier power level of +10 dBm. Assume that the spectrum analyzer measures -106 dBm with a 10 Hz bandwidth setting at a particular Fourier frequency. The value of spectral density which will be plotted or stored is calculated as,

$$\mathcal{L}(f) \text{ in decibels relative to } 1\text{ Hz}^{-1} = [-106\text{ dBm} - 10\text{ dBm} - 6\text{ dB} + 2.5\text{ dB} - 10\log 10 - 3\text{ dB}] = -132.5\text{ dBc/Hz} \quad (24)$$

The -3 dB correction is included since it is assumed that the two units under test are similar and produce the same amount of noise.

In this example, the value of the spectral density of phase fluctuations in dB relative to one square radian per hertz is calculated as,

$$S_{\delta\phi}(f) \text{ in dB(rad}^2\text{/Hz)} = -132.5\text{ dBc/Hz} + 3\text{ dB(rad}^2) = -129.5\text{ dB(rad}^2\text{/Hz)} \quad (25)$$

If one desires to plot the spectral density of frequency fluctuations, it is necessary to recall that

$$S_{\delta\phi}(f) \text{ in dB(rad}^2\text{/Hz)} = 10\log S_{\delta\phi}(f) \quad [\text{dB(rad}^2\text{/Hz)}] \quad (26)$$

and that the spectral density of frequency fluctuations is

$$S_{\delta\nu}(f) = f^2 \, S_{\delta\phi}(f) \qquad [Hz^2/Hz] \qquad (27)$$

System Noise Floor Verification

A plot of the noise floor is obtained by repeating the automated measurements with Attenuator No. 1 set to maximum or by disconnecting the unit under test and terminating the input of the mixer with a matched load.

We use the following equation to correct for noise floor.

$\mathcal{L}(f)$ (corrected) in decibels relative to 1 Hz^{-1} =

$$\mathcal{L}(f) + \left[\frac{P_{\mathcal{L}(f)} - P_{noise\ floor}}{P_{\mathcal{L}(f)}}\right] \qquad [dBc/Hz] \qquad (28)$$

If adequate calculator memory is available, a set of measurements of the oscillator plus floor can be performed at selected Fourier frequencies and stored. The calculations of corrections can be made when the noise floor is measured.

The correction for noise floor can also be obtained at an earlier point in the program at Equation (13). Measurement of $S_{\delta v}(f)$ of the oscillator plus floor is obtained then $S_{\delta v}(f)$ is obtained for the noise noise floor only. Then,

$$S_{\delta v}(f)\Big|_{corrected} =$$

$$S_{\delta v}(f)\Big|_{osc.\ +\ floor} - S_{\delta v}(f)\Big|_{floor} \qquad [V^2/Hz] \qquad (29)$$

Basics of the TRW Metrology Automated

Phase Noise Measurement Program

1. The HP 3330B Synthesizer serves as the local oscillator for the HP 3571A Tracking Spectrum Analyzer. The calculator program controls the switching of the synthesizer to the desired Fourier frequencies.
2. The Fourier frequency increments are chosen to be equal to the selected IF noise bandwidth in order to obtain a continuous spectrum plot.
3. The minimum delay time for a measurement is determined by the IF filter build-up in the spectrum analyzer. The range is from 2.5 seconds for the 3-Hz bandwidth, decreasing to 70 milliseconds for the 10-kHz bandwidth setting.
4. Video smoothing is used in order to obtain a better approximation of the mean. The program can be designed so that a large number of measurements can be taken for better estimation of the mean value. From statistical theory the confidence in an average is improved by the square root of the number of samples.
5. The IF bandwidth settings for the Fourier (offset) frequency range selections are as follows:

IF Bandwidth (Hz)	Fourier Frequency	IF Bandwidth (kHz)	Fourier Frequency (kHz)
3	10 - 400 Hz	1	40 - 100
10	400 Hz - 1 kHz	3	100 - 400
30	1 - 4 kHz	10	400 - 13 MHz
100	4 - 10 kHz		
200	10 - 40 kHz		

Program running time is 27 minutes when using 100 measurements at each Fourier measurement frequency out to 200 Hz.

6. The 60 Hz line frequency interference appears smaller than the actual amplitude if the noise corrections are applied as set forth in the noise measurement program. The corrections for the log amplifier and detection, bandwidth, and equal oscillator contribution should be removed for a plot of discrete frequencies.
7. Amplitude auto-ranging is used in the program to select the most sensitive range that does not result in overload conditions.

The low-pass filter prevents local oscillator leakage power from overloading the spectrum analyzer when baseband measurements are performed at the Fourier (offset) frequencies of interest. Leakage signals will interfere with autoranging and the dynamic range of the spectrum analyzer.

The low-noise high-gain preamplifier provides additional system sensitivity by amplifying the noise signals to be measured. Also, since spectrum analyzers usually have high values of noise figure, this amplifier is highly desirable. As an example, if the high-gain preamplifier had a noise figure of 3 dB and the spectrum analyzer had a noise figure of 18 dB, the system sensitivity at this point has been improved by 15 dB. The overall system sensitivity would not necessarily be improved 15 dB in all cases because the limiting sensitivity could have been imposed by a noisy mixer.

Figure 5 illustrates the type of plot obtained using our program and the two-oscillator measurement system. Each plot is the Normalized Phase Noise Sideband Power Spectral Density in dBc/Hz. However, any defined spectral density can be plotted as illustrated by Figure 1.

Figure 5a measurements were performed on two F.E.I. Model 1050A Disciplined Time Frequency Standards. The sample plots beneath the continuous plot are the noise floor.

Figure 5b characteristics of the Austron 1120 S Oscillator were obtained using the Austron 1125 S Multiplier to enhance the noise by 13.98 dB. Direct measurement at 5 MHz was slightly beyond the system capability. Direct measurement may be possible by incorporating a high level mixer in our system.

Figure 5c represents the characteristics of a Hewlett-Packard Model 105B Quartz Oscillator. The F.E.I. DTF Standard was used as the reference. The lowest plot is for the HP 105B and the upper plot illustrates increase in phase noise when the signal was amplified using a Hewlett-Packard 5087A Distribution Amplifier.

Figure 5d is the phase noise plot obtained using two identical 250 MHz Surface Acoustic Wave (SAW) Oscillators developed at TRW. The roll-off in the range from 10 Hz out to about 200 Hz was due to inadequate isolation between the two oscillators. The slight injection locking was also noted when fine tuning one oscillator. This illustrates a problem of measuring in close to the carrier since the effective locking loop bandwidth will depend upon the isolation of the oscillators.

Figure 5e shows measurements performed on two HP 8660C Synthesizers, at 5 MHz. The reference oscillator of one was locked to the reference oscillator in the other unit. Therefore, the phase noise plot does not slope upward from about 500 Hz to 10 Hz on the plot. The slope upward was verified using the single-oscillator measurement technique which will be discussed later. The phase noise reduction by the phase-lock loop of the synthesizer is shown and one notes where the loop bandwidth ends and where the characteristics of the source oscillator are being measured.

Figure 5f shows measurements performed on a Fluke 6160B Synthesizer with the F.E.I. 7507A-1 Frequency Standard as the reference. The measurements were performed at 100 MHz. The characteristics discussed previously are now shown. The top plot at the lower Fourier frequencies was obtained with the oscillator "free-running". The lower plot was performed when the reference oscillator of the synthesizer was phase-locked to the F.E.I. Frequency Standard Reference Oscillator.

Phase Noise Measurements Using Delay Line FM Discriminators

Frequency fluctuations are measured directly using FM discriminator techniques (References 5, 6, and 7). One of the important advantages of this type system is that the phase noise characteristics of a single oscillator can be measured without the requirement of a similar or better source as a reference.

The delay line yields a phase shift by the time the signal arrives at the balanced mixer. The phase shift depends upon the instantaneous frequency of the signal. The presence of frequency modulation (FM) on the signal gives rise to differential phase modulation (PM) at the output of the differential delay and its associated (non-delay) reference line. This relationship is linear if the delay (τ_d) is non-dispersive. This is the property which allows the delay line to be used as an FM discriminator. In general, the conversion factors are a function of the delay (τ_d) and the Fourier frequency (f) but not the carrier frequency. The delay of the transmission line and variable length airline as a function of phase noise measurement at various offset frequencies will be discussed later. However, the maximum sensitivity of the transmission line discriminator depends upon the attenuation value of the delay line at the carrier frequency.[5,7] For maximum sensitivity, the optimum length of the delay line for the system shown in Figure 6 occurs when the total attenuation is approximately one neper (one neper = 8.686 dB). It represents the one-way loss of the delay line in Figure 6a and the round-trip loss of the delay line shown in Figure 6b. The optimum delay line length is determined at a particular selectable carrier frequency. However, since the attenuation varies slowly (approximately proportional to the square root of frequency), this characteristic allows near optimum operation over a considerable frequency range without appreciable degradation in the measurements.

General Theory

The measurement system is shown in Figure 6. Two delay line configurations are illustrated. The signals experience the one-way delay of the delay line in Figure 6a.[5,6] The FM discriminator of Figure 6b is a reflective type used by Ashley et al[7] and it is noted that the signals experience the round-trip and resultant two-way loss in the delay line.

With adequate drive power we have not limited the delay line in this system to one neper. Delay lines with attenuation greater than 11 dB were used in some of our measurements. The resulting noise floor was more than 10 dB below the measured values of phase noise.

The calculator program performs the same functions as outlined for the measurement system of Figure 3.

The system calibration and measurement procedure is outlined as follows:

- o Measure IF Noise Power Bandwidths

  The IF noise power bandwidth is measured for each IF bandwidth setting of the spectrum analyzer as set forth in a previous section of this paper.

- o Establish System Power Levels

  The oscillator under test is connected and Attenuators 1, 2, and 3 are used to set the equal power levels at the mixer inputs. The inputs were from +7 to +10 dBm for the particular mixer used in our system.

- o Quadrature Adjustment

  The phase shifter is adjusted for Zero volts DC at the output of the mixer as indicated on the oscilloscope connected as shown. This establishes the quadrature condition for the two inputs to the mixer. This quadrature condition is continuosly monitored and is adjusted if necessary.

- o Discriminator Calibration

  The calibration factor of the discriminator should be established at selected Fourier frequencies over the range of measurements. If the calibration factor is not constant, appropriate measured values should be used in the automated measurement program. The measurement and calibration will be described in terms of a 20 kHz modulation frequency.

The calibration factor is defined as:

$$CF = \Delta\nu_{rms}/V_{rms} \quad [Hz/V] \qquad (30)$$

Where $\Delta\nu_{rms}$ is the rms frequency deviation of the carrier due to the intentional modulation and V_{rms} is the spectrum analyzer voltage measurement of the modulation sideband.

$$\Delta\nu_{rms} = \Delta\nu_{peak}/\sqrt{2} = m\,(f_m)/\sqrt{2} \quad [Hz] \qquad (31)$$

The calibration factor of the discriminator is calculated as:

$$CF = m\,(f_m)/\sqrt{2}\,V_{rms} \quad [Hz/V] \qquad (32)$$

The calibration factor of this system has been found to be constant to about 400 kHz which was the limit due to resolution in setting the carrier null.

The calibration factor is established as follows:

a) Replace the oscillator under test with a signal generator or oscillator that can be frequency modulated.

   The power output and operating frequency of the generator must be set to the same precise frequency and amplitude values as the oscillator under test.

b) Select a modulation frequency of 20 kHz and increase the modulation until the carrier is reduced to the first Bessel null of the carrier as indicated on the spectrum analyzer connected to Coupler No. 1. One can choose any other convenient setting that will produce a known modulation index (m).

c) Tune the HP 3571A Tracking Spectrum Analyzer to the modulation sideband frequency (20 kHz). Maintain the phase quadrature of the two signals applied to the balanced mixer.

d) The Tracking Spectrum Analyzer is now displaying a power reading that corresponds to the

d) modulation frequency. This reading must be corrected for the dB setting of Attenuator No. 4 in order to calculate the calibration factor of the discriminator. Attenuator No. 4 will be set to Zero dB reference indication during the measurements of Fourier frequency noise power. This results in an increase of sideband power at the mixer which must be accounted for in the calculation of the calibration factor.

As an example, if the attenuator was set to 50 dB as proposed. and the spectrum analyzer reading of baseband power is -35 dBm, the corrected power is then

$$P\ (dBm) = 50\ dB - 35\ dBm = +15\ dBm \qquad (33)$$

e) The discriminator calibration factor can now be calculated, since this power in dBm can be converted to the corresponding rms voltage using the following equation.

$$V_{rms} = \sqrt{10^{P(dBm)/10}/1000 \times R} \quad [V] \qquad (34)$$

where R is 50 onms in this system.

f) The discriminator calibration factor is calculated as,

$$CF = m \cdot f_m/\sqrt{2}\ V_{rms}$$

$$= 2.405 \cdot f_m/\sqrt{2}\ V_{rms} \quad [Hz/V] \qquad (35)$$

since 2.405 is the modulation index (m) for the first Bessel carrier null as used in this technique. The modulation frequency is f_m.

Measurement and Data Plotting

a) Connect the unit under test in place of the modulated signal generator and readjust quadrature at the signals if necessary.

b) Set Attenuator No. 4 to its Zero dB indication.

The measurements, calculations, and data plotting are completely automated. The calculator program selects the Fourier frequency, auto-ranging is performed, bandwidth is set, and measurements of Fourier frequency power are performed by the tracking spectrum analyzer.

Each Fourier frequency noise power reading P_n (dBm) is converted to the corresponding rms voltage designated as δv_{1rms}.

$$\delta v_{1rms} = \sqrt{\frac{10^{(P_n(dBm)\ +\ 2.5)/10}}{1000} \times R} \quad [V] \qquad (36)$$

the rms frequency fluctuations are calculated as,

$$\delta\nu_{rms} = \delta v_{1rms} \times CF \quad [Hz] \qquad (37)$$

The spectral density of frequency fluctuations is calculated as,

$$S_{\delta\nu}(f) = (\delta\nu_{rms})^2/B \quad [Hz^2/Hz] \qquad (38)$$

where B is the measured IF noise power bandwidth of the spectrum analyzer.

The spectral density of phase fluctuations is calculated as,

$$S_{\delta\phi}(f) = S_{\delta\nu}(f)/f^2 \quad [rad^2/Hz] \qquad (39)$$

The normalized phase noise sideband power spectral density is calculated as,

$$\mathcal{L}(f) = 1/2\ rad^2)\ \ S_{\delta\phi}(f) \quad [Hz^{-1}] \qquad (40)$$

Script $\mathcal{L}(f)$ expressed in decibels relative to 1 Hz^{-1} is plotted in real time in our program. However, the data can be stored and the desired spectral density can be plotted in other forms.

Noise Floor Measurements

The system noise floor can be plotted by setting Attenuator No. 3 to maximum and repeating the automated measurements. We increase the LO power to obtain essentially the same mixer output impedance which exists when both signals are applied. When using the 12 mw and 3 mw levels, there is no difference in the noise floor with and without the 3 mw signal. However, in systems which require different impedance levels, it is necessary that the total input power (Signal and LO) be adjusted to produce the required mixer output impedance. This will require a method of measuring the mixer output impedance and re-establishing the particular value when measuring the noise floor. In the calculation process in the program, the rms voltage corresponding to the noise floor is designated δv_{2rms}.

In our program the noise floor is checked only at selected Fourier frequencies over the range of measurement.

A correction for the noise floor requires a measurement of the rms voltage of the oscillator and floor δv_{1rms} and a measurement of the noise floor rms voltage δv_{2rms}. These voltages are used in the following equation to obtain the corrected value.

$$\delta v_{rms} = \sqrt{(\delta v_{1rms})^2 - (\delta v_{2rms})^2} \quad [V] \qquad (41)$$

The corrected value δv_{rms} is then used in the calculation of frequency fluctuations. If adequate memory is available, each value of δv_{1rms} can be stored and used after the other set of measurements is performed at the same Fourier frequencies.

Figure 6b illustrates the FM discriminator used by Ashley et al[7]. The circulator, double stub tuner, delay line and adjustable length airline form the FM discriminator. The double stub tuner and adjustable line are used to null out the carrier.

The carrier nulling procedure makes the discriminator relatively immune to incidental AM on the signal being tested. The system is not restricted by burnout characteristics of the crystals in the mixer.

The quadrature adjustment differs from the system of Figure 6a which uses the one-way length of the delay line. During the calibration process the phase shifter is adjusted for maximum output indication on the Tracking Spectrum Analyzer at the 20 kHz modulation frequency. This sets the phase quadrature of the two signals applied to the balanced mixer.

The phase noise plots of Figure 7 were obtained with the TRW Metrolgoy program and the measurement system shown in Figure 6a. The phase noise plot follows the power law process to about 700 kHz. Beyond this point the delay line discriminator is not calibrated. A plot of $S_{\delta\nu}(f)$ has nulls at every multiple of f_d, where f_d is the Fourier frequency corresponding to $1/t_d$, and t_d is the total differential delay of the FM discriminator.[5]

Figure 7a measurements were performed using two different delay lines. The delay line used in measurements of the 600 MHz fundamental frequency was about 500 ns as indicated by the first null at about 2 MHz. The delay line used in the measurements of the 2.4 GHz (multiplied output) had about 250 ns delay as indicated by the first null at about 4 MHz. Note that the theoritical enhancement and measured data agree within the resolution of the plot.

APPENDIX A

Basic Modulation Theory and Spectral Density

Applying a sinusoidal frequency modulation (f_m) to a sinusoidal carrier frequency (ν_o) produces a wave that is sinusoidally advanced and retarded in phase as a function of time. The instantaneous voltage is expressed as,

$$V(t) = V_o \operatorname{Sin}[2\pi\nu_o t + \Delta\phi \operatorname{Sin} 2\pi f_m t] \quad (1)$$

where $\Delta\phi$ is the peak phase deviation caused by the modulation signal.

The first term inside the brackets represents the linearly progressing phase of the carrier. The second term is the phase variation (advancing and retarded) from the linearly progressing wave.

The effects of modulation can be expressed as residual FM noise in which modulation by a single sinusoidal signal produces a peak frequency deviation of the carrier (ν_o).

$$\Delta\nu_o = \Delta\phi \cdot f_m \quad (2)$$

$$\Delta\phi = \Delta\nu_o / f_m \quad \text{(peak)} \quad (3)$$

This ratio of peak frequency deviation to modulation frequency is called modulation index (m) so that $\Delta\phi = m$ and

$$m = \Delta\nu_o / f_m \quad (4)$$

The frequency spectrum of the modulated carrier contains frequency components (side-bands) other than the carrier. For small values of modulation index (m<<1), as is the case with random phase noise, only the carrier and first upper and lower sidebands are significantly high in energy.

The effects of modulation can also be expressed as single sideband phase noise in which the peak phase deviation is $\Delta\phi = m$. The ratio of the amplitude of either sideband (single sideband) to the amplitude of the carrier, V_o is:

$$V_{sb}/V_o = m/2 \quad (5)$$

This ratio is expressed in dB and referred to as dB below the carrier for the given bandwidth.

$$\left.\frac{V_{sb}}{V_o}\right|_{dB} = 20\log(m/2) = 20\log(\Delta\nu_o/2f_m)$$
$$= 10\log\left[\frac{m}{2}\right]^2 = 10\log(\Delta\nu_o/2f_m)^2 \quad [\text{dBc/B}] \quad (6)$$

If the frequency deviation is given in terms of its rms value then,

$$\Delta\nu_{rms} = \Delta\nu_o/\sqrt{2} \quad (7)$$

Equation 6 now becomes

$$V_{sb}/V_o \Big|_{db} = \mathcal{L}(f) = 20\log(\Delta\nu_{rms}/\sqrt{2}\, f_m)$$
$$= 10\log(\Delta\nu_{rms}/\sqrt{2}\, f_m) \quad (8)$$

The ratio of single sideband to carrier power in dBc/Hz is,

$$\mathcal{L}(f) = 20\log\left[\frac{\Delta\nu_{rms}}{f_m}\right] - 3\text{ dB} \quad [\text{dBc/Hz}] \quad (9)$$

and

$$S_{\delta\phi}(f) = 10\log\left[\frac{\Delta\nu_{rms}}{f_m}\right] \quad [\text{dB}(\text{rad}^2/\text{Hz})] \quad (10)$$

i.e., in decibels relative to 1 rad^2/Hz.

The interrelationships of modulation index, peak frequency deviation, rms frequency deviation and spectral density of phase fluctuations can be found from

$$\left[\frac{m}{2}\right]^2 = \left[\frac{\Delta\nu_o}{2\,f_m}\right]^2 = \left[\frac{\Delta\nu_{rms}}{\sqrt{2}\,f_m}\right]^2$$

or

$$= 10^{\mathcal{L}(f)/10} = S_{\delta\phi}(f)/2 \quad (11)$$

$$\frac{m}{2} = \frac{\Delta\nu_{rms}}{\sqrt{2}\,f_m} = \sqrt{10^{\mathcal{L}(f)/10}} = \sqrt{S_{\delta\phi}(f)/2} \quad (12)$$

and

$$m = \frac{\Delta\nu_o}{f_m} = \frac{2\Delta\nu_{rms}}{\sqrt{2}\,f_m} = 2\sqrt{10^{\mathcal{L}(f)/10}}$$
$$= 2\sqrt{S_{\delta\phi}(f)/2} \quad (13)$$

The basic relationships are plotted in Figure 1.

Discussion

Identical phase noise plots were obtained using the two discriminator systems. Measurements performed with the single-oscillator technique and the two-oscillator technique agree over the range out to about 40 percent of the Fourier frequency of the first null of the plot obtained with the single-oscillator (FM discriminator) system. The FM discriminator is uncalibrated beyond the described 40 percent area as indicated in Figure 7.

When using the two-oscillator technique accurate measurements close to the carrier can be limited depending upon the isolation between the two sources. This was clearly indicated in the data plots of Figure 5.

In order to obtain the lowest noise floor, the mixers used in these systems were driven with equal signals to the RF and LO ports. The nominal IF output impedance of the mixers was 50 ohms and we did not experience system sensitivity problems with a broad range of total input power to the mixer. Mismatch problems were not critical when the mixers were driven for IF output impedance values in the range of 30 to 70 ohms.

If the mixer IF output impedance is driven to a low value for use with an amplifier such as the Ortec 9431, which has a high input impedance, the total input power to the mixer can become critical when one desires optimum measurement results. In this case the mixer is driven just short of burnout of the diodes. A special impdeance monitoring circuit may be required in this particular type application.

Dr. J. Robert Ashley and Gustaf J. Rast, Jr.,[7] performed measurements on a 250 MHz TRW Surface Acoustic Wave Oscillator using their system at Redstone Arsenal. The measurements agreed within one dB with measurements performed with the three systems described in this paper.

Measurements performed using the Hewlett-Packard program in Application Note AN 207[3] fell precisely on the continuous plot obtained with the TRW Metrology program. The measurements were performed in TRW Metrology by Chuck Reynolds of Hewlett-Packard, Loveland Division.

Joe Oliverio of the Hewlett-Packard Company performed phase noise measurements in TRW Metrology using the HP 5420 Digital Signal Analyzer which measures over the frequency range from sub-millihertz to 25 kHz. The measurements were also in precise agreement with measurements performed using our program.

Holly Cole of the Hewlett-Packard Company performed verification measurements out to a few hundred hertz using the HP 5390A Frequency Stability Analyzer which performs measurements in the time domain and the data is transformed into the frequency domain.

We have performed phase noise and AM noise measurements on Impatt and Gunn oscillators at 34 GHz. The measurement technique requires two sources. The two sources are applied to a balanced mixer and are offset to obtain a difference frequency in the range of the single-oscillator (FM delay line discriminator) measurement system. Measurements are performed using the system in Figure 6 as described in this presentation.

Acknowledgements

We thank Marlyn Hed, Manager of TRW Metrology, for his support and interest in this work. We appreciate the evaluations and suggestions of Dale Dryer of TRW, the review and suggestions of Dr. William Lindsey of the University of Southern California, and the valuable cross-checks performed by Chuck Reynolds, Joe Oliverio, and Holly Cole of the Hewlett-Packard Company. We thank Dr. J. Robert Ashley of the University of Colorado, Colorado Springs, and Gustaf J. Rast of the U. S. Army Missile Command, for performing the measurements at Redstone Arsenal and for supplying complete documentation on their system. We are very grateful to Dr. Donald Halford of NBS in Boulder, Colorado, for his interest, consultations, valuable suggestions during the development of the systems, review of the final paper, and for suggesting the technique for measuring phase noise of millimeter sources. The efforts of Don Leavitt of TRW were vital in initiating this measurement program. Also, Don Leavitt, Jim Kotow, Ron Hilberg, and Dale Dryer arranged the loan of project equipment which permitted us to develop techniques and measurement systems several months prior to receiving the present owned Metrology equipment (HP 3045 system).

References

1. Barnes, J. A., Chi, A. R., Cutler, L. S., et al., "Characterization of Frequency Stability", NBS Technical Note 394, October 1970.

2. Shoaf, John H., Halford, D., and Risley, A.S., "Frequency Stability Specifications and Measurement", NBS Technical Note 632, January 1973.

3. Hewlett-Packard Application Note 207, "Understanding and Measuring Phase Noise in the Frequency Domain", October 1976.

4. Halford, Donald, Shoaf, John H., and Risley, A.S., "Frequency Domain Specification and Measurement of Signal Stability", Proc. Annual Symposium on Frequency Control at Cherry Hill, New Jersey, 1973

5. NBS Notebook, F 10, Pages 19-38, Dr. Donald Halford's notes on "The Delay Line Discriminator", April 1975.

6. Tykulsky, Alexander, "Spectral Measurements of Oscillators", Proc. IEEE, Vol 54, No. 2, Feb. 1966

7. Ashley, J. R., Barley, T. A., and Rast, G. J., The Measurement of Noise in Microwave Transmitters", IEEE Trans. on Microwave Theory and Techniques, Special Issue on Low Noise Technology, April 1977.

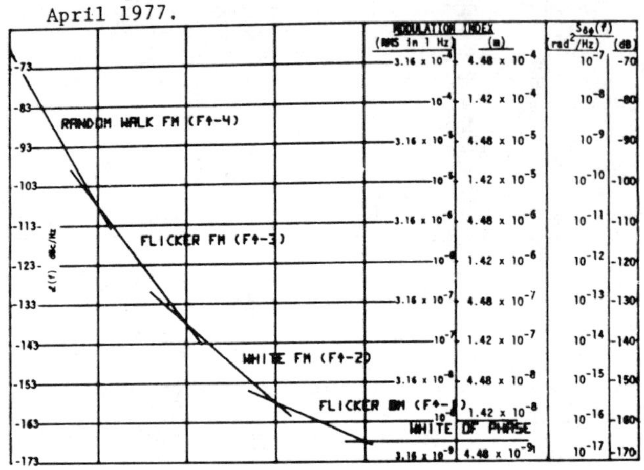

FIGURE 1. RELATIVE SCALE PLOTS OF SPECTRAL DENSITIES.

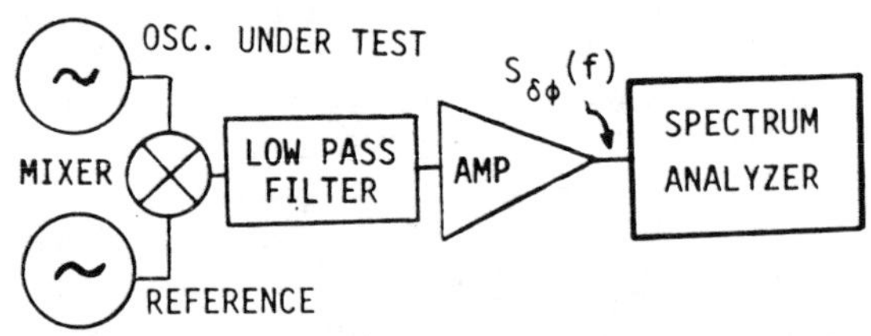

FIGURE 2. FUNCTIONAL BLOCK DIAGRAM OF A PHASE NOISE MEASUREMENT SYSTEM.

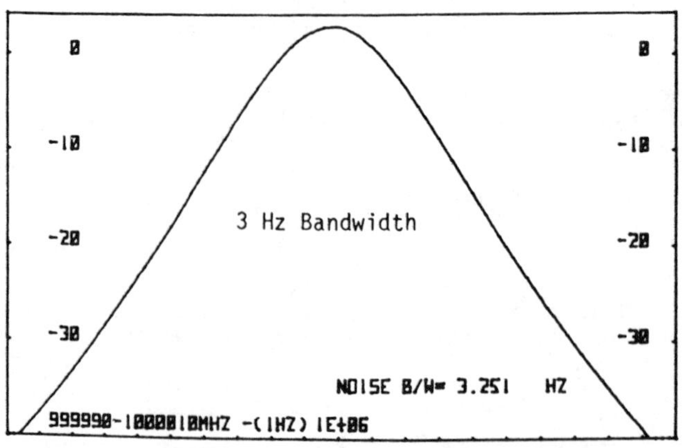

FIGURE 4. AUTOMATED PLOT OF NOISE POWER BANDWIDTH.

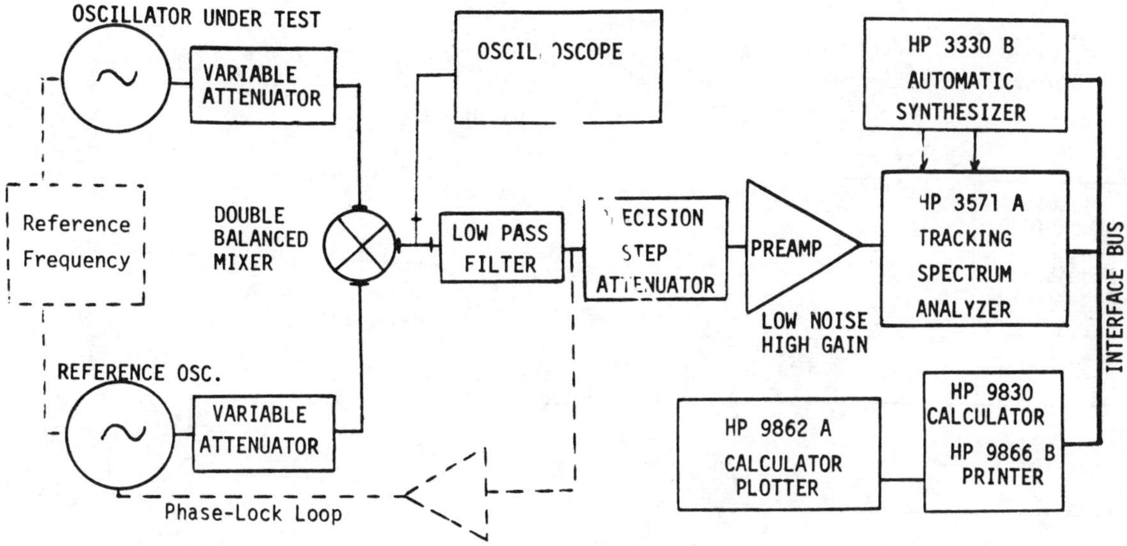

FIGURE 3. PHASE NOISE MEASUREMENT SYSTEM USING THE TWO - OSCILLATOR TECHNIQUE.

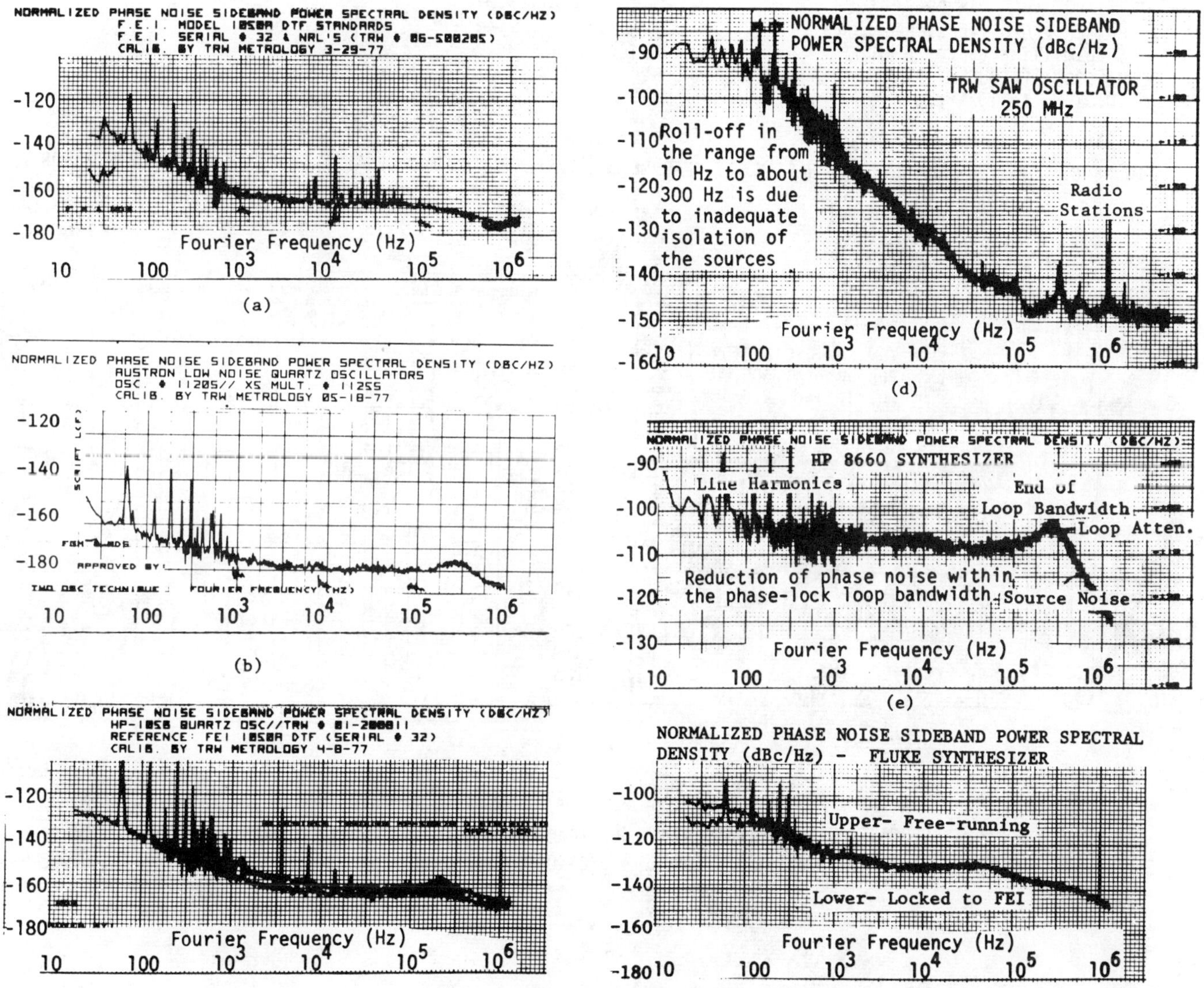

Figure 5. PHASE NOISE PLOTS

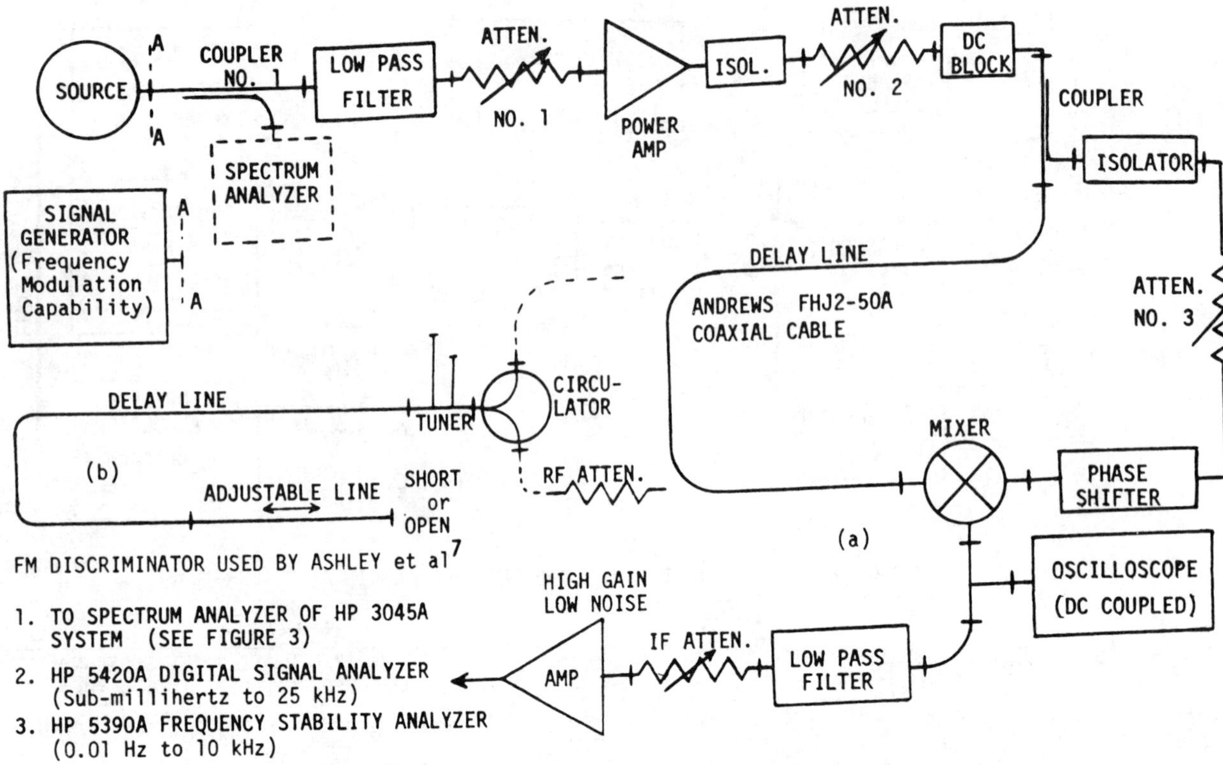

FIGURE 6. SINGLE - OSCILLATOR PHASE NOISE MEASUREMENT SYSTEM USING THE DELAY LINE AS AN FM DISCRIMINATOR.

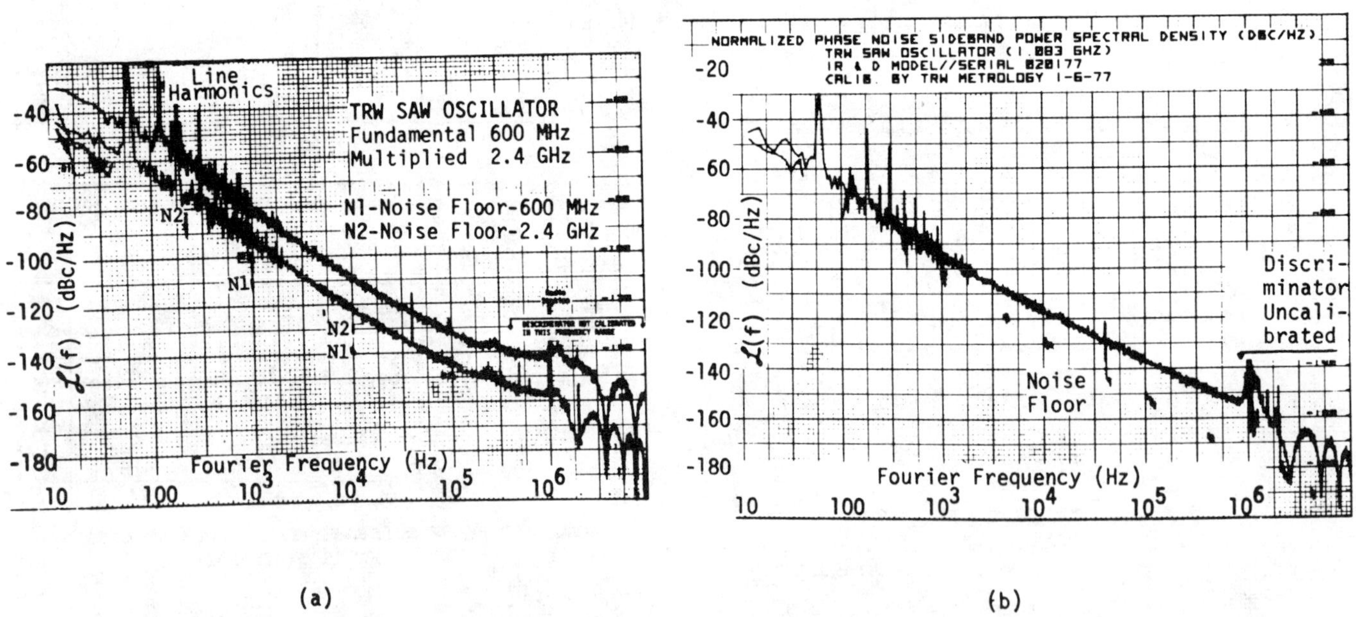

FIGURE 7. NORMALIZED PHASE NOISE SIDEBAND POWER SPECTRAL DENSITY OF TRW DEVELOPED SURFACE ACOUSTIC WAVE OSCILLATORS MEASURED USING THE SINGLE - OSCILLATOR TECHNIQUE.

PHASE NOISE MEASUREMENT USING A HIGH RESOLUTION COUNTER WITH ON-LINE DATA PROCESSING

Luiz Peregrino
David W. Ricci
Hewlett-Packard Company
Santa Clara, California

Summary

The measurement of close-in phase noise using time domain techniques will be discussed. A review of the theoretical basis for the measurements and some extensions will be presented. A system for making the measurements using a high resolution reciprocal counter and a desk top calculator will be described. The capabilities of the system will be reviewed.

Introduction

The measurement of phase noise can be made by a variety of methods which are described in the current literature. However, if one desires to measure the spectral characteristics close into the carrier, the task becomes difficult and time consuming. Measurements down to the 10 to 100 Hz range can be accomplished with spectrum or wave analyzers depending on the noise level. Measurements much below these offset frequencies are virtually impossible with these techniques.

The techniques to make measurements closer to the carrier usually resort to time domain analysis which is also described in the current literature. The primary impediment to the utilization of these techniques has been the lack of equipment to make it easy to use. Recent equipment advances such as programmable instruments and versatile calculators, for example, have provided the means to implement these techniques.

Before describing the measurement system, a review of the theoretical basis for the measurement will be presented. Most of this review will follow the standard literature with some exceptions that will tend to put more emphasis on the calculator utilization.

A general transfer function will be derived for the counter-calculator system which should cover a standard frequency measurement or comparison of the phase between two oscillators or the phase of an oscillator compared to itself through a delay line.

For a constant gate time (or constant delay), the transfer function can be written as the product of two functions, one controlled by the counter (or delay line) and the other by the calculator.

We prove that a linear combination of frequency measurements is equivalent to a filter which can be adjusted by proper choice of the linear combination coefficients. A particular choice of coefficients is assumed for which we derive closed formulas for the continuous and the delta function (bright line) spectral densities. The minimum spectral level due to counter quantization is derived.

Derivation of Time Domain Relationships to Phase Spectra Characteristics

Definitions and Notation

Let us begin with a brief review of random variables, linear systems, notations and definitions.

Let $x(t)$ be a real random variable. The auto correlation function of $x(t)$ is defined as:

$$R_x(t,\tau) = \langle x(t) \cdot x(t+\tau) \rangle \tag{1}$$

where $\langle\ \rangle$ means statistical or ensemble average.

If $x(t)$ is stationary in the wide sense, then the auto correlation does not depend on t and we can write:

$$R_x(\tau) = \langle x(t) \cdot x(t+\tau) \rangle \tag{2}$$

If $x(t)$ is ergodic, then the statistical average can be replaced by time average.

We will assume without any loss of generality that the mean value of $x(t)$ is zero. In this case we have:

$$R_x(0) = \langle x^2(t) \rangle = \sigma_x^2 \tag{3}$$

The spectral density of $x(t)$ is defined as the Fourier transform of its auto correlation function:

$$S_x(\omega) = \int_{-\infty}^{\infty} R_x(\tau)\, e^{-j\omega\tau} dt \tag{4}$$

The inversion formula gives:

$$R_x(\tau) = \frac{1}{2\pi} \int_{-\infty}^{\infty} S_x(\omega)\, e^{+j\omega\tau} d\omega \tag{5}$$

Let a linear time invariant system with transfer function $H(\omega)$ be driven by the random variable $x(t)$. The output of the system represented by $m(t)$ will be also a random variable as in Fig. 1.

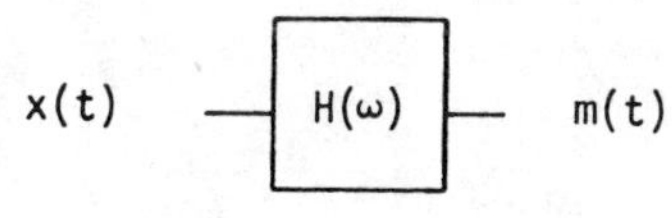

Figure 1

It can be shown that:

$$S_m(\omega) = \left| H(\omega) \right|^2 S_x(\omega) \tag{6}$$

Equation 6 will be used to determine $S_x(\omega)$ in the next section. For more details see references 1,2 and 3.

Reprinted with permission from *Proc. 30th Annu. Symp. Frequency Contr.*, Atlantic City, NJ, June 2–4, 1976, pp. 309–317.

Measuring Spectral Density

Normally a narrow band filter represented by $H(\omega)$ is used to determine the spectra of the random variable $x(t)$ as explained below. Two special cases are considered.

A. Continuous Spectra - The spectral density can be considered approximately constant over the filter bandwidth and we can write:

$$\sigma_m^2 = R_m(0) \cong S_x(\omega_o) \frac{1}{\pi} \int_{\omega_1}^{\omega_2} |H(\omega)|^2 d\omega$$

$$\cong \frac{1}{\pi} |H(\omega_o)|^2 \cdot (\omega_2 - \omega_1) \cdot S_x(\omega_o) \tag{7}$$

where ω_o is the filter center frequency and ω_1, ω_2 define the filter band. (We have used the fact that $|H(\omega)|$ and $S_x(\omega)$ are even functions of ω.)

Solving for the spectral density we get:

$$S_x(\omega_o) \cong \frac{\sigma_m^2}{\frac{1}{\pi}\int_{\omega_1}^{\omega_2} |H(\omega)|^2 d\omega} \cong \frac{\sigma_m^2}{\frac{1}{\pi}|H(\omega_o)|^2 (\omega_2-\omega_1)} \tag{8}$$

Where σ_m^2 can be determined with an average power meter or using samples of K measurements of m as:[1,4,5]

$$\sigma_m^2 = \left\langle \frac{1}{K-1} \sum_{i=1}^{K} (m_i - \frac{1}{K} \sum_{j=1}^{K} m_j)^2 \right\rangle$$

$$= \left\langle \frac{1}{K-1} \left[(\sum_{i=1}^{K} m_i^2) - \frac{1}{K} (\sum_{j=1}^{K} m_j)^2 \right] \right\rangle^* \tag{9}$$

If we use only one sample of K measurements of $m(t)$, we get an estimation of σ_m^2 which is satisfactory for most cases.

B. The Spectra Contains Delta Functions. This is the case when $x(t)$ contains periodic terms, then the result of the integration is no longer dependent on the filter bandwidth.

For example let $x(t)$ be given by:

$$x(t) = x_p \cdot \cos \omega_o t = \frac{1}{2} x_p [e^{j\omega_o t} + e^{-j\omega_o t}] \tag{10}$$

---

*For complex $m(t)$, σ_m^2 is given by:

$$\sigma_m^2 = \left\langle \frac{1}{K-1} \sum_{i=1}^{K} |m_i - \frac{1}{K} \sum_{j=1}^{K} m_j|^2 \right\rangle$$

$$= \left\langle \frac{1}{K-1} \left[(\sum_{i=1}^{K} |m_i|^2) - \frac{1}{K} |\sum_{j=1}^{K} m_j|^2 \right] \right\rangle$$

We have:

$$S_x(\omega) = 2\pi (\frac{x_p}{2})^2 [\delta(\omega-\omega_o) + \delta(\omega+\omega_o)] \tag{11}$$

Which gives:

$$\sigma_m^2 = |H(\omega_o)|^2 \frac{1}{2} x_p^2 \tag{12}$$

Note that $\frac{1}{2} x_p^2$ is the average power of $x(t)$.

It should be pointed out that during all the derivations we will use the "two-sided" spectral density, that is the integrations will be from $-\infty$ to $+\infty$. This will keep our Fourier transform pairs in the standard notation used by most EE's. Fourier frequency will be in radians per seconds.

Frequency Counter as a Linear System

Consider the signal

$$V(t) = V_o \cdot \cos(2\pi\gamma_o t + \theta_o + \theta(t)) \tag{13}$$

where V_o and γ_o are the nominal amplitude and frequency and $\theta(t)$ represents a random phase variation. θ_o is chosen such that $\langle\theta(t)\rangle = 0$.

Any amplitude variation is assumed to be eliminated by limiters or some other method and for this reason is disregarded.

The signal phase is defined as:

$$\phi(t) = 2\pi\gamma_o t + \theta_o + \theta(t) \tag{14}$$

The signal frequency $\nu(t)$ and the angular frequency $\Omega(t)$ are related by:

$$2\pi\nu(t) = \Omega(t) = \frac{d\phi}{dt} \tag{15}$$

and are assumed to be positive for all t, which is equivalent to $\phi(t)$ be a monotonic increasing function of time, that is, $2\pi\nu_o - \theta(t) > 0$.

Counter Model

An ideal counter can be modeled as a system that measures phase variation over an interval of time τ, called gate time, and divides the result by $2\pi\tau$.

Let $\nu(t)$ represent the result of a counter measurement, then have:

$$\nu(t) = \frac{1}{2\pi\tau} \cdot [\phi(t) - \phi(t-\tau)] \tag{16}$$

Taking the Fourier transform from both sides and using the shifting theorem, we have:

$$\Gamma(\omega) = \frac{1}{2\pi\tau} [\Phi(\omega) - \Phi(\omega) e^{-j\omega\tau}] \tag{17}$$

Here the upper case letters are used to represent the Fourier transform.

Which can be reduced to:

$$\Gamma(\omega) = \frac{\omega}{2\pi} \cdot \frac{\sin \omega \tau/2}{\omega \tau/2} \cdot e^{-j\omega\tau/2} \cdot \Phi(\omega) \quad (18)$$

We may conclude that the counter is a linear system with transfer function (or gain) given by:

$$H_\phi(\omega) = \frac{\omega}{2\pi} \frac{\sin \omega \tau/2}{\omega \tau/2} \cdot e^{-j\omega\tau/2} \quad (19)$$

Where the index ϕ in transfer function means that it applies for the phase as input.

The transfer function when the angular frequency $\Omega(t)$ is considered as input is given by:

$$H_\Omega(\omega) = \frac{1}{j\omega} H_\phi(\omega) = \frac{1}{2\pi j} \cdot \frac{\sin \omega\tau/2}{\omega \tau/2} e^{-j\omega\tau/2} \quad (20)$$

Practical Counters

Below are some problems that we might incur due to practical considerations.

In reality a counter detects and counts zero crossings of the signal being measured. If the phase of the input signal is a monotonic increasing function of time, as it was assumed, then the model is quite satisfactory.

On the other hand if the phase of the input signal can decrease, which is equivalent to a negative frequency, then we may get extra counts, for example in the vector diagram representation of the signal, Fig. 2. Every time that the resultant vector crosses the Y axis from right to left, we get a count; if the vector goes back we will get an extra count.

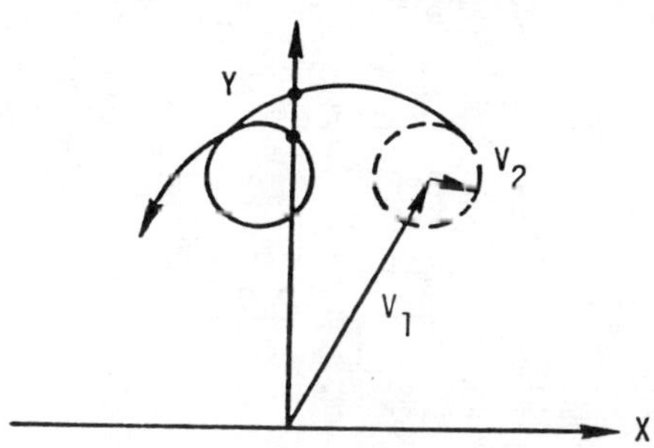

Figure 2

In Fig. 2, V_1 represents a sine wave and V_2 a interference that causes phase modulation.

Practical counters also have quantization problems, that is the number of counts is an integer and any fraction of 2π in the phase variation is disregarded. This problem is reduced by reciprocal counters with a high frequency clock. In a reciprocal counter, the actual gate time is a multiple of the signal period and if the phase fluctuation is not too large ($\dot{\Theta}(t) \ll 2\pi\nu_o$), then the gate time can be assumed constant as far as the counter transfer function is concerned.

The last problem in a real counter is the sampling, that is the counter output is not a continuous variable but a sequence of numbers obtained at the end of each gate time. A special case solution is presented in reference 5, Appendix 1.

Modifying the Counter Transfer Function

In order to measure spectral density, we want the counter to look like a very narrow band filter, ideally we want a delta function.

A weighted combination of a sequence of measurements taken at different times will give:

$$m(t) = \sum_{i=0}^{M} \alpha_i \cdot \nu_i \quad (21)$$

where $\nu_i = \nu(t_i)$

Let Fig. 3 represent the phase as a function of time, τ_i the gate time and t_i the time at the end of the i reading. The last reading is taken at the time which is considered to be the current time, that is $t_o = t$.

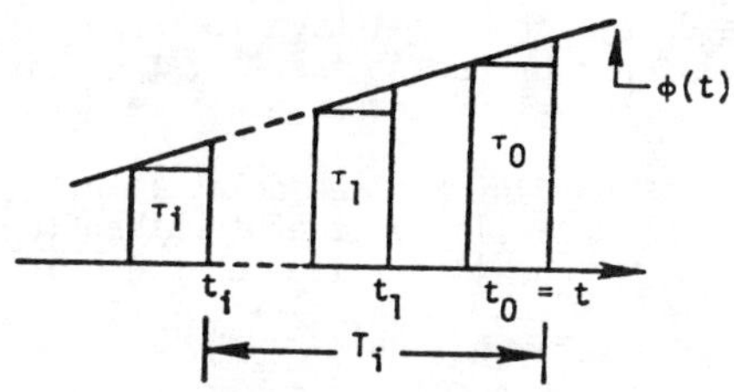

Figure 3

Replacing Y(t) as function of $\phi(t)$ and defining $T_i = t_o - t_i$ we have:

$$m(t) = \frac{\alpha_o}{2\pi\tau_o}\left[\phi(t) - \phi(t-\tau_o)\right] + \dots + \frac{\alpha_M}{2\pi\tau_M}\left[\phi(t-T_M) - \phi(t-T_M-\tau_M)\right] \quad (22)$$

Taking the Fourier transform we get the transfer function:

$$H_\phi(\omega) = \left[\frac{\alpha_o}{2\pi\tau_o}(1-e^{-j\omega\tau_o}) + \dots + \frac{\alpha_M}{2\pi\tau_M} e^{-j\omega T_M}(1-e^{-j\omega\tau_M})\right] \quad (23)$$

The counter calculator equivalent transfer function can be adjusted to approximate the desired filter transfer function by proper choice of α_i, τ_i and T_i. The weights α_i can be complex and their phase is equivalent to a change in the time the measurements are made, which can be used to simulate a variable time between measurements.

If we make the gate time τ_i a constant τ and the time between consecutive measurements, defined as dead time, a constant τ_d we have:

$$H_\phi(\omega) = \left[\alpha_0+\alpha_1 e^{-j\omega(\tau+\tau_d)}+\ldots+\alpha_M e^{-jM\omega(\tau+\tau_d)}\right] \cdot \left[\frac{\omega}{2\pi}\cdot\frac{\sin\omega\tau/2}{\omega\tau/2}\cdot e^{-j\omega\tau/2}\right] \tag{24}$$

The first bracket, which is mainly determined by the calculator via the α_s, can be interpreted as a truncated complex Fourier series and its terms can be adjusted as a series representation of the desired filter function[6,7]. The second bracket can be considered as a fixed filter and for large M the first bracket completely determines H_ϕ.

Other types of measurement such as comparing the phase of two oscillators of same frequency or comparing the phase of an oscillator to itself, using a delay line, will have the same type of transfer function due to the fact that time differences are equivalent to phase differences.

In the case of a fixed delay line, we use the calculator controlled part of the transfer function to achieve the desired filter characteristic.

A particular but very useful choice[7] is $\alpha_i = (-1)^i$ which reduces m(t) to:

$$m(t) = (y_0 - y_1) + (y_2 - y_3) + \ldots + (y_{M-1} - y_M) \tag{25}$$

Let $N = (M+1)/2$ be the number of pairs of measurements represented by the terms in parenthesis, then we can reduce H_ϕ to:

$$H_\phi(\omega) = \left[1+e^{-j2\omega(\tau+\tau_d)} + \ldots e^{-j(N-1)2\omega(\tau+\tau_d)}\right] \cdot 2\sin\left(\omega\frac{\tau+\tau_d}{2}\right)\cdot\frac{\omega}{2\pi}\,\frac{\sin\omega\tau/2}{\omega\tau/2}\,e^{-j\omega(\tau+\tau_{d/2})} \tag{26}$$

The bracket can be recognized as a geometric series which gives:

$$H_\phi(\omega) = \frac{\omega}{\pi}\cdot\frac{\sin\omega\tau/2}{\omega\tau/2}\cdot\sin\left(\omega\frac{\tau+\tau_d}{2}\right)\cdot\frac{\sin N\omega(\tau+\tau_d)}{\sin\omega(\tau+\tau_d)}\cdot e^{-j\omega[N(\tau+\tau_d)-\tau_{d/2}]} \tag{27}$$

A very important point to be kept in mind is that in general this corresponds to a filter with many pass bands.

For most practical cases, the phase noise is a fast decreasing function of ω and only the first pass band of H_ϕ has to be considered, as in references 6 and 7, which have plots of $|H_\Omega(\omega)|$ which is equal to $|H_\phi(\omega)/\omega|$. Error will occur depending on how fast the phase noise decreases as a function of ω such as ω^{-3}, ω^{-2}, ω^0 = constant, and we may have to consider more than one pass band.[6,7]

Calculation of Spectra Density

So far we have proven that the linear combination of the frequency measurements m(t) can be considered as the output of a filter whose input is $\phi(t)$ and transfer function H_ϕ or input $\Omega(t)$ and transfer function $H_\Omega(\omega)$. This is represented in the block diagram, Fig. 4.

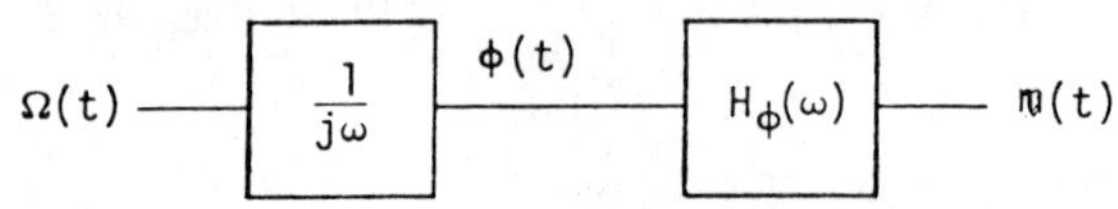

Figure 4

As in the linear system we have two cases.

A. Continuous Spectra - The spectra can be considered as constant over the filter bandwidth. Using the fact that $\alpha_m^2 = R_m(0)$ and eq. 5, we have:

$$\sigma_m^2 = \frac{1}{2\pi}\int_{-\infty}^{+\infty}\left|H_\phi(\omega)\right|^2 S_\phi(\omega)d\omega \tag{28}$$

If we use $H_\Omega(\omega)$ instead of $H_\phi(\omega)$, we have:

$$\sigma_m^2 = \frac{1}{2\pi}\int_{-\infty}^{+\infty}\left|H_\Omega(\omega)\right|^2 \omega^2 S_\phi(\omega)\,d\omega \tag{29}$$

A normalized expression in ω for $|H_\phi|$ is given as:

$$|H_\phi(\omega)| = \left|\frac{N\omega}{\pi}\cdot\frac{\sin r\frac{\pi}{2}\frac{\omega}{\omega_0}}{r\frac{\pi}{2}\frac{\omega}{\omega_0}}\cdot\sin\pi/2\frac{\omega}{\omega_0}\cdot\frac{\sin N\pi\frac{\omega\pm\omega_0}{\omega_0}}{N\sin\pi\frac{\omega\pm\omega_0}{\omega_0}}\right| \tag{30}$$

Where r and ω_0 are:

$$r = \frac{\tau}{\tau+\tau_d} \tag{31}$$

$$\omega_0 = \frac{\pi}{\tau+\tau_d} \tag{32}$$

We have used the fact that sin(x) only changes the sign when we add or subtract π to x to obtain the term $(\omega \pm \omega_o)/\omega_o$. For large N and $|\omega \pm \omega_o| << \omega_o$ we can approximate $|H_\phi|$ as:

$$|H_\phi| \cong \left| \frac{N\omega_o}{\pi} \cdot \frac{\sin r\pi/2}{r\pi/2} \cdot \frac{\sin N\pi \frac{\omega \pm \omega_o}{\omega_o}}{N\pi \frac{\omega \pm \omega_o}{\omega_o}} \right| \qquad (33)$$

Note that $|H_\phi|$ is equal to a constant times $|\sin(x)/x|$ centered at $+\omega_o$ and $-\omega_o$. Solving the integral (28), we have:

$$S_\phi(\omega_0) \cong \frac{\sigma_m^2}{\frac{1}{\pi}\left|H_\phi(\omega_0)\right|^2 \frac{\omega_0}{N}} = \frac{1}{8}\left(\frac{r\frac{\pi}{2}}{\sin r\frac{\pi}{2}}\right)^2 \frac{\sigma_m^2}{N f_0^3} \qquad (34)$$

Where ω_o/N can be interpreted as the equivalent bandwidth in radians per second (see Fig. 5) and $f_o = \omega_o/2\pi$.

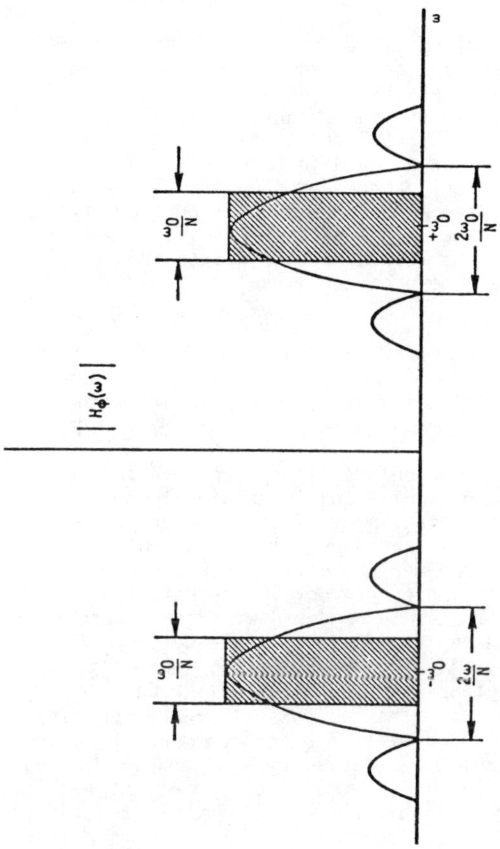

Figure 5

The shaded area in Fig. 5 represents the equivalent ideal filter.

If we select $\tau_d = \frac{1}{2}\tau$, then the multiple responses of $H\phi(\omega)$ do not begin until $5\omega_o$ is reached. See figure 6.

$$S_\phi(\omega_o) \cong \frac{1}{8}\left(\frac{\pi/3}{\sin\pi/3}\right)^2 \frac{\sigma_m^2}{N f_o^3} \qquad (35)$$

$$S_\phi(\omega_0) \cong .183 \frac{\sigma_m^2}{N f_0^3} \qquad (36)$$

The phase spectral density due to the pass band centered at ω_0 can be determined by equation (36) where σ_m^2 can be determined by numerical methods.

B. <u>Power Spectra Contains Delta Function (Bright Line)</u>. Let us consider the special case of a phase modulated signal such as

$$V = V_o \cos(\Omega_o t + \theta_p \cos \omega_o t) \qquad (37)$$

The phase spectra for the modulation is:

$$S_\theta(\omega) = 2\pi \left(\tfrac{1}{2}\theta_p\right)^2 [\delta(\omega-\omega_o) + \delta(\omega+\omega_o)]$$

$$= \left(\tfrac{1}{2}\theta_p\right)^2 [\delta(f-f_o) + \delta(f+f_o)] \qquad (38)$$

The signal can be expanded in the usual form using Bessel functions and for small θ_p we have:

$$J_o(\theta_p) \cong 1$$

$$J_1(\theta_p) \cong 1/2\, \theta_p$$

$$\mathcal{L}(f) \cong \left(\tfrac{1}{2}\theta_p\right)^2 \delta(f-f_o) \qquad (39)$$

Where $\mathcal{L}(f)$ is defined as the energy at $\Omega_o + \omega_o$ per Hz divided by the total energy.

From the previous equations (37, 38), we conclude that $\mathcal{L}(f)$ and the "two-sided" phase noise S_ϕ are approximately equal for small θ_p.

The relation between σ_m^2 and θ_p is obtained by solving the integral (28) giving:

$$\left(\tfrac{1}{2}\theta_p\right)^2 = \frac{\sigma_m^2}{2\left|H_\phi(\omega_0)\right|^2} = \left[\frac{\sigma_m^2}{\frac{1}{\pi}\left|H_\phi(\omega_0)\right|^2 \frac{\omega_0}{N}}\right]\left(\frac{f_0}{N}\right) \qquad (40)$$

From the previous equation, we conclude that we can determine the phase spectrum as for the continuous case and multiply the result by f_o/N, to obtain $(\frac{1}{2}\theta_p)^2$.

The presence of delta functions (bright lines) can easily be detected due to the fact that σ_m^2 is not dependent on N, as is indicated by the equation relating σ_m^2 and $(\frac{1}{2}\theta_p)^2$.

So far we have relations to determine the phase spectral density $S_\phi(\omega)$ and the bright lines intensity $(\frac{1}{2}\theta_p)^2$. In the next section, we will derive the system resolution using a statistical approach to the quantization problem.

Estimation of the System Resolution

The system resolution is determined by the minimum time variation that the reciprocal counter can resolve, which is the clock period.

Let τ_c be the clock period, τ the gate time and ν_i the frequency of the input signal.

We will assume that the noise of the input signal will cause a maximum time variation of $\ell\tau_c$ during the gate time. Let $\delta\nu$ represent the resultant frequency variation then we have:

$$\delta\nu = \frac{\tau_c}{\tau}\nu_i \tag{41}$$

If we assume that ℓ is uniformly distributed from -1 to +1 and that the resultant $\delta\nu$ in each frequency measurement is independent of all previous ones, than m(t) is the sum of independent uniformly distributed random variables.

Using the central limit theorem, we concluded that for large N the random variable m(t) is approximately Gaussian with variance equal to the sum of the variances for each frequency measurement resulting in:

$$\sigma_m^2 = 2N\left[\frac{1}{3}\left(\frac{\tau_c}{\tau}\nu_i\right)^2\right] \tag{42}$$

where 2N is the number of frequency measurements and the term in brackets is the variance of $\delta\nu$.

This implies in a resolution given by:

$$S_\phi(\omega_0) \cong \left(\frac{r\pi/2}{\sin r\pi/2}\right)^2 \frac{\tau_c}{3r^2}\frac{\nu_i^2}{f_0} \tag{43}$$

The Measurement System

System Description

The system consists of five major components: a high resolution reciprocal counter (HP 5345A) with external gating capabilities, a measurement storage plug-on unit (HP 5358A), a mixer/amplifier unit (HP 10830A), a desk top calculator (HP 9825A) and printer/plotter output device (HP 9871A) as shown in figure 7. Communication and control between the various instruments is provided by a digital interface system (HP-Interface Bus). In addition, a test tone generator (HP 10831A) and time of day clock are included to enhance the system's capabilities. A functional block diagram is shown in figure 8.

The counter provides the system with the ability to make high resolution (2 nsec) period or time interval measurement or frequency (by the reciprocal technique). This determines the system's sensitivity floor limit as will be shown later. In addition, the counter has the ability to be gated from an external sample time signal. This is necessary in order to utilize various sampling functions and thus determine the offset frequency at which a spectra measurement is made.

The counter by itself does not have sufficient control or data output capacity, so a measurement storage plug in unit is included. It extends several capabilities of the counter. The front panel gate times are (as in most counters) in decade steps. This does not allow enough flexibility for spectra characterization so a gate generator is incorporated in the plug-in which generates a measurement time (τ) signal in the range of 1 to 999 x 10^6 µsec. The dead time between measurements (τ_d) is also controllable via the plug-in. The plug-in also stores the measurement data in a buffer memory for output over the interface bus to the calculator.

The measurement cycle of the counter is also controlled by the plug in by-passing such things as the display cycle to minimize the dead time between measurements. Dead time as low as 15 µs can be achieved.

The mixer/amplifier unit provides a means to further increase the system's resolution by heterodyning the test signal down to a low frequency signal by mixing it with an offset reference oscillator. It includes the necessary filtering, bandwidth control amd amplifiers to properly condition the input signals for application to the counter.

The control of all the previously described instruments is provided by the calculator via the interface bus. Measurement data is also sent to the calculator by the same means. The calculator is programmed from its keyboard to perform the various measurement operations and to process the data received from the counter and plug-in. All the aspects of system behavior are under program control of the calculator. The operator specifies measurement parameters at the keyboard. The printer device is used to output the processed results either in numeric or plotted form.

System Operation

For the following discussion,refer to figures 7 and 8.

Signal Conditioning. There are a variety of ways of preconditioning the input signal before applying them to the input of the counter. The major objective of these techniques is to increase the resolution of the measurement. The one shown in Figure 8 is a simple heterodyning technique where the test oscillator is compared with a reference oscillator which is offset in frequency from the test oscillator by ν_i. The output of the two oscillators is mixed together to produce an audio range beat note. The output of the mixer is passed thru appropriate filtering to eliminate the undesired mixer products. The signal is then amplified by a high gain, low noise, limiting amplifier. The main purpose of the amplifier is to provide reliable detection of the zero crossings of the beat note. The output of the amplifier is essentially a square wave which is used to drive the input of the counter. This approach is necessitated by the wide bandwidth (500 MHz) of the counter input and the resulting input noise which makes it impossible to detect a low frequency zero crossing with a 2 nsec resolution. As such the amplifier is provided with bandwidth control so that the input noise bandwidth can be adjusted to be compatible with the signal being measured.

The technique of comparing two oscillators, while it has some disadvantages, is still the best overall method of performing this measurement. It is not possible to measure the carrier directly, especially at microwave frequencies, with sufficient resolution, thus some down conversion technique is required. The problem is the noise contributed by the local (reference) oscillator. In order for it to be negligible its spectra must be 10 to 20 dB below the test oscillator in which case the error is small (.4 to .2 dB respectively). When it is not possible to obtain a reference oscillator which is better than the test oscillator, then two oscillators of assumed identical spectral characteristics can be used and the measurement result taken as the average of the two. This approach will always measure the worst of the two with a maximum error of less than 3 dB. The uncertainty of which oscillator is the poorest of the pair can be resolved by taking 3 or more and comparing them in all combinations (A vs B, A vs C, B vs C, etc.). This approach will identify both the best and the worst of the group.

The remaining problem with the heterodyning technique is the requirement that the reference oscillator be offset from the test oscillator. In certain cases (cesium, rubidium or crystal oscillators for example), it is difficult or impractical to offset them, and as such, a different signal conditioning approach is required. These have been discussed in the literature[8]. The main requirement in terms of the current approach is to present the counter with a signal which is representative of the phase of the test oscillator and whose period is measurable by the counter with sufficient resolution to be meaningful.

Counter Operation. The counter counts the number of cycles of the input signal that occurs within the gate time and the number of cycles of the 500 MHz time base that occurred from the first zero crossing after the opening of the counter's main gate to the last zero crossing after the closing of the gate. Thus, the measurement consists of two numbers: 1) the number of cycles of the input and, 2) the elapsed time in 2 nsec steps that it took the n cycles to occur.

The operation of the counter and plug-in unit are directed by commands received over the interface bus from the calculator. These are determined by the stored program in the calculator. Each series of measurement is made by programming the counter to the desired function (period, frequency, time interval), setting up the desired gating function (measurement and dead time) and the number of measurements to be made. The resulting data is stored in the plug-in in order to reduce the dead time between measurements and subsequently transferred to the calculator via the interface bus for processing.

Sampling Functions.

The frequency selective characteristics of the counter is determined by the way the measurement data is acquired by the counter (i.e. the measurement and dead time) and the processing algorithm used in the calculator (i.e. the choice of α's) as given by eq. 24. A variety of resulting transfer functions have been discussed in the literature[6,7]. The most attractive of these from the point of view of an on line process is the so called modified Hadamard variance or 50% dead time sampling function proposed by Baugh (see Reference 7, p 225, Figure 6). The reason for this choice is the sampling function has two α coefficients which are zero and the counter can be allowed to reset and transfer data during these intervals. Hence the counter is essentially free from dead time constraints when using this sampling function. Secondly, the sampling function is a short sequence suitable to on line processing & allows measurements to be made out to a reasonable distance away from the carrier.

Software.

The software (the stored program in the calculator) determines a significant portion of the system's behavior and performance characteristics. The set-up of the operating modes of the instrument and the method of data reaction is determined by the program written for the calculator. As such the system has a great deal of flexibility in executing various sampling and processing methods as a function of writing the appropriate program.

System Performance

Sensitivity: The system's sensitivity using the simple heterodyning method and the 50% dead time sampling function is given by eq. 43. Evaluating this gives a family of curves as shown in figure 9. Eq. 43 assumes no other sources of noise in the system. This is valid as long as the mixer and amplifier noise are designed to be below this limit. As can be seen, the sensitivity increases with a decrease in the beat frequency (ν_i) and as the offset frequency is increased.

In the primary region of interest (10 Hz and below), the sensitivity is quite good compared to most oscillators available today. In the region above 10 Hz, which is primarily for comparison with other techniques, the sensitivity can be inadequate. In these cases other resolution enhancement techniques, such as multiplying the input signals to microwave frequencies, can be used.

Maximum Offset Frequency: The maximum offset frequency is limited by the counter's dead time and number of measurements required per cycle of the sampling function. In the case of the 50% dead time sampling, the upper offset frequency is given by:

$$f_{max} = \frac{1}{6\tau_d}$$

Since the counter is only able to measure an integral number of cycles of the input, one of these is equal to the dead time, this can also be related to the beat note frequency, thus

$$f_{max} = \frac{\nu_i}{6}$$

Filter Bandwidth: The approximate equivalent filter bandwidth as given in figure 5 is

$$B = \frac{f_o}{N}$$

It is important to note two characteristics: 1) As the filter fundamental response f_o is moved closer to the carrier, the bandwidth of the filter becomes proportionally smaller, and 2) By increasing N, the bandwidth can be made arbitrarily small. Both of these facts are what allows this method to measure phase noise arbitrarily (in theory) close to the carrier, whereas traditional analog methods are limited by the skirts of analog filters.

Digital Filter Harmonic Responses

As shown in figure 6, the harmonics of the digital filter have the same response as the fundamental response. The usual assumption is that the spectra in the region of interest is declining at a rate of f^{-2} or greater, and that the number of harmonic responses is limited to a finite number by the selectable IF filter and thus the error is negligible. However, this assumption must be verified each time a different class of oscillator is measured as white phase noise can contribute significantly. Further, so called bright lines occurring at any of the harmonic responses will produce erroneous results. The problem of white phase noise is discussed by Lesage and Audoin[6]. The problem of bright lines (usually 60 and 120 Hz and harmonics thereof) can be coped with by judicious choice of the offset frequencies and/or bandwidths.

Performance Verification

Several methods were used to verify the results obtained by this measurement technique.

FM Modulation: By applying a FM modulated signal of a known index of modulation and using the relationship given by eq. 40, the measured value of the sidebands can be compared with those values predicted by the modulation index.

The index of modulation is given by

$$\beta = \frac{\Delta f}{f_m} = \theta_p$$

where f_m is the modulation frequency. If the modulation index is small, $\beta << \frac{\pi}{2}$, then the value of the sideband at $f_c + f_m = \beta/2$.

Sensitivity Floor Verification: The system's resolution limit as given by eq. 43 was verified experimentally. A test tone was generated by dividing a good crystal oscillator's output frequency down to a variety of frequencies suitable as a beat note input to the system. The frequency dividers were digital circuits (74LS161's) operated in a manner so as to introduce a minimum of phase noise. The resultant signals could be calculated to be much below the sensitivity limit of the system. If the square wave outputs of the dividers were applied to the amplifier, the results were much better than the model predicted. This is explained by the fact that the square wave eliminated trigger errors and both the counter's time base and the test tone were derived from very stable crystal oscillators and a degree of synchronization occurred. If the outputs of the dividers were filtered to create sine waves, a $\pm$ one count error due to trigger noise could be observed and good correlation to the resolution floor model occurred.

Correlation with Other Techniques: In the region of offset frequencies where other techniques can be used, measurements were made on various oscillators and the results compared. Results correlated within 3 to 6 dB. One difficulty with correlating these results is estimating the mean value of the spectra using spectrum or wave analyzers. This tends to be subject to operator judgment.

Sample Results

A typical printout of the system is shown in figure 10. A series of measurements are made at the specified frequencies and digital filter bandwidths. The measurements are repeated at each frequency for specified numbers of "sweeps" to provide some statistical information about the measurements since the measured value represents an estimation of the mean value of the spectra. The results can also be presented in graphical form as shown in figure 11. Here the "X" represents the average of the values measured and the "-", the one sigma value of the variations.

Conclusions

The use of time domain techniques for close-in phase noise measurements can now be performed in a practical manner. Use of a high resolution reciprocal counter gives good sensitivity for offset frequencies below 100 Hz. By combining a programmable calculator with the programmable instrumentation, results can be obtained in the time it takes to collect the data.

Acknowledgement

The authors wish to thank Mohamed Sayed, David Chu and Holly Cole for their comments and proof reading and Sharon Gantman for her diligency in preparing the manuscript.

References

1. Athanasios Papoulis, Probability, Random Variables, and Stochastic Process, McGraw Hill.

2. Laning and Batlin, Control Systems Engineering, McGraw Hill.

3. W.B. Davenport and W.L. Post, Random Signals and Noise, McGraw Hill.

4. N.B.S. Technical Note 632, U.S. Department of Commerce, National Bureau of Standards.

5. "Characterization of Frequency Stability," IEEE Transactions on Instrumentation and Measurement, pp 105-120, May 1971.

6. P. Lesage and C. Audoin, "A Time Domain Method for Measurement of Spectral Density of Frequency Fluctuations at Low Fourier Frequencies," Laboratorie do L'Horloge Atomique, Bat. 221, Universite' Paris, Sud 91405, Orsay, France.

7. Richard A. Baugh, "Frequency Modulation Analysis with Hadamard Variance," Proc. Freq. Ctl. Sump., pp 222-225, April 1971.

8. David W. Allen, "The Measurement of Frequency and Frequency Stability of Precision Oscillators", NBS Technical Note 669, May 1975.

FIGURE 6

Equivalent Digital Filter

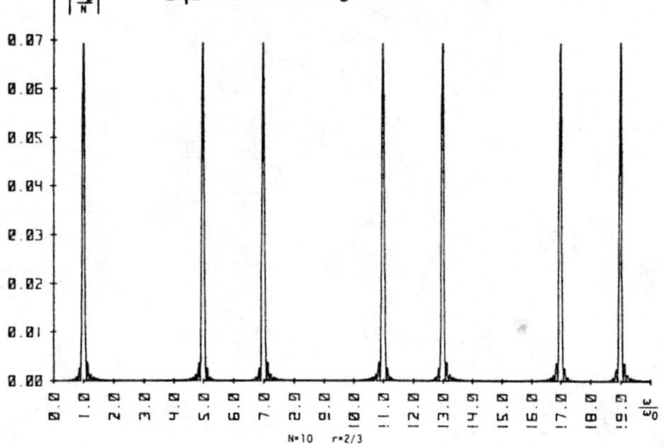

Figure 7 HP 5390A Frequency Stability Analyzer System Configuration

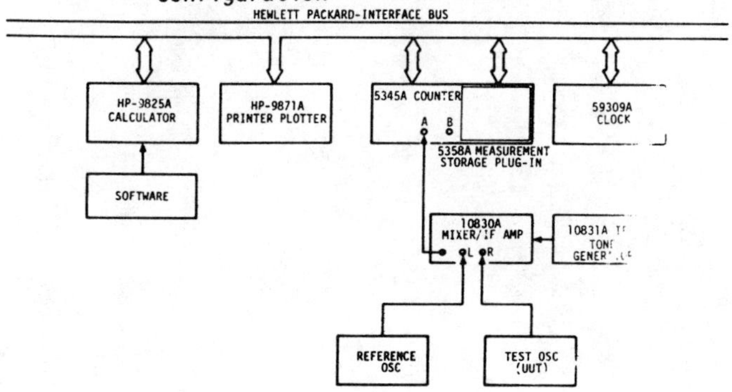

FIGURE 8

Phase Noise Measuring System Functional Block Diagram

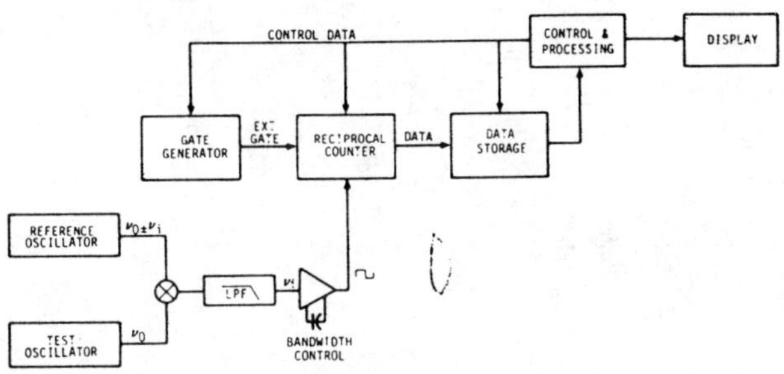

FIGURE 9

Phase Noise Sensitivity Limit

(- 73 + 20 log ν_i -10 log f)

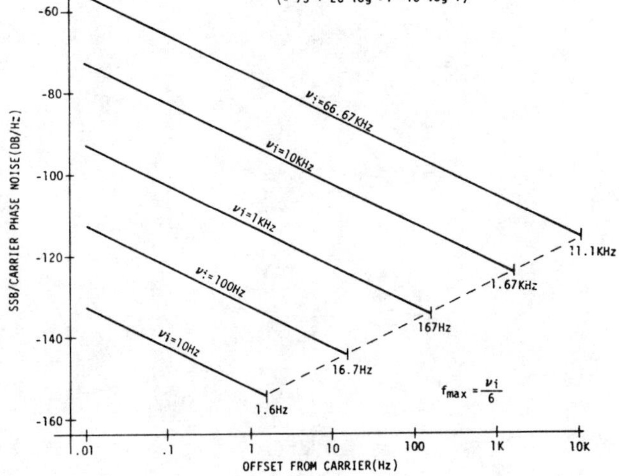

FIGURE 10

Sample Printout

*** P H A S E   N O I S E   A N A L Y S I S ***

PHASE NOISE PROGRAM (9825) rev 05-26-76-2310

MEASUREMENT DESCRIPTION:

10544A-H36 VS 10544A-H36
DATE 05/26/76 TIME 23:23:00

MEASUREMENT PARAMETERS:

K= 10
IF FREQUENCY 1000 HZ

SSB/CARRIER PHASE NOISE (DB/HZ)

FREQ=	10.4	2.5	1.0	0.5	0.2	HZ
BW =	0.52	0.12	0.05	0.03	0.01	HZ
FLOOR	-123.2	-117.0	-113.0	-110.0	-106.0	DB
SWEEP						
1	-125.1	-116.1	-103.5	-91.8	-79.9	
2	-126.0	-114.4	-104.6	-95.2	-80.8	
3	-125.8	-112.7	-103.7	-93.6	-79.7	
4	-124.0	-117.7	-104.7	-91.5	-77.7	
5	-122.3	-118.9	-103.0	-92.3	-77.2	
	------	------	------	------	------	
AVE=	-124.4	-115.4	-103.9	-92.7	-78.8	
SIG=	1.5	2.5	0.6	1.5	1.6	
MIN=	-126.0	-118.9	-104.7	-95.2	-80.8	
MAX=	-122.3	-112.7	-103.0	-91.5	-77.2	

END TIME:01:52:03

FIGURE 11

Sample Plot

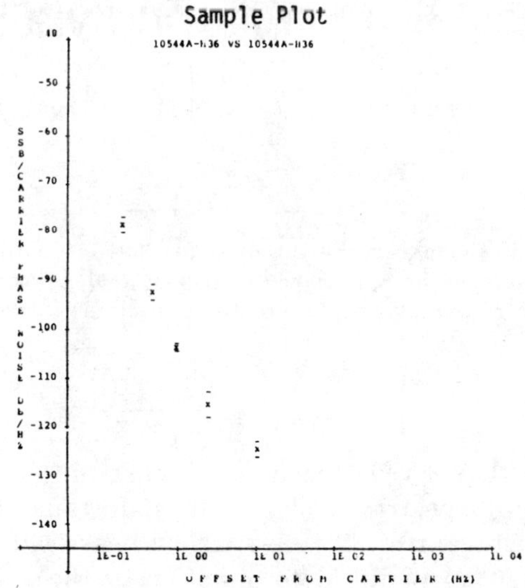

A New Spectrum Analyzer Using Both Analog and Digital Filtering via Hadamard Variance

JACQUES GROSLAMBERT

Abstract—**In this paper, the author shows how to simplify the complex response of the Hadamard variance by analog filtering to design a spectrum analyzer. The realized spectrum analyzer is described.**

Introduction

THE HADAMARD variance is useful for low-frequency spectral analysis but unfortunately its transfer function presents some spurious peaks which can give some errors in the determination of the spectral density. A simple analog filter solves this problem. A spectrum analyzer using both analog and Hadamard variance filtering have been realized.

Manuscript received July 1, 1976.
The author is with the Laboratoire de Physique et Métrologie des Oscillateurs du C.N.R.S., Besançon, France.

The Hadamard Variance

The Hadamard variance is characterized by many ensembles of $2N$ frequency measurement samples averaged over successive time intervals and alternately added and subtracted,

$$\sigma_H^2(N,\tau,T_M) = \langle (\bar{\nu}_{k+1} - \bar{\nu}_{k+2} + \cdots - \bar{\nu}_{k+2N})^2 \rangle \quad (1)$$

where ν_{k+i} is the ith measurement in the kth ensemble of

Reprinted from *IEEE Trans. Instrum. Meas.*, vol. IM-25, pp. 480–483, Dec. 1976.

$2N$ measurements and T_M is the dead time between two measurements.

The relationship between the Hadamard variance and the power spectral density of the frequency fluctuations is given by

$$\sigma_H^2(N,\tau,T_M) = \int_0^\infty |H(f)|^2 S_{\delta\nu}(f)df \quad (2)$$

where $S_{\delta\nu}(f)$ is the power spectral density of frequency fluctuations.

The transfer function of the Hadamard variance presents a narrow response at a frequency f_1 corresponding to $1/2(\tau + T_M)$ and spurious responses at odd harmonics

$$H(f_n) = 2N \frac{\sin \pi\tau f_n}{\pi\tau f_n}, \qquad f_n = \frac{n}{2(\tau + T_M)}. \quad (3)$$

This property can be used to realize a spectrum analyzer, but the spurious response at odd harmonics can give many errors if the spectral density of frequency fluctuations is of the form

$$S_{\delta\nu}(f) = K_m f^m, \qquad m = 2,1,0,\cdots. \quad (4)$$

Consistant errors occur especially with $m = 2$ (white noise of phase). In fact, all the p harmonic peaks have the same magnitude. The result is p time the exact value. For a bandlimited frequency spectrum $S_{\delta\nu}(f)$ the number of the p peaks depends on the time τ used in the Hadamard variance. Consequently the slope value of the measured $S_{\delta\nu}(f)$ cannot take the value of 2. With $m = 1$, the contribution of each peak p is $1/p$. The error will depend on the value of p.

Many authors [1]–[3] have given some methods to reduce the influence of the spurious response of odd harmonics. The first method [1] uses a modified Hadamard variance with binomial envelope to eliminate the contribution of the $3f_1$ peak. The second method [2] uses a pseudosinusoïdal weighting which takes away the $3f_1$, $5f_1$, $7f_1$, $9f_1$ peaks. The third method [3] uses the nth difference of the phase which consists of weighting the ith measurement by the binomial coefficient C_{k-1}^i. This method eliminates the spurious peaks up to the eleventh.

These methods require a great number of measurements and, therefore, are difficult to achieve mainly because of the time they need.

THE NEW METHOD

We can eliminate the spurious peaks of the Hadamard variance by combining analog and digital techniques. In fact, we can obtain a voltage proportional to the frequency fluctuations of an oscillator with a phase-lock technique. This signal feeds through a low pass filter with a f_c cutoff frequency. The output signal of the filter is then converted into frequency by a voltage to frequency converter of conversion factor L(Hz/V). If we take the Hadamard variance of the output signal of the converter we obtain

$$\sigma_{H1}^2(N,\tau,T_M) = K^2L^2 \int_0^\infty |H(f)|^2|H_1(f)|^2 S_{\delta\nu}(f)df$$

where $H_1(f)$ is the transfer function of the filter and K the sensitivity in volt/hertz of the signal proportional to the frequency fluctuations.

The number of the poles of the low-pass filter and the cutoff frequency are chosen to reject at 30 dB the first spurious harmonic without attenuation of the first peak of the $H(f)$ response. Therefore, the higher order spurious harmonics are eliminated.

For all types of $S_{\delta\nu}$ spectra the relation (5) can be written as

$$\sigma_H^2(N,\tau,T_M) = K^2L^2H(f_1)^2 S_{\delta\nu}(f_1)\Delta f_1 \quad (6)$$

where f_1 is the frequency of the fundamental peak and Δf_1 is the equivalent bandwidth

$$f_1 = \frac{1}{2(\tau + T_M)} \qquad \Delta f_1 = \frac{f_1}{N}.$$

The spectral density of the frequency fluctuations is obtained by calculating the quantity

$$S_{\delta\nu}(f_1) = \frac{\sigma^2(N,\tau,T_M)}{K^2L^2H(f_1)^2\Delta f_1}. \quad (7)$$

THE SPECTRUM ANALYZER

With this principle a spectrum analyzer has been realized. This analyzer can be divided in four parts: the phase-lock acquisition of the frequency fluctuations, the filter section, the time base and logic circuits, and the computing section (Fig. 1).

A. The Phase-Lock Acquisition of the Frequency Fluctuations

This section is a classical phase lock loop using a good double balanced mixer and a very low-noise amplifier. The oscillator under test is fed into one side of a low noise double balanced mixer which uses Schottky barrier diodes. The reference oscillator is fed into the other input of the mixer. The mixer operates as a phase detector when the two signals have the same frequency and are in phase quadrature. The output, which in this condition is approximately 0 V, fed through an amplifier and goes back to the electronic tuning of the reference oscillator. The voltage at the amplifier output can be written in terms of $\delta\nu_1(p)$ and $\delta\nu_r(p)$ which are respectively the frequency fluctuations of the free under test oscillator and the free reference oscillator:

$$V_1(p) = \frac{K}{1 + Kp/G\mu}(\delta\nu_1(p) + \delta\nu_r(p)) \quad (8)$$

where $p = j\omega$, K is the sensitivity of the electronic tuning of the reference oscillator (V/Hz), μ is the sensitivity of the mixer (rad/V), and G is the low-noise amplifier gain.

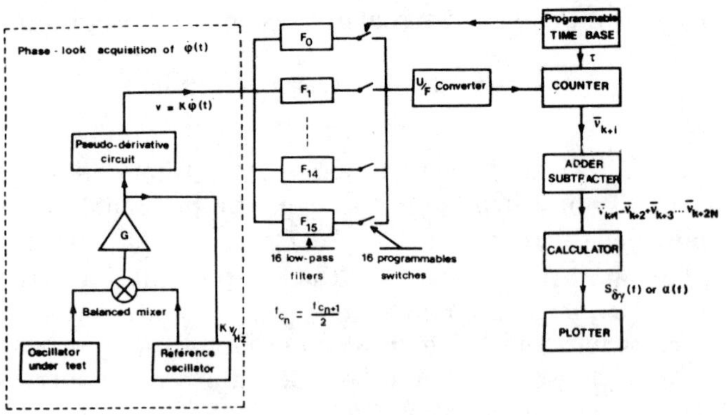

Fig. 1. Schematic diagram of the spectrum analyzer.

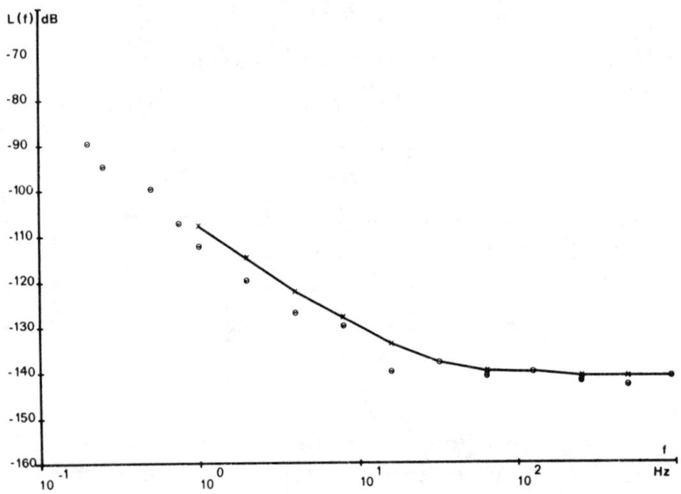

Fig. 2. Spectral density of the phase fluctuations of an oscillator which presents white phase noise obtained with a conventional spectrum analyzer (solid line) and an Hadamard spectrum analyzer (sample points).

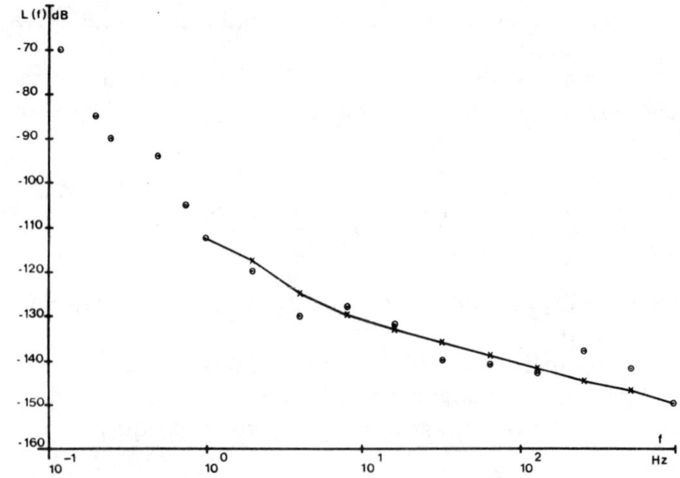

Fig. 3. Spectral density of the phase fluctuations of an oscillator which presents a phase flicker noise obtained with a conventional spectrum analyzer (solid line) and an Hadamard spectrum analyzer (sample points).

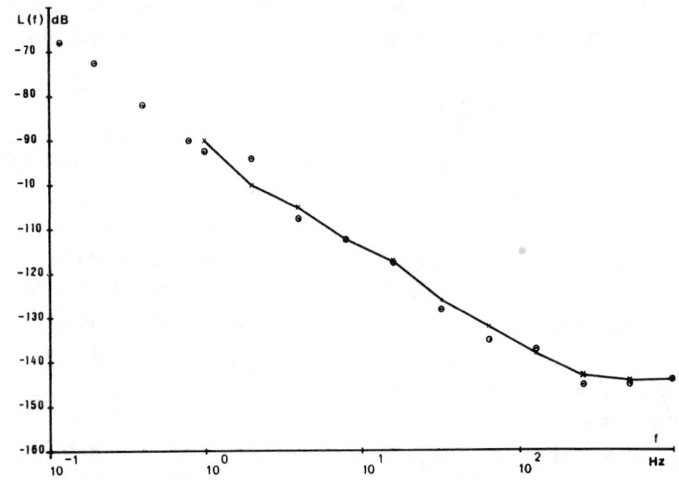

Fig. 4. Spectral density of the phase fluctuations of an oscillator which presents a frequency flicker noise obtained with a conventional spectrum analyzer (solid line) and an Hadamard spectrum analyzer (sample points).

If this voltage $V_1(p)$ fed through a selectable pseudoderivative circuit with a transfert function given by

$$G_1 = G_0(1 + Kp/G\mu). \tag{9}$$

The output voltage $V_2(p)$ can be written as

$$V_2(p) = K(\delta\nu_1(p) + \delta\nu_r(p)). \tag{10}$$

This voltage $V_2(p)$ is proportional to the frequency fluctuations of the undertest oscillators up to 2 kHz whatever is the phase-loop attack time. This limit is due to the pseudo-derivative circuit.

B. The Filter Section

This section consists of sixteen filters. These low-pass filters are three poles Butterworth type. The cutoff frequencies of these filters are given by

$$f_{n_c} = \frac{f_{n+1_c}}{2}. \tag{11}$$

All the inputs of these filters are connected together but the output of each filter is applied to a voltage to frequency converter via a programmable switch. One megahertz is the output frequency of the U/F converter which exhibits a transfert coefficient L of 30 kHz/V.

C. The Time Base and Logic Circuit

Sixteen time bases are generated from the 5-MHz reference oscillator. The first time base, obtained by a 1200 divider, gives a 240-μs square wave. With this first time base the analyzing frequency is given by:

$$f = \frac{1}{4\tau} = \frac{1}{4 \times 240 \times 15^6} \simeq 1040 \text{ Hz}. \tag{12}$$

The other time bases are obtained by successive dividers of 2. When the first time base is on, the U/F converter receives the signal from the filters via the first switch. When the nth time base is used the nth switch is turned on automatically.

D. The Computing Section

This computing section consists of a counter, an adder–subtracter and a calculator. A fast adder–subtracter[1] is necessary to take the Hadamard variance. The adder–subtracter acts during the dead time of the counter. The acquisition time of the calculator, typically 0.2 s, is too long for this measurement, but this calculator takes the contents of the accumulator of the adder–subtracter, squares it and computes the mean value of the $m \times 2N$ measurements. We can choose $10 \leqslant m \times 2N \leqslant 160$.

The program of the calculator can be established to obtain the spectral density of the frequency fluctuations $S_{\delta\nu}(f)$ or the one sided spectral density of the phase fluctuations $\mathcal{L}(f)$:

$$S_{\delta\nu}(f) = \frac{\sigma^2(N,\tau,T_M)}{N\tau K^2 L^2}, \qquad \tau = T_M, \quad f = \frac{1}{4\tau} \tag{13}$$

$$\mathcal{L}(f) = 10 \log \frac{8\tau\sigma^2(N,\tau,T_M)}{NK^2L^2} \tag{14}$$

where $\sigma^2(N,\tau,T_M)$ is the variance of the samples actually given by the counter, $2N$ the number of samples, τ the integration time, f the Fourier frequency, K the sensitivity of the electronic tuning of the reference oscillator, and L the transfer coefficient of the U/F converter.

[1] Actually, we study the replacement of the fast adder–subtracter and of the calculator by a microprocessor.

Some comparisons have been made with oscillators, exhibiting different types of noise (white phase noise, frequency flicker noise, . . .) by using a conventional spectrum analyzer and our Hadamard analyzer. Good agreement between the results of both methods has been found, as shown by Figs. 2–4.

Conclusion

Without complicated weighting of the Hadamard variance, but with the use of analog and Hadamard variance filtering, a spectrum analyzer have been realized. This spectrum analyzer can measure the spectral density of the frequency fluctuations $S_{\delta\nu}(f)$ or the spectral density of the phase fluctuations $\mathcal{L}(f)$ of the best oscillator presently on the commercial market.

References

[1] R. A. Baugh, "Frequency modulation analysis with the Hadamard variance," in *Proc. 25th Frequency Control Symp.* (Atlantic-City, NJ), pp. 222–225, 1971.

[2] G. Sauvage and J. Rutman, "Analyse spectrale du bruit de fréquence des oscillateurs par la variance de Hadamard," *Ann. Télécommun.*, vol. 28, no. 7–8, 1973.

[3] P. Lesage and C. Audoin, "Instabilité de fréquence des oscillateurs, mesure dans le domaine fréquence par une méthode temporelle," *Onde Electrique*, vol. 55, no. 2, 1975.

[4] J. H. Shoaf and A. S. Risley, "Frequency stability specification and measurement: High frequency and microwave signals," NBS Tech. Rep., Note 632.

[5] P. Lesage and C. Audoin, "A time domain method for measurement of the spectral density of frequency fluctuations at low Fourier frequencies," in *Proc. 29th Symp. on Frequency Control* (USAEC, Ft. Monmouth, NJ), pp. 394–403, 1975.

PICOSECOND TIME DIFFERENCE MEASUREMENT SYSTEM

David W. Allan, National Bureau of Standards, Boulder, Colorado
Herman Daams, National Research Council, Ottowa, Canada

Abstract

Recently a time difference measurement system was developed in the Frequency & Time Standards Section of the National Bureau of Standards and independently at the National Research Council which shows more than three orders of magnitude improvement over that which is currently available commercially. Current state-of-the-art time difference measurement devices have specified accuracies of about 1 nanosecond. Measurement precision and potential accuracy of better than 1 picosecond has been demonstrated in this new time difference measuring device. This has significant implications in frequency and time metrology using: state-of-the-art frequency standards and clocks. A brief report was given on this measurement system at the PTTI Planning Meeting at the Naval Research Laboratory on the 5th of December 1974. This paper will give more detailed circuit diagrams necessary to build up such a measurement system.

This particular measurement system has the advantage that it can measure time differences with accuracies of a few picoseconds and with repetition rates ranging from a few milliseconds to as slow a repetition rate as would be desirable, thus expanding convenient measurement of time domain stabilities of frequency and time standards over several decades with only one measurement system. The system is also very amenable to self-calibration and self-noise analysis. Specifically, a fractional frequency stability of about 10^{-16} was measured for the noise of this measurement system at a sample time of 10^3 s.

Introduction & Perspective

When making measurements between a pair of frequency standards or clocks, it is often desirable to have less noise in the measurement system than the composite noise in the pair of standards being measured. This places stringent requirements on measurement systems as the state-of-the-art of precision frequency & time standards has advanced to its current level. [1] As will be shown, perhaps one of the greatest areas of disparity between measurement system noise and the noise in current standards is in the area of time difference measurements. Commercial equipment can measure time differences to about 10^{-10}s, but the time fluctuations - second to second - of state-of-the-art standards is as good as 10^{-13}s.

The disparity is unfortunate because if time differences between two standards could be measured with adequate precision then one may also know the time fluctuations, the frequency differences, and the frequency fluctuations. In fact, one can set up an interesting hierarchy of kinds of measurement systems: 1) those that can measure time; $x(t)$; 2) those that can measure changes in time or time fluctuations $\delta x(t)$; 3) those that can measure frequency, ν ($y \equiv (\nu-\nu_0)/\nu_0$); and 4) those that can measure changes in frequency or frequency fluctuations, $\delta\nu$ ($\delta y \equiv \delta\nu/\nu_0$). As depicted in Table 1 if a measurement system is of status 1 in this hierarchy, i.e., it can measure time, then time fluctuations, frequency and frequency fluctuations can be deduced. However, if a measurement system is only capable of measuring time fluctuations (Status 2 - Table 1), then time cannot be deduced, but frequency and frequency fluctuations can. If frequency is being measured (Status 3 - Table 1), then neither time nor time fluctuations may be deduced with fidelity because essentially all commercial frequency measuring devices have "dead time" (technology is at a point where that may soon change with data processing speeds that are now available). Dead time in a frequency measurement destroys the opportunity of integrating the fractional frequency to get to "true" time fluctuations. Of course, if frequency can be measured, then trivially one may deduce the frequency fluctuations. Finally, if a system can only measure frequency fluctuations (Status 4 - Table 1), then neither time, nor time fluctuations, nor frequency can be deduced from the data. If the frequency stability is the primary concern then one may be perfectly happy to employ such a measurement system, and similarly for the other statuses in this measurement hierarchy. Obviously, if a measurement method of Status 1 could be employed with state-of-the-art precision, this would provide the greatest flexibility in data processing. The dual mixer time difference system set forth in this paper is purported to be such a method. Before discussing this method in detail, we will briefly review the other measurement method examples of Table 1.

The measurement method example of Status 2 in Table 1 is shown in block diagram in Figure 1. This method - as described elsewhere [2-6] - has proved very useful in analyzing the spectral density, $S_\phi(f)$, of the phase (time) fluctuations in precision oscillators; ϕ is the phase in radians and f is the Fourier frequency. We use "phase (time)" because they are linearly and simply related by:

$$x(t) = \frac{\phi(t)}{2\pi\nu_0}, \tag{1}$$

where ν_0 is the oscillator's nominal carrier frequency. The "keystone" in all the state-of-art measurement systems reviewed in this paper is the low-noise Schottky barrier diode mixer. The signal from an oscillator under test is fed into one port of a mixer. The signal from a reference oscillator is fed into the other port of this mixer. The signals are in quadrature, that is, they are 90 degrees out of phase so that the average voltage out of the mixer is nominally zero, and the instantaneous voltage corresponds to phase fluctuations rather than to the amplitude fluctuations between the two signals. The output of this mixer is fed through a low pass filter and them amplified in a feedback loop, causing the voltage controlled oscillator (reference) to be phase locked to the test oscillator. The gain is adjusted to obtain a very loose phase lock condition. For times shorter than the attack time of the loop a voltage fluctuation will be proportional to a phase or time fluctuation which is equivalent to the condition where the oscillators are "free running" (the attack time is the inverse of 2π times the unity gain bandwidth of the loop). In turn the output of the low noise amplifier may be fed to a spectrum analyzer, for example, to measure the Fourier components of the phase fluctuations. Alternatively, the output of the low-noise amplifier may be fed to a voltage to frequency converter which in turn is fed to a counter. A frequency fluctuation, $\delta\nu_c$, as read on the counter is proportional to a phase (time) fluctuation, $\delta\nu_c \sim \delta x(t,\tau)$ allowing one to also analyze the fluctuations

Reprinted with permission from *Proc. 29th Annu. Symp. Frequency Contr.*, Atlantic City, NJ, May 28–30, 1975, pp. 404–411.

in the time-domain. With typical attack times this measurement method is useful for measuring the noise for Fourier frequencies equal to or greater than 1 Hz and for sample times of the order of 2π seconds and shorter. It is also specifically very useful if discrete sidebands such as 60 Hz or detailed structure exist in the spectrum. The measurement method example of Status 3 in the hierarchy of Table 1 is shown in block diagram in Figure 2a. As illustrated, the signals from two independent oscillators are fed into the two ports of a double balanced mixer. The difference frequency or the beat frequency out, ν_b, is obtained as the output of a low pass filter which follows the mixer. This beat frequency is then amplified and fed to a frequency counter and printer or some recording device. The fractional frequency can simply be obtained by dividing ν_b, by the nominal carrier frequency ν_0, $|y(t,\tau)| = \nu_b/\nu_0$. This heterodyne or beat frequency method and slight variations of it are commonly used for measuring precision oscillators. This method has the advantage of being simple and inexpensive; and state-of-the-art measurement precision is achievable. It has the disadvantages that only the magnitude of the frequency difference can be measured; i.e., without additional information one cannot ascertain which of the two oscillators is higher or lower than the other in frequency. Also, they must be at different frequencies in order to get a beat frequency; and typically, for common oscillators it is difficult without special synthesis to make this beat frequency even as high as 1 Hz, hence, frequency sample times are usually limited to about 1 second and longer. Commercial frequency counters cause a dead time in a data sequence equal to or greater than the period of the beat frequency. Dead time causes three fundamental problems: 1) One cannot precisely deduce the time fluctuations by integrating the frequency; 2) Increased complications often occur in the stability analysis of the data causing in some instances misleading conclusions; and 3) The time required to take the data in order to do a frequency stability characterization of a pair of oscillators is often significantly increased over that necessary with no dead time. Some have used a method of dating the zero-crossings of the beat frequency to successfully avoid the dead time problem.

The measurement method example of Status 4 in the hierarchy of Table 1 is shown in block diagram in Figure 2b. It is essentially the same as in Figure 1, except in this case the loop is in a tight phase lock condition; i.e., the attack time of the loop should be of the order of a few milliseconds. In such a case, the phase fluctuations are being integrated so that the voltage output is proportional to the frequency fluctuations relative to the "unlocked" reference oscillator and is no longer proportional to the "free running" phase fluctuations for sample times longer than the attack time of the loop. The bias box is used to adjust the voltage on a voltage variable frequency tuning capacitor (varicap) to a convenient tuning point. The voltage fluctuations prior to the bias box (biased slightly away from zero) are fed to a voltage to frequency converter which in turn is fed to a frequency counter where the frequency fluctuations can be read with great amplification of the instabilities between this pair of oscillators. The frequency counter data are logged with a printer or some other data logging device. The coefficient of the varicap and the coefficient of the voltage to frequency converter are used to determine the fractional frequency fluctuations, y_i, between the oscillators, where i denotes the ith measurement as shown in Figure 2b. A system sensitivity of a part in 10^{14} per Hz resolution of the frequency counter is easily achievable. This method is useful for making time-domain measurements of the frequency fluctuations for sample times of the order of one second and longer (usually limited in long-term by the maximum gate time of the frequency counter employed).

Dual Mixer Time Difference System

A block diagram of the dual mixer time difference system is shown in Figure 3. Let us restate that if the time or the time fluctuations can be measured directly an advantage is obtained over just measuring the frequency, because one can calculate the frequency from the time without dead time as well as know the time behavior. Special laboratory test-sets similar to that shown in Figure 3 have been developed in the past with equivalent time-difference precisions as good as 0.6 picosecond. [7, 8]

The system described in this section demonstrated a precision of 0.1 picosecond and with the potential of achieving 0.01 picoseconds (10^{-14}s). Such precisions open the door to making time measurements as well as frequency and frequency stability measurements for sample times as short as a few milliseconds as well as for longer sample times and all without dead time. In Figure 3, Oscillator 1 could be considered under test and Oscillator 2 could be considered the reference oscillator. These signals go to the ports of a pair of double balanced mixers. Another oscillator with separate symmetric buffered outputs is fed to the remaining other two ports of the pair of double balanced mixers. This common oscillator's frequency is offset by a desired amount from the other two oscillators producing approximately the same beat frequencies with Oscillator 1 and with Oscillator 2 as illustrated in Figure 3. These beat frequencies will be out of phase by an amount proportional to the time difference between Oscillator 1 and 2 when running as clocks - excluding the differential phase shift that my be inserted; and will differ in frequency by an amount equal to the frequency difference between Oscillators 1 and 2. If Oscillator 1 and Oscillator 2 are on the same frequency the time difference remains constant. In contrast the heterodyne or beat frequency method is not useful if the oscillators are near zero beat frequency which is often the case with atomic standards (cesium, rubidium and hydrogen frequency standards).

A phase shifter may be inserted as illustrated in Figure 3 to adjust the phase so that the two beat frequencies are nominally in phase; this adjustment sets up the nice condition that the noise of the common oscillator tends to cancel when the time difference is determined in the next step - depending on the level and the type of noise as well as the sample time involved as described below. After amplifying these beat signals, the start port of a time interval counter is triggered with the zero crossing of one beat and the stop port with the zero crossing of the other beat. If the phase fluctuations of the common oscillator are small during this time interval as compared to the phase fluctuations between Oscillators 1 and 2 over a full period of the nominal beat frequencies, the noise of the common oscillator is insignificant in the measurement noise error budget, which means the noise of the common oscillator can be worse than that of either Oscillator 1 or 2 and still not contribute significantly. The above condition will exist if the following equation is satisfied:

$$\Delta t \cdot \left\langle \sigma^2_{y_{common}} (2, \tau, \Delta t, f_h) \right\rangle^{1/2} << \tau \cdot \sigma_{y_{12}}(\tau) \qquad (2)$$

where the variances (or their square roots) are as defined by the IEEE subcommittee on frequency stability [9, 10]. The left side of Eq. (2) is representative of the phase (time) noise of the common oscillator over Δt and the right side is of the combined phase (time) noise of Oscillators 1 and 2 over τ.

By taking the time difference between the zero crossings of these beat frequencies, what effectively is being measured is the time difference between Oscillator 1 and Oscillator 2, but with a precision which has been amplified by the ratio of the carrier frequency to the beat frequency over that normally achievable with this same time interval counter. The time difference $x(i)$, for the ith measurement between Oscillators 1 and 2 is given by Eq. (3):

$$x(i) = \frac{\Delta t(i)}{\tau\nu} - \frac{\phi}{2\pi\nu} + \frac{n}{\nu} \qquad (3)$$

where $\Delta t(i)$ is the ith time difference as read on the counter, τ is the beat period, ν is the nominal carrier frequency, ϕ is the phase delay in radians added to the signal of Oscillator 1, and n is an integer to be determined in order to remove the cycle ambiguity. It is only important to know n if the absolute time difference is desired; for frequency and frequency stability measurements and for time fluctuation measurements, n may be assumed zero unless a cycle or more elapses during a set of measurements. If the frequency difference between Oscillators 1 and 2 is large enough to cause n to accumulate significantly over the duration of an experiment, one could easily build a divider (counter) to determine n as a function of time.

The fractional frequency can be derived in the normal way from the time fluctuations.

$$Y_{1,2}(i,\tau) = \left\{ \begin{array}{c} \dfrac{\nu_1(i,\tau) - \nu_2(i,\tau)}{} \\[2ex] \dfrac{x(i+1) - x(i)}{\tau} \\[2ex] \dfrac{\Delta t(i+1) - \Delta\tau(i)}{\tau^2\nu} \end{array} \right. \qquad (4)$$

The last equality in Eq. (4) is convenient for using the direct readings of the time interval counter but is not valid if n changes between the i and i + 1 readings. In Eqs. (3) and (4), the assumptions are made that the transfer or common oscillator is set at a lower frequency than Oscillators 1 and 2, and that the beat $\nu_1 - \nu_0$ starts and $\nu_2 - \nu_0$ stops the time interval counter. The sample time by appropriate calculation can be any integer multiple of τ:

$$Y_{1,2}(i, m\tau) = \frac{x(i+m) - x(i)}{m\,\tau} \qquad (5)$$

where m is any positive integer. If needed, τ can be made to be very small by having very large beat frequencies. One may replace the common or transfer oscillator with a low phase noise synthesizer, which derives its basic reference frequency from Oscillator 2, for example. In such a set up the nominal beat frequencies are simply given by the amount the output frequency of the synthesizer is offset from ν_2. Sample times as short as a few milliseconds were easily obtained. Logging the data at such a rate can be a problem without special equipment, e.g., magnetic tape. The system developed at the National Research Council (NRC) differs from that of the National Bureau of Standards (NBS) in this respect: that the transfer oscillator is always offset 500 Hz. To obtain, e.g., a 1 second averaging time, pulses from a 10 MHz clock are gated by the 500 time intervals between leading edges of the 500 Hz beats before being fed into a counter preceded by a divide by 5 stage. The counter has two counting registers which are used on alternate seconds at the end of each second the accumulated count is transferred to the storage and display register which is so arranged that it always displays 200,000.0 picosecond full scale. The use of the dual counter eliminates dead time between measurements while the use of 500 Hz although it widens the bandwidth eliminates problems with stabilization of dc levels in stages after the first squarer. At NBS a computing counter was used to achieve sample time stabilities as short as a few milliseconds and with no dead time (see appendix for computing counter program possibilities). A circuit diagram for this dual mixer time difference system is shown in Figure 4.

A review of some of the advantages of the dual mixer time difference system are: If the oscillators, including the transfer oscillator, and a time interval counter are available, the component cost is fairly inexpensive ~ $500, most of which is the cost of the phase shifter (the phase shifter is not necessary if the noise of the common oscillator is sufficiently low). The measurement system bandwidth is easily controlled (note that this should be done in tandem with both low pass filters being symmetrical). The measurement precision is such that one can measure essentially all state-of-the-art oscillators. For example, if the oscillators are at 5 MHz, the beat frequencies are 0.5 Hz and the time interval counter employed has a precision of 0.1 μs, then the potential measurement precision is 10^{-14}s (10 femto seconds) for τ = 2s; other things may limit the precision such as noise in the amplifiers. This precision does not imply that absolute time differences can be measured this well; calibrating all the relevant phase delays limits the time difference accuracy to the order of a picosecond. As has been stated above, there is no dead time in the frequency averages (see ref. 10), which is a distinct advantage for sample times shorter than a second - a region of sample times where dead time is difficult to avoid in most other measurement methods. The sample time is conveniently selectable by adjusting the frequency of the common oscillator and is given by the nominal beat period or multiple of the same. If one replaces the common oscillator by a synthesizer then the beat period may be selected very conveniently. The synthesizer should have fairly low phase noise to obtain the maximum precision from the system. If discrete sidebands exist in the spectrum, having a synthesizer as the common oscillator is also very convenient because by setting the beat frequency equal to the sideband one can estimate the stability as if the sideband weren't present (the sideband has been filtered out). The system measures time difference putting it in Status 1 of the measurement hierarchy. One may calculate from the data both the magnitude and the sign of the frequency difference. This system, therefore, allows the measurement of time fluctuations as well as time difference, and the calculation of frequency fluctuations as well as frequency differences between the two oscillators in question. The system may be calibrated and the system noise be measured by simply feeding the same signal symmetrically split from one oscillator in place of Oscillators 1 and 2.

A review of the disadvantages are: The system is somewhat more complex than the others. Because of the low frequency beats involved, precautions must be taken to avoid ground loop problems; there are some straightforward solutions; e.g., a saturated amplifier followed by a differentiator and isolation transformer worked very well in avoiding ground loops as shown in Figure 4. Buffering is needed on the split output signals from the common oscillator because the mixers present a dynamic load to the oscillator - allowing the possibility of cross-talk. The time difference reading is modulo the beat period. For example, at 5 MHz there is a 200 nanosecond per cycle ambiguity that must be resolved if the absolute time difference is desired: this ambiguity is usually a minor problem to resolve for precision oscillators.

It is instructive to compare the measurement methods discussed to this point in terms of the nominal region of sample times over which eacy typically finds usefulness. This is illustrated in Figure 5. One sees again a significant advantage using the dual mixer time difference approach because of the significant number of decades of sample time covered. To estimate $\sigma_y(\tau)$ using the beat frequency method it is necessary to use a set of bias functions as explained in Ref. [2 and 10], since this method as described in the text generates data having dead time. However, it should be noted that by precisely logging the dates of zero crossings of a beat frequency, the problem of dead time may be avoided. This method has proven useful in the Frequency & Time Standards Section of the NBS. An estimate of $\sigma_y(\tau)$ can be obtained from the frequency-domain method suggested in Figure 1 by using Table 2 or Ref. [1, 2, 8 or 9]. Table 2 is restricted to integer power-law spectral densities for the noise models.

Some Measurements Using the Dual Mixer Time Difference (DMTD) System

By using a low-noise synthesizer as the local or common oscillator and by feeding the signal from a quartz crystal oscillator symmetrically split into the ports for Oscillators 1 and 2 of the DMTD system we estimated the measurement noise, and this is shown in Figure 6 as a $\sigma_y(\tau)$ vs τ plot. The curve plotted may be reasonably modeled as: $\sigma_y(\tau) \simeq 10^{-13}\ \tau^{-1}$ (τ in seconds) with slight exceptions at sample times of milliseconds, where the measurement system bandwidth was opened from 1.8 Hz up to 300 Hz, and at sample times of $\tau > 10^3$s where $\sigma_y(\tau)$ flattened at several parts in 10^{17}; apparently some diurnal effects were perturbing the system. Figure 7a is a plot using the DMTD system of a strip chart recording of a digital to analog output of the significant digits from the time interval counter between a quartz oscillator and a high performance commercial cesium oscillator. In other words this is a plot of the time difference between these two oscillators as a function of time. The high frequency fluctuations (over fractions of a second) would most probably be those between the quartz oscillator and the quartz oscillator in the cesium servo system. The low frequency fluctuations (over seconds) would most probably be those induced by the cesium servo in its effort to move the frequency of its quartz oscillator to the natural resonance of the cesium atom - causing a random walk of the time fluctuations for sample times longer than the servo attack time. Figure 7b shows a similar plot obtained with the NRC system for the phase difference between the NRC standard Cs V and a similar cesium oscillator as in Figure 7a. The averaging times of 1 second causes, of course, much coarses steps although the general shape of the curves is similar.

Figure 8 is a $\sigma_y(\tau)$ vs τ plot of the same pair of oscillators as for Figure 7a using the DMTD system. The plot contains a lot of information: The measurement noise of the dual mixer system is indicated. One can see the short term stability performance of the quartz oscillators for $\tau \leq 1$ s. One can see a little bit of 60 Hz present as indicated by slightly increased values of $\sigma_y(\tau)$ at 1/2 and 3/2 of τ = 1/60 Hz. One observes the attack time of the servo in the cesium electronics perturbing the short term stability of the quartz oscillator and degrading it to the level of the shot noise of the cesium resonance for $10\ s \leq \tau \leq 10^2 s$. The white noise frequency modulation characteristic then becomes the predominant power law causing $\sigma_y(\tau)$ to improve as $\tau^{-1/2}$ until the flicker floor of the quartz crystal oscillator, in this case 6 parts in 10^{13}, becomes the predominant noise source for $\tau \geq 10^2$ s. Thus, using this particular measurement system on this pair of precision frequency standards, their stabilities were well characterized for sample times of a few milliseconds all the way out to 1000 seconds. Longer sample times are of course easily achievable - patience is the main ingredient!

Figure 9 is an example of where the common or transfer oscillator was known to be about 5 times worse in frequency instability than the two oscillators being measured. The common oscillator was a high performance commercial cesium beam frequency standard and the oscillators being measured were primary frequency standards at NBS (NBS-4 and NBS-5). Using the DMTD method and having the phase relationship between the beat frequencies near enough coincident to satisfy Eq. (2), one may deduce that the long term fractional frequency stability of NBS-4 or NBS-5 was better than 1×10^{-14} for $\sigma_y(\tau)$, τ = 1 day.

Conclusions

Using low noise techniques one can measure the time and/or frequency performance characteristics of state-of-the-art frequency standards and clocks. The low level of measurement noise is achieved using Schottky diode mixers with low noise amplifiers following.

It has been shown that in practice when making measurements between a pair of clocks or oscillators, the greatest degree of flexibility is achieved in data analysis if the time-difference can be measured with adequate precision. A symmetric dual mixer time difference system was demonstrated which fulfilled this maximum flexibility. A significant variety of alternatives exist which could fruitfully employ this system. For example, if the common oscillator involved could be synthesized internal to the system, and the time difference device could be also constructed internal to the system, these coupled with a micro-processor and appropriate peripherals could make a very powerful state-of-the-art frequency and time standards data acquisition and analysis system.

Also a common transfer oscillator followed by a suitable distribution amplifier could drive several balanced mixers simultaneously. The beat signals then could be multiplexed into a time interval counter or into a precise dating device and the data thus obtained at regular intervals would be useful in a timekeeping system with several contributing clocks.

Acknowledgements

The authors wish to express appreciation to James A. Barnes, Helmut Hellwig, Fred L. Walls, and Sam R. Stein for very helpful discussions and critiquing of the manuscript.

Appendix: Computing Counter Program

The following program may be used in a computing counter, is useful in determining the fractional frequency stability, $\sigma_y(\tau)$, and is unique as compared with other similar types of programs to determine stability in that it does so with no dead time. The following program actually determines the root-mean square second difference, $(\Delta^2\ (\Delta t))$ rms, of the time difference readings between a pair of clocks or oscillators, and therefore complements very nicely the dual mixer time difference measurement system described in the text. The fractional frequency stability may be calculated from computer program results as follows:

$$\sigma_y(\tau) = \frac{1}{\sqrt{2}\,\tau^2\nu} (\Delta^2(\Delta t))_{rms} \qquad (A1)$$

If additional programming steps were available, of course one could program the computing counter to calculate an estimate of $\sigma_y(\tau)$ directly. Following is the program procedure to generate $(\Delta^2(\Delta t))_{rms}$:

1. clear x
2. c ⇄ x
3. Plug-in
4. a ⇄ x
5. Plug-in
6. a ⇄ x
7. $\overrightarrow{a\ x\ y}$
8. - (subtract)
9. b ⇄ x
10. Xfer Program
11. Plug-in
12. a ⇄ x
13. $\overrightarrow{a\ x\ y}$
14. - (subtract)
15. b ⇄ x
16. $\overrightarrow{b\ x\ y}$
17. - (subtract)
18. $\overrightarrow{xy}$
19. x (multiply)
20. $\overrightarrow{c\ x\ y}$
21. + (add)
22. c ⇄ x
23. Repeat
24. Xfer Program
25. c ⇄ x
26. $\overrightarrow{N\ x\ y}$
27. ÷ (divide)
28. $\sqrt{x}$
29. Display x
30. Pause

The confidence of the estimate will improve approximately as the square root of the number of times (N) the sub-loop is repeated as preset by the programmer [11].

References

[1] Hellwig, H., Allan, D. W., Walls, F. L., Proc 5th Int. Conf. on Atomic Masses & Fundamental Constants; Paris, France, 2-6 June 1975.

[2] Shoaf, J. H., Halford, D., Risley, A. S., NBS Technical Note 632 (1973).

[3] Allan, D. W., Shoaf, J. H., Halford, D., Chapter 8 of NBS Monograph 140, pp. 151-204 (1974).

[4] Rutman, J., IEEE Trans. on Instr. & Meas., IM-23, pp. 40-48 (1974).

[5] Rutman, J. and Savage, G., IEEE Trans. on Instr. & Meas., IM-23, pp. 515-518 (1974).

[6] Meyer, D.G., IEEE Trans. on Instr. & Meas., IM-19, pp. 215-227 (1970).

[7] Maxwell, D.E., IEEE Trans. on Instr. & Meas., IM-15, pp. 304-310 (1966).

[8] Mungall, A. G., Morris, D., Can. J. Phys. Vol. 37, pp. 1349-1356 (1959).

[9] Barnes, J. A., et al., NBS Technical Note 394 (1970); IEEE Trans. on Instr. & Meas., IM-20, pp. 105-120 (1971).

[10] Barnes, J. A., NBS Technical Note 375 (1969).

[11] Lesage, P., Audoin, C., IEEE Trans. on Instr. & Meas., IM-22, pp. 157-161 (1973).

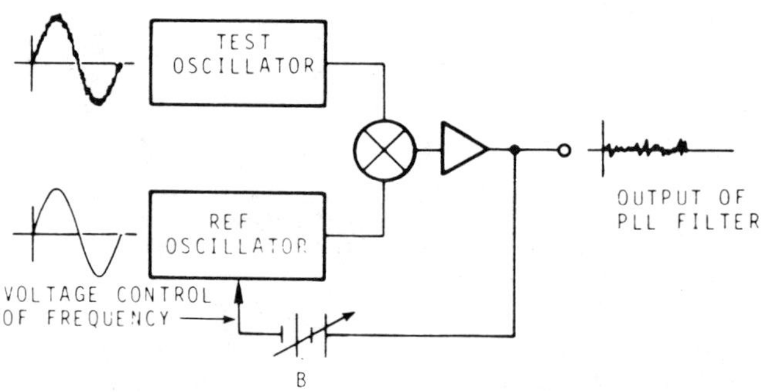

FIGURE 1.

A phase (or time) fluctuation measurement system. The reference oscillator is loosely phase-locked to the test oscillator - attack time is about 1 second. The reference and test oscillators are fed into the two ports of a Schottky barrier diode double balanced mixer whose output is fed through a low pass filter and low noise amplifier, a battery bias box and to the varicap of the reference oscillator. The instantaneous output voltage of the phase locked loop (PLL) following the low noise amplifier will be proportional to the phase or time fluctuations between the two oscillators which is useful for times short compared to the attack time of the servo.

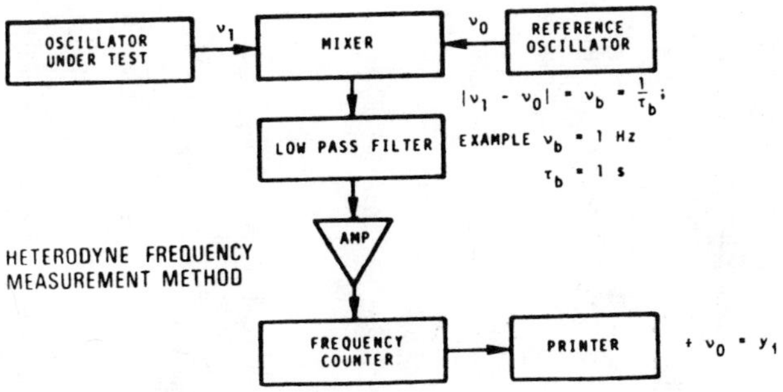

FIGURE 2a.

A frequency and frequency fluctuation measurement system. The difference frequency, $|\nu_1 - \nu_0|$ is measured with a frequency counter. A counter measuring the period (or multiple period) of the beat (difference) frequency could equivalently be used.

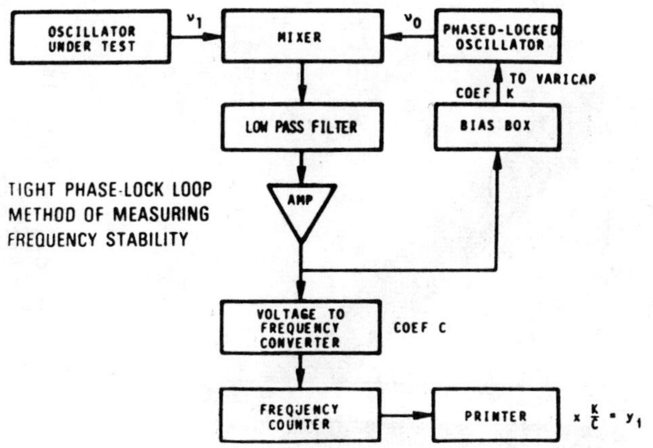

FIGURE 2b.

A frequency fluctuation measurement system. The attack time of the phase lock loop in this case is much less than a second. The amplifier (AMP) output voltage fluctuations for sample times significantly longer than the servo loop attack time will be proportional to the frequency fluctuations which would have existed between the "free running" oscillators.

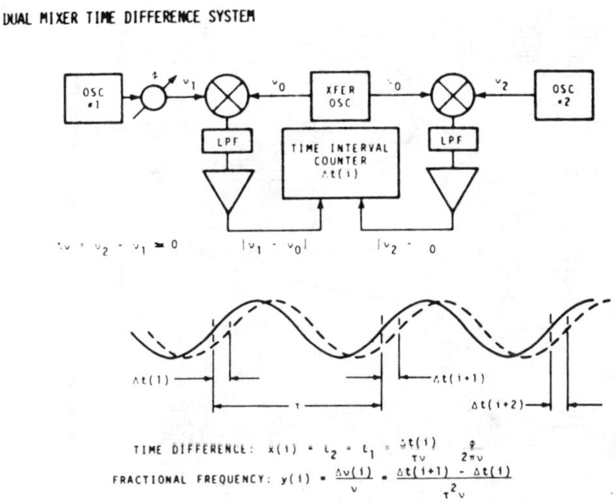

FIGURE 3.

A time difference and time fluctuation measurement system. The low pass filters (LPF) determine the measurement system bandwidth and must pass the difference frequencies which are depicted by the solid-line and dashed-line sinusoids at the bottom of the figure. The positive going zero volts crossing of these difference (beat) frequencies are used to start and stop a time interval counter after suitable low noise amplification. The ith time difference between Oscillator 1 and 2 is the $\Delta t(i)$ reading of the counter divided by $\tau\nu$ and plus any phase shift added, ϕ, where $\nu \simeq \nu_1 \simeq \nu_2$ is the nominal carrier frequency. The frequency difference is straight forwardly calculated from the time difference values.

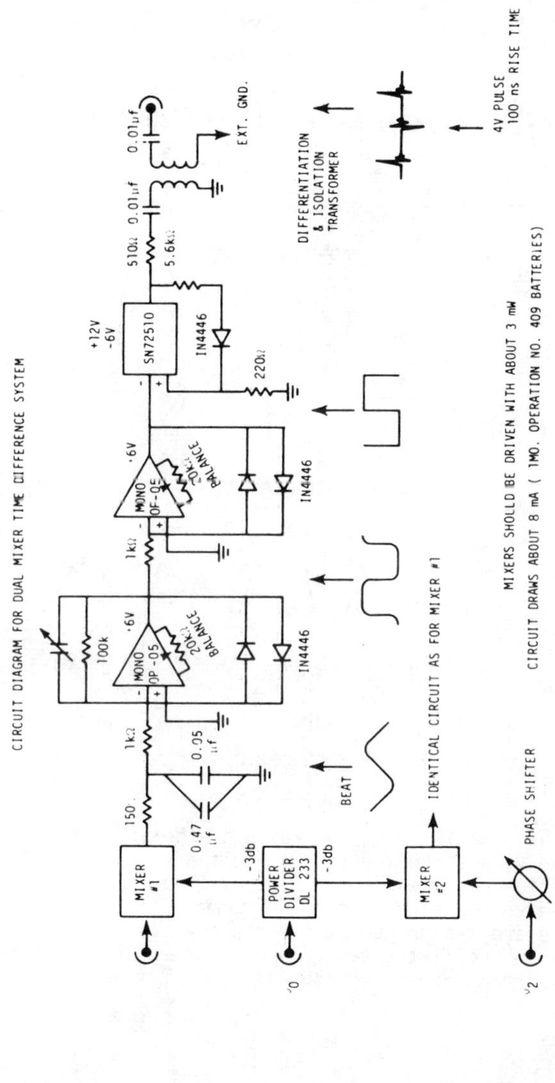

FIGURE 4. An example circuit diagram for a Dual-Mixer-Time-Difference System

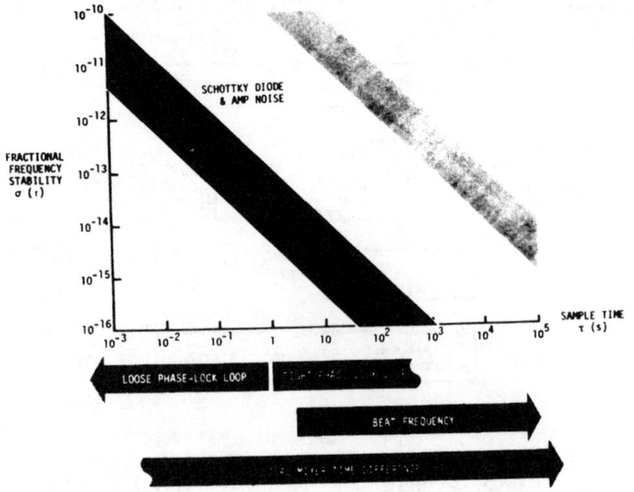

Figure 5. Comparison of the nominal sample time range of some state-of-the-art methods of measuring frequency stability.

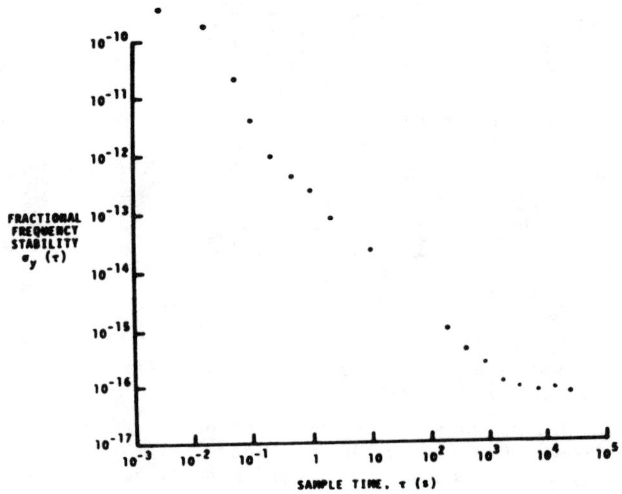

Fig. 6 The fractional frequency stability plot as a function of sample time, τ of the measurement noise of the dual mixer time difference system. The first 3 points are with system bandwidth of 300 Hz, then 2 at 18 Hz, 1 at 8 Hz, 1 at 3 Hz, and the rest of the stability points are at 1.8 Hz. The flatening as the stability reaches a few parts in 10^{17} is believed due to diurnal temperature effects.

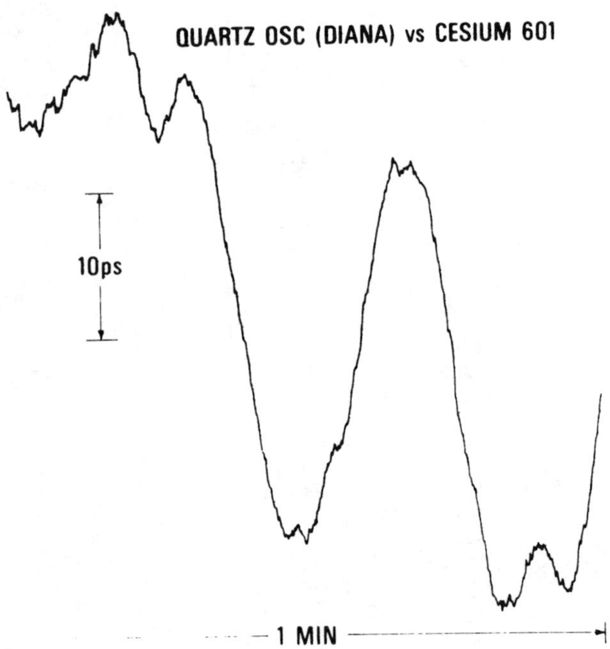

FIGURE 7a.

A copy of a strip chart recording of the time fluctuations versus running time using the dual mixer time difference measurement system. The oscillators involved were a high performance commercial cesium standard and a high quality quartz crystal oscillator. The common oscillator employed was a low noise synthesizer. The measurement system noise was about 0.1 ps.

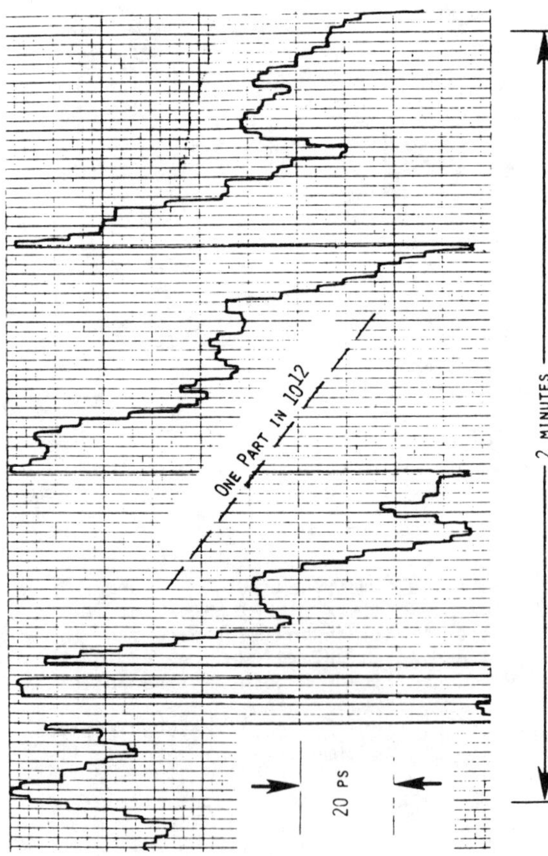

FIGURE 7b. Time Difference between NRC CsV and High Performance Commercial Cesium using NRC Dual Mixer Time Difference System.

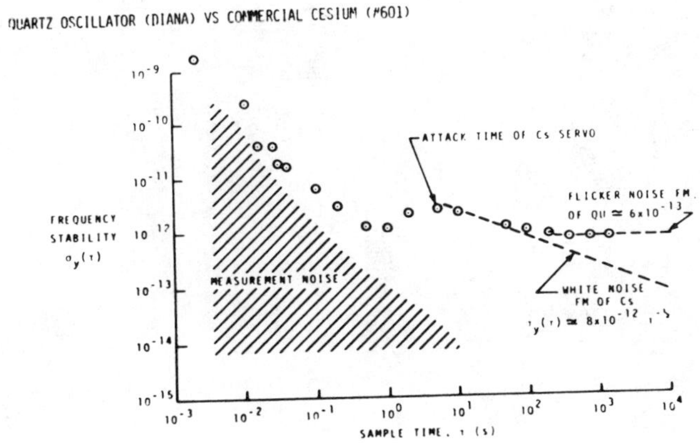

FIGURE 8.

A $\sigma_y(\tau)$ versus τ plot of the fractional frequency fluctuations, y(t) between a high performance commercial cesium beam frequency standard and a commercial quartz crystal oscillator.

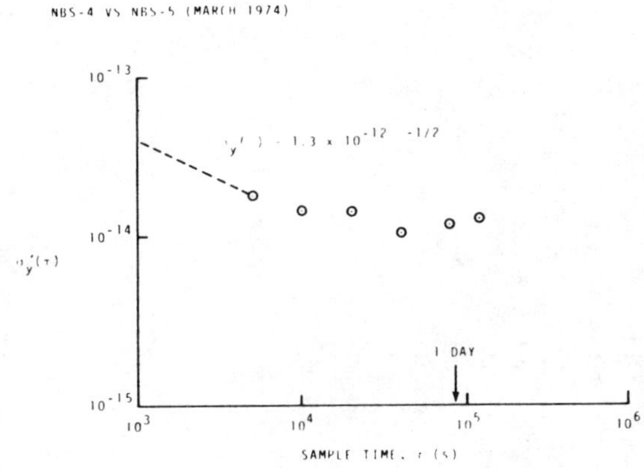

Figure 9. A measurement of the fractional frequency stability $\sigma_y(\tau)$ of the two primary frequency standards NBS-4 & NBS-5 versus sampling time τ using the dual mixer time difference method of measurement. The frequency instability of the common oscillator employed was about a factor of 5 greater than that of either NBS-4 or NBS-5.

Hierarchy Status	Example of Measurement Method	Data deducible from the measurement			
		Time	δ Time	Freq.'	δ Freq.
1	Dual Mixer Time Diff.	$x(t) = \frac{\Delta t}{T\nu}$	$\delta x(t,\tau) = x(t+\tau)-x(t)$	$y(t,\tau) = \frac{\delta x(t,\tau)}{\tau}$	$\delta y(t,\tau) = y(t+\tau,\tau)-y(t,\tau)$
2	loose phase-locked reference oscillator	can't measure	$\delta x(t,\tau) = \frac{\delta \phi(t,\tau)}{2\pi\nu}$	"	"
3	Heterodyne or beat frequency	can't measure	can't measure	$\lvert y(t,\tau)\rvert = \frac{\nu_{beat}}{\nu_o}$	"
4	tight phase-locked reference oscillator	can't measure	can't measure	can't measure	$\delta y(t,\tau) = c \cdot \delta V$

TABLE 1. Hierarchy of measurement methods: showing that the measurement of time differences between a pair of oscillators (clocks) gives the greatest data processing capability.

$S_y(f) = h_\alpha f^\alpha$ $\alpha =$	$S_y(f) = a\,\sigma_y^2(\tau)$ $a =$	$\sigma_y^2(\tau) = b\,S_\phi(f)$ $b =$
2 (white phase)	$\frac{(2\pi)^2 \tau^2 f^2}{3 f_h}$	$\frac{3 f_h}{(2\pi)^2 \tau^2 \nu_o^2}$
1 (flicker phase)	$\frac{(2\pi)^2 \tau^2 f}{3.81 + 3 \ln(\omega_h \tau)}$	$\frac{[3.81 + 3 \ln(\omega_h \tau)] f}{(2\pi)^2 \tau^2 \nu_o^2}$
0 (white frequency)	2τ	$\frac{f^2}{2\tau \nu_o^2}$
-1 (flicker frequency)	$\frac{1}{2 \ln(2) \cdot f}$	$\frac{2 \ln(2) \cdot f^3}{\nu_o}$
-2 (random walk frequency)	$\frac{6}{(2\pi)^2 \tau f^2}$	$\frac{(2\pi)^2 \tau f^4}{6 \nu_o^2}$

TABLE 2. Conversion table from time-domain to frequency-domain and from frequency-domain to time-domain for common kinds of integer power law spectral densities; f_h ($= \omega_h/2\pi$) is the measurement system band-width.

$$S_\phi(f) = \frac{\nu_o^2}{f^2} S_y(f)$$

A Direct Time-Domain Measure of Frequency Stability: The Modified Allan Variance

RICHARD G. WILEY, MEMBER, IEEE

Abstract—A simple filter is shown to closely approximate the procedure used to compute the "Allan variance," an often used measure of the short term frequency stability of oscillators. This modified Allan variance is obtained by measuring the rms value of the instantaneous frequency waveform after filtering. Some advantages of the modified Allan variance include ease of implementation for short averaging times and better behavior when the instabilities include periodic components. The effect of the bandwidth of the measuring process on the minimum usable averaging time is also discussed.

I. Introduction

THE "ALLAN VARIANCE" for an oscillator is defined in terms of the fractional variation of the average frequency over intervals of length τ [1]–[3]. It is often estimated using frequency counters and digital computation. A direct method for estimating a modified Allan variance is described here.

Manuscript received February 27, 1976.
The author is with the Syracuse Research Corporation, Merrill Lane, Syracuse, NY 13210.

At present, the Allan variance is estimated by computing the average of the instantaneous frequency of an oscillator over τ-seconds, and then computing the mean square value of the variation of this average frequency. This is obtained by taking the difference of the averaged frequency at times τ-seconds apart; and computing the mean square of these differences. Finally, this is divided by the average or "carrier" frequency (usually estimated by averaging the frequency over a time much longer than τ).

Following Barnes [3], the instantaneous fractional frequency is defined as

$$y(t) \equiv \frac{\dot{\psi}(t)}{2\pi f_s} \tag{1}$$

where $\dot{\psi}(t)$ is the derivative of the total instantaneous phase, and f_s is the average frequency of the signal from the oscillator

$$f_s = \lim_{T\to\infty} \left[\frac{\psi(T/2) - \psi(-T/2)}{2\pi T}\right].$$

Let $S_y(f)$ denote the one-sided spectral density on a per Hertz basis of the real function $y(t)$. The spectral density

Reprinted from *IEEE Trans. Instrum. Meas.*, vol. IM-26, pp. 38–41, Mar. 1977.

of $y(t)$ can be computed from the spectral density of $\psi(t)$ as follows:

$$S_y(f) = \frac{1}{2\pi f_s} S_{\dot\psi}(f) = \frac{f^2}{f_s} S_\psi(f). \tag{2}$$

Barnes shows that the Allan variance, denoted by $\sigma_y^2(\tau)$, is related to the frequency domain frequency stability measure $S_y(f)$ by

$$\sigma_y^2(\tau) = 2 \int_0^\infty S_y(f) \frac{\sin^2(\pi\tau f)}{(\pi\tau f)^2} \left\{1 - \frac{\sin^2(2\pi\tau f)}{4\sin^2(\pi\tau f)}\right\} df. \tag{3}$$

Equation (3) states that the Allan variance can be estimated by integrating the power spectrum of the instantaneous fractional frequency fluctuation after passing it through a filter with transfer function given by the expression under the integral in (3). The integration simply computes the total power, which can also be done by integrating in the time domain after filtering the signal $y(t)$ using the transfer function in (3).

Examination of this Allan variance filter shows that it consists of a $(\sin x)/x$ "low-pass" stage followed by a periodic transfer function which has a high-pass character at low frequencies. A plot of this transfer function is shown as the solid curve in Fig. 1. The parameter τ is often referred to as the "averaging time" and is of central importance since the frequency variation over some specified interval usually corresponds to some performance attribute of the system in which the oscillator is used. For example, in FMOP radar, the frequency variation from linear within the pulse determines the target resolution which can be achieved. Therefore, the frequency variations over time intervals corresponding to the radar's resolution cell width are of major interest. Compare this to the pulse-Doppler radar, where coherent integration time (over several pulse intervals) determines the velocity resolution which can be achieved. In this case, frequency variation over time intervals corresponding to the reciprocal of the velocity resolution cell width is of major importance. The point is that, for some applications, averaging time intervals of nanoseconds or microseconds are important, for others variations over milliseconds or seconds is important, while for still others, variation over hours or even years is important. Short-term stability measurements generally deal with fractional frequency variation over averaging times of a few seconds or less.

II. Approximating the Allan Variance Filter

The dashed curve in Fig. 1 shows the response of a filter which is easily implemented and which closely approximates the Allan variance filter transfer function. It consists of a simple C-R high-pass stage and a simple R-C low-pass stage with identical break points at $RC = \tau/2$. If the instantaneous frequency waveform is fed through such a filter and the rms value of the resulting voltage is measured, the result is an estimate of the square root of a modified Allan variance (after division by the nominal or

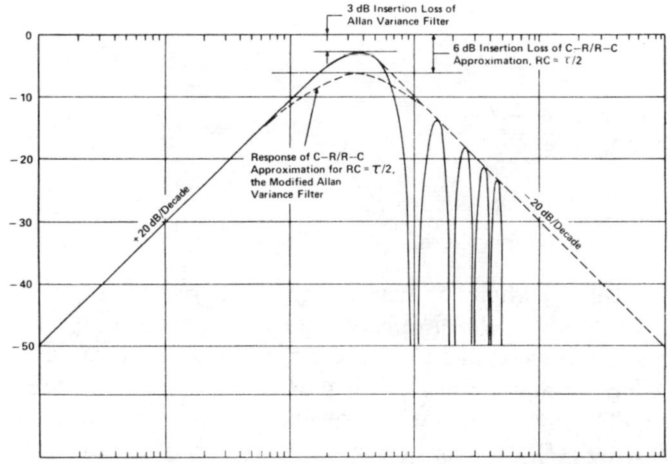

Fig. 1. Allan variance filter transfer function (τ = averaging time).

average frequency of the oscillator). The estimate is conservative since it includes all of the spectrum under the dotted portion of the curve in Fig. 1, without the deep notches in the Allan variance filter transfer function. However, it does have a 3 dB greater insertion loss at τ.

If the instantaneous total phase waveform is available (instead of the instantaneous frequency) the filter characteristic shown in Fig. 1 must be modified by a factor proportional to $1/f^2$ to account for the differentiation which occurs naturally in the discrimination process. To achieve the same result as obtained by filtering the instantaneous frequency waveform, the phase waveform must be filtered using a high pass slope characteristic of +40 dB/decade at low frequencies, and no low-pass characteristic to diminish the response at the higher frequencies. The problem of approximating the filter for processing the phase data has been addressed by Rutman [4]. However, if the instantaneous frequency waveform is available, these filters are not needed. Consider the case when $S_y(f)$ is constant (i.e., the frequency instability is white noise). Then the ideal Allan variance estimate from (3) would be

$$\sigma^2(\tau) = 2\int_0^\infty \frac{\sin^2(\pi\tau f)}{(\pi\tau f)^2}\, dt - 2\int_0^\infty \frac{\sin^2(2\pi\tau f)}{(2\pi\tau f)^2}\, df \tag{4}$$

$$= \frac{1}{2\tau}. \tag{5}$$

Using simple R-C high-pass and low-pass filters would yield a modified Allan variance (for white noise) of

$$\hat\sigma^2(\tau) = 2\int_0^\infty \left[\frac{1}{1 + \dfrac{1}{(\pi f\tau)^2}}\right]\left[\frac{1}{1 + (\pi f\tau)^2}\right] df \tag{6}$$

$$= 2\int_0^\infty \frac{(\pi f\tau)^2}{[1 + (\pi f\tau)^2]^2}\, df. \tag{7}$$

Let

$$u = 1 + (\pi\tau f)^2 \text{ and } du = 2(\pi\tau)(\pi\tau f)\, df \tag{8}$$

$$\hat{\sigma}^2(\tau) = \frac{2}{\pi\tau}\int_1^{\infty} \frac{(u-1)^{1/2}}{2u^2}\,du \tag{9}$$

$$= \frac{2}{\pi\tau}\left[\frac{(u-1)^{3/2}}{u} - (u-1)^{1/2} + \tan^{-1}(u-1)^{1/2}\right]_1^{\infty} \tag{10}$$

$$= 1/\tau \text{ (for } RC = \tau/2). \tag{11}$$

Therefore, when the instantaneous frequency waveform is white noise, the simple cascade C-R high-pass and R-C low-pass gives exactly twice the mean square output as the Allan variance filter. Although the RC approximation has a 3 dB larger insertion loss, this only partially compensates for the lack of attenuation "notches" at frequencies $f = N/\tau$; $N = 1,2,3,\cdots$. Of course, if the spectrum of the frequency fluctuation is not white, the approximate filter may provide an estimate for the Allan variance which is greater or smaller than the actual Allan variance. By simply incorporating an attenuation of 0.5, the modified Allan variance as measured at the output of the RC filter will be equal to the unmodified Allan variance when the frequency disturbance is white noise.

The notches in the bandpass characteristics of the Allan variance filter are an obvious consequence of the definition: frequency variations which are periodic such that the period is a submultiple of the averaging time τ contribute nothing to the Allan variance. For instance, if $\tau = 1/60$ second, then any 60 Hz (or multiples of 60 Hz) variations in frequency will not affect the Allan variance for $\tau = 1/60$. Also large changes in the Allan variance for small changes in averaging time are possible if periodic frequency variations are present. For most applications, this appears to be a disadvantage to using the Allan variance. Certainly it should be carefully measured over a range of averaging times to avoid problems with periodic frequency fluctuations. Indeed, the modified Allan variance computed by the simple C-R/R-C filter combination appears to be the more desirable measure of oscillator stability, with the presently defined Allan variance being a second choice which is easy to calculate for sampling measurement systems (e.g., frequency counters). Certainly, for short averaging times, the modified Allan variance suggested here has many advantages.

III. Limitations Due to Bandlimited Measurement Systems

In any real measurement system, the instantaneous frequency (or phase) is determined over a limited bandwidth. Thus, the Allan variance cannot be estimated for arbitrarily short averaging times (τ). In fact, for bandlimited signals, the variation in a time τ is bounded. Papoulis [5] has shown that a random ergodic signal, bandlimited to f_b having an average power P has a mean square variance over time τ which is bounded by

$$\sigma_y^2(\tau) = \lim_{T\to\infty}\frac{1}{T}\int_{-T/2}^{+T/2}(\overline{y}(t) - \overline{y}(t+\tau))^2 dt < (2\pi f_b \tau)^2 P \tag{12}$$

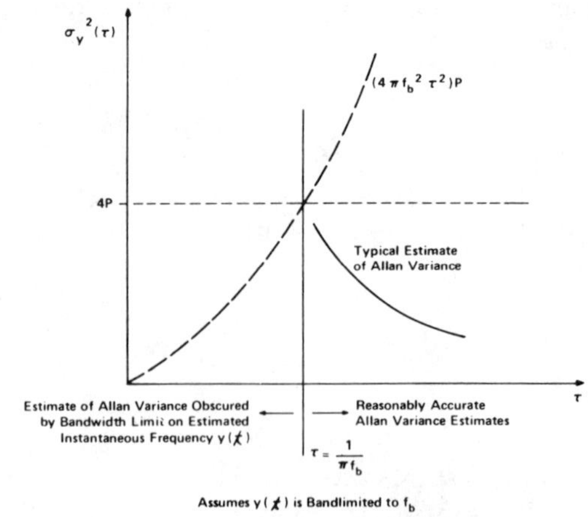

Fig. 2. Bounds on Allan variance estimate for a bandlimited measurement system.

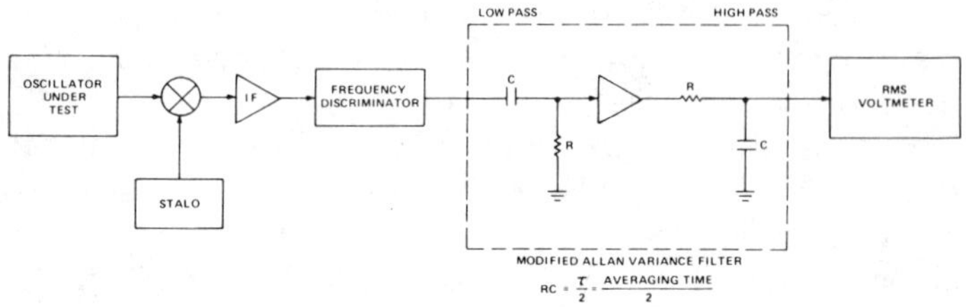

Fig. 3. Instrumentation for modified Allan variance measurement.

where $\overline{y}(t)$ is the instantaneous frequency averaged over time τ

$$\overline{y}(t) = \int_{-\infty}^{\infty} y(t-\tau)p(\tau)\,d\tau$$

$p(t)$ is a unit pulse extending from $t = -\tau/2$ to $t = +\tau/2$, and

$$P = \lim_{T\to\infty}\frac{1}{T}\int_{-T/2}^{+T/2}\overline{y}^2(t)dt$$

= average power of the τ-averaged fractional frequency.

Expanding the integrand in (12) gives another bound on $\sigma_y^2(\tau)$.

$$\sigma_y^2(\tau) = \lim_{T\to\infty}\frac{1}{T}\left[\int_{-T/2}^{+T/2}(\overline{y}^2(t) - 2\overline{y}(t)\overline{y}(t+\tau) + \overline{y}^2(t+\tau))\,dt\right]. \tag{13}$$

The center term in the integrand of (13) is the correlation $R_{yy}(\tau)$, which can be no larger than $R_{yy}(0) = P$. Thus

$$\sigma_y^2(\tau) \le 4P. \tag{14}$$

Assuming that the estimate of $y(t)$ is subject to an upper frequency cutoff f_B, the Allan variance estimated by such a measurement system is subject to the bounds shown in Fig. 2, which graphically illustrates (12) and (14). Clearly, reasonably accurate Allan variance estimates can be made only for

$$\tau > 1/\pi f_b. \quad (15)$$

IV. Summary

There are advantages to slightly modifying the Allan variance definition to eliminate the effects of periodic sampling. When this is done, the resulting modified Allan variance is easy to estimate using the output of a frequency discriminator as shown in Fig. 3. For short averaging times (τ) of less than one second, the modified Allan variance is a useful measure of oscillator quality which is easier to estimate than the presently defined Allan variance.

References

[1] T. F. Curry, "Specification of short-term frequency stability by maximum likelihood estimates," in *Proc. IEEE-NASA Symp. on Short-Term Stability,* NASA Rep. SP-80, pp. 101–109, Nov. 1964.

[2] D. W. Allan, "Statistics of atomic frequency standards," *Proc. IEEE,* vol. 54, pp. 221–230, Feb. 1966.

[3] J. A. Barnes *et al.,* "Characterization of frequency stability," *IEEE Trans. Instrum. Meas.,* vol. IM-20, pp. 105–120, May 1971.

[4] J. Rutman, "Characterization of frequency stability: A transfer function approach and its application to measurements via filtering of phase noise," *IEEE Trans. Instrum. Meas.,* vol. IM-23, pp. 40–48, Mar. 1974.

[5] A. Papoulis, "Limits on bandlimited signals," *Proc. IEEE,* vol. 55, pp. 1677–1686, Oct. 1967.

Active Filters for Measuring Time-Domain Frequency Stability

LIAO YOULIN AND YU JUEBANG

***Abstract*—The instrumentation for measuring the filter variance is simple and low cost. An assembly of the test set known as PCC-1 for evaluating the time-domain frequency stability of signal generators was developed in the Chengdu Institute of Radio Engineering. It also includes specially designed active filters for measuring the filter variance. Experimental results are presented which show that the high-pass filter variance is a good estimation of the Allan variance.**

I. INTRODUCTION

THE ALLAN variance has been recognized internationally for a long time as a commonly used specification for evaluating the time-domain frequency stability of signal generators [1], [2]. When using two electronic counters to realize the Allan sampling (i.e., two adjacent zero deadtime sampling), a subordinate controlling circuitry is necessary to alternatively switch the two counters. The recording and processing of the experimental data are rather tedious. One may use, of course, a computing counter (e.g., HP model 5360A [3]) to perform these measurements automatically. However, because of its high cost, the computing counter is not yet universal equipment beyond metrology centers or large laboratories. On the other hand, the designing and manufacturing of high-stability oscillators and other low-noise circuits (such as amplifiers, multipliers, dividers, mixers, etc.) make the measurement of frequency stability or phase noise a common practice. Therefore, it is necessary to develop a simpler and cheaper technique for measuring the frequency stability of signal generators. In this respect, Rutman has proposed a filter variance as a useful characterization of the time-domain frequency stability [4]. If the oscillator's noise behavior conforms to the power law model's spectral density, there is one-to-one correspondence between the filter variance and the Allan variance.

The filter variance may be measured with the well-known phase-noise demodulation technique, and the required test set contains only comparatively simple and low-cost measurement apparatus. In this case, the data processing procedure is rather straightforward. In the measurement of filter variance, it is desirable to design a filter according to given specifications derived from the original definition. For instance, when the bandpass variance $\sigma^2_{\Pi B_2}$ is to be measured, one should design a second-order Butterworth bandpass filter with an equivalent

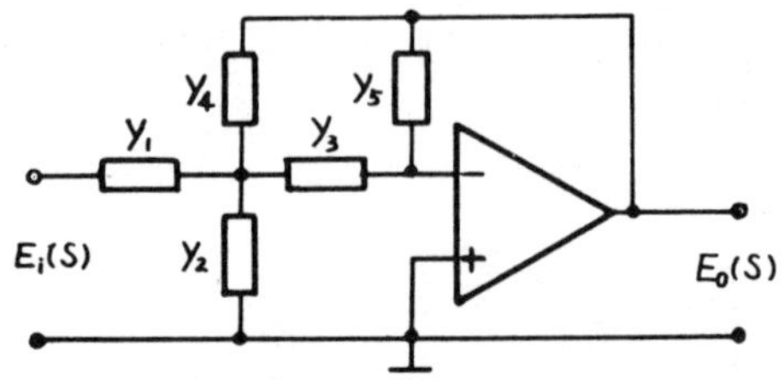

Fig. 1. Multipath feedback active filter.

figure of merit, $Q = 1$. When this is done, the comparative experimental results among the filter variance, Allan variance, and the phase-noise spectral density can be checked in terms of available data [4]. Based on this consideration, we have designed two types of adjustable active filters for measuring the spectral purity of signal generators. The design details, the derivation of the design equations, and the range-switching considerations are described later in this paper. We have measured two types of filter variance: the high-pass variance and the bandpass variance with respect to a 100-MHz crystal oscillator. Using the same oscillator, the Allan variance was also measured. Based on experimental results we concluded that the filter variance is a good estimation of the Allan variance and can serve as a useful characterization of the time-domain frequency stability of signal generators.

II. THE DESIGN OF ADJUSTABLE ACTIVE FILTERS

To make the circuit simple and easy to adjust, we chose the scheme of the multipath feedback and infinitive gain model shown in Fig. 1, for both the high-pass and bandpass active filters.

The circuit shown in Fig. 1 consists of an *RC* network and an operational amplifier (OP), upon which we imposed conventional assumptions in our theoretical analysis. Under these assumptions we can write the transfer function as follows:

$$H(s) = \frac{E_0(s)}{E_i(s)} = \frac{\begin{vmatrix} y_1 & y_1 + y_2 + y_3 + y_4 \\ 0 & y_3 \end{vmatrix}}{\begin{vmatrix} y_1 + y_2 + y_3 + y_4 & -y_4 \\ -y_3 & -y_5 \end{vmatrix}} \tag{1}$$

or

$$H(s) = \frac{-y_1 y_3}{y_5(y_1 + y_2 + y_3 + y_4) + y_3 y_4}. \tag{2}$$

Following the synthesis procedure suggested by Bridgman and Brennen [5], we rewrite (2) by replacing its admittances

Manuscript received September 25, 1979; revised November 6, 1979 and January 26, 1980.

The authors are with the Chengdu Institute of Radio Engineering (CIRE), Chengdu, Sichuan, People's Republic of China.

Reprinted from *IEEE Trans. Instrum. Meas.*, vol. IM-29, pp. 89–93, June 1980.

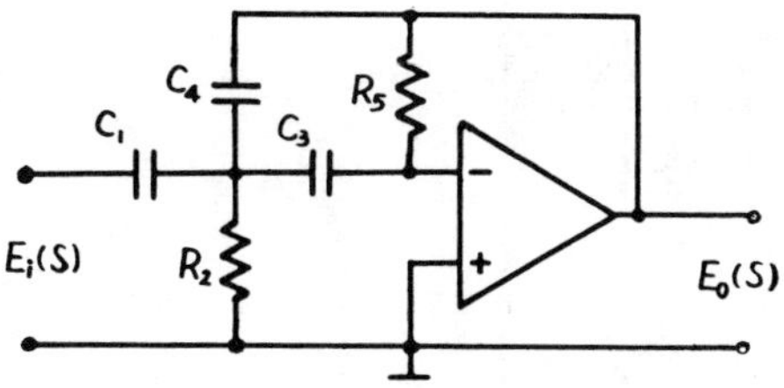

Fig. 2. Circuit of the high-pass filter.

by the corresponding impedances as follows:

$$H(s) = \frac{-z_2 z_4 z_5}{z_1 z_2 z_3 + z_2 z_3 z_4 + z_1 z_2 z_4 + z_1 z_3 z_4 + z_1 z_2 z_5}. \tag{3}$$

A. Active High-Pass Filter

The circuit of the high-pass filter is shown in Fig. 2.

Substituting the corresponding impedances of the components shown in Fig. 2 into (3), we have

$$H(s) = \frac{-s^2 C_1 C_3 R_2 R_5}{s^2 R_2 R_5 C_3 C_4 + S R_2 (C_1 + C_3 + C_4) + 1}. \tag{4}$$

On the other hand, according to the definition of the high-pass filter variance given by Rutman [4], the corresponding filter should be the second-order Butterworth type, whose transfer function can be written as

$$H(s) = \frac{K S^2}{S^2 + \frac{\omega c}{Q} S + \omega_c^2} \tag{5}$$

or

$$H(s) = \frac{\frac{K}{\omega_c^2} S^2}{\frac{1}{\omega_c^2} S^2 + \frac{1}{\omega_c Q} S + 1} \tag{6}$$

where K is the maximum transmission coefficient, ω_c the cutoff angular frequency, and Q the selective factor.

By comparing the coefficients of (4) with that of (6), we obtain

$$\left.\begin{aligned} C_1 C_3 R_2 R_5 &= -K/\omega_c^2 \\ R_2 R_5 C_3 C_4 &= 1/\omega_c^2 \\ R_2(C_1 + C_3 + C_4) &= 1/(\omega_c Q). \end{aligned}\right\} \tag{7}$$

Let $C_3 = C_4$, $K' = -K$, and from (7) we can get the formulas for calculating the components of the second-order Butterworth high-pass filter as follows:

$$\left.\begin{aligned} C_3 &= C_4 \\ C_1 &= K' C_3 \\ R_2 &= \frac{1}{(2 + K')\omega_c Q C_3} \\ R_5 &= \frac{1}{\omega_c^2 C_3^2 R_2}. \end{aligned}\right\} \tag{8}$$

With respect to the values of K', Q, and $\omega_c = 2\pi f_c$, it is known [4] that for the prototype second-order Butterworth high-pass filter used for measuring the filter variance

$$\left.\begin{aligned} K' &= 1 \\ Q &= 1/\sqrt{2} \\ f_c &= 1/\pi\tau \end{aligned}\right\} \tag{9}$$

and for the modified-type

$$\left.\begin{aligned} K' &= 0.6125 \\ Q &= 1/\sqrt{2} \\ f_c &= 1/4\tau \end{aligned}\right\} \tag{10}$$

where τ is the equivalent sampling time.

Using 1:3 division of τ from 1 s to 10 μs (e.g., 1 s and 0.33 s, 0.1 s, and 33 ms, etc.), the values of f_c for the prototype and the modified-type filters calculated from (9) and (10) are 0.3183 Hz, 1.067 Hz, 3.183 Hz, $\cdots$, 31.83 kHz, and 0.25 Hz, 0.833 Hz, 2.5 Hz, $\cdots$, 25 kHz, respectively.

All component values in (8)–(10) can be calculated by assigning an appropriate values to C_3.

The circuit of the high-pass filter is shown in Fig. 3. The use of a five-pole eleven throw switch in this circuit is self-evident. The OP is a BG305 type (made in China) whose performance is somewhat like that of μA741. The values of the circuit component are labeled beside each component, the upper row corresponds to the prototype, and the lower row corresponds to the modified-type filter.

The theoretical and experimental frequency responses of the modified-type high-pass filter are plotted in Fig. 4, in which $\omega_n = \omega/\omega_c$. The experimental curves are measured with respect to the switch positions 2, 7, and 10. From Fig. 4, it is clear that the maximum error of ω_c and K' are both less than ± 5 percent.

B. Active Bandpass Filter

The circuit of the bandpass filter is shown in Fig. 5. Substituting the corresponding impedances of the components shown in Fig. 5 into (3), we have

$$H(s) = \frac{-S R_2 R_5 C_3}{S^2 R_1 R_2 R_5 C_3 C_4 + S R_1 R_2 (C_3 + C_4) + (R_1 + R_2)} \tag{11}$$

or

$$H(S) = \frac{-\frac{R_2 R_5 C_3}{R_1 + R_2} S}{\frac{R_1 R_2 R_5 C_3 C_4}{R_1 + R_2} S^2 + \frac{R_1 R_2 (C_3 + C_4)}{R_1 + R_2} S + 1}. \tag{12}$$

On the other hand, to measure the bandpass filter variance defined by Rutman [4], the filter now discussed should be the second-order Butterworth type, whose transfer function can be written as

$$H(S) = \frac{K \omega_0 S}{S^2 + \frac{\omega_0 S}{Q} + \omega_0^2} \tag{13}$$

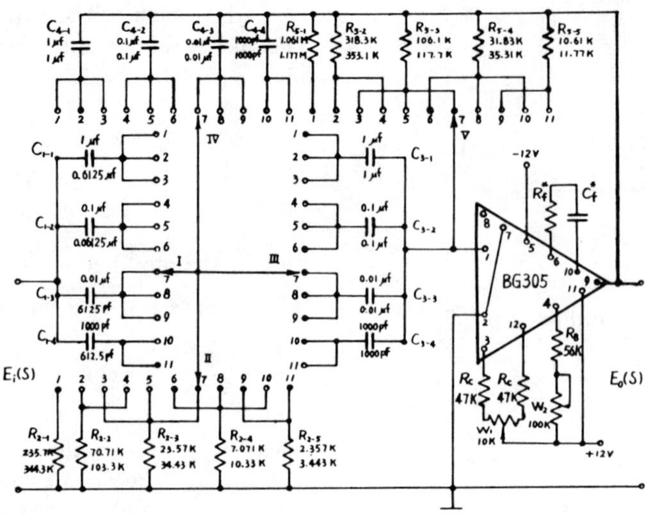

Fig. 3. Circuitry detail of the high-pass filter.

or

$$H(S) = \frac{\dfrac{K}{\omega_0} S}{\dfrac{S^2}{\omega_0^2} + \dfrac{S}{\omega_0 Q} + 1} \tag{14}$$

where K is the maximum transmission coefficient, ω_0 is the resonant angular frequency, and Q is the selective factor.

By comparing the coefficients of (12) with those of (14), we obtain

$$\left.\begin{aligned} \frac{R_2 R_5 C_3}{R_1 + R_2} &= \frac{-K}{\omega_0} \\ \frac{R_1 R_2 R_5 C_3 C_4}{R_1 + R_2} &= \frac{1}{\omega_0^2} \\ \frac{R_1 R_2 (C_3 + C_4)}{R_1 + R_2} &= \frac{1}{\omega_0 Q}. \end{aligned}\right\} \tag{15}$$

Let $R_1 = R_2$, $C_3 = C_4$, $K' = -K$; then we can get the formulas for calculating the components of the second-order Butterworth bandpass filter as follows:

$$\left.\begin{aligned} R_1 = R_2 &= \frac{1}{Q\omega_0 C_3} \\ R_5 &= 2K'QR_1 \\ C_4 &= C_3. \end{aligned}\right\} \tag{16}$$

It is known that for the second-order Butterworth bandpass filter used for measuring the filter variance

$$\left.\begin{aligned} K' &= 1 \\ Q &= 1 \\ f_0 = \omega_0/2\pi &= \frac{1}{2\tau}. \end{aligned}\right\} \tag{17}$$

To evaluate if the experimental data agree with the theoretical results, we substitute the values of K and Q into (13),

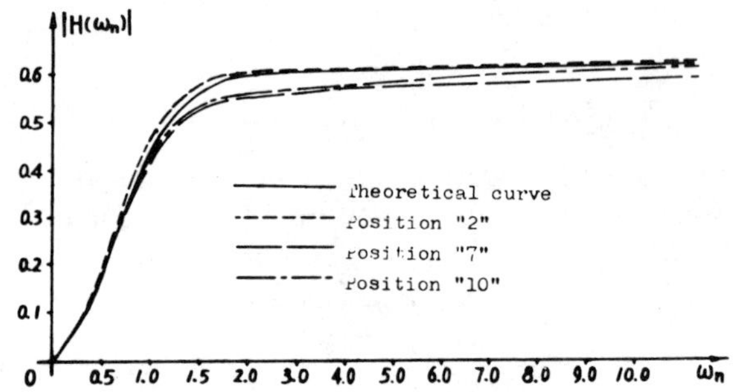

Fig. 4. Theoretical and experimental frequency responses of the modified-type high-pass filter.

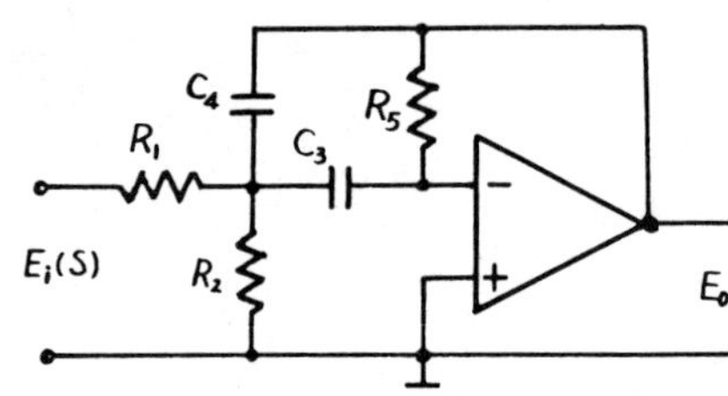

Fig. 5. Circuit of the bandpass filter.

and obtain the 3-dB angular frequencies as follows:

$$\omega_{3\text{ dB}} = \sqrt{\frac{3 \pm \sqrt{5}}{2}}\, \omega_0 \tag{18}$$

which yields the upper half-power point frequency

$$f_{3\text{ dB } u} = \sqrt{\frac{3 + \sqrt{5}}{2}}\, f_0 = 1.618 f_0. \tag{19}$$

and low half-power point frequency

$$f_{3\text{ dB } l} = \sqrt{\frac{3 - \sqrt{5}}{2}}\, f_0 = 0.618 f_0. \tag{20}$$

The difference combination of (19) and (20) gives the bandwidth $BW = f_{3\text{ dB } u} - f_{3\text{ dB } l} = f_0$, which is as expected since $Q = 1$. Using the 1:3 division of τ again from 10 μs to 1 s, it is easy to calculate, according to (17)–(20), the values

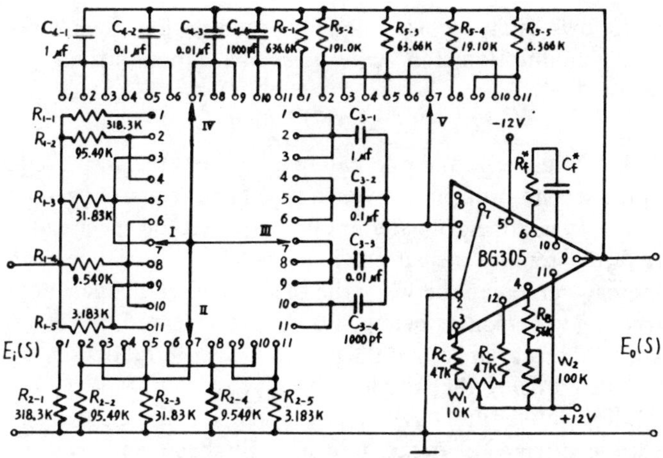

Fig. 6. Circuitry detail of the bandpass filter.

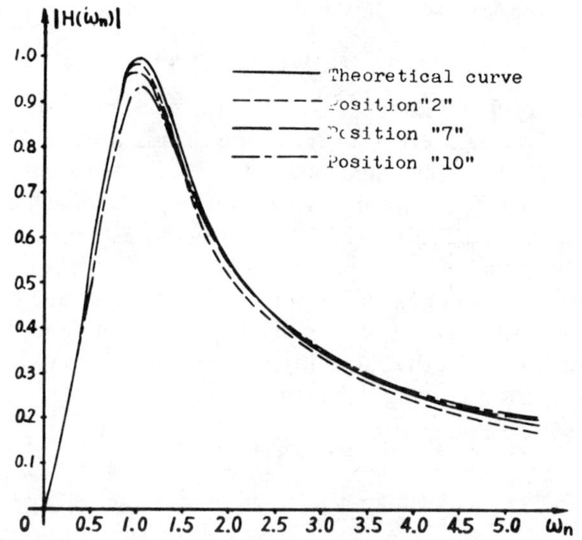

Fig. 7. Theoretical and experimental frequency responses of the bandpass filter.

of $f_0 = BW = 0.5$ Hz, 1.667 Hz, 5 Hz, $\cdots$, 50 kHz; $f_{3\ \mathrm{dB}\ u} = 0.809$ Hz, 2.697 Hz, 8.09 Hz, $\cdots$, 80.9 kHz; $f_{3\ \mathrm{dB}\ l} = 0.309$ Hz, 1.031 Hz, 3.09 Hz, $\cdots$, 30.9 kHz, respectively.

All component values in (16) and (17) can be calculated by assigning an appropriate value to C_3.

The circuit detail of the bandpass filter and the component values are shown in Fig. 6.

The theoretical and experimental frequency responses of the bandpass filter are plotted in Fig. 7, in which $\omega_n = \omega/\omega_0$. The experimental curves are measured with respect to the switch positions 2, 7, and 10. From Fig. 7 it is clear that both the maximum error of f_0 and BW are less than ±5 percent and that the maximum error of K' is less than ±10 percent.

III. The Application of Active Filters Used for Measuring the Frequency Stability and Some Experimental Results

The block diagram of PCC-1 is shown in Fig. 8. The signal sources under test are two identical 100-MHz crystal oscillators.

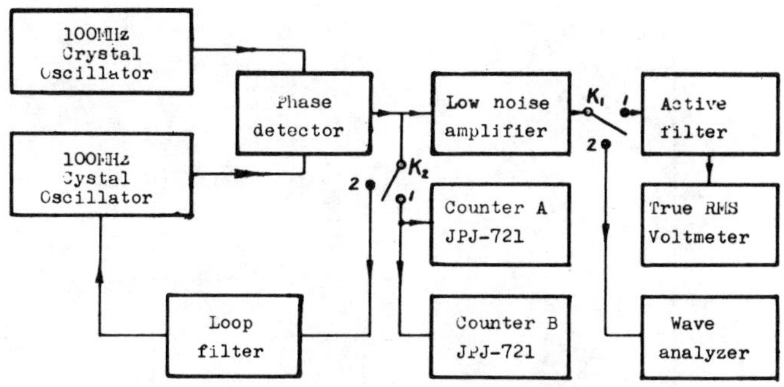

Fig. 8. Block diagram of PCC-1.

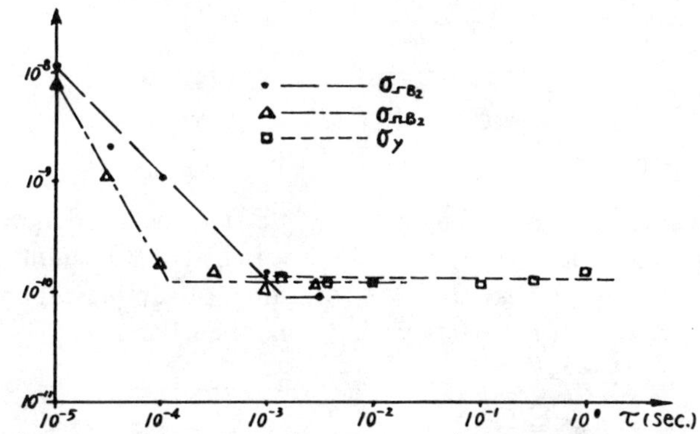

Fig. 9. Experimental data of frequency stability measurement of 100-MHz crystal oscillator using PCC-1 test set. $\sigma_{\int B_2}$, $\sigma_{\sqcap B_2}$, and σ_y are standard deviations calculated, respectively, from high-pass filter, bandpass filter, and Allan variances.

When the filter variance is to be measured, the switch K_1 is set to position 1, and the switch K_2 is set to position 2, the output voltage of the phase detector is proportional to the phase fluctuation of its input signal when the phase control system is kept in a very loose locking state [6], [7]. This output voltage is amplified by the low-noise amplifier, then filtered by the high-pass or bandpass filter. The readings of the true rms voltage meter V_{rms} give the corresponding square root of the high-pass variance $\sigma_{\int B_2}$, or bandpass variance $\sigma_{\sqcap B_2}$, according to the following expression: [4]

$$\sigma_F(\tau) = \frac{\sqrt{8}\, V_{rms}}{2\pi K K_m K_y \nu_c \tau} \tag{21}$$

where $\sigma_F = \sigma_{\int B_2}$ or $\sigma_{\sqcap B_2}$, K is phase-detecting sensitivity in volts per radian of the phase detector, K_m is voltage gain of the low-noise amplifier, K_y is transmission coefficient of the active filter, ν_0 is the average frequency of the crystal oscillator under test, and τ is the sampling time.

The experimental data of $\sigma_{\int B_2}$ and $\sigma_{\sqcap B_2}$ versus τ are plotted in Fig. 9. Due to the limitation of the rms voltmeter's low and high cutoff frequency, only the filter variances for τ between the range of 10 μs to a few milliseconds were measured.

When the Allan variance is to be measured, the switch K_2 is set to position 1, and the phase detector, whose structures is in the form of a doubly balanced mixer, now operates as a mixer. The reference 100-MHz crystal oscillator whose frequency is offset from the test oscillator, e.g., 1 kHz, is fed to the two digital electronic counters attached to each other by

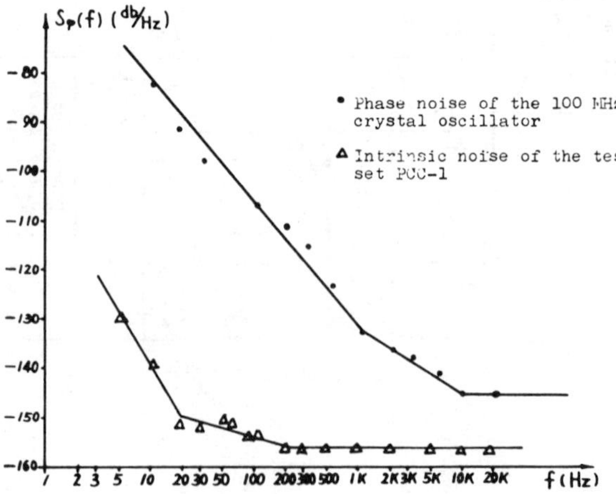

Fig. 10. Experimental curve of $S_\phi(f)$.

a subordinate controlling circuitry. Let T_{Ai} and T_{Bi} denote the period readings displayed alternatively by the two counters in a sampling process, and N the number of samples; then the square root of the Allan variance can be written as [7]

$$\sigma_y = \frac{1}{\sqrt{2T_p^2\nu_0}}\frac{1}{N}\sqrt{\sum_{i=1}^{N}(T_{Ai}-T_{Bi})^2} \tag{22}$$

where T_p is the average period of the offset signal and ν_0 is the average carrier frequency of the 100-MHz crystal oscillator. In our experiment we took $N = 100$. The experimental curve of σ_y versus τ is also plotted in Fig. 9. It is clear from this figure that there is good agreement between the Allan variance and the filter variance when τ is in the milliseconds order. Furthermore, when τ varies from milliseconds to hundreds of milliseconds, σ_y is constant to the order of $(2\text{-}3) \times 10^{-10}$. This fact implies that $\sigma_y^2(\tau) \propto \tau^0$ or, in other words, in a certain range of the phase-noise spectral density of the crystal oscillator under test there exists a relationship of $s(f) \propto f^{-3}$, [4] which is also confirmed by the spectral density measurement.

When the phase-noise spectral density $S_\phi(f)$ is to be measured, the operating principle is the same as that of measuring the filter variance. In this case the switch K_2 is set to position 2, and K_1 to position 2. The output voltage of the low-noise amplifier, which is proportional to the phase fluctuation of the input signal, is fed into the wave analyzer for spectral density measurement. Fig. 10 gives the experimental results in which $S_\phi(f)$ is plotted versus Fourier frequency f. From Fig. 10 we can identify the noise models of the 100-MHz crystal oscillator as follows: flicker-noise frequency modulation, flicker-noise phase modulation, and white-noise phase modulation.

IV. Conclusion

Two high-pass and one bandpass adjustable active filters of the second-order Butterworth type have been designed with conventional operational amplifiers to meet the specifications required for the measurement of the filter variance. Satisfactory agreement between the theoretical and experimental frequency responses below 200 kHz is observed. Filter variance, Allan variance, and the phase-noise spectral density were measured using the simple and low cost test set PCC-1, including active filters as one of its assembly. Experimental results confirmed that the high-pass filter variance is a good estimation of the Allan variance. As Rutman [4] has pointed out the bandpass filter variance can be used to distinguish the white phase noise from the flicker phase noise.

Acknowledgment

The authors wish to thank Z. Shi-ji of the Chengdu Institute of Radio Engineering for reviewing the manuscript, encouraging study of this topic, and helpful comments and discussions, and W. Yu-ming, Z. Chao-wei, W. Zhu-hong, X. Cheng-xi, and C. Jie-mei for developing the test set PCC-1 and for their use of the experimental results on measuring the phase-noise spectral density of the 100-MHz crystal oscillator, as well as Dr. J. Rutman for helpful discussions, and Prof. F. Binquan of South China Institute of Technology, Canton, China, for reviewing this manuscript.

References

[1] D. W. Allan, "Statistics of atomic frequency standards," *Proc. IEEE*, vol. 54, pp. 221–230, Feb. 1966.

[2] J. A. Barnes *et al.*, "Characterization of frequency stability," *IEEE Trans. Instrum. Meas.*, vol. IM-20, pp. 105–120, May, 1971.

[3] D. Martin, "Frequency stability measurements by computing counter system," Hewlett-Packard J., vol. 11, pp. 9–14, Nov. 1971.

[4] J. Rutman, "Characterization of frequency stability: A transfer function approach and its application to measurements via filtering of phase noise," *IEEE Trans. Instrum. Meas.*, vol. IM-23, pp. 40–48, Mar. 1974.

[5] A. Bridgman, R. Brennen, "Simulation of transfer functions using only one operational amplifier," in *IRE WESCON Conv. Rec.*, pt. 4, pp. 273–278, 1957.

[6] J. Rutman, "Measurement of frequency stability in time and frequency domain via filtering of phase noise," *IEEE Trans. Instrum. Meas.*, vol. IM-23, pp. 515–518, Dec. 1974.

[7] J. Groslambert *et al.*, "Méthodes expérimentales de caractérisation des oscillateurs," *Onde Élec.*, vol. 55, no. 2, pp. 74–81, Feb. 1975.

[8] Z. Shi-ji, "Characterization and measurement of frequency stability," published as Lecture Notes in the Chengdu Institute of Radio Engineering, Chengdu, Sichuan, China, July, 1975.

PHASE NOISE OF VARIOUS OSCILLATORS AT VERY LOW FOURIER FREQUENCIES

D. Babitch and J. Oliverio
Hewlett-Packard Company
Santa Clara, California

Summary

Measurements of very low Fourier frequency (.003 to 100 Hz) phase noise of various high stability oscillators are reported. The objective of this experiment is to provide a check of the theoretical relations between time domain and frequency domain spectra for some real oscillators, particularly in the very low Fourier frequency region where no frequency domain measurements have been made. These measurements were made directly with a double multiplier chain and Fourier Analyzer. The results are then mathematically converted to time domain and compared with direct time domain stability measurements as an experimental check of theory. To our knowledge, these are the first phase noise measurements of high stability oscillators at such low Fourier frequencies.

Key Words

Frequency stability, phase noise, Fourier analysis, flicker noise, atomic clock, crystal oscillator.

Introduction

The oscillators measured include a quartz crystal oscillator, two types of cesium beam atomic clocks and a rubidium vapor frequency reference. In each case the reference oscillator was the same type as the unit under measurement. A Hewlett-Packard 5451B Fourier Analyzer with a Band Selectable Fourier Analysis (also known as zoom) transform is used to analyze and display the phase noise spectra of a low frequency beat of two frequency multiplied sources. The BSFA transform allows analysis in a selectable bandwidth about a selectable center frequency, rather than the usual analysis of Fourier frequencies from zero to some f max. This has several advantages: non-zero beat frequency eliminates the requirement of phase locking the unknown to the reference; the resolution is increased by increasing absolute resolution limit, reducing processor gain; increased dynamic range and resistance to out-of-band signals also result. The Fourier Analyzer allows real or near real time operation without which the analysis of very low rates may be impractical. Of course, averaging of several measurements is still required to reduce the variance of the measured noise levels; i.e., to get a precise measurement.

Frequency Domain Measurements

Test Setup

A block diagram of the system used for frequency domain measurements is shown in Figure 1. This type of system has been described in several publications.[1,2] As a calibration check of the Fourier Analyzer, the signal-to-phase noise ratio measurement for the crystal oscillator was duplicated with a conventional wave analyzer in the region 10 to 100 Hz. The signal-to-noise ratios measured with these different instruments agree within 2 dB which is within the error bracket for the 10 average measurement chosen in the Fourier Analyzer for that frequency range.

The Fourier Analyzer

Hardware and Software

The 5451B Fourier Analyzer is used to solve problems in the areas of electronics, filter design, servo testing, communications and mechanical vibration, to name a few.

The system (Figure 2) includes an HP 2100S Microprogrammable Systems Computer, two-channel Analog-to-Digital Converter, keyboard control unit, CRT display, Teleprinter for hard copy output and a plotter.

Figure 3 shows a block diagram of the system.

The ADC converts analog data to digital data by sampling the analog signals at discrete points, then converting each sample into a digital word. Controls on the front panel allow the operator to set sampling rate, number of inputs, attenuation, etc. An internal quartz crystal clock (accuracy ±0.01%) provides the sampling pulses at a rate of twice the selected maximum analysis frequency.

The user can select single or dual channel input. Each channel is simultaneously sampled for cross measurements; e.g., crosscorrelation.

The hardware operation is based on a pushbutton keyboard with function related nomenclature on the keys. The user pushes a short sequence of keys to collect and analyze data, or control other system operations, such as power spectrum, correlation, convolution, histogram and transfer function (see Figure 4). Pushbutton key sequences can be stored and automatically executed any number of times. Programs written by the user can be incorporated into the system and executed from the keyboard. Programs and data are output to the teleprinter for hardcopy.

A CRT display and display control unit provide a means of viewing signals both during acquisition and after processing. Signals or measurement results can be viewed in either the time or frequency domain, linear or log coordinates, and rectangular or polar modes; i.e., real and imaginary, or magnitude and phase components. The display is calibrated with vertical calibration shown digitally on the display control unit.

The system uses an HP 2100S Microprogrammable Systems Computer as the central processor. The processor includes extended arithmetic instructions, 16K memory, dual channel direct memory access, floating point hardware and writable control store.

The system software executive makes extensive use of microprogramming to speed up the execution times of the various routines.

Fourier Transform

The Fourier Analyzer utilizes a digital computer to calculate Fourier transforms of time-varying voltage signals. The Fourier Transform of a time-varying function x(t) is defined as:

Reprinted with permission from *Proc. 28th Annu. Symp. Frequency Contr.*, Atlantic City, NJ, May 29–31, 1974, pp. 150–159.

$$X(f) = \int_{-\infty}^{\infty} x(t)\, e^{-i2\pi ft}\, dt \qquad \text{(Forward Transform)} \quad (1)$$

where $e^{\pm 2\pi ft} = \cos(2\pi ft) \pm i \sin(2\pi ft)$.

X(f) is the Fourier Transform of x(t) and contains the amplitude and phase information at every frequency present in x(t).

Discrete Finite Transform (DFT)

In order to implement the Fourier Transform digitally, the ADC first converts the continuous input signal into a series of discrete data samples. This is accomplished by sampling the input waveform, x(t), at discrete intervals of time, Δt. As a result, Eq. 1 is modified to

$$X'(f) = \Delta t \sum_{n=-\infty}^{n=+\infty} x(n\Delta t)\, e^{-i2\pi fn\Delta t} \qquad (2)$$

where x(nΔt) are the measured values of the input function.

Equation (2), even though dealing with a sampled version of x(t), still calculates a valid Fourier Transform. However, the Fourier Transform as calculated by (2) no longer contains accurate magnitude and phase information at all of the frequencies contained in X(f). Rather, X(f) accurately describes the linear spectrum of x(t) up to some maximum frequency (F_{max}) which is dependent upon the sample spacing, Δt. The determination of F_{max}, from the Nyquist criteria, is $F_{max} = 1/2\Delta t$.

In order to calculate X'(f), an infinite number of samples of the input waveform must be taken. As each sample must be separated by a finite amount of time, one would have to wait forever for the calculation of X'(f) to be completed. Clearly the observation time must be limited. Assuming that the input signal is "observed" (sampled) from some zero time reference to time T seconds, we have

$$T/\Delta t = N \qquad (3)$$

where N is the number of samples, and T is the "time window" (total observation time).

Restricting the observation time to T seconds is equivalent to truncating equation (2). Since there are no longer an infinite number of time points, we cannot expect to calculate magnitude and phase values at an infinite number of frequencies between zero Hz and F_{max}. Equivalently, the truncated version of equation (2) does not produce a continuous spectrum. This discrete finite transform (DFT) is given by Eq. (4).

$$X'(m\Delta f) = \Delta t \sum_{n=0}^{N-1} x(n\Delta t)\, e^{-i2\pi m\Delta fn\Delta t} \quad \text{volts/Hz} \qquad (4)$$

Only periodic functions have such a discrete frequency spectra. Therefore, equation (4) requires that the input function be periodic with period T. Conversely, equation (4) assumes that the function observed between zero and T seconds repeats itself with period T for all time. This assumption is made whether or not x(t) is actually periodic. The discrete finite transform, as calculated by Eq. (4), is actually a sampled Fourier Series.

Fast Fourier Transform (FFT)

The Fast Fourier Transform is a collection of algorithms developed by Cooley and Tookey which make use of the symmetrical nature of the DFT matrix in order to simplify and reduce the number of calculations (multiplies and adds) required. To fully describe a frequency in the spectrum, two values must be calculated (i.e., the magnitude and the phase, or the real and imaginary part at the given frequency). As a result, N points in the time domain allow only N/2 complex quantities in the frequency domain.

If F_{max} is the maximum frequency present in the spectrum, then

$$\Delta f = \frac{2F_{max}}{N} \qquad (5)$$

where Δf is the separation of frequencies (referred to as resolution) in the frequency domain.

Δf is also related to the observation time, T, by

$$\Delta f = \frac{1}{T} \qquad (6)$$

Since x(t) is assumed to be a real-valued function the linear transform of x(t), X(f) is Hermitian and only the positive half of the spectrum is displayed on the analyzer. If A_n denotes the peak amplitude of an input sinusoid f_n in volts, the DFT is implemented such that the amplitude calculated for this f_n is $A_n/2$ volts. Thus, the amplitudes calculated by the Fourier Analyzer must be multiplied by a factor of 2 in order to display peak amplitudes.

Power Spectrum

The power spectrum of the function x(t) is obtained by multiplying the Fourier Spectrum X(f) by its complex conjugate, X*(f), and normalizing by 1/T the measurement time. The resultant function, called $S_x(f)$ is real and positive and has the units volts²/Hz.

$$S_x(f) = \frac{X(f)X^*(f)}{T} \qquad (7)$$

If A_n represents the peak amplitude of an input sinusoid, as above, the spectrum $S_x(f)$ as calculated by the analyzer has the value $A_n{}^2/4$ volts squared. $S_x(f)$ must be multiplied by 4 to display peak amplitude. The analyzer calculates the log amplitude of the spectrum in dBv where

$$\text{dBv} = 10 \log \frac{S_x(f)}{1\ \text{volt}^2} \qquad (8)$$

Averaging

Equation (7) neglects the effect of noise on the input signal x(t). If $X_n(f)$ represents the spectrum of random, independent noise on the signal, then Equation (7) becomes,

$$S_x'(f) = S_x(f)+S_n(f)+X(f)X_n^*(f)+X_n(f)X^*(f) \qquad (9)$$

The last two terms in equation (9) reduce to zero as the number of averages increase, and can be neglected. $S_n(f)>0$ and ensemble averaging must be done to allow precise measurement of its value.

Corrections to the DFT

The Discrete Finite Transform assumes the sampled signal is bandlimited with a bandwidth of F_{max} and that the signal is periodic in the observation time, T. Generally this is not the case and two errors occur in the frequency domain representation of the signal. They are frequency aliasing and amplitude leakage.

Aliasing

Frequency aliasing is caused by frequencies in the signal that are higher than F_{max} "folding over" and appearing as "aliased" lower frequencies in the spectrum. This happens in exactly the same way that a conventional mixer will give the same output frequency for inputs equally above or below the local oscillator rate and its harmonics. Once the signal has been sampled, aliasing cannot be corrected. The usual solution is to employ low-pass "anti-aliasing" filters in the signal path between the source and the ADC. The filter is set such that it attenuates frequencies above some cut-off frequency which is less than F_{max}. Frequencies that do alias are sufficiently reduced in amplitude such that their contribution to the spectrum is negligible.

Leakage

The DFT of an input sinusoid, f_n, which does not have an integral number of periods in the time domain will appear at more than one frequency (although the actual location of f_n is uncertain by less than Δf). The amplitude of the sine wave will be reduced from its true value. This occurs because the DFT thinks it is operating on a function that looks like Figure 5a, which is clearly not a sine wave.

The sinusoid has been multiplied by the gate function of the analyzer $\pi(t)$. In the frequency domain, this is equivalent to convolving the spectrum of the sinusoid, a delta function at f_n with the spectrum of the gate function, sinc T. The area of the gate function is normalized.

$$X'(f) = \delta(f-f_n)*\text{sinc } T$$

$$\text{sinc } T \equiv \frac{\sin \pi T}{\pi T} \qquad (10)$$

The DFT of this function is shown in Figure 5b where the solid lines represent what the analyzer "sees" and the dash lines represent the actual signals.

In effect, the information, or energy, in the sinusoid has "leaked" out to adjacent frequencies and smeared the spectrum. Leakage cannot be eliminated entirely; however, one can reduce many of the leakage effects and gain accurate amplitude information at the expense of less precise frequency resolution. This is done by using a Hanning window. The Hanning window is chosen because it is easily implemented on the Fourier Analyzer keyboard. The Hanning window modifies the effective shape of the time window by multiplying the window by the function $(1/2-1/2 \cos (2\pi t/T))$. The effective window then takes on the shape of this Hanning function. The new window is shown in Figure 5c. Note that this window eliminates discontinuities at the ends of the time record.

To implement this Hanning window in the Fourier Analyzer, the Hanning window and the gated input wave are multiplied. This multiplication of two time domain functions results in convolution in the frequency domain. The convolution of the two frequency functions is illustrated in Figure 5d. The result of the convolution of the two window functions is shown in Figure 5e. We see that the main lobe of our modified window function is widened and flattened, while the side lobes are greatly reduced in amplitude. The convolution of this modified window transform and a sinusoid which is not periodic in the window results in Figure 5f.

While one Hanning attenuates the amplitude by 6.02 dB, the uncertainty (unknown amplitude error) of the measurement is greatly reduced due to the relative flatness of the main lobe of the modified window function. It is also clear that the price paid for reducing the amplitude uncertainty is degradation of frequency resolution (exactly where the frequency f_n is located is now less certain).

Each time the window is Hanned, the side lobes of its transform are attenuated by 12 dB/octave (40 dB/decade), while the main lobe of the transform is widened by $2\Delta f$. Each Hanning also has a corresponding correction factor which must be added to the results (or which the results must be multiplied by, if the dBs are converted to their equivalent power ratio) in order to display true amplitude. This factor is 6.02 dB (a multiplication factor of 2) for one Hanning. In the case of two, three and four Hannings, the narrow band correction factors (for discrete lines) are 9.52 dB (2.67), 10.10 dB (3.2) and 11.26 dB (3.66), respectively. (These are total correction factors for Hanning two, three and four times, not the incremental correction factor for each Hanning. These numbers are squared for a power spectrum.) The uncertainty then between the resulting values and the correct amplitude is as shown in Figure 6. As worst case amplitude error occurs when the sine wave is precisely in the middle of two channels, it is seen that two Hannings reduce this error to less than 1 dB. Two Hannings were used in our computations.

When the spectrum of interest contains two or more sinusoids, the increased side lobe roll off, the result of Hanning, enhances the Analyzer's ability to separate these frequencies.

The following table gives the narrowband and wideband power spectrum correction factors for one, two and three Hanns.

Window Corrections — Hanning

Number of Hannings	Amplitude Correction for Power Spectrum*: Wide Band** Signal (e.g., noise)	Narrow Band** Signal
1	8/3	4
2	128/35	64/9
3	1024/231	256/25

* A linear spectrum should be corrected by the square root of these values.
**The bandwidth is relative to Δf.

Effective Noise Bandwidth

The effective noise bandwidth of the Fourier Analyzer is equal to the resolution, Δf, of the measurement for a rectangular window. The table below gives the effective noise bandwidth corrections for one, two and three Hanns.

Number of Hanns	Equivalent Noise Bandwidth
1	1.5 Δf
2	1.92 Δf
3	2.31 Δf

Band Selectable Fourier Analysis

Band Selectable Fourier Analysis is a technique for performing Fourier-based spectrum analysis over a band of frequencies whose minimum and maximum frequency are independently selectable.

The resulting frequency resolution (Δf) is bandwidth ÷ number of transform points. Since the measurement time is the reciprocal of the frequency resolution (this is true for any measurement technique), this time increases for decreasing bandwidths at a fixed number of transform points.

BSFA results in increased resolution over a band of frequencies, reduced block sizes to achieve this increase, more efficient use of core memory, faster processing and increased processor gain. BSFA is arrived at by combining (digital) filtering with standard Fourier Analysis processing. A functional block diagram of the processing involved is shown in Figure 7. The same corrections that apply for a baseband DFT also must be applied to a Band Selectable Fourier Analysis. For an in-depth study of these matters, see Reference 3, 4, 5 and 6.

Double Multiplier Chain

This system is shown in Figure 7. Each chain consists of a 5 to 100 MHz (or 5 to 90) multiplier feeding a step recovery diode harmonic generator which multiplies the frequency by 92 (102) to 9.20 GHz (9.18). The signals are mixed down to 20 MHz, or, for 100 and 90 MHz inputs, down to 10 MHz. The 20 MHz signal mixed down from X band is rather weak and is amplified in the IF Amp before reaching the final mixer where it is beat against an HP 5100B Synthesizer down to some suitable audio rate. The 10 MHz mixed down from 100 MHz is filtered and mixed down again with the synthesizer. Thus the total multiplication is either 1840 or 20 selected by a switch. The lower multiplication is required when analyzing at Fourier frequencies less than 0.1 Hz in order to avoid carrier supression. The synthesizer was modified to make its noise adequate (less than 26 dB worse than the expected input signals) for mixing down with a multiplication of only 20. The synthesizer driver used an external reference frequency from one of the signal sources under test to insure adequate stability in the millhertz region so as not to perturb the noise of the instruments under test.

Results

The single sideband signal-to-phase-noise ratio data, $\mathcal{L}(f)$, are displayed in Figures 8-11. Several plots of raw data are shown in Figures 12-14. Ten averages of data were used in measuring the noise down to 0.1 Hz and three averages below that. This gives 1 sigma error brackets of about ±3 dB down to 0.1 Hz and ±6 dB below that, assuming the noise is approximately white within a measurement bandwidth.

As shown in Figure 14, the crystal oscillators were subjected to the vibration of a fan which produced significant sidebands. These show up in the time domain spectrum, Figure 18. The theoretical conversion from frequency domain to time domain, discussed later, neglected these lines. Note that the phase noise of the cesium standard with standard beam tube on long time constant is not as much worse as expected than the cesium standard with high performance beam tube in the range .003 to .1 Hz. This is because the standard units happened to have unusually good crystal oscillator stability (7×10^{-13} for 1 sec.) compared to the high performance standard (3×10^{-12} for 1 sec.) This situation is not typical.

The straight line approximations on each curve are used to calculate the time domain stability as explained in following sections.

Time Domain Measurements

The test setup for time domain measurements was the same as for frequency domain measurements except an HP 5360A computing counter and 5055A printer are used in place of the Fourier Analyzer. Since the computing counter has a maximum internally generated gate time of 90 sec. the data for 180, 360 and 720 sec. were computed from 90 sec. data by hand. The results are shown in Figures 15-18. The computed variance was for N=2 (the "Allan" variance) and the high frequency cutoff was 50 kHz.

Theoretical Conversion of Data from Frequency Domain to Time Domain

As a check on the overall agreement of measurements and theory, the data in frequency domain was converted to time domain using the formula [7]

$$\left\langle \sigma^2_y(2,T,\tau)\right\rangle = 2\int_0^\infty df\, S_y(f)\, \frac{\sin^2(\pi f\tau)}{(\pi f\tau)^2} \cdot \left\{1 - \frac{\sin^2(2\pi r f\tau)}{4\sin^2(\pi r f\tau)}\right\} \quad (11)$$

where T = time interval between measurements

τ = averaging time

$r = \frac{T}{\tau}$

For the computing counter the minimum dead time for counting a sufficiently high input frequency is 3.3 ms. We therefore restrict $\tau \geq .01$ so that $r \approx 1$ and for the experimental frequency domain measurement limits the formula reduces to

$$\sigma^2(2,\tau,\tau) \approx 2\int_{.003}^{100} df S_y(f)\, \frac{\sin^4(\pi f\tau)}{(\pi f\tau)^2} \quad (12)$$

except for the crystal oscillator where the lower limit is .007 Hz. It is interesting to find the limits of τ which are experimentally correct for the frequency domain limits of .002 to 100 Hz, so we have deliberately computed for $.01 < \tau < 1000$ sec. to verify where the formula will become incorrect due to an excessively narrow frequency range for $S_y(f)$.

To simplify the computations we fitted straight line segments of the form kf^α, α an integer from -1 to -4, to the (f) curves. This allows precalculated integrals[7] to be used to find $\sigma_y(\tau)$. The computations are summarized in Figure 19. The results are shown on the time domain stability curves, Figures 15-18. Agreement with experimental data is generally fair or better. The worst fit was for the very complicated spectrum of the 5061A cesium standard where it is clear that $\sigma_y(\tau)$ is very sensitive to minor variations in the amplitude of each f^α section due to the presence of large f^{-3} and f^{-4} components from .003 to 10 Hz. There are also some systematic effects on the stability of the crystal oscillator in the region 40 to 700 sec. where $\sigma_y(\tau)$ measured about 50% worse than predicted. This could be explained by temperature effects.

The question of the range of validity of $\sigma_y(\tau)$ computed from $S_y(f)$ can now be answered by observing from the data (such as Figure 16 for $\tau>200$ sec) that if the frequency limits of measured S_y are f_L to f_H then the valid limits of σ_τ will be about $\tau_L \leq 2/f_H$ to $\tau_H \geq 1/2f_L$ if $S_y \propto f^\alpha$ is extrapolated at its ends by the best fit power law spectrum.

Conclusion

It has been shown that the Fourier Analyzer is a useful tool for direct measurement of frequency domain spectra for frequencies much lower than can be measured with pure analog techniques. We have also shown that the theoretical conversion of frequency domain to time domain spectra must be done with great care, accuracy and strictly limited range of τ if the spectra are complicated or have f^α components with $|\alpha| \geq 3$ because the sidelobe supression of the weighting function $\sin^2\pi f\tau/(\pi f\tau)^2$ in the frequency domain to time domain conversion formula is insufficient to prevent sensitivity to noise much removed from $f=1/\tau$, τ the measure ment time of interest, for steep spectra or discrete sideband perturbations.

Acknowledgments

Many thanks to Webb McKinney and Pete Roth of the Digital Signal Analysis group for their patient explanations and assistance in using the HP5451B Fourier Analyzer.

References

1. Shoaf, John H., "Specification and Measurement of Frequency Stability," NBS Report 9794.

2. Cutler, L.S., and Searle, C., "Some Aspects of the Theory and Measurement of Frequency Fluctuations in Frequency Standards," IEEE Proceedings, 54, 2, Feb. 1966.

3. Bracewell, "The Fourier Transform and its Applications," McGraw-Hill, 1965.

4. Bendat and Piersol, "Measurement and Analysis of Random Data," John Wiley and Sons, 1966.

5. Gold and Rader, "Digital Processing of Signals," McGraw-Hill, 1969.

6. Carlson, "Communication Systems: An Introduction to Signals and Noise in Electrical Communication," McGraw-Hill, 1968.

7. Barnes, et.al., "Characterization of Frequency Stability," IEEE Trans. on Instrumentation and Measurement, 20, 2, May 1971.

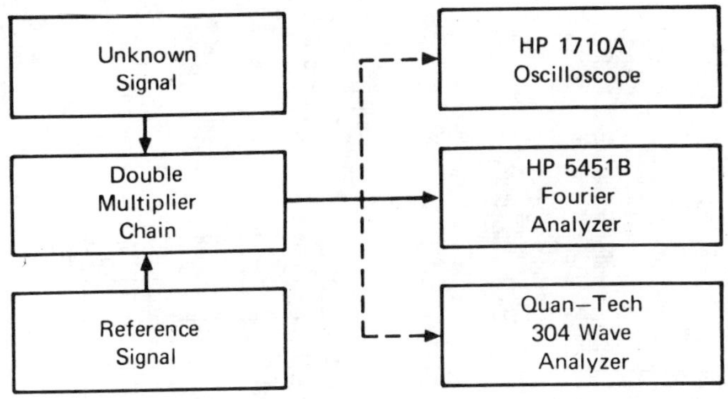

1. Frequency Domain Measurement System

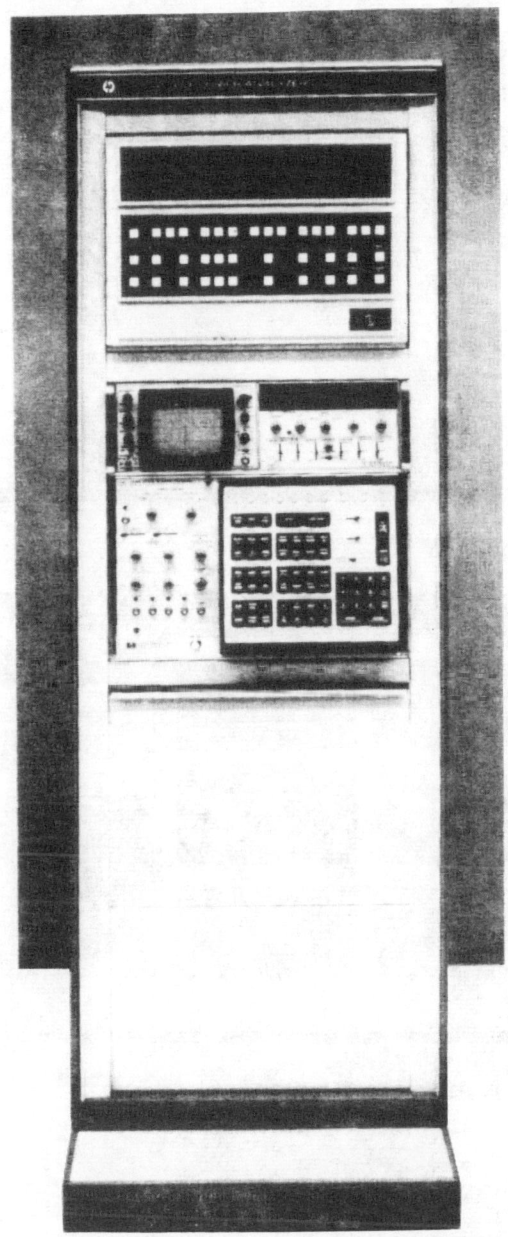

2. Fourier Analyzer

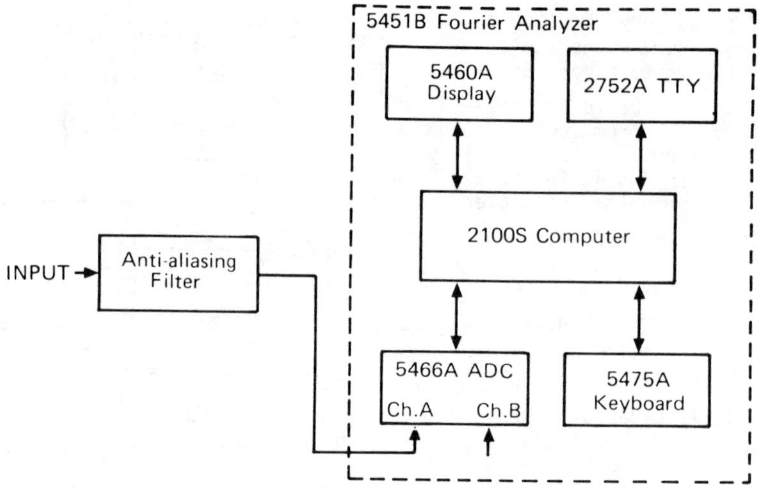

3. Fourier Analyzer Block Diagram

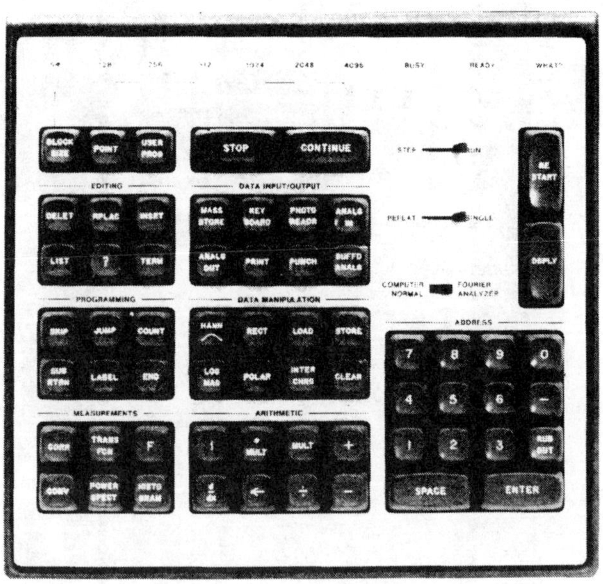

4. Fourier Analyzer Keyboard

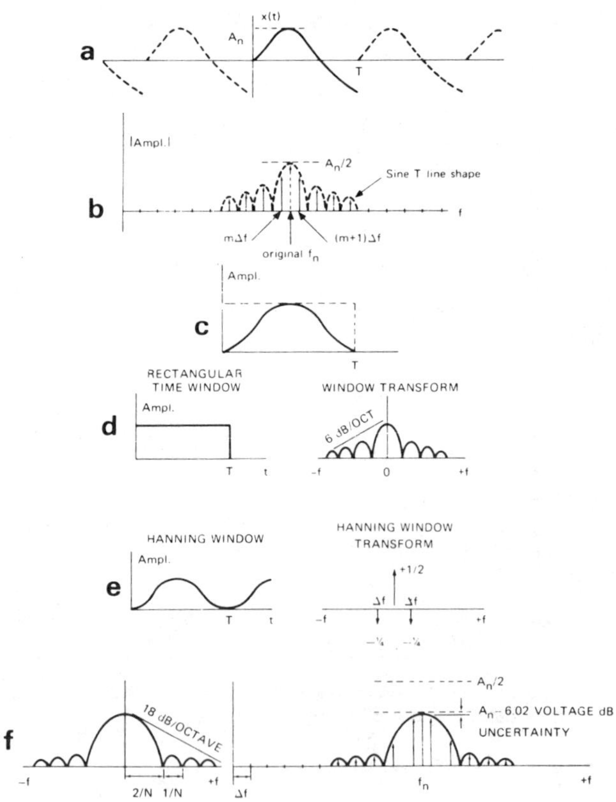

5. Leakage Error and the Hanning Window.

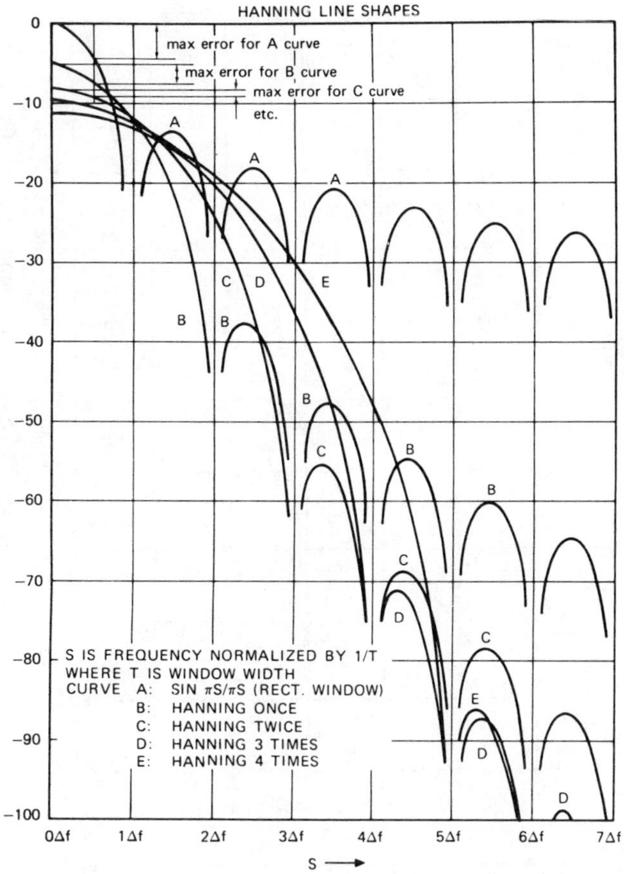

6. Hanning Line Shapes

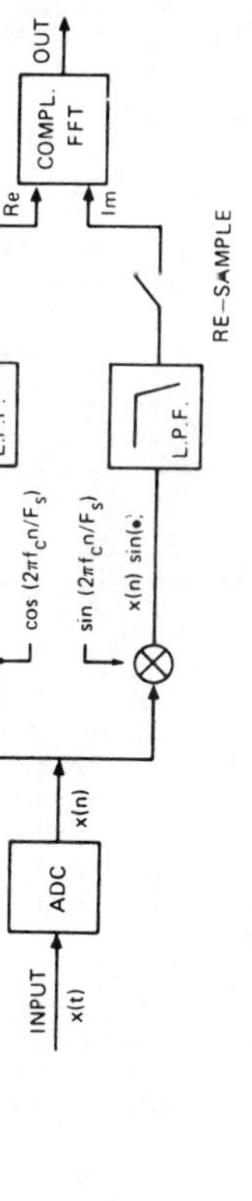

7. Band Selectable Fourier Analysis

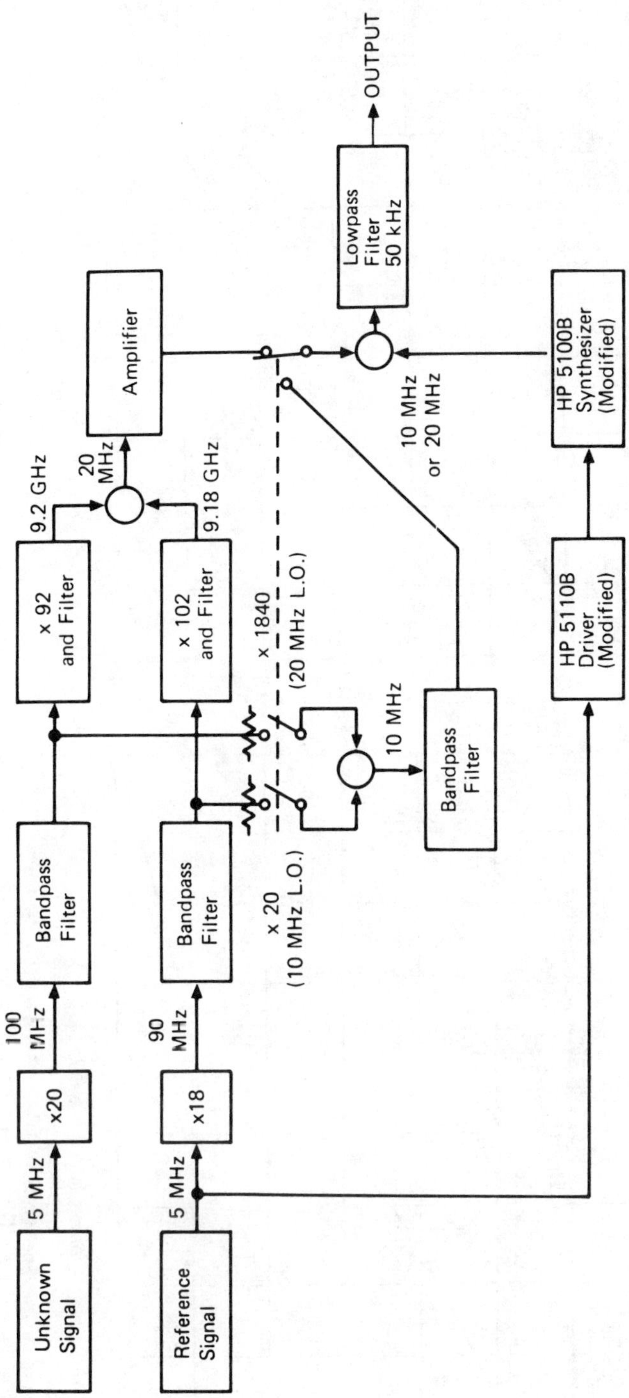

8. Double Multiplier Chain

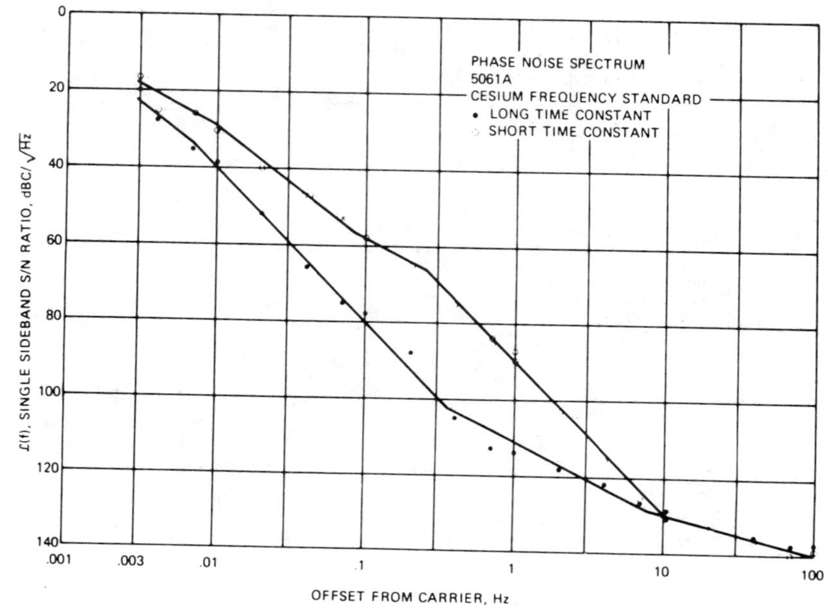

9. Phase Noise Spectrum, 5061A Cesium Frequency Standard

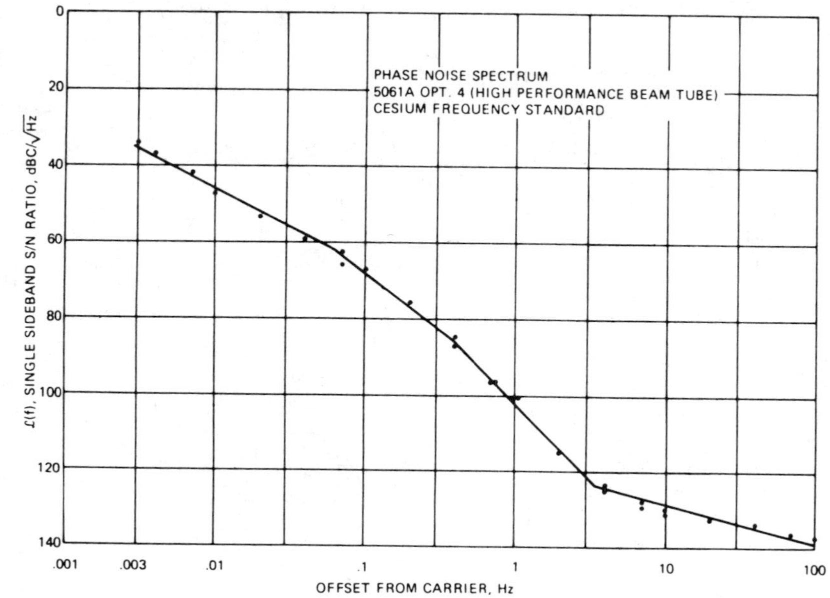

10. Phase Noise Spectrum, 5061A Opt. 4 (High Performance Beam Tube) Cesium Frequency Standard

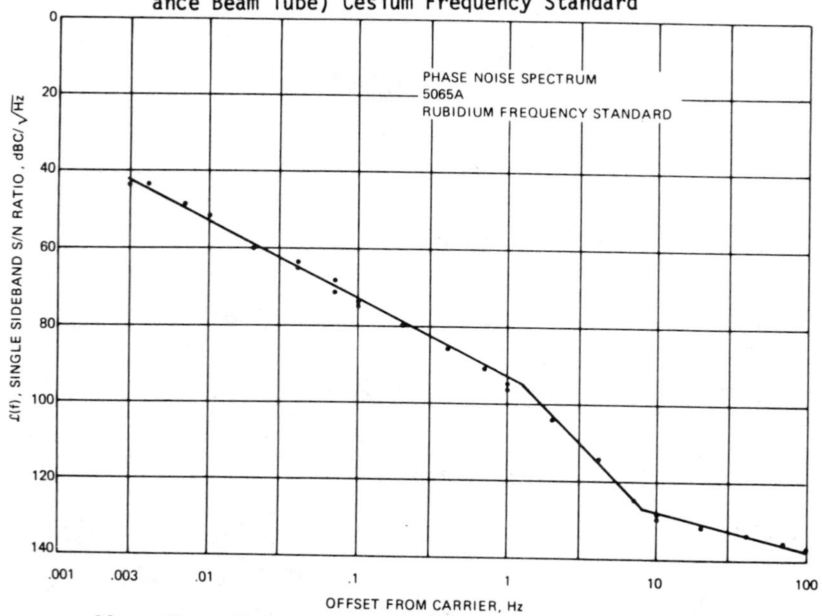

11. Phase Noise Spectrum, 5065A Rubidium Frequency Standard

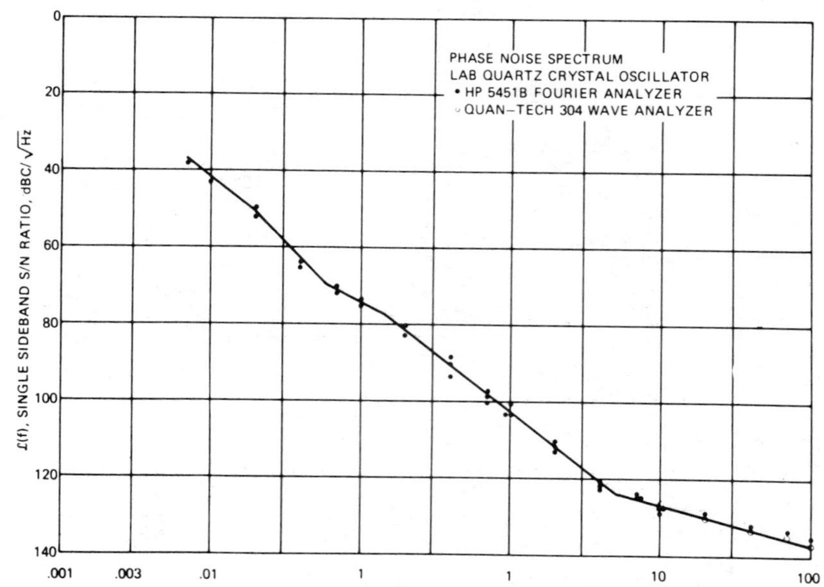

12. Phase Noise Spectrum, Lab Quartz Crystal Oscillator.

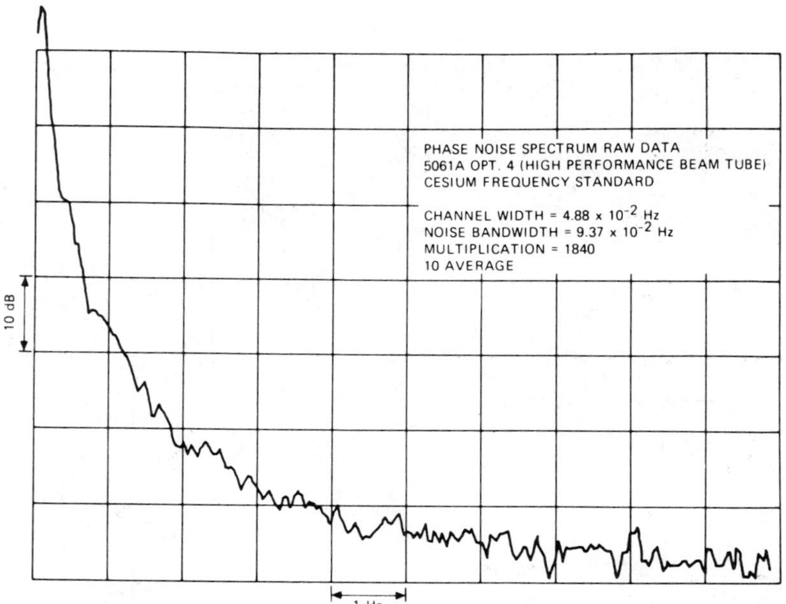

13. Phase Noise Spectrum Raw Data, 5061A Opt. 4 (High Performance Beam Tube) Cesium Frequency Standard

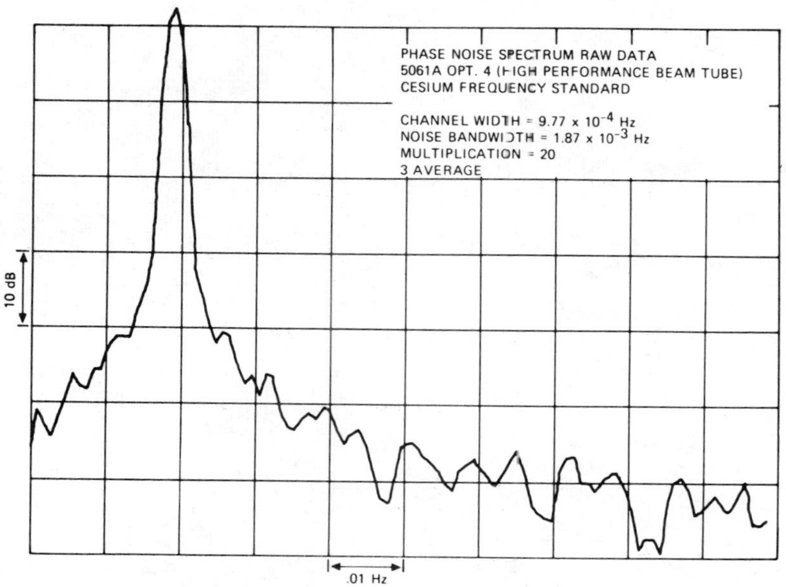

14. Phase Noise Spectrum Raw Data, 5061A Opt. 4 (High Performance Beam Tube) Cesium Frequency Standard

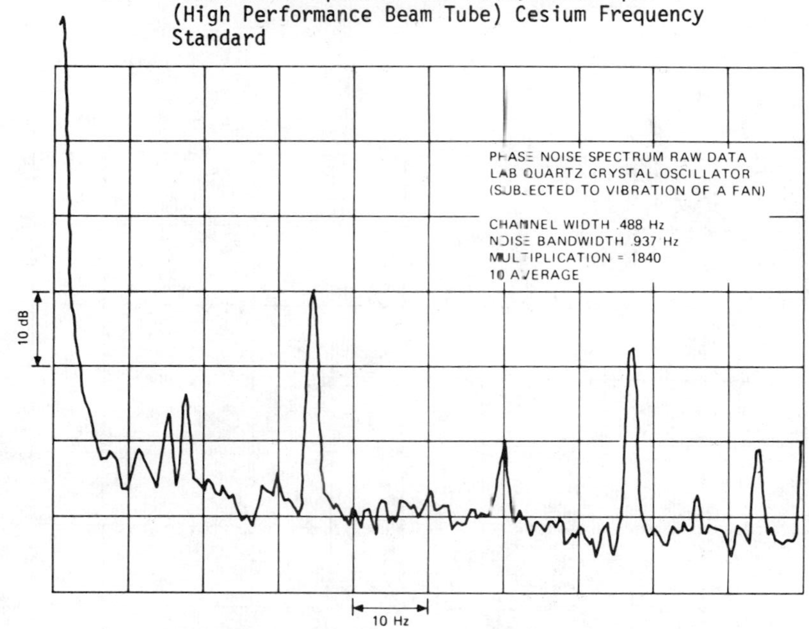

15. Phase Noise Spectrum Raw Data, Lab Quartz Crystal Oscillator (subjected to vibration of a fan)

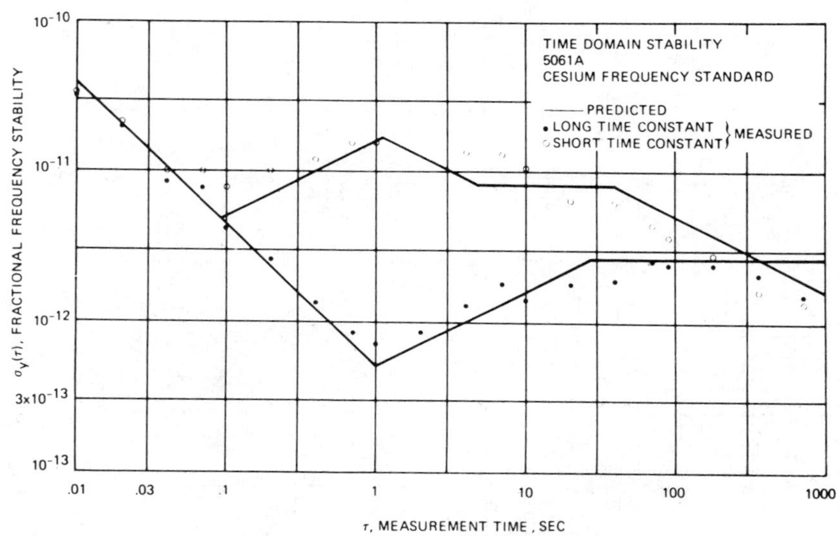

16. Time Domain Stability, 5061A Cesium Frequency Standard

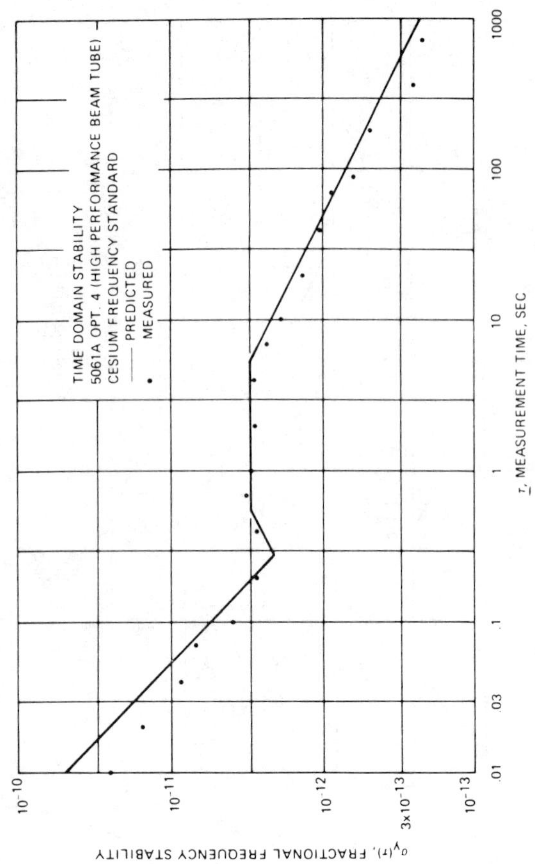

17. Time Domain Stability, 5061A Opt. 4 (High Performance Beam Tube) Cesium Frequency Standard

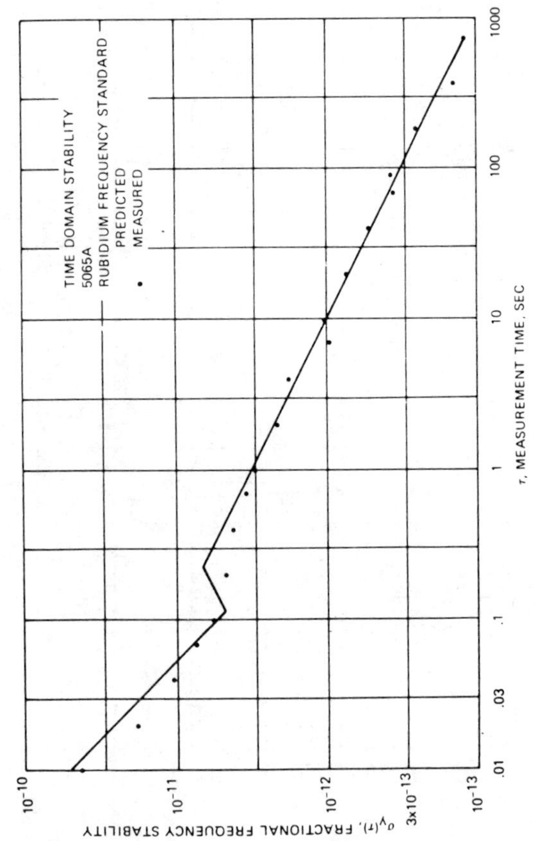

18. Time Domain Stability, 5065A Rubidium Frequency Standard

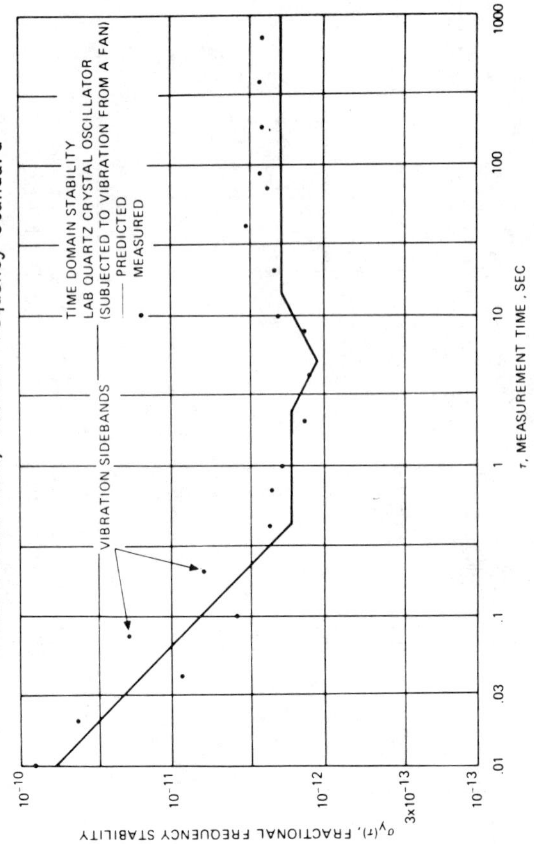

19. Time Domain Stability, Lab Quartz Crystal Oscillator (subjected to vibration of a fan)

SUMMARY OF COMPUTATIONS FOR FREQUENCY DOMAIN TO TIME DOMAIN CONVERSION

$$S_y(f) = \frac{f^2}{f_o^2}\mathcal{L}(f);\ f_o = 5\ \text{MHz}$$

$$f_h = 50\ \text{kHz}$$

$S_y(f)$	$\sigma_y(\tau)$
$h_2 f^2$	$61.64\, h_2^{1/2}\, \tau^{-1}$
$h_1 f$	$[h_1 (6.65 + .477\, \ln \tau)]^{1/2}\, \tau^{-1}$
h_0	$.707\, h_0^{1/2}\, \tau^{-1/2}$
$h_{-1} f^{-1}$	$1.177\, h_{-1}^{1/2}$
$h_{-2} f^{-2}$	$2.57\, h_{-2}^{1/2}\, \tau^{1/2}$

Region 1, 2, 3 etc. refers to the straight line segment approximation counting from the left (lowest frequency) for the phase noise spectra, or counting from the right (longest τ) for the time domain stability spectra.

5061A CESIUM, LONG T.C.

	$\mathcal{L}(f)$	$S_y(f)$	$\sigma_y(\tau)$
1	$1.26 \times 10^{-10} f^{-3}$	$5.04 \times 10^{-24} f^{-1}$	2.64×10^{-12}
2	$10^{-12} f^{-4}$	$4 \times 10^{-26} f^{-2}$	$5.14 \times 10^{-13}\sqrt{\tau}$
3	$6.31 \times 10^{-12} f^{-2}$	2.52×10^{-25}	$\frac{3.55 \times 10^{-13}}{\sqrt{\tau}}$
4	$10^{-12} f^{-1}$	$4 \times 10^{-26} f$	$(6.65 + .477\, \ln \tau)^{1/2}\, \frac{2 \times 10^{-13}}{\tau}$

5061A CESIUM, SHORT T.C.

	$\mathcal{L}(f)$	$S_y(f)$	$\sigma_y(\tau)$
1	$1.25 \times 10^{-7} f^{-2}$	5.04×10^{-21}	$\frac{5.02 \times 10^{-11}}{\sqrt{\tau}}$
2	$1.26 \times 10^{-9} f^{-3}$	$5.04 \times 10^{-23} f^{-1}$	8.36×10^{-12}
3	$1.58 \times 10^{-8} f^{-2}$	6.34×10^{-22}	$\frac{1.78 \times 10^{-11}}{\sqrt{\tau}}$
4	$10^{-9} f^{-4}$	$4 \times 10^{-23} f^{-2}$	$1.63 \times 10^{-11}\sqrt{\tau}$
5	$10^{-12} f^{-1}$	$4 \times 10^{-26} f$	$(6.65 + .477\, \ln \tau)^{1/2}\, \frac{2 \times 10^{-13}}{\tau}$

LAB CRYSTAL OSCILLATOR

	$\mathcal{L}(f)$	$S_y(f)$	$\sigma_y(\tau)$
1	$6.31 \times 10^{-11} f^{-3}$	$2.52 \times 10^{-24} f^{-1}$	1.87×10^{-12}
2	$10^{-12} f^{-4}$	$4 \times 10^{-26} f^{-2}$	$5.14 \times 10^{-13}\sqrt{\tau}$
3	$3.16 \times 10^{-10} f^{-2}$	1.26×10^{-23}	$\frac{2.51 \times 10^{-12}}{\sqrt{\tau}}$
4	$5.01 \times 10^{-11} f^{-3}$	$2 \times 10^{-24} f^{-1}$	1.66×10^{-12}
5	$2 \times 10^{-12} f^{-1}$	$8 \times 10^{-26} f$	$(6.65 + .477\, \ln \tau)^{1/2}\, \frac{2.83 \times 10^{-13}}{\tau}$

5065A RUBIDIUM

	$\mathcal{L}(f)$	$S_y(f)$	$\sigma_y(\tau)$
1.	$5.0 \times 10^{-10} f^{-2}$	2.0×10^{-23}	$\frac{3.17 \times 10^{-12}}{\sqrt{\tau}}$
2	$7.94 \times 10^{-10} f^{-4}$	$3.18 \times 10^{-23} f^{-2}$	$1.45 \times 10^{-11}\sqrt{\tau}$
3.	$1.78 \times 10^{-12} f^{-1}$	$7.11 \times 10^{-26} f$	$(6.65 + .477\, \ln \tau)^{1/2}\, \frac{2.67 \times 10^{-13}}{\tau}$

5061A OPT. 4 CESIUM, HIGH PERFORMANCE BEAM TUBE

	$\mathcal{L}(f)$	$S_y(f)$	$\sigma_y(\tau)$
1.	$2.51 \times 10^{-9} f^{-2}$	1.004×10^{-22}	$\frac{7.09 \times 10^{-12}}{\sqrt{\tau}}$
2	$1.58 \times 10^{-10} f^{-3}$	$6.34 \times 10^{-24} f^{-1}$	2.96×10^{-12}
3.	$6.31 \times 10^{-11} f^{-4}$	$2.52 \times 10^{-24} f^{-2}$	$4.08 \times 10^{-12}\sqrt{\tau}$
4	$1.41 \times 10^{-12} f^{-1}$	$5.65 \times 10^{-26} f$	$(6.65 + .477\, \ln \tau)^{1/2}\, \frac{2.38 \times 10^{-13}}{\tau}$

20. Summary of Computations for Frequency Domain to Time Domain Conversion

Identification of Power-Law Type Oscillator Phase Noise Spectra from Measurements

WILLIAM C. LINDSEY, FELLOW, IEEE, AND CHAK MING CHIE, MEMBER, IEEE

Abstract—**In this paper we demonstrate how structure functions can be used to identify power-law type spectral density components $1/f^{\nu}$ associated with oscillator phase noise measurements when polynomial type systematic drifts are present. The technique discussed is new and novel in that it avoids the usual errors that are introduced in the measurement due to removing systematic trends in the observed data by the usual estimation detrending techniques. Several applications of the theory are given and a convenient and expedient look-up table is provided for those who perform measurements. The paper begins by summarizing accepted frequency stability measures and places these into perspective with structure functions of oscillator instability. The paper ends by showing how the rms fractional frequency deviation differs from the Allan variance; numerical examples are given.**

I. Introduction

We begin by setting the notation to be used in this paper. This also provides us with a short review of oscillator instability measures. The instantaneous output voltage of a high-quality signal generator may be written as [1]

$$V(t) = [V_0 + \epsilon(t)] \sin [2\pi\nu_0 t + \varphi(t)] \tag{1}$$

V_0 and ν_0 are the nominal amplitude and frequency, respectively, $(\omega_0 = 2\pi\nu_0)$. $\epsilon(t)$ and $\varphi(t)$ are random processes representing amplitude noise and phase noise, respectively.

The random process $\dot{\varphi}(t) = d\varphi(t)/dt$ denotes angular frequency fluctuations and is often assumed to have a zero mean value. Provided $\epsilon(t)$ and $\dot{\varphi}(t)$ are sufficiently small for all time t, the fractional instantaneous frequency deviation from nominal is defined as

$$y(t) = \frac{1}{2\pi\nu_0}\frac{d\varphi(t)}{dt} = \frac{\dot{\varphi}(t)}{\omega_0} \triangleq \dot{x}(t). \tag{2}$$

The recommended definition for the frequency stability measure in the Fourier frequency-domain is the one-sided spectral density on a per hertz basis $S_y(f)$ of the pure real random process $y(t)$. Denoting $S_\varphi(f)$ the spectral density of phase fluctuations and $S_{\dot{\varphi}}(f)$ the spectral density of angular frequency fluctuations, then

$$S_y(f) = \frac{1}{\omega_0^2} S_{\dot{\varphi}}(f) \tag{3}$$

$$S_y(f) = \frac{1}{\nu_0^2} f^2 S_\varphi(f). \tag{4}$$

The recommended definition for the frequency stability measure in the time domain is the square root of the two-sample zero-deadtime Allan variance [1], i.e., with $t_{k+1} = t_k + \tau$

$$\sigma_y^2(\tau) = \langle \sigma_y^2(2,\tau,\tau)\rangle = \frac{\langle(\bar{y}_{k+1} - \bar{y}_k)^2\rangle}{2} \tag{5}$$

where

$$\bar{y}_k = \frac{1}{\tau}\int_{t_k}^{t_k+\tau} y(\theta)d\theta$$

is the kth individual sample averaged over a sampling time τ and where the angular brackets indicate an average of the quantity over infinite time. Making the usual assumption that the random process $y(t)$ is ergodic, (5) can be written as

$$\sigma_y^2(\tau) = E\{[\Delta^2 x(t;\tau)]^2\}/2\tau^2 \tag{6}$$

where

$$\Delta^2 x(t;\tau) \triangleq x(t + 2\tau) - 2x(t + \tau) + x(t) \tag{7}$$

is the usual second difference of $x(t)$ and $E\{\cdot\}$ is the (mathematical) expectation operator.

The relationship between $\sigma_y^2(\tau)$ and $S_y(f)$ may be demonstrated very briefly [2] and reads

$$\sigma_y^2(\tau) = \int_0^\infty S_y(f)\frac{2\sin^4 \pi\tau f}{(\pi\tau f)^2}df. \tag{8}$$

This relation allows one to calculate the Allan variance from the spectral density $S_y(f)$, and hence from $S_\varphi(f)$.

Another useful quantity, the rms fractional frequency deviation [2], $\Delta f(\tau)/f_0$ is defined by

$$\left[\frac{\Delta f(\tau)}{f_0}\right]^2 = \frac{E\{[\Delta x(t;\tau)]^2\}}{\tau^2} \tag{9}$$

where $\Delta x(t;\tau)$ is the first difference $\Delta x(t;\tau) \triangleq \Delta(x + \tau) - x(t)$. The rms fractional frequency deviation is related to $S_y(f)$ via

$$\left[\frac{\Delta f(\tau)}{f_0}\right]^2 = \int_0^\infty S_y(f)\frac{\sin^2 \pi\tau f}{(\pi\tau f)^2}df. \tag{10}$$

Lindsey and Chie [2], [3], have recently generalized the rms fractional frequency deviation and the Allan variance in the sense of providing a larger class of τ-domain oscillator instability measures. They have characterized the

Manuscript received March 10, 1977. This work was supported by LinCom Corporation under IR and D funds.
W. C. Lindsey is with the Department of Electrical Engineering, University of Southern California, Los Angeles, CA 90007.
C. M. Chie is with LinCom Corporation, Los Angeles, CA 90015.

Reprinted from *IEEE Trans. Instrum. Meas.*, vol. IM-27, pp. 46–53, Mar. 1978.

short term τ-domain phase instability of the oscillator phase noise process by the set of stability functions $\{D_x^{(k)}(\tau^k), k = 1,2,\cdots,N\}$. In particular, they show that [2]

$$D_x^{(1)}(\tau) = \tau^2 \left[\frac{\Delta f(\tau)}{f_0}\right]^2 \tag{11}$$

and

$$D_x^{(2)}(\tau) = 2\tau^2\sigma_y^2(\tau). \tag{12}$$

They also characterize the short-term frequency stability of the oscillator frequency noise process by the set of stability functions $\{D_y^{(k)}(\tau^k), k = 1,2,\cdots,N\}$. These two sets of stability functions are related in a known way [2], [3] and are connected to the f-domain stability measure $S_y(f)$ via the Mellin transform [2].

In this paper, we first consider an important practical application of higher order structure functions, namely, in the identification of power-law type spectrum $S_y(f) = h_{-\nu}f^{-\nu}$ when a polynomial type systematic drift is present in the fractional frequency process $y(t)$, e.g., due to "aging" in crystal oscillators [4, ch. 2], "cavity pulling" in hydrogen masers [4, ch. 4], etc. In yet another example, we shall demonstrate, using the theory of structure functions [2], how to translate between $\Delta f(\tau)/f_0$ and $\sigma_y^2(\tau)$ without resorting to the power spectral density (PSD) $S_y(f)$.

II. Utility of Higher Order Structure Functions

A. Structure Functions

For most oscillators, the assumption that $y(t)$ has a zero mean value is not always valid. As an example, state of the art crystal oscillators exhibit a typical frequency drift component $E[y(t)] = d_1 t$ where d_1 is of the order of 5×10^{-10}/day [5]. Due to this reason, it is often suggested [1] that this drift be estimated and removed before determining, say, $\sigma_y^2(\tau)$. Usually, this drift term is estimated by a minimum mean square fit. Let $\hat{d}_1$ be such an estimate. Unless the estimated value is exactly d_1, the error introduced shows up as an additional term $(d_1 - \hat{d}_1)^2\tau^2/2$ in the value of $\sigma_y^2(\tau)$. In what follows, we develop a method of annihilating the effect of a polynomial type systematic drift in $y(t)$ using the higher order structure functions $D_x^{(m)}(\tau)$, $m > 2$. In addition, we show how a power-law PSD $S_y(f)$ can be identified from a plot of $D_x^{(m)}(\tau)$ versus τ.

From this point on we assume that the frequency process $y(t)$ has a time-dependent mean drift, viz.

$$y(t) = E[y(t)] + \tilde{y}(t) \tag{13}$$

where $\tilde{y}(t)$ is a zero mean random process and

$$E[y(t)] = \sum_{k=1}^{N-1} \frac{d_k}{k!} t^k. \tag{14}$$

Let us denote the PSD of the process $\tilde{y}(t)$ by $S_y(f)$ (instead of $S_{\tilde{y}}(f)$) to comply with the "standard" notation [1]. Notice that if $d_1 = \cdots = d_{N-1} = 0$ in (14), then $\tilde{y}(t) = y(t)$ and our generalization reduces to the conventional definition.

It is well known that taking higher-order increments reduces the degree of a polynomial. In particular, the $(N-1)$th increment of $y(t)$ reduces to [2]

$$\Delta^{N-1}y(t;\tau) = d_{N-1}\tau^{N-1} + \Delta^{N-1}\tilde{y}(t;\tau) \tag{15}$$

and the Nth increment of $y(t)$ reduces to

$$\Delta^N y(t;\tau) = \Delta^N \tilde{y}(t;\tau) \tag{16}$$

which is independent of the systematic drift. Let us define the mth structure function of a process $\xi(t)$ by [2] (assuming stationarity)

$$D_\xi^{(m)}(\tau) \triangleq E\{[\Delta^m \xi(t;\tau)]^2\}. \tag{17}$$

For $m = N$, we observe that $D_y^{(m)}(\tau) = D_{\tilde{y}}^{(m)}(\tau)$ which is independent of the systematic drift. In [2], it was shown that $D_y^{(N)}(\tau)$ and the PSD $S_y(f)$ are related via

$$D_y^{(M)}(\tau) = 4^M \int_0^\infty \sin^{2M} \pi\tau f S_y(f)df, \qquad M \geq N. \tag{18}$$

Similar results are developed [2] for the fractional phase process $x(t)$. In particular, it was shown that

$$\Delta^N x(t) = d_{N-1}\tau^N + \Delta^{N-1}\left[\int_t^{t+\tau} \tilde{y}(\xi)d\xi\right] \tag{19}$$

$$\Delta^{M+1}x(t) = \Delta^M\left[\int_t^{t+\tau} \tilde{y}(\xi)d\xi\right]; \qquad M \geq N \tag{20}$$

and

$$D_x^{(M+1)}(\tau) = 4^M \int_0^\infty \frac{\sin^{2(M+1)}\pi\tau f}{(2\pi f)^2} S_y(f)df, \qquad M \geq N \tag{21}$$

Notice also that $D_x^{(N+1)}(\tau)$ is independent of the systematic drift. We now use these theoretical results to show how power-law type spectral density components can be identified in τ-domain measurements without removing the systematic drift by first estimating it and then subtracting it out of the measurement.

B. Power-Law PSD $S_y(f)$

In practice, the PSD $S_y(f)$ is usually modeled by a linear combination of power-law PSD's having the form

$$S_y(f) = \begin{cases} h_\nu f^\nu, & 0 \leq f \leq f_h \\ 0 & f > f_h. \end{cases} \tag{22}$$

For various reasons, it is also assumed that $-2 \leq \nu \leq 2$ in the literature [1]. Using this model of $S_y(f)$ in (18) and (21), we arrive at two important relationships, namely,

$$D_x^{(M+1)}(\tau) = \frac{4^M h_\nu}{\pi^{\nu+1}} F(\nu - 2, M+1, \pi\tau f_h)\tau^{1-\nu}, \qquad M \geq N \tag{23}$$

and

$$D_y^{(M)}(\tau) = \frac{4^M h_\nu}{\pi^{\nu+1}} F(\nu, M, \pi\tau f_h)\tau^{-\nu-1}, \qquad M \geq N \tag{24}$$

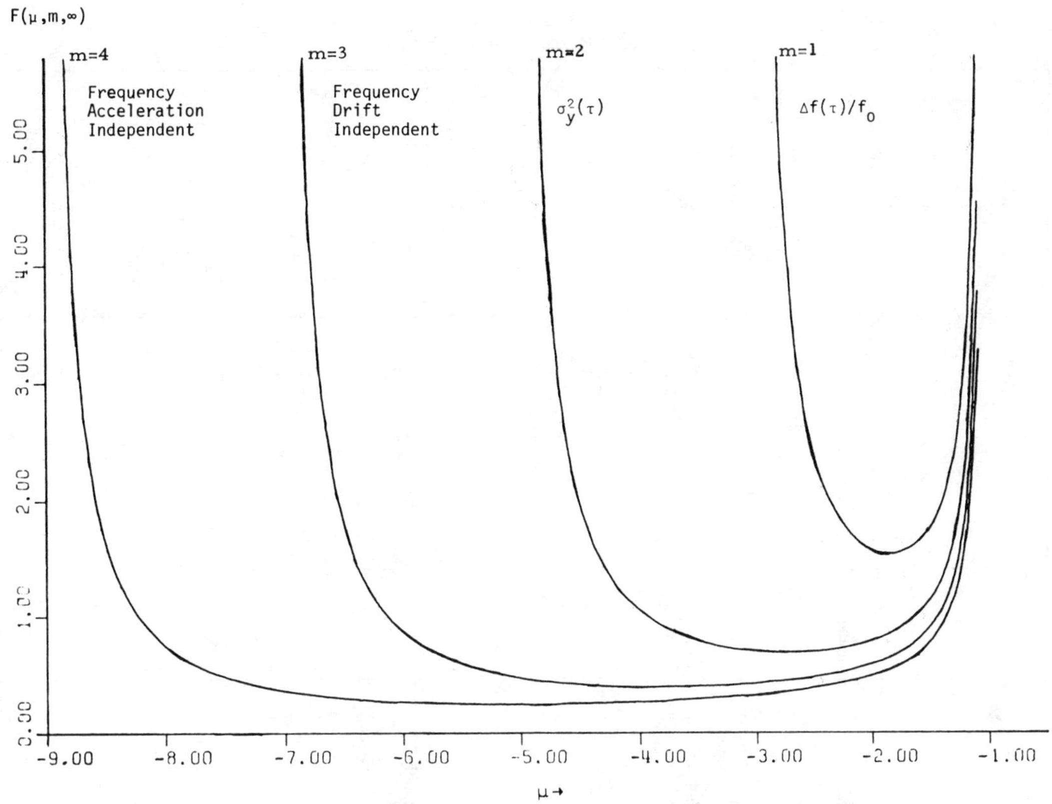

Fig. 1. A plot of $F(\mu,m,\infty)$ for $m = 1,2,3,4$ (universal curves for identifying power-law type components).

where (μ not necessarily an integer)

$$F(\mu,m,B) \triangleq \int_0^B \xi^\mu \sin^{2m}\xi d\xi, \qquad \mu > -(2m+1). \tag{25}$$

In Appendix I, the function $F(\mu,m,B)$ is investigated and evaluated for $B \gg 1$. For $\mu > -1$, $F(\mu,m,B)$ is dependent on B. However, for $-(2m+1) < \mu < -1$, $F(\mu,m,B)$ is essentially independent of B for $B \gg 1$. The function $F(\mu,n,\infty)$ for $-(2m+1) < \mu < -1$ is plotted in Fig. 1. Using this curve and the relations (23) and (24), one can determine $S_y(f)$ from $D_x^{(N+1)}(\tau)$ or $D_y^{(N)}(\tau)$ which are readily measured experimentally. We shall dwell on this point when we given an example in the next section. Other commonly encountered quantities of interest are tabulated in Tables I, II. We have checked our results with those given in [4, ch. 8] using the conversion

$$\sigma_y^2(\tau) = \frac{1}{2\tau^2} D_x^{(2)}(\tau). \tag{26}$$

They are found to be identical except when $S_y(f) = h_1 f$. In that case, the constant term added to the ln $(\pi\tau f_h)$ term is slightly different. Our experience with direct numerical integration shows that our expression is in closer agreement. This may result from a better approximation technique employed.

Another interesting point is worth noting here. From observing these tables, we see that higher order structure functions basically have the same τ dependence as the lower order one. The only difference lies in the multiplicative constants which increase with the order of the structure functions.

C. *Applications of the Theory Presented*

1. *Interconnection with the Allan Variance:* We shall now consider two specific examples demonstrating the use of higher order structure functions and various power-law PSD/τ-domain translations. We first consider an example of a typical crystal oscillator where the aging effect has to be reckoned with. In the other example, we shall apply Fig. 1 and (21) to determine the PSD $S_y(f)$ from a $\sigma_y^2(\tau)$ versus τ plot in which the power-law dependence is not an integer.

Consider a crystal oscillator with an aging rate of 10^{-10}/day, i.e., $N = 2$, $d_1 = 1.2 \times 10^{-15}$/s in (14). Notice that from (19) and (21), the systematic drift gives rise to a τ^2 dependence of $\sigma_y^2(\tau)$, namely,

$$\sigma_y^2(\tau) = \frac{1}{2} d_1^2 \tau^2 + \frac{4}{\pi^2}\int_0^\infty \frac{\sin^4(\pi\tau f)}{f^2} S_y(f) df. \tag{27}$$

If $S_y(f)$ is modeled by a flicker frequency PSD $S_y(f) = h_{-1} f^{-1}$ with $h_{-1} = 7.2 \times 10^{-25}$, $\sigma_y(\tau)$ can be depicted as in Fig. 2 [6, fig. 16]. Notice, however, that the third difference of $x(t)$ is independent of the drift and from Table I, we find that $D_x^{(2)}(\tau) = 4.86 \times 10^{-24}\tau^2$. The normalized quantity $\sqrt{D_x^3(\tau)/\tau^2} = 2.2 \times 10^{-12}$ is a constant independent of τ. For comparison purposes, it is plotted alongside with $\sigma_y(\tau)$ in Fig. 2 (the dashed line).

TABLE I
Higher Order Phase Structure Functions

Process	$D_x^{(1)}(\tau)$ $\left(\tau^2\left[\frac{\Delta f(\tau)}{f_0}\right]^2\right)$	$D_x^{(2)}(\tau)$ $(2\tau^2\sigma_y^2(\tau))$	$D_x^{(3)}(\tau)$	$D_x^{(4)}(\tau)$
White x $S_y = h_2 f^2$	$\frac{1}{2\pi^2} h_2 f_h$	$\frac{3}{2\pi^2} h_2 f_h$	$\frac{5}{\pi^2} h_2 f_h$	$\frac{35}{2\pi^2} h_2 f_h$
Flicker x $S_y = h_1 f$	$\frac{1}{2\pi^2} h_1 (\ln \pi\tau f_h + 1.27)$	$\frac{3}{2\pi^2} (\ln \pi\tau f_h + 1.04)$	$\frac{5}{\pi^2} (\ln \pi\tau f_h + 0.96)$	$\frac{35}{2\pi^2} (\ln \pi\tau f_h + 0.93)$
White y $S_y = h_0$	$\frac{1}{2} h_0 \tau$	$h_0 \tau$	$3 h_0 \tau$	$10 h_0 \tau$
Flicker y $S_y = h\text{-}1/f$	Low cutoff frequency dependent	$4 \ln 2 h_{-1} \tau^2$	$6.75 h_{-1} \tau^2$	$20.7 h_{-1} \tau^2$
Random Walk y $S_y = h_{-2}/f^2$	Low cutoff frequency dependent	$\frac{4}{3} \pi^2 h_{-2} \tau^3$	$2\pi^2 h_{-2} \tau^3$	$\frac{16}{3} \pi^2 h_{-2} \tau^3$

TABLE II
Higher Order Frequency Structure Functions

Process	$D_y^{(1)}(\tau)$	$D_y^{(2)}(\tau)$	$D_y^{(3)}(\tau)$	$D_y^{(4)}(\tau)$
White x $S_y = h_2 f^2$	$\frac{2}{3} h_2 f_h^3$	$2 h_2 f_h^3$	$\frac{20}{3} h_2 f_h^3$	$\frac{70}{3} h_2 f_h^3$
Flicker x $S_y = h_1 f$	$h_1 f_h^2$	$3 h_1 f_h^2$	$10 h_1 f_h^2$	$35 h_1 f_h^2$
White y $S_y = h_0$	$2 h_0 f_h$	$6 h_0 f_h$	$20 h_0 f_h$	$70 h_0 f_h$
Flicker y $S_y = h_{-1}/f$	$2h_{-1}(\ln \pi\tau f_h + 1.27)$	$6h_{-1}(\ln \pi\tau f_h + 1.04)$	$20h_{-1}(\ln \pi\tau f_h + 0.96)$	$70h_{-1}(\ln \pi\tau f_h + 0.93)$
Random Walk y $S_y = h_{-2}/f^2$	$2\pi^2 h_{-2} \tau$	$4\pi^2 h_{-2} \tau$	$12\pi^2 h_{-2} \tau$	$40\pi^2 h_{-2} \tau$

Let us turn the problem around. If we have an experimental plot of $D_x^{(3)}(\tau)$ versus τ, we can infer h_{-1} from Table I. Since $D_x^{(3)}(\tau)$ is not a function of the systematic trend in this case (linear frequency shift), detrending of experimental data (and hence the error associated with it) is no longer necessary! This is one immediate advantage over that of the Allan variance. Obviously, for other (higher order) types of polynomial drift, higher order structure

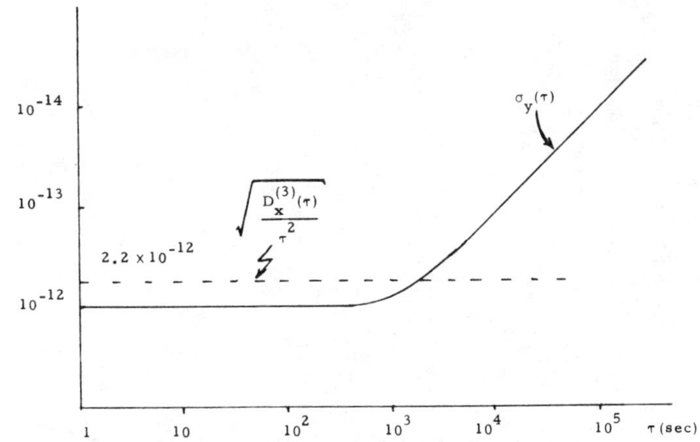

Fig. 2. An oscillator with both random fluctuations of flicker noise FM and a nonrandom linear fractional frequency drift of 10^{-10} per day.

functions are necessary. Furthermore, if the PSD has an additional independent component, say, $S_{y_1}(f) = h_{-4}f^{-4}$. From (23) and (I.14) in Appendix I, we conclude that

$$D_x^{(3)}(\tau) = \frac{22}{5}\pi^4 h_{-4}\tau^5 \tag{28}$$

for this particular component. However, the Allan variance fails to exist. For this reason, we maintain that higher order structure functions provide a more powerful and a more general characterization of oscillator stability. The *frequency* structure functions [2] play a similar role and for the purpose of conciseness, we will not comment on them.

2. *Utility of Tables I and II and the Universal Curves of Fig. 1:* The next example serves to illustrate the use of the curves in Fig. 1. From (23), we find that if $\sigma_y^2(\tau) = a_\mu \tau^\mu$, $S_y(f) = h_{-\mu-1}f^{-\mu-1}$ where

$$h_{-\mu-1} = \frac{\pi^{-\mu} a_\mu}{2F(-\mu - 3, 2, \pi\tau f_h)}. \tag{29}$$

Now consider an oscillator that has an Allan variance

$$\sigma_y^2(\tau) = 10^{-28}\tau^{-1.2} + 10^{-24}\tau^{0.2} + 10^{-32}\tau^{0.8} \tag{30}$$

with a corresponding PSD

$$S_y(f) = h_{0.2}f^{0.2} + h_{-1.2}f^{-1.2} + h_{-1.8}f^{-1.8} \tag{31}$$

containing nonintegral powers of f. From Fig. 1, we find that $F(-1.8,2,\infty) = 0.87$, $F(-3.2,2,\infty) = 0.72$, and $F(-3.8,2,\infty) = 0.91$. Finally, using (29), we conclude that $h_{0.2} \cong 2.3 \times 10^{-28}$, $h_{1.2} \cong 5.6 \times 10^{-25}$ and $h_{-1.8} \cong 7.0 \times 10^{-34}$.

III. The Allan Variance and RMS Fractional Frequency Deviation

In time applications of frequency generators, it is of interest to interrelate the current standard stability measure, i.e., the Allan variance and the rms fractional frequency deviation which is important in answering questions like: What is the rms deviation on a clock time interval of say, 1 hour? One obvious way is to translate one measure to $S_y(f)$ and then evaluate the other one from $S_y(f)$. There are problems associated with this approach. First of all, for simple τ-domain to f-domain translations, a power-law PSD has to be assumed (yet see [2] for other advanced methods which do not assume this). Unfortunately, for some cases, power-law spectrum is not totally satisfactory as the next example will show. Secondly, even if a power-law PSD is a good assumption, the rms fractional frequency deviation may not exist unless a low cutoff frequency f_c is assumed. However, in that case $\Delta f(\tau)/f_0$ is usually highly f_c dependent, e.g., for $S_y(f) \sim f^{-\nu}, \nu > 1$. One immediate question arises: What should we choose for f_c?

In [2], Lindsey and Chie have developed a different approach. Under the reasonable assumption that the variance of the stationary part of $y(t)$, $\tilde{y}(t)$ is bounded (it could be very large), it was shown that $\sigma_y(\tau)$ and $\Delta f(\tau)/f_0$ are interrelated via

$$\frac{\Delta f(\tau)}{f_0} = \sqrt{\frac{1}{2}\sigma_y^2(\tau) + \frac{1}{2}\sum_{k=1}^{\infty}\sigma_y^2(2^k\tau)} \tag{32}$$

and

$$\sigma_y^2(\tau) = 2\left\{\left[\frac{\Delta f(\tau)}{f_0}\right]^2 - \left[\frac{\Delta f(2\tau)}{f_0}\right]^2\right\} \tag{33}$$

directly in the τ-domain. In this section, we shall give an example to demonstrate the use of the two formulas. Before doing so, we need some preliminary preparations.

A. *Asymptotic Behavior of $\sigma_y(\tau)$ and $\Delta f(\tau)/f_0$*

For all real-world oscillators, we can safely assume that the phase and frequency noise variance are bounded. A convenient model is that both $S_y(f)$ and $S_y(f)/f^2$ diminishes for $f < f_l$ and $f > f_h$. Assuming this in (21), we

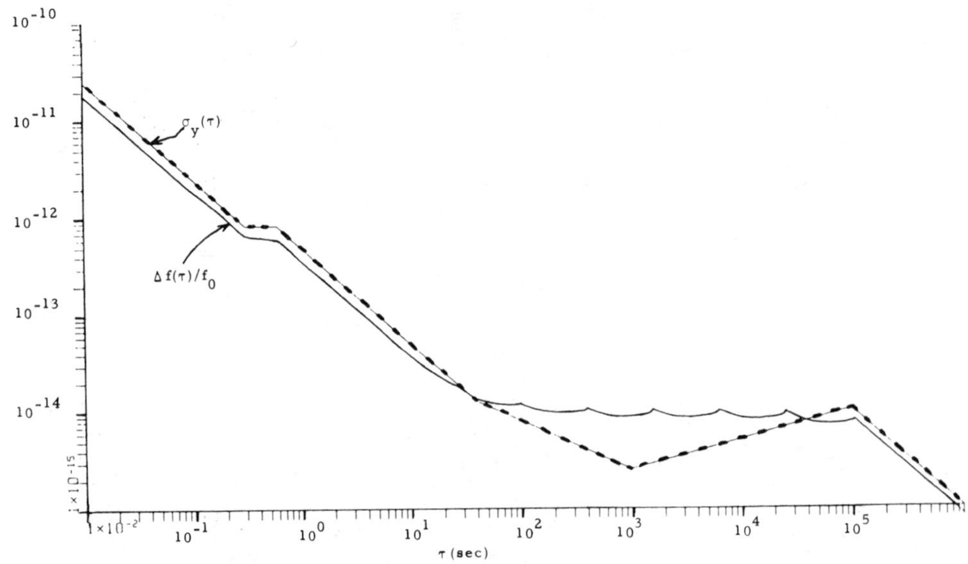

Fig. 3. $\sigma_y(\tau)$ for nonautotuned NASA hydrogen maser.

have

$$D_x^{(1)}(\tau) = 4\int_{fl}^{fh} \sin^2(\pi\tau f)\frac{S_y(f)}{(2\pi f)^2}\,df$$

$$= 2\int_{fl}^{fh}\frac{S_y(f)}{(2\pi f)^2}\,df - 2\int_{fl}^{fh}\cos(2\pi\tau f)\frac{S_y(f)}{(2\pi f)^2}\,df. \quad (34)$$

As a consequence of the Riemann–Lebesque lemma, the second term in (34) tends to 0 as $\tau \to \infty$. Now from (11), we see that

$$\left[\frac{\Delta f(\tau)}{f_0}\right]^2 \sim 2\sigma_x^2\tau^{-2}; \qquad \tau \gg 1 \quad (35)$$

where σ_x^2 is the phase noise variance. Similarly, from (21) and (26) we see that

$$\sigma_y^2(\tau) \sim 3\sigma_x^2\tau^{-2}, \qquad \tau \gg 1 \quad (36)$$

B. Example

Let us consider the following NASA hydrogen maser [7, fig. 9] having an Allan variance given by

$$\sigma_y(\tau) = \begin{cases} 2.5\times 10^{-13}\tau^{-1}, & 10^{-12}\le\tau\le 3\times 10^{-1}\\ 8.3\times 10^{-13}, & 3\times 10^{-1}\le\tau\le 6\times 10^{-1}\\ 5.0\times 10^{-13}\tau^{-1}, & 6\times 10^{-1}\le\tau\le 4\times 10^{-1}\\ 7.9\times 10^{-14}\tau^{-0.5}, & 4\times 10^{-1}\le\tau\le 10^{3}\\ 3.2\times 10^{-16}\tau^{0.3}, & 10^{3}\le\tau\le 10^{5}.\end{cases} \quad (37)$$

Following the argument given in Section II-A we shall assume that

$$\sigma_y(\tau) = 10^{-9}\tau^{-1}, \qquad \tau \ge 10^5 \quad (38)$$

This example is depicted in Fig. 3. Notice that a power-law series approximation of (37) would probably miss the white frequency noise region where $3\times 10^{-1}\le\tau\le 6\times 10^{-1}$ and hence not a very good model. Besides, if we try the table look-up approach for power-law PSD's, the resultant PSD consist of too many terms. Also in order to obtain $\Delta f(\tau)/f_0$, an arbitrary low cutoff frequency f_c has to be assumed.

The τ-domain translation using (32) is performed and the result is indicated in Fig. 3. Conversely, if the curve for $\Delta f(\tau)/f_0$ is given, we can simply use (33) to obtain $\sigma_y(\tau)$.

APPENDIX I

It is of interest to derive approximate expressions for the function

$$F(\mu,m,B) = \int_0^B \xi^\mu \sin^{2m}\xi\,d\xi, \qquad \mu > -(2m+1) \quad (\text{I.1})$$

for $B \gg 1$.

Case (i): $\mu > -1$. We rewrite (I.1) into the form

$$F(\mu,m,B) = B^{\mu+1}\int_0^1 z^\mu \sin^{2m} Bz\,dz \quad (\text{I.2})$$

When $B \gg 1$,

$$B^{M+1}\int_0^1 z^\mu \sin^{2m} Bz\,dz \cong \frac{\binom{2m}{m}}{2^{2m}}\frac{B^{\mu+1}}{\mu+1} \quad (\text{I.3})$$

as the dc term in the $\sin^{2m} Bz$ expression is $\binom{2m}{m}/2^{2m}$ [8, eq. 1.320(1)].

Case (ii): $\mu = -1$. For $\mu = -1$, we have to interpret (I.1) as

$$F(\mu,m,B) = \lim_{\epsilon\to 0}\int_\epsilon^B \xi^\mu \sin^{2m}\xi\,d\xi \quad (\text{I.4})$$

Then from [8, eq. 2.642(1)]

$$F(\mu,m,B) = \lim_{\epsilon\to 0}\left\{\binom{2m}{m}\frac{\ln\xi}{2^{2m}} + \frac{(-1)^m}{2^{2m-1}}\sum_{k=0}^{m-1}(-1)^k\binom{2m}{m}\operatorname{ci}[2(m-k)\xi]\right\}_\epsilon^B \quad (\text{I.5}a)$$

where the cosine–integral ci (z) is given by

$$\mathrm{ci}(z) = C + \ln z + \sum_{k=1}^{\infty} (-1)^k \frac{z^{2k}}{2k(2k)!} \qquad \text{(I.5b)}$$

and C is the Euler's constant, $C = 0.577 \cdots$. Using (I.5b) in (I.5a) simplifying and taking limits, we obtain

$$F(\mu,m,B) = \frac{\binom{2m}{m}}{2^{2m}} (\ln B + \ln 2 + C) - \frac{(-1)^m}{2^{2m-1}} \sum_{k=0}^{m-1} (-1)^k \binom{2m}{k} \ln (m-k) + \delta_m(B) \qquad \text{(I.6)}$$

where the error term

$$\delta_m(B) = \frac{(-1)^m}{2^{2m-1}} \sum_{k=0}^{m-1} (-1)^k \binom{2m}{k} \mathrm{ci}\,[2(m-k)B] \qquad \text{(I.7)}$$

goes to zero as $B \to \infty$.

Case (iii): $-(2m+1) < \mu < -1$. First of all, we derive an expression for $-3 < \mu < -1$. We then extend our result to noninteger μ, $-(2m+1) < \mu < -1$. Finally, by taking limits, we obtain the general case.

Using the expansion [2, eq. II.14]

$$2^{2(m-1)} \sin^{2m}\alpha = \sum_{k=1}^{m} (-1)^{k-1} \binom{2m}{m-k} \sin^2 k\alpha \qquad \text{(I.8)}$$

and [8, eq. 3.823] we simplify to get

$$F(\mu,m,\infty) = -\frac{\pi \sum_{k=0}^{m} (-1)^{k-1} \binom{2m}{m-k} k^{-\mu-1}}{2^{m+1+\mu}\Gamma(\mu) \cos\left(-\frac{\mu\pi}{2}\right)}, \quad -3 < \mu < -1. \qquad \text{(I.9)}$$

Let (I.9) hold for arbitrary nonintegral μ, $-(2m-1) < \mu < -1$. We shall show that (I.9) also holds for nonintegral μ, $-(2m-1) < \mu - 2 < -1$. Using [8, eq. 3.821(12)] and (I.9), we simplify to yield

$$F(\mu-2,m,\infty) = \frac{-\pi \left\{ m^2 \sum_{k=1}^{m} (-1)^{k-1} \binom{2m}{m-k} k^{-\mu-1} - 2m(2m-1) \sum_{k=1}^{m-1} (-1)^{k-1} \binom{2(m-1)}{(m-1)k} k^{-\mu-2} \right\}}{\Gamma(-\mu+2) \cos\left[\frac{(-\mu+2)}{2}\right] 2^{2m+1+\mu-2}}. \qquad \text{(I.10)}$$

However, since (can be verified from definition)

$$2m(2m-1) \binom{2(m-1)}{(m-1)k} = \binom{2m}{m-k} (m^2 - k^2) \qquad \text{(I.11)}$$

it follows that

$$F(\mu-2,m,\infty) = \frac{-\pi \sum_{k=1}^{m} (-1)^{k-1} \binom{2m}{m-k} k^{-(\mu-2)-1}}{\Gamma[-(\mu-2)] \cos\left[-\frac{(\mu-2)\pi}{2}\right] 2^{2m+1+(\mu-2)}}. \qquad \text{(I.12)}$$

Equations (I.9) and (I.12) extend our result for nonintegral μ, $-(2m+1) < \mu < -1$.

To handle the general case, notice that $F(\mu,m,\infty)$ converges for $-(2m+1) < \mu < -1$ whether μ is an integer or not. Hence we can take the limit of (I.9) to evaluate $F(\mu,m,\infty)$ at the integer values of μ. For μ odd, using the l'Hospital rule,

$$F(\mu,m,\infty) = \frac{\sum_{k=2}^{m} (-1)^k \binom{2m}{m-k} k^{-\mu-1} \ln k}{2^{2m+\mu}\Gamma(-\mu) \sin\left(-\frac{\mu\pi}{2}\right)}. \qquad \text{(I.13)}$$

When μ is even, the value remains unchanged.

To summarize, for $-(2m+1) < \mu < -1$ where μ is not an odd integer

$$F(\mu,m,\infty) = \frac{-\pi \sum_{k=0}^{m} (-1)^{k-1} \binom{2m}{m-k} k^{-\mu-1}}{2^{2m+1+\mu}\Gamma(-\mu) \cos\left(-\frac{\mu\pi}{2}\right)} \qquad \text{(I.14)}$$

For odd μ, $-(2m+1) < \mu < -1$,

$$F(\mu,m,\infty) = \frac{-\sum_{k=2}^{m} (-1)^k \binom{2m}{m-k} k^{-\mu-1} \ln k}{2^{2m+\mu}\Gamma(-\mu) \sin\left(-\frac{\mu\pi}{2}\right)} \qquad \text{(I.15)}$$

Finally, notice that

$$F(\mu,m,B) = F(\mu,m,\infty) - \delta_m(B) \qquad \text{(I.16)}$$

where for $\mu < -1$

$$\delta_m(B) = \int_B^{\infty} \xi^{\mu} \sin^{2m} \xi \, d\xi < \frac{B^{\mu+1}}{\mu+1} \qquad \text{(I.17)}$$

which tends to zero as $B \to \infty$.

References

[1] J. A. Barnes *et al.*, "Characterization of frequency stability," *IEEE Trans. Instrum. Meas.*, vol. IM-20, pp. 105–120, May 1971.

[2] W. C. Lindsey and C. M. Chie, "Theory of oscillator instability based upon structure functions," *Proc. IEEE*, vol. 64, pp. 1652–1666, Dec. 1976.

[3] ——, "Interpretation and application of oscillator instability measures based upon structure functions," in *Proc. 8th Annu. Precise Time and Time Interval (PTTI) Applications and Planning Meeting* (Nov. 30–Dec. 2, 1976, Naval Research Laboratory, Washington, DC), 1976.

[4] B. E. Blair, Ed., "Time and frequency: Theory and fundamentals," NBS Monograph 140, CODEN: NBSMA 6, U.S. Superintendent of Documents, Washington, DC, 1974.

[5] E. Hafner, "Outlook for precision frequency control in the 1980's,"

in *Proc. 7th Annu. Precise Time and Time Interval (PTTI) Applications and Planning Meeting* (Dec. 2–4, 1975, Goddard Space Flight Center, Greenbelt, MD), pp. 101–119, 1975.

[6] D. W. Allan, "The measurement of frequency and frequency stability of precision oscillators," NBS Tech. Note 669, CODEN: NBTNAE, U.S. Superintendent of Documents, Washington, DC, 1975.

[7] V. Reinhardt, "Frequency stability requirements for two way range tracking," in *Proc. 7th Annu. Precise Time and Time Interval (PTTI) Applications and Planning Meeting* (Dec. 2–4, 1975, Goddard Space Flight Center, Greenbelt, MD), pp. 265–283, 1975.

[8] I. S. Gradshteyn and I. M. Ryzhik, *Table of Integrals, Series, and Products*, 4th ed. New York: Academic Press, 1965.

Part V
Noise Measurements at Microwave Frequencies

IN instances where only the frequency stability, i.e., the phase noise, of microwave generators is of concern, the "two-oscillator" measurement techniques, discussed in the previous part, can be used throughout the whole microwave and millimeter-wave ranges up to optical frequencies [V-1], [V-2].

On the other hand, in many modern radar and communications systems the knowledge of both the spurious amplitude and phase (or frequency) modulation of microwave sources and the added close-to-carrier noise in some "passive" components, such as amplifiers, modulators, etc., is of great importance. The possibility of analyzing AM and FM noise contributions with the assistance of the measurement generator only led to the development and perfection of "one-oscillator" methods, the properties of which are discussed in depth in the reprint papers by Ondria and Ashley *et al.* Both papers are tutorial reviews reflecting the state of the art in the mid-60's and about ten years later. What is interesting is that the principle of the measurement has not effectively changed [V-3]; The major contribution of Ashley's paper is the introduction of automation into an otherwise tiresome measurement procedure. To those interested in the perfection of the automation procedure, we recommend for further study the article by McClintock *et al.* [V-4] and the paper by Lance *et al.* reprinted in Part IV in which commercial FFT spectrum analyzers are incorporated into the measuring equipment.

Recently, Simmons and Smith [V-5] reported extension of direct FM noise measurements into the *W*-band (75–100 GHz), whereas Harth *et al.* [V-6] suggested an innovation of the "one-oscillator" measurement method based on the use of a quasioptical resonator.

One important feature of the "one-oscillator methods," not mentioned in either of the reprint papers, is a much larger background noise at very low Fourier frequencies compared to two-oscillator measurement systems.

The theoretical explanation of the phenomenon is a simple one. By referring to Part VI of this book, we find out that the power spectral density of the output noise voltage of mixers and amplifiers approximately follows the law

$$S_{V_n}(f) = \frac{a_1}{f} + a_0. \qquad \text{(V-1)}$$

In instances where "two-oscillator methods" are used, the measured quantity is the phase noise; to get to the respective power spectral density both sides of (V-1) must be divided by K_d^2, K_d being the phase detector constant [V/rad]. Evidently,

$$S_{\varphi n} = \frac{a_1/K_d^2}{f} + a_0/K_d^2. \qquad \text{(V-2)}$$

On the other hand, in all one-oscillator methods we refer to the frequency noise. Consequently, the transition to the phase noise is made via integration which in Fourier transform symbology requires division by f^2. However, the baseband output voltage v_d of the balanced mixer is proportional to the product of the frequency deviation Δf and the time delay τ:

$$v_d = \text{Constant} \cdot \Delta f \cdot \tau \qquad \text{(V-3)}$$

Evidently, the power spectral density of the background phase noise is proportional to

$$S_{\varphi n}(f) \sim \left[\frac{1}{f^3} \cdot \frac{a_1}{\tau^2} + \frac{1}{f^2} \cdot \frac{a_0}{\tau^2}\right]. \qquad \text{(V-4)}$$

The experimental verification of this property has been provided by Lance *et al.* [V-7], by McClintock [V-4], by Labaar [V-8], and can be deduced from other publications too.

The consequence is that the one-oscillator measurement systems are unsuitable for the evaluation of the close-to-carrier phase noise of many modern microwave sources (e.g., crystal oscillators followed by frequency multipliers or low-noise frequency multipliers) because of the masking phenomenon. Some remedy can be provided with the assistance of the cross-correlation technique [V-7], but even this solution is inferior to the two-oscillator systems. Another possibility is the application of the delay-line RF bridge [V-9] with low noise amplification of its output [V-8].

Investigation of the relation (V-3) reveals that the constant depends on the input power and does not change too much. Evidently, the sensitivity improves with τ, i.e., the effective threshold is lowered, since the additive mixer noise voltage remains constant.

Now to the actual background noise of one-oscillator systems based on published experimental data; the Δf_{rms} at 1 Hz in 1 Hz bandwidth for the flicker noise constant is 0.05–0.5 Hz/$\sqrt{\text{Hz}}$ and for the white noise constant is 0.001–0.01 Hz/$\sqrt{\text{Hz}}$. Since microwave cavities have, generally, a very large Q they exhibit a large time delay

$$\tau = \frac{Q_{\text{load}}}{\pi f_0} \qquad \text{(V-5)}$$

and as a consequence the sensitivity of devices employing cavities is at the lower ends of the above indicated intervals.

In the microwave frequency ranges the noise properties of generators are usually given in terms of the rms frequency deviation Δf_{rms} in a certain bandwidth B_w. This practice is based on applications in communications on one hand and on the widespread use of one-oscillator measurement systems on the other hand. However, the noise properties of modern low-noise frequency sources are generally specified in terms of the phase noise power spectral densities $S_\varphi(f)$. Consequently, we feel that the conversion formulas can be of a great help for our readers.

According to Cuttler and Searle (see the first paper in Part III), the mean-square frequency error or variance of frequency in a given bandwidth is

$$(\Delta f_{rms})^2 = \sigma^2(\varphi) = \int_{f_L}^{f_H} S_{\dot{\varphi}}(f)\, df. \qquad \text{(V-6)}$$

For a small bandwidth

$$B_w = f_H - f_L \qquad \text{(Hz)}$$

we can approximate relation (V-6) with

$$(\Delta f_{rms})^2 \approx B_w \cdot S_{\dot{\varphi}}(f) = B_w \cdot f^2 \cdot S_\varphi(f) \qquad \text{(V-7)}$$

and vice versa:

$$S_\varphi(f) \approx \frac{1}{B_w}\left(\frac{\Delta f_{rms}}{f}\right)^2 \quad (\text{rad}^2/\text{Hz}). \qquad \text{(V-8)}$$

By remembering the piecewise linearization of the noise characteristics

$$S_{\dot{\varphi}}(f) = S_y(f) \cdot f_0^2 \approx f_0^2\left[\frac{h_{-1}}{f} + h_0 + h_1 f + h_2 f^2\right] \qquad \text{(V-9)}$$

we can compute the basic coefficients h_k invariant with respect to the noiseless multiplication and division,

$$h_k = \left(\frac{\Delta f_{rms}}{f_0}\right)^2 \cdot \frac{1}{f^k B_w} \quad (\text{Hz}^{-k-1}) \qquad \text{(V-10)}$$

or

$$\Delta f_{rms,k}(f) = f_0 \sqrt{h_k f^k \cdot B_w} \quad (\text{Hz}). \qquad \text{(V-11)}$$

Note that in the flicker frequency noise range Δf_{rms} is proportional to $1/\sqrt{f}$, in the white frequency noise range it is constant, and in the white phase noise range it is proportional to the Fourier frequency f.

References

[V-1] M. Têtu, R. Brousseau, and J. Vanier, "Frequency-domain measurement of the frequency stability of a maser oscillator," *IEEE Trans. Instrum. Meas.*, vol. IM-29, pp. 94–97, June 1980.

[V-2] A. Godone, C. O. Weiss, and G. Kramer, "FM-noise measurements on an optically pumped FIR laser," *IEEE J. Quantum Electron.*, vol. QE-14, pp. 339–342, May 1978.

[V-3] C. H. Grauling and D. J. Healey, "Instrumentation for measurement of the short-term frequency stability of microwave sources," *Proc. IEEE*, vol. 54, pp. 249–257, Feb. 1966.

[V-4] W. J. McClintock, A. R. Faulkner, and M. J. B. Scanlan, "An automated close-to-carrier noise measurement system at *S*-band," *Marconi Rev.*, vol. XLIII, no. 219, pp. 185–202, 1980.

[V-5] T. J. Simmons and D. C. Smith, "Direct measurement of FM noise in 80 GHz solid-state oscillators," *Electron. Lett.*, vol. 18, pp. 308–309, Apr. 1982.

[V-6] W. Harth, D. Leistner, and J. Freyer, "FM noise measurement of *W*-band impatt diodes with a quasi-optical direct detection system," *Electron. Lett.*, vol. 18, pp. 355–356, Apr. 1982.

[V-7] A. Lance *et al.*, "Phase-noise measurements using cross-spectrum analysis," presented at Conf. Precision Electromag. Meas., Ottawa, Canada, June 1978.

[V-8] F. Labaar, "New discriminator boosts phase-noise testing," *Microwaves*, vol. 21, no. 3, pp. 65–66, 68–69.

[V-9] K. H. Sann, "The measurement of near-carrier noise in microwave amplifiers," *IEEE Trans. Microwave Theory Tech.*, vol. MTT-16, pp. 761–766, Sept. 1968.

A Microwave System for Measurements of AM and FM Noise Spectra

JOHN G. ONDRIA

Abstract—This paper describes a noise measuring system with a high degree of flexibility in regard to video frequency coverage as well as radio frequency coverage. Noise systems for *S*, *X*, and Ku-bands were constructed to obtain high sensitivity at spectral frequencies from several Hz to 30 MHz off the carrier. The measured limiting sensitivities [1] of the *X*-band set over the video range 20 Hz to 325 kHz with an input power of 10 mW are −145 dB for AM noise and 0.02 Hz rms for FM noise in a 100-Hz measuring bandwidth.

Basic Measuring Principles

At THE PRESENT time, the most commonly used commercial equipment for measuring noise in microwave sources is the Allscott [2] set, made in England, or the equivalent Spectra Electronics [3] set made in the U.S.A. Commercially available instrumentation and measurement sets are limited to a spectral range of 1 kHz to 140 kHz off the carrier. Further, the radio frequency range is narrow; therefore, full coverage of one band, such as *X*-band, may require several RF sections that comprise the whole of the microwave bridge and mixer circuits.

Present-day sources such as crystal-controlled varactor multiplier sources have noise close to the carrier frequency considerably lower than other microwave sources, therefore, the system measuring sensitivity must be correspondingly improved. These sources are particularly effective where precise frequency control is required, such as in coherent radar and for those communication systems where frequency allocations are precise and closely spaced. Recent incorporation of such sources in Doppler radar and other systems make use of the region from several Hz to 10 MHz off the carrier. In these applications, very pure sinusoidal oscillations are desired and the output spectrum would consist of a discrete line in the frequency domain. Generally, in addition to the desired component termed the carrier, there is a concentration of power surrounding the carrier as shown in Fig. 1. These spectral distributions on opposite sides of the carrier are termed *noise sidebands*. The distribution located below the carrier frequency is referred to as the *lower sideband*, and that part above the carrier frequency, the *upper sideband*. Usually there is some correlation between the spectral components on opposite sides of the carrier resulting in unsymmetrical [4] sideband distributions. When noise modulation indices are small, correlation noise can be neglected.

Manuscript received March 19, 1968; revised May 29, 1968. This work was supported in part by the U. S. Navy Bureau of Ships, under Contract NObsr-95009.

The author was with Microwave Associates, Inc., Burlington, Mass. He is now with the Department of Electrical Engineering, Lehigh University, Bethlehem, Pa. 18015

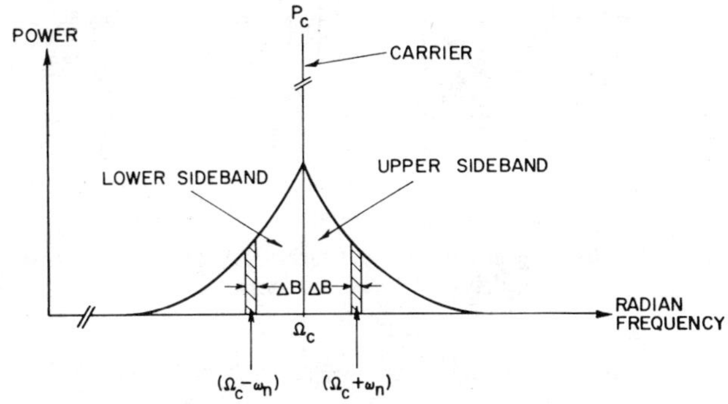

Fig. 1. Typical power-frequency spectrum of a microwave source. Ω_c is the carrier radian frequency. The carrier power P_c is of much greater magnitude than the mean sideband noise power.

It is useful to separate the noise sidebands into AM and FM components, and to discuss noise as a modulation phenomenon. Noise sidebands can be thought of as arising from a composite of low-frequency signals. Each of these signals modulate the carrier producing components in both sidebands separated from the carrier by the modulation frequency. The intensity of the sidebands generally limits the sensitivity of equipment used, i.e., in Doppler radar applications. It will be shown that the carrier has fluctuations in amplitude and phase or frequency. Amplitude fluctuations can be measured with a simple square law detector and phase or frequency fluctuations can be detected with a discriminator.

Small Signal AM Measuring Principles

Noise power spectra encountered in microwave sources are of very low magnitude compared with the carrier power; therefore, noise modulation indices are small and small signal principles can be used because the correlation noise is neglected. It is mathematically instructive to subdivide the noise continuum of total mean noise power P_{ni} into a large number of strips of bandwidth ΔB [5]. The noise power within each strip may then be replaced by an equivalent sinusoidal noise voltage source which has the same rms voltage and mean power as the replaced strip or band of noise, as shown in Fig. 2. Each equivalent source is assumed to have a mean frequency located at the center of its noise band. When all the equivalent sinusoidal noise voltages in the spectrum are added on a power or root-sum-square (rss) basis, the same total mean power as the actual noise spectrum is produced. Any type of interfering signal, such as the one located at $\Omega_c+\omega_n$ (upper sideband) in Fig. 2, can be

Reprinted from *IEEE Trans. Microwave Theory Tech.*, vol. MTT-16, pp. 767–781, Sept. 1968.

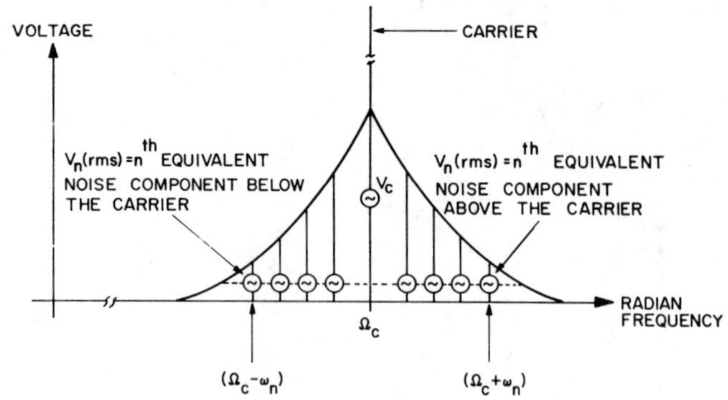

Fig. 2. Noise spectrum of Fig. 1 decomposed into an equivalent spectrum of sinusoidal noise voltages.

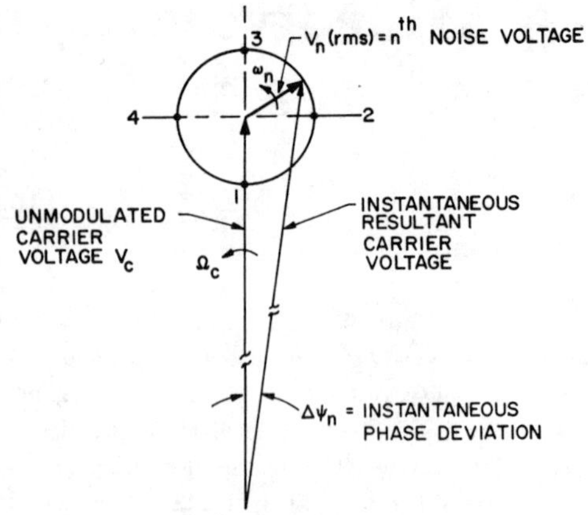

Fig. 4. Vector diagram of a carrier simultaneously phase and amplitude modulated by the nth noise voltage interference shown in Fig. 3(a) and expressed by (1).

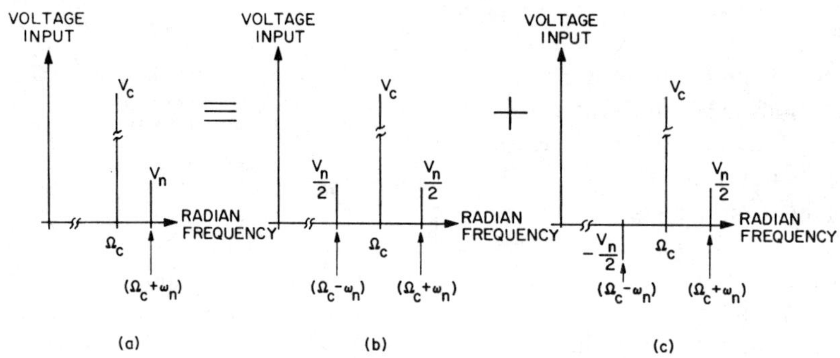

Fig. 3. Alternate representations of a carrier plus a single small noise sideband interference. (a) Carrier plus single interference signal. (b) Pair of symmetrical sidebands (pure AM). (c) Pair of antisymmetrical sidebands (pure FM).

represented as a pair of symmetrical sidebands plus a pair of antisymmetrical sidebands as shown in Fig. 3 [6], [7].

Continuing this procedure, another set of components is found, due to the presence of the equivalent noise voltage source at $\Omega_c-\omega_n$ (lower sideband), situated at the same frequencies shown in Fig. 3(b) and (c). When noise measurements are made, the noise power is actually measured at video frequencies and then referred to radio frequency.

Restricting our discussion to the case where the carrier power is much larger than the noise power, modulation of the carrier voltage by the noise voltage causes the only significant detected noise voltage at the video (baseband) frequency, as the effect of noise voltages modulating each other is negligible. The voltage expression for the carrier plus interfering sinusoidal signal [see Fig. 3(a)],

$$v(t) = \mathrm{Re}\left\{V_c\epsilon^{j\Omega_c t} + V_n\epsilon^{j}[(\Omega_c+\omega_n)t+\theta_n]\right\} \tag{1}$$

is represented in the vector diagram Fig. 4. The points marked 1 and 3 represent the positions of maximum and minimum carrier amplitude, respectively, for the single interference located at ω_n above the carrier frequency. The first term on the right-hand side of (1) is the carrier, while the second term is the interfering signal. For the small signal case, $V_n \ll V_c$, the instantaneous voltage $v(t)$ can be represented as [1]

$$v(t) \approx V_c\left[1+\frac{V_n}{V_c}\cos(\omega_n t+\theta_n)\right] \cdot \cos\left[\Omega_c t+\frac{V_n}{V_c}\sin(\omega_n t+\theta_n)\right]. \tag{2}$$

Equation (2) represents a carrier at radian frequency Ω_c that has been simultaneously amplitude and phase modulated at radian frequency ω_n. The amplitude of the unmodulated carrier is V_c. The interfering sinusoid has amplitude V_n and an arbitrary initial phase angle θ_n. Replacing the single sinusoidal interference by a summation of N uncorrelated disturbances [see Fig. 2], (1) becomes

$$v(t) = \mathrm{Re}\left\{V_c\epsilon^{j\Omega_c t} + \sum_{n=1}^{N} V_n\epsilon^{j[(\Omega_c+\omega_n)t+\theta_n]}\right\}. \tag{3}$$

The small signal assumption $V_n \ll V_c$ leads to

$$v(t) \approx V_c\left[1+\frac{1}{V_c}\sum_{n=1}^{N} V_n\cos(\omega_n t+\theta_n)\right] \cdot \cos\left[\Omega_c t+\frac{1}{V_c}\sum_{n=1}^{N} V_n\sin(\omega_n t+\theta_n)\right]. \tag{4}$$

For every equivalent noise voltage located at ω_n above the carrier, there is a corresponding equivalent noise voltage

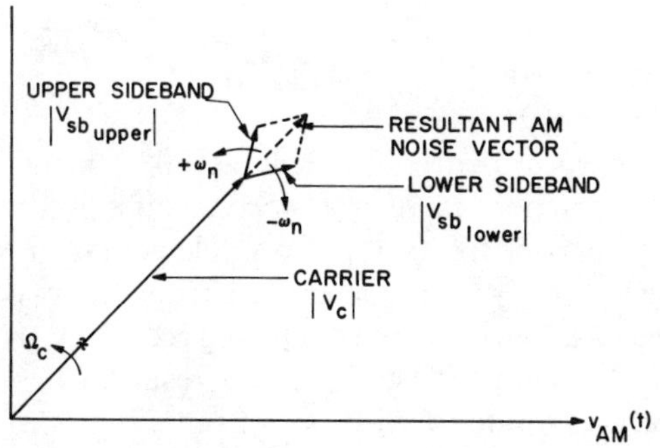

Fig. 5. Amplitude modulation of a carrier by two noise components located, respectively, above and below the carrier frequency Ω_c by radian frequency ω_n. The noise sidebands rotate in opposite directions relative to the carrier vector, causing the resultant vector amplitude to vary between a maximum and minimum value.

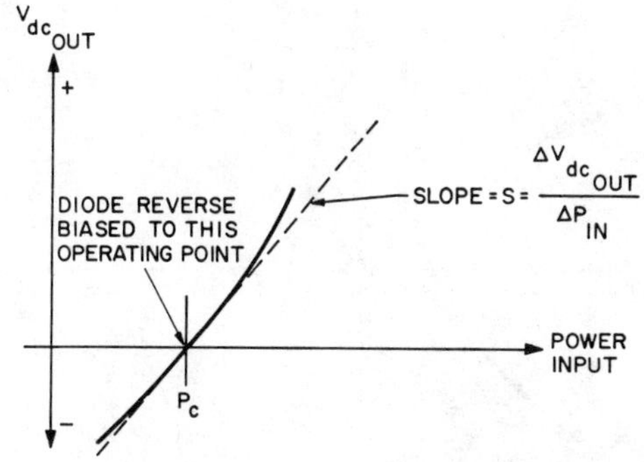

Fig. 6. Variation of dc output voltage as a function of input power after reverse biasing the diode to zero volts.

component situated at ω_n below the carrier frequency. The AM components for these two noise voltages can be represented in vector form as shown in Fig. 5 [6]. When these two sideband vectors rotate to the point where they are in the same direction as the carrier, the maximum amplitude condition results.

$$|V|_{\max} = |V_c| + |V_{sb_{\text{upper}}}| + |V_{sb_{\text{lower}}}|. \tag{5}$$

When the two sidebands rotate to the point 180° from this position, the minimum voltage results.

$$|V|_{\min} = |V_c| - |V_{sb_{\text{upper}}}| - |V_{sb_{\text{lower}}}|. \tag{6}$$

As negligible correlation is assumed

$$|V_{sb_{\text{upper}}}| = |V_{sb_{\text{lower}}}| = |V_{sb}| \tag{7}$$

and (5) and (6) become

$$|V|_{\max} = |V_c| + 2|V_{sb}| \tag{8}$$

$$|V|_{\min} = |V_c| - 2|V_{sb}|. \tag{9}$$

In terms of power

$$P_{\max} = \frac{|V|^2_{\min}}{R_0} = \frac{[|V_c| + 2|V_{sb}|]^2}{R_0}$$

$$= P_c\left[1 + 4\frac{V_{sb}}{V_c} + 4\frac{V_{sb}^2}{V_c^2}\right], \tag{10}$$

$$P_{\min} = \frac{|V|^2_{\min}}{R_0} = \frac{[|V_c| - 2|V_{sb}|]^2}{R_0}$$

$$= P_c\left[1 - 4\frac{V_{sb}}{V_c} + 4\frac{V_{sb}^2}{V_c^2}\right]$$

and

$$P_{\max} - P_{\min} = 8P_c\frac{V_{sb}}{V_c}. \tag{11}$$

In these equations

$$P_c = \frac{V_c^2}{R_0} \tag{12}$$

is the carrier power developed across the RF impedance R_0 of the diode, V_c the carrier voltage, and V_{sb} the voltage due to *one* sideband.

When the power spectrum of Fig. 1 is fed into a diode detector, its output voltage, as monitored with a high-impedance dc voltmeter, depends on the input power. If now, the diode is externally biased until the output voltage is zero and then, the input power is varied about the nominal position, the dc voltage output versus power input curve Fig. 6 is generated. The diode detector responding only to amplitude variation of the input signal is producing a demodulated output at video frequency ω_n due to the presence of the two noise sidebands equally located above and below the carrier frequency.

The value of the slope S at the operating point is an important system parameter that can be easily obtained by noting the incremental change in dc output voltage for an incremental change in input power, i.e.,

$$S = \frac{\Delta V_{\text{dc}_{\text{out}}}}{\Delta P_{\text{in}}}. \tag{13}$$

Variations of the signal, due to modulation by noise, produce extremely small perturbations about the operating point. This is shown magnified in Fig. 7. The detected peak-to-peak output voltage is

$$\Delta V = V_{\max} - V_{\min} = 2\sqrt{2}\,\Delta V_{\text{rms}} = S\Delta P$$
$$= S(P_{\max} - P_{\min}). \tag{14}$$

From (11) and (14)

$$\frac{V_{sb}}{V_c} = \frac{\sqrt{2}\,\Delta V_{\text{rms}}}{4SP_c} \tag{15}$$

and

$$\frac{\text{Power in One Sideband}}{\text{Power in Carrier}} = \frac{P_{sb}}{P_c} = \left\{\frac{V_{sb}}{V_c}\right\}^2$$
$$= \left\{\frac{\sqrt{2}\,\Delta V_{\text{rms}}}{4SP_c}\right\}. \tag{16}$$

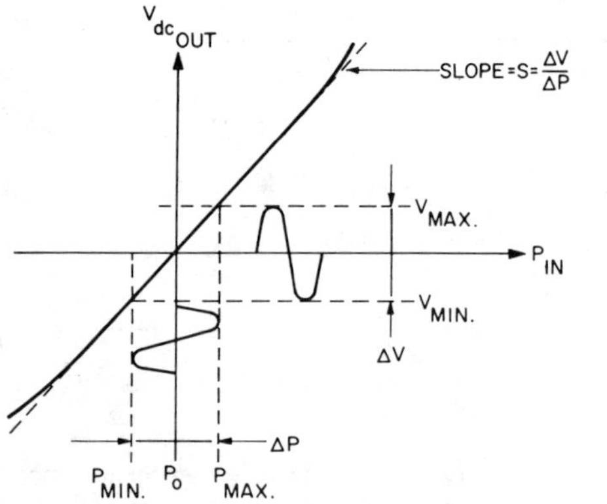

Fig. 7. Operating point of Fig. 6 shown magnified.

The power in two sidebands relative to the carrier power is

$$\frac{P_{AM}}{P_c} = \frac{\text{Power in AM Sidebands}}{\text{Power in Carrier}} = \frac{2P_{sb}}{P_c} = \frac{1}{4}\left\{\frac{\Delta V_{rms}}{SP_c}\right\}^2. \quad (17)$$

The quantity ΔV_{rms} is the root-mean-square output noise voltage developed across the diode at the video frequency ω_n for a given bandwidth window ΔB.

Small Signal FM Measuring Principles

Fig. 4 shows that the degree of amplitude and phase modulation of the carrier depends only on the magnitudes of the noise vectors. The amplitude and phase of the resultant carrier vector change with a frequency dependent on the rotational speed of the noise vector. When the noise modulation indices are very small, only the first-order sidebands of the FM spectrum need be considered [6], [8], [9]. The desired small-signal FM measuring principles can be obtained from the vector diagram Fig. 4 and the voltage-frequency spectrum Fig. 3(c). The time-dependent voltage (2)

$$v(t) = V(t)\cos\phi(t)$$

has, by definition, instantaneous phase

$$\phi(t) = \Omega_c t + \frac{V_n}{V_c}\sin(\omega_n t + \theta_n), \quad (18)$$

and instantaneous phase deviation

$$\frac{V_n}{V_c}\sin(\omega_n t + \theta_n) = \Delta\phi_n(t). \quad (19)$$

The maximum or peak value of the phase deviation is V_n/V_c, defined as the FM modulation index $\Delta\phi_{max}$. The instantaneous radian frequency deviation $\Delta\omega(t)$ of voltage $v(t)$ is by definition the first time derivative of the instantaneous phase deviation. For one noise disturbance located ω_n above the carrier

$$\Delta\omega_1(t) = \frac{d[\Delta\phi_n(t)]}{dt} = \frac{V_n}{V_c}\omega_n\cos(\omega_n t + \theta_n), \quad (20)$$

$$|\Delta\omega_1|_{max} = \left|\frac{V_n}{V_c}\omega_n\right| = \left|\frac{V_{sb}}{V_c}\omega_n\right|. \quad (21)$$

For every equivalent noise voltage component located ω_n above the carrier, there is a corresponding component ω_n below the carrier frequency. From elementary FM modulation theory [6], the frequency spectrum for two sidebands [see Fig. 3(c)] can be represented vectorially as shown in Fig. 8. Maximum phase deviation results when *both* sideband vectors rotate to positions of right angles to the carrier vector. When this occurs,

$$\tan|\Delta\phi_n|_{max} \approx |\Delta\phi_n|_{max} = \frac{2V_{sb}}{V_c} \quad (22)$$

and

$$|\Delta\omega|_{max} = \frac{2V_{sb}}{V_c}\omega_n.$$

In terms of frequency deviation

$$\Delta f_{peak} = \sqrt{2}\,\Delta f_{rms} = \frac{2V_{sb}}{V_c}f_n. \quad (23)$$

In terms of power

$$\frac{\text{Power in One Sideband}}{\text{Power in Carrier}} = \frac{P_{sb}}{P_c} = \left(\frac{V_{sb}}{V_c}\right)^2 = \frac{\overline{\Delta f^2_{rms}}}{2f_n^2}, \quad (24)$$

the power in two sidebands relative to the carrier power is

$$\frac{P_{FM}}{P_c} = \frac{\text{Power in FM Sidebands}}{\text{Power in Carrier}} = \frac{2P_{sb}}{P_c} = \overline{\Delta\phi^2} = \frac{\overline{\Delta f^2_{rms}}}{f_{n2}}. \quad (25)$$

The usual definition of rms frequency deviation obtained from (24) is

$$\Delta f_{rms} = \sqrt{2}f_n\sqrt{\frac{P_{sb}}{P_c}}. \quad (26)$$

This definition may be further illustrated by the following example. Assuming that a microwave source has an FM noise of Δf_{rms} = 10 Hz as measured in a 1-kHz bandwidth at 100 kHz from the carrier frequency, then in this bandwidth the noise power is the same as the power that would be measured in one sideband if a noise-free source of the same carrier level and same frequency was frequency modulated at 100 kHz with a carrier frequency deviation of Δf_{rms} = 10 Hz.

Frequency modulation noise or rms frequency deviation, can be measured by an AM detection system after it is converted to amplitude modulation noise. In practice, this conversion can be achieved with a discriminator or by changing

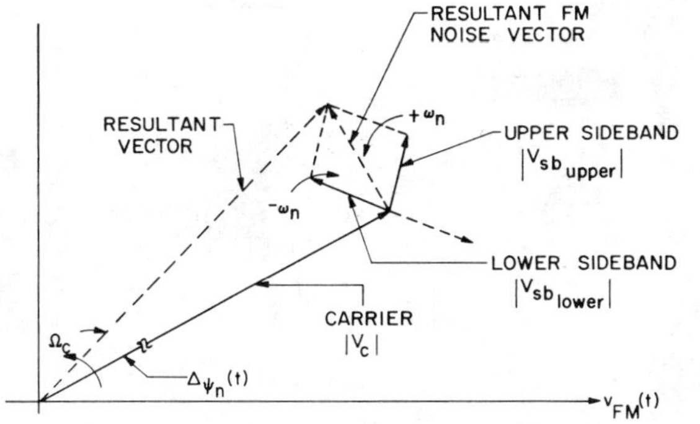

Fig. 8. Frequency modulation of a carrier by two noise components located, respectively, above and below the carrier frequency Ω_c by radian frequency ω_n. The significance of the negatively directed lower sideband component in Fig. 3(c) is shown vectorially in this figure. The resultant FM noise vector lies at right angles to the carrier vector.

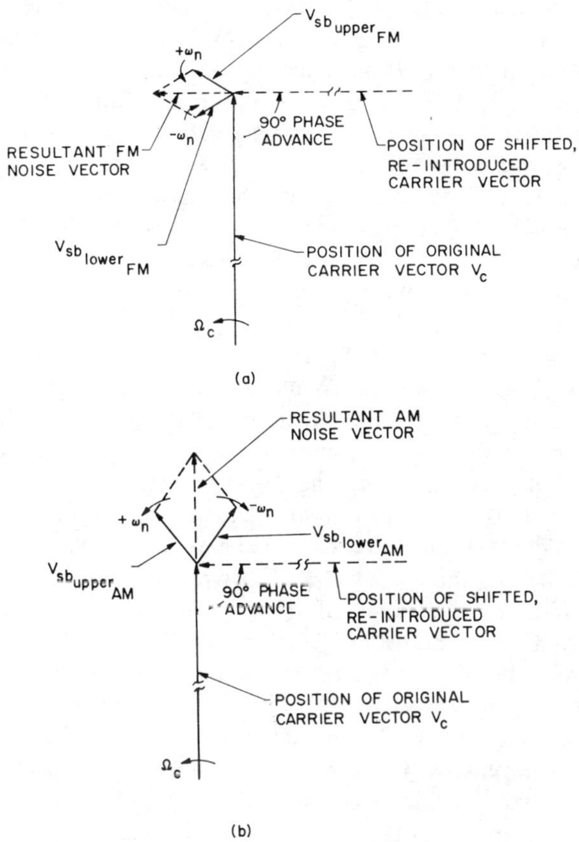

Fig. 9. (a) Vector diagram representation of conversion of FM into AM by suppressing the carrier and reintroducing it phase advanced by 90 degrees. (b) Vector diagram showing simultaneous conversion of the AM into FM.

the phase of the carrier, relative to the quadrature noise components, by 90 degrees. At the same time, the AM noise components are converted to FM and are effectively removed. Fig. 9(a) and (b) show that for a 90° phase advanced (retarded) carrier, frequency modulation is converted to amplitude modulation, while the amplitude modulation is converted to frequency modulation. The physical realization of this conversion will be shown when the noise measuring system is explained.

Deleterious Environment Effect

Noise contributions from extraneous effects such as radio frequency interference, mechanical vibrations, power supply instabilities, and power-line pickup may sometimes exceed the quantity of noise generated within the device under test.

In order to minimize radio frequency interference effects, noise measurements must be performed within an electromagnetically shielded room. Power-line pickup is effectively reduced with the aid of filters located in series with each power line feed at the point of entry to the room. Power-line instabilities can be minimized with harmonically neutralized voltage regulators. Storage cells should be used to power the measuring equipment, diodes, amplifiers, and, where possible, the spectrum analyzers as well as the devices under test. Shielding of cables, use of shock-absorbing material, and interconnection of source and measuring system with flexible waveguide are essential in minimizing extraneous noise pickup.

Noise Measuring System

Fig. 10(a) illustrates the basic method used for AM noise measurement and is generally referred to as a direct-detection system. When the power spectrum, cf. Fig. 1, is fed into the detector, the upper and lower AM noise sidebands beat with the carrier to produce a video power output spectrum of the form shown in Fig. 10(b). When the modulation indices are small, the first-order FM noise sidebands are in antiphase and cancel [1], [10].

In order to measure FM noise with a direct-detection system, some form of FM-to-AM conversion, and preferably the reverse process as well, is required. Fig. 11(a) illustrates the simplest method of conversion using a discriminator and AM detector. The system sensitivity is low when the bandwidth is broad and high when the bandwidth is narrow (high Q), but the measuring range is essentially restricted to the linear portion of the discriminator response curve.

FM-to-AM conversion, and the reverse process as well, can also be achieved by suppressing the carrier with a microwave filter and then reintroducing it into the system shifted by the required 90 degrees, [see Fig. 9(a) and (b)]. Of course, the direct-detection system also indicates, besides the AM or converted FM noise, its own noise; however, that can be accounted for by application of a balanced mixer in place of the detector [1].

Basic Measuring System

A block diagram of the complete noise measuring system is shown in Fig. 12. For detection, low-noise *backward diodes* were used in a balanced mixer configuration with provisions for electrically switching diodes between the unbalanced and balanced conditions. In both modes of operation, the diodes were externally back biased to the zero voltage point of the voltage-output versus power-input curve [see Fig. 6].

After detection, the video signals are introduced into a broad-band high-gain low-noise amplifier. In order to prevent overloading of the amplifier, particularly by 60-Hz

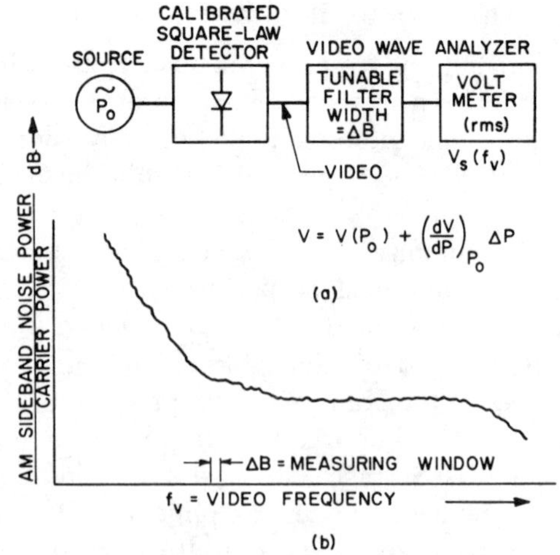

Fig. 10. (a) Block diagram illustrating the basic direct detection system for the measurement of AM noise fluctuations of oscillators and other sources. (b) Typical measured AM power spectra for a varactor microwave source measured at video frequencies with a tunable wave analyzer.

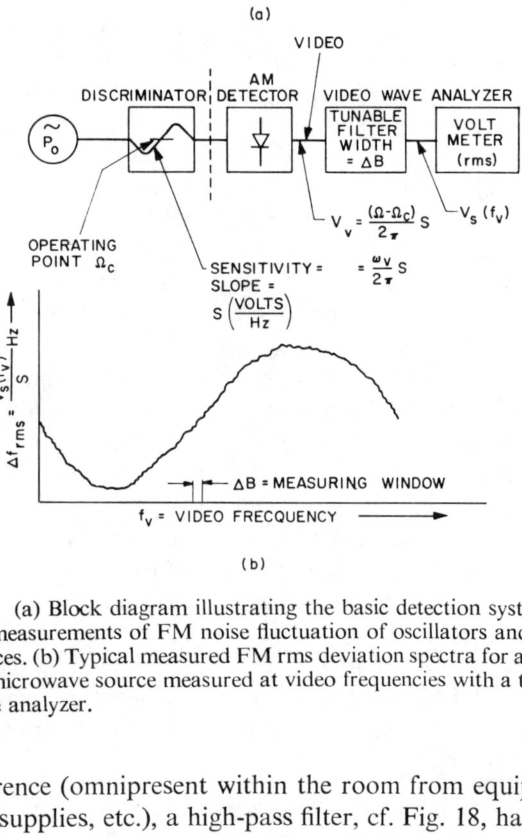

Fig. 11. (a) Block diagram illustrating the basic detection system for the measurements of FM noise fluctuation of oscillators and other sources. (b) Typical measured FM rms deviation spectra for a varactor microwave source measured at video frequencies with a tunable wave analyzer.

interference (omnipresent within the room from equipment power supplies, etc.), a high-pass filter, cf. Fig. 18, having a measured suppression of 52 dB at 60 Hz and essentially 0 dB above 1 kHz, was designed [1] and inserted between balanced mixer and low-noise amplifier. For accurate noise measurements at video frequencies below 1 kHz, this filter must be switched out since it greatly suppresses the relatively low-noise components to be measured. For measurements below 100 kHz, the measuring system was converted to total battery operation and all ac power supplies to the shielded room disconnected. Lighting was subsequently derived from storage cells.

Before describing the methods for measuring AM and FM noise, the operation of a balanced mixer and a special low-

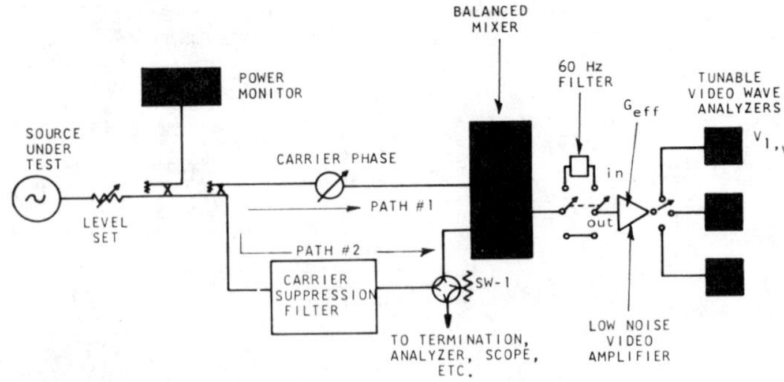

Fig. 12. Block diagram of AM and FM noise measuring system.

loss carrier suppression filter will be first considered since these form the basis of the measuring method used here.

Operation of a Balanced Mixer

A balanced mixer [11]–[14] consisting of two diode detectors fed from a hybrid junction is often employed at microwave frequencies. It is often used in a superheterodyne receiver to cancel the local oscillator AM noise. Although we are dealing here with a *direct-detection* system, the term *balanced mixer* is still used.

The qualitative operation of a balanced mixer, in relation to the hybrid-T configuration is illustrated in Fig. 13. AM noise modulation of the local oscillator is cancelled, whereas the received signal is retained.

As shown in Fig. 13, the mixer diodes are positioned in the coplanar arms 1 and 2. The received signals enter the E-plane, arm 3, the local oscillator power (carrier plus oscillator noise components) through the H-plane, arm 4, and have the instantaneous polarities in the directions indicated in Fig. 13. Looking into the H-plane arm [see Fig. 13(b)], the local oscillator power at frequency Ω_{LO} and its noise components Ω_n are equally divided between the two coplanar arms arriving at diodes A and B with the same polarities. The received carrier signal Ω_c, amplitude modulated at ω_n, entering the E-plane arm [see Fig. 13(c)] divides equally between the two coplanar arms and arrives at diodes A and B with opposite polarities. If the output terminals of the diodes are connected together and then to an IF (video) amplifier, only the various difference frequencies due to nonlinear mixing will be selected. In Table I, the relative polarity of the components at each diode is deduced for the arbitrary choice of instantaneous polarities of the received and local oscillator signals chosen in Fig. 13(a). The polarities of the first three components of Table I are obvious from Fig. 13(b) and (c). The directions of the last three components can be written down directly recalling that the *difference frequencies* in the mixing process actually result from the *product* of two exponential terms. Thus, for example, the fifth set of components listed in Table I are shown positive (upward) at diode A but negative (downward) at diode B. Since both Ω_c and Ω_{LO} are positive at diode A, their product is positive, whereas Ω_{LO} is positive at diode B but Ω_c is negative at diode B; thus, their product is negative. The remaining frequency components are obtained in a similar manner.

Figs. 13(b) and (c) show that the two diodes appear *balanced* (in antiparallel) to the local oscillator signals Ω_{LO}

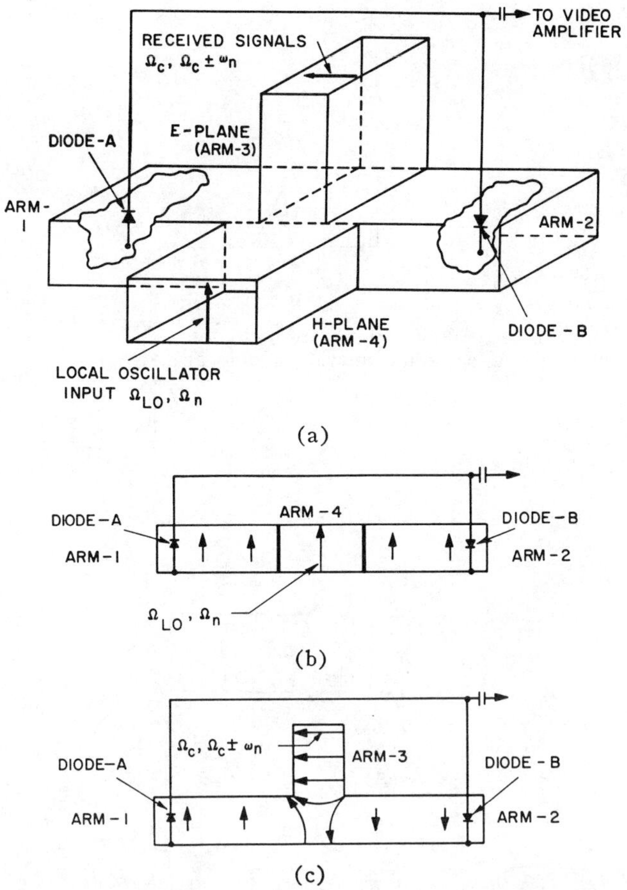

Fig. 13. (a) Hybrid-T balanced mixer shown with the diodes arranged to cancel local oscillator noise. The sense of direction chosen for the received and local oscillator signals is purely arbitrary. Interchanging the local oscillator and signal ports produces the same results. Ω_c is the received carrier frequency that is amplitude modulated at $\omega_n \ll \Omega_c$. (b) Relative sense of polarity of Ω_{LO} and Ω_n at diodes A and B for the assumed H-plane input polarity. (c) Relative sense of polarity of Ω_c and $\Omega_c \pm \omega_n$ at diodes A and B for the assumed E-plane input polarity.

and Ω_n, but are *unbalanced* (in parallel) to the received signals Ω_c and $\Omega_c \pm \omega_n$. This means that the local oscillator noise (fourth set of components in Table I) will cancel at the IF; whereas, the received signal (fifth set of components in Table I) will be detected.

The same results (local oscillator AM noise cancellation) are obtained if the received signal and local oscillator ports are reversed.

Carrier Suppression Filter

A typical power-frequency spectrum of a microwave oscillator is shown in Fig. 1. Most of the available power is contained in the carrier since the sideband power is generally more than 60 dB below the carrier level. When crystal diodes are employed in a direct-detection system, the carrier may be thought of as supplying the local oscillator power, the sidebands being the signal. In the FM noise measuring system employed, the local oscillator signal is derived from the same source as the noise to be measured. This signal, local oscillator plus noise, enters the H-plane arm of a balanced mixer where, it was seen, the noise entering this arm is effectively suppressed, resulting in a pure L.O. signal.

TABLE I

Frequencies of Interest at Diodes A and B of Fig. 13 and Their Relative Polarities

Signal Component	Frequency Designation	Relative Polarity at Diode A	Relative Polarity at Diode B
1) Local oscillator power (RF)	Ω_{LO}	↑	↑
2) Local oscillator noise (RF)	Ω_n	↑	↑
3) Received signal (RF)	Ω_c	↑	↓
4) Local oscillator noise (IF)	$\Omega_{LO}-\Omega_n$	↑	↑
5) Received signal (IF)	$\Omega_c-\Omega_{LO}$	↑	↓
6) Received signal and local oscillator noise (IF)	$\Omega_c-\Omega_n$	↑	↓

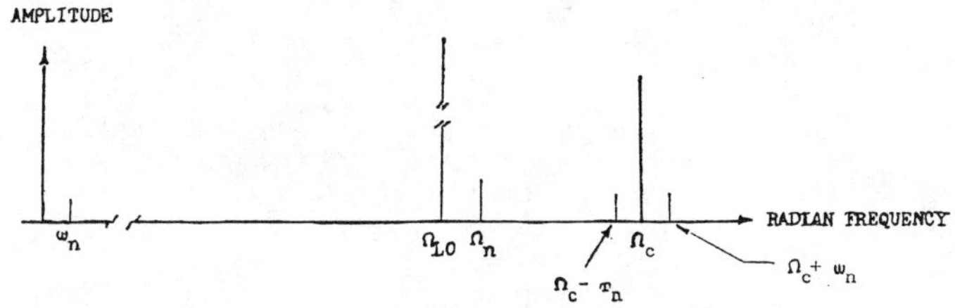

Relative position of radian frequency components listed above.

The noise which is to be measured is introduced into the E-plane arm. If the carrier power entering the E-plane arm is *somehow removed*, the relative strength of the noise may be raised and the resulting measuring sensitivity will be correspondingly increased. When FM noise measurements are made, FM is converted to AM while the carrier is removed before applying the noise to the E-plane arm of the balanced mixer. The circuitry that effects the removal of the carrier is termed a *carrier suppression filter*. An ideal filter would provide infinite suppression at the carrier frequency [see Fig. 14(b)] leaving the noise sidebands unattenuated (i.e., zero suppression at all other frequencies). An actual filter always exhibits a finite bandwidth that suppresses, in addition to the carrier, noise components on both sides of the carrier [see Fig. 14(c)]. The Q of the circuit determines the degree to which these sidebands are attenuated. In addition, suppression filters inherently possess a constant insertion loss which, for the case of the infinite-Q filter, is just the difference between the noise power level P_{n_a} that enters the filter and the corresponding noise power level P_{n_b}, which emerges from the filter [see Fig. 14], i.e., insertion loss $=P_{n_a}-P_{n_b}$.

Several microwave carrier suppression filters have been described in the literature [15] and range from the very simple [16] to the highly sophisticated loop filters [10]. Sometimes a Pound discriminator [11], [14] or a simple magic-T carrier suppression filter [10], [17] that bears a superficial resemblance to the Pound stabilizer is employed, but in practice it is found to have an insertion loss of about 10 dB. Even the best loop-type filter is reported to have, at best, an insertion loss of 4.5 dB. The filter described here is extremely simple in construction and exhibits an insertion loss of, at most, a few tenths of a dB.

The filter shown in Fig. 15(a), requires, in principle, only

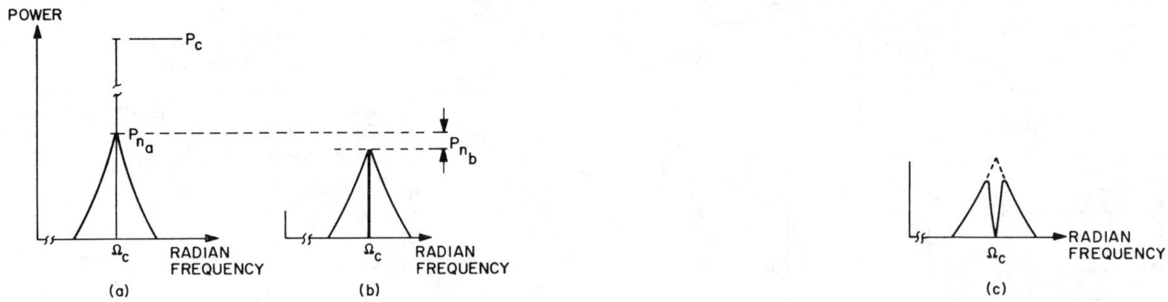

Fig. 14. (a) Carrier plus noise sidebands. (b) Infinite Q filter with carrier alone suppressed and noise sidebands decreased by the insertion loss of the filter. (c) Actual filter showing additional sideband attenuation close to the carrier.

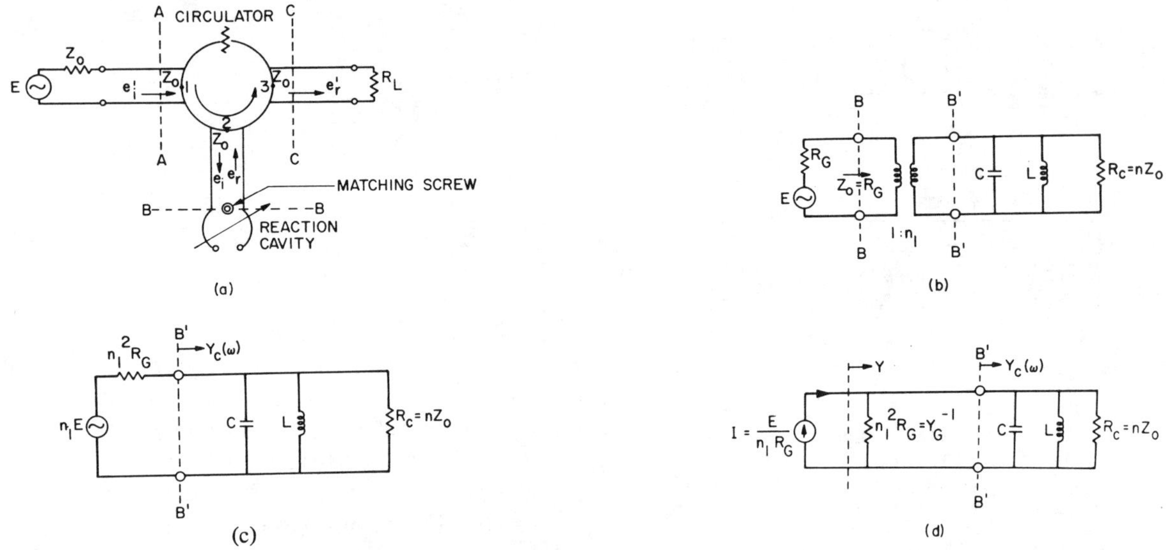

Fig. 15. (a) A carrier suppression filter employing a ferrite circulator. The typical insertion loss between points 1 and 3 is approximately 0.3 dB. (b) Equivalent circuit of the input circuit plus cavity for the carrier suppression filter in (a). (c) Thevenin voltage source to the left of plane B-B in (b), transformed into the cavity circuit assuming an ideal transformer. $Y_c(\omega)$ is the cavity admittance. (d) Equivalent Norton current source of (c). Y is the total circuit admittance.

two components, a tunable cavity, and a ferrite circulator. Provision for matching the cavity is also necessary but this is usually available as a simple matching screw at the coupling iris of the cavity input. Since the initial cavity did not possess a matching screw, a slide screw tuner was employed. Slide screw tuners, while effectively matching the cavity to the guide at the resonant frequency, also tend to distort the power spectrum around resonance. This must be avoided for noise measurements close to the carrier; therefore, a cavity was used with a matching screw located at the coupling iris so that sideband distortion was eliminated. The cavity had a vernier micrometer located at the opposite end from the coarse tuning micrometer. This construction can be recommended whenever high-Q cavities are employed in any microwave filter to effect precise frequency tuning.

The magnitude and phase of the transfer function of the carrier suppression filter employing a ferrite circulator are derived in the Appendix using equivalent circuits [18]–[20] [Fig. 15(b)–(d)]. Suppression and phase [see (52) and (51), respectively] of the suppression filter are plotted [see Figs. 16 and 17] for a resonant frequency of 9.530 GHz with Q_L as a parameter. The insertion loss for the circulator-type carrier suppression filter depends mainly on the insertion loss of the circulator, which is typically on the order of 0.3 dB. Thus, for all practical purposes, the ordinate of Fig. 16 may be taken as the total suppression.

The Complete Measuring System

Noise measurements are performed using a combined AM and FM noise measuring system shown in block diagram form in Fig. 18. For AM noise measurements path 2 is interrupted by way of waveguide switch SW-1, placing a termination on the E-plane arm of the balanced mixer hybrid-T. The remaining system is a basic direct-detection system already shown in its elementary form in Fig. 10(a). When the output power spectrum of the oscillator under test [see Fig. 1] is fed into the H-plane arm of the balanced mixer, the carrier signal serves as the local oscillator power and the

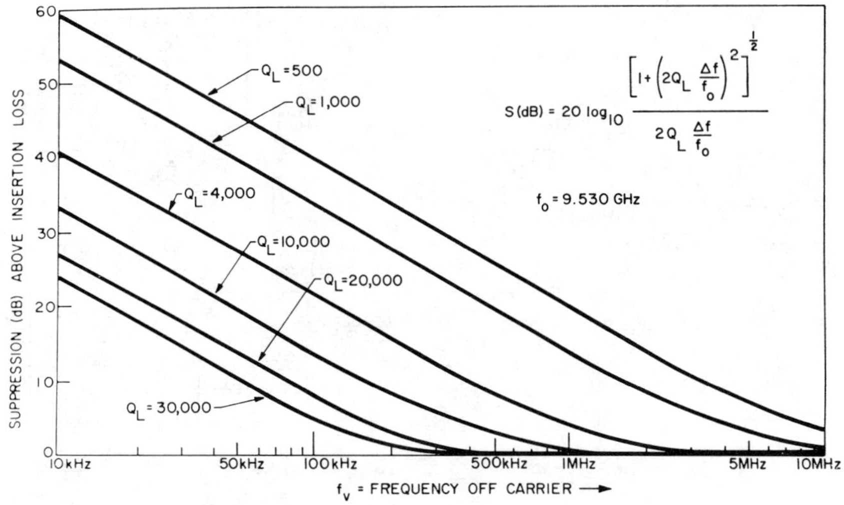

Fig. 16. Calculated suppression curves for circulator type C. S. filters.

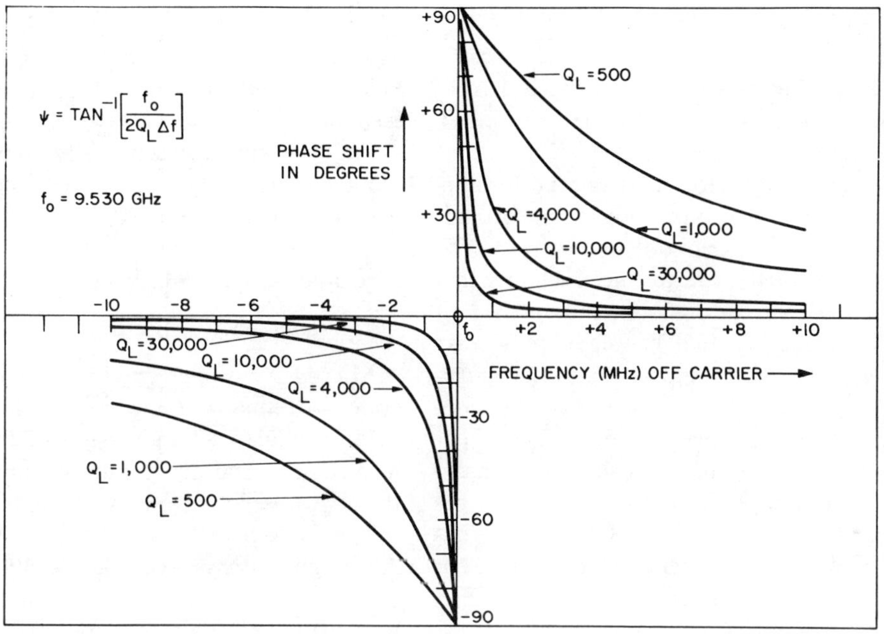

Fig. 17. Frequency dependence of insertion phase shift of circulator type carrier suppression filter.

sidebands as the signal to be detected. The upper and lower AM noise sidebands beat with the carrier to produce an output; whereas, the beat products of carrier and FM sidebands for small modulation indices are in antiphase and cancel. The relatively high carrier power is converted to dc and effectively blocked from the succeeding video amplifier.

One diode is dc isolated from the system to permit electrical switching of the mixer between the unbalanced and balanced conditions. In the unbalanced condition, the diodes appear in parallel to the H-plane arm and yield a demodulated output voltage v_1, consisting of AM noise from the source under test plus system noise [see Fig. 7]. When the isolated diode is electrically reversed, the mixer is placed into the balanced condition; the two diodes appear in antiparallel to the noise sidebands of the source under test. If the diodes are carefully matched, mixer suppression factors greater than 60 dB can be achieved and the slope of the output voltage versus power input curve [see Fig. 7], is essentially zero. Any demodulated output voltage v_2 that results when the mixer is in the balanced position, is due to system noise only. Since source and system noise voltages are uncorrelated, the true source noise voltage $\Delta V_{\rm rms}$ used to calculate the noise-to-signal ratios [see (17)] is the square root of the difference of the squares of these two readings [1].

$$\Delta V_{\rm rms} = \sqrt{v_1^2 - v_2^2}. \tag{27}$$

Since the accuracy of the readings is greatly dependent upon proper mixer adjustments, the characterization, biasing, and matching of diodes will be discussed in a subsequent section.

The demodulated video output voltage is introduced into a low-noise high-gain amplifier and then into a series of

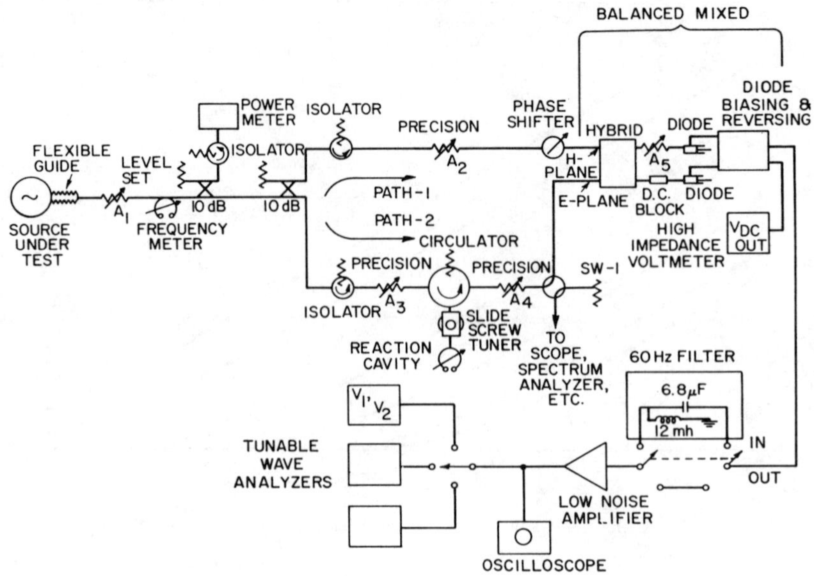

Fig. 18. Complete AM and FM noise measuring system.

tunable wave analyzers. For measurements above 1 kHz from the carrier, a 60-Hz filter is switched in between mixer and amplifier [see Fig. 18].

FM noise measurements are performed using both paths shown in Fig. 18. A portion of the source power fed through path 1 enters the H-plane arm of the hybrid and serves as the local oscillator signal. The diodes are balanced against this signal to suppress the local oscillator noise, thus providing the required clean carrier [see Fig. 13]. The signal plus noise entering path 2 arrive at the carrier suppression filter where the carrier is effectively removed. The phases of the sidebands are also changed by the suppression filter. The upper sidebands are phase-advanced and the lower sidebands phase-retarded by the same amount. This effect is derived in the Appendix and the results shown in Fig. 17. FM to AM conversion occurs if the phase of the reintroduced carrier is shifted 90 degrees with respect to the sidebands [see Fig. 9(b)]. The relatively low noise power emerging from the carrier suppression filter enters the E-plane arm of the hybrid and mixes with the clean *shifted carrier* entering the H-plane arm. The signal is demodulated since the diodes now appear unbalanced (in parallel) to signals entering the E-plane arm. The resultant video (converted FM) output is processed by the same video circuits used in the AM noise measurements. For FM noise measurements, the mixer is in the balanced condition to obtain v_1. The system noise voltage v_2 is obtained by interrupting path 2 via SW-1 and keeping the mixer in the balanced condition.

Further improvement in AM sensitivity is possible by using both paths of the measuring system as in the case of FM noise measurements. If the carrier is suppressed in one path of the bridge and the phase shifter in the other path adjusted to a position exactly 90 degrees away from the setting required for the FM measurement, full noise power can be applied to the diodes. Depending upon the magnitude of the available source power, the bridge input coupler may be increased to a value larger than the 10 dB coupler shown in Fig. 18 and the system sensitivity will increase by a corresponding amount.

The limiting AM sensitivity (noise-to-signal ratio) using the direct-detection system was measured to be −145 dB over the video range 20 Hz to 325 kHz for 1.2 mW of carrier power into the mixer, the noise power having been measured in a 100-Hz bandwidth.

Selection and Characterization of Diodes

Microwave diodes employed in the balanced mixer generate additional noise, exceeding thermal or Johnson noise [21], over a fairly wide band of frequencies. This noise is measured at the diode output along with the source noise that is to be determined when v_1 is obtained. The effects of this noise contribution can be accounted for through v_2. Although it is not necessary to know the absolute amount of this noise, too excessive an amount will reduce the system measuring sensitivity. A special class of detector diodes, known as *backward diodes*, exhibit a reduced noise spectrum at the low video or intermediate frequencies compared to the spectrum of conventional mixer diodes such as the 1N23G. At 100 Hz, the noise figure improvement is better than 30 dB and at 10 kHz it is better than 20 dB [22].

Experiments prove that the use of a noise lamp as a noise power reference is not practical for measurements at video frequencies less than a few MHz. In this region the excess mixer diode noise usually exceeds the available noise power of the lamp. In addition, the low-frequency flicker noise of lamp beam current yields erratic and nonrepeatable results. Various methods [10] have been proposed to extend the measurable noise range by about 20 dB above the noise level of the standard waveguide lamp, but this is still not sufficient for measurements at spectral frequencies below a few hundred kHz off carrier. Even at frequencies of 1 MHz removed from the carrier, where source noise is relatively low, the noise lamp substitution method exhibits anomalous behavior and the measured results vary widely. For these reasons, the signal generator method [21] is to be preferred, demanding a complete knowledge of the mixer characteristics and the following circuitry parameters. This requires initial measurements of diode video impedance, conversion loss, noise temperature ratio, dc rectified output voltage as a function of input power, dc current required to bias the diodes to the optimum operating point [see Fig. 6], etc. As it is generally not possible to obtain diode pairs that are identical in all respects, a batch of crystals is first characterized independently and a selection of nearly matched pairs [1] made after analysis of the tabulated resultant data. The impedances of backward diodes are low; the IF impedance is typically 50 to 150 ohms.

Diode Matching and Conversion Loss Results

The matched diode pairs are examined in the balanced mixer. When the diodes are inserted in the holders, RF drive power is applied and the waveguide mount adjusted for maximum rectified dc output voltage. Each diode is initially RF matched separately; the alternate diode being disconnected from the biasing circuit as the other diode is adjusted. The diodes are then placed in the unbalanced (parallel) condition and the total video impedance measured. Next they are switched to the balanced (antiparallel) condition and the total video impedance measured again. A plot of these impedance values as a function of RF drive power is made. As the RF drive level is increased from about 0.3 mW per diode to 0.8 mW per diode, the video impedance for the unbalanced diode pair decreases from about 65 ohms to 41 ohms, respectively, and decreases from about 35 ohms to 28 ohms in the balanced condition. An optimum drive level is established to yield an unbalanced video impedance of approximately 50 ohms. This value is chosen since the noise figure of the subsequent amplifiers can conveniently be measured with a 50-ohm load. The amplifier noise figure at this impedance level is approximately 4.0 dB over its useful range.

Conversion loss (L_c) measurements are made using the dc incremental method [21], [23]. (This method was chosen since it is commonly used to establish an absolute calibration of standard crystals.) The measurement is relatively simple to perform in that it resembles the method required to obtain the slope S of the voltage output versus power input curve shown in Fig. 6. This method is also convenient since conversion loss is independent of the intermediate frequency chosen (i.e., L_c does not vary over the video range). The typical value of conversion loss for the two diodes in parallel, under optimum drive conditions and proper biasing, is found to be on the order of 6.5 dB. Overall receiver noise figure as calculated from Frii's [24] relationship is typically 9 dB [1].

System Calibration

Most microwave noise measuring systems usually require expensive auxiliary equipment for calibration purposes. In addition, it is desirable to periodically check the calibration during extensive noise measurements. In most cases this requires removai of the source under test, reactivation of the auxiliary calibrating equipment, and finally reconnection of the source under test to the measuring system. Calibration of the noise measuring system described, as shown in Fig. 18, avoids these difficulties since calibration can be accomplished by using the source under test. This flexibility permits rapid checks on system performance and insures accurate calibration at each measuring frequency. For this reason, narrow-band tunable diode mounts can be used and the mixer easily balanced to guarantee the highest degree of sensitivity and suppression for each source under test.

AM Calibration

For AM noise measurements usually only path 1 [see Fig. 18] is used. With diodes in the unbalanced (parallel) position, the largest value of SP_c obtainable is desirable to achieve maximum sensitivity [see (17)]. The notation S_uP_c will be adopted for this condition where S_u is the slope of the voltage output versus power input curve for the parallel diodes at the operating point [Figs. 6 and 7]. When the diodes are switched to the balanced position with respect to the H-plane arm, the value of the slope S_{Bal} should be made as close to zero as possible; i.e., the mixer diodes should be so chosen and so matched that an incremental change (amplitude variation) in local oscillator power input to the H-plane arm produces essentially no change in dc output voltage. When this condition is achieved, the balanced mixer exhibits a high degree of AM suppression. Ideally

$$\text{Suppression (dB)} = 20 \log_{10}\left(\frac{S_u}{S_{\text{Bal}}}\right) \to \infty. \qquad (28)$$

If the diodes differ somewhat in their characteristics, it is usually possible to slightly detune one diode to yield a high degree of suppression in the balanced position (i.e., $S_{\text{Bal}} \to 0$) and yet not reduce the sensitivity in the unbalanced position. This detuning, however, must be done with discretion. If the detuning is too severe, among other things, the system noise figure will degrade rapidly. In all cases, there is no substitute for a well matched pair of diodes.

After the dc suppression factor has been established, the suppression factor as a function of video frequency is found. By applying an amplitude modulated CW signal to the balanced mixer at the required sideband frequency, the demodulated output voltage is measured with a tunablewave analyzer set to the corresponding video frequency in the unbalanced and then the balanced conditions. To ensure accuracy, the modulation sidebands should be well above the level of the noise that the mixer introduces. The value of the suppression factor at each frequency is

$$\text{Suppression (dB)} = 20 \log_{10} \frac{V_{\text{out}}\ (\text{Unbalanced})}{V_{\text{out}}\ (\text{Balanced})}. \qquad (29)$$

With properly matched diode pairs, the measured suppression factor of the balanced mixer as a function of video frequency was greater than 60 dB over the video range extending from essentially 20 Hz to 30 MHz.

The output from the mixer is passed through a high-pass filter, a low-noise amplifier, and then to a series of tunable wave analyzers [Fig. 18]. The noise voltages v_1 and v_2 read at the analyzers must be referred back to the mixer terminals. Therefore it is necessary to measure the effective gain G_{eff} between mixer and analyzer as a function of video frequency under the same conditions as the corresponding noise measurements; i.e., under consistent amplifier gain settings and analyzer bandwidths. The rms source output noise voltage measured at the analyzers in a given bandwidth ΔB is [see (27)],

$$\Delta V_{out}(\Delta B) = \sqrt{v_1^2 - v_2^2}$$

and referred to the mixer terminals

$$\Delta V_{rms}(\Delta B) = \frac{\Delta V_{out}(\Delta B)}{G_{eff}} = \frac{\sqrt{v_1^2 - v_2^2}}{G_{eff}} \tag{30}$$

where v_1 is the rms output voltage measured at the analyzer in a given bandwidth ΔB at a given video frequency when the mixer is unbalanced and represents AM source plus system AM noise voltage; v_2 is the same as v_1 except for the case when the mixer is balanced and represents system AM noise voltage when the mixer suppression is high. The noise-to-signal ratio [see (17)], can be written

$$\frac{P_{AM}}{P_c} = \frac{1}{4}\left[\frac{\Delta V_{out}(\Delta B)}{G_{eff}S_u P_c}\right] = \frac{v_1^2 - v_2^2}{[2G_{eff}S_u P_c]^2} \tag{31}$$

or, expressed in dB,

$$\frac{P_{AM}}{P_c}(\text{dB}) = 10 \log_{10}(v_1^2 - v_2^2) - 20 \log_{10} G_{eff} - 20 \log_{10}(2S_u P_c). \tag{32}$$

FM Calibration Below Cavity 3 dB Points

FM noise measurements are performed using both paths of the bridge [see Fig. 18], path 1 supplying local oscillator power and path 2 the noise power. The FM calibration [1] below the 3 dB points of the cavity is accomplished by introducing a known amount of frequency deviation into the source under test and measuring the corresponding output voltage.

The cavity suppression filter is first matched at the resonant frequency of the source under test by applying a swept leveled signal at the input to the bridge and adjusting the cavity matching screw [Fig. 15(a)] while sampling the output with a detector and scope located at waveguide switch SW-1 [see Fig. 18]. This can also be done using a microwave spectrum analyzer; however, care must be taken to assure that sideband distortion is avoided. The source under test is then attached to the system and an audio oscillator signal is used to frequency modulate its carrier. The modulating oscillator is set to some convenient video frequency f_v and as its amplitude is increased the X-band output is frequency modulated at f_v. The peak frequency deviation is dependent upon the magnitude of the applied modulating voltage. The Crosby zero crossing method [25] is used to complete the calibration.

As the amount of deviation increases, the carrier power decreases and distributes its energy in the sidebands [6], [26]. At discrete frequencies the carrier power reduces to zero. The first zero produces a peak deviation.

$$\Delta f_{peak} = 2.405 f_v. \tag{33}$$

When the carrier suppression filter is tuned to the source frequency, a demodulated output voltage at the frequency f_v is obtained as the phase shifter in path 1 is adjusted. Maximum output occurs when the carrier is shifted 90 degrees relative to the FM noise sidebands. The resulting sensitivity is calculated from the voltage reading of the analyzer and from the known amount of deviation.

$$\text{Sensitivity (rms)} = S_c(\text{rms}) = \frac{\text{Output Voltage of Analyzer}}{\Delta f_{rms}}$$

$$S_c(\text{rms}) = \frac{V_{\text{analyzer output}}}{\frac{1}{\sqrt{2}}\Delta f_{peak}} = \frac{\sqrt{2}\, V_{\text{analyzer output}}}{2.405 f_v}\left[\frac{V}{\text{Hz}}\right]. \tag{34}$$

For this part of the calibration the low-noise video amplifier is not used.

The system limiting sensitivity can be measured by removing the modulation, diverting path 2 into a termination, and placing the high-gain broad-band amplifier after the balanced mixer. The output voltage at frequency f_v is measured at the wave analyzer and referred back to the mixer terminals.

$$\text{Limiting Sensitivity}_{(\text{Hz})} = \frac{\text{System Noise Voltage Referred to Mixer Terminals}}{S_c(\text{rms})} = \frac{v_2(f_v)}{G_{eff}S_c(\text{rms})}. \tag{35}$$

When the available X-band source power entering the bridge is 10 mW, the measured value of rms sensitivity at the mixer output is typically 1.25 μV/Hz while the corresponding limiting sensitivity is 0.02 Hz. These values are essentially independent of measuring bandwidth ($\Delta B \leq 500$ Hz).

FM noise measurement is similar to AM noise measurement. The FM noise components (minus carrier) entering the E-plane arm of the balanced mixer combine with the *shifted local oscillator* carrier entering the H-plane arm. The video output v_1 consists of the sum of the *converted* noise (FM to AM) plus system noise. When path 2 is diverted the output voltage v_2 is just system noise. The source noise voltage referred to the mixer terminals is identical in form to (30). The rms frequency deviation in a given bandwidth ΔB and at a given frequency off carrier can be computed as

$$\Delta f_{rms}(\Delta B) = \frac{\Delta V_{out}(\text{rms})}{G_{eff}S_c(\text{rms})} = \frac{\sqrt{v_1^2 - v_2^2}}{G_{eff}\sqrt{2}\, S_c(\text{peak})}. \tag{36}$$

The power spectrum can easily be obtained from the analysis leading to (25), i.e.,

$$\frac{\text{Power in Two FM Sidebands}}{\text{Power in Carrier}} = \frac{P_{FM}}{P_c} = \frac{\Delta f^2{}_{rms}}{f_v{}^2}.$$

FM Calibration Above Cavity 3 dB Points

When the frequencies away from carrier exceed approximately the loaded Q half-power frequency points of the cavity, (36) is no longer valid in the form shown. From physical reasoning it is clear that the noise power sidebands are totally reflected at frequencies far removed from the filter center frequency. For a loaded cavity Q of 10 000, the 3 dB points at X-band occur at about ± 500 kHz from center frequency. Hence, at about 700 kHz, we can assume that total noise power reflection occurs. These reflected FM sidebands are still detected in the balanced mixer since the phase shifter was pre-set to the position required to convert FM to AM. In this part of the video region, we must invoke (25).

$$\Delta f_{rms} = f_v \sqrt{\frac{P_{FM}}{P}}.$$

The ratio P_{FM}/P_c has the same quantitative expression as that given by (31) for the AM case; therefore,

$$\Delta f_{rms} = f_v \frac{\sqrt{v_1{}^2 - v_2{}^2}}{2G_{eff}S_u P_c}. \tag{37}$$

Equation (37) can now be used to calculate Δf_{rms} for video frequencies above the half-power frequency points *provided* the directional coupler at the input to the phase bridge of Fig. 18 is a 3 dB coupler. Generally this coupler is 10 dB; hence, if we remain perfectly general we can define a factor A for the directional coupler and multiply (37) by $\sqrt{A}$. If the path losses differ between path 1 and path 2, this can also be lumped with the factor A. For a 3-dB coupler and equal path losses, $A=1$; for a 10-dB coupler, $A=10^{-1}$; for a 20-dB coupler, $A=10^{-2}$; etc. Thus,

$$\Delta f_{rms} = f_v \sqrt{A} \frac{\sqrt{v_1{}^2 - v_2{}^2}}{2G_{eff}S_u P_c}. \tag{38}$$

A computer program was established to handle the large quantity of data and various circuit arrangements. The values of Δf_{rms} for both (36) snd (38) are computed. When the data is then displayed in graphic form, the crossover point between these two equations is found and several points omitted from the plot within the transition region. The printed results can be examined for anomalous effects and if none are noted, a smooth curve is drawn in the transition range. Equations (36) and (38) encompass the entire video range from 20 Hz to 30 MHz.

Correction Factors for Various Measuring Bandwidth Shapes

Noise measurements are usually made with a series of analyzers, each having different bandwidths and band shapes. Noise measurements below 1 kHz are made using a 10-Hz slot and the results referred to a 100-Hz slot. Conversion of measured data from one measuring window to another is accomplished by multiplying the rms frequency deviation by the square root of the ratios of the associated bandwidths. If the measuring band shape is not rectangular, this must be accounted for by a suitable correction factor. Thus, for example, if an FM noise measurement is made in a 10-Hz window that is *round top* rather than *rectangular*, the rms frequency deviation referred to a 100-Hz bandwidth will be

$$\Delta f_{rms}(100 \text{ Hz}) = \Delta f_{rms}(10 \text{ Hz}) \sqrt{\frac{100 \text{ Hz}}{10 \text{ Hz}}} \times (\text{CF}). \tag{39}$$

The correction factor (CF) that accounts for the difference in response characteristics of the 10-Hz slot with respect to a rectangular 100-Hz slot is easily deduced from plotted data. For example, if data measured in a 10-Hz slot and referred to a 100-Hz slot without a correction factor is compared with corresponding data actually measured in a rectangular 100-Hz slot, the correction factor will just be the constant amount of separation between the two curves. This factor can be expressed conveniently as a ratio between the two values of deviation at any given frequency within the range of measuring overlap and is denoted CF in (39).

Sample AM Noise-to-Signal Ratio Calculation and Interpretation

By way of example, assume the analyzer is set to 5 kHz with the bandwidth $\Delta B = 200$ Hz. Typical measured values at this frequency might be $v_1 = 0.14$ mV, $v_2 = 0.105$ mV, $S_u = 100$ V/W, $P_c = 1.2$ mW (i.e., 0.6 mW per diode), and $G_{eff} = 0.94 \times 10^3$. Substituting these values into (32),

$$\frac{P_{AM}}{P_c}(\text{dB}) = -127.7 \text{ dB}.$$

This example indicates that the AM noise power of two noise sidebands measured in a bandwidth $\Delta B = 200$ Hz at 5 kHz from the carrier is 127.7 dB below the carrier power of 1.2 mW.

Sample FM rms Frequency Deviation Calculation and Interpretation

By way of example, assume the analyzer is set to 5 kHz with bandwidth $\Delta B = 200$ Hz. Typical measured values at this frequency might be $v_1 = 0.24$ mV, $v_2 = 0.091$ mV, $G_{eff} = 0.94 \times 10^3$, and $S_c(\text{rms}) = 1.25$ μV/Hz. Substituting these values into (36),

$$\Delta f_{rms}(200 \text{ Hz}) = 0.19 \text{ Hz}.$$

This example shows that the source has an FM noise of 0.19 Hz rms as referred to a 200-Hz bandwidth 5 kHz away from the carrier frequency. Thus, in the reference bandwidth, the noise power is the same as the power that would be measured in one sideband if a noise-free source of the same carrier level was frequency modulated at a 5-kHz rate with a carrier deviation of 0.19 Hz rms.

Appendix

Derivation of the Transfer Function for a Carrier Suppression Filter Employing a Ferrite Circulator

The voltage transfer function of the carrier suppression filter [see Fig. 15(a)] between planes A–A and C–C is

$$T(\omega) = \frac{e_r'(\omega)}{e_i'(\omega)} \cdot \tag{40}$$

where $e_r'(\omega)$ and $e_i'(\omega)$ are the voltages at the output and input ports of the circulator. Neglecting circulator insertion losses, which are usually less than 0.2 dB from port to port, the transfer function is equal to the reflection coefficient at plane B–B.

$$T(\omega) \approx \Gamma(\omega) = \frac{e_r(\omega)}{e_i(\omega)} \cdot \tag{41}$$

The circulator isolates the input generator from the output load; thus, the equivalent circuit [see Fig. 15(b)] looking into plane A–A (B–B) is just that of the cavity. When the source is matched to the line $R_G = Z_0$ and if the transformer is assumed to be ideal, the Thevenin voltage source can be transformed into the cavity circuit [see Fig. 15(c)] and then into its equivalent Norton current source [see Fig. 15(d)]. The reflection coefficient at plane B–B can also be written

$$\Gamma(\omega) = \frac{Y_G - Y_c(\omega)}{Y_G + Y_c(\omega)} \tag{42}$$

where $Y_G = (n_1^2 R_G)^{-1}$ is the transformed load admittance. The cavity admittance

$$Y_c(\omega) = \frac{1}{nZ_0} + j\left(\omega C - \frac{1}{\omega L}\right) \approx \frac{1 + jQ_u \dfrac{2\Delta\omega}{\omega_0}}{nZ_0} \tag{43}$$

where $Q_u = nZ_0/\omega_0 L$ is the unloaded cavity Q, $\omega_0 = 1/\sqrt{LC}$ the resonant frequency, and $\Delta\omega = \omega - \omega_0$. Combining (42) and (43)

$$\Gamma(\omega) = \frac{\left[\dfrac{nZ_0}{n_1^2 R_G} - 1 - jQ_u \dfrac{2\Delta\omega}{\omega_0}\right]}{\left[\dfrac{nZ_0}{n_1^2 R_G} + 1 + jQ_u \dfrac{2\Delta\omega}{\omega_0}\right]} \tag{44}$$

results. When maximum cavity absorption (maximum power transfer) occurs, $nZ_0 = n_1^2 R_G$, then

$$\Gamma(\omega) = \frac{-jQ_u \dfrac{\Delta\omega}{\omega_0}}{1 + jQ_u \dfrac{\Delta\omega}{\omega_0}} = \left(Q_u \frac{\Delta\omega}{\omega}\right) \frac{\left[-Q_u \dfrac{\Delta\omega}{\omega_0} - j\right]}{\left[1 + \left(Q_u \dfrac{\Delta\omega}{\omega_0}\right)^2\right]} \tag{45}$$

which has magnitude

$$|\Gamma(\omega)| = \frac{Q_u \dfrac{\Delta\omega}{\omega_0}}{\left[1 + \left(Q_u \dfrac{\Delta\omega}{\omega_0}\right)^2\right]^{\frac{1}{2}}} = \frac{Q_u \dfrac{\Delta f}{f_0}}{\left[1 + \left(Q_u \dfrac{\Delta f}{f_0}\right)^2\right]^{\frac{1}{2}}} \tag{46}$$

and phase

$$\operatorname{Tan} \phi = \frac{-1}{-Q_u \dfrac{\Delta\omega}{\omega_0}} = \frac{f_0}{Q_u \Delta f}; \quad \phi = \operatorname{Tan}^{-1}\left(\frac{f_0}{Q_u \Delta f}\right). \tag{47}$$

The total circuit conductance

$$G_T = Y_G + R_c^{-1} = \frac{1}{n_1^2 R_G} + \frac{1}{nZ_0} = \frac{2}{nZ_0} \tag{48}$$

and the loaded Q of the cavity

$$Q_L = \frac{1}{G_T \omega_0 L} = \frac{1}{2} \frac{nZ_0}{\omega_0 L} = \frac{Q_u}{2} \cdot \tag{49}$$

Combining (46), (47), and (49)

$$|\Gamma(\omega)| = \frac{2Q_L \dfrac{\Delta f}{f_0}}{\left[1 + \left(2Q_L \dfrac{\Delta f}{f_0}\right)^2\right]^{\frac{1}{2}}} \tag{50}$$

$$\phi = \operatorname{Tan}^{-1}\left(\frac{f_0}{2Q_L \Delta f}\right). \tag{51}$$

The suppression, above insertion loss, is

$$S(\text{dB}) = 20 \log_{10} \frac{1}{|\Gamma(\omega)|} = 20 \log_{10} \frac{\left[1 + \left(2Q_L \dfrac{\Delta f}{f_0}\right)^2\right]^{\frac{1}{2}}}{2Q_L \dfrac{\Delta f}{f_0}} \cdot \tag{52}$$

Equations (52) and (51) are plotted in Figs. 16 and 17, respectively, for a resonant frequency of 9.530 GHz with Q_L as a parameter.

Acknowledgment

The author would like to acknowledge Dean J. J. Karakash for his personal assistance in helping to set up the initial facilities in the Electrical Engineering Department at Lehigh University, and to thank Microwave Associates for providing the full facilities. The author is grateful for the invaluable technical suggestions of Prof. W. E. Dahlke and M. E. Hines.

References

[1] J. G. Ondria, "Study of noise reduction techniques for varactoharmonic generator type microwave sources Navy Bureau or Ships, 1st Quart. Rept. Contract NObsr-95009, August–November 1965.

[2] A. L. Whitwell and N. Williams, "A new microwave technique for determining noise spectra at frequencies close to the carrier," *Microwave J.*, p. 27, November 1959.

[3] Spectra Electronics, Inc., Los Altos, Calif., Spectra Electronics Appl. Note AN-1.

[4] L. A. MacKenzie, "General theory of noise in oscillators," U. S. Signal Corps, School of Elec. Engrg., Cornell University, Ithaca, N. Y., Res. Rept. E.E. 548.

[5] M. Schwartz, *Information Transmission, Modulation, and Noise*. New York: McGraw Hill, 1959.

[6] S. Goldman, *Frequency Analysis, Modulation and Noise*. New York: McGraw-Hill, 1948.

[7] J. Fagot, and P. Magne, *Frequency Modulation Theory*. New York: Pergamon, 1961.

[8] C. P. Walsh, "Frequency modulation principles," *Electro-Technology*, p. 53, January 1965.

[9] H. S. Black, *Modulation Theory*. New York: Van Nostrand, 1962.

[10] B. G. Bosch, "An investigation of noise in microwave oscillators," Ph.D. dissertation, University of Southampton, England, 1960.

[11] R. V. Pound, *Microwave Mixers*, M.I.T. Rad. Lab. Ser., vol. 16. New York: McGraw-Hill, 1948, p. 257.

[12] A. F. Harvey, *Microwave Engineering*. New York: Academic Press, 1963, p. 767.

[13] T. S. Saad, "The microwave mixer," Sage Labs., Natick, Mass., 1965.

[14] J. L. Altman, *Microwave Circuits*. New York: Van Nostrand, 1964, p. 332.

[15] A. M. Stone and J. L. Lawson, "Infinite-rejection filters," *J. Appl. Phys.*, vol. 18, p. 691, August 1947.

[16] G. C. Dalman and A. S. Rhoads, Jr., "Microwave oscillator noise spectrum measurements," *IRE Trans. Electron Devices*, vol. ED-1, pp. 51–55, December 1954.

[17] W. P. N. Court, "A tunable filter for use in the measurement of excess noise from local oscillators," *Electron. Engrg.*, p. 208, April 1958.

[18] R. Beringer, *Resonant Cavities as Microwave Circuit Elements; Principles of Microwave Circuits*, M.I.T. Rad. Lab. Ser., vol. 8. New York: McGraw-Hill, p. 207.

[19] R. Beringer, *The Measurement of Wavelength; Technique of Microwave Measurement*, M.I.T. Rad. Lab. Ser., vol. 2. New York: McGraw-Hill, p. 285.

[20] E. A. Guillemin, *Communication Networks*, vol. II. New York: Wiley, 1935, ch. 5.

[21] H. C. Torry and C. A. Whitmer, *Crystal Rectifiers*, M.I.T. Rad. Lab. Ser., vol. 15. New York: McGraw-Hill, 1964.

[22] S. T. Eng, "Low-noise properties of microwave backward diodes," *IRE Trans. Microwave Theory and Technique*, vol. MTT-9, pp. 419–423, September 1961.

[23] W. E. Doherty, "A study of $1/f$ noise in point contact silicon and germanium microwave mixer diodes," M.S. thesis, University of Pennsylvania, Philadelphia, May 1962.

[24] H. T. Friis, "Noise figures of radio receivers," *Proc. IRE*, vol. 32, pp. 419–422, July 1944.

[25] M. G. Crosby, "Carrier and side-frequency relations with multitone frequency or phase modulation," *RCA Rev.*, p. 103, July 1938.

[26] J. P. Carson, "Notes on the theory of modulation," *Proc. IRE*, vol. 10, pp. 57–64, February 1922.

The Measurement of Noise in Microwave Transmitters

J. ROBERT ASHLEY, SENIOR MEMBER, IEEE, THOMAS A. BARLEY, MEMBER, IEEE, AND GUSTAF J. RAST, JR.

Invited Paper

***Abstract*—A tutorial review of the basis for transmitter noise measurements shows that noise is best described and measured as AM and FM noise. The determination of RF spectrum is done by calculation after the AM and FM noise are known. The contribution of AM noise to RF spectrum shape is determined by the power spectral density shape of the AM noise. The contribution of FM noise to RF spectrum is to make the shape that of an *RLC* circuit resonant response rather than a δ function with a sideband structure. The measurement of AM noise is done with a direct detector diode. The measurement of FM noise for frequencies above 5 GHz is done with a discriminator based on a one-port cavity resonator. The measurement of FM noise below 5 GHz is done with an improved transmission line discriminator which is described in detail. Measurement of low-power low-noise signal sources is made possible with an injection-locked oscillator for a preamplifier to the discriminator. The most widely used baseband analyzer is the constant bandwidth superheterodyne wave or spectrum analyzer. Most differences in measurement results are resolved by understanding the baseband analyzers. At least the baseband spectrum analysis of transmitter noise measurements can be automated with worthwhile savings in time and improvement of documentation.**

Manuscript received September 1, 1976; revised September 20, 1976.
J. R. Ashley is with the Department of Electrical Engineering, University of Colorado, Colorado Springs, CO 80907.
T. A. Barley and G. J. Rast, Jr., are with DRDMI-TER, U.S. Army Missile Research and Development Command, Redstone Arsenal, AL 35809.

I. Background Concepts

NOISE in a microwave transmitter is an unwanted kind of signal which degrades the ability of the system to transmit communications or other information. We must include both stochastic signals and coherent spurious signals—usually related to bias supply ripple or mechanical vibration—in our thinking about transmitter noise. In the first stages of receiving systems, we are usually concerned with a ratio of desired signal to noise in the order of unity. In a transmitter, the ratio of carrier to noise is on the order of 10^3 or more. Most of the amplifying or oscillating devices in a transmitter are operating in a saturated mode; thus some of the linear system theory ideas that are so useful in the study of receiving systems must be used with care in studying a transmitter system.

The user of a microwave transmitter also will be using a detection system which may respond to either amplitude modulation or angle modulation (an FM discriminator), but not often both. His essential question is, "How much of the noise in a transmitter will come out of my detection system?" The kind of a detector being used will determine if the user is concerned with what has come to be called "FM noise"

Reprinted from *IEEE Trans. Microwave Theory Tech.*, vol. MTT-25, pp. 294–318, Apr. 1977.

or "AM noise." Fortunately, the study of oscillator theory as pioneered by Middleton [1], Mullen [2], and others yields answers in terms of FM noise or AM noise; and even more fortunately, the measurement techniques which yield the best data are based on separately detecting AM noise and FM noise [3], [4].

The measurement of transmitter noise is accomplished by applying a suitable sample of the output to detectors which respond to either AM or FM. It is easy to visualize an AM detector which rejects a moderate amount of incidental FM on the signal being tested. It is not so easy to visualize a microwave discriminator without a limiter stage rejecting incidental AM on the signal being tested; yet, the cavity resonator discriminators of Marsh and Clare [5], Ashley *et al.* [3], and the new transmission line discriminators of Ashley *et al.* [6], do indeed reject incidental AM. Any competitive methods must reject the unwanted form of modulation in measuring the desired form.

Current microwave transmitters include those with a simple high-power oscillator (such as a magnetron) as the only device, those which include direct microwave oscillators (reflex klystrons, TEO, IMPATT), driving amplifiers (amplitron, klystron, TWT, locked IMPATT oscillator, transistor), and those which synthesize the microwave signal from sources controlled by VHF or UHF crystal oscillators and then increase the power with amplifiers. To measure noise in transmitters, we must be able to study direct oscillators, amplifiers, and the VHF and UHF oscillators, amplifiers, and frequency multipliers.

The range where the demodulated noise must be studied varies with the applications. Doppler radar systems are degraded by excessive noise in the modulation range 10 Hz–100 kHz, and the usual noise specification covers a significant portion of this range. Communications systems using FM multiplex the subcarriers into a band typically 10 MHz wide, and this sets the upper limit on the measurement of demodulated noise. These modulation frequencies are sometimes described as distance from the carrier with an implied idea of the RF spectrum spread.

In FM communications, the nonlinearity in amplifiers can cause cross modulation of multiplexed subcarriers which is a form of distortion. This will appear in the system output as unwanted signals in the individual channels. This meets the definition of noise as being unwanted, but we will not treat the problem further in this work.

Understanding noise and the mathematics for studying noise requires some careful definitions and explanations. Some of the intuitive ideas about noise based on modulation theory using sinusoids only are imprecise to say the least, and we advise you to not avoid the next section which gives some of the concepts needed.

II. Modulation Noise Relationships

Most of the relationships to be given here are tersely abstracted from [7] and [8]. Naturally, we leave the proofs and many details to the References.

For amplitude modulation transmitters, the output can be written

$$v(t) = A_0[1 + \lambda V_s(t)] \cos (\omega_c t + \phi) \tag{1}$$

where

- t time;
- A_0 unmodulated amplitude (peak volts);
- ω_c $2\pi\nu_c$ = carrier frequency (radians per second);
- ν_c carrier frequency (hertz);
- ϕ phase constant usually set to zero by proper choice of coordinates;

and $\lambda V_s(t) \leq 1$ is the information to be communicated. We are interested in the noise floor which exists when $V_s(t) = 0$. Thus we replace $\lambda V_s(t)$ by $N_A(t)$ where $N_A(t)$ is the AM noise and must be treated as a random variable.

To distinguish the frequency variable near the carrier from frequency in the baseband region where modulation noise must be described, we shall use the lower case Greek nu (ν) for frequency (hertz) near the carrier frequency ν_c. In the baseband region from zero to a few megahertz, we shall use f to designate frequency (hertz).

Similarly, for angle modulated transmitters, the output can be written

$$v(t) = A_0 \cos (\omega_c t + \Phi(t) + \phi) \tag{2}$$

where the information to be transmitted is carried in the instantaneous phase $\Phi(t)$. The instantaneous frequency ν_i of this signal is

$$\nu_i = \frac{d}{dt} (\omega_c t + \Phi(t) + \phi) = \omega_c + \dot{\Phi}(t). \tag{3}$$

We are interested in the residual $\Phi(t)$ or $\dot{\Phi}(t)$ for the unmodulated state which we will call $\Phi_N(t)$ or $\dot{\Phi}_N(t)$. Again, $\Phi_N(t)$ must be considered as a random variable. From a measurement standpoint, it is fortuitous that we usually measure $\dot{\Phi}_N(t)$ with a discriminator and can integrate the result to find $\Phi_N(t)$.

Combining these equations, the signal from an unmodulated transmitter can be written

$$v(t) = A_0(1 + N_A(t)) \cos (\omega_0 t + \Phi_N(t) + \phi) \tag{4}$$

where

$$|N_A(t)| \ll 1 \tag{5}$$

and

$$|\Phi_N(t)| \ll 1. \tag{6}$$

Our task is to learn something about $N_A(t)$ and $\Phi_N(t)$ by measurements made on $v(t)$.

Except for components caused by bias supply ripple or deterministic mechanical vibrations, the signals $N_A(t)$ and $\dot{\Phi}_N(t)$ are not deterministic and must be studied from a statistical point of view. Any dc average is absorbed in either A_0 or ω_c; so, it is appropriate to consider that $N_A(t)$ and $\Phi_N(t)$ have zero mean and study the density function, variance, autocorrelation function, and band-limited spectrum.

The probability density function for $N_A(t)$ or $\Phi_N(t)$ is difficult to measure to say the least. It is usually assumed by implication to be Gaussian for the simple reason that this is the mathematically best known function. In terms of physical processes in oscillators and amplifiers, this is not an unreasonable assumption.

The usual definition of an autocorrelation function can be applied to $A_N(t)$ and $\Phi_N(t)$.

$$R_A(\tau) = \lim_{T \to \infty} \frac{1}{2T} \int_{-T}^{T} N_A(t) N_A(t + \tau)\, dt \tag{7}$$

$$R_\phi(\tau) = \lim_{T \to \infty} \frac{1}{2T} \int_{-T}^{T} \Phi_N(t) \Phi_N(t + \tau)\, dt \tag{8}$$

$$\tau = t_2 - t_1 \tag{9}$$

and for the real random variables $N_A(t)$ and $\Phi_N(t)$

$$R(-\tau) = R(\tau). \tag{10}$$

The autocorrelation function is used as the mathematical link between the time domain and frequency domain descriptions of random variables. The Wiener–Khintchine theorem states that for ergodic and well-behaved real processes the power spectral density $S(f)$ and the autocorrelation functions $R(\tau)$ are a Fourier transform pair

$$S(f) = \int_{-\infty}^{\infty} R(\tau) \cos (2\pi f \tau)\, d\tau \tag{11}$$

$$R(\tau) = \int_{-\infty}^{\infty} S(f) \cos (2\pi f \tau)\, df. \tag{12}$$

The power spectral density is an even, real function and nonnegative

$$S(-f) = S(f) \tag{13}$$

$$S(f) \geq 0. \tag{14}$$

The power spectral density of the noise modulations is band limited to a small fraction of the carrier frequency by resonant circuits in the signal generation source.[1] This limited band is usually termed the "baseband."

In terms of the measurement problem, applying the source to demodulators as shown in Fig. 1 yields voltages proportional to $N_A(t)$, $\Phi_N(t)$, and $\dot{\Phi}_N(t)$ which are functions of time. The simplest data processing in the time domain is to use a cathode ray or mechanical oscillograph to record instantaneous amplitude, phase, or frequency versus time. The records are often called "stability" although the word instability might be more appropriate. In terms of frequency or phase instability, the time domain processing can be quite sophisticated if long-term stability (time period of days or more) is important—as in timekeeping applications. The use of computing counters to determine the sigma versus tau characteristic is well described by Shoaf *et al.* [15].

In theory, it should be possible to apply the output of the demodulators to autocorrelation computing equipment.

[1] This is not true for traveling-wave tube amplifiers or backward wave tube oscillators.

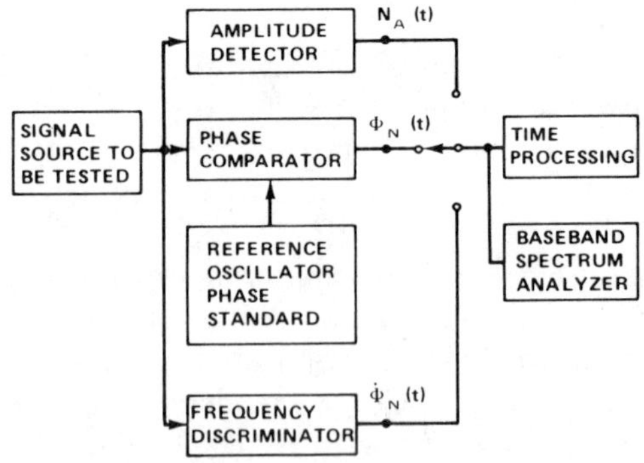

Fig. 1. Schematic representation of modulation noise measurements.

We do not know of published results of such an experiment. One experiment which would be of great significance in oscillator noise theory would be to determine the cross correlation between amplitude and phase noise modulation.

Since most microwave transmitters are used in either radar or communication systems, the most useful analysis of the modulation noise is in the frequency domain. The baseband analysis equipment is usually called a spectrum analyzer if frequency is automatically swept through and a wave analyzer if the frequency is manually set. These equipments are of sufficient importance that we devote a later section to their discussion.

You may wonder why nothing has been said to this point about the microwave spectrum of $v(t)$. The main reason is that the measurement of the signal spectrum will not predict the effects of transmitter noise on the amplitude or angle demodulators in the receiver—from the signal spectrum alone it is not possible to determine if a particular sideband is caused by N_A or $\Phi_N(t)$. If one needs to know the spectrum structure in the region of the carrier, he will usually find it well hidden under the noise floor of the spectrum analyzer. Thus $N_A(t)$ and $\Phi_N(t)$ must be measured and used to compute the near carrier spectrum.

This is not to deny the usefulness of a microwave spectrum analyzer in studying transmitter noise. Observation of spurious signals at subharmonic, harmonic, and unwanted mixing product frequencies is best done with these powerful tools. We will also make much use of the microwave spectrum analyzer in the alignment and calibration of AM and FM noise measuring equipment.

The theory required to understand the near carrier spectrum structure requires the most difficult mathematics of this paper but should not be avoided.

III. The Spectrum of Modulated Signals

Our approach to this topic paraphrases that given by Middleton [7] by considering low index AM, ϕM, and FM separately and then combining results for simultaneous AM and FM. Throughout, the use of autocorrelation functions and the Wiener–Khintchine theorem makes the results apply to both deterministic (usually sinusoidal) and stochastic modulation signals. First, for amplitude modulation only,

$$v(t) = A_0(1 + \lambda V_M(t)) \cos (2\pi v_c t) \tag{15}$$

where $N_A(t) = \lambda V_M(t)$. With just a little bit of work, the spectrum $S_{RF}(\nu)$ is found from ([7], eq. 12.10b)

$$S_{RF}(\nu) = A_0{}^2\delta(\nu - \nu_c) + \lambda^2 A_0{}^2 \int_0^\infty R_M(\tau)\cos 2\pi(\nu - \nu_c)\tau \, d\tau \tag{16}$$

where $R_M(\tau)$ is the autocorrelation function of the modulation $V_M(t)$. The δ-function $\delta(\nu - \nu_c)$ indicates a spectral line at ν_c. If $V_M(t) = \cos 2\pi f_a t$ and $\lambda < 1$, then

$$S_{RF}(\nu) = \frac{A_0{}^2}{2}\,\delta(\nu - \nu_c) + \frac{A_0{}^2\lambda^2}{8}\,\delta(\nu - (\nu_c + f_a)) + \frac{A_0{}^2\lambda^2}{8}\,\delta(\nu - (\nu_c - f_a)) \tag{17}$$

which is the familiar carrier with a pair of side frequencies located at $\nu_c \pm f_a$. If V_M is a more complex signal which can be expressed as a Fourier series of sinusoids, then pairs of side frequencies will appear for each Fourier component. There will be no cross products as long as the total modulation index is kept less than 1 (i.e., no overmodulation).

A most useful result can be deduced for the case of any form of low index amplitude modulation. If we were to find the spectrum of V_M from its autocorrelation function, we would evaluate

$$S_{AM}(f) = 2\int_0^\infty R_M(\tau)\cos 2\pi f\tau \, d\tau \tag{18}$$

while the sideband portion $S_{SB}(\nu)$ of (16) requires evaluation of

$$S_{SB}(\nu) = \int_0^\infty R_M(\tau)\cos 2\pi(\nu - \nu_c)\tau \, d\tau. \tag{19}$$

An obvious change of variable yields

$$S_{RF}(\nu) = \frac{A_0{}^2}{2}\,\delta(\nu - \nu_c) + \frac{\lambda^2 A_0{}^2}{4}\,S_{AM}(\nu - \nu_c) \tag{20}$$

which tells us that the spectrum of the modulating signal is shifted from being symmetrical about $f = 0$ to being symmetrical about $\nu = \nu_c$. Therefore, if we can measure the spectrum of the AM noise $S_{AM}(f)$, we can make a prediction of the contribution of AM noise to the RF spectrum using (20).

This important concept is illustrated by Fig. 2. The measured noise spectrum of the AM, $S_{MAM}(f)$, is shown as it would be measured in Fig. 2(a). This is a power spectral density for the baseband range. Mathematically, we must carry the negative frequency portion of the spectrum which is the mirror image of the positive portion as shown in Fig. 2(b). The fact that $S_{AM}(f)$ is the AM spectrum means that the RF spectrum can be found by translating $S_{AM}(f)$ to the regions near $\pm\nu_c$ and adding the δ-function representing the carrier. This is illustrated in Fig. 2(c). Our example AM noise spectrum has both δ-functions to represent sinusoids and a continuous portion to represent random noise.

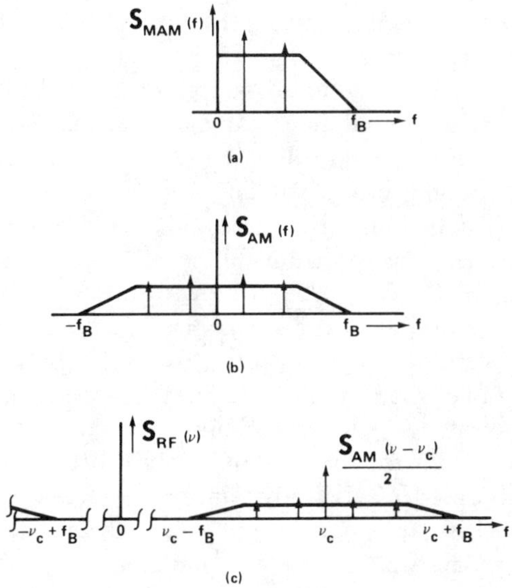

Fig. 2. (a) Measured spectrum of AM noise. (b) Mathematical representation of AM noise. (c) RF spectrum caused by AM noise.

For the case of angle modulation only, the RF signal can be written

$$v(t) = A_0 \cos(2\pi\nu_c t - \Phi(t)) \tag{21}$$

where $\Phi(t)$ is chosen for phase or frequency modulation as

$$\Phi(t)_{\phi M} = D_\phi V_M(t) \tag{22}$$

$$\Phi(t)_{FM} = D_F \int^t V_M(t')\,dt'. \tag{23}$$

Here D_ϕ has units of radian/volt and D_F has units of angular frequency (radian/second) per volt. For the case where V_M is sinusoidal, let

$$\Phi(t) = \mu_{\phi M}\cos 2\pi f_a t \tag{24}$$

for phase modulation and

$$\Phi(t) = \mu_{FM}\sin 2\pi f_a t \tag{25}$$

where $\mu_{\phi M} = \Delta\phi$, the magnitude of the phase deviation and $\mu_{FM} = \Delta\nu/f_a$, the frequency deviation divided by the modulation frequency are the appropriate modulation indices for ΦM and FM. With more than a little bit of work and Bessel function manipulation, Middleton ([7], eq. 14.10b f.f.) finds the autocorrelation function and the spectrum

$$R_V(\tau) = \frac{A_0{}^2}{2} J_0\left(2\mu\cos\frac{2\pi f_a\tau}{2}\right)\cos 2\pi\nu_c\tau \tag{26}$$

$$S_{RF}(\nu) = \frac{A_0{}^2}{2} \cdot \sum_{m=0}^{\infty} J_m{}^2(u)\{\delta(\nu - (\nu_c + mfa)) + \delta(\nu - (\nu_c - mfa))\} \tag{27}$$

where J_m is an mth order Bessel function. This is the familiar result with the multitude of side frequencies for large μ. If $\mu < 0.01$, only the $m = 1$ side frequencies are significant and the approximation $J_1(x) = x/2$ is useful. This theory is the basis for calibration of the discriminators used for FM noise measurements.

If more than one sinusoid is present in the modulation signal, then additional side frequencies are added for low values of μ. For large μ, the nonlinearity of angle modulation causes some cross coupling of the modulation. For example, if two sinusoids at 300 and 1000 Hz angle modulate a carrier, the spectrum will have components located at ± 1300, 1600, 2300, 2900 Hz, etc., away from the center frequency as well as at the harmonies of 300 and 1000 Hz. These facts and the expression $\mu_{\mathrm{FM}} = \Delta \nu / f_a$ should make us expect trouble when the modulation is not deterministic.

The theory does not disappoint our expectation. The limiting behavior for small modulation is fortunately all we need to consider in this study. Since we are concerned with the case where the modulation is just the residual noise in the source, we can take Middleton's equations [7, sec. 14.2-1] and abandon higher order terms with great vigor because Φ is quite small. For phase modulation,

$$R_V(\tau) \approx \frac{A_0^2}{2} \cos 2\pi\nu_c\tau\{1 + R_{\phi\mathrm{M}}(\tau)\} \tag{28}$$

which is Fourier transformed to find the RF spectrum

$$S_{\mathrm{RF}}(\nu) \approx \frac{A_0^2}{2} \delta(\nu - \nu_c) + \frac{A_0^2}{2} \int_0^{\infty} R_{\phi\mathrm{M}}(\tau) \cos 2\pi\nu_c\tau \cos 2\pi\nu\tau \, d\tau. \tag{29}$$

In the integral, the identity for $\cos A \cos B$ can be used, a term in $\cos 2\pi(\nu_c + \nu)$ abandoned by a stationary phase argument, and the same clever substitution used in the AM theory invoked to show

$$S_{\mathrm{RF}}(\nu) = \frac{A_0^2}{2} \delta(\nu - \nu_c) + \frac{A_0^2}{4} S_{\phi\mathrm{M}}(\nu - \nu_c). \tag{30}$$

We should be overjoyed with this result because it shows the same sort of limiting behavior as we found in the AM case. A carrier line is the center frequency for the symmetrical phase modulation spectrum which is unchanged in spectral shape. Thus we might measure $S_{\phi\mathrm{M}}(f)$ and very quickly translate it to the RF spectrum.

Our joy begins to fade when we look at typical measurements of $S_{\phi\mathrm{M}}(f)$ and study oscillator theory. Leeson [17] shows general forms for $S_{\phi\mathrm{M}}(f)$ which show a behavior below about 1 kHz which is of the order $1/f^2$ or $1/f^3$. Going back to the theory with this knowledge of $S_{\phi\mathrm{M}}(f)$ shows that all our approximations were not very well justified. Those who use the "theorem" implied in (30) would be in for a rude shock if they could *measure* $S_{\mathrm{RF}}(\nu)$ for a signal where $S_{\phi\mathrm{M}}(f)$ is known.

The true understanding of the RF spectrum for small angle modulation comes from the study of limiting forms for FM noise modulation. The classic reference is Mullen and Middleton [18] although the work is well documented in Middleton's book [7, sec. 14.2-1]. The root of the mathematical trouble is in the value of an integral

$$L = \int_0^{\infty} \frac{S_{\mathrm{FM}}(f)}{4\pi^2 f^2} \, df \tag{31}$$

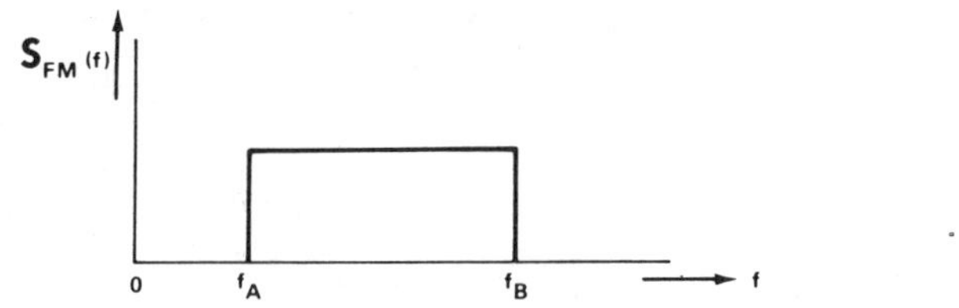

Fig. 3. An idealized FM noise spectrum.

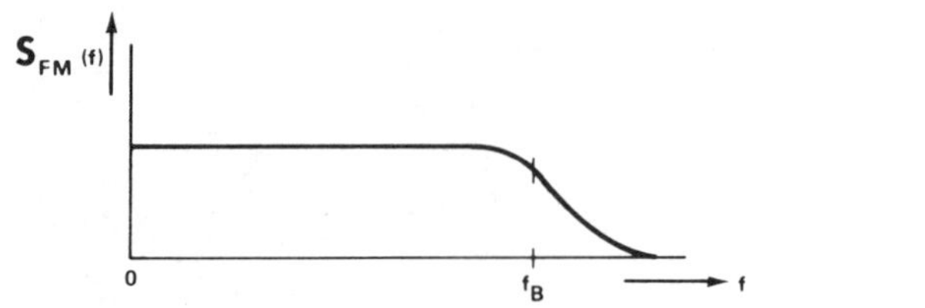

Fig. 4. A more typical FM noise spectrum.

needed in finding the autocorrelation function. There are two cases which have been worked through. The first case is for an idealized spectrum for $S_{\mathrm{FM}}(f)$ sketched in Fig. 3. When the noise spectrum does not extend to zero frequency, $L < \infty$ and the limiting form for the RF spectrum is [7, eq. 14.42b]

$$S_{\mathrm{RF}}(\nu) = \frac{A_0}{2} \left[\delta(\nu - \nu_c) + \frac{1}{2} \frac{S_{\mathrm{FM}}(\nu - \nu_c)}{(\nu - \nu_c)^2} \right] \tag{32}$$

which is valid only for

$$|\nu - \nu_c| < f_b. \tag{33}$$

This is the kind of a result we would intuitively expect from using sinusoidal modulation theory for the small deviation case. Since this idealized spectrum S_{FM} does not go to zero frequency, the analogous sinusoidal modulation index $\Delta\nu/f_A$ does not go to infinity and cause our intuition trouble.

The real world case corresponds to the FM noise spectrum sketched in Fig. 4. Even this sketch is idealized because the oscillator should show a $1/f$ spectral shape near $f = 0$. Still, if the FM noise spectrum is constant as $f \to 0$, then the integral L of (31) diverges. By a different mathematical route Middleton [7, eq. 14.45] obtains the RF spectrum as

$$S_{\mathrm{RF}}(\nu) = \frac{\gamma A_0^2}{\gamma^2 + 4\pi^2(\nu - \nu_c)^2} + \text{correction term.} \tag{34}$$

The correction term is significant only in the "tails" of the spectrum, where

$$|\nu - \nu_c| > f_b. \tag{35}$$

There is a surprising physical significance in this development. In the region

$$|\nu - \nu_c| < f_b \tag{36}$$

the shape of $S_{\mathrm{FM}}(\nu)$ does not appear in the leading term for the spectrum. Also, there is not a δ-function in (34) so we cannot talk of a carrier for computing carrier-to-noise sideband ratios. Instead of the discrete carrier and sideband structure of AM or overly idealized angle modulation, the real world signal sources have a spectral shape resembling the resonant response of a series L–R–C circuit shown in Fig. 5. The width of this spectral "line," is γ/π where γ is a

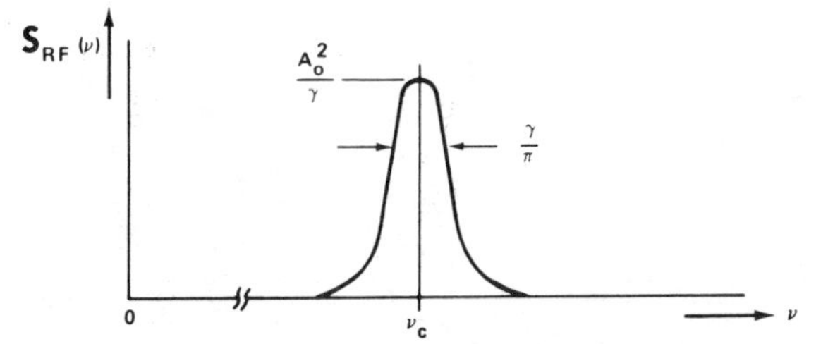

Fig. 5. The limiting form for the spectrum with low FM noise of the shape given in Fig. 4.

measure of the total power in the FM noise spectrum, that is, a factor proportional to the area under the $S_{FM}(f)$ curve. In practice, the linewidth is of the order of a few hertz at the most and cannot be resolved with commercially available RF spectrum analyzers which might have a minimum bandwidth of 10 Hz. (Also, the noise floor in the spectrum analyzer is too high to see anything important.) This is why the failure of (34) to match the intuitive extrapolation of sinusoidal FM theory has not been detected by measurement.

The theory just presented is not the conventional wisdom widely practiced for computing spectrum.[2] One cannot help but wonder why someone has not discovered the error through measurement. With 20-20 hindsight, we can see one area where the correct spectrum computation explains a good result. As documented by Risley *et al.* [13] and McDonald *et al.* [26], a research program to beat the output of *X*-band signal sources with various laser radiations in a Josephson junction has been carried out at the National Bureau of Standards, Boulder, CO. The first experiment used a reflex klystron electronic phase locked to a UHF crystal oscillator. Only a beat with the 0.964-THz line from an HCN laser had been detected. Substitution of a transmission cavity stabilized klystron made possible detection of a beat of the 401-st harmonic of the klystron frequency with a 3.82-THz water-vapor laser. Misuse of (32) would predict similar spectrums for the two different klystron systems within the 8-kHz bandwidth of the receiving system used. Correct computation of the RF spectrum explains why the lower FM noise above 10 kHz in the cavity stabilized klystron made possible the detection of the higher harmonic.

The case when FM and AM are present simultaneously deserves only brief comment. For the sinusoidal case, the result of simultaneous amplitude and angle modulation is to destroy the symmetry of the RF spectrum. For our concern with transmitter noise measurements, this asymmetrical spectrum when seen while looking at calibration side frequencies means that the pure amplitude or angle modulation needed for calibration is not present and the calibration equipment must be realigned.

The unmodulated RF output of a microwave transmitter or of any RF signal source is going to have both AM and FM noise simultaneously present. This will give basically three terms to compute to find the RF spectrum: 1) a component caused by the FM noise, 2) a component caused by the AM noise, and 3) a component caused by the cross correlation of the AM and the FM noise. For frequencies within a few kilohertz of the carrier region, the FM noise term dominates. For frequencies more than a few tens of kilohertz from the carrier region, the AM term dominates. The cross-correlation term would be found by applying the Wiener–Khintchine theorem to the cross-correlation function. Middleton [7, sec. 14.3-1] shows that the computed effect is to skew the spectrum. Thus the result is of diagnostic value for transmitters (such as counter-measures jammers) which use large index noise modulation to produce wideband noise as the primary output. If the spectrum is asymmetrical, then the modulation is not pure amplitude or angle. For the low residual level of the AM and FM noise in an unmodulated transmitter, a measurement of the cross-correlation function would be required before anything definitive could be said about the importance of this term in the RF spectrum computation.

What conclusions can we reach from this diversion into communications theory? In relation to our topic of transmitter noise measurements, these conclusions must include the following.

1) Both the nature of noise in transmitters and the measurement equipment available dictate the measurement of AM noise, and ϕM or FM noise.

2) The theory of amplitude or angle modulation by sinusoids gives the basis for alignment and calibration of noise measurement equipment.

3) A microwave spectrum analyzer is useful for detecting spurious outputs several megahertz from the carrier region in a transmitter but cannot yield sufficient information about the noise modulation of the output.

4) A microwave spectrum analyzer is needed for alignment and calibration of AM and angle modulation noise measurement equipment.

5) Estimation of the contribution of AM noise to the RF spectrum is straightforward [see (20)] and has no booby traps.

6) Estimation of the contribution of angle modulation noise to the RF spectrum is not straightforward [see (34)] and many unknowingly fall into a serious booby trap [using (32)].

7) FM or ϕM noise specifications based on a few numbers such as carrier-to-noise ratio at specified frequency (or frequencies) from the carrier are not sufficiently precise. A curve showing maximum limits of the angle modulation noise must be specified for completeness.

8) If you are asked to find the RF spectrum in the region near the carrier, you should ask why.

9) If there is a good reason for determining the spectrum in the region near the carrier, carefully measure AM noise and ϕM or FM noise and then compute. (You will not find enough detail in this paper to make the computations. See Middleton [7].)

10) One must be careful about extrapolating theory based on sinusoids only to a situation where random noise is involved.

[2] The majority of spectrum computations described in books, journals, and specifications correspond to the unjustified use of (30) and (32).

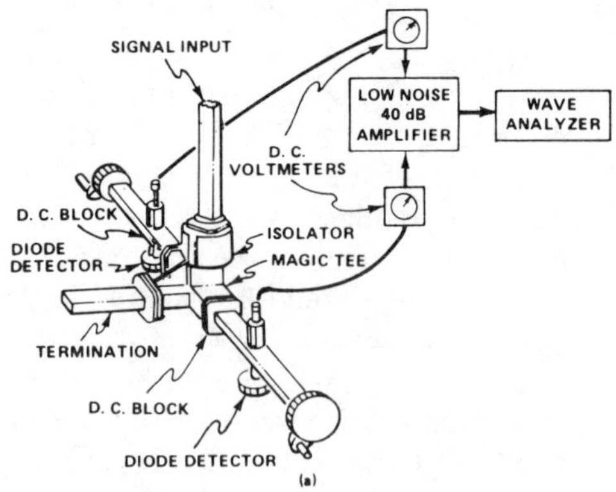

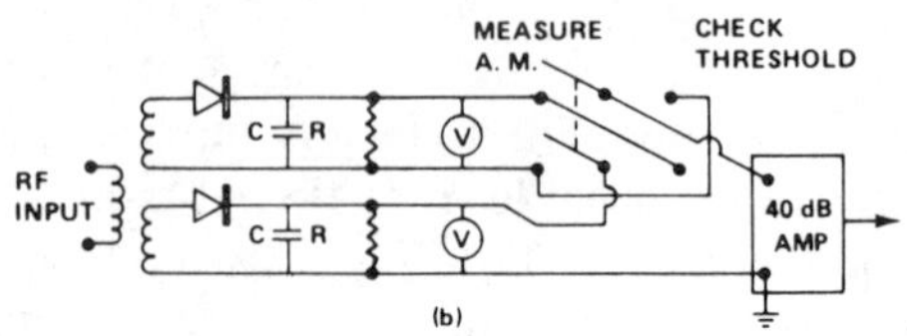

Fig. 6. (a) The balanced AM detector. (b) Its low-frequency equivalent circuit.

With those ten conclusions, we must now get back to our main topic of measuring transmitter noise.

IV. Measurement of AM Noise

The first measurement to consider is determining $N_A(t)$ in (5). In practice, the measurement yields the spectral description $S_{AM}(f)$. There has been no significant advance in this technique since the reports by Ashley *et al.* [3] and Ondria [4] in the September 1968 Special Issue on Noise of this Transactions. As reported there, a direct detector diode is simpler than a superheterodyne demodulator and has a lower threshold. For measuring noise and determining the threshold or noise floor of the measurement, a double diode circuit such as the one shown in Fig. 6 is convenient. The two detector mounts are insulated for dc and baseband frequencies up to several megahertz by plastic shims between the mounts and the main portion of the waveguide. This allows the diodes to be connected in series aiding or opposing as indicated in the low-frequency equivalent circuit of Fig. 6(b). The dc blocks account for the two secondary windings on the equivalent circuit. The two capacitors labeled C are actually cable capacity and the resistors R are 1–10 kΩ each. At this impedance level, we have not been troubled with interference at the power line frequency either in or out of shielded rooms.

Silicon point contact diodes can be operated at about 1 V dc without exceeding the peak inverse voltage limit. Since this is a high-level detection application and the noise from the diodes sets the power limit, we can improve the threshold by increasing the power applied to the diodes. Here the higher peak inverse voltage that can be applied to a Schottky-barrier diode makes it a superior device for AM measurements. (Note that detector mounts designed for point contact diodes will not function properly with Schottky-barrier diodes because of the difference in *I–V* characteristics of the two kinds of diodes. This can be remedied over a limited but adequate microwave bandwidth either with a slide screw tuner or by inserting an adjustable capacitive post in the diode mount at a point $0.3\lambda_g$ in front of the diode.)

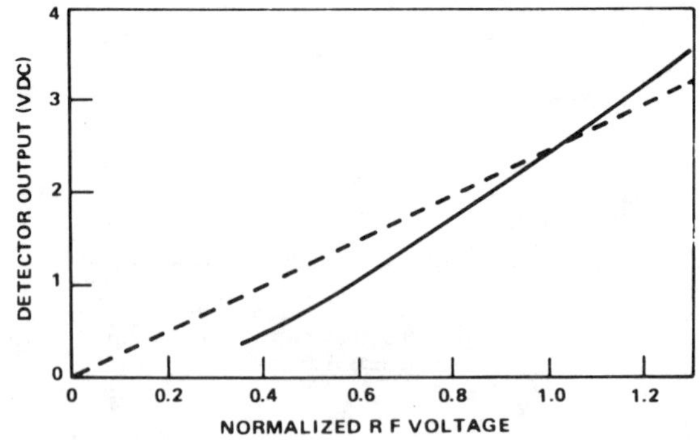

Fig. 7. Schottky-barrier diode detection characteristics (HPA2627 diodes at 10 GHz).

Depending on the type of detector used, a signal level of +1 to 5 dBm is applied to the detector input and the two mounts are tuned for maximum dc voltage output. Both detectors should be adjusted as nearly as possible for identical operation. Then data are taken to determine the detector characteristics of Fig. 7. The abscissa of this curve is obtained by normalizing the RF voltage with respect to the voltage at the operating point. The RF input to the detector is controlled by a precision waveguide attenuator. The ratio of the slope at the operating point to the slope of a line from the origin to the operating point gives a correction factor C which is used in (37). The correction factor for the diode characteristic of Fig. 7 amounts to 3.5 dB at an operating level of 2.45 V.

In stating the correction factor in this simple form, we have made two assumptions about the equipment. First, the microwave bandwidth of the detector mounts is much wider than either the maximum deviation or the highest amplitude modulation frequency to be measured. Second, the ac input impedance of the video amplifier is greater than $20R$.

Once the slope correction has been determined, data are taken from the wave analyzer to determine the AM noise power spectral density:

$$\mathrm{AM}(f) = 20 \log_{10} V_{AM}(f) - 20 \log_{10} V_{REF} - 20 \log_{10} C - G_v - 6 - 10 \log_{10} B_N \quad (37)$$

where

$\mathrm{AM}(f)$	amplitude modulation noise-to-carrier ratio in a 1-Hz bandwidth (decibels);
$V_{AM}(f)$	reading of wave analyzer (rms volts);
V_{REF}	dc voltage for each detector in Fig. 6;
B_N	noise bandwidth of the wave analyzer (hertz);
G_v	gain of the video amplifier (decibels);
C	correction factor derived from Fig. 7;
f	modulation frequency, read from wave analyzer (hertz).

The 6-dB factor is used to account for the two detectors operating in series. If a single diode detector is used, this 6-dB factor is omitted.

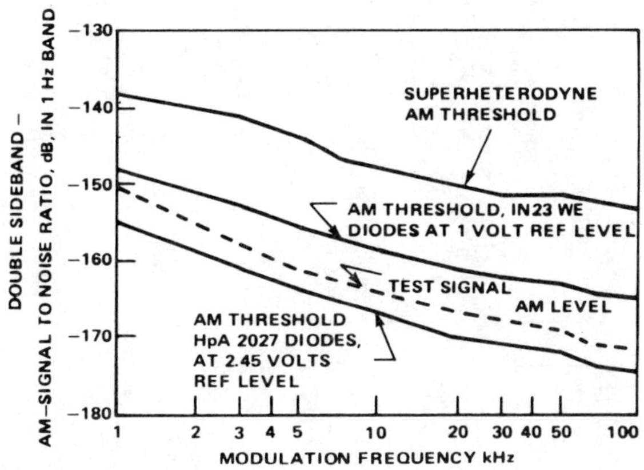

Fig. 8. The AM threshold for a superheterodyne receiver, silicon point contact diodes (IN23WE), and Schottky-barrier diodes (HPA2627).

The quantity AM(f) is the ratio of $S_{AM}(f)$ to carrier power expressed in decibels. This is often described as the double-sideband noise-to-carrier ratio because the use of (20) means that in the RF spectrum, this AM noise will appear one-half on each side of the carrier.

In practice, we have observed that the most frequent failing is neglecting to determine the measurement threshold.[3] Thus, if one attempts an AM noise measurement with a single-ended diode AM demodulator, he should check the typical threshold curves for his demodulator device as given by Fig. 8. If measured data are within 6 dB of expected threshold, then a balanced detector such as the one illustrated for waveguide in Fig. 6 needs to be constructed. The test for threshold shown in Fig. 6(b) consists of operating the individual diodes in series opposition so that the AM on the test signal is mostly canceled. Here, the balance of the diodes is most important and must be tested under RF excitation. One easy method is to apply intentional sinusoidal AM on the test signal and match the detectors to yield minimum output at the AM frequency. The first intuition would be to balance the detectors by tuning adjustment. This tends to cause unbalance for modulation frequencies above 100 kHz, and a better way is to adjust the sizes of the load resistors R to balance the detector system. A plot of noise spectrum from the detector is a good indication of threshold. Reconnecting the diodes to series aiding then gives an output which is the sum of threshold noise and AM noise. If the output is within 2–10 dB of threshold, then a correction

$$V_{AM}(f) = \sqrt{V_{AMM}^2(f) - V_{ATH}^2(f)} \tag{38}$$

where

$V_{AM}(f)$ corrected noise voltage;
$V_{AMM}(f)$ uncorrected AM noise data;
$V_{ATH}(f)$ threshold data;

improves accuracy. Note that the correction must be done on a power spectral density basis before expressing the answer as an AM noise-to-carrier ratio in decibels.

[3] In a conversation at the 1976 IEEE-MTT-S International Symposium, Dr. J. Ondria made a similar observation.

Some of our practical experience with AM noise measurements may help you save time in the laboratory. If you are measuring a source which has a known relatively high AM noise, such as IMPATT amplifiers and oscillators or transmitter TWT amplifiers, then the simple single-ended silicon point contact diode is adequate, available, and reliable. The least expensive versions of 1N21 or 1N23 sorts of diodes are just as useful in this application as those which are sorted for best superheterodyne noise figure.

If you are measuring a source with relatively low AM noise, such as klystron oscillators or transferred electron oscillators (TEO's) using GaAs diodes (Gunn diodes), then the full balanced detector system with Schottky-barrier diodes must be employed and the threshold check made every time. Certainly, if measuring a completely new and unknown source, then a careful and complete calibration and determination of measurement threshold must be done.

We have experienced a degradation of Schottky-barrier diodes which we do not pretend to understand.[4] The peak inverse voltage allowable decreases with operating life and some strange noises appear in the demodulator output. This requires replacement of the diodes before taking more data.

A simple operational check for an adequate measurement threshold is worth knowing about. In the calibration procedure, a precision signal frequency attenuator is needed to take the slope data. We leave this attenuator in the system for the balance of the measurement. Regardless of the kind of detection system, increasing attenuation on the input of the demodulator by 3 dB should make the wave analyzer output indicator go down by the same 3 dB. If there is less than 1 dB of change on the wave analyzer output, the system threshold is inadequate. If there is a 2-dB change, taking threshold data and correcting will give a reliable answer.

One method of reducing AM threshold [4], [5] has been thoroughly evaluated by Fikart *et al.* [9]–[11] and found to be of dubious utility. Essentially, the cavity discriminator [3]–[5] can be used for AM noise measurements by inserting a 90° phase shift in the reference line going to the output phase detector. Component imperfections and drift of the source and/or the discriminator cavity tuning cause sufficient errors to make this a method of last resort. Since only the two-resonator klystron oscillators have approached the threshold of the dual diode system with Schottky-barrier diodes, the method of last resort is not often needed.

When measuring an amplifier or frequency multiplier, AM noise at input and output is measured separately. Since these components usually operate in a highly saturated mode, there is little correlation between input and output AM noise. Furthermore, the AM noise at the output is quite dependent on relative carrier power output. We have observed a tendency to neglect making the AM noise

[4] In a conversation with Dr. John Ondria at the 1976 IEEE-MTT-S International Symposium, he reported similar experiences. The diode manufacturers are aware of the problem and recently produced diodes are better in this regard.

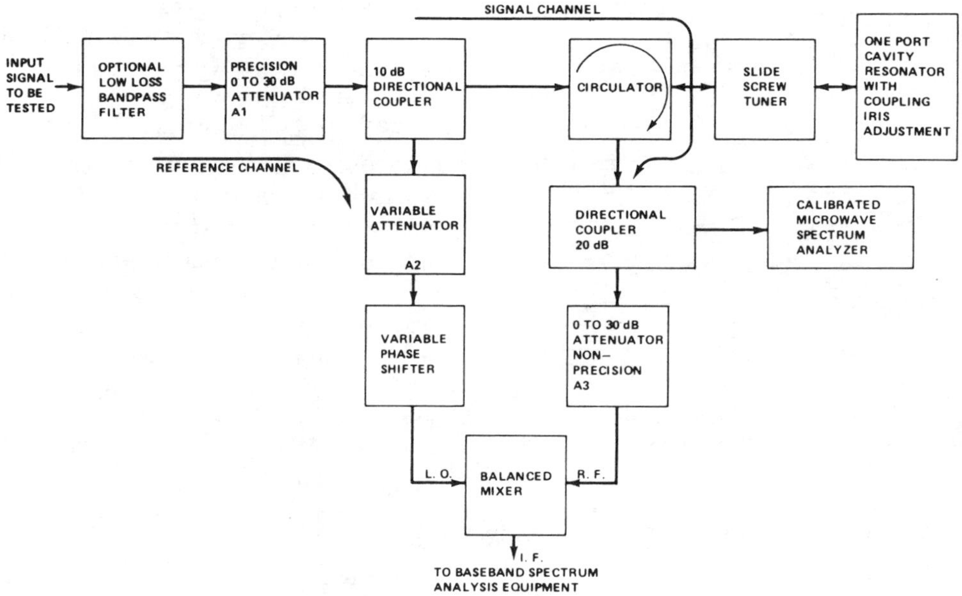

Fig. 9. A microwave frequency discriminator for FM noise measurement.

measurement which is not at all justified by transmitter device performance.

V. Microwave FM Noise Measurements

We will follow the historical development of accepted noise measurement circuits in describing FM or phase noise measurements; thus the discussion in this section will concentrate on the microwave region above about 5 GHz, and the following section will discuss FM noise measurements for carrier frequencies below 5 GHz.

The history is difficult to document because much of the work was classified for security reasons in the era following World War II and during the Korean War. This work has now been declassified, but the archival journal papers were not written at the time, and one would have to delve into company and military reports to adequately give credit to those who made the first contributions. With regard to equipments available in the USA, we can trace a clear path back to the Royal Signals and Radar Establishment (RSRE), Great Malvern, Worcester, England. During the 1950's the RSRE supported the development of klystron oscillators and amplifiers for Doppler radar use. These klystrons quickly presented a measurement challenge, and the apparent result was the development of the cavity discriminator which Whitwell and Williams [12] attribute to "S. B. Marsh, J. D. Clare, and S. A. Drage of RSRE who evolved the frequency discriminator and constructed the first experimental models . . .". This equipment was marketed in the United States by James Scott & Company, Ltd., Glasgow, Scotland, as the Allscott Model 123 Noise Measuring Equipment, and the first author of this paper made his first believable FM noise measurement with Allscott equipment in the early 1960's.

Aside from the fact that it did represent a completely designed equipment with a set of operating instructions, the RSRE discriminator bridge had two superiorities over the prior art: first, it would reject a good amount of incidental AM noise on the test signal; and, second, input power could be increased (with proper adjustment of the equipment) to override the threshold noise in the equipment. Thus the 10-W sort of two-resonator klystron oscillators which still hold the world record for low FM noise could be measured.

Those who use the RSRE cavity discriminator quickly tire of tuning the microwave bridge based on a magic tee or 3-dB four-port directional coupler, and of tuning the klystron used as a local oscillator (LO) for the superheterodyne equipment. Many microwave sources put out 10 mW or more, and the input power sensitivity is not needed. Also, the 3-dB hybrid in the bridge costs a total of 6 dB in threshold or detection slope because the signal goes through the hybrid twice. These factors and the need to measure lower noise sources motivated Ondria [4] first and, working independently, Ashley *et al.* [3] to make significant improvements in this equipment. As shown in Fig. 9, a circulator has replaced the magic tee, and the cavity has been modified to include an adjusting screw on the coupling iris. This simplifies tuning and improves the threshold by nearly 6 dB. Since the intermediate frequency (IF) amplification is not often needed, the LO and IF equipment have been left out in the interest of simplicity. Although the final phase detector could be the magic tee circuit of Fig. 6, we have found it more convenient to use a standard balanced mixer[5] in the discriminator as illustrated in Fig. 10.

There are two approaches to deriving a theory for demodulator circuits such as this frequency discriminator: 1) the quasi-stationary frequency assumption [3]; or, 2) the sideband phasor method [4]. The first has the advantage

[5] Such as a Varian Orthomode® mixer designed as a downconverter for superheterodyne receivers.

Fig. 10. Photograph of the *C*-band implementation of Fig. 9.

of yielding a direct answer for the discriminator slope and showing that the answer is good for modulation frequency approaching zero; thus we can use the discriminator for short-term and even medium-term stability indications for comparison with other measurement techniques. The second yields the upper modulation frequency correction factors more directly. The first method is the one used in communication theory study of noise problems in FM systems [7].

Either method will yield an equation for the slope S of the form

$$S = \frac{\Delta v}{\Delta \nu} = \frac{K\sqrt{P_i}10^{-(G_H/20)}Q_0}{\nu_c} \tag{39}$$

where

- $\Delta\nu$ small deviation from the carrier frequency (peak hertz);
- Δv small dc change in output (volts);
- P_i input power (watts);
- G_H loss in the system between input and phase detector (decibels);
- Q_0 internal Q of the microwave resonator;
- ν_c carrier frequency (hertz);
- K a factor depending mainly on the conversion loss in the phase detector.

This equation is again one that is mainly for study rather than for actual calculation. The measurement of the factor K has some conceptual difficulties, and we do not need to tell you that measurement of a resonator Q_0 which is probably greater than 20 000 is a challenging problem. Thus most laboratories use a technique based on applying a known angle modulation and measuring Δv to determine S. Such a technique will be described along with the tuning and operating procedure. For now, it is important to draw some conclusions from (39) that influence operation.

First, we note that the slope is proportional to $\sqrt{P_i}$. Therefore, if a calibration is done to determine S, this result is valid only as long as P_i is not changed. All the other factors of (39) are quite stable, and as long as the power is maintained constant, the slope is very stable with time. Therefore, it is quite reasonable to use one source (which might be easy to angle modulate) for calibration to measure another source; all that must be done is to either hold input power constant, or to measure the power ratio and make the appropriate correction.

Second, measurement of a small deviation $\Delta\nu$ (corresponding to a quiet oscillator) requires the highest obtainable value of S to override the noise in the crystals of the phase detector. Since carrier frequency ν_c and K are fixed, this means that P_i must be as high as possible, Q_0 high, and G_H minimized. If the circulator shown in Fig. 9 is used, G_H is of the order of 1 dB; whereas the use of a magic tee or 3-dB coupler will make G_H about 7 dB. This 6-dB difference directly translates to a 6-dB lower measurement threshold. If these measures do not result in an adequately low threshold, then methods to be discussed in Section VII must be employed.

In practice, the determination of S is done as part of the tuning and operating procedure, and this will be the organization of our discussion. You will notice that the evolution of signal sources and spectrum analyzers has caused appreciable change in calibration procedure from those reported earlier [3], [4]. In the 1950's and early 1960's, most of the sources being measured were direct microwave oscillators (klystrons, TEO's, IMPATT's, etc.) which could be easily frequency modulated via a bias port. Also, the deviation in these devices could be proved from theory and by measurement to be linear with respect to voltage applied. Microwave spectrum analyzers of that era were not stable and calibration accuracy was low; thus the spectrum analyzer was used to determine the first carrier null and the corresponding modulation index, 2.405. Then the modulating voltage was attenuated by a measured amount and the linearity of the oscillator depended on to calculate the amount of angle modulation used in the calibration. Now, a good fraction of sources to be tested are controlled by VHF or UHF crystal oscillators and usually do not have angle modulation facility. Also, microwave spectrum analyzers have experienced continuous improvement, and now several manufacturers offer equipment with excellent stability and reliable calibration. In particular, the IF attenuator is a stable device that can be accurately calibrated and used as a secondary standard in making the slope calibration. Dynamic range and internal noise in the spectrum analyzer make it possible to measure calibration sidebands 50–60 dB below the carrier with adequate accuracy.

For the discussion of the discriminator tuning, calibration, and operation, we will assume that angle modulation side frequency can be applied to the sources being measured that are 50–60 dB down from the carrier. If the source has internal phase or frequency modulation capability, then the microwave spectrum analyzer is used to measure the side frequency to carrier ratio. If the source cannot be modulated, then a circuit such as the one of Fig. 11 can be connected between source and the discriminator. The idea is simple—amplitude modulation side frequencies with suppressed carrier are generated with a crystal modulator

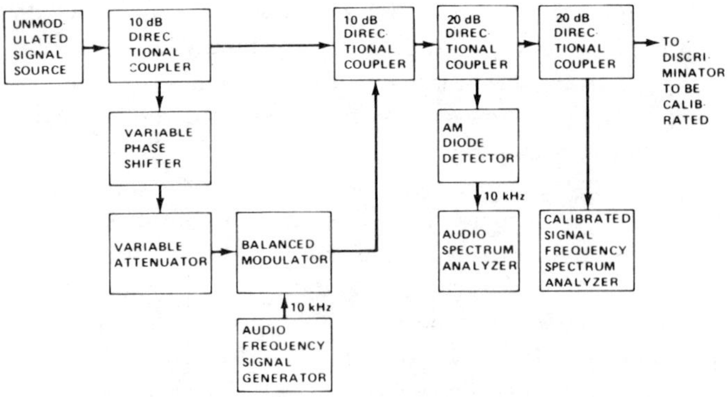

Fig. 11. Generation of angle modulated side frequencies for discriminator calibration.

and reinserted with the needed 90° phase shift to appear as angle modulation side frequencies. The difficulty is to know that the needed 90° phase shift has been obtained and to know if carrier leakage through the balanced modulator will confuse the calibration. The amplitude detector tells us that the phase shifter is set correctly because a simple phasor diagram indicates that minimizing amplitude modulation on the recombined signal is an accurate indication of proper side frequency phasing, even if the balanced modulator has carrier leakage on the order of the side frequencies generated (as has proved the case in modulators we have tested).

During the process of setting up and measuring the calibration side frequencies, the precision input attenuator of Fig. 9 (or Fig. 13) should be set for maximum attenuation to protect the crystals in the balanced mixer.

The operation procedure for the circuit of Fig. 11 is as follows.

1) Apply as much signal frequency power to the input of the balanced modulator as the device will stand without burnout.

2) Set the audio signal generator at the desired side frequency spacing (say, 10 kHz), and adjust the output level while viewing the signal frequency spectrum analyzer. Do not exceed the level which causes second-order side frequencies (at, say, ± 20 kHz) to appear.

3) Tune the audio spectrum analyzer to the frequency of the audio signal generator (say, 10 kHz), and then adjust the variable phase shifter to minimize the indication on the audio spectrum analyzer.

4) Check the level of the angle modulation side frequencies on the calibrated spectrum analyzer.

In operating the circuit, be careful to not burn out the diodes in the balanced modulator with too much RF input or to cause generation of higher order sidebands by the application of excessive audio input power. Usually, it is possible to obtain calibration side frequencies in the range 50–60 dB below the carrier. Side frequencies more than 60 dB below the carrier are getting too close to the internal noise level of the microwave spectrum analyzer to measure accurately enough for calibration purposes. Side frequencies less than 50 dB below the carrier require a modification of Eq. (40).

One clever idea which might occur to you is to use this calibration technique to do a swept calibration of the output indicator; for instance, putting a line of constant side frequency level on the output graph. We have tried this idea with reasonable success. For side frequency spacing greater than 5 kHz, the microwave spectrum analyzer can resolve the sideband structure and prove that the calibration side frequency amplitude does remain constant. On several balanced modulators we have used in Fig. 11, the calibration level line did not behave as expected below 5 kHz, and we were able to isolate the trouble to the balanced modulator. Perhaps a thermal time constant or an unsuspected filter inside the balanced modulator is the cause of the trouble, but our advice is to not to assume the side frequency level is constant if you cannot prove it with the microwave spectrum analyzer.

With the side frequency level set and measured, the discriminator of Fig. 9 is tuned and calibrated as follows.

5) With the reference channel attenuator (A2) set at 0 dB and the threshold check attenuator (A3) set at maximum, decrease the input attenuator (A1) until the reference power at the LO input of the balanced mixer is as high as the mixer will stand without burnout. (A crystal current indicator is a useful way to make this determination.)

6) Replace the cavity resonator with a well-matched termination, and adjust the slide screw tuner for minimum carrier indication on the microwave spectrum analyzer. This cancels undesired leakage in the three-port circulator.

7) Reinstall the cavity resonator and tune to resonance. Check the mode chart and cavity tuning calibration to make certain that the desired mode[6] is being used. The indication of resonance is a dip in carrier level on the microwave spectrum analyzer. Adjust the coupling iris to minimize carrier amplitude, recheck the tuning, etc. Within just a few iterations, the carrier level can be reduced some 60 dB which is more than sufficient. A more important goal to strive for at this stage is to make certain that the resonator is tuned as precisely as possible to the carrier frequency. With the calibration side frequencies present, this is simple to determine because the side frequencies' amplitudes will be the same (the picture is symmetrical) if the resonator is properly tuned. At this stage, it is normal to see the side frequencies' amplitudes within 10 dB of the residual carrier.

8) Reduce attenuator A3 to zero while observing mixer current which should not change. Tune the baseband spectrum analyzer to the modulation frequency of the calibration side frequencies (say, 10 kHz), and adjust the reference channel phase shifter for maximum indication on the baseband spectrum analyzer. (Watch for spectrum analyzer overload.)

9) If the input attenuator A1 is not at 0 dB, increase A2, then decrease A1 until A1 is 0 dB. Recheck the reference phase adjustment. Check for overload in the baseband spectrum analyzer, and insert input attenuation if needed.

[6] These cavities using the TE_{01n} modes are an order of magnitude more difficult to protect from spurious mode troubles than the TE_{111} mode used in cavity wavemeters.

(Here, increasing the precision attenuator A1 by 3 dB should reduce the output indication by 3 dB.)

10) Calculate the discriminator slope S, using

$$S = \frac{\sqrt{2}V_{ac}}{2f_m 10^{-D/20}} \tag{40}$$

where

- V_{ac} discriminator ac output voltage at f_m (rms volts);
- f_m modulation frequency (hertz);
- D side frequency amplitude with respect to carrier frequency amplitude (decibels);

and store this value for later use in data processing.

11) Insert the threshold check attenuator A3, turn off the calibration side frequencies and operate the wave or spectrum analyzer at all frequencies where FM noise data are desired. (The input attenuator of the wave or spectrum analyzer should be set for minimum at this step. An overload light on the input amplifier of the wave analyzer is most useful here.) Record or store these threshold voltages $V_{\mathrm{FTH}}(f)$ and the measurement bandwidth B_N.

12) Reduce the threshold check attenuator A3 to 0 dB, check overload of the baseband equipment, and take FM noise data at the desired baseband frequencies $V_{\mathrm{FMM}}(f)$. (Use the same values of B_N as in step 11.)

13) Calculate corrected deviation FM(f) at each modulation frequency using:

$$\mathrm{FM}(f) = \sqrt{V_{\mathrm{FMM}}^2(f) - V_{\mathrm{FTH}}^2(f)}/S. \tag{41}$$

Note that the bandwidth for the measurement B_N was set in the baseband wave analyzer and must be stated along with the values for the FM noise.

14) Convert the FM noise data to other bandwidths, forms, etc.

Several items from our laboratory experience may be helpful to you if you are working on your first or second FM noise measurement. As in the case of AM measurements, we have observed that the most consistent failing is to neglect determining measuring system threshold. We consider the development of threshold checking routines as our most important contribution to the art of noise measurement. If you are attempting to measure a state-of-the-art source with just a few milliwatts output, you will probably find the threshold is too high, and our suggestions for this problem can be found in Sec. VII.

The usual problem is finding a cavity suitable for this use. Cavities intended for use as wavemeters do not have a sufficiently large coupling iris to achieve the critical coupling needed here. Modification is usually not a trivial task. To achieve the high Q_0 needed, the cavity mode is normally from the TE_{01n} family. The TE_{011} will have a $Q_0 \approx 22\,000$ at 10 GHz with a 10-percent tuning range which is moderately free of spurious modes. As the higher order modes are used to obtain higher Q, the spurious mode problem becomes more serious. The vendors most familiar with this kind of cavity resonator are those who develop transmission cavities for oscillator stabilization [14].

Most of the mysteries are found in the baseband equipment. All the good wave and spectrum analyzers have both input and IF attenuators to allow optimization of the level at the first upconverter. These attenuators must be set correctly (as noted in steps 8), 9), 11), and 12)). Some older wave analyzers which were not intended for noise spectrum analysis do not have enough gain in the input stages and a preamplifier is needed.

Another baseband problem is ground loops. When the baseband spectrum analyzer indicates at 60 Hz and harmonics thereof, the question is whether the indication means trouble in the signal source or the measurement apparatus. The threshold check does partially indicate if the indication is in the measurement apparatus because FM from the source will be attenuated by the amount of the threshold check attenuation inserted. Eliminating ground loops seems to be more a matter of experience than anything that is capable of being written down. One source of trouble is the safety ground in the power cords of electronic instruments. All the baseband equipment should be plugged into one box of outlets, and the cords run as a bundle as far as possible. The RF input to the final amplitude or phase detector is best supplied through dc blocks to break up the worst ground loop. Keep in mind that it is the magnetic field that couples into a loop; thus minimizing loop area and placing low-frequency breaks in the loops will minimize current flow.

Minimizing external interference with the measurements is sometimes accomplished by using a shielded room. Our experience has been that this is not as effective as one might think. In general, we have been able to take data in R&D laboratory environments and not find a significant difference if the whole apparatus is moved inside a shielded room. In fact, we have sometimes seen a degradation because the shielded rooms often have ventilation blowers with appreciable acoustic noise output. This acoustic noise vibrates waveguide walls and cavity resonators (as well as the signal source) to cause a messy spectrum below 2 kHz. The threshold check does not distinguish whether this is in the signal source or in the measurement apparatus. If you see a messy spectrum in the FM noise below 2 kHz, turn off all possible blowers and acoustic noise generators to estimate how much of the mess is acoustic. Unfortunately, shielded rooms get hot in just a few minutes after the blower is turned off, and this acoustic noise must be tolerated. One instance where outside interference could not be eliminated without a shielded room was in trying a measurement out on the production floor near a test station for high-power pulsed klystrons. The magnetic radiation from the pulse transformers clearly cluttered the noise measurement on a CW klystron at a nearby test station.

There does seem to be a learning curve associated with making transmitter noise measurements. About a man-week is required to really understand the specification and concept of transmitter noise. After equipment is assembled, another man-week is required to learn the ritual of all the tuning, calibrating, and data taking that must be done. At the end of this second week, you can expect to have some faith in eventually having adequate confidence in the

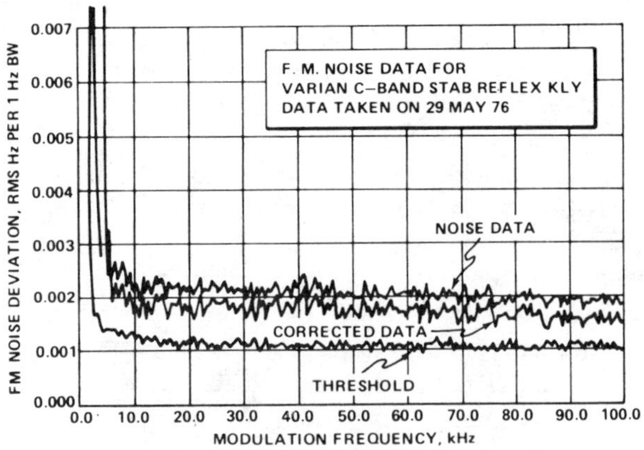

Fig. 12. Typical *C*-band klystron data plot using the HP-3045A automated spectrum analysis system.

measurements. At the end of three to four man-weeks, you can expect enough confidence to tell your superior that the answers are right, and you have mastered the art of transmitter noise measurements. Results such as the typical measurement of Fig. 12 are a pleasure to report.

The use of a frequency discriminator for the measurement of amplifier phase noise might seem inappropriate at first glance. A phase bridge has been used to compare output and input of amplifiers to determine the noise added by the amplifiers. There are several problems with this latter idea. First, an adequately low-noise drive source must be used and the simple phase bridge will not determine adequacy. Second, the threshold will be inadequate if carrier injection is not added. Third, after the measurement results are available, the amplifier noise and driving signal noise have to be combined to find output noise [24].

The frequency discriminator solves all of these problems. Since it is needed for the study of drive source noise, the discriminator is available for studying an appropriate sample of the amplifier output.

The measurement of both AM and FM noise at the output of a 10-W or higher power TWT or klystron will normally show power supply ripple and grounding problems (as seen in Fig. 25). We have reported [25] that the output noise cannot be predicted on the basis of noise figure measurements made at low level and using an IF amplifier at 30 MHz. In the range from 3 dB below to saturated power output, the output noise significantly increases, probably because of RF defocusing of the electron beam and interception on the RF structure.

The complete noise study of a microwave transmitter requires measurement of both AM noise and FM noise at all significant intermediate and output ports. The cavity resonator discriminator just discussed will make the FM noise measurement at the output ports but is not practical for some of the intermediate stages. The next section describes a newly developed technique that is practical for the submicrowave portions of a transmitter.

VI. Submicrowave FM Noise Measurements

The need to make transmitter noise measurements in the lower microwave frequency range, VHF, and UHF ranges has grown because of the evolution of frequency synthesis techniques for microwave transmitters. When a microwave transmitter noise measurement of a synthesized signal yields an unacceptable noise, then one must go back and measure the components used in the synthesis. This was shown to be the case with the Radio Frequency Simulation System developed and installed at the McMorrow Laboratories of the U.S. Army Missile Command, Redstone Arsenal, AL. We found it necessary to make many of our noise measurements below 2 GHz.

The method recommended by Shoaf *et al.* [15] for this frequency range is to electronically phase lock the oscillator under test to a second supposedly identical unit. Then a spectral analysis of the correction voltage in the phase lock loop is the basis for determining the phase noise. To achieve a minimal proof that the oscillators are identical (and that the measured noise can be attributed to each oscillator on a 50–50 basis), a third oscillator is needed so that the three oscillators can be compared two at a time. The need to develop the equipment for the electronic phase locking is a strong disadvantage of this method.

What is needed is a method similar to the cavity discriminator methods described in the previous section. As frequency gets lower, the TE_{011} cavities get larger, and somewhere below 2 GHz attain unreasonable size and weight. (We have not been able to purchase cavities for the region below 5 GHz.) Such a new method must share the following advantages with the RSRE cavity discriminator if the method is to attain bench-mark status.

1) It must be relatively easy to adjust and calibrate.

2) It must be adequately stable in frequency and time.

3) An operational method for determining the measurement threshold must be included.

4) If additional signal power is available, the threshold must be reducible.

5) When properly tuned, the discriminator must reject incidental AM noise on the signal under test.

This need for a new method led us to study transmission line discriminators. These have been known for a long time, but the conventional wisdom for dismissing them was well expressed by Campbell [16] in 1964. "The simplest detector—with the exception of the slope detector—is the interferometer, where the dispersive element is a long line. This is usually used for wide-band applications, with its accompanying low sensitivity. More sensitive versions require an excessively long line and are characterized by multiple responses." The more important faults of prior art transmission line discriminators were that additional input signal power could not be used to reduce threshold and that AM on the test signal was not rejected.

After all the years of operating the microwave discriminator of Fig. 9, it suddenly became obvious that replacing the cavity with a long, shorted transmission line and slide screw tuner, as shown in Fig. 13, gives a discriminator which meets all of the aforementioned requirements.[7] In the VHF range, circulators and slide screw tuners may be hard to obtain while 3-dB four-port hybrid junctions are readily

[7] Three patents [27]–[29] have been issued.

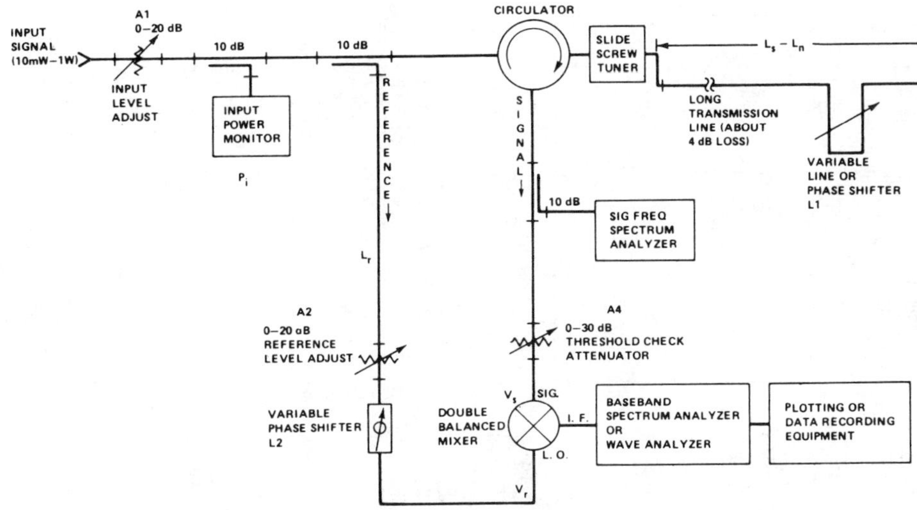

Fig. 13. An optimum sensitivity transmission line discriminator.

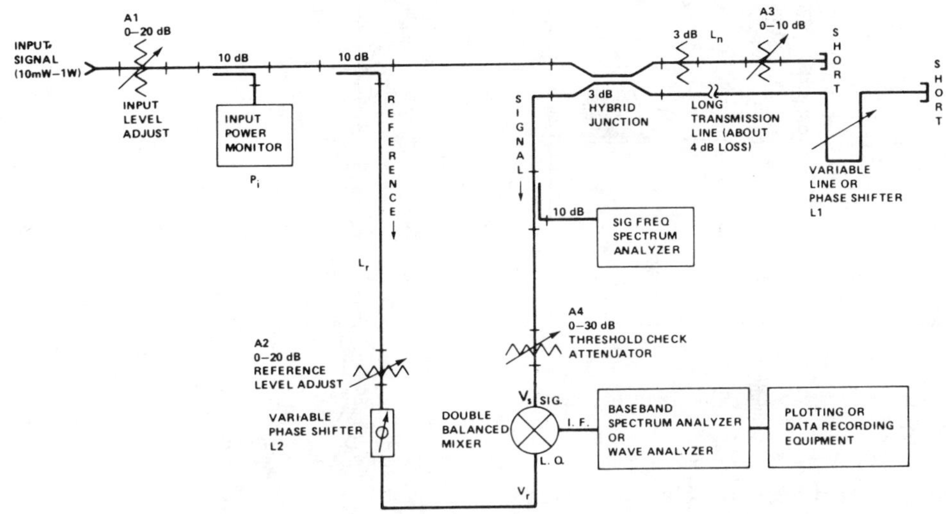

Fig. 14. A wide operating frequency range transmission line discriminator.

available; thus the adaptation shown in Fig. 14 is useful although the threshold and slope are 6 dB poorer than the circulator version.

Although we cannot claim to have looked much further than the ends of our noses, we did not find a published theory to adequately illuminate the use of the transmission line discriminator. Thus we developed some theory of our own which quickly disclosed that all transmission line discriminators have an optimum length for the transmission line. At this optimum length, the signal attenuation is 1 Np.

We have used the quasi-stationary frequency approach to analyze these discriminators. Two transmission line lengths, L_s and L_n, are defined as the mechanical lengths of the signal line and nulling line in either of the discriminators. L_s is the mechanical length of long transmission line and phase shifter in the signal path (*m*), and L_n is the mechanical length of transmission line in the null circuit path (m). For the discriminator using a circulator and slide screw tuner (Fig. 13) L_n is the distance to the tuner probe. The derivation [22] determines the magnitude and phase angle of the signal applied to the SIG port of the balanced mixer used as a phase detector. Both magnitude and phase angle depends on the depth of the tuning null at ν_c as illustrated by Figs. 15 and 16. Using this signal input to the phase detector, we obtain the slope equation

$$S = \frac{\Delta v}{\Delta \nu} = 8\pi\sqrt{2Z_0P_i}[10^{-(G_M+G_H)/20}](L_s - L_n)e^{-2\alpha L_s}/U_0 \tag{42}$$

where

- Z_0 characteristic impedance of transmission line components (ohms);
- P_ι input power to discriminator (watts);
- G_M conversion loss of the balanced mixer (decibels);
- G_H loss in the hybrid junctions between input and SIG port of the phase detector (decibels);
- α transmission line attenuation factor (nepers per meter);
- U_0 phase velocity in the transmission line (meters per second).

This expression for slope does not contain any factors which depend on the depth of the tuning null, which is

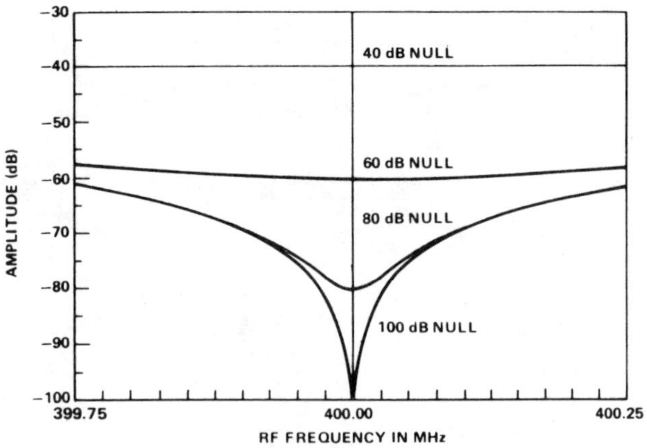

Fig. 15. Amplitude of the signal applied to the SIG input of the phase detector of Fig. 14.

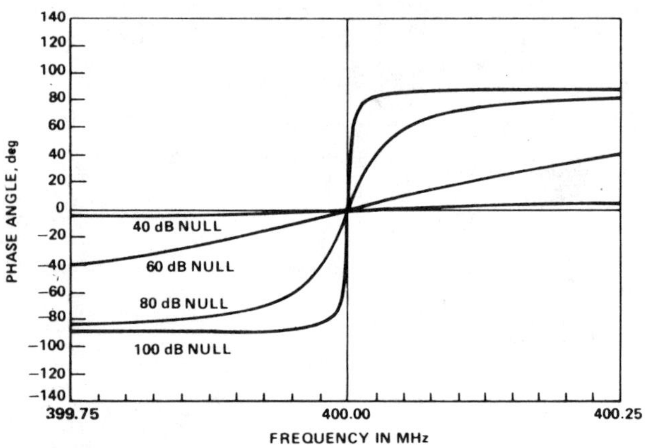

Fig. 16. Phase of the signal applied to the SIG input of the phase detector of Fig. 14.

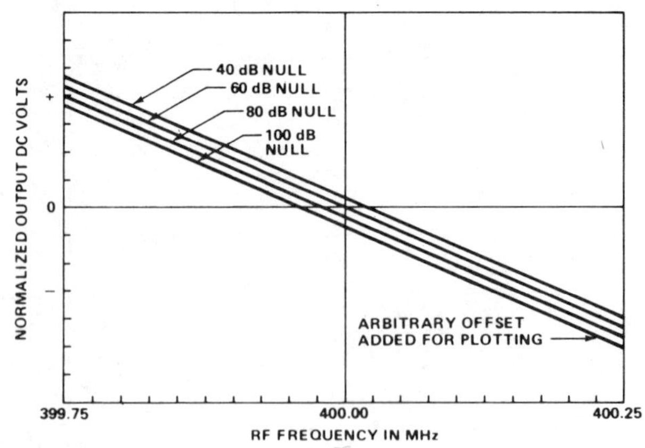

Fig. 17. Output voltage versus input RF frequency for the discriminator of Fig. 14.

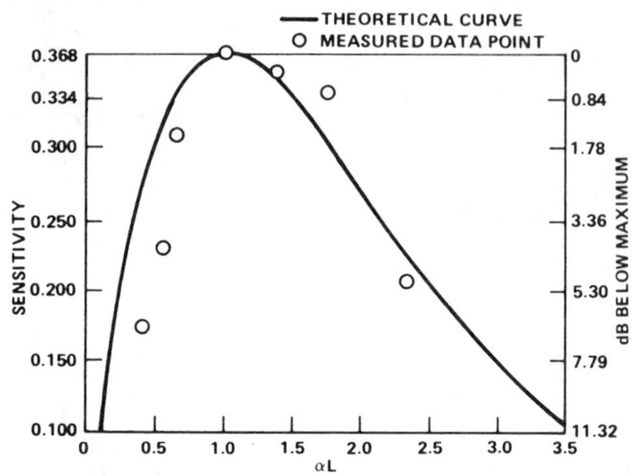

Fig. 18. Sensitivity versus total transmission line length.

surprising in the light of Figs. 15 and 16. Thus we took the expressions which are plotted in Figs. 15 and 16, applied the equation for the phase detector response, and computed the output voltage versus quasi-stationary input frequency in the neighborhood of v_c. Plotting this response yielded curves which fell one on top of the other. An artificial offset was added to make the plots shown in Fig. 17, which clearly shows that the slope does not depend on the depth of the null. This conclusion has been borne out by detailed laboratory measurements.

Any transmission line discriminator has multiple responses. If there are n wavelengths in the long transmission line for the desired tuning, additional nearly equal responses occur for $n + m$ wavelengths where m is an integer. The first adjacent responses are found at

$$v_A = v_c \pm U_0/(2L_s). \tag{43}$$

If this expression is evaluated for the use of an optimum length of low-loss coaxial cable, the adjacent modes are spaced 1 MHz for operation at 100 MHz and spaced 3 MHz for operation at 1 GHz. This spacing is sufficient that FM noise data can be taken through the baseband of interest for Doppler radar (below 100 kHz) without applying correction factors. When measuring TWT's BWO's and other devices without an output filter circuit, a bandpass filter must be used between these sources and any kind of a discriminator.

Notice the analogy between the equations for the slope of the cavity resonator discriminator [see (39)] and the slope of the transmission line discriminator [see (42)]. Both are portiortional to $\sqrt{P_i}$ and energy stored in the system. Both show that amplitude modulation on the input signal is rejected in the output. Perhaps the only surprise is that (42) does not have the factor v_c. Therefore, the transmission line discriminator can be calibrated at one carrier frequency and then used over a $\pm$5-percent carrier frequency band (with retuning at each value of v_c within the band) without recalibration if P_i is held constant or corrected for.

The slope equation can be normalized and plotted as shown in Fig. 18. Notice that this curve has a maximum at $\alpha L_s = 0.5$ corresponding to a round-trip attenuation of 1 Np (8.68 dB) in the transmission line. Thus we know there is an optimum length for the transmission line which depends only on α, the loss per unit length. The lower the line loss, the longer can be the line and the higher the slope. Our intuition that lower loss line should be better is indeed the truth!

This maximum shown in Fig. 18 is broad enough that a transmission line cut to be optimum at one frequency can be used over a surprisingly wide carrier frequency range. For coaxial cable below 5 GHz, α is solely dependent on skin depth and changes as $\sqrt{\nu_c}$. Thus for a 3-dB limit on sensitivity (and threshold) degradation, a line optimum at ν_1 can be used from two octaves below ν_1 to three octaves above ν_1.

To verify the aforementioned theory, we have made detailed measurements in the MICOM Laboratories at Redstone Arsenal to confirm Figs. 15–17. Even more gratifying has been the experience in using these transmission line discriminators to make FM noise measurements on sources from 10 MHz to 5 GHz. At 5 GHz, the slope is about 5 dB less than a cavity resonator discriminator with $Q_0 \approx 33\,000$ corresponding to a TE_{013} mode.

The theory and our operating procedure with this transmission line discriminator can be summarized as an alignment, calibration, and operating procedure. Obtaining calibration side frequencies is essentially the same as discussed in steps 1)–4) of Section V, and the discussion need not be repeated here except to indicate it is a required operation. The typical transmission line discriminator shown in Fig. 13 is normally implemented with coaxial components and can be operated over several octaves of input signal frequency range without changing mixer, directional coupler, and the hybrid junction. The alignment and operating procedure can be described in the following steps.

1) Adjust the input attenuator (A1) for approximately 20 mW going into the discriminator. Then adjust the threshold attenuator (A4) for maximum attenuation.

2) Adjust the reference level attenuator (A2) for the proper power level at the input LO port of the mixer. (Normal LO power requirements for mixers of the coaxial type in these frequency ranges vary from 2 to 10 mW). Power level adjustment for these mixers differs from the waveguide mixer in that most coaxial mixers in the 2-GHz and below frequency range do not have a current monitoring capability. For this reason, it has been found that an easy way to set up the mixer LO port power is to disconnect the coaxial cable at the mixer LO port and connect in a power meter. After attenuator (A2) has been set, then reconnect the coaxial cable to the mixer.

3) Adjust phase shifter (L1) at the end of the long transmission line and attenuator (A3) in the other arm of the hybrid for a carrier null as observed on the signal frequency spectrum analyzer. Normally, it is easy to obtain a null on the order of 75 dB. However, nulls as good as 110 dB have been observed. When adjusting for a carrier null, after about 40 dB of nulling has been attained, if the calibration side frequencies have been placed on the carrier, these side frequencies can be used to advantage in determining which direction to adjust the attenuator (A3) and phase shifter (L1). The calibration side frequencies will tend to exhibit an unbalance in amplitude which can be used to adjust the nulling elements. After a relatively few nulling operations it becomes very easy to watch the side frequency amplitude changes and then be able to tell which element should be adjusted and in which direction (i.e., increase or decrease attenuation and more or less phase shift). A properly tuned null will exhibit a carrier with side frequency amplitude not more than 10 dB below the carrier (70 dB null) and both side frequencies will have the same amplitude. The display will be completely symmetrical on the spectrum analyzer used to monitor the nulling operation.

4) Reduce threshold check attenuator (A4) to zero. Reduce input level attenuator (A1) to zero while readjusting reference level adjust attenuator (A2) (remember to use the power meter) to maintain proper mixer LO port power level. Apply angle modulation (if this was not done in the nulling procedure) with an amplitude of 50–60 dB below the incoming carrier and offset from 10 to 20 kHz from the carrier.

5) Locate the modulation frequency on the baseband spectrum analyzer. Adjust variable phase shifter (L2) for a maximum signal output on the baseband spectrum analyzer. Remove the angle modulation calibration signal. (The system is now aligned, calibrated, and ready to take data.)

6) Calibrate the discriminator slope S using (40) presented in step 10) of Section V and store

$$S = \frac{\sqrt{2}V_{ac}}{2f_m 10^{-D/20}} \tag{40}$$

for later use in data processing.

7) Insert maximum threshold check attenuation (A4) and operate the wave analyzer to record the equipment threshold or noise floor.

8) Reduce threshold check attenuator (A4) to zero, check for baseband overload, and operate the wave analyzer, recording the FM noise as the system is operating.

9) If the FM noise measurement is less than 10 dB above the equipment threshold, calculate the corrected deviation at each modulation frequency, using (41).

$$\mathrm{FM}(f) = \sqrt{V_{\mathrm{FMM}}^2(f) - V_{\mathrm{FTH}}^2(f)}/S. \tag{41}$$

Note that the bandwidth for the measurement was set in the baseband wave analyzer and must be stated along with the values for the FM noise.

10) Convert the FM noise data to other bandwidths, forms, etc., as required for the format of presentation desired.

As in the case of the cavity discriminator, the most vexing problem is inadequate threshold. A typical example would be measuring a crystal oscillator directly at 100 MHz or lower frequency. After finding the lowest loss coaxial line that one can afford, the threshold is still too high and the only remedy is to use some form of amplification. (Or, use the two-oscillator method being equally careful to determine threshold.) We need about 30 dB of amplification with power output in the order of 1 W to overcome the threshold problem. This brings up the question "Will the amplifier noise confuse the measurement?" This is the topic of the next section but we will admit for now that our experience indicates that such an oscillator can be measured with an amplifier and transmission line discriminator.

For the total microwave transmitter, we evade this issue by noting that the amplifier is needed to drive the frequency

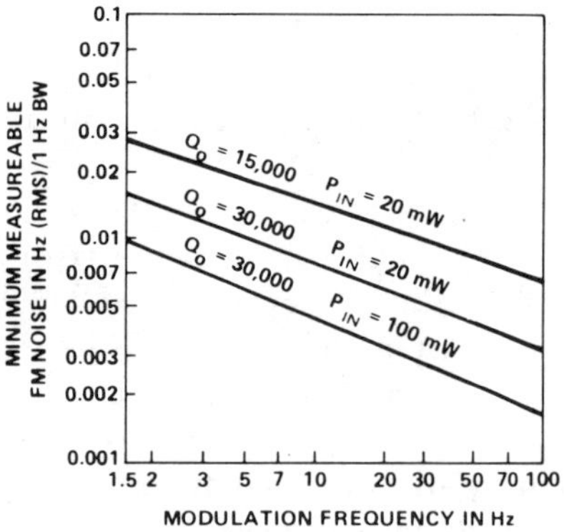

Fig. 19. Threshold level of a 16-GHz FM discriminator. This threshold is set by noise in the phase detector crystals.

multiplier. It is the output of this amplifier which is important, and this can be measured with the discriminator. Once the power is increased and then the deviation increased by the frequency multiplier, we can measure with ease and confidence in the submicrowave and lower microwave regions.

VII. Reduction of FM Noise Measurement Threshold

The problem which has caused the most trouble in the laboratory is the measurement of FM noise on low-power signal sources. At microwave frequencies, noise reduction by transmission cavity stabilization reduces output power by typically 6 dB as well as significantly reducing the FM noise. At submicrowave frequencies, the basic crystal oscillators are operated at relatively low output power, and even after amplification the following frequency multiplier loses power in the process. Yet these sources for microwave signals represent the state-of-the-art in low noise and must be measured.

The discriminator slope equations for both the cavity resonator discriminator [see (39)] and the transmission line discriminator [see (57)] both show that the slope is proportional to $\sqrt{P_i}$ and energy storage (Q of the cavity, length of the line) in the system. The noise generated in the crystals of the phase detector is constant because the LO port input power is held constant. Thus the threshold depends on $\sqrt{P_i}$ and energy storage. For the cavity discriminator, this is illustrated by Fig. 19. After one has used the highest possible cavity Q_0 or the optimum length of lowest loss transmission line, the only way to reduce the measurement threshold is to add amplification to the system. This amplification must be added with care so that the verification of the system threshold is not lost. One useful method for adding this amplification was developed by Ashley and Palka [19].

The usual noise properties of an amplifier (expressed as the noise figure) are not important here because this amplifier must operate at a high carrier level. The important factor is that the amplifier not add FM noise to the signal test. Also, since the discriminator rejects AM, the amplifier can add some AM noise without degrading the measurement. The data published by Ashley and Palka [20] show

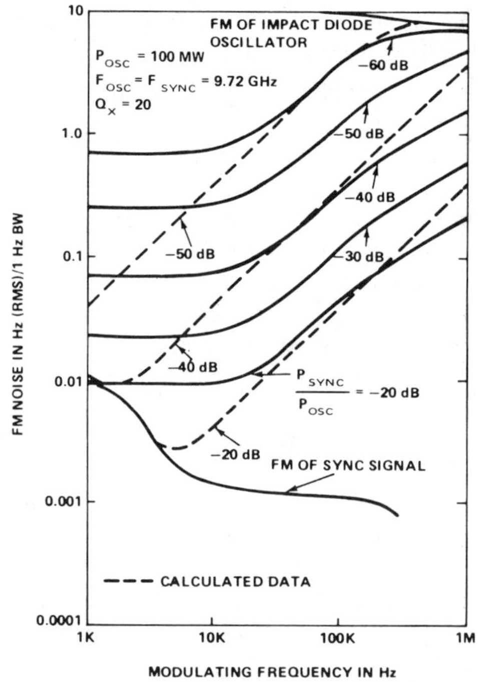

Fig. 20. FM noise in an injection synchronized phase avalanche diode oscillator.

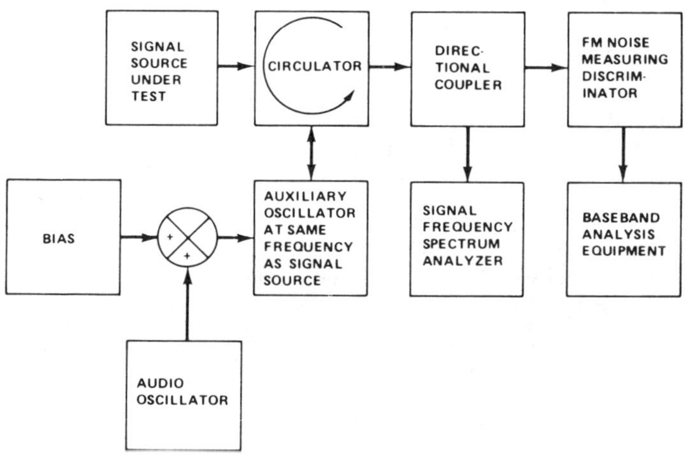

Fig. 21. Use of an injection-locked oscillator to lower the threshold of an FM measurement.

that for a locking gain of about 20 dB, an injection-locked oscillator will reproduce below 100 kHz the FM noise of the locking signal for any but the most sophisticated of signal sources. These data are shown in Fig. 20.

Thus the addition of an injection phase-locked oscillator as an input stage for the FM noise discriminator as shown in Fig. 21 can improve the measurement threshold some 10–20 dB, an improvement which is directly translated to a measurable 10–20-dB lower power signal. As a typical example, consider the use of an auxiliary 100-mW avalanche diode oscillator such as the one in Fig. 20. If we use this oscillator as the input to a discriminator with a cavity Q of 20 000–30 000, comparison of Fig. 19 with Fig. 20 will show that the measurement threshold will be set by the 20-dB curve of Fig. 20. Using an auxiliary oscillator with

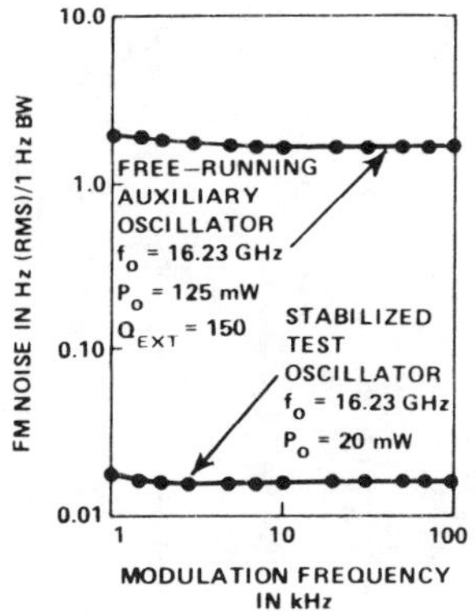

Fig. 22. Typical measurement of a low-noise low-power oscillator.

lower FM noise (higher Q_{ext}) would lower the threshold even more. Most important is the fact that this threshold is achieved with an input power of 1 mW, a level that would be insufficient to operate the discriminator without the auxiliary oscillator. This is a significant improvement.

In addition to improving the measurement threshold, the auxiliary oscillator also makes it much easier to calibrate the discriminator. The usual oscillator can be frequency modulated by applying ac to the bias port. While locked, the result will not be FM but ϕM; but this will be quite satisfactory for calibration purposes because a side frequency level of 50–60 dB below the carrier can be achieved with a symmetrical spectrum indicating pure angle modulation.

A typical result reported by Ashley and Palka [19] is shown in Fig. 22. This oscillator could not have been measured without added amplification of some kind.

We have noted some fear of attempting any kind of an injection phase-locking experiment when we first propose this technique to someone with a threshold problem. This is one of those fears of the unknown which promptly recedes after the experiment is started. Viewing the output of the circulator of Fig. 21 with a signal frequency spectrum analyzer will show the locking process most vividly and quickly will give the needed confidence in the method. We have used the technique often (as has F. M. Palka) and have never experienced any difficulty caused by injection phase locking.

The key idea that made this amplification method work was the availability of theory and measurements for injection phase locking. If another amplification method is to be used as a carrier frequency preamplifier, then it must be understood as well as the injection phase locking. First, we note that this is a high-level output and the usual noise figure measurements and concepts must be used with caution if any degree of output saturation exists. Second, this is a chicken and egg problem because verification of the amplifier contribution requires a known low-noise drive signal. In the case of the injection-lock noise experiment [20], we had available the TE_{015} transmission cavity stabilized two-resonator klystron oscillator to provide the 600 mW needed to make a good measurement of driver source noise.

Going back to the problem postulated at the end of Section VI, the method of attack would be to obtain two 30-dB amplifiers, each capable of about 1-W output power. The crystal oscillator would then drive one amplifier which would in turn drive the second amplifier via a 30-dB attenuator. Comparison of the two amplifier outputs with the transmission line discriminator should tell something about the FM noise added (or hopefully not added) by the second amplifier.

VIII. Baseband Noise Analysis Equipment

The most frequent question asked about transmitter noise measurements is, "Are these measurements repeatable from one laboratory to another?" Our answer to this can be only a qualified "yes" with the qualification being that the personnel making the measurements to be compared have at least several man-months of experience in what is often regarded as a topic with more art than science. When we do run down differences, the usual diagnosis is that the baseband equipment is different. On many occasions in our own laboratories, we have used two different analyzers to study the noise output of a common discriminator and found the answers to be several decibels different on first evaluation. Only careful attention to detail in understanding and calibrating the baseband equipment will resolve these differences.

The signal coming out of either the amplitude or angle demodulator is a function of time; yet the use of time domain analysis equipment is not very common. The use of a storage oscilloscope [3] has proved useful for both medium-term instability display and for the display of pathological misbehavior of signal sources.

Since the usual microwave radar or communications system employs frequency domain processing of the demodulated signals, it is the frequency domain analysis of the residual modulation noise which is most significant to transmitter evaluation and specification. The equipment used has evolved from the wave analyzers of the 1930's developed for the study of audio frequency signals, harmonic distortion in amplifiers, etc. In current terminology, an analyzer which is manually tuned (or perhaps sweep tuned with a motor drive) is called a wave analyzer, while an analyzer which has an electronic sweep of frequency is called a spectrum analyzer. The obvious conflict of terminology is that a sweeping analyzer for displaying the spectrum RF or microwave frequencies is also called a spectrum analyzer. We try to give an idea of the frequency range of the input signal (baseband, RF) whenever using the name spectrum analyzer.

Baseband analyzers operating in the frequency domain can be further classified as either constant bandwidth (say 100 Hz) or constant percentage bandwidth (say 1/3 octave.) Since the theory of noise is rooted in the idea of constant measuring bandwidth and since system noise performance depends on constant incremental bandwidth (i.e., multiplexing a number of 3-kHz channels into a broad baseband signal for transmission), the constant bandwidth analyzers are in far more frequent use although some

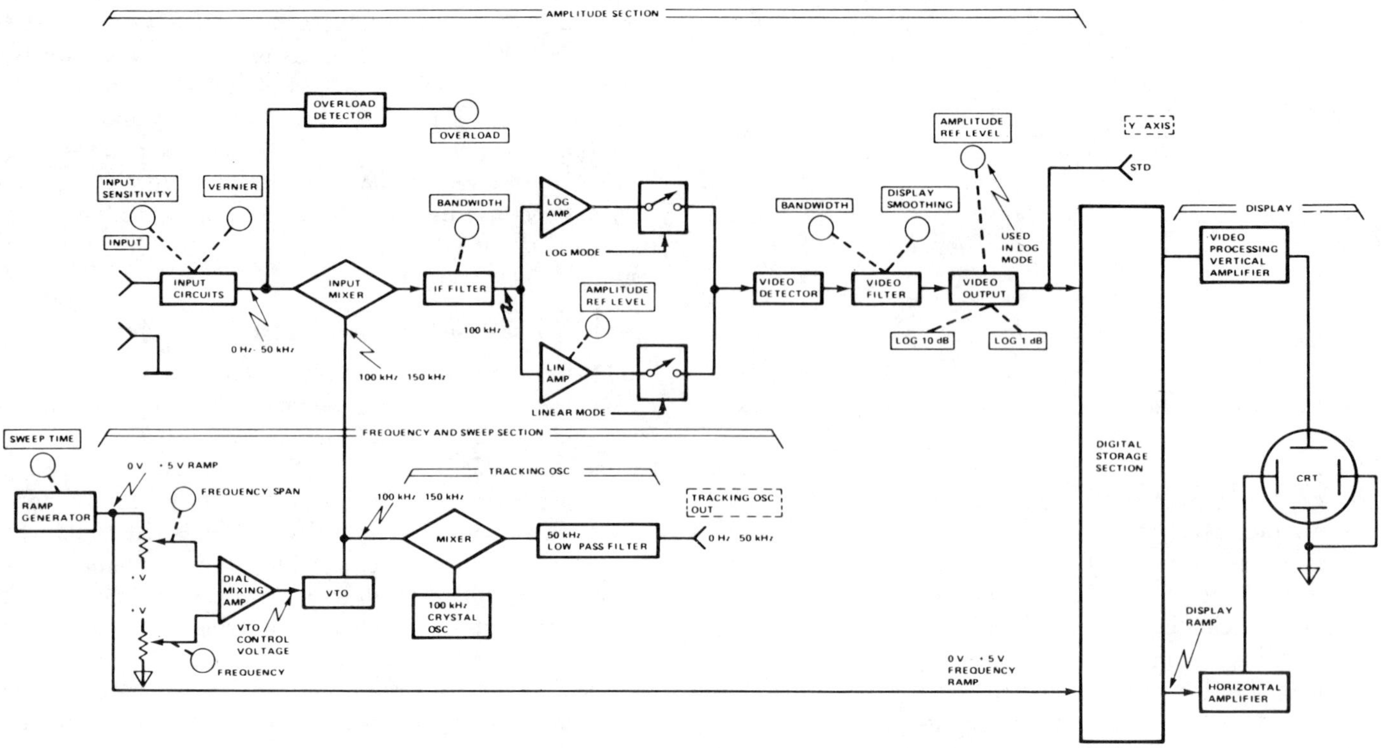

Fig. 23. Simplified block diagram of HP-3580A spectrum analyzer.

operating time convenience could be achieved by correct use of constant percentage bandwidth equipment. In this last category, measurement time can be tremendously reduced by operating parallel banks of overlapping constant bandwidth filters, and this equipment is usually called a real time analyzer.

Digital filtering and processing of noise data to obtain frequency domain information (by use of correlation computation and fast Fourier transform methods) is gaining wide acceptance in environmental studies and audio applications, but has not been used, to our knowledge, for analyzing modulation noise.

The constant bandwidth wave or spectrum analyzers are superheterodyne receivers such as the one shown in the block diagram of Fig. 23. This particular equipment has been selected for discussion because it is typical of all constant bandwidth analyzers and illustrates the basic concepts. Other equipments may use higher IF to accommodate wider input frequency range, but the block diagrams will be remarkably similar.

The input circuits box contains three essential items: 1) the input attenuator system, 2) amplification, and 3) a low-pass filter. The input attenuator is needed when the analyzer is used for relatively high-level input such as during the calibrate or alignment stage of a noise measurement. To minimize the noise contribution of the internal stages of the analyzer, this attenuation must be adjusted to the minimum possible loss for a measurement.

The amplifier is needed to allow measurement of lower level signals such as the noise output from a demodulator. Since the noise figure for an amplifier can be made several decibels better than the noise figure for the mixer, the noise performance of the analyzer can be improved with at least 10 dB of gain in the input amplifier. The amplifier must be used with care because it can be over driven (as can the following mixer) to make a measurement invalid. An input stage overload indicator is a tremendous operational convenience.

The low-pass filter is needed to reject signal components at IF (100 kHz) and in the image band (200–250 kHz).

The mixer converts the input spectrum (0–50 kHz) to sidebands above and below the LO frequency (100–150 kHz). The mixer is balanced to minimize the LO component in the output. The objective is to move the input spectrum past the fixed frequency and relatively narrow bandpass of the IF amplifier and filter. The dynamic range of the analyzer is set in the mixer. The mixer normally works so well that we are not appreciative of the design skill required.

The stability and operational features designed into the LO system, a swept voltage tuned oscillator in our example, are the biggest factors in determining the stability and operational features of the analyzer. We can appreciate these features without detailed study of the circuits.

The IF filter stage sets the measurement bandwidth which is crucial to the accurate analysis of the noise signal. In this analyzer, a five-stage synchronously tuned crystal filter has a bandwidth (between 3-dB points) of 1 Hz. By adding resistive losses and compensating the interstage amplifier gains, the bandwidth can be switched in a 1-3-10 sequence to a maximum of 300 Hz. For this synchronous tuning, the frequency response has the familiar bell shape of the Gaussian function.

The choice of filter shape is usually made on the basis of

measuring a spectrum consisting of a collection of discrete frequencies (rather than noise.) The first decision is the choice of Gaussian or rectangular shape. In sweeping analyzers, the Gaussian shape is preferred for nearly optimum buildup as a sinusoidal signal is swept past the filter. For manually tuned analyzers, the rectangular shape is preferred to allow a minor amount of frequency drift between the source generating the spectrum and the LO in the analyzer.

For noise analysis, the choice is not important. What is important is to know the noise bandwidth of the response shape—a number not normally in the instruction manual. If the frequency response of the amplifier $H(jf)$ is centered at f_0 and is down 40 dB at f_{LO} and f_{HI}, then the noise bandwidth is

$$B_N = \frac{1}{|H(jf_0)|^2} \int_{f_{\mathrm{LO}}}^{f_{HI}} |H(jf)|^2 \, df. \tag{44}$$

We have found [21] that about 30 data points defining $H(jf)$ are sufficient for a very simple rectangular rule integration to determine B_N. This must be done for each bandwidth of the analyzer.

Typically, the noise bandwidth is wider than the 3-dB bandwidth. For this analyzer, it is typically 12-percent wider than the measured 3-dB bandwidth.

Another specification of the IF filter shape relates the frequencies at which the filter attenuation is 60 and 3 dB. The ratio of 60-dB bandwidth to 3-dB bandwidth is called the shape factor and is about 10 to 1 for Gaussian-shape filters. This factor determines the spacing between discrete frequencies which can be resolved by the analyzer and the lowest frequency which can be measured without interference from the zero response. As a good rule of thumb, discrete frequencies spaced 15 times the bandwidth can be accurately measured. The minimum frequency which can be measured is of the order of 8 times the IF bandwidth. The cause of this limitation is the leakage of the LO output through the balanced mixer. As the analyzer is tuned to low input frequency, the leaked LO signal rides up the skirts of the IF filter response and registers as an output of the filter which must be classed as an error.

The output of the IF filter is amplified by a bandpass amplifier to increase the voltage enough to adequately override dc drift in the detector circuitry. To increase the dynamic range of the analyzer, attenuators are again used to optimize signal levels in the amplifier.

The logarithmic compression of a decibel vertical scale is most useful for first looks at a signal. The most stable and smooth way to accomplish the logarithmic compression is to perform this operation in the IF amplifier. Thus the spectrum analyzer we have been describing (Fig. 23) has a second IF amplifier, the LOG AMP, which has a dynamic compression range of 120 dB. This amplifier causes us no extra thinking for sinusoidal signal measurement but does exercise the brain cells for noise measurement.

If the input to the spectrum analyzer has a Gaussian probability density function, the output of the narrow-band IF filter also has a Gaussian density function. The nonlinearity of the logarithmic compression changes this density function and thereby changes the power relationship. The result of the compression of the peaks is that a true rms detection circuit will read 1.45 dB less than the correct value. If the LOG AMP is used, then noise readings must be corrected by adding 1.45 dB while sinusoidal readings do not require correction.

After IF amplification, the detector stage produces a dc output proportional to the IF output voltage (or the logarithm for the decibel output mode).

In most analyzers, the detector responds to the time average of the IF output voltage. This is taken care of in the calibration by the well-known ratio of average to rms for a sinusoid. But when used as a noise analyzer, this is in error by 1.05 dB which must be added to the output indication to obtain the effective time average (rms) of the noise. A true rms detector would be better for a noise analyzer.

For the measurement of sinusoidal signals, the dc from the detector is a nice stable value while for noise measurement, it bounces around with the bigger bounces on the narrower IF bandwidths. Post detection filtering (the video filter of Fig. 23) reduces the bounce of the output at the expense of measurement time.

The detected output is used to drive some form of output indicator. In the usual wave analyzer, the indicator is a simple dc voltmeter calibrated to read rms volts and possibly decibels with respect to a stated reference. For the spectrum analyzer, the output indicator is often an oscilloscope with some form of memory. For the analyzer of Fig. 23 the memory is digital with two sections; thus two different frequency sweeps of data can be compared.

All of this digression into nonmicrowave instrumentation must be understood to make effective microwave transmitter noise measurements. Before using a baseband spectrum analyzer, spend a few hours reading the instruction book—it will be time well invested.

IX. Equipment Operation

Once you have decided to make a transmitter noise measurement, decided on the methodology, acquired the necessary equipment, and located a working space, a very significant question arises: What is the sequence of operations? Intuition might guide you through the maze but it is doubtful that the path will be close to optimum; furthermore, if you hope to automate the measurement, the sequence of operations must be thought out in advance and probably worked through on a manual basis to understand the development of automatic routines such as those of the next section.

Our experience is outlined in the operations flow chart of Fig. 24. A similar operations sequence can be developed for AM noise measurements. Discussion of this flow chart will be the basis for sharing our laboratory experience with you.

Once you have the equipment assembled and operating, you have probably found all of the defective equipment and had it repaired. Next, take the time to verify the calibration

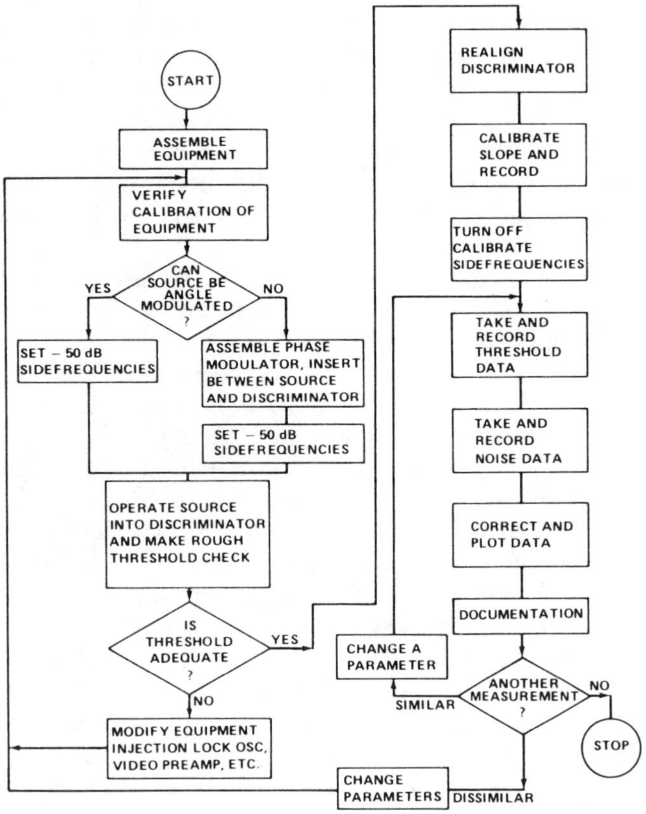

Fig. 24. Operations flow chart for making FM noise measurements.

of the RF spectrum analyzer, power meters, audio oscillators, and precision attenuators. For the RF spectrum analyzer, determine that the IF attenuator calibration is correct. The calibration of the baseband spectrum analyzer will not normally include the noise bandwidth so a separate project needs to be initiated to take the data and do the numeric integration to determine this bandwidth for each position of the bandwidth selector switch. While considering the baseband analyzer, check to see if the IF amplifier is linear or logarithmic and if the detector is average or rms responding. Make note of the necessary correction factors.

Once the source is working and the discriminator has been aligned, a few checks will tell if it is worthwhile to take detailed data. With the calibrate side frequencies on, tune the baseband analyzer to this modulation frequency and check for overload. Operate the input attenuator for maximum sensitivity without overload. If the analyzer does not have an overload indicator, take 10 dB out of the IF attenuator and put in 10 dB with the input attenuator. The output indicator should remain unchanged. If a preamplifier is used ahead of the baseband analyzer, use an oscilloscope to look for clipping at the output. Insertion of 3 dB at the input of the discriminator should make the output indicator fall 3 dB if nothing is overloading.

Turn off the calibration signal and reduce the input attenuator of the baseband analyzer as much as possible without overload. If the output indication is not useful, then remove IF attenuation. If you cannot achieve at least midscale indication at the analyzer output with the minimum bandwidth to be used, then a preamplifier is needed for the input of the baseband analyzer. After this check is passed, drop the input power by 3 dB and note whether the output indicator also drops 3 dB. If it does, the threshold check routine to be done later will yield a negligible correction. If it drops at least 1 dB, there is hope that the threshold check routine will allow correction of the final data to reasonable accuracy. If it drops less than 1 dB, then the threshold is not adequate and either a video preamplifier or the injection-locked oscillator of Section VII must be added.

After you know that the threshold is adequate, then it is worthwhile to take detailed data. If the threshold is more than adequate and you are taking and recording data manually, the slope calibration can be manipulated to make the full scale of the output indicator be a convenient deviation. For instance, the 10-μV full scale might be made equivalent to 10-Hz deviation by clever manipulation of input power and the variable gain knob on the analyzer. At each point where noise data are to be taken, the threshold equivalent deviation or actual voltage, the analyzer bandwidth, and the modulation frequency must be recorded. The same data are taken to determine the noise. Several important factors were written into the step-by-step alignment and operation instructions for the discriminators (steps 8)–13) of Section V and steps 4)–9) of Section VI).

The preceding recording of data is made most convenient if an X–Y recorder can be attached to the wave or spectrum analyzer. This raises the question of how fast the frequency can be swept in taking the data. For noise spectra which are dominated by discrete frequencies (often caused by power supply ripple), the rule is

$$\text{sweep rate} = 0.5\,(\text{bandwidth})^2\ \text{Hz/s} \tag{45}$$

where the bandwidth is also in hertz. The reason for this rule is that time is required for the high-Q filter in the IF section to build up to the steady state. If the noise spectrum is random and white or pink in shape, then this sweep rate can be multiplied by a factor of 5, which is an appreciable saving in data taking time.

Notice that the sweep rate goes as the square of bandwidth so the largest possible bandwidth should be used. Here is where judgment in taking the data is required. If modulation frequencies below 1000 Hz must be studied, a 10-Hz or smaller bandwidth is needed to resolve power supply ripple components spaced at 50 or 60 Hz and to approach zero frequency as closely as the baseband analyzer will allow. For higher modulation frequency ranges, make a first sweep with a wider bandwidth, say, 100 or 300 Hz. If the result is a white or pink spectrum with no significant humps, the data can be given the final processing and used. If a hump is seen, make a second sweep with the narrowest possible bandwidth and at a sweep rate appropriate for sinusoids. The reason for this advice is shown in Fig. 25. The lump in the curve taken with the 300-Hz bandwidth at 3700 Hz might normally be disregarded as something mysterious in the TWT. Use of the 10-Hz bandwidth and a much slower sweep rate shows that a fine structure related to the 60-Hz power

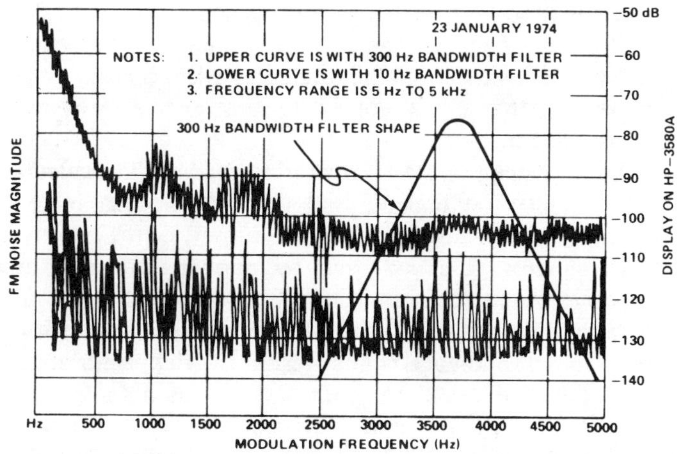

Fig. 25. Modulation frequency (hertz). These data were taken on an HP-3580A using the stabilized klystron as a source to drive a 20-W TWTA.

supply is the cause of this hump. Additional data [22] show that the same spectrum analyzer can detect the ripple components in the TWTA power supply, and a look in the time domain showed switching spikes from the silicon rectifiers were getting past the regulator stage.

Fig. 25 also illustrates several other important facts about spectrum analyzer operation and data interpretation. For the 300-Hz bandwidth data, notice the similarity of the curve below the 500 Hz and the shape of the IF filter response. This is the zero response (helped with a big spectral component at 120 Hz) of the analyzer. In the range above 1500 Hz, a pure random noise spectrum can be measured in one bandwidth and mathematically transformed to another. For these two bandwidths, the curve should be separated by 30 dB which is certainly not true in Fig. 25. Thus, before mathematically transforming bandwidth, take a narrow-band sweep to insure that the spectrum is that of random noise only.

The rules for mathematically transforming noise data to different bandwidths are based on the fact that noise power through a linear filter is proportional to the bandwidth. Thus, if data such as AM noise are taken in decibels with a measurement bandwidth B_N, transformation to another band B_A is accomplished by adding $10 \log_{10} (B_A/B_N)$. An rms noise deviation is transformed by multiplying by $\sqrt{B_A/B_N}$. We repeat, these transformations are valid only when the noise spectrum is flat over the larger bandwidth.

Measuring an amplifier with a discriminator yields the FM noise $S_{FM}(f)$ at the input and output. To understand the phase noise sources within the amplifier, it is useful to plot the input and output phase modulation noise $S_{\phi M}(f)$. This can be obtained by the use of

$$S_{\phi M}(f) = f^2 S_{FM}(f). \tag{46}$$

The proof of this equation is constructed by remembering the definition of instantaneous frequency [see (3)], and using a well-known theorem of communications theory quoted by Papoulis [8, table 10-1].

One block of Fig. 24 which is often neglected now and lamented later is documentation. Especially when taking data with a sweeping analyzer, record bandwidths, sweep rates, sweep ranges, and something to indicate if the noise measurement threshold is sufficiently low or must be corrected for. Mark the curves for the deviation scale *now*.

The decision block after completing one data run shows a path around the data taking loop which does not include the calibration block. When taking this shortcut path, remember that the calibration remains constant for 5-percent carrier frequency changes and for constant input power.

All the preceding applies to well-behaved portions of a transmitter. While working with klystrons, BWO's, IMPATT oscillators, TEO's, and crystal oscillator controlled sources, we have seen pathological behavior typically in the form of big jumps in the output of the baseband analyzer. The jumps seem to occur at randomly related bands of frequency, but, in truth, are occurring over a wide band of baseband frequencies and coming and going with time. A better indicator is to use a preamplifier and storage oscilloscope to show the time of occurrence of these noise bursts.

X. Automation of FM Noise Measurements

After you have traversed the major and minor loops of Fig. 24 for a few weeks, you will find that you are spending most of your time taking, documenting, and processing data. It is difficult to visualize the complete automation of the noise measurement process but not so difficult to visualize computer control of the noise data taking and processing to significantly reduce operating time.

The first decision is the choice of calculator, minicomputer, microprocessor, or time-shared computer as the controller for the automatic system. For this automation, we have only a few instruments to control and the measurement time is set in the equipment; therefore, blinding speed and large size are not needed in the controller. The calculator has enough size and speed for this job. Our task is similar in size and speed to many other measurement systems, and this class of systems brought about the development of the IEEE Standard #488-1975 Interface Bus for calculator or computer control of measurements and data processing. Using the IEEE Standard #488-1975, the Hewlett-Packard Loveland Instrument Division has developed a spectrum analyzer system for the 10 Hz to 13 MHz range, the HP 3045A system. The controller is the HP 9830A calculator which is programmed in the computing language BASIC. We have been able to write the software to use this system in partially automating transmitter noise measurements.

The organization first divides the tasks into manual or calculator control as shown in Fig. 26. Next, the programming is subdivided on the basis of the operations sequence (Fig. 24). In this sequence, there are branching paths and even the possibility of branching within blocks. Several of the blocks have enough work to be subprograms. One simple way to implement the calculator control is to use the 20 function keys to control the operation of various portions of the program; then much of the branching is accomplished by pressing the appropriate function key.

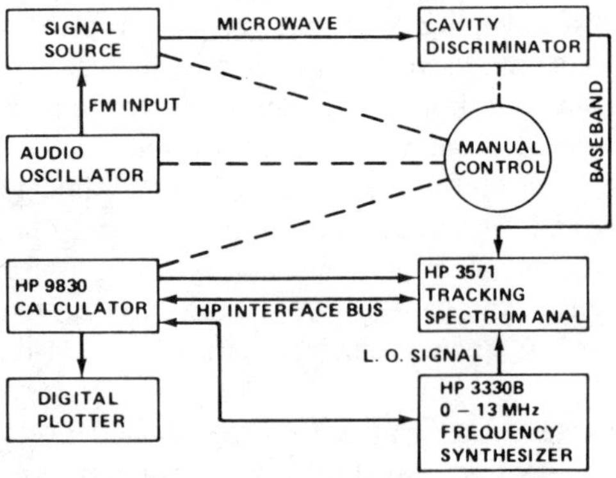

Fig. 26. Automated measurement of near carrier noise.

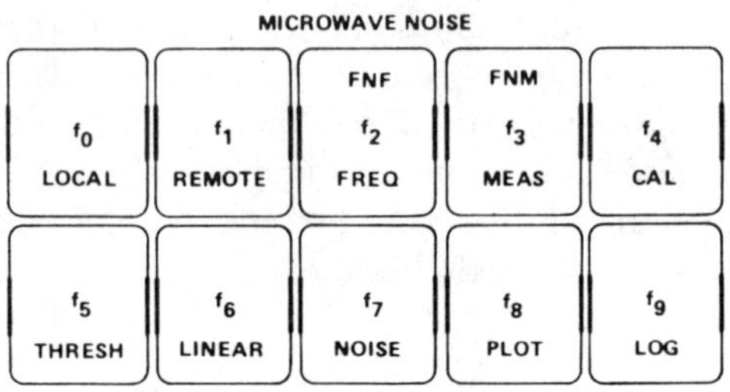

Fig. 27. HP-9830 function keys used to sequence the automated noise measurement.

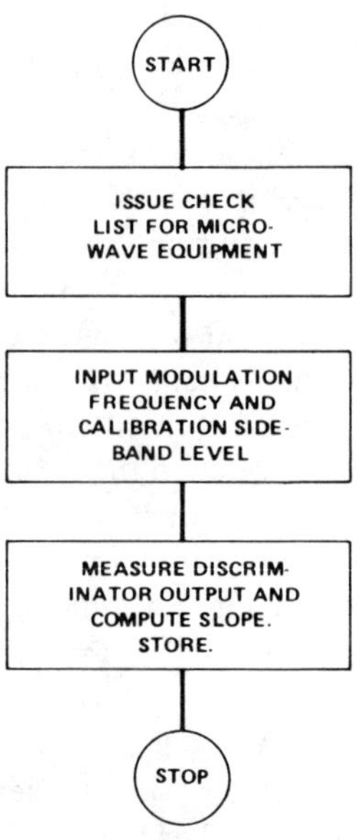

Fig. 28. Flow chart subprogram CAL.

The following subprograms were written and stored on the function keys.

Name	Function
CAL	Calibrates all equipment.
THRESH	Stores threshold data.
NOISE	Takes and stores noise data.
PLOT	Processes and plots noise data.
LINEAR GRID	Draws a linear grid for plotted data.
LOG GRID	Draws a logarithmic grid for plotted data.

The function keys with the appropriate overlay are shown in Fig. 27. Each of these subprograms can be operated independently as far as the calculator is concerned, and this makes possible several noise data runs for one calibrate and threshold run.

Usually, the first step is operation of CAL. The calculator issues control statements and takes data to calibrate the slope (hertz deviation per volt of output) of the microwave discriminator. This is done by applying a small angle modulation to the source under test and adjusting for a known deviation. (The equations to use were given with the discriminator theory.) The result of this calibration is stored in the calculator memory and used in later data processing. The flow chart for subprogram CAL is shown in Fig. 28 and the BASIC program used by the HP 9830A calculator is given in List 1.

```
10  OUTPUT  (13,20)1024;
20  FORMAT B
30  DIM A$[3],B$[36],C$[36],TS[251],NS[251],FS[251],BS[251]
40  DIM D$[10],I$[30],CS[251]
50  DISP "WHAT IS TODAY'S DATE";
60  INPUT D$
70  DISP "WHAT SOURCE ARE U MEASURING";
80  INPUT I$
90  DISP "HAS THE SIGNAL SOURCE WARMED UP";
100  INPUT A$
110  IF A$="NO" THEN 90
120  DISP "IS THE REFERENCE POWER CORRECT";
130  INPUT A$
140  IF A$="Y" THEN 180
150  DISP "SET THE REFERENCE POWER, DIMBULB"
160  WAIT 5000
170  GO TO 120
180  DISP "IS THE REFERENCE PHASE CORRECT";
190  INPUT A$
200  IF A$="Y" THEN 230
210  DISP "TWIDDLE THE PHASE SHIFTER"
220  GO TO 180
230  DISP "HAVE YOU CHECKED INPUT OVERLOAD";
240  INPUT A$
250  IF A$="NO" THEN 90
260  OUTPUT (13,20)768;
270  DISP "WHAT IS CALIBRATE MOD FREQ";
280  INPUT A
290  B=30
300  F=A
310  GOSUB FNF(F) OF 310
320  GOSUB FNB(0) OF 320
330  DISP "HOW MANY DB DOWN ARE SIDEBANDS";
340  INPUT D
350  S=A*2*(10↑(-D/20))/(SQR(2)*Y)
360  B$="THE DISCRIMINATOR SLOPE IS"
370  C$="PK HZ / MICROVOLT"
380  S7=INT(S*1E-04)/100
390  PRINT B$,S7,C$
400  PRINT LIN(3)
410  DISP B$
420  WAIT 2000
430  DISP 97;C$
440  WAIT 5000
450  K1=S/(SQR(2)*1.13)
460  DISP "CALIBRATION IS COMPLETE"
470  END
```

The second step in the measurement sequence is operation of THRESH to determine the noise floor or threshold of the discriminator. The theory is given in Sections V and VI, but the practice is very time consuming because data for the threshold must be taken at each frequency where noise is to be measured. Now the calculator is given the frequency limits for which measurements are desired and the number of data points; then it takes the threshold data and stores the results in an array variable which will be used later to correct actual noise data. Thus all of the measurements can be corrected for threshold before plotting.

The third step is to use NOISE to take data at the same

frequencies where the threshold is known from the previous step. This is completely under the control of the calculator which again stores the results in its memory (as an array variable.) During this step, the human operator can observe the operation of the source under test and the microwave discriminator tuning. Those of us who have spent many hours of tuning a wave analyzer and recording data with pencil and paper especially appreciate the ease of this step.

Probably the most enjoyable step is pushing the calculator key labeled PLOT. This subprogram takes the arrays for threshold and noise, calibration factors, and other corrections to compute threshold corrected noise deviation while the calculator driven plotter plots complete results. Documentation is made painless by answering "yes" to a displayed question, "Do you want to label the axes?" The result is a plot such as the one in Fig. 12 where everything but the caption was put there by the plotter. The linear grid portion was plotted by activating the proper function key. After years of doing noise measurements the hard way, it is pure pleasure to watch the plotter deliver the results of automated noise measurements.

Far more significant than the pleasure of the engineers is the time savings and improved documentation that result from automation. We have proved by personal example that someone familiar with making FM noise measurements manually can learn to operate the automated system in less than an hour. For production line testing, test personnel can be taught to operate this automated equipment much easier than they can be taught to make manual controlled noise measurements and *to do the data processing calculations*. If significant numbers of noise measurements need to be made, then this extra equipment to automate the system will quickly pay for itself.

At the time of this writing, we are at perhaps the 90-percent point on the automation learning curve. It is premature to formally document all of the subprograms and listings; however, we have written this work into a less formal but accessible format [22] which includes all listings ([23] is also helpful).

XI. Concluding Remarks

We listed ten conclusions at the end of the theoretical part of this paper in Section III. We continue with conclusions regarding the experimental technique of making transmitter poise measurements.

1) The measurement of AM noise in all stages of a microwave transmitter is best done with direct diode detectors. The worst pitfall is failing to determine the measurement threshold.

2) The measurement of FM noise for carrier frequencies above 5 GHz is now a repeatable and reliable procedure using the discriminator of Section VI. Be careful in calibrating and determining threshold.

3) The measurement of FM noise for carrier frequencies below 5 GHz can now be accomplished with the improved transmission line discriminator of Section VII. After a few months of operational experience at frequencies between 30 MHz and 5 GHz, we believe the technique is as repeatable and reliable as the methods above 5 GHz.

4) The most difficult FM noise measurement problem is a low-power low-noise source. The use of an injection phase-locked oscillator as a discriminator preamplifier (as described in Section VII) has proved a viable solution for the problem.

5) Most of the problems concerning lack of repeatability from one laboratory to another can be resolved by understanding the baseband analysis equipment.

6) The art of a transmitter noise measurement is in the optimum use of various bandwidths in the wave analyzer.

7) Automation of the measurement is fun (and time consuming) for those doing the work.

8) Not many transmitter noise measurements per month are required to justify the cost of automation.

You will notice two topics we did not discuss. First, the use of two-oscillator techniques for phase modulation noise measurement, and second, pulsed noise measurements. Two-oscillator methods are well described by Shoaf *et al.* [15] and apply more to time and frequency applications than to microwaves. We believe that for submicrowave portions of transmitters, the transmission line discriminator is a more useful tool. With regard to pulsed transmitter noise measurements, we have not yet implemented these and have nothing to add to the work of Sann [24].

Acknowledgment

The work reported in this paper is the result of 26 years of experience by the authors, which makes the task of proper acknowledgment difficult. Whenever we have encountered a theoretical problem, we have either found the answer or a method of attack in the writings of Dr. J. Mullen and Dr. D. Middleton; they deserve commendation for their excellent research. The measurement work which inspired our contributions to this subject has been done in the Gainesville, FL, and Clearwater, FL, laboratories of the Sperry Rand Corporation and the Redstone Arsenal, AL, laboratories of the U.S. Army Missile Command. Without the extensive physical resources of these laboratories, our accomplishments would not be significant. Correspondence and conversations with colleagues too numerous to call by name have added significantly to our knowledge of this topic and equally deserve our heartfelt appreciation.

References

[1] D. Middleton, "Theory of phenomenological models and measurements of fluctuating output of CW magnetrons," *IRE Trans. Electron Devices*, vol. ED-1, pp. 56–89, Feb. 1954.

[2] J. A. Mullen, "Background noise in nonlinear oscillators," *Proc. IRE*, vol. 48, pp. 1467–1473, Aug. 1960.

[3] J. R. Ashley, C. B. Searles, and F. M. Palka, "The measurement of oscillator noise at microwave frequencies," *IEEE Trans. Microwave Theory Tech.*, vol. MTT-16, pp. 753–760, Sept. 1968.

[4] J. G. Ondria, "A microwave system for measurements of AM and FM noise spectra," *IEEE Trans. Microwave Theory Tech.*, vol. MTT-16, pp. 767–781, Sept. 1968.

[5] S. B. Marsh and J. D. Clare, "Microwave frequency discriminator," U.S. Patent 3 079 563, Feb. 1963, assigned to James Scott and Company (Electrical Engineers) Limited, Glasgow, Scotland.

[6] J. R. Ashley, T. A. Barley, and G. J. Rast, Jr., "Transmission line discriminators for FM noise measurement," *Proc. IEEE* (Letters), vol. 65, pp. 578–580, Apr. 1977.

[7] D. Middleton, *Introduction to Statistical Communication Theory*. New York: McGraw-Hill, 1960.
[8] A. Papoulis, *Probability, Random Variables and Stochastic Processes*. New York: McGraw-Hill, 1965.
[9] J. L. Fikart, "A theory of oscillator noise and it's application to IMPATT diode oscillators," Ph.D. dissertation, University of Alberta, Edmonton, Canada, Spring 1973.
[10] J. L. Fikart and P. A. Goud, "The direct detection noise measuring system and its threshold," *IEEE Trans. Instrum. Meas.*, vol. IM-21, pp. 219–224, Aug. 1972.
[11] J. L. Fikart, J. Nigrin, and P. A. Goud, "The accuracy of AM and FM noise measurements employing a carrier suppression filter and phase detector," *IEEE Trans. Microwave Theory Tech.*, vol. MTT-20, pp. 702–703, Oct. 1972.
[12] A. L. Whitwell and N. Williams, "A new microwave technique for determining noise spectra at frequencies close to the carrier," *Microwave J.*, vol. 2, pp. 27–32, Nov. 1959.
[13] A. S. Risley, J. H. Shoaf, and J. R. Ashley, "Frequency stabilization of *X*-band sources for use in frequency synthesis into the infrared," *IEEE Trans. Instrum. Meas.*, vol. IM-23, pp. 187–195, Sept. 1974.
[14] J. R. Ashley and C. B. Searles, "Microwave oscillator noise reduction by a transmission stabilizing cavity," *IEEE Trans. Microwave Theory Tech.*, vol. MTT-16, pp. 743–748, Sept. 1968.
[15] J. H. Shoaf, D. Halford, and A. S. Risley, "Frequency stability specification and measurement: High frequency and microwave signals," NBS Tech. Note 632, Jan. 1973.
[16] R. A. Campbell, "Stability measurement techniques in the frequency domain," *Proc. IEEE-NASA Symp. Short Term Frequency Stability*, NASA SP-80, pp. 231, Nov. 1964.
[17] D. B. Leeson, "Short term stable microwave sources," *Microwave J.*, pp. 59–69, June 1970.
[18] J. A. Mullen and D. Middleton, "Limiting forms of FM noise spectra," *Proc. IRE* (Letters), pp. 874–877, June 1957.
[19] J. R. Ashley and F. M. Palka, "Improvement of a microwave discriminator by an injection phase locked oscillator," *IEEE Trans. Microwave Theory Tech.*, vol. MTT-18, pp. 100–101, Nov. 1970.
[20] J. R. Ashley and F. M. Palka, "Measured FM noise reduction by injection phase locking," *Proc. IEEE*, vol. 58, pp. 155–157, Jan. 1970.
[21] T. A. Barley, G. J. Rast, Jr., and J. R. Ashley, "Wave analyzer dynamic range and bandwidth requirements for signal noise analysis," U.S. Army Missile Command Rep. TR-RE-76-26, Mar. 26, 1976.
[22] G. J. Rast, Jr., T. A. Barley, and J. R. Ashley, "Automated measurement of transmitter noise at HF through microwave frequencies," U.S. Army Missile Command Rep. TR-RE-7T-8, in publication.
[23] "Understanding and measuring phase noise in the frequency domain," Appl. Note 207, Hewlett-Packard Instruments Division, Loveland, CO.
[24] K. H. Sann, "The measurement of near carrier noise in microwave amplifiers," *IEEE Trans. Microwave Theory Tech.*, vol. MTT-16, pp. 761–766, Sept. 1968.
[25] J. R. Ashley, T. A. Barley, and G. J. Rast, Jr., "Near carrier noise in TWT amplifiers," *IEEE Int. Electronic Devices Meeting Technical Digest*, p. 599, Washington DC, Dec. 1974.
[26] D. G. McDonald, A. S. Risley, J. D. Cupp, K. M. Evenson, and J. R. Ashley, "Four-hundredth-order harmonic mixing of microwave and infrared laser radiation using a Josephson junction and a mixer," *Appl. Phys. Lett.*, vol 20, p. 296, Apr. 15, 1972.
[27] T. A. Barley, G. J. Rast, Jr., and J. R. Ashley, "Optimum length transmission line discriminator with low noise detector," U.S. Patent 4 002 969, Jan. 11, 1977.
[28] J. R. Ashley, G. J. Rast, Jr., and T. A. Barley, "Optimum threshold transmission line discriminator," U.S. Patent 4 002 970, Jan. 11, 1977.
[29] G. J. Rast, Jr., T. A. Barley, and J. R. Ashley, "Wide operating frequency range transmission line discriminator," U.S. Patent 4 002 971, Jan. 11, 1977.

Part VI
Noise Properties of the Measurement System Building Blocks

By measurements of frequency stability one often finds that the background noise may spoil all of the efforts the experimentalist has taken, particularly with high-grade frequency sources. In order not to waste time and money, the information about noise performance of all major building blocks encountered in frequency stability measurement systems must be taken into account as early as the planning stage.

The first of the papers in this part supplies a great deal of needed information about low-noise amplifiers, double balanced mixers, and simple phase-lock loops. However, for more information about low-noise amplifiers, the designer should further consult the book by Motchenbacher *et al.* [VI-1], and the surveying papers by Netzer [VI-2], Letzter *et al.* [VI-3], or Skritek [VI-4]. In cases where one is looking for actual low-noise low-frequency amplifiers, one will find them presented in the first reprint paper in Part IV, in the fourth paper of Part VI, in the book by Kartaschoff [VI-5], and in the book edited by Blaire [IV-2, p. 169]. Modern low-noise operational amplifiers are discussed in the papers by Erdi *et al.* [VI-6] and Bernardi [VI-7]. The phase noise properties of RF and isolation amplifiers are reviewed in the first, second, and seventh of the reprint papers in this part.

It has been found that double balanced mixers, in the ring configuration—with two Schottky barrier diodes in the arm—are the best for applications as phase detectors in frequency stability measurement systems (see Part IV, paper 1 and Part VI, paper 1 and [VI-8]). On the other hand, there is some information that logic circuits phase detectors are noisier [VI-8], [VI-9].

One of the circuits often encountered in frequency stability measurement devices is the frequency multiplier which is used for increasing the phase noise power spectral density by the square of the multiplication factor (N^2). Evidently, the low-noise design is imperative, and it is discussed in the second, third, and fourth of the reprint papers. The consequence of direct frequency multiplication is that the output frequency generally lies somewhere in the microwave range (see Part IV, paper 8), which is often undesirable. To overcome this difficulty, one can take recourse to the "error multiplication technique" [VI-18], the principle of which is shown in Fig. 1. With both input power spectral densities, $S_{\varphi 1}(f)$ and $S_{\varphi 2}(f)$, of the same order, i.e.,

$$S_{\varphi 1}(f) \approx S_{\varphi 2}(f) \approx S_{\varphi \mathrm{in}}(f) \qquad \text{(VI-1)}$$

the output power spectral density is approximately

$$S_{\varphi \mathrm{out}}(f) \approx 2 \times 10^{2M} S_{\varphi \mathrm{in}}(f) \qquad \text{(VI-2)}$$

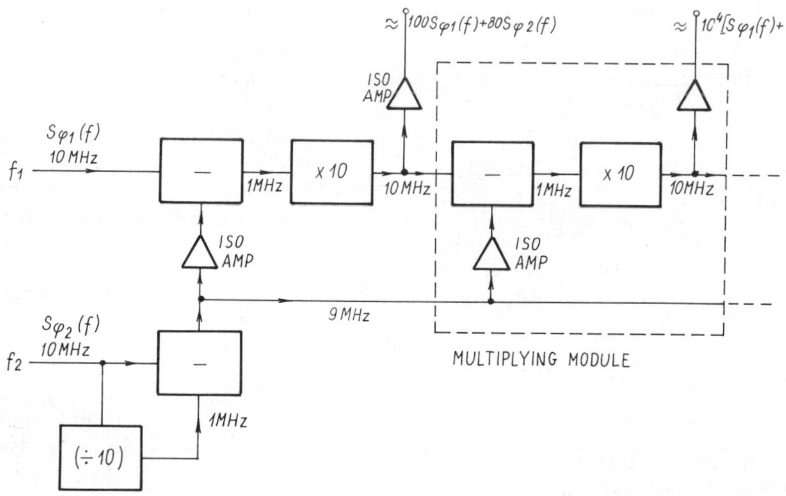

Fig. 1. Special-purpose frequency synthesis for phase or frequency error multiplication technique.

where M is number of the "multiplying modules" used. One of the problems encountered with this technique is the danger of spurious couplings, so as a consequence the perfect isolation of individual stages is imperative.

Another difficulty met with a high-order frequency multiplication is the distortion of the output RF spectrum which, eventually, ceases to truly reproduce the spurious input FM or PM modulation. Some insight is provided by the fifth reprint paper, which is by Lindsey and Chie. The problem is closely connected with the timely topic of the direct measurement of laser frequencies where the maximum multiplication factor is of paramount importance, as it follows from the next reprint paper by Bava *et al.* (for further information see e.g., [VI-10]–[VI-12]). Both the flicker frequency noise and white noise pedestal should be kept at a minimum.

In accordance with the Leeson model, the oscillator noise (see Part II, paper 4) is given by the power law

$$S_{\varphi\text{osc}}(f) = \left(\frac{f_0}{f}\right)^2 \frac{a_{-1}}{f \cdot 4Q_L^2} + \left(\frac{f_0}{f}\right)^2 \frac{a_0}{4Q_L^2} + \frac{a_{-1}}{f} + a_0. \qquad \text{(VI-3)}$$

In instances where oscillators with different output frequencies should be compared, the following normalization is of advantage

$$\frac{S_{\varphi\text{osc}}(f)}{f_0^2} = \frac{1}{f^3} \cdot \frac{a_{-1}}{4Q_L^2} + \frac{1}{f^2} \cdot \frac{a_0}{4Q_L^2} + \frac{1}{f_0^2} \cdot \frac{a_{-1}}{f} + \frac{a_0}{f_0^2} \qquad \text{(VI-4a)}$$

$$= \frac{1}{f^3} h_{-1} + \frac{1}{f^2} h_0 + \frac{1}{f} h_1 + h_2 \qquad \text{(VI-4b)}$$

since the h_k coefficients are invariant with respect to a noiseless frequency multiplication or division. Some theoretical support for the Leeson model is provided by Healey (see the seventh reprint paper), whereas the editor has investigated noise characteristics of actual oscillators published by different authors all over the world, and thus arrived at the practical values of the h coefficients in (VI-4b); the results are summarized in the last reprint paper. However, one problem remains.

Many crystal oscillator designers during the 1970's eventually arrived at the conclusion that the ultimate level of the white and flicker frequency noises was not due to the maintaining electronic circuits but was associated with the resonator itself. Some qualitative information is provided by the eighth reprint paper. One sees that in the passband the power spectral density of the fractional frequency obeys the flicker frequency law and at larger Fourier frequencies the slope changes to $1/f^2$. The same results were published later by French authors [VI-13]. However, the problem has not yet been satisfactorally solved. On one hand, there are experimenters who deny the limiting role of the resonator noise [VI-14], and on the other hand there are researchers who try to find physical backgrounds of the phenomenon [VI-15]–[VI-17]; cf. also [I-4].

In many frequency stability measurement systems, we encounter frequency synthesizers either as "local oscillators" or as parts of the system itself. Evidently, knowledge of the overall noise behavior and the noise levels to be expected is of the highest importance. Since for the derivation of a new frequency from a given reference frequency (i.e., frequency synthesis) combinations of several additions, subtractions, multiplications, and divisions are necessary [VI-18]–[VI-20], evaluation of the noise properties of the output frequency or frequencies seems to be a formidable task.

The PLL technique is a favored technique in frequency synthesis systems since it makes possible a substantial hardware simplification; it is here that the difficulties start since nearly all PLL building blocks may add (sometimes substantial) noise power to the useful signal. The problem is not yet generally understood and, in addition, is often underestimated. However, by following all major paths of the reference signal we can establish the sources of additive noises or spots of noise enhancements, and thus establish the expected output noise level. The leading ideas are summarized in the last reprint paper and in [VI-21].

References

[VI-1] C. D. Motchenbacher and F. C. Fitchen, *Low Noise Electronic Design*. New York: Wiley, 1973.

[VI-2] Y. Netzer, "Design of low-noise amplifiers," *Proc. IEEE*, vol. 69, pp. 728–741, June 1981.

[VI-3] S. Letzter and N. Webster, "Noise in amplifiers," *IEEE Spectrum*, pp. 67–75, Aug. 1970.

[VI-4] P. Skritek, "Noise figures of differential-pair input-stage equalizer preamplifiers," *IEEE Trans. Acoust., Speech, Signal Processing*, vol. ASSP-30, pp. 167–174, Apr. 1982.

[VI-5] P. Kartashoff, *Frequency and Time*. London: Academic, 1978, p. 174, ch. 7.

[VI-6] G. Erdi *et al.*, "Op amps tackle noise—And for once, noise loses," *Electron. Des.*, vol. 28, pp. 65–71, Dec. 20, 1980.

[VI-7] T. S. Bernardi, "Designing low noise op-amp based systems, part one," *Electron. Eng.*, pp. 43–44, 47, July 1982, and "Designing low noise op-amp based systems, part two," *Electron. Eng.*, pp. 33, 35–36, 39, 41, Aug. 1982.

[VI-8] L. Šojdr, "Phase noise in oscillators and synthesizers" (in Czech), Ph.D. dissertation, Czech Tech. Univ., Prague, Czechoslovakia, June 1979.

[VI-9] M. J. Uderhill, P. A. Jordan, M. A. G. Clark, and R. I. H. Scott, "A general purpose LSI frequency synthesizer system," in *Proc. 32nd Annu. Symp. Frequency Contr.*, Atlantic City, NJ, May–June 1978, pp. 365–372.

[VI-10] A. Godone, A. De Marchi, and E. Bava, "An improved multiplier chain for precise frequency measurements up to 20 THz," in *Proc. 33rd Annu. Symp. Frequency Contr.*, Atlantic City, NJ, May–June 1979, pp. 498–503.

[VI-11] J. J. Jimenez, "Microwave to visible frequency synthesis," *Radio Sci.*, vol. 14, pp. 541–560, July–Aug. 1979.

[VI-12] V. F. Kroupa and L. Šojdr, "Close-to-carrier noise of precision oscillators after multiplication," in *Conf. Precision Electromag. Meas. CPEM-82 Dig.*, June–July 1982, pp. N10-N13.

[VI-13] G. Marianneau, J. J. Gagnpain, and J. Uebersfeld, "Bruit de phase et d'amplitude des resonateurs a quartz," in *Proc. Int. Symp. Telecommun.*, Lanion, France, Oct. 1977, pp. 240–244.

[VI-14] L. Penavaire *et al.*, "A 120 MHz SAW resonator stabilized oscillator with high spectral purity," presented 1980 IEEE Ultrasonics Symp., Boston, MA, Nov. 1980.

[VI-15] J. J. Gagnpain, "Fundamental noise studies of quartz crystal resonators," in *Proc. 30th Annu. Symp. Frequency Contr.*, June 1976, pp. 84–89.

[VI-16] T. Musha, "$1/f$ resonant frequency fluctuation of a quartz crystal," in *Proc. 29th Annu. Symp. Frequency Contr.*, May 1975, pp. 308–310.

[VI-17] J. J. Gagnpain *et al.*, "Relation between $1/f$ noise and Q-factor in quartz resonators at room and low temperatures, first theoretical interpretation," in *Proc. 35th Annu. Frequency Contr. Symp.*, May 1981, pp. 476–483.

[VI-18] V. F. Kroupa, *Frequency Synthesis, Theory, Design and Applications*. London: Griffin, 1973; New York: Wiley, 1973, p. 243.

[VI-19] V. Manassewitsch, *Frequency Synthesizers, Theory and Design.* New York: Wiley, 1980.

[VI-20] W. F. Egan, *Frequency Synthesis by Phase Lock.* New York: Wiley, 1981.

[VI-21] V. F. Kroupa, "Low-noise microwave frequency synthesizers: Design principles," *Proc. IEE*, Pt. H, to be published.

DESIGN CONSIDERATIONS IN STATE-OF-THE-ART SIGNAL PROCESSING AND PHASE NOISE MEASUREMENT SYSTEMS

by

F. L. Walls, S. R. Stein, James E. Gray
and David J. Glaze
Frequency and Time Standards Section
National Bureau of Standards
Boulder, Colorado 80302

Introduction

The recent rapid improvement of oscillator phase noise has resulted in significantly more stringent requirements for signal handling equipment. However, information concerning the phase noise performance of the two most important types of circuits - amplifiers and mixers - is often difficult to find. Some general principles are presented which allow one to estimate the phase noise performance of an amplifier. Also, techniques are described which permit one to obtain the best possible results from the traditional double balanced mixer. A measurement set-up which has 15 to 25 dB improvement in its noise floor is shown in detail to illustrate proper mixer drive and termination. Although traditional circuits can with extreme care achieve S_ϕ = -175 dB or slightly better, this is not sufficient for all present requirements. One technique to obtain an additional improvement of 10 to 40 dB in measurement system noise is to reduce the mixer and amplifier contributions to the noise floor by the use of correlation techniques. A circuit to accomplish this is discussed along with some preliminary results.

One of the most frequently needed systems in the study and use of oscillators is the phase-lock loop. However, since the performance of this system is often incidental to the ultimate goal of the experimenter, e.g. the measurement of phase noise, the design of such a system is sometimes given too little consideration, resulting in unanticipated difficulties and wasted time. The design of an extremely simple phase-lock loop which is suitable for almost all high stability oscillator applications is discussed with particular attention to the advantages over more traditional circuits.

I. PHILOSOPHY OF LOW NOISE AMPLIFIER DESIGN

If an amplifier were driven from a noiseless oscillator, then the output phase spectrum would typically have a flicker noise region at low frequencies and a white noise region at higher frequencies. The break between the two is usually between one and one hundred Hz and the white noise extends out to the bandwidth of the amplifier. The source of the white noise modulation can be identified and the magnitude estimated, but similar generalizations can not be made for the flicker noise. Nevertheless, empirical guidelines can be established which should ensure against unnecessarily poor flicker noise performance.

Provided that the integrated noise of the amplifier over its entire bandwidth is small compared to the signal power, half the thermal noise power contributes to the phase modulation of the signal. Thus the spectral density of phase fluctuations due to noise of the amplifier is

$$S_\phi(f) = S_o/2P_s \qquad (1)$$

where S_o is the spectral density of the noise power and P_s is the power available to a matched load. For Johnson noise, the most common situation,

$$S_o = kT = 4 \times 10^{-21}\ \text{J}$$

and the achievable phase noise performance is -184 dB below a 1 rad^2/Hz for a 1 V_{rms} signal from a 50Ω source.

In contrast to the white phase noise which is added to the carrier by the amplifier, the flicker phase noise is produced by direct phase modulation in the active element. It has been found empirically that a transistor stage which does not use emitter degeneration typically has phase noise given by $S_\phi(f) \approx 10^{-11}$ rad^2/f. However the use of local RF negative feedback can reduce this noise power by as much as a factor of 10^4.[1] Passive elements can also contribute to the flicker phase noise. Electrolytic, ceramic, and silver mica capacitors and carbon composition resistors can give excessive flicker noise and should only be used in non-critical locations.

Three design requirements for low phase noise amplifier design follow from the above discussion. Firstly, each stage of an amplifier must incorporate emitter degeneration to minimize the flicker phase modulation. Secondly, critical passive components should be examined for excessive phase noise. Finally, the signal level must be always maintained at a high enough level to achieve the desired white phase noise level.

In order to illustrate the influence of this philosophy on the design of an amplifier, a new isolation amplifier is described. This amplifier was developed to provide a high degree of isolation between very low phase noise RF signals which are used to compare atomic and other frequency standards.

II. WIDE-BAND LOW-NOISE ISOLATION AMPLIFIER

The amplifier shown in Figure 1 is designed to operate from one to several hundred megahertz. In order to minimize current drain a method of achieving high isolation which used a small number of stages was needed. This requirement was satisfied by a cascaded pair of common base transistor stages, Q_1 and Q_2 A signal applied to the output port propagates towards the input through the collector-base capacitance of Q_1. The 2N3904 was selected because of its small output capacitance, 4 pF. Since the base of Q_1 is grounded through a capacitor and the emitter looks into the high output impedance of the preceding stage, the signal is low pass filtered. A second stage of filtering is performed by the transistor Q_2 in the same way. It is also possible for a signal to propagate from the output to input through the bias chain. Transmission through this path is reduced to the same level as transmission through the transistor chain by the cascaded low

Reprinted with permission from *Proc. 30th Annu. Symp. Frequency Contr.*, Atlantic City, NJ, June 2-4, 1976, pp. 269-274.

pass filters. Typical isolation which is achieved is greater than 120 dB at 5 MHz degrading to 100 dB at 50 MHz.

The common emitter input stage determines the collector current of the transistors. The 27Ω dc emitter resistance produces an average collector current of 40 mA. Noise performance is generally best when the amplifier operates well within the class A region. With a 50Ω load this amplifier can produce an output of nearly 1 V_{rms} (13 dBm) with minimum distortion. The gain of the amplifier is determined by the load resistance and the unbypassed portion, 27Ω, of the emitter resistance. With the values shown, the full output swing occurs for an input of approximately 1.5 V_{ptp}.

The white noise floor which one would estimate for this amplifier is S_ϕ = -184 dB. The measured noise floor is shown as curve A of Figure 2. The noise floor appears to be only -174 dB, but since this level corresponds to the measurement system noise it can only be said that the amplifier is not worse than this. The measured flicker phase noise of the amplifier is $S_\phi = 10^{-14}$ rad^2/f. This performance level is reached because each transistor has a reasonable amount of local RF negative feedback. The emitters of Q_1 and Q_2 both look into the high dynamic impedance of the preceding stage while the emitter of Q_3 has the unbypassed 27Ω resistor. For a given application, this unbypassed resistor should be made as large as possible, limited only by the necessity of having full output voltage swing.

In addition to achieving low noise levels it is necessary to minimize microphonics and pickup of powerline frequencies and other signals. For this reason, no use has been made of filter inductors or coupling transformers. It is also possible that temperature changes could cause sufficient collector current variation to degrade the flicker performance. As a result, the diode has been included in the bias chain to further stabilize the collector current. It should be placed in physical contact with transistor Q_3. The amplifier has been constructed on a double sided printed circuit board measuring 3.25 cm x 9 cm. The art work for this circuit board is available from the authors.

III. DOUBLE BALANCED MIXERS AND PHASE NOISE MEASUREMENT SYSTEMS

The most common and also most sensitive method of measuring phase noise is to use a double balanced mixer. If the input ports are driven by quadrature signals, then the output voltage is proportional to the phase deviation of the input signals from the quadrature condition. The spectral density of the phase noise can be calculated from the very simple expression

$$S_\phi(f) = \left[\frac{V_n(f)}{V_s}\right]^2 \qquad (2)$$

where V_n^2 (f) is the noise density in units V^2/Hz at the output of the mixer and V_s is the sensitivity of the mixer in V/rad.

A variety of circuits for the measurement of phase noise have been discussed extensively in the literature. Here we will look closely at specific problem areas common to all circuits using double balanced mixers: components, input drive levels and output termination. Data are presented which show that proper treatment of these details results in a 15 to 25 dB improvement in the performance of phase noise measurement systems.

Figure 3 shows a typical double balanced mixer. The best performance has been obtained with units which use hot carrier diodes in the ring. Some differences may also result from the type of transformers in the coupling circuits. The noise observed at the output of the mixer, consisting of mixer and amplifier contributions, is nearly constant over a range of input power level. However, the output signal, proportional to the phase fluctuations, increases with the drive power. The best signal-to-noise ratio for Fourier frequencies in the white noise region is obtained at very high drive levels. For one type of mixer, using a single diode in each arm, the best noise floor was obtained with approximately 30 mA rms current at each input. This drive level exceeded the manufacturer's maximum drive current specification. The optimum drive is not necessarily the same for all Fourier frequencies. The same mixer performed best below 40 Hz at lower drive level. Since such a double balanced mixer is a dynamic impedance the average drive current does not sufficiently describe the operating conditions. The optimum method of coupling to the mixer also depends upon the output impedance of the signal source. Although the use of 50Ω pads to attenuate the drive level is traditional, a series resistor whose value is chosen to set the desired current often gives superior performance. The improvements which are observed may be due to reduced ringing of the drive currents.

The signal-to-noise ratio at the mixer output is also affected by the type of termination used. Since the mixer has a low output impedance, near 50Ω, the dc termination must be high impedance compared to 50Ω. Failure to observe this may result in 6 dB or more loss in signal level. However, it has been determined empirically that the mixer must be terminated differently at RF. In the circuit shown in Figure 4 the impedance to ground at the output of the mixer is 1 kΩ at dc and approximately 50Ω at 10 MHz. The net result of the high drive level and the output termination is illustrated in Figure 5, which shows the beat frequency between the two oscillators in Figure 4 at low drive level (sinewave) and high drive level (clipped waveform). The slope of the clipped waveform at the zero crossings is more than twice the slope of the sinewave. It follows from Eq. (2) that the noise floor is improved more than 6 dB by this technique. The increase in slope is not realized without appropriate termination, but the optimum circuit has not been determined. The noise floor achieved with the circuit of Figure 4 is shown in curve C of Figure 2. The spectral density of phase is -150 dB at 1 Hz and drops to a floor of -176 dB.

Several other special circumstances may occur. One may wish to measure the signal from a device which has insufficient output power to drive a double balanced mixer. Figure 6 shows a simple buffer amplifier which may be used under the circumstances. In keeping with the stated philosophy of amplifier design, this circuit can drive a mixer with a nominal 50Ω input impedance with a 1 V_{rms} signal in class A operation. The mixer input impedance appears as an unbypassed 200Ω in the emitter circuit which results in excellent flicker noise performance. As shown in curve B of Figure 2, the spectral density of phase is -149 dB at 1 Hz and falls off to the noise floor of the measurement system. The 10 dB improvement in flicker noise over the previously described isolation amplifier is probably due to the greater emitter degeneration or lower intrinsic flicker noise in the 2N5943 transistor compared to the 2N3904 or both. The 2N5943 was suggested for use in this circuit by Charles Stone of Austron Inc.

If the device being tested is capable of more output power than a standard double balanced mixer can accept, then it is possible to achieve even lower noise floors. Provided the driving voltage exceeds about 1 V_{rms}, it is possible to use a high level mixer. Such a device has more than one diode in each leg of the ring and is therefore able to achieve higher output voltage without a corresponding increase in the noise. The circuit diagram of the mixer in Figure 3 shows two diodes in each leg. Using a mixer with three diodes per leg a noise floor of -184 dB was achieved with a drive level of 1.6 V_{rms}. If the oscillator's output impedance is low but the voltage is insufficient to drive a high level mixer a step-up transformer can be used to obtain the appropriate drive voltage. Since the signal and noise power increase by the same ratio, the spectral density of phase of the device under test is unchanged but the noise floor of the measurement system is reduced.

IV. CORRELATION TECHNIQUE

With all of the improvements described the traditional double balanced mixer phase noise measurement system is unable to resolve the noise floor of the best oscillators and amplifiers.

If times 20 or more frequency multiplier chains with noise levels 20 dB below that of the measurement system shown in Figure 2 were available, then that would solve the present problem. So far we are unaware of such multiplier chains, although some prototype multiplier chains show white noise floors 5 to 10 dB below Figure 2. It would also be convenient if the measurement system were broadband so as to accept carrier frequencies from approximately 1 to 100 MHz. Figure 7 shows the block diagram of a phase noise measurement system which is inherently very broadband and also has the capability of improving the measurement system noise by at least 20 dB. It consists primarily of two equivalent traditional phase noise measurement systems.

At the output of each double balanced mixer there is a signal which is proportional to the phase difference, $\Delta\phi$, between the two oscillators and a noise term, V_N, due to contributions from the mixer and amplifier. The voltages at the input of each bandpass filter are

$$\begin{aligned} V_1(\text{BP filter input}) &= A_1\,\Delta\phi(t) + C_1 V_{N1}(t) \\ V_2(\text{BP filter input}) &= A_2\,\Delta\phi(t) + C_2 V_{N2}(t) \end{aligned} \tag{3}$$

where $V_{N1}(t)$ and $V_{N2}(t)$ are substantially uncorrelated. Each bandpass filter produces a narrow band noise function around its center frequency f:

$$\begin{aligned} V_1(\text{BP filter output}) &= A_1[S_\phi(f)]^{\frac{1}{2}} B_1^{\frac{1}{2}} \cos\,[2\pi ft + \Psi(t)] \\ &+ C_1[S_{VN1}(f)]^{\frac{1}{2}}\; B_1^{\frac{1}{2}} \cos\,[2\pi ft + \eta_1(t)] \\ V_2(\text{BP filter output}) &= A_2[S_\phi(f)]^{\frac{1}{2}} B_2^{\frac{1}{2}} \cos\,[2\pi ft + \Psi(t)] \\ &+ C_2[S_{VN2}(f)]^{\frac{1}{2}}\; B_2^{\frac{1}{2}} \cos\,[2\pi ft + \eta_1(t)] \end{aligned} \tag{4}$$

where B_1 and B_2 are the equivalent noise bandwidths of filters 1 and 2 respectfully. Both channels are bandpass filtered in order to help eliminate aliasing and dynamic range problems. The phases $\Psi(t)$, $\eta_1(t)$ and $\eta_2(t)$ take on all values between 0 and 2π with equal likelihood. They vary slowly compared to 1/f and are substantially uncorrelated. When these two voltages are multiplied together and low pass filtered only one term has finite average value. The output voltage is

$$\begin{aligned} V^2_{out} &= 1/2\, A_1A_2\, S_\phi(f)\, B_1^{\frac{1}{2}}B_2^{\frac{1}{2}} + D_1\langle\cos[\Psi(t) - \eta_1(t)]\rangle \\ &+ D_2\langle\cos[\Psi(t) - \eta_2(t)]\rangle + D_3\langle\cos[\eta_1(t) - \eta_2(t)]\rangle. \end{aligned} \tag{5}$$

For times long compared to $B_1^{-\frac{1}{2}}B_2^{-\frac{1}{2}}$ the noise terms D_1, D_2 and D_3 tend towards zero as $\sqrt{\tau}$. Limits in the reduction of these terms are usually associated with harmonics of 60 Hz pickup, dc offset drifts, and nonlinearities in the multiplier. Also if the isolation amplifiers have input current noise then they will pump current through the source resistance. The resulting noise voltage will appear coherently on both channels and can't be distinguished from real phase noise between the two oscillators. One half of the noise power appears in amplitude and one half in phase modulation.

Curve A of Figure 8 shows $S_\phi(f)$ for the mixer and dc amplifiers in Channel 1 and 2 when used separately. The mixers in this case have three diodes in each leg instead of the two shown in Figure 3 and are driven with approximately 5 V_{ptp} at 5 MHz from 10Ω source impedance using 33Ω series resistors. Curve B Figure 8 indicates the correlated component of this noise between the two channels. In order to predict performance in a specific measurement using this scheme, the noise level of the isolation amplifiers used would have to be added to Curve A and proportionately to Curve B.

The above data clearly indicate that significant improvements over any presently existing phase noise measurement system can be obtained using correlation techniques. Such improvements are vitally necessary in order to measure present state-of-the-art signal processing equipment and to test future components and circuits. The simple single frequency correlator used in this experiment could be replaced by a fast digital system which would simultaneously compute the correlated phase noise for a large band of Fourier frequencies. Ultimate noise floors could probably be reduced 40 dB below the noise level of a single channel.

V. PHASE-LOCK TECHNIQUES

One of the most ubiquitous elements of phase noise measurement systems is the phase-lock loop.[2,3] When the phase noise of a pair of oscillators is measured a phase-lock loop is normally used to maintain a condition of approximate phase quadrature. For accurate measurements it is necessary to keep the phase error less than about 1/6 rad despite any initial frequency offset between the two oscillators or any frequency drift during the course of the measurement. The phase-lock loop is usually the most neglected element of the measurement system because its purpose is only tangential to the measurement requirements. As a consequence it often performs marginally. In this section the requirements for a phase-lock system are discussed and an extremely simple yet elegant circuit is presented which more than meets these requirements.

A specific example illustrates the problem: Design a feedback loop to lock a 5 MHz VCO to a reference oscillator with a unity gain frequency of .16 Hz and calculate the open loop frequency difference for phase error of 1/6 rad. The VCO has a tuning rate of 5×10^{-3} Hz/V and the phase deviation from quadrature. One solution, the first order phase-lock loop, is shown in Figure 9. The 50Ω resistor and 0.1 μF capacitor are for proper termination of the mixer and do not contribute appreciably to the frequency response of the phase-lock loop. The open loop gain of this servo is

$$\begin{aligned} G_{servo}(\omega) &= \frac{2\;(5\text{x}10^{-3}\ \text{Hz/V})\;(0.17\ \text{V/rad})\; G_{amp}(\omega)}{j\omega} \\ &= \frac{5.3 \times 10^{-3}\; G_{amp}(\omega)}{j\omega} \end{aligned} \tag{6}$$

where $G_{amp}(\omega)$ is the frequency response function of the amplifier. For the first order loop the amplifier has constant gain and the open loop gain of the servo system falls off at 6 dB/octave at all frequencies as shown in the dashed curve of Figure 10. This type of response results from the fact that a phase error measured at the mixer is corrected by changing the frequency of the VCO, i.e., the feedback loop contains one inherent integration.

To produce a unity gain frequency of 0.16 Hz, or an attack time of one second, requires amplifier gain of 185. The loop produces negligible correction for Fourier frequencies greater than 1 Hz and may therefore be used to make phase noise measurements in this frequency range. The open-loop frequency offset which corresponds to a 1/6 rad phase error is

$$\Delta\nu = (1/6\ \text{rad})\ (0.17\ \text{V/rad})(5 \times 10^{-3}\text{Hz/V})(185) = 2.5 \times 10^{-2}\ \text{Hz} \quad (7)$$

or only 5 parts in 10^9 at 5 MHz. It is often difficult to achieve and maintain such a small frequency offset. However, the following circuit increases the dc gain of the feedback loop by 10^5 and decreases the phase error proportionally. The improved performance is achieved by adding a stage of quasi-integration which makes a second order loop as shown in Figure 11. The first stage of amplification is the same as in the first order loop. The second stage has gain equal to R_1/R_2 for Fourier frequencies larger than $1/(2\pi R_1C_1)$. If R_1C_1 exceeds the attack time of the first order loop the new phase-lock loop is stable. It is critically damped when R_1C_1 is approximately four times the attack time and has good step response for R_1C_1 between 1 and one and five times the attack time. The solid curve in Figure 10 is the magnitude of the open loop gain of the second order loop assuming $R_1 = R_2$. Figure 12 compares the step response of a first order loop to that of a second order loop with R_1C_1 equal to the attack time.

The second order loop increases the long-term gain by the open loop gain of the second operational amplifier provided that the leakage resistance, R_L, of the capacitor is sufficiently large. The open loop frequency offset between the two oscillators which can be tolerated is therefore increased to a limit determined either by the maximum voltage swing of the second amplifier or the maximum tuning available in the VCO. The second order loop can be implemented with a single operational amplifier rather than the two which were shown for clarity. In this case the attack time is adjusted by varying the input resistor, R_2, of the operational amplifier, A_2 and omitting amplifier A_1 in Figure 11.

ACKNOWLEDGMENTS

The authors would like to thank Charles Stone of Austron Inc. for many helpful conversations and in particular for his advice concerning low noise transistors and other components.

REFERENCES

[1] Donald Halford, A.E. Wainwright, and James A. Barnes, in Proc. 22nd Annual Symposium on Frequency Control (NTIS Accession No. AD 844911), 1968, p. 340.

[2] F. M. Gardner, Phaselock Techniques (John Wiley, New York, 1966).

[3] Don Kesner, EDN 5 Jan 1973, p. 54.

ISOLATION AMPLIFIER

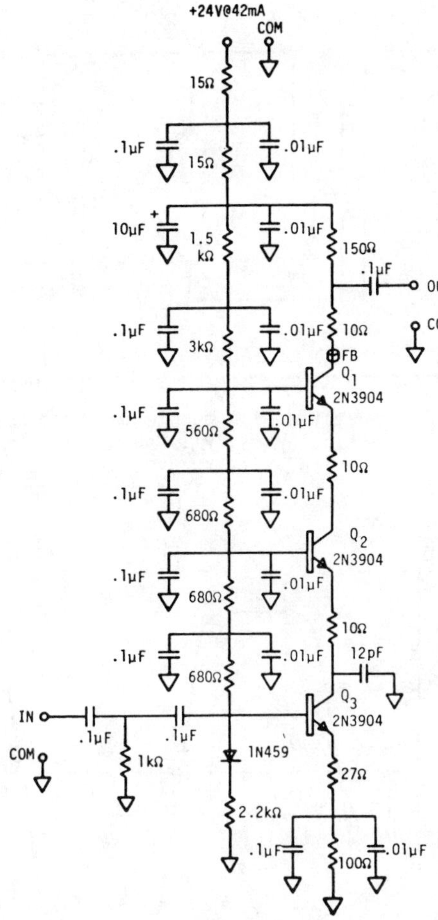

Fig. 1 Schematic of 1 MHz to 200 MHz isolation amplifier. The isolation is -120 dB at 5 MHz. Art work for this amplifier is available from the authors.

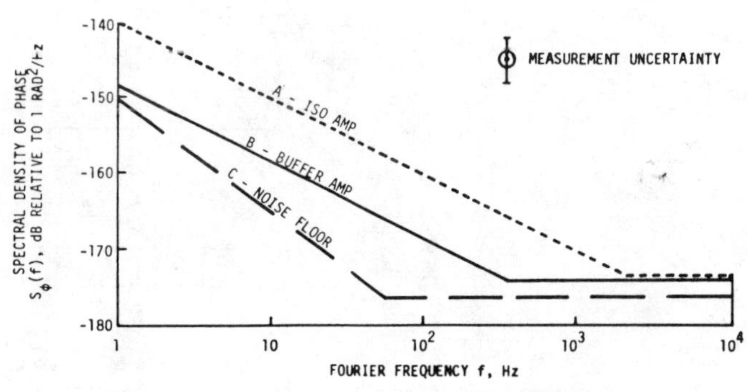

Fig. 2 Spectral density of phase, $S_\phi(f)$, for A) the isolation amplifier of Figure 1 at 5 MHz, B) the buffer amplifier of Figure 6 at 5 MHz, C) noise floor of the measurement systems of Figure 4 with a single diode mixer.

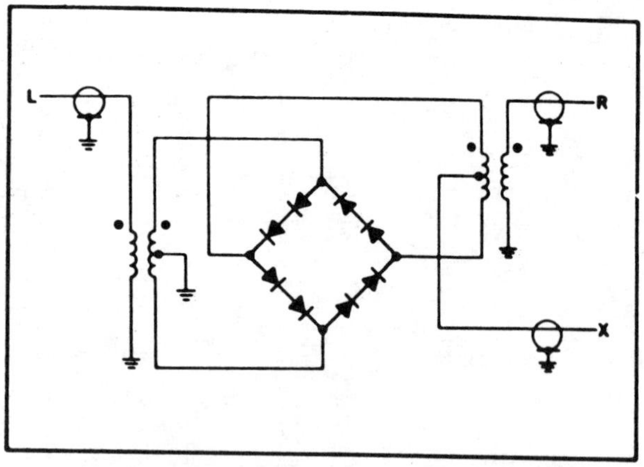

Fig. 3 Schematic of a double balanced mixer. This particular mixer is shown with two diodes in each leg.

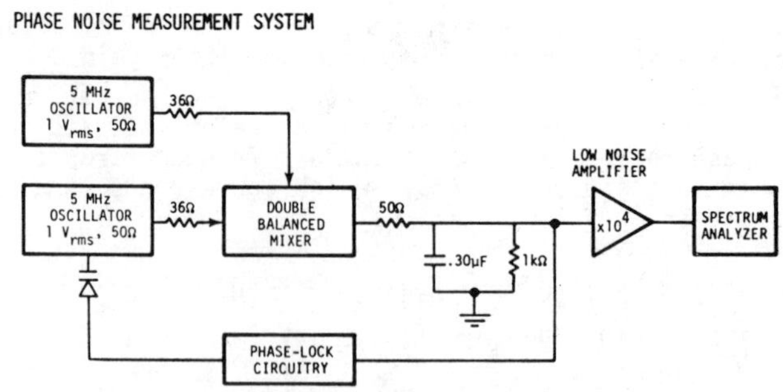

Fig. 4 Typical system for measuring $S_\phi(f)$ of a pair of equal frequency oscillators. Noise floor for this system is shown in Figure 2, curve C.

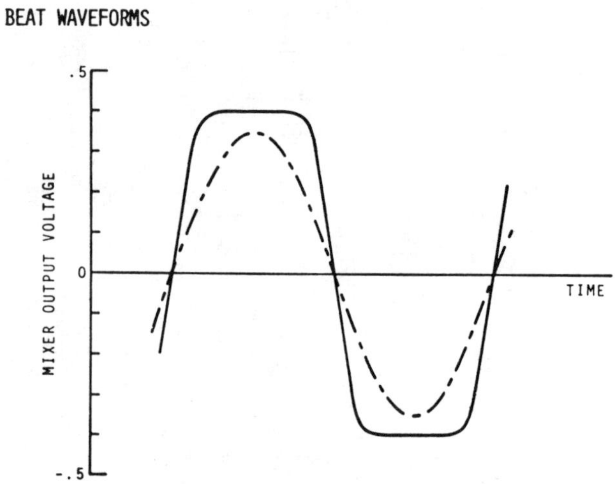

Fig. 5 Filtered waveform at the output (x port) of double balanced mixer due to frequency difference between signals at R and L ports. The solid curve is obtained at high drive levels while the dashed curve is obtained at low drive levels.

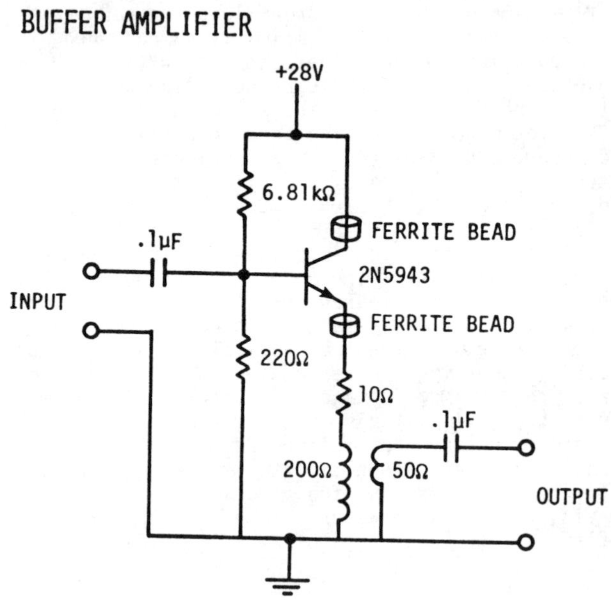

Fig. 6 Very low noise buffer amplifier which can be used to drive mixers. The noise performance is shown in Figure 2 curve B.

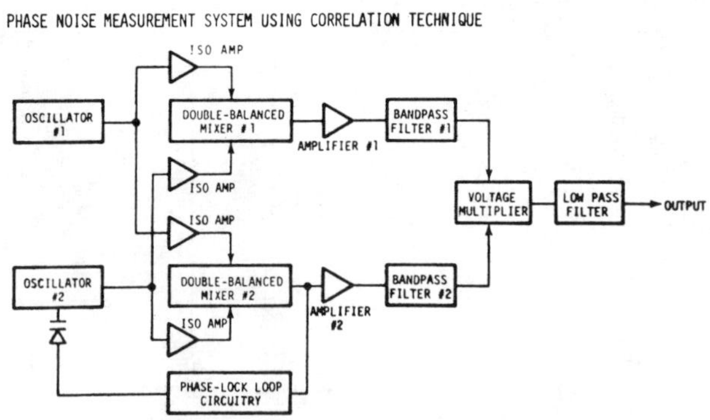

Fig. 7 Block diagram of a new phase noise measurement system which is broadband and features low noise.

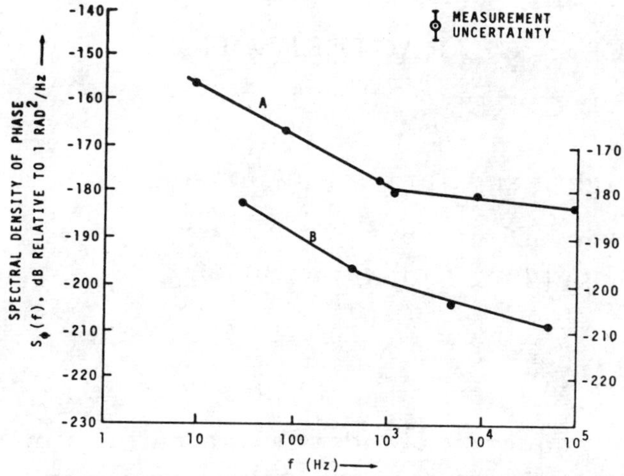

Fig. 8 Curve A shows $S_\phi(f)$ for each channel of the measurement system of Figure 7 excluding the isolation amplifiers. Curve B shows the correlated component of $S_\phi(f)$ between the two channels.

OPEN-LOOP GAIN

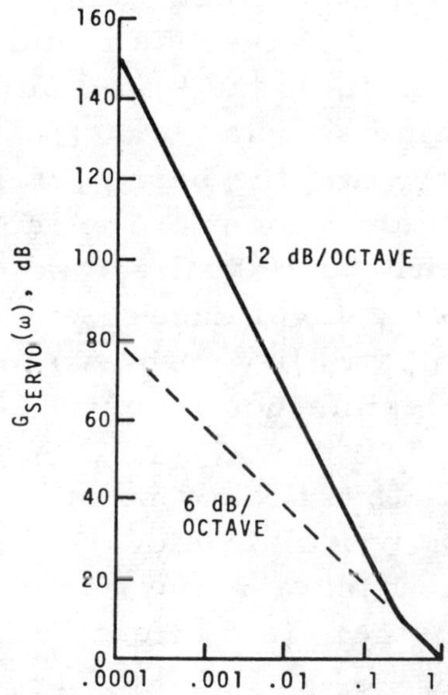

Fig. 10

The dashed curve shows the open-loop gain of the first order phase-lock loop of Figure 9, while the solid curve shows the open-loop gain of the second order phase-lock loop of Figure 11.

FIRST ORDER PHASE-LOCK LOOP

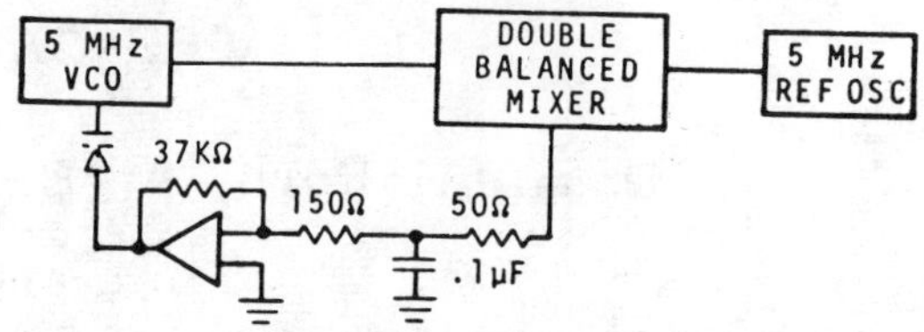

Fig. 9 Schematic of typical first order phase-lock loop. Loop attack time is adjusted by changing amplifier gain.

SECOND ORDER PHASE-LOCK LOOP

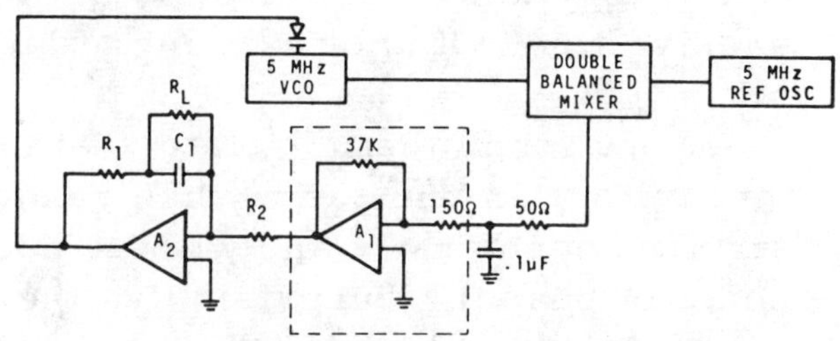

Fig. 11 Second order phase-lock loop. The attack time is adjusted by changing gain as in Figure 9. The time constant R_1C_1 is adjusted to be 1 to 5 times longer than the attack time.

PHASE-LOCK LOOP STEP RESPONSE

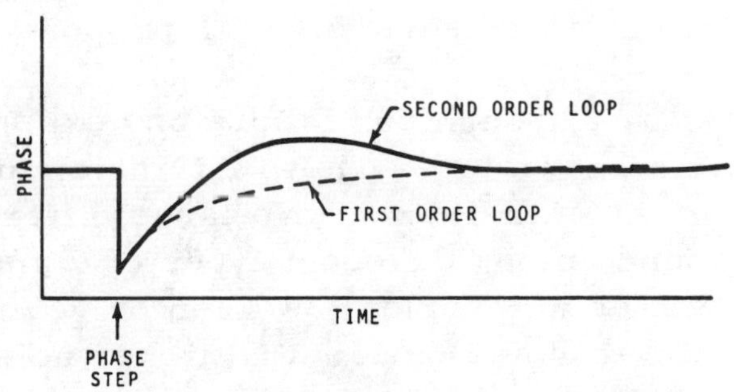

Fig. 12 The dashed curve shows the response of the first order phase-lock loop of Figure 9 to a phase step, while the solid curve shows the response of the second order phase-lock loop of Figure 11, with R_1C_1= the attack time.

FLICKER NOISE OF PHASE IN RF AMPLIFIERS AND FREQUENCY MULTIPLIERS: CHARACTERIZATION, CAUSE, AND CURE

Donald Halford, A. E. Wainwright, and James A. Barnes

Time and Frequency Division
National Bureau of Standards, Boulder, Colorado 80302

Summary

The high phase stability of atomic frequency standards has called for the development of associated electronic equipment of equivalent or superior stability. We have surveyed the performance of existing high quality frequency multipliers and RF amplifiers which operate in the range of 5 MHz to microwave frequencies. We were most interested in the phase noise in the range of Fourier frequencies, f, of about 10^{-2} Hz to 10^{+3} Hz, since most electronic servo systems in existing atomic frequency standards use modulation frequencies which fall within this range. This range also includes the passive linewidths of existing atomic frequency standards.

We found that all state-of-the-art, solid state amplifiers and frequency multipliers in our survey had random phase fluctuations (phase noise) of significant intensity and with a spectral density proportional to 1/f (flicker noise of phase). Surprisingly, the flicker noise of phase was found to be approximately the same in most of the apparatus included in the survey. The typical one-sided spectral density of the phase noise was about $(10^{-11.2}$ radian$^2)/f$ when referred to the input frequency, and the best performance was only 6 dB better. This performance was independent of the input frequency over at least the surveyed range of 5 MHz to 100 MHz, and did not depend upon the multiplication factor. This noise level exceeds by about 20 dB the stability which is needed for full compatibility with existing hydrogen atom masers and with proposed high flux cesium beam designs.

Through our laboratory experimentation, this flicker noise of phase was shown to be due to intrinsic, direct, phase modulation of the RF carrier by the transistors. No large differences in the flicker noise intensity were found among different types of transistors, whether field effect or bipolar (including overlay), silicon or germanium, high f_T or medium f_T, high DC flicker noise or low DC flicker noise, and hermetic can or plastic. With the knowledge that the active elements invariably cause phase noise via intrinsic, direct, phase modulation, we could predict that negative feedback, and only negative feedback, could reduce the phase noise.

By systematically applying local RF negative feedback (emitter degeneration) to each of the transistors in an apparatus, we have realized typically more than 30 dB (up to 40 dB in some cases) reduction of the flicker noise of phase in amplifiers and frequency multipliers. No other circuit

Reprinted with permission from *Proc. 22nd Annu. Symp. Frequency Contr.*, Atlantic City, NJ, Apr. 22–24, 1968, pp. 340–341.

improvements were found which yielded significant additional reduction of the flicker noise of phase. Individual silver mica capacitors (80% of our stock) gave excessive flicker noise of phase, but, by testing and selecting capacitors, this source of noise could always be avoided. Selection of individual transistors had relatively little effect, typically no more than 4 dB.

This improved performance of amplifiers and frequency multipliers comfortably exceeds the requirements for compatibility with contemporary and proposed atomic frequency standard systems. At the Frequency Control Symposium we will discuss

1. our survey of the state-of-the-art,
2. the measurements which proved that the active elements (transistors) were the unavoidable source of the flicker noise of phase,
3. the use of local RF negative feedback as the only practical means of achieving extremely low flicker noise of phase,
4. the measured performance of our new amplifier and frequency multiplier designs, and
5. the outlook for possible similar improvements in oscillators, mixers, and other phase-processing electronic equipment.

LOW NOISE FREQUENCY MULTIPLICATION

Richard A. Baugh
Hewlett-Packard Company
Palo Alto, California

Summary

The phase of the output of an ideal frequency multiplier is an integer number times the phase of the input. Practical frequency multipliers depart from this ideal situation in several respects. Variations in reactive components and charge storage times in active devices cause a random phase error with an F^{-1} spectral density (flicker noise); shot noise and thermal noise cause a phase noise with a white spectral density and imperfect filtering of adjacent harmonics cause a phase error which varies sinusoidally at the input frequency.

Since the phase information in a bandpass-filtered sinewave is contained in the zero crossings, a frequency multiplier which tightly couples the output phase to the zero crossings of the input waveform should have very low phase noise. This can be done with an emitter-coupled current switch acting as a hard limiter. The output is a squarewave current which changes polarity when the input voltage crosses through zero. The squarewave is differentiated and rectified to give a grain of impulses of the appropriate polarity. Since the position of each impulse is accurately established by the input zero crossing, it should have very little jitter or phase noise. The impulses can either be put through a bandpass filter tuned to the desired harmonic or they can be used to drive a sampler for phase locking a voltage controlled oscillator to the desired harmonic.

Hard limiting can also be used to remove undesired adjacent harmonics. If the train of impulses is passed through a bandpass filter centered on the desired harmonic, then the adjacent harmonics will be attenuated and phase shifted in such a manner that the output will have no phase modulation sidebands, only amplitude modulation. If this signal is hard limited, then the amplitude modulation sidebands will be removed. Thus a simple one or two pole bandpass filter plus a limiter will provide the same sideband suppression as a more elaborate bandpass filter.

Introduction

The output of a frequency multiplier is an integral multiple of the input frequency. In one type of frequency multiplier the input is distorted with some nonlinear device such as a transistor, rectifier, varactor or step recovery diode and the desired harmonic is selected with a bandpass filter. Another type of multiplier locks an oscillator operating at the desired harmonic frequency to the input by means of a sampling phase lock loop. A defect which is common to all frequency multipliers is random phase noise on the output. If the input is

$$v_o = V_o \sin \omega_o t$$

then the output of the multiplier is

$$v_n = V_N \sin [N \omega_o t + \varphi(t)]$$

where the phase, $\varphi(t)$, is a random variable. The phase noise can be described by $S_\varphi(f)$, the spectral density of $\varphi(t)$ where

$$S_\varphi(f) = 4 \int_0^\infty R_\varphi(\tau) \cos 2\pi f \tau \, d\tau$$

and $R_\varphi(\tau)$ is the autocorrelation function of $\varphi(t)$ or $\mathcal{L}(f)$, the single sideband noise power/carrier power ratio in a one Hz noise bandwidth expressed as a function of f, the offset from the carrier. It can be shown[1] that

$$\mathcal{L}(f) = 1/2 \, S_\varphi(f)$$

for small phase deviations.

In general $\mathcal{L}(f)$ for a frequency multiplier has two components, a white phase noise term extending from dc out to the passband of the multiplier, and a "flicker" phase noise term with a spectral density which varies inversely with frequency and extends down to very low frequencies. The white noise is due to additive thermal and shot noise and the flicker noise is due to random variations in the reactive elements and random variations in delay times and minority carrier lifetimes of active devices.

The theoretical lower limit on the white phase noise that a frequency multiplier should add to the output depends on the ratio of thermal noise power to signal power at the input. The lower

Reprinted with permission from *Proc. 26th Annu. Symp. Frequency Contr.*, Atlantic City, NJ, June 6–8, 1972, pp. 50–54.

limit is

$$\mathcal{L}\ \text{min} = N^2 \left(\frac{2\,k\,T\,R_s}{V_s^2} \right)$$

where V_s is the rms source voltage and R_s is the source resistance. The noise figure of a frequency multiplier, analogous to the noise figure of an amplifier, is the ratio of measured phase noise on the output relative to the theoretical minimum.

$$F_N = 10 \log_{10} \left(\frac{V_s^2\, \mathcal{L}}{2\,k\,T\,R_s\,N^2} \right)$$

At present there is no theoretical lower bound on the flicker noise in frequency multipliers.

Design Approach

The phase information of a signal is contained in the positions of the zero crossings. Therefore, to minimize phase noise, the phase of the multiplier output should somehow be closely coupled to the zero crossings of the input. Thus, the class C frequency multiplier was rejected as a possible circuit because the output current pulses occur on the peaks of the input waveform. Similarly varactor multipliers were not considered because of their tendency to convert small amplitude variations into phase noise and because of their dependence on minority carrier lifetime, a quantity which seems to have a great deal of flicker noise. Reasoning such as this eliminated all possible candidates except the emitter coupled current switch[2], a circuit which is used in very high speed logic circuitry.

For a slowly varying input the collector currents of the emitter coupled switch, Fig. 1, are

$$i_1 = \left(I_{EE}/2 \right) \left[1 + \tanh \left({-qV_B}/{2kT} \right) \right]$$

$$i_2 = \left(I_{EE}/2 \right) \left[1 - \tanh \left({-qV_B}/{2kT} \right) \right]$$

The switching speed is very great because the transistors are never saturated.

The Circuit

The collector current of Q1, Figure 2, which is a square wave, is differentiated to produce a train of positive and negative pulses which occur on the positive and negative zero crossings of the input waveform.

The positive pulses are amplified by Q2 and filtered with a single resonant circuit tuned to the desired harmonic. Experimentally it was observed that the phase noise was 3 db smaller when the second transistor in the standard emitter-coupled switch was replaced with a switching diode.

Another emitter coupled current switch, Q3 and Q4, is used as a "sideband skimmer" to remove the undesired harmonics of the input frequency. When tuned circuit L2C1 is tuned exactly to the desired harmonic the output (Fig. 3a) has no phase modulation, only amplitude modulation. Consequently when the signal is passed through a symmetric limiter the am sidebands are removed and a square wave at the desired frequency is obtained. If the circuit is not tuned exactly the output will have am and pm sidebands (Fig. 3b). By using a single tuned circuit and a symmetric limiter the undesired sidebands can be reduced substantially. This is, in many situations, less expensive than a multi-pole bandpass filter.

Measurement System

The system used to measure phase noise in frequency multipliers, Fig. 4, includes a double balanced mixer, lowpass filter, low-noise preamplifier, audio wave analyzer with a 7.1 Hz noise bandwidth, and a strip chart recorder. Systems similar to this have been described by Cutler and Searle[3], and Meyer[4]. The only unusual features are the use of a strip chart recorder to average the meter reading for a long period of time and indicate anomolous noise bursts and a calibration procedure which compensates for a non-sinusoidal output from the double balanced mixer.

Normally the double balanced mixer is operated with a high level local oscillator input and a low level signal input so that the output is proportional to the cosine of the phase angle between the two inputs. Reduction of signal level to obtain a known relationship between phase angle and output voltage means that sensitivity to very low level phase noise is lost. This sacrifice of sensitivity is not necessary if the proper calibration procedure is used. With two high level input signals, slightly offset in frequency, the output of the double balanced mixer looks like Fig. 5. The slope of the voltage versus phase plot is obtained from the Fourier coefficients of the output, Fig. 6. The magnitude of these coefficients can be obtained by examining the beat note with an audio wave analyzer. Their phase can be obtained from the symmetry properties of the waveform. The results are that, for the two inputs in quadrature, the slope is

$$\frac{dV}{d\varphi} = \sqrt{2} \left[V_1 - 3\,V_3 + 5\,V_5 \cdots \right]$$

where V_N is the rms value of the N'th harmonic in the waveform.

Experimental Results

Two X35 frequency multipliers using the circuit of Fig. 2 were made. No unusual construction techniques or non-standard components were used. All Motorola 2N3906 transistors tried in the input, Ql, gave similar results. A common 1 MHz signal source was used to drive both multipliers and the phase noise on the 35 MHz outputs was measured with the system described earlier. With an input of 1.75 Vrms and source resistance of 50Ω, the $\mathcal{L}(f)$ was

$$\mathcal{L}(f) = 10^{-17.4} + \left(10^{-14.7}\right)/f$$

referred to the input. This is an improvement of 11 db in the white noise region and 7 db in the 1/f region over the best results published to date by Halford, Wainwright and Barnes[5]. The phase noise and noise figure for several input power levels are given in Table 1.

TABLE 1. The Noise and Noise Figure of a Low Noise Frequency Multiplier

Drive Level on base of Ql	f	$\mathcal{L}(f)$	Noise Figure
3.5 V - 200 Ω	5 kHz	-173.6 db	15 db
	1 kHz	-172.3 db	
	90 Hz	-167.2 db	
3.5 V - 400 Ω	5 kHz	-172.6 db	13 db
	1 kHz	-171.2 db	
	90 Hz	-164.9 db	
3.5 V -1600 Ω	5 kHz	-170.0 db	9.7 db
	1 kHz	-168.0 db	
	90 Hz	-159.8 db	

Conclusions

It has been demonstrated that the output of a current-mode switch has very little phase noise. Although the experimental work was done with a tuned filter frequency multiplier the results should also be applicable to the generation of narrow pulses for sampling phase lock loops.

References

1 Some Mathematics of Frequency/Phase Noise, Don Halford, NBS Atomic Frequency and Time Standards Section, 16 May 1969.

2 W.D.Roehr, editor, Motorola Switching Transistor Handbook, 253-265 (1963).

3 L.S.Cutler and C.L.Searle, "Some Aspects of the Theory and Measurement of Frequency Fluctuations in Frequency Standards", Proceedings IEEE, Vol 54, pp 136-154, February 1966.

4 D.G.Meyer, "An Ultra Low-Noise Direct Frequency Synthesizer, Proceedings of the 29 Annual Symposium on Frequency Control, pp 209-232, 1970.

5 Donald Halford, A.E.Wainwright, and James A. Barnes, "Flicker Noise of Phase in RF Amplifiers and Frequency Multipliers: Characterization, Cause, & Cure", Proceedings of the 22 Annual Symposium on Frequency Control, pp 340-341, 1968.

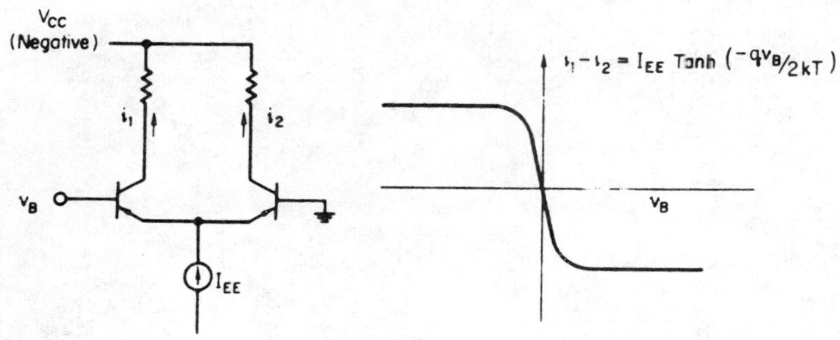

Figure 1

The Emitter-Coupled Current Switch

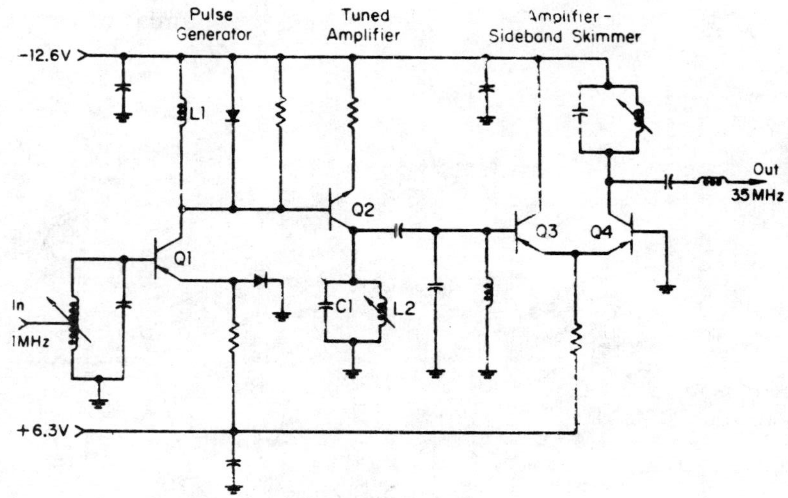

Figure 2

A Low Noise Frequency Multiplier

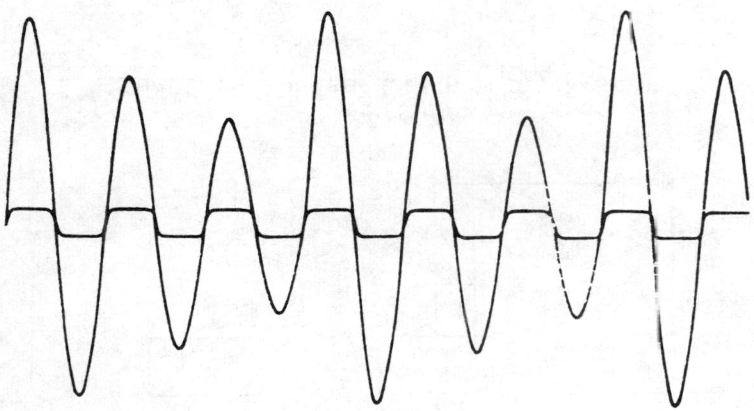

Figure 3a

The voltage waveform on the collector of Q2 when L2 and C1 are tuned exactly to the third harmonic of the input. The third harmonic square wave is superimposed to show that although there is considerable amplitude modulation the phase is constant.

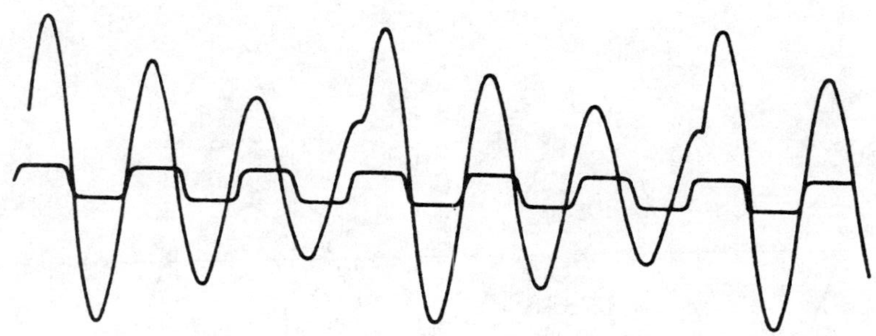

Figure 3b

Same as Figure 3a, except that L2 and C1 are tuned slightly above the third harmonic. The phase deviates in a sawtooth manner.

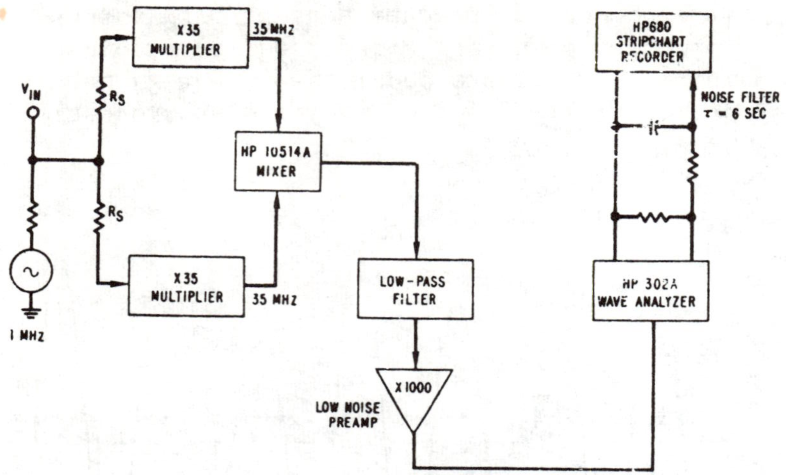

Figure 4

The Phase Noise Measurement System

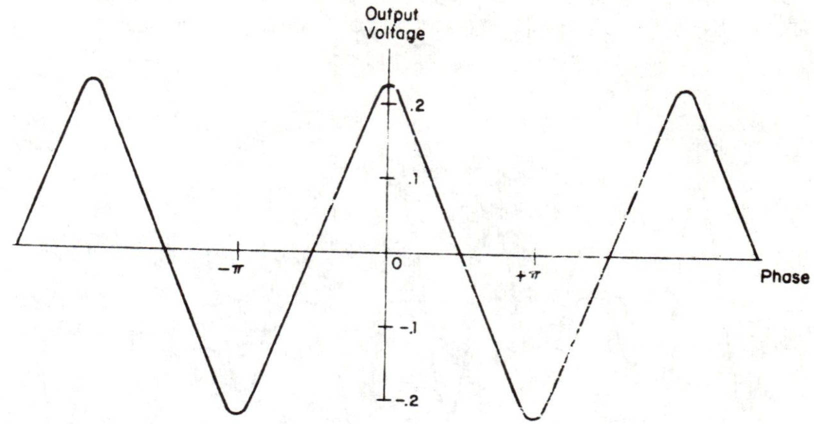

Figure 5

The output of a double balanced mixer with high drive level.

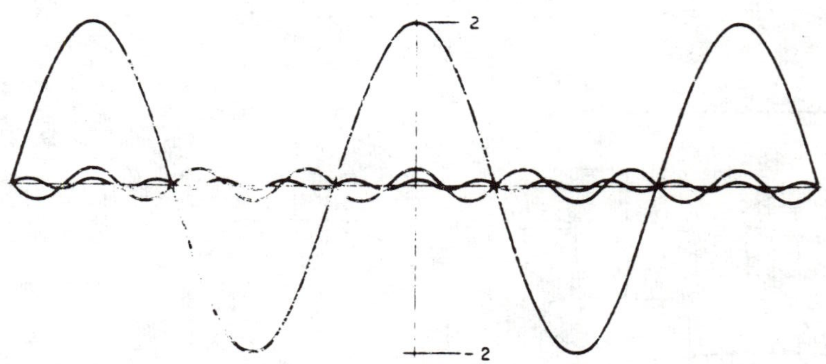

Figure 6

The Fourier components of the mixer output

Phase Noise of Various Frequency Doublers

SVEIN G. ANDRESEN AND JOEL K. NESHEIM

Abstract–Previous studies of phase noise contributed by amplifiers and frequency multipliers have indicated the existence of a so-called "flicker of phase noise law of p-n junction." Contradictions of this law were obtained using low-noise transistors in frequency doublers. Schottky barrier diode and vacuum-tube multipliers were compared to the p-n junction devices. The tube doublers exhibited an anomaly, since their phase noise did not change with the application of negative feedback, indicating the possibility that the tubes themselves may be much better than the observed noise. The noise of the transistor doublers was progressively reduced by application of increasing amounts of negative feedback until a lower limit was reached. The noise then remained constant in spite of further increases in the amount of negative feedback. Measurements were obtained on all doublers from 2 Hz to 5 kHz with extrapolation to 1 Hz. During the investigation, a low-flicker-noise low-drift dc preamplifier was developed.

INTRODUCTION

In the fields of time and frequency standards as well as in communication and radar, it is necessary to know the phase noise contributions by various amplifiers and frequency multipliers. Previous work [1] indicates the existence of a "flicker of phase noise law of p-n junction" amounting to $S_\phi(f) = (10^{-11.2}\ \text{rad}^2)/f$, where $S_\phi(f)$ is the spectral density of the phase fluctuations [2] in rad^2/Hz referred to the input of the device. For comparison purposes, the value of $S_\phi(f)$ at 1 Hz in dB is usually quoted, thus for the law, one gets −112 dB at 1 Hz. This value applies to amplifiers or frequency multipliers without feedback. With negative feedback, a reduction of about 15 dB or more is usually possible. In establishing the phase noise contribution of a given device, it is usual practice to compare it to a similar device in a measuring setup such as that shown in Fig. 1, in which the total phase noise due to both devices is measured, and on reporting the results assume that each device of the same kind contributes equally to the noise. Alternately, the total spectral density measured may be stated as an upper limit for either device. The difference between the two methods is a factor of 1/2 or 3 dB in $S_\phi(f)$. The work reported here is based on the first method.

RESULTS

Two types of transistors, Motorola 2N3904 and Philips BC109, were tested. In both cases, an improvement of approximately 7 dB in the noise level was obtained by comparing one doubler to a Schottky barrier diode doubler and rejecting the noisiest transistors. The results shown in Figs. 2 and 3 were obtained after such sorting. The introduction of emitter degeneration improved the noise until the emitter resistor reached a value of 34 Ω for the 2N3904 transistors or a value of 99 Ω for the BC109 transistors. Increasing the resistors beyond those values caused no further improvement in the noise levels. The maximum improvement with emitter degeneration was 18 dB ± 3 dB for the 2N3904 and 16 dB ± 3 dB for the BC109. In both cases, the noise was proportional to $1/f$.

Two types of tubes were used in the frequency doublers. The 2C51 tubes resulted in a value of $S_\phi(1) = -132$ dB ± 3 dB. When 12AT7 tubes were substituted, the results changed by less than 3 dB. In both cases, the introduction of cathode degeneration caused no improvement in the noise level. The noise was again proportional to $1/f$.

In order to get data on some other types of doublers, a transistor doubler using 2N3904 transistors and an unbypassed 34-Ω emitter resistor was compared to two passive doublers: one a double balanced mixer (HP 10514A) connected as a doubler, the other a diode circuit designed for frequency doubling (HP 10515A). This resulted in $S_\phi(1) \simeq$

Manuscript received December 27, 1971; revised September 25, 1972. This work was supported in part by the Frequency and Time Section, National Bureau of Standards, Boulder, Colo.
The authors are with the Department of Electrical Engineering, University of Colorado, Boulder, Colo. 80302.

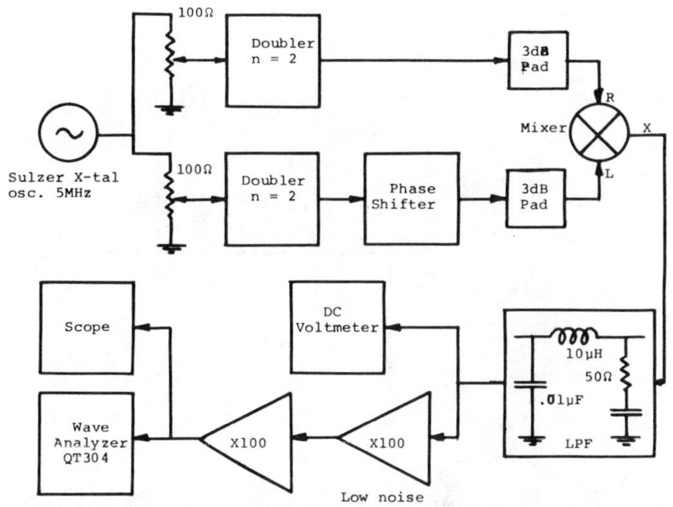

Fig. 1. Typical phase noise measurement system. Phase shifter consists of switchable sections of RG-58A/U coaxial cable plus 20-cm air line stretcher. Low-noise amplifier is of special design (see Fig. 5). Wave analyzer tunes from 1 Hz to 5 kHz (QuanTech model 304).

−128 dB for both the mixer/transistor and the doubler/transistor. Thus these diode multipliers have a phase noise contribution of not more than −128 dB since the transistor doubler is better than that figure.

The wide noise peak around 1 kHz is due to the 5-MHz Sulzer oscillator and the peak in the system noise at 20 Hz appears to be due to building vibration and is simply deleted in plotting $S_\phi(f)$ (similarly for 60 Hz and its harmonics). The system noise level is measured by replacing the devices under test by a short coaxial cable and operating with the same peak voltage output V_p (see the Appendix) from the phase detector. Since the noise contributions of the devices are referred to their inputs (5 MHz), this establishes a lower limit on the sensitivity of the measuring system and may be plotted along with the $S_\phi(f)$ versus f curves.

Fig. 4 is a circuit diagram of the transistor doublers used, while Fig. 5 shows the low noise × 100 preamplifier with its noise characteristics shown in Fig. 6.

CONCLUSIONS AND DISCUSSION

From the experimental results, it follows that some types of transistors contribute less phase noise than other types when used in the same circuit. A further reduction in phase noise can be achieved by sorting transistors within the same type number.

The reduction of the noise by emitter degeneration implies that the phase noise is being generated within the transistor. The noise plateau that is reached when using sufficient feedback is probably due to noise generated by the passive components of the doubler circuits. By choosing low-noise transistors and using degeneration, active frequency multipliers may be built having less phase noise than some passive multipliers employing Schottky barrier diodes.

The lack of improvement in the phase noise contribution of the tube doublers indicates that the noise contributed by the tubes is less than that of the passive components in their doubling circuits.

The flicker of phase noise law of p-n junctions, while correctly predicting the slope of the noise spectral density curve, does not accurately predict the value of $S_\phi(1)$ for all transistors. If the noise plateau that is reached when using sufficient degeneration is due to the passive components, then it is mainly due to their flicker noise and is difficult to estimate. However, it is possible to measure the flicker of phase noise due to capacitors and resistors in a setup similar to that used for the doublers. Since a doubler contains several passive components, it is difficult to estimate the combined effect. However, the measurement of the phase noise of single components provide guidelines for their selection. Our experience and discussions with other workers in this field indicate that piston and variable disc trimmers are quite stable in this respect, while butterfly capacitors are not [3]. The commonly

Reprinted from *IEEE Trans. Instrum. Meas.*, vol. IM-22, pp. 185–188, June 1973.

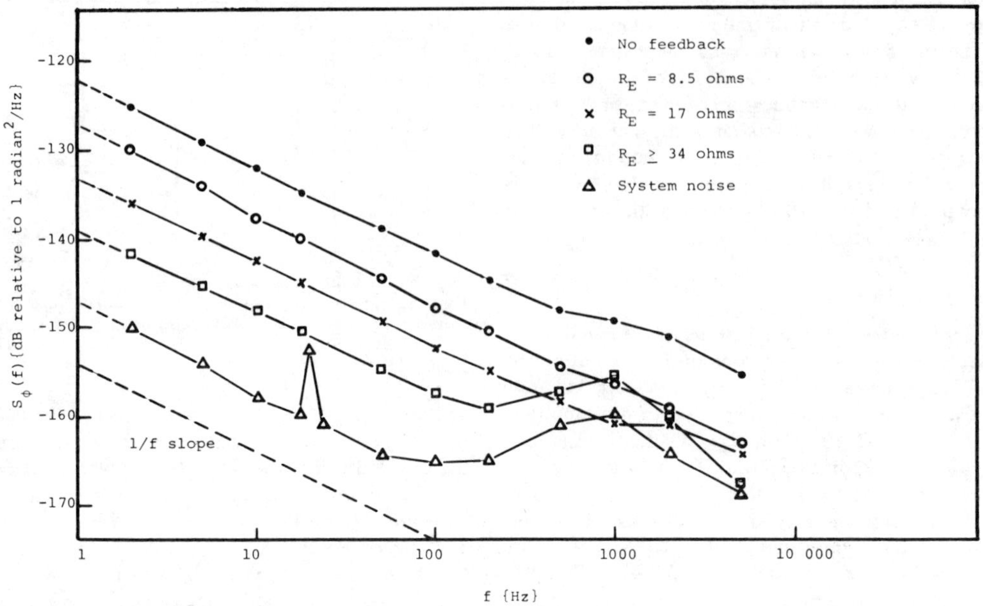

Fig. 2. Spectral density of phase fluctuations referred to input for each of pair of transistor frequency doublers, 5–10 MHz, using Motorola 2N3904 transistors. Results are shown for varying amounts of emitter degeneration. Also shown is system noise level and $1/f$ slope line. V_p = 180 mV.

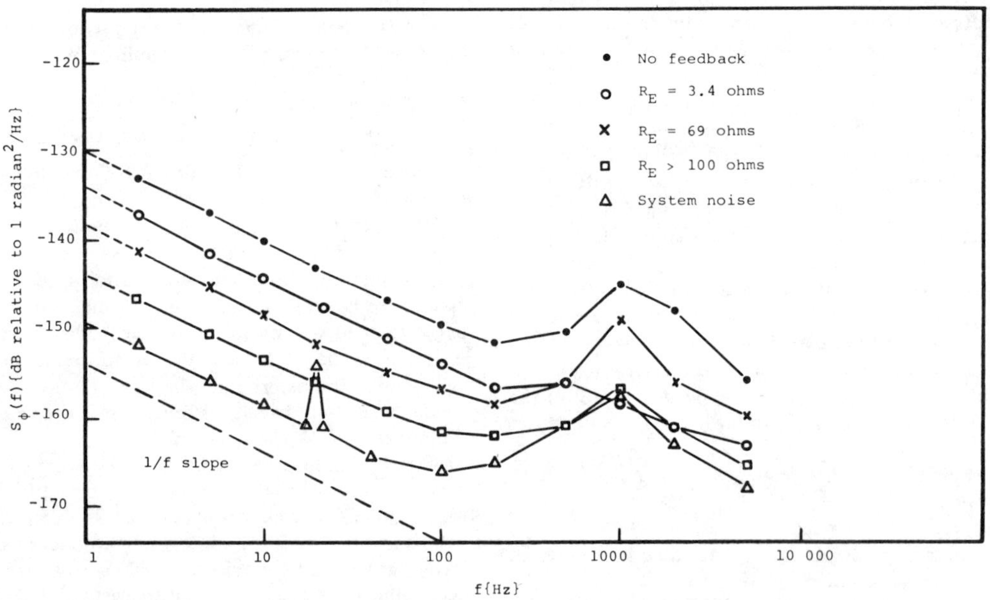

Fig. 3. Spectral density of phase fluctuations referred to input for each of pair of transistor frequency doublers, 5–10 MHz, using Philips BC109 transistors. Results are shown for varying amounts of emitter degeneration. Also shown is system noise level and $1/f$ slope line. V_p = 180 mV.

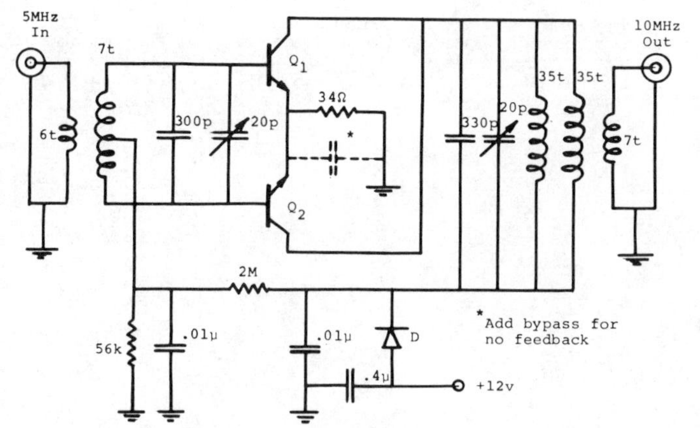

Fig. 4. Transistor frequency doubler circuit diagram. Coils are both toroidal air cores with OD = 0.9 in, ID = 0.6 in, and $h = 0.25$ in wound with no. 22 copper wire.

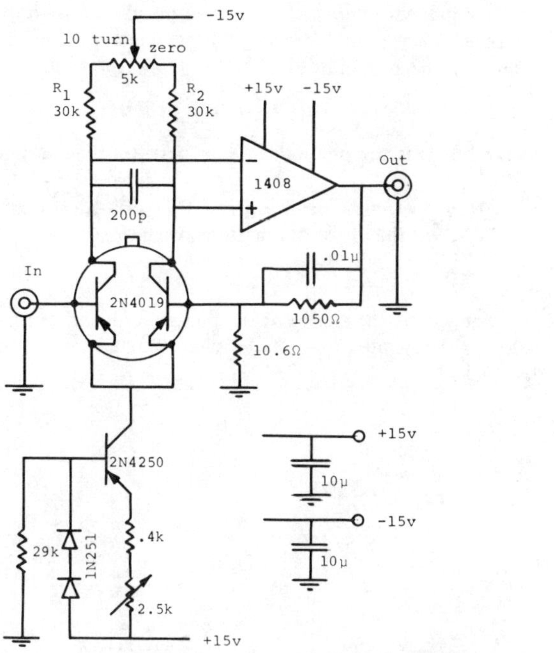

Fig. 5. Low-flicker-noise low-drift × 100 preamplifier circuit diagram. All resistors are metal-film type. Op-amp is Philbrick type 1408 with FET input stage. R_1 and R_2 are optimum for low noise. Collector current per input transistor is 220 μA.

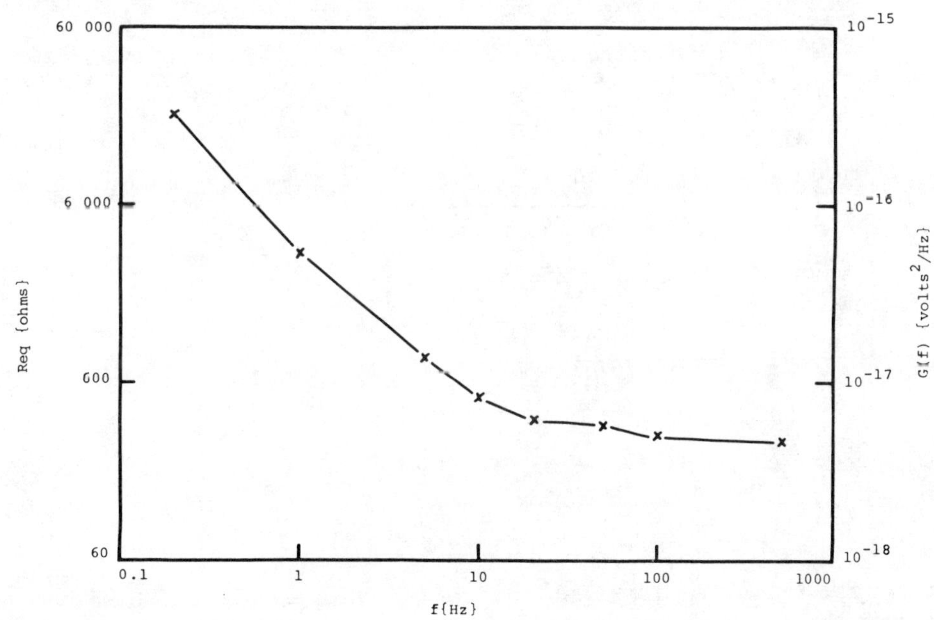

Fig. 6. Input noise power spectral density and equivalent input noise resistor for low-noise preamplifier with input shorted. Data points at 0.2 and 1 Hz were obtained by specially constructed wave analyzer.

used ceramic disc capacitor is a particularly poor choice, while silver-mica, glass, or mylar are generally good. Metal-film resistors appear to be superior to carbon resistors. Components sensitive to vibration should be avoided and the multiplier or amplifier should be of rigid construction.

Appendix

This appendix derives $S_\phi(f)$ in terms of the multiplication factor of the frequency multipliers and the parameters of the measuring system. Consider the system in Fig. 1. The output of the low-pass filter is

$$v_0(t) = V_p \cos \phi(t) \tag{1}$$

where $\phi(t)$ is the phase difference between the two input signals to the phase detector.

Setting $\phi(t) = 0$ yields $v_0(t) = V_p$, and for $\phi(t) = \pi/2 + \epsilon(t)$ with $\epsilon(t) << \pi/2$ the phase fluctuations, we have

$$v_0(t) \simeq -V_p \epsilon(t). \tag{2}$$

Thus $v_0(t)$ is the analog of the phase noise, and its rms value is proportional to the rms value of the phase fluctuations at the output of the multipliers. $v_0(t)$ is amplified by the preamplifier (of voltage gain A) and read by the wave analyzer as a rms voltage V in a noise equivalent bandwidth Δf centered at Fourier frequency f. By referring this reading back to the input of the frequency multipliers with the use of (2) and the multiplication factor n, and assuming that each multiplier contributes equally to the phase noise, $S_\phi(f)$ may be obtained:

$$S_\phi(f) = \frac{V^2}{2n^2 V_p^2 A^2 \Delta f} \text{ Hz}^{-1} \tag{3}$$

where $n = 2$ for a doubler and $n = 1$ for measuring the system noise.

References

[1] D. Halford, A. E. Wainwright, and J. A. Barnes, "Flicker noise of phase in RF amplifiers and frequency multipliers: characterization, cause, and cure," in *Proc. 22nd Annu. Symp. Frequency Control*, 1968, pp. 340–341.

[2] J. A. Barnes *et al.*, "Characterization of frequency stability," Nat. Bur. Stand., Boulder, Colo., Note 394, Oct. 1970; also *IEEE Trans. Instrum. Meas.*, vol. IM-20, pp. 105–120, May 1971.

[3] D. Halford and J. Shoaf, Nat. Bur. Stand., Boulder, Colo., private correspondence.

Frequency Multiplication Effects on Oscillator Instability

WILLIAM C. LINDSEY, FELLOW, IEEE, AND CHAK MING CHIE, MEMBER, IEEE

***Abstract*—The effects which frequency multiplication produces on the power spectral density of an oscillator is examined as a function of the multiplication ratio and the power spectral density of the phase noise process. In addition, the power spectral density (PSD) of the multiplied-up oscillation is interconnected with the frequency standard $\mathcal{L}(f)$. It is explicitly demonstrated how errors are introduced when one attempts to define the rms fractional frequency deviation $\Delta f(\tau)/f_0$ and the Allan variance $\sigma_y^2(\tau)$ using the measurement of $\mathcal{L}(f)$. Finally, the genesis of spectral spreading of the PSD due to frequency multiplication is demonstrated in such a way that this interesting and important phenomena can be grasped by the practicing engineer.**

I. Introduction

ALMOST ALL modern telecommunication, navigation, ranging, and radar systems require frequency control; hence, such systems involve frequency multiplication. When a signal, which possesses a small amount of phase noise, is passed through a frequency multiplier the power spectral density of the phase noise process is increased. The corresponding effect on the power spectral density (PSD) of the multiplied-up oscillation changes in a way which is highly dependent on the phase noise level prior to multiplication and the multiplication ratio. Frequently, the PSD of the multiplied-up oscillation is used as a means of obtaining the PSD of the phase noise process. Sometimes this approach is accurate yet there are other times when it is not. Moreover, direct measurement of the PSD of an RF oscillation near the carrier frequency (1 to 10 kHz) is usually not possible and the PSD in this region is of great interest in determining systems performance.

Manuscript received April 14, 1977. This work was supported by LinCom Corporation under IR and D funds.

W. C. Lindsey is with the University of Southern California, Los Angeles, CA 90007.

C. M. Chie is with LinCom Corporation, Pasadena, CA 91105.

In this paper we give the necessary theoretical background for assessing the effects of frequency multiplication on a noisy oscillator signal and interconnect the multiplied-up measurement with other direct phase noise measurements, how they differ and how they interface with current accepted frequency stability standards.

II. Mathematical Model

Let us consider the oscillation model

$$r(t) = \sqrt{2P} \cos (\omega_0 t + \psi(t) - \psi(0) + \Phi(0)) \qquad (1)$$

where $\Phi(0)$ is the initial phase offset and

$$\psi(t) \triangleq \int_0^t \dot{\psi}(\xi) d\xi + \psi(0) \qquad (2)$$

is the phase noise generated from integrating the zero-mean stationary process $\dot{\psi}(t)$. Equation (1) can be rewritten as

$$r(t) = \sqrt{2p} \cos (\omega_0 t + \tilde{\psi}(t) + \theta_0) \qquad (3)$$

where

$$\tilde{\psi}(t) = \int_{-\infty}^t \dot{\psi}(\xi) d\xi \qquad (4)$$

and

$$\theta_0 = \Phi(0) + \psi(-\infty) - \psi(0). \qquad (5)$$

Let us assume further that $\dot{\psi}(t)$ is Gaussian with a PSD $S_{\dot{\psi}}(\omega)$. Since $\tilde{\psi}(t)$ is a linear transformation of a stationary Gaussian process, it is also stationary, Gaussian with a PSD equal to $S_{\dot{\psi}}(\omega)/\omega^2$. If we make the usual assumption that θ_0 is uniformly distributed over $[0,2\pi]$, then $r(t)$ is wide-sense stationary zero-mean with a correlation function

Reprinted from *IEEE Trans. Instrum. Meas.*, vol. IM-27, pp. 26–28, Mar. 1978.

$$R_r(\tau) = P \cos \omega_0 \tau \exp\left[-D_{\dot\psi}^{(1)}(\tau)/2\right] \tag{6}$$

where

$$D_{\dot\psi}^{(1)}(\tau) \triangleq E\{[\psi(t+\tau) - \psi(t)]^2\}$$

is the first structure function of the process $\psi(t)$.[1] Now the PSD of oscillation $r(t)$ is given by

$$S_r(\omega) = P[S_{r_0}(\omega - \omega_0) + S_{r_0}(\omega + \omega_0)] \tag{7}$$

where (the symbol $*$ denotes convolution)

$$S_{r_0}(\omega) = \exp(-\sigma_{\dot\psi}^2)\left[\delta(\omega) + \frac{S_{\dot\psi}(\omega)}{\omega^2} + \frac{1}{2!}\frac{S_{\dot\psi}(\omega)}{\omega^2} * \frac{S_{\dot\psi}(\omega)}{\omega^2} + \cdots\right] \tag{8}$$

and

$$\sigma_{\dot\psi}^2 \triangleq \int_{-\infty}^{\infty} \frac{S_{\dot\psi}(\omega)}{\omega^2}\frac{d\omega}{2\pi} \tag{9}$$

is assumed to be bounded. Obviously, the delta function in (8) corresponds to the unmodulated carrier and the remaining terms are due to oscillator instability. In fact, in specifying the rms fractional frequency deviation, the Allan variance [2] or any other oscillator instability standard [3], only the term $S_{\dot\psi}(\omega)/\omega^2$ which exists in the sum on the right-hand side of (8) is required. The other terms present in any direct measurement at the RF level will introduce error in the frequency stability measure [1], [2], or [3] chosen. The amount of error introduced is dependent on the size of the remaining terms (excluding the carrier). We will return to this later in the section on frequency multiplication.

In practice, the approximation

$$S_{r_0}(\omega) \cong \exp(-\sigma_{\dot\psi}^2)\left[\delta(\omega) + \frac{S_{\dot\psi}(\omega)}{\omega^2}\right] \tag{10}$$

is often used [4]. This is a valid approximation as long as $\sigma_{\dot\psi}^2 \ll 1$. However, if $\sigma_{\dot\psi}^2$ is not small (for example, see our later discussion on frequency multiplication), the phase noise spectrum spreading due to higher order convolution terms in (8) cannot be neglected. To examine this phenomenon quantitatively, notice that if we integrate (8) over all frequencies and divide the result by 2π, we obtain

$$\sigma_{r0}^2 = \exp(-\sigma_{\dot\psi}^2)\left[1 + \sigma_{\dot\psi}^2 + \frac{\sigma_{\dot\psi}^4}{2!} + \cdots\right] = 1. \tag{11}$$

Hence the residual energy $\rho(\sigma_{\dot\psi})$ in the convolved phase noise spectrum is given by

$$\rho(\sigma_{\dot\psi}) = 1 - \exp(-\sigma_{\dot\psi}^2)(1 + \sigma_{\dot\psi}^2) \tag{12}$$

which is depicted in Fig. 1. Notice that for $\sigma_{\dot\psi}^2 \leq 0.5$, the portion of the energy due to *spectrum spreading* is at least 10 dB down. As an example, Fig. 2 shows a band-limited (high-frequency cutoff = f_h) white phase noise PSD and the resulting two convolution products. Fig. 3 shows the PSD $S_{r_0}(\omega)$ (for $\omega \neq 0$) obtained from (8) for various values of $\sigma_{\dot\psi}^2$ [5]. In this particular example, it is obvious that spectrum spreading is noticeable for $\sigma_{\dot\psi}^2 \geq 1$, which confirms our earlier argument. It was pointed out [6] that as $\sigma_{\dot\psi}^2 \to \infty$, $S_{r_0}(\omega)$ tends to be Gaussian, namely,

$$S_{r_0}(\omega) \sim \frac{\exp\{-\omega^2/2\sigma_{\dot\psi}^2\}}{\sigma_{\dot\psi}\sqrt{2\pi}} \tag{13}$$

as the case for $\sigma_{\dot\psi}^2 = 10$ in Fig. 3 shows.

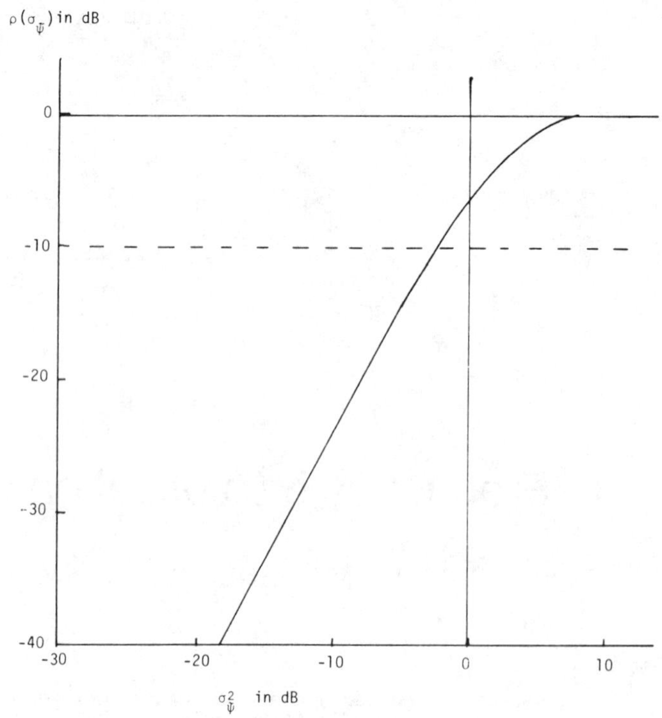

Fig. 1. Residual Energy $\rho(\sigma_{\dot\psi})$ versus $\sigma_{\dot\psi}^2$.

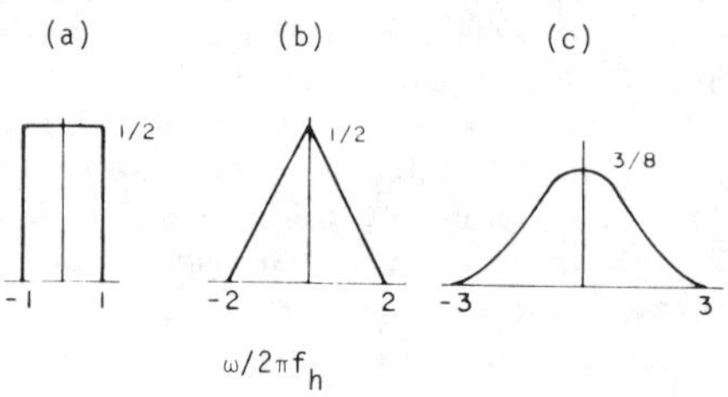

Fig. 2. Bandlimited white phase noise PSD and first two convolution products. (a) Normalized spectrum $S_{\dot\psi}(\omega)f_h/\omega^2\sigma_{\dot\psi}^2$. (b) First convolution product. (c) Second convolution product.

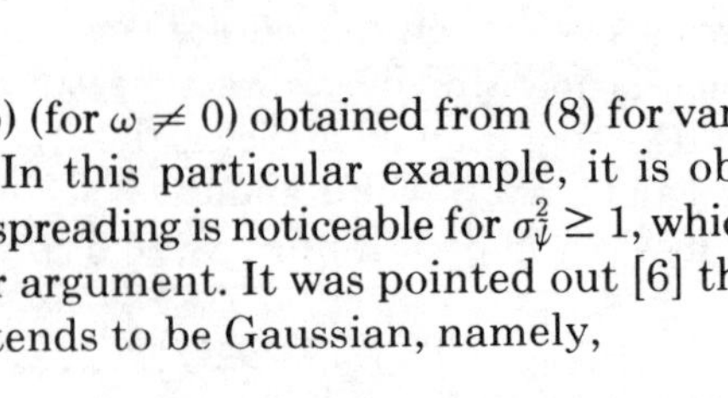

Fig. 3. PSD of $S_{r_0}(\omega)$ with suppressed spectral line.

[1] We have used the fact that $D_{\dot\psi}^{(1)}(\tau) = D_{\psi}^{(1)}(\tau)$.

III. Effect of Frequency Multiplication on RF Spectrum in the Presence of Oscillator Phase Noise

Frequency multiplication is actually a phase angle multiplication by a factor of, say, M. The PSD of the resultant oscillation $\hat{r}(t) = \sqrt{2P}\cos[M(\omega_0 t + \dot{\psi}(t) - \dot{\psi}(0) - \Phi(0)]$ is

$$S_{\hat{r}}(\omega) = P[S_{\hat{r}_0}(\omega - M\omega_0) + S_{\hat{r}_0}(\omega + M\omega_0)] \quad (14)$$

where

$$S_{\hat{r}_0}(\omega) = \exp(-M^2\sigma_{\dot{\psi}}^2)\left[\delta(\omega) + \frac{M^2 S_{\dot{\psi}}(\omega)}{\omega^2} + \frac{M^4}{2!}\frac{S_{\dot{\psi}}(\omega)}{\omega^2} * \frac{S_{\dot{\psi}}(\omega)}{\omega^2} + \cdots\right]. \quad (15)$$

If we take the first two terms in (15) for $S_{\hat{r}_0}(\omega)$ the residual energy $\hat{\rho}(\sigma_{\dot{\psi}})$ in the convolved phase noise spectrum is amplified to

$$\hat{\rho}(\sigma_{\dot{\psi}}) = 1 - \exp(-M^2\sigma_{\dot{\psi}}^2)(1 + M^2\sigma_{\dot{\psi}}^2). \quad (16)$$

For example, a multiplier gain of $M = 1000$ shifts the curve in Fig. 1 to the left by 60 dB. Hence the first-order approximation is no longer accurate and spectrum spreading (which tends to a Gaussian shape as discussed earlier for high gains) is always present.

IV. Relationship with the Standard $\mathcal{L}(f)$

The one-sided RF standard measure $\mathcal{L}(f)$, [4], is related to the two-sided PSD $S_{r_0}(\omega)$ via

$$\mathcal{L}(f) \triangleq \mathcal{L}(\omega)|_{\omega=2\pi f} = 4 \cdot S_{r_0}(\omega), \qquad \omega \neq 0 \quad (17)$$

when the phase noise can be considered as a narrow-band process modulating the carrier. Frequently, to obtain $\mathcal{L}(f)$, one measures the normalized RF PSD of the multiplied-up process

$$\mathcal{L}_M(\omega) = 4 \cdot S_{\hat{r}_0}(\omega), \qquad \omega \neq 0 \quad (18)$$

and approximates $\mathcal{L}(f)$ by $\mathcal{L}_M(f)/M^2$. Using $\mathcal{L}_M(f)/M^2$ can introduce an error in defining stability as we shall now demonstrate.

The rms fractional frequency deviation, $\Delta f(\tau)/f_0$, is related to $S_{\dot{\psi}}(\omega)/\omega^2$ via

$$\left(\frac{\Delta f(\tau)}{f_0}\right)^2 = 4\int_0^\infty \sin^2\left(\frac{\omega\tau}{2}\right)\frac{S_{\dot{\psi}}(\omega)}{\omega^2}\frac{d\omega}{2\pi} \quad (19)$$

while the Allan variance $\sigma_y^2(\tau)$ is related to $S_{\dot{\psi}}(\omega)/\omega^2$ via [3]

$$\sigma_y^2(\tau) = 16\int_0^\infty \sin^4\left(\frac{\omega\tau}{2}\right)\frac{S_{\dot{\psi}}(\omega)}{\omega^2}\frac{d\omega}{2\pi}. \quad (20)$$

In general, if the approximation $\mathcal{L}_M(\omega)/M^2$ is used, the magnitude of the error is dependent on M and the particular fluctuations in the phase error process, i.e.,

$$\frac{\mathcal{L}_M(\omega)}{M^2} - \frac{S_{\dot{\psi}}(\omega)}{\omega^2} = [\exp(-M^2\sigma_{\dot{\psi}}^2) - 1]\frac{S_{\dot{\psi}}(\omega)}{\omega^2} + \exp(-M^2\sigma_{\dot{\psi}}^2)\left[\frac{M^2}{2!}\frac{S_{\dot{\psi}}(\omega)}{\omega^2} * \frac{S_{\dot{\psi}}(\omega)}{\omega^2} + \frac{M^4}{3!}\frac{S_{\dot{\psi}}(\omega)}{\omega^2} * \frac{S_{\dot{\psi}}(\omega)}{\omega^2} * \frac{S_{\dot{\psi}}(\omega)}{\omega^2} + \cdots\right]. \quad (21)$$

When $M^2\sigma_{\dot{\psi}}^2 \ll 1$, then (21) simplifies to

$$\frac{\mathcal{L}_M(\omega)}{M^2} - \frac{S_{\dot{\psi}}(\omega)}{\omega^2} \cong \frac{M^2}{2!}\frac{S_{\dot{\psi}}(\omega)}{\omega^2} * \frac{S_{\dot{\psi}}(\omega)}{\omega^2} + \frac{M^4}{3!}\frac{S_{\dot{\psi}}(\omega)}{\omega^2} * \frac{S_{\dot{\psi}}(\omega)}{\omega^2} * \frac{S_{\dot{\psi}}(\omega)}{\omega^2} + \cdots. \quad (22)$$

Thus any attempt to measure $\mathcal{L}(f)$ directly and use this as $S_{\dot{\psi}}(\omega)/\omega^2$ in (19) or (20) to evaluate the τ-domain stability measures gives rise to an error!

References

[1] L. S. Cutler and C. L. Searle, "Some aspects of the theory and measurement of frequency fluctuations in frequency standards," *Proc. IEEE*, vol. 54, pp. 136–154, Feb. 1966.

[2] J. A. Barnes *et al.*, "Characterization of frequency stability," *IEEE Trans. Instrum. Meas.*, vol. IM-20, pp. 105–120, May 1971.

[3] W. C. Lindsey and C. M. Chie, "Theory of oscillator instability based upon structure functions," *Proc. IEEE*, vol. 64, pp. 1652–1666, Dec. 1976.

[4] B. E. Blair, Ed., *Time and Frequency: Theory and Fundamentals* (NBS Monograph 140). Washington, DC: U.S. Government Printing Office, 1974, ch. 8.

[5] N. Abramson, "Bandwidth and spectra of phase- and frequency-modulated waves," *IEEE Trans. Commun. Systems*, vol. CS-11, pp. 407–414, Dec. 1963.

[6] A. J. Viterbi, *Principles of Coherent Communication.* New York: McGraw-Hill, 1966, pp. 169–174.

Spectral Analysis of Synthesized Signals in the mm Wavelength Region

ELIO BAVA, ANDREA DE MARCHI, AND ALDO GODONE

***Abstract*—In this paper some experimental results on the spectral purity along a multiplier chain from 5 MHz up to 600 GHz are reported. The measured points are compared to the numerical data which can be obtained from simple calculations considering the phase noise spectrum of the crystal oscillators used to drive the multiplier chains.**

I. INTRODUCTION

THE GENERATION of electromagnetic signals with high stability and good spectral properties in the submillimeter and far-infrared regions has been a field of impressive development in this last decade aiming at frequency synthesis and measurements for spectroscopic and metrological purposes [1]–[3]. As a matter of fact, new beam frequency standards in the submillimeter wave region have been proposed [4]. The most important characteristics, concerning the use of a FIR signal obtained by multiplication, are its power and its spectral purity. The first must be strong enough either to excite transitions suitable for frequency standards or to allow frequency measurement of sources with high signal-to-noise ratio (SNR). The efficiency of point contact, Schottky barrier, metal-insulator-metal diodes and Josephson junctions have an important role in coping with this problem. The second depends directly on the spectral purity of the driving source (usually a quartz oscillator), on the phase stability of the first stages of the multiplier chain, and on its AM/PM conversion if the oscillator exhibits an appreciable amplitude modulation. (Theoretical study of the fast linewidth broadening by the multiplication process was introduced by Halford [5].)

A rigorous analytical treatment of the phase-noise increment in the harmonic generation, even if possible, requires the aid of a computer when the power in the sidebands becomes comparable or higher than the power in the carrier. However, it seems interesting to foresee with simple calculations the RF spectrum of a signal multiplied from a quartz oscillator. For this purpose a simplified theory was presented in [6] and supported by a simulation experiment.

Following the outlines of this theory, the effective phase-noise spectrum at the input of a multiplier, is given by the formula:

$$S_\phi^{\text{eff}}(\nu_0,f) = S_\phi^{\text{sig}}(\nu_0,f)G^2(f) + S_\alpha^{\text{sig}}(\nu_0,f)K^2(f) + S_\phi^{\text{ch}}(\nu_0,f)$$

Manuscript received October 6, 1976. This research was performed with support of the Consiglio Nazionale delle Ricerche, Roma, Italy.

The authors are with the Istituto Elettrotecnico Nazionale Galileo Ferraris, Torino, Italy.

where $S_\phi^{\text{sig}}(\nu_0,f)$ and $S_\alpha^{\text{sig}}(\nu_0,f)$ are respectively the phase and amplitude noise spectra of the signal, fed to the multiplier frequency ν_0, as a function of the Fourier frequency f; $G^2(f)$ describes the filtering of the chain for high values of f; $K^2(f)$ is the AM/PM conversion factor for the chain [7]; $S_\phi^{\text{ch}}(\nu_0,f)$ is the phase instability spectrum of the chain. The quantities K and S_ϕ^{ch} are to be referred to the frequency ν_0.

In this paper, experimental results on the spectral purity at different multiplied frequencies are compared to the numerical data obtained from the simplified theory corrected for the roll-off law of the phase-noise pedestal of the effective input spectrum. No measurement was made on the AM of the sources used, in the assumption that it were so small as to be unimportant. Disagreements between the phase noise spectra reported to the input from different multiplied frequencies may well be imputed to AM/PM conversion in the chain if they occur inside the multiplier bandwidth.

The comparison is carried out also in the frequency region where the spectral purity is largely spoiled, in order to analyze the increase of the pedestal linewidth.

II. EXPERIMENTAL SETUP

The block diagram of the multiplier chain built for frequency measurements in the millimeter and submillimeter wavelength region is shown in Fig. 1. The values of frequencies along the chain were chosen for measurements close to the $^3P_0 \leftrightarrow {}^3P_1$ Mg transition. The master oscillator, a 5 MHz quartz, is followed by a few stages of symmetric transistor doublers and triplers up to 120 MHz. Then a step-recovery diode yields a microwave signal, at 91.2 GHz, which stabilizes, by injection locking, a Gunn oscillator. A millimetric klystron at 100.30 GHz (VRB-2111A) is phase-locked to the eleventh harmonic of the X band signal via a harmonic mixer and a synchronizer. The klystron output drives a frequency multiplier equipped with an Au-GaAs Schottky barrier diode. The different harmonics of the millimetric source, selected by a series of below-cutoff filters are sent to a Golay detector. The relative power levels of each harmonic are reported in Fig. 2. At the sixth harmonic (near the Mg resonance) a power of about 100 nW was detected.

For the spectral purity measurements discussed in next section, two equal multiplier chains were built. Both multiplier and harmonic mixer driven by the 100-GHz stabilized sources are equipped with Schottky barrier diodes which, according to the literature [8], [9] and our

Reprinted from *IEEE Trans. Instrum. Meas.*, vol. IM-26, pp. 128–132, June 1977.

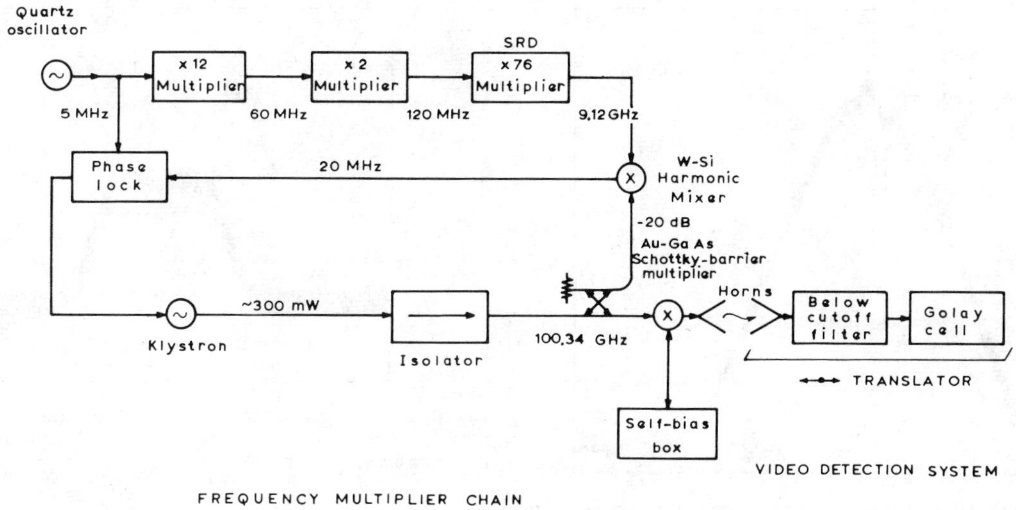

Fig. 1. Block diagram of the multiplier chain and experimental setup for power measurements and wavelength check of each harmonic of the klystron.

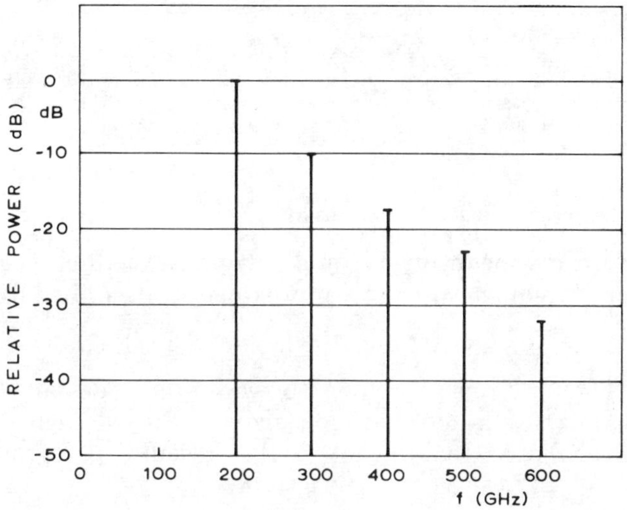

Fig. 2. Power levels of the klystron harmonics, from the 2nd up to the 6th, relative to the level at 200 GHz, as detected by a Golay cell.

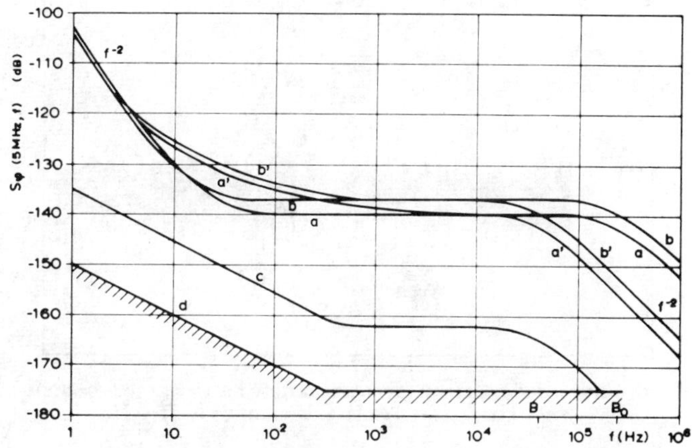

Fig. 3. Spectral densities of phase fluctuations. *a*) and *b*) 5-MHz crystal controlled oscillators. *a'*) and *b'*) Outputs of the chains at 9.12 GHz referred to the input. *c*) 60-MHz noise of the chains referred to the input. *d*) Phase noise sensitivity of the measurement circuit.

previous experimental results, appear the most convenient devices for this purpose.

III. Results

The purpose of this section is to compare the results of the spectral analysis carried out on the multiplied signal at different frequencies along the chain with the simplified theory which has been exposed in [6]. Although few experimental points could be taken, their agreement with the theory is satisfactory both for the relative height of the carrier compared to the noise pedestal and for the 3 dB linewidth of the pedestal itself.

A. Spectral Analysis of the Signals Along the Chain

The phase noise spectral density $S_{\phi 1,2}(\nu_0,f)$ of the two quartz crystal oscillators used as sources for the two chains was measured at $\nu_0 = 5$ MHz by beating directly their output signals in a double balanced mixer and spectrum analyzing the zero frequency beat at the Fourier frequency f with 1-Hz bandwidth.

Curves a and b in Fig. 3 show $S_\phi{}^{\text{sig}}$ (5 MHz, f) respectively for oscillator 1 and oscillator 2; since the two white noise levels were found different by analyzing the multiplied signal directly at X band, these two curves have been corrected for this difference. Furthermore, the white phase noise is shaped as a Lorentzian with a cutoff frequency B_0 of 250 kHz.

Curves a' and b' on the same graph are the plot of $S_{\phi 1,2}$ (5 MHz, f) of the two oscillators (reported to the input) as they were measured at X band by beating the two signals in a low noise double balanced mixer, the above mentioned correction being introduced. These are the affective phase noise spectra at the input of the two chains.

In the same figure curve c is $S_\phi{}^{\text{ch}}$ (5 MHz, f) as it was measured at 60 MHz by driving the first section of the two chains with the same source. It is shown that the multiplier does not increase the phase noise of the multiplied signal more that the theoretical n^2 factor, moreover the $G^2(f)$

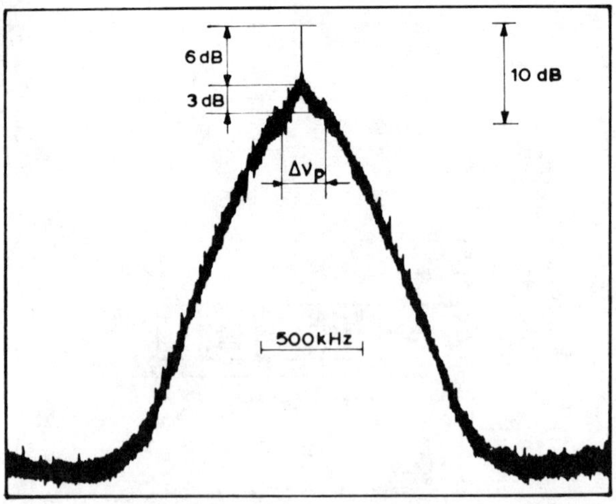

Fig. 4. Recorded spectrum of the beat note between the two chains at 200 GHz: f_{beat} = 42 MHz, IF bandwidth 1 kHz.

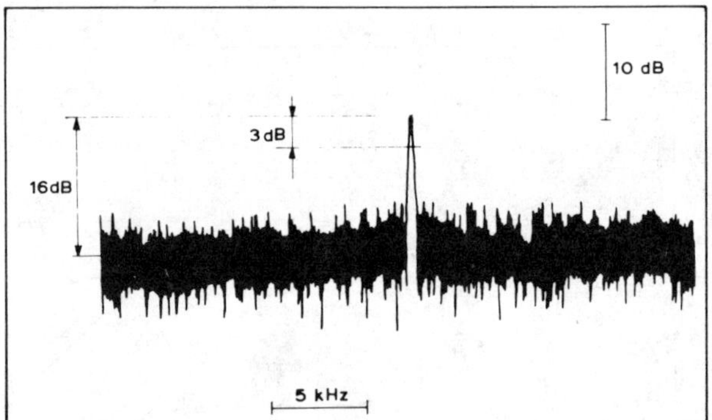

Fig. 5. Recorded spectrum of the beat note between the two chains at 200 GHz: f_{beat} = 42 MHz, IF bandwidth 100 Hz.

factor has a high frequency cutoff B of about 42 kHz and rolls off as f^{-2} for higher Fourier frequencies, so that it shapes in this way the effective input spectrum to be used in calculations.

The RF spectrum of the signal was then recorded at 5 MHz and 9.12 GHz by direct spectral analysis: no noise pedestal was expected to be visible at 5 MHz, while a carrier-to-pedestal ratio of respectively 48 and 45 dB with 1-kHz bandwidth was measured at X band for the two oscillators. In both cases the 3-dB linewidth of the pedestal turned out to be 84 kHz.

A beat note of 42 MHz between the two chains at 100 and 200 GHz was recorded. The spectrum of the 200 GHz beat note is displayed in Figs. 4 and 5 with different scan widths and analyzer bandwidths. A carrier-to-pedestal ratio of 6 dB (1 kHz bandwidth) and 16 dB (100-Hz bandwidth) can be seen. The 3 dB pedestal linewidth is 220 kHz.

At 600 GHz the beat note between the multiplied signal of chain 1 and a CH_3F optically pumped laser was analyzed. The spectrum is shown in Fig. 6 with 30-kHz bandwidth. No carrier was observed either in this case or with a 100-Hz bandwidth. The 3-dB pedestal linewidth is estimated to be 650 kHz.

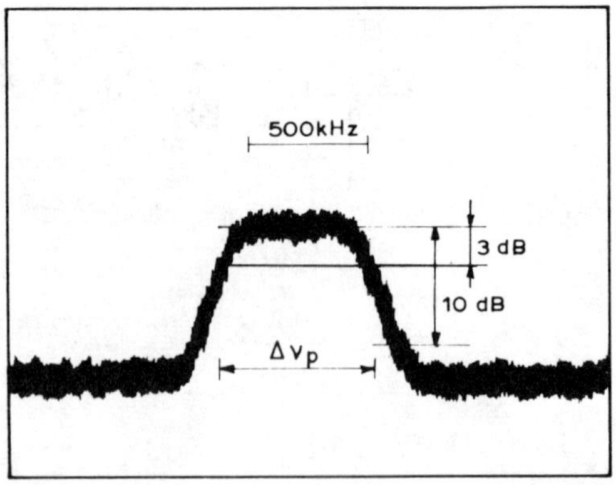

Fig. 6. Beat note between a CH_3F laser and the sixth harmonic of phase-locked klystron at 100.34 GHz. The pattern is obtained from a photograph on the spectrum analyzer.

TABLE I
Values of Total Phase Noise in the Pedestal Φ^p at Different Multiplied Frequencies.

	5 MHz	9.12 GHz	100 GHz	200 GHz	600 GHz
n	1	1824	2×10^4	4×10^4	1.2×10^5
P_1 (rad^2)	6.6×10^{-10}	2.2×10^{-3}	2.5×10^{-1}	1	9
P_2 (rad^2)	1.3×10^{-9}	4.4×10^{-3}	$5 \times 10^{-}$	2	19

B. Agreement with the Theory

The RF spectrum of a randomly phase modulated carrier at frequency ν can be approximately described by

$$S_{RF}(\nu,f) = e^{-\Phi^p(\nu)}[\delta(f) + S_\phi(\nu,f)]. \quad (1)$$

The total rms phase noise power in the pedestal $\Phi^p{}_{1,2}$ (5 MHz) for the effective input signal of the two chains can be calculated from the data of the preceding paragraph.

$$\Phi^p{}_{1,2}(5\text{ MHz}) = \int_0^\infty \frac{S^w{}_{\phi 1,2}(5\text{ MHz})}{1 + (f/B)^2}\, df = \frac{\pi}{2} B S^w{}_{\phi 1,2}\,(5\text{ MHz}) \quad (2)$$

where $S^w{}_{\phi 1,2}$ (5 MHz) is the white phase noise level at 5 MHz for the two effective input signals.

The validity of the approximations used here is shown by the agreement with the results reported hereafter. Being

$$\Phi^p(n\nu) = n^2 \Phi^p(\nu) \quad (3)$$

provided the pedestal maintains the same roll off shape along the chain, Table I can be calculated, which resumes the values of Φ^p at different points of the two chains with multiplier bandwidth B = 42 kHz.

For a beat note between two signals it can be easily shown that the RF spectrum is described, in the same approximation used in [6], by

$$S^{beat}{}_{RF}(\nu_{beat},f) = e^{-(\Phi^p{}_1 + \Phi^p{}_2)} \cdot [\delta(f) + S_{\phi 1}(f) + S_{\phi 2}(f)]. \quad (4)$$

As a consequence the beat note of the two 200-GHz signals must be interpreted correctly as having a total ped-

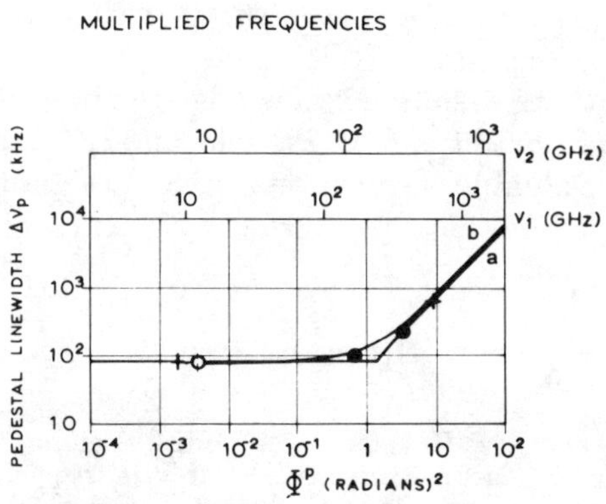

Fig. 7. 3 dB linewidth of the noise pedestal as a function of Φ^p. Curves a and b are calculated from equations (6) and (8), respectively. The crosses are points relative to chain 1 and the circles are points relative to chain 2; the circles with the cross are relative to a beat note between the two chains.

estal phase noise $\Phi^p{}_b = \Phi^p{}_1(200\text{ GHz}) + \Phi^p{}_2\,(200\text{ GHz}) = 1 + 2 = 3\text{ rad}^2$ and a phase noise spectrum $S_{\phi b}(f) = S_{\phi 1}(f) + S_{\phi 2}(f)$. For the beat note at 600 GHz on the contrary, although little experimental data is available up to now on the spectral purity of an optically pumped FIR organic laser, it will be assumed that both S_ϕ and Φ^p be very small for the laser in comparison with those of the multiplied signal; that is, the analyzed beat note is supposed to reproduce the spectrum of the synthesized 600 GHz.

In Figs. 7 and 8 the experimental results are compared with this simplified theory.

In Fig. 7 two curves are plotted versus ϕ^p which show the results of two different approximations for the linewidth of a Lorentzian-shaped pedestal. Curve a is the complete solution for the formula proposed in [6]

$$\int_{\Delta\nu_p/2}^{\infty} S^p{}_\phi(\nu,f)\,df = \ln 2 \tag{5}$$

which gives, assuming that $S^p{}_\phi(f)$ is a Lorentzian

$$\Delta\nu_p = 2B\cot\left(\frac{\pi}{2}\,\frac{\ln 2}{\Phi^p}\right). \tag{6}$$

Curve b is obtained by simply imposing that $S^p{}_\phi$ and $S^p{}_{\mathrm{RF}}$ have the same shape and that they are asymptotically equal. If their common shape is of the form $[1 + (f/(\Delta\nu/2))^\alpha]^{-1}$, then we have

$$\Delta\nu_p = 2B\left(\frac{\Phi^p}{1-e^{-\Phi^p}}\right)^{1/\alpha-1} \tag{7}$$

which gives, for a Lorentzian ($\alpha = 2$)

$$\Delta\nu_p = 2B\,\frac{\Phi^p}{1-e^{-\Phi^p}}. \tag{8}$$

In Fig. 8 two curves are drawn representing, as a function of Φ^p, the total power in the carrier: $P_c = e^{-\phi^p}$ (curve c), and the noise power per Hz in the white section of the pedestal of the RF signal in a double band representation: $h_p = S^w{}_{\mathrm{RF}}/2$ (curve p). Both curves are normalized with respect to the total signal power.

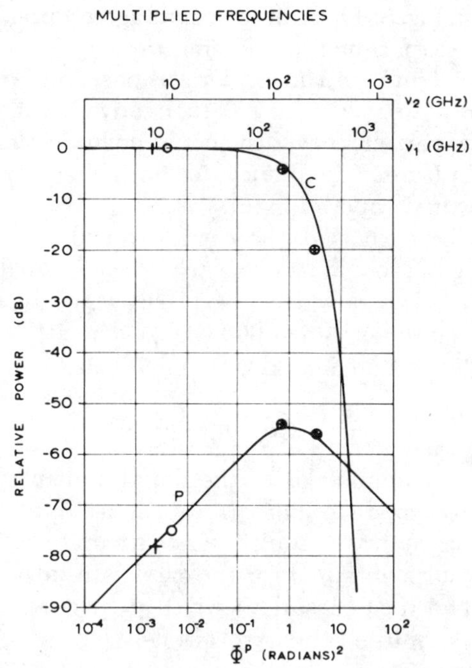

Fig. 8. Relative levels of carrier power (curve c) and pedestal power per Hz (curve p) as a function of Φ^p. Crosses, circles, and circles with the cross have the same meaning as in Fig. 8.

For a Lorentzian h_p can be calculated from (7)

$$h_p = \frac{2P_p}{\pi\Delta\nu_p} = \frac{P^2{}_p}{\pi B\Phi^p} = \begin{cases} \dfrac{\Phi^p}{\pi B}, & \Phi^p \ll 1 \\[2ex] \dfrac{1}{\pi B\Phi^p}, & \Phi^p \gg 1. \end{cases} \tag{9}$$

Two scales are added, on the top of Figs. 7 and 8, representing the frequencies ν_1 and ν_2, multiplied from the two oscillators, that yield the corresponding Φ^p. Being $S^w{}_\phi\,(\nu_0)$ the white noise level at frequency ν_0, from (2) and (3) we obtain

$$\nu = n\nu_0 = \frac{\nu_0}{\left[\dfrac{\pi}{2}BS^w{}_\phi(\nu_0)\right]^{1/2}}[\Phi^p(\nu)]^{1/2}$$

$$= \nu^{\mathrm{col}}[\Phi^p(\nu)]^{1/2} \tag{10}$$

where

$$\nu^{\mathrm{col}} \equiv \nu_0[\Phi^p(\nu_0)]^{-1/2} \tag{11}$$

is the carrier collapse frequency of the synthesizer (complete of driving oscillator); this means that, at the frequency ν^{col}, the fractional power of the carrier has decreased to e^{-1}.

For the two chains we have

$$\nu_1{}^{\mathrm{col}} = 200\text{ GHz};\quad \nu_2{}^{\mathrm{col}} = 140\text{ GHz}. \tag{12}$$

The measured pedestal linewidth for the X band signals of the two chains, for the beat note between them at 100 and 200 GHz, and for chain 1 at 600 GHz are compared with the theoretical curves in Fig. 7. Also, the measured carrier to pedestal ratio at X band at 100 and 200 GHz are reported in Fig. 8. Thus, all the points agree with the theory within the experimental uncertainties.

No broadening of the carrier was observed in the 200-GHz beat note by using a 100-Hz bandwidth; as a matter of fact the carrier linewidth should be lower than 10 Hz, at that multiplied frequency, for both chains, according to the theory proposed in [6].

It can be seen that the carrier-to-pedestal ratio for chains 1 and 2 should be respectively around 20 dB and 15 dB with 1-kHz bandwidth at 200 GHz, while no carrier can be expected to be seen at 600 GHz, even with the best of our oscillators and a bandwidth of 100 Hz.

IV. Conclusions

The spectral analysis of the signal multiplied from 5 MHz quartz crystal oscillators was carried out at different frequencies along the multiplier chain up to 600 GHz.

The results obtained for the carrier-to-noise-pedestal ratio and for the pedestal linewidth are in good agreement with the simplified theory previously proposed for their calculation.

Unfortunately the oscillators we could employ, although they were the best available at the time of the purchase, do not provide a carrier-to-pedestal ratio suitable for application to the excitation and a precise frequency measurement of atomic transitions around 600 GHz, which is our immediate goal. For this and other purposes in the same range of frequency, such as the phase lock of molecular lasers in the THz region, further improvement of the spectral purity of the sources and (possibly) of the multiplier chains will be required. The use of a superconducting cavity as a filter at X band or at 100 GHz may be considered, if necessary, in the future.

Acknowledgment

The authors wish to acknowledge the helpful cooperation of H. R. Fetterman, B. Fiscella, and L. Canarelli along with the valuable suggestions of S. Leschiutta and F. Strumia.

References

[1] D. A. Jennings, F. R. Petersen, and K. M. Evenson "Extension of absolute frequency measurements to 148 THz: frequencies of the 2.0 and 3.5 μm Xe laser," *Appl. Phys. Lett.*, vol. 26, no. 9, May 1975.

[2] K. M. Evenson, J. S. Wells, F. R. Petersen, B. L. Danielson, and D. W. Day, "Accurate frequencies of molecular transitions used in laser stabilization: the 3.39 μm transition in CH_4 and the 9.33 and 10.18 μm transitions in CO_2," *Appl. Phys. Lett.*, vol. 22, no. 4, Feb. 1973.

[3] F. R. Petersen, K. M. Evenson, D. A. Jennings, J. S. Wells, K. Goto, and J. J. Jimenez, "Far infrared frequency synthesis with stabilized CO_2 lasers: accurate measurements of the water vapor and methyl alcohol laser frequencies," *IEEE J. Quantum Electron.*, vol. QE-11, no. 10, Oct. 1975.

[4] F. Strumia, "A proposal for a new absolute frequency standard using a Mg or Ca atomic beam," *Metrologia*, no. 8, 1972.

[5] D. Halford, "Infrared-microwave frequency synthesis design: some relevant conceptual noise aspects," presented at the Proc. Frequency Standards and Metrology Seminar, 1971.

[6] F. L. Walls and A. De Marchi, "RF spectrum of a signal after frequency multiplication; measurement and comparison with a simple calculation," *IEEE Trans. Instrum. Meas.*, vol. IM-24, Sept. 1975.

[7] E. Bava, G. P. Bava, A. De Marchi, and A. Godone, "Measurements of Static AM-PM conversion in frequency multipliers," *IEEE Trans. Instrum. Meas.*, vol. IM-26, pp. 33–38, Mar. 1977.

[8] H. R. Fetterman, B. J. Clifton, P. E. Tannenwald, C. D. Parker, and H. Penfield: "Submillimeter heterodyne detection and harmonic mixing using Schottky diodes," *IEEE Trans. Microwave Theory Tech.*, vol. MTT-22, no. 12, Dec. 1974.

[9] H. R. Fetterman, B. J. Clifton, P. E. Tannenwald, C. D. Parker, "Submillimeter detection and mixing using Schottky diodes," *Appl. Phys. Lett.*, vol. 24, no. 2, Jan. 1974.

FLICKER OF FREQUENCY AND PHASE AND WHITE FREQUENCY AND PHASE FLUCTUATIONS IN FREQUENCY SOURCES

D.J. Healey III

Westinghouse Electric Corp. Baltimore, Md.

Introduction

The frequency stability of a signal generator can be characterized by the single-sided spectral density of y(t), which is the fractional frequency offset of V(t), the signal generator output voltage from the nominal frequency of the signal generator output. Thus, letting

$$V(t) = \left[V_o + \epsilon(t)\right] \sin\left[2\pi\nu_o + \phi(t)\right] \tag{1}$$

$$y(t) = \dot{\phi}(t)/2\pi\nu_o \tag{2}$$

$$S_y(f) = \left(\frac{1}{2\pi\nu_o}\right)^2 S_{\dot{\phi}}(f) \tag{3}$$

$$S_y(f) = \left(\frac{1}{\nu_o}\right)^2 f^2 S_\phi(f) \tag{4}$$

Although $S_y(f)$ is a fundamental definition for the frequency stability of a frequency source and is the definition recommended by the IEEE Subcommittee on Frequency Stability of the Technical Committee on Frequency and Time, it is somewhat more convenient to describe frequency sources used for radar and communications applications by $S_V(f)$, the RF power spectrum or by $S_{V\phi}(f)$, the phase portion of the RF power spectrum.

The frequency sources with which this paper is concerned will be confined to harmonic oscillators that employ resonators utilizing quartz crystal units or LC circuits, and description of the frequency stability will be given by the single-sided $S_{V\phi}(f)$ which we will simply denote in this paper as $S_\phi(f)$ expressed in units $10 \log_{10} S_\phi(f)/2$ (dB/Hz).

The Harmonic Oscillator

A harmonic oscillator comprises an amplifying device, a limiter, and a frequency selective network arranged in a positive feedback configuration. This paper will limit itself to simple single-pole resonators for the frequency selective network that is the principal component determining the frequency of oscillation.

Figure 1 shows the arrangement of a harmonic oscillator.

In Figure 1(a), A is a linear gain function, and β is the frequency selective network. If we simply let the β network consist of a series resonant circuit and terminate the input of the amplifier and output of the limiter with capacitors (input impedance of amplifier and output impedance of limiter being very large), the harmonic oscillator that results is the Colpitts oscillator utilizing the Clapp resonator or more commonly the Clapp oscillator when the series resonant circuit consists of an inductor and a capacitance or the Pierce oscillator when the series resonant circuit is provided by a quartz crystal unit (with or without simple reactive impedance transformer).

A more convenient method of examining the behavior of the harmonic oscillator is by means of the circuit shown in Figure 1b. This corresponds to disconnecting the series resonant circuit from terminals 1 and 2 of Figure 1(a) and considering the impedance exhibited between terminals 1' and 2'.

It can be seen that, when 1' and 2' are terminated by capacitors, the impedance exhibits a negative resistance in series with a capacitive reactance. If the small signal transconductance (g_m) results in a negative resistance having magnitude greater than the series resistance of the series resonant circuit (which must include any loading of the resonant circuit or Z_1 or Z_2 to extract useful power from the harmonic oscillation), harmonic oscillation will occur, and the oscillation amplitude will build up to a level at which the g_m becomes exactly that value necessary to cause the negative resistance of $Z_{1'2'}$ to be exactly the magnitude of the positive resistance of the oscillating resonator. The oscillating resonator includes the impedance $Z_1 + Z_2$ as part of the resonator so that oscillation is slightly above the resonant frequency of the series resonant β network when Z_1 and Z_2 are capacitive reactances.

The reduction of g_m necessary to produce sustained oscillation occurs due to the limiter. The limiter may be a symmetrical hot carrier diode limiter, regulation of A by means of an automatic level control (ALC) circuit that adjusts the negative feedback of A by comparing the output voltage to a fixed dc reference and using the difference as the signal to control the negative feedback, or simply due to the nonlinear current versus voltage characteristic of the amplifier. The highest short-term frequency stability is obtained using the first two methods for providing the limiting function provided that extreme care is taken in the detailed design of the components. In many applications, however, the last method is adequate and this will be the principal type of circuit considered. The circuit that produces the amplifier-limiter function is called the maintaining circuit.

The current versus voltage characteristic which yields g_m can be described by a power series

$$i = a_o + a_1 V + a_2 V^2 + a_3 V^3 + \dots\dots \tag{5}$$

If we let V be a single large sinusoidal voltage as will be the case in frequency source designs.

$$i = [a_1 V + 3/4 a_3 V^3 + 5/8 a_5 V^5 + \dots] \sin\omega_o t \tag{6}$$

$$+ K \sin 2\omega_o t + \dots\dots\dots$$

Reprinted with permission from *Proc. 26th Annu. Symp. Frequency Contr.*, Atlantic City, NJ, June 6–8 1972, pp. 29–42.

The fundamental frequency component is the most important in determining frequency stability. (Harmonics, however, are also of importance, but will not be considered in this discussion[1]).

When a_3, a_5, etc. terms of the power series expansion of i in terms of V are negative coefficients, the required reduction of g_m with increase of the oscillation amplitude is obtained.

The solution for the oscillation current is difficult because it involves a nonlinear differential equation. The only case that is conveniently solved in closed form is the case where the transconductance can be described by

$$i = a_1 V + b_1 V^3 \tag{7}$$

The solution of the harmonic oscillator using this function was first made by B. van der Pol[2].

When the maintaining circuit is provided by a self-limiting transistor, however, we need to consider

$$i = I_{co}\left[\exp\left(\frac{qV}{\{1+m\}kT}\right) - 1\right] \tag{8}$$

for the current voltage relationship. Here, V is the base-emitter voltage exceeding the contact potential ϕ of approximately 0.6 volt, q is the electronic charge, k is Boltzmann's constant, T is the junction temperature in degrees Kelvin, and m is a constant-typically between 0.4 and 0.45 for silicon. When eq (8) is expanded in a power series involving typical values for V that are encountered in oscillator circuits, the series diverges so that the required limiting function does not occur unless the negative portion of the waveform V swings below V=0, and a dc bias shift occurs. Since forward biasing is necessary to ensure starting of oscillation, the negative swing will exceed 0.6 volt before the current i becomes 0. The effective transconductance is then obtained by Fourier analysis of the current produced by the transistor when the base-emitter is driven below ϕ during a portion of the cycle.

Short-Term Frequency Stability of the Harmonic Oscillator

The frequency stability of a harmonic oscillator depends on the following.

a. The stability of the basic oscillating resonator as a function of time, pressure, humidity, temperature, oscillation amplitude.

b. The Q of the oscillating resonator.

c. The perturbation of the maintaining circuit transconductance by noise sources in the maintaining circuit and resonator, and the resultant perturbation of maintaining circuit reactance and perturbation of the phase of the signal current.

We will only consider b. and c. in this paper.

To consider c., the equivalent signal plus noise voltage appearing across the terminals of the resonator is examined. The oscillator signal voltage and the noise voltage can be represented by the phasors shown in Figure 2.

Although nonlinear operation is involved the total noise resulting from various intermodulation terms involving V_o and its harmonics is still very small relative to V_o so that the noise can be considered the superposition of a number of individual frequency components whose amplitude is determined by considering the effects of intermodulation when the wide-band noise is added to the signal V_o. Noise frequency components separated from the frequency of V_o by typically 1 to 10^6 Hz are usually of concern. The effect of a single noise component, (examined through a 1-Hz bandwidth filter and separated from the frequency of V_o which we will denote f_o by an amount f which we denote as the Fourier frequency), on the frequency of oscillation can be determined by consideration of Adler's theory on the injection locking of oscillators[3].

It can be shown that Adler's equations are also obtained from analysis of a simple first order phase lock loop in which the open loop-transfer function is

$$K/_s = \frac{1}{s}\left(\frac{\omega_o}{2Q}\right)\left(\frac{E_1}{E_o}\right) \tag{9}$$

where E_1 is the injection voltage and E_o is the oscillation voltage. Letting E_o be V_o and letting E_1 be a noise voltage frequency component $V_n \sin 2\pi ft$ it can be shown that the phase perturbation which is simply $\Delta\phi = \frac{V_n}{V_o}$ $(V_r/V_o << 1)$ will result in a frequency change given by

$$\Delta f = \Delta\phi\left(\frac{f_o}{2Q}\right) \tag{10}$$

If we simply consider the "white" noise sources in the maintaining circuit, $\Delta\phi(f)$ will exhibit a "white" spectral density over the Fourier frequency range of concern even when the effects of intermodulation between the wide-band noise and various harmonics of f_o resulting from the non-linearity of i versus V is considered.

We would then expect that $S_\phi(f)$ of a harmonic oscillator would simply have a $1/f^2$ power spectral density. However, the observed power spectral density for $S_\phi(f)$ has typical shapes as shown in Figure 3.

Halford first reported his findings that the characteristic of Figure 3a result from a fundamental limitation of semiconducting devices[4].

Using Halfords definition for the spectral density of $\mathcal{L}(f)$, the observed $S_\phi(f)$ is given simply by

$$S_\phi(f) = \mathcal{L}(f)\left[1 + \left(\frac{f_o}{2Qf}\right)^2\right] \tag{11}$$

where $\mathcal{L}(f)$ is a power spectral density of the form

$$\mathcal{L}(f) = C\left[1 + Af^{-\alpha}\right] \tag{12}$$

with $\alpha \simeq 1$. This type of spectral density for $\mathcal{L}(f)$ is observed when any effects on $S_\phi(f)$ due to large drive levels affecting the properties of the resonator are removed, and burst noise is not present. The constants C and A depend on the characteristics of the amplifier. We will examine the device and circuitry to determine what controls $\mathcal{L}(f)$ and how circuit and device designs can be modified to reduce it. Since the harmonic oscillator maintaining can be considered as amplifier and limiter or as a two terminal impedance, it is sufficient that we first simply examine the case of the transistor amplifier under large signal input and then use the result in connection with maintaining circuit applications.

Noise in Bipolar Transistors

Noise in bipolar transistors consists of full-shot noise in the emitter current and collector current, thermal noise in the resistance of the junctions, and flicker noise or 1/f noise characterized by $1/f^\alpha$ power spectral density with $\alpha \simeq 1$. There is also generation recombination noise and partition noise, but these need not be considered[5]. There is significant correlation between the shot noise in the emitter and collector currents at frequencies $f \ll f_T$, (common emitter unity current gain frequency) so that at HF when typical transistors such as the 2N918, 2N2857 class having f_T greater than 0.9 GHz are employed, the transistor equivalent circuit of a single-stage transistor amplifier incorporating current negative feedback can be represented by the circuit of Figure 4.

In Figure 4, combining the noise voltages and noise currents the transistor noise can simply be represented by two noise generators at the input, one a voltage source and the other a current source. The noise voltage $\sqrt{4kTr_{bb}'\,df}$ is the thermal noise in the base spreading resistance which is the most significant series resistance. Since r_c is typically less than 1 ohm and r_{bb}' greater than 15 ohms, the noise contributed by r_c is neglected. The voltage generator $\sqrt{4kT\frac{r_e}{2}\,df}$ represents the noise voltage introduced in the base-emitter circuit loop due to the potential developed by the emitter shot noise current flowing through the emitter impedance of the transistor which is simply r_e at HF. The noise current generator accounts for the correlation between collector and emitter shot noise and at low frequencies and HF when $\omega \ll \omega_T$ is essentially just full shot noise in the dc base current. The generators V_f and I_f account for flicker noise. The placement of these generators in the circuit is not exact, but it is found that placement as shown is adequate[6]. It is found by experiment that we need only be concerned with the source I_f when r_e is small.

The spectral density of I_f^2 varies as $1/f^\alpha$ where α is frequency dependent being 1 at very low frequency, but between 3/2 and 2 in some transistors in the frequency range 10 to 1000 Hz and nearly 1 in others in this range. Presumably this departure is caused by Burst noise.

In analyzing the performance of frequency source circuits, we need to know the value of I_f. This parameter is not immediately obvious from an inspection of noise figure versus frequency curves for the transistor.

The equivalent circuit shown in Figure 5 must be considered. Then from spot noise figure data at frequencies where the noise figure becomes constant with frequency, the value of r_{bb}' can be obtained by analysis of the circuit of Figure 5.

From Siemen's data for the BC109C transistor, r_{bb}' is found to be approximately 450 ohms at I_c = 0.2 mA. Using this value of r_{bb}' and noise figure data at 0.1 Hz, the frequency at which I_f becomes equal to the current i_{NT} is then determined to be 57 Hz. The noise figure data, however, shows that flat noise occurs at $f \cong 1$ Hz. This example is to show that the 1/f noise that needs to be considered in connection with flicker of phase is somewhat higher than the value one normally associates with the transistor on basis of published noise figure curves.

The flicker noise I_f is believed to be a function of h_{FE}. This is the result of van der Ziel's[5] interpretation of Guttkov's measurements[7]. The relationship between h_{FE} and the surface recombination velocity is

$$\frac{1}{h_{FE}} = \frac{1}{h_{FEI}} + ks \tag{13}$$

where h_{FE} is the measured dc common emitter current gain of the transistor h_{FEI} is the intrinsic common emitter current gain, k is a constant, and s is the surface recombination velocity. Reducing s reduces the 1/f noise so it is expected that super h_{FE} transistors such as the BC109C will have lower 1/f noise. From this relationship we expect the 1/f noise of transistors of the 2N2857 type to be significant up to f = 50 to 100 kHz. Equation (8) shows that the bipolar transistor is an inherently nonlinear device. As a result, the various intermodulation components that occur between the signal current and its harmonics is of no concern where the signal applied to the transistor circuit develops less than 10 or 20 millivolts across the base-emitter junction as in the case of receiving amplifiers but can not be neglected in the frequency source application. When the voltage V does not swing below the contact potential during any part of its time function, equation 8 can adequately describe the performance. We are concerned with the case of

$$\frac{qV}{(1+m)\,kT} = A\sin\omega_1 t + \sum_{\omega_n = 0}^{\omega_n = \infty} B\sin\omega_n t \qquad (14)$$

The total voltage due to the noise will be very small compare to A, and we can simple concern ourselves with single term of the noise $B \sin \omega_2 t$. The current is then of the form

$$i_c = k\left[(\exp A\sin\omega_1 t \times \exp B\sin\omega_2 t) - 1\right] \qquad (15)$$

Using well-known expression for $e^{iz\,8}$ we get for the case of B = 0,

$$i_c = k\Big[I_o(A) - 1 + 2I_2(A)\cos 2\omega_1 t + 2I_4(A)\cos 4\omega_1 t - 2I_6(A)\cos 6\omega_1 t + 2I_1(A)\sin\omega_1 t - 2I_3(A)\sin 3\omega_1 t + 2I_5(A)\sin 5\omega_1 t + \ldots\ldots\ldots \Big] \qquad (16)$$

In eq. 16, I_n (A) is the modified bessel function of the first kind. When A = 0 and B is considered, the approximation

$$I_n(Z) \approx (1/2Z)^n/\Gamma(n+1);\ (n \neq -1, -2, -3\ldots.)$$

is valid. The i_c due to B is then simply B/2K at the frequency $\omega_2 t$. The multiplication of the two time functions simply amounts to convolution of the frequency functions. We then get for the case of concern where B<<A, output currents of the form

$$2I_1(A)\sin\omega_1 t \qquad (17)$$

$$2I_1(A)\,I_1(B)\sin(\omega_1 \pm \omega_2)t \qquad (18)$$

$$2I_2(A)\,I_1(B)\sin(2\omega_1 - \omega_2)t \qquad (19)$$

$$2I_3(A)\,I_1(B)\sin(3\omega_1 - \omega_2)t \qquad (20)$$

$$2I_4(A)\,I_1(B)\sin(4\omega_1 - \omega_2)t \qquad (21)$$

Expressions 19, 20, and 21 yield the increase in white noise for Fourier frequencies of 0 to typically 1 MHz that results from the signal harmonics heterodyning with wide-band noise. Note that when the operating frequency approaches f_T of the transistor the value of B that must be considered increases for the harmonics due to the decorrelation between emitter and collector shot noise current, which causes I_{N_1} of Figure 4 to to increase.

A typically large signal operation might involve a voltage of 100 millivolts peak for base-emitter with sufficient zero signal biasing that the -100 millivolts does not cutoff the collector current. For such condition we obtain

$$\frac{2I_1(A)\,I_1(B)}{2I_1(A)} = I_1(B) \qquad (22)$$

$$\frac{2I_2(A)\,I_1(B)}{2I_1(A)} = 0.56\, I_1(B) \qquad (23)$$

$$\frac{2I_3(A)\,I_1(B)}{2I_1(A)} = 0.24\, I_1(B) \qquad (24)$$

$$\frac{2I_4(A)\,I_1(B)}{2I_1(A)} = 0.08\, I_1(B) \qquad (25)$$

These relationships show that for this condition if B were constant for all frequencies, the increase in direct white phase noise due to intermodulation effects will only be about 0.8 dB. Higher values, will of course be observed due to the change in I_{N_1} with frequency as noted, and the external circuit impedance which affects the magnitude of noise potential across b'e will usually be different at the harmonics of the signal frequency. Typical parameters usually lead to white S/N_ϕ of +160 dB/Hz with 50 to 100 millivolts peak base-emitter HF signal.

The other effect that is observed is that the low-frequency noise will produce amplitude fluctation of the current, and the percentage of amplitude modulation is independent of A. Thus, we expect the S/N in the output current that is dominated by low-frequency noise to improve as the signal voltage is increased.

The amplitude modulation spectrum will exhibit 1/f noise when I_f dominates the low-frequency noise developed across b'e. In some cases, the amplitude noise can be reduced by proper attention to the impedance of the circuit external to the transistor that exists at zero and low frequencies.

The effect of the amplitude modulation by low frequency noise is to:

a. Modulate the phase of the signal current.

b. Modulate the input reactance of the amplifier stage by an amount dependent on the feedback impedance of the circuit that are involved.

To show how the modulation of the current by low frequency signals modulates the reactance or the current we need simply to consider the circuit of Figure 6. Straightforward circuit analysis yields the effective transconductance $i_c/V = gm'$ as

$$g_m' = \frac{h_{FE}}{r_{bb}' + \frac{26}{I_e}(1+h_{FE}) + 2Z_E + h_{FE}Z_E + j\,(r_{bb}' + Z_E)h_{FE}\,f/f_T} \qquad (26)$$

The phase angle of the transconductance

$$\phi = \tan^{-1} \frac{(r_{bb}' + Z_E) h_{FE}\, f/f_T}{r_{bb}' + Z_E (2 + h_{FE}) + \frac{26}{I_e} (1 + h_{FE})} \qquad (27)$$

and

$$d\phi/dI = \frac{26\, h_{FE}\, (f/f_T)\, (1 + h_{FE})}{I_e^2 \left\{ [h_{FE}\, f/f_T\, (r_{bb}' + Z_E)]^2 + [r_{bb}' + Z_E (2 + h_{FE})]^2 \right\} + I_e (52) (r_{bb}' + Z_E) (1 + h_{FE}) [r_{bb}' + Z_E (2 + h_{FE})] + [26 (1 + h_{FE})]^2} \qquad (28)$$

A plot of equation 28 for some typical transistor parameters is shown in Figure 7.

It should be noted that the presence of negative feedback at the frequency of the excitation signal significantly decrease the AM to PM conversion efficiency. Without negative feedback, it is observed that the conversion of AM to PM is actually larger when using the super h_{FE} transistor such as the BC109 than when using a good VHF transistor such as the 2N2857, 2N918 or 2N2222. This fact is offset, however, by the fact that the frequency at which flicker noise becomes dominant is much lower for the BC109 than for the 2N2857 types. To estimate $\mathcal{L}(f)$ one needs to consider the external circuit in some detail. In the majority of oscillator circuits the low-frequency impedance between base and emitter is much larger than $(1 + h_{FE}) \frac{26}{I_e}$ so that I_f essentially dominates the noise voltage for low Fourier frequencies by the voltage $I_f (1 + h_{FE}) \frac{26}{I_e}$ that is the approximate base-emitter voltage magnitude. Because of the relationship in equation 13, we expect the $\mathcal{L}$(1Hz) to be much less when using the BC109C rather than the 2N2857. However, in many applications where $S_\phi(f)$ at frequencies above 5000 Hz is important, the performance provided by the BC109C will be poorer than with the 2N2857 owing to the larger value of r_{bb}'. This performance can possibly be improved then by paralleling several transistors as is done when very low noise amplifiers are required.

The analysis of the model that is proposed to explain the nature of $\mathcal{L}(f)$ indicates that both AM and PM noise will occur, and that the AM should be approximately 16 dB above the PM. Careful laboratory measurements were made in which caution was taken that no AM suppression was introduced by circuitry for the instrumentation that followed the stage under observation. The measurement results showed the PM to be typically 15 to 20 dB below the AM.

When $\mathcal{L}(f)$ is the only factor involved, performance is described by equation 11. However, in many oscillator circuits, the low-frequency noise modulation of the current results in reactance modulation of the input impedance of the circuit that cannot be neglected. The apparent $\mathcal{L}(f)$ is then found to be larger than the theory would predict. This problem is encountered with single transistor self-limiting maintaining circuits and also in some buffer amplifiers and frequency multipliers. To avoid such problems a two-stage transistor maintaining circuit is always employed in which the transconductance of the first stage is kept constant[9 10].

In buffer amplifiers, neutralization using a transistor in which $I_e = 0$ for the neutralizing impedance is employed. In frequency multipliers, the neutralization is provided by using a push-push transistor doubler circuit as has been done by D. Halford.[11]

The effects of the reactance modulation are predicted by simply solving the circuit of Figure 6 for $d(V/I)/dI_e$ in similar manner as done for the phase variation. The expression in closed form is rather complicated and will not be given. In practice, one does not need to obtain an equation for $dV/I/dI_e$ for the circuit since it is a trivial matter to analyze it numerically using today's time-shared computers.

Self-Limiting Oscillator

In the self-limiting oscillator, either voltage limiting by collector-base voltage bottoming or current-limiting due to reduction in the fundamental Fourier coefficient of the collector current as a result of swinging the base emitter below the contact potential for more than half cycle of the input can occur. At HF both can be used, but only current limiting will be considered. Since the Bessel function expansion previously given would necessitate expressing V as a Fourier series to account for V becoming less than ϕ, the results for the intermodulation coefficients will be different than in the sinusoidal V case. Rather than examine the transistor using the Bessel expansion and convolution, the actual V_{IN} versus i_c characteristic with HF impedance Z_E, taken as a resistance of 50 ohms was obtained. Then a fifth degree polynominal was fitted to this function for two levels of input voltages, for case of a large dc resistance for Z_E so that the biasing point shifts due to the effect of the average value of i_c that is a function of V. The coefficients of the polynominal were then used to compute the intermodulation terms. The results of these computation showed that the $\omega_1 \pm \omega_2$ term increases approximately 2 dB for variation from 0.2 to 0.6 volt peak base-emitter voltage. However, due to the approximation error using only a fifth degree polynominal it is not believed such variation would actually occur so that equation 22 is considered applicable even in the self limiting case i.e., $\mathcal{L}(f)$ at low frequency should not be affected whether transistor is essentially Class A or Class C.

The computation did, however, show the nulls in the intermodulation terms involving high-frequency noise and various harmonics of the signal frequency. These nulls have been reported previously[12, 13].

Owing to the fact that the high order signal harmonic - fundamental noise frequency contribution to the white phase noise is small, there is no noticeable benefit by operation at a signal and current level that produces a particular null.

Experimental Work

Amplifier

According to the theory for $\mathcal{L}(f)$ that was presented, the fractional modulation caused by the 1/f noise should decrease as the signal level increases.

$\mathcal{L}(f)$ should therefore be expected to be a function of signal level. Measurements of $\mathcal{L}(f)$ on a single stage transistor amplifier were made for different signal levels and the results are shown in Figure 8. Extrapolation of the 79 mV curve to f = 1 Hz yields $\mathcal{L}(1) \cong -114$ dB/Hz. With larger signal levels $\mathcal{L}(1)$ is reduced to -128 dB/Hz. The measurements also showed that the AM noise was 10 to 15 dB larger than the PM noise in the Fourier frequency range of 100 to 400 Hz.

When negative feedback was applied using R_E between 51 and 82 ohms, $\mathcal{L}(f)$ was reduced to the level of the instrumentation noise.

Oscillators

The single transistor self-limiting circuit and the two transistor maintaining circuits of the type shown in Figure 9 were investigated using an LC Clapp resonator, and a quartz crystal unit. The use of an LC resonator permits investigation of the effects of negative feedback since $\mathcal{L}(f)$ produces frequency noise which is well above the instrumentation noise level. The modified Butler circuit (bridged Tee) was also examined. Figure 10 shows the typical measured performance for the single stage circuit using various transistors. The results are in agreements with the theory showing $1/f^2$ rather than $1/f^3$ power spectral density when using the BC108C, (Note BC108C is same as the BC109C but is not guaranteed noise characteristic) but show also an increase in white noise in the BC108C due primarily to the larger base spreading resistance.

Figure 11 shows the reduction in 1/f frequency noise resulting from incorporation of negative feedback in the self-limiting oscillator circuit. Figure 12 shows measurement of the single-stage Clapp oscillator with various transistors and with and without negative feedback. The results are as expected from the theory that has been given.

Figure 13 shows the results of measurement of the single-stage Pierce circuit. For this circuit, the oscillating resonator Q was increased 120 times that of the Clapp oscillating resonator. (Note: Q of the oscillating resonator was determined by calculation from measurements of the frequency shift of the oscillator when the maintaining circuit phase was perturbed by 0.1 radian, by measuring injection locking range, and by a dynamic phase modulation technique.)

However, it will be observed that $S_\phi(f)$ decreased by only 20 dB. According to equation 11, the reduction should have been 41 dB. The principal reason this improvement was not obtained is quite simple. It is due to the fact that because of the larger resonator resistance of the composite quartz crystal-capacitor resonator, the reactance of Z_M, the maintaining circuit impedance seen by the crystal unit is a much larger part of the composite oscillating resonator than in the case of the LC Clapp circuit. Computations of Figure 6 to obtain $dXm/_{dI}$ were made as previously discussed, and the expected frequency deviation due to the reactance fluctuation caused by I_f was computed. The results are shown in Figure 13 and indicate that this was a plausible explanation for the observed $S_\phi(f)$. Further measurements were then made using the two crystal maintaining circuit. (With the LC Clapp resonator, negligible difference in $S_\phi(f)$ was observed using single transistor or two transistor maintaining circuits.) Figure 14 shows the results of the measurements. It will be observed that $S_\phi(f)$ is 36 dB lower than for the LC Clapp oscillator. The expected reduction was 41.6 dB.

Also shown in Figure 14 is the $S_\phi(f)$ when a Westinghouse integrated circuit was used in place of the two 2N2857. This device is the same circuit configuration as for the discrete component circuit. It will be observed that the flicker noise is somewhat larger when using the integrated circuit than when using discrete components. This is possibly the result of not having included the R_E of 51 ohms in the integrated circuit that was employed in the discrete component circuit.

Figure 15 shows the results of measurements made at 100 MHz. These measurements provides phase expansion of three to one, allowing more readily made measurements at high Fourier frequencies since the instrumentation noise is effectively reduced by 10 dB.

The measurement of $S_\phi(f)$ was made using active filters and the HP302A wave analyzer with an IR602A integrator. The direct analogue filtering and observation was made only to frequency of 10 Hz. An observation was made to obtain $S_\phi(f)$ at 1 Hz by computation of the spectral density from the time function because of difficulty and uncertainty in measurement through the active filters below 20 Hz. The output of the phase detector was recorded on a chart recorder and by photographing an oscilloscope trace. The signal observed in this manner was band limited to 10 Hz. Data points from the recording and photograph were used as input to an FFT program on a time-shared computer to obtain the spectral density of the phase. The results were corrected by the measured error response of the phase locked loop used for the measurement. (Bandwidth of the loop was nominally 0.3 Hz but fluctuates depending on the static error due to variation in $df/_{dV}$ of the tuneable oscillator which employed a simple weakly coupled varactor in order not to degrade the oscillating resonator Q.) The final result for $S_\phi(f)$ when due account for all parameters was made was -78 dB/Hz. Extrapolation of the higher frequency measurements of $S_\phi(f)$ indicated $S_\phi(1)$ of -72 dB/Hz. The discrepancy is typical for measurements of this kind made at low Fourier frequency. Using the conservative value of $S_\phi(1)$ = -72 dB/Hz we can determine $\mathcal{L}(1)$.

$$\mathcal{L}(1) = S_\phi(1)\ \text{dB/Hz} - 20\log_{10}\left(\frac{f_o}{2Q}\right)$$

$$= -72 - 55 = -127\ \text{dB/Hz}$$

From the FFT of the phase data we get $\mathcal{L}(1)$ = = -133 dB/Hz.

The single-transistor modified Butler circuit (bridged Tee) was also examined. Figure 16 shows the results. These results show only small differences in $S_\phi(f)$ for two widely different crystal unit drive levels.

The oscillating resonator Q for this circuit, however, is only about 16000 whereas the crystal unit Q is 160000. In the two transistor Pierce circuit, on the other hand, it is possible to obtain oscillating resonator Q approaching the crystal unit Q within a few percent. As a result lower FM noise can be achieved using the fundamental mode crystal unit in the Pierce configuration than can be achieved using the third overtone crystal unit in the bridged tee configuration.

Conclusions

A model has been presented for the purpose of explaining the observed behaviour of $\mathcal{L}(f)$ in frequency sources. $\mathcal{L}(f)$ at low Fourier frequencies is simply dominated by the noise resulting from conversion of AM noise to PM noise. The AM noise is result of noise modulation of the transconductance by the low-frequency transistor noise. The modulation occurs due to the inherent nonlinear current versus voltage characteristic of the bipolar transistor and is independent of whether operation is class A, B, or C. Simple negative feedback reduces the AM to PM conversion efficiency, and it becomes essentially the same for all transistor types when a small amount of negative feedback exists. For Fourier frequencies below 1000 Hz, however, the reduced 1/f noise of super h_{FE} transistors becomes important and they can in some circuits provide improved performance at f = 1 Hz. However, the thermal noise in such transistors resulting from the larger base resistance becomes a limiting factor in typical frequency source circuit designs so that VHF transistors having high h_{FE} and low r_{bb}' are preferable for the majority of radar and communications applications of frequency sources where low FM noise is required. With careful design of the HF crystal controlled source, $\mathcal{L}(1)$ of -125 to -130 dB/Hz is achieved; and $\mathcal{L}(5000) \approx -160$ dB/Hz; and $\mathcal{L}(50000) \approx -163$ dB/Hz.

When care is not taken, the effects of positive feedback can sometimes cause the apparent $\mathcal{L}(f)$ to be much worse. The apparent $\mathcal{L}(f)$ can typically be 15 to 20 dB worse than the actual $\mathcal{L}(f)$.

Negligible change in 1/f frequency noise of moderate Q HF oscillators was observed at Fourier frequencies greater than 10 Hz when crystal unit power dissipation was varied from 500×10^{-6} to 1500×10^{-6} watts. All observation were at 25 to 30°C.

References

1. Janus Groszkowski, "The Interdependence of Frequency Variation and Harmonic Content and the Problem of Constant Frequency Oscillators", Proc. IRE v. 21 p. 958 - 981, 1933.

2. Balthasar van der Pol, "Non Linear Theory of Electric Oscillations", Proc. IRE 22, 1051, 1934.

3. R. Adler, "A Study of Locking Phenomena in Oscillators", Proc. IRE June 1946, pp. 351-357.

4. D. Halford et al. "Flicker of Phase in RF Amplifiers and Frequency Multipliers: Characterization, Cause and Cure", Proc. 22d Annual Symposium on Frequency Control", pp. 340 - 341, April 1968.

5. A. van der Ziel "Noise in Solid-State Devices and Lasers", Proc. IEEE August 1970 pp 1178 - 1206.

6. W. Fonger, "Noise in Electron Devices", John Wiley and Sons, 1959.

7. I. D. Guttkov, "Investigation of the 1/f Low Frequency Noise of Backbiased p-n Junctions", Radio Electron Eng. Physics, pp 880 - 882, May 1967.

8. "Handbook of Mathematical Functions" National Bureau of Standards Applied Mathematics Series 55, 1964.

9. U.S. patent No. 3,569,865.

10. U.S. patent No. 3,571,754.

11. Letter from D. Halford NBS to D. Healey, 22 Apr. 1969 with copies of laboratory notebook pages of A. Wainwright.

12. H. Lotsch "Third-Order Distortion and Cross Modulation in a Grounded Emitter Transistor Amplifier", IEEE Trans. on Audio pp. 49 - 58 March-April 1961.

13. H. Lotsch "The Non-Linear Theory for Semiconductor Diodes", IEEE Trans on Electron Devices, May 1968.

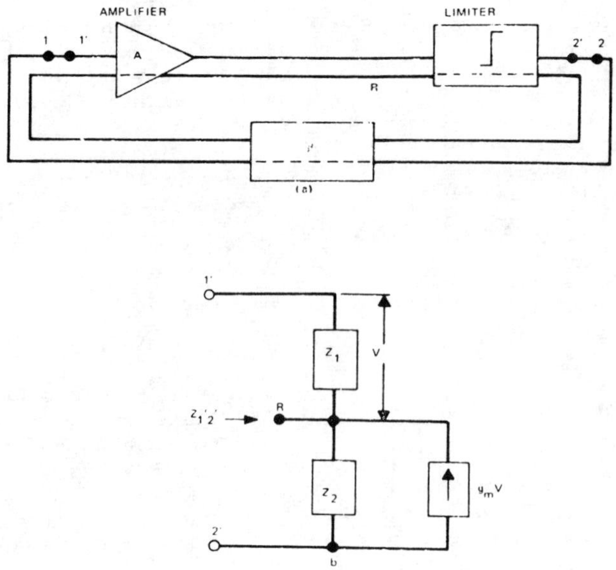

$$Z_{1'2'} = Z_1 + Z_2 + g_m Z_1 Z_2$$

$$Z_{1'2'} = \frac{1}{j\omega C_1} + \frac{1}{j\omega C_2} - \frac{g_m}{\omega^2 C_1 C_2}$$

72-0389 V-25

Figure 1

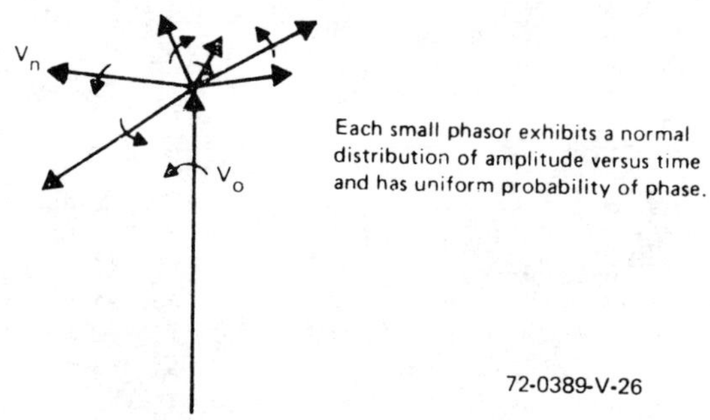

72-0389-V-26

Figure 2

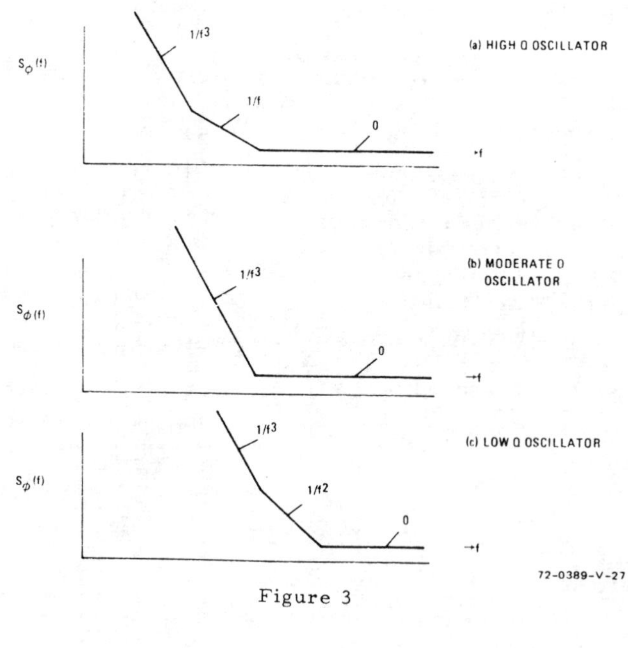

72-0389-V-27

Figure 3

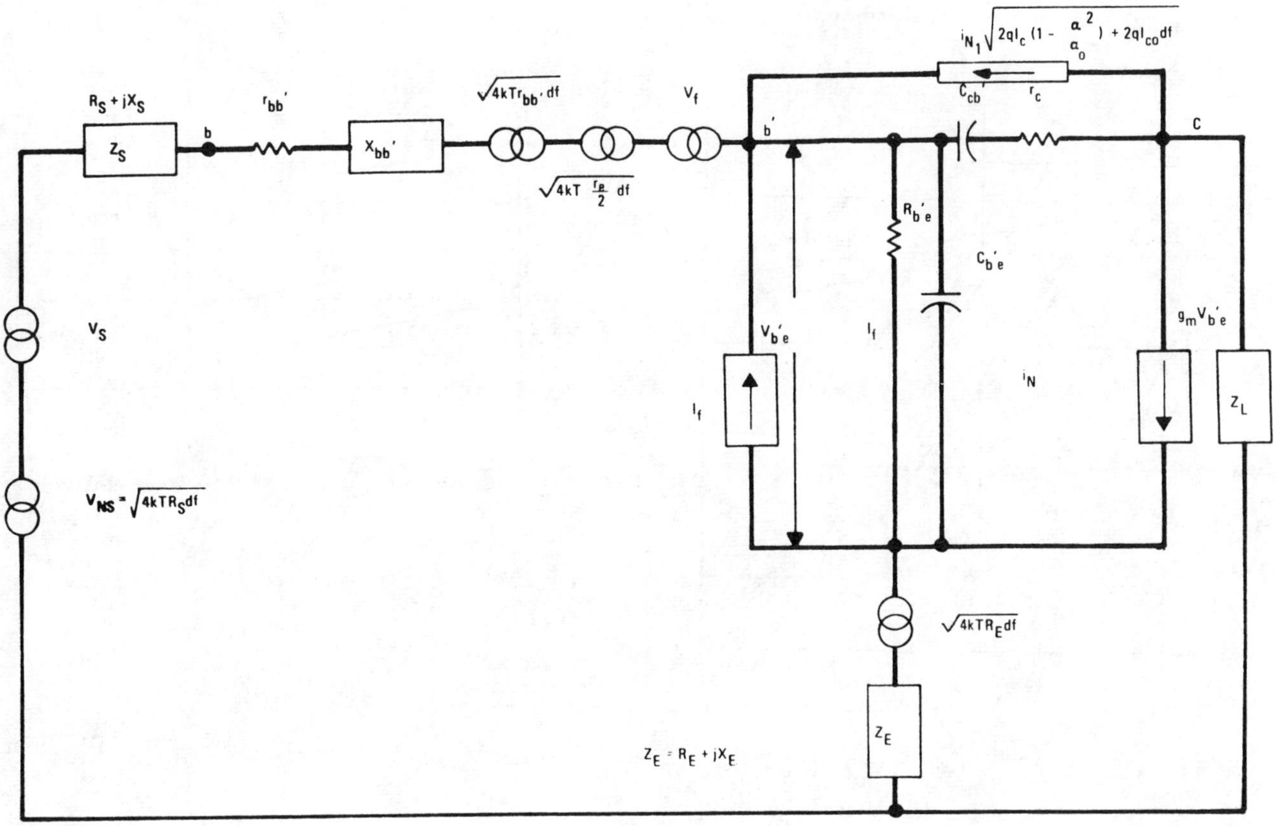

$C_{b'e} = \frac{1}{\omega_T r_e} \simeq \frac{I_e\,(mA)}{26\,\omega_T}$ $R_{b'e} = (1 + h_{FE})$

$g_m \simeq I_e\,(mA)/26$ for $\omega \ll \omega_T$

72-0389-V-28

Figure 4

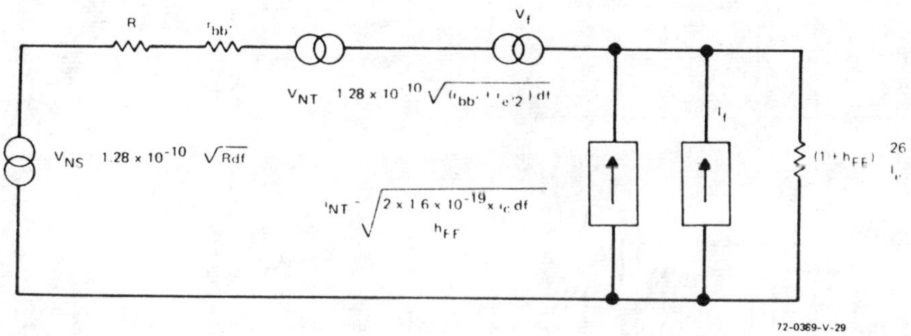

72-0389-V-29

Figure 5

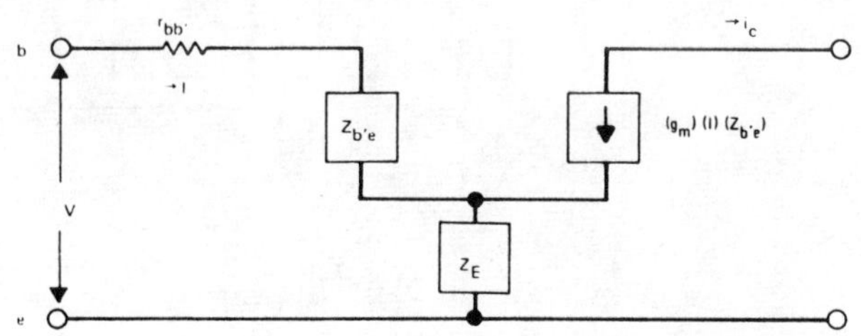

72-0389-V-30

Figure 6

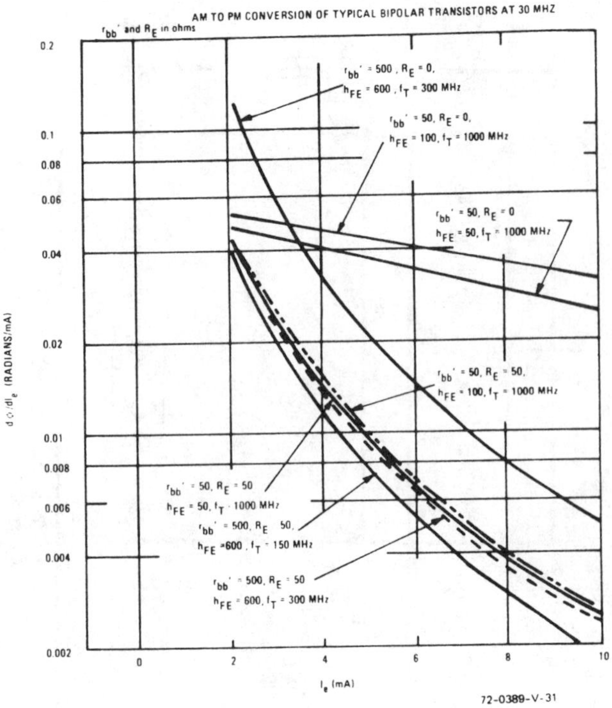

Figure 7

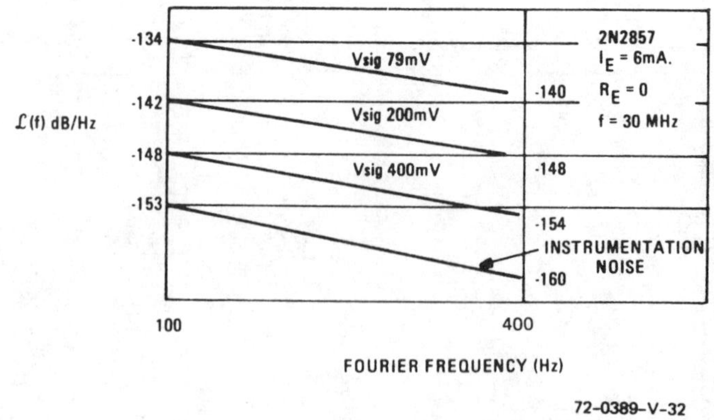

Figure 8

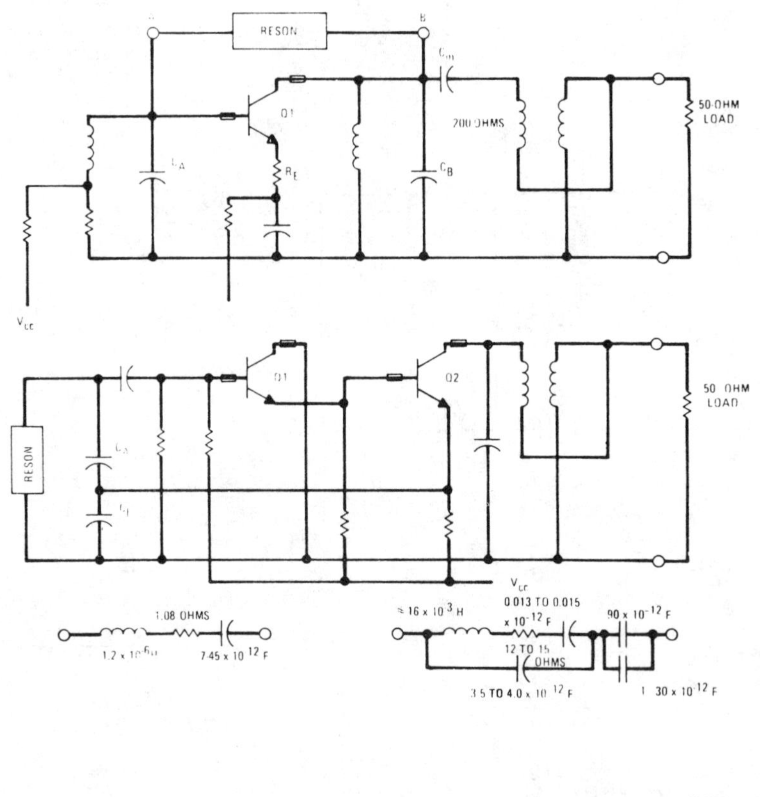

Figure 9

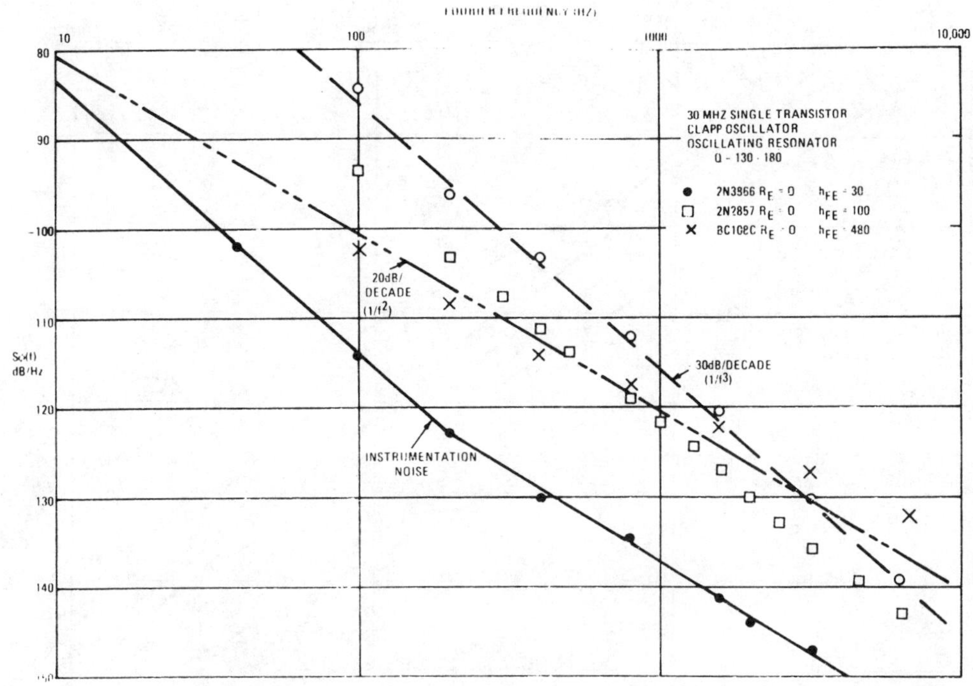

Figure 10

72-0389-V-34

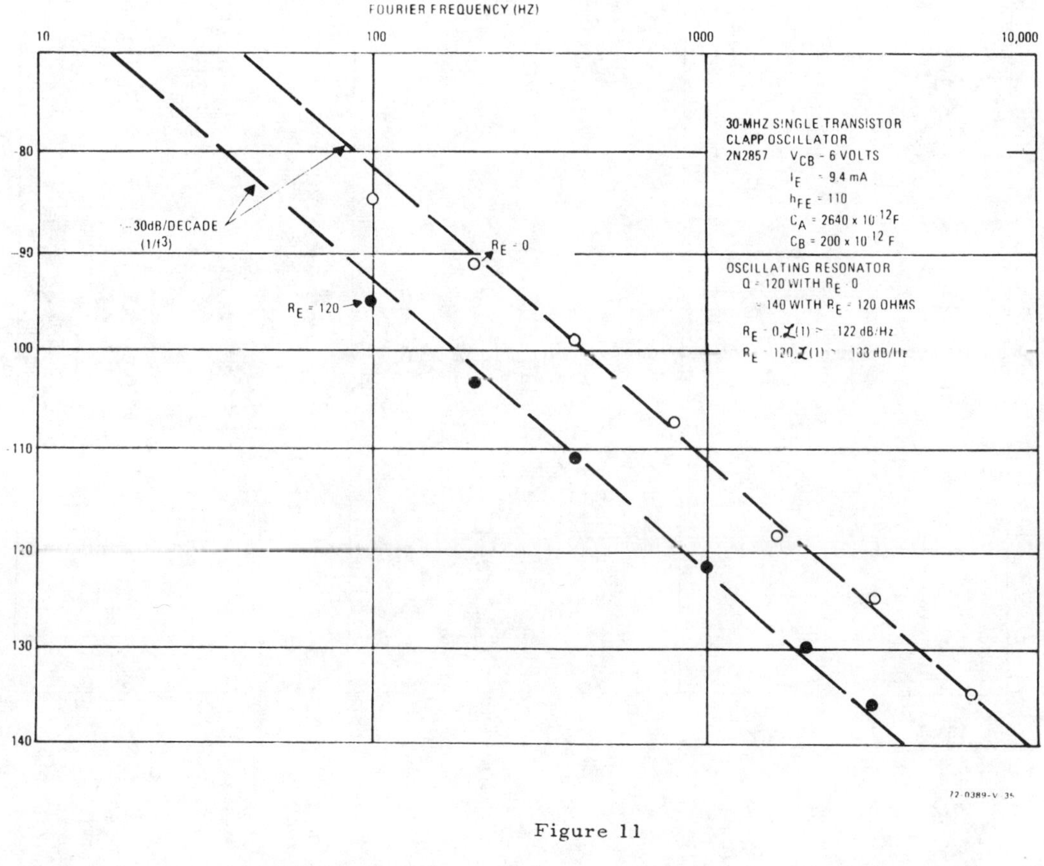

72-0389-V-35

Figure 11

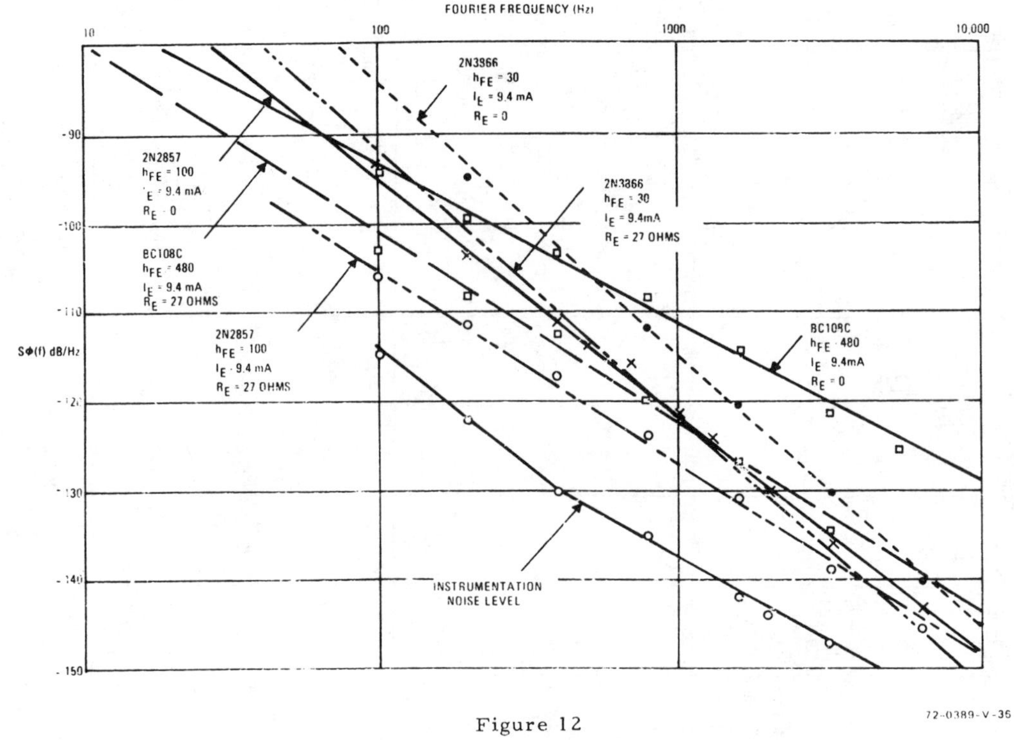

Figure 12

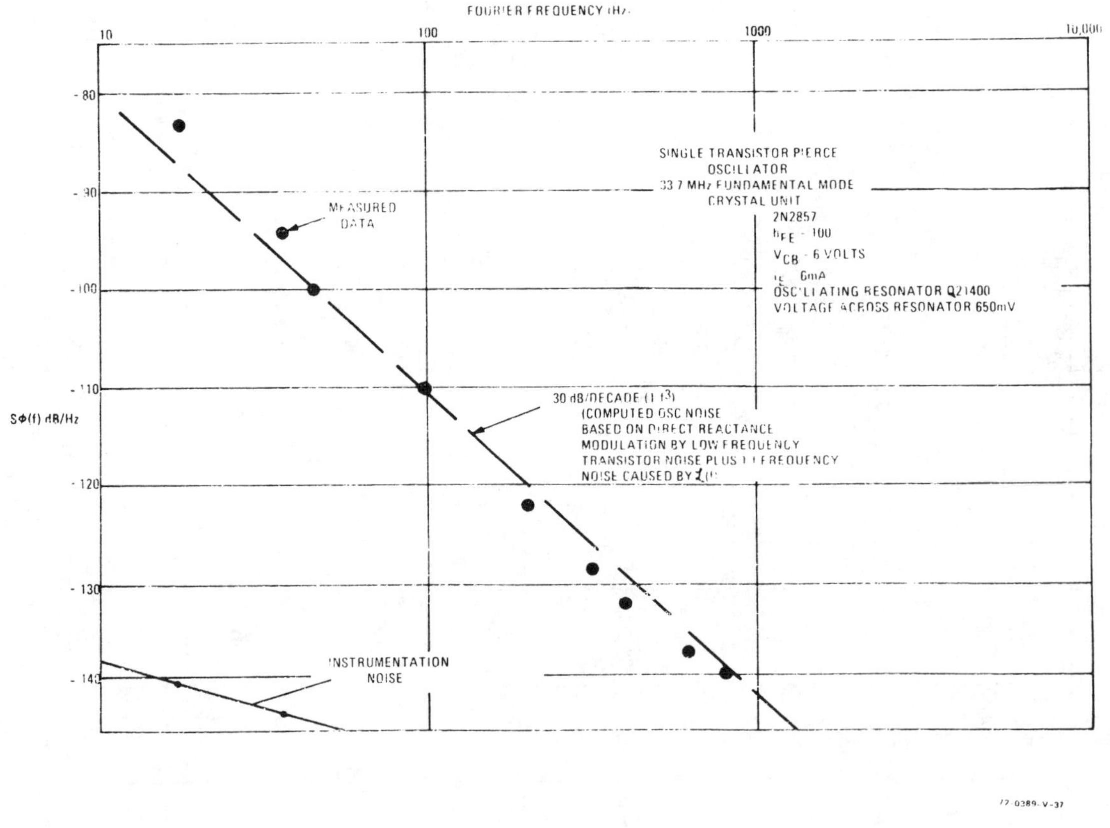

Figure 13

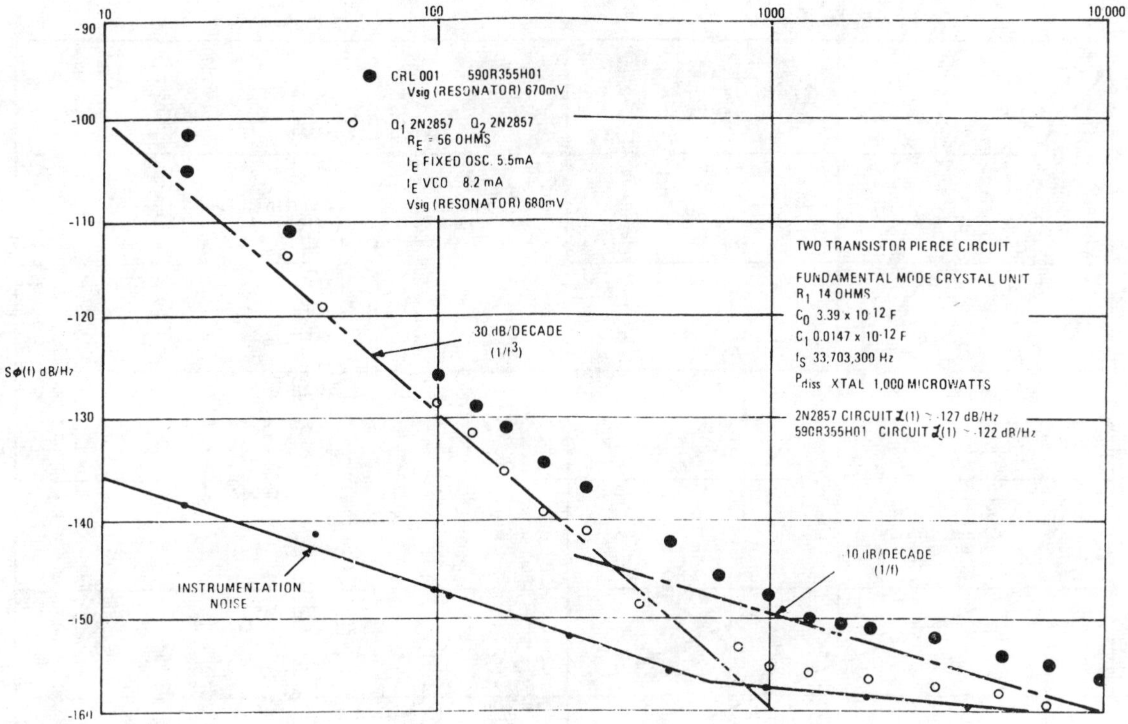

Figure 14

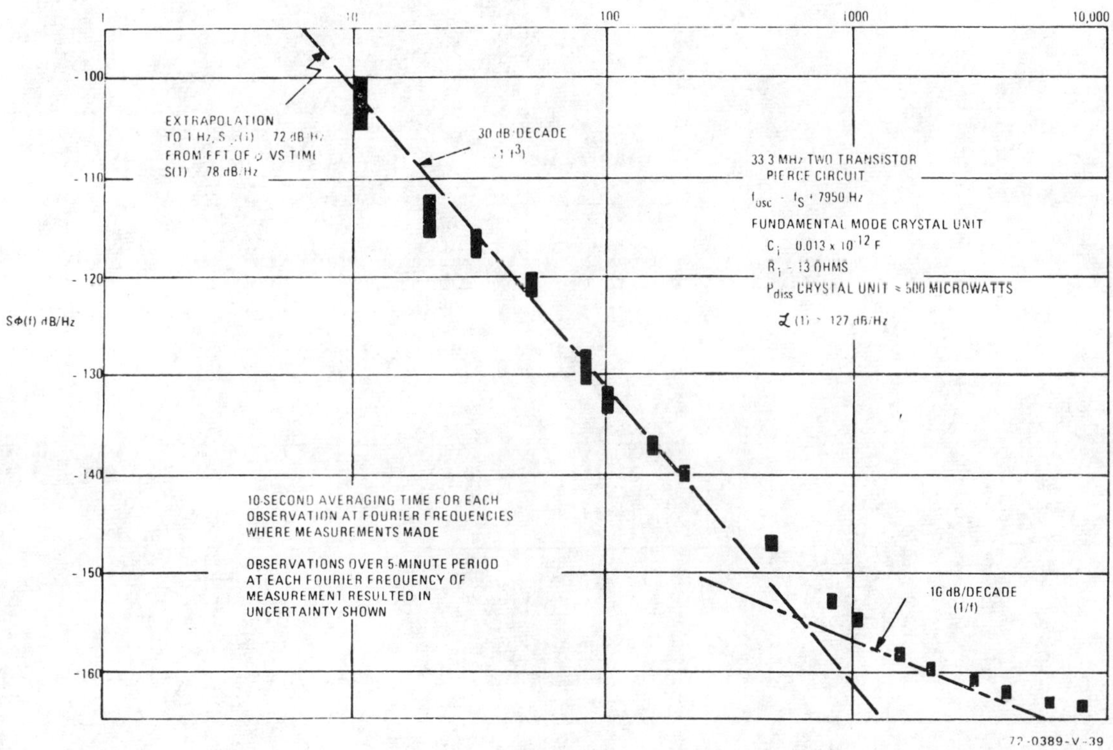

Figure 15

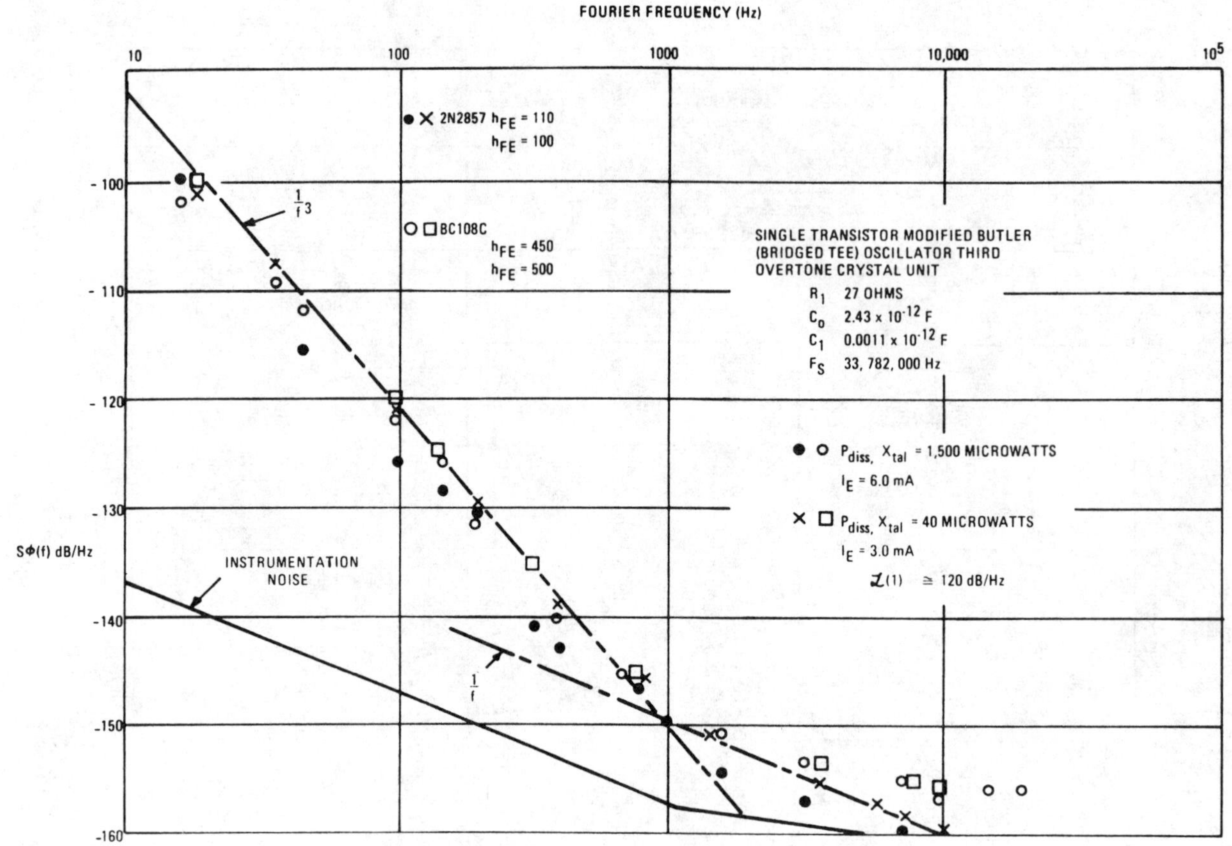

72-0389-V-40

Figure 16

Measurement of the Short-Term Stability of Quartz Crystal Resonators and the Implications for Crystal Oscillator Design and Applications

FRED L. WALLS AND ARTHUR E. WAINWRIGHT

***Abstract*—A new technique is presented which makes it possible to measure the inherent short-term stability of quartz crystal resonators in a passive circuit. Comparisons with stability measurements made on crystal controlled oscillators indicate that noise in the electronics of the oscillators very seriously degrades the inherent stability of the quartz resonators for times less than 1 s. A simple model appears to describe the noise mechanism in crystal controlled oscillators and points the way to design changes which should improve their short-term stability by two orders of magnitude. Calculations are outlined which show that with this improved short-term stability it should be feasible to multiply a crystal controlled source to 1 THz and obtain a linewidth of less than 1 Hz. In many cases, this improved short-term stability should also permit a factor of 100 reduction in the length of time necessary to achieve a given level of accuracy in frequency measurements.**

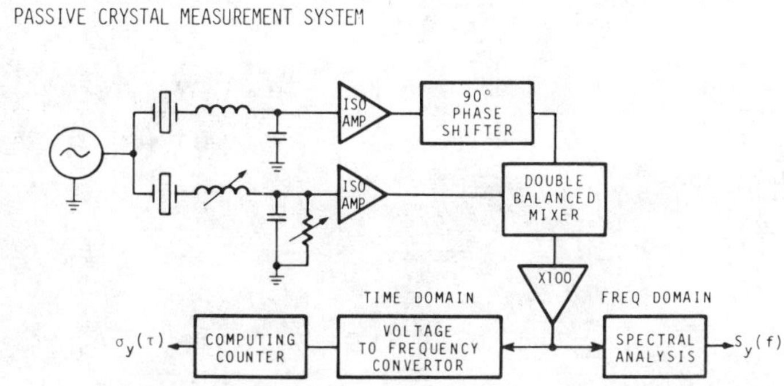

Fig. 1. Passive crystal measurement system.

INTRODUCTION

CRYSTAL controlled oscillators play a key role in frequency metrology in that they are used in nearly all precision frequency measurement and generation devices. Measurements of the inherent frequency stability of the quartz crystal resonators used in the frequency control of oscillators are therefore of particular interest.

The technique described here permits one to measure the inherent frequency stability of quartz crystal resonators in a passive circuit without the noise usually associated with an active oscillator. Comparisons with stability measurements made on crystal controlled oscillators indicate that noise in the electronics of the active oscillators very seriously degrades the inherent stability of the quartz resonators for times less than 1 s. Hence, in addition to making possible the evaluation and modeling of the frequency stability of crystal resonators and the dependence on various manufacturing processes, these measurement techniques also make it possible to evaluate directly the degradation of the frequency stability due to different oscillator circuits. A simple model is developed which appears to explain this degradation in stability and which also points the way to design changes in crystal oscillators which should greatly improve their short-term stability.

Calculations indicate that with this improved short-term stability, it should be feasible to frequency multiply a crystal controlled source to 1 THz and obtain a linewidth of less than 1 Hz. This would make possible a much more precise transfer of frequency information and stability between the infrared and RF regions of the spectrum than is possible now. In many cases, this improved short-term stability would also permit a substantial reduction in the length of time necessary to reach a given level of accuracy in frequency measurements.

MEASUREMENT TECHNIQUES

The frequency stability of the crystal resonators has been measured in the system schematically shown in Fig. 1. Two crystals which are as identical as possible are driven from the same low-noise source. By careful adjustment of crystal tuning and the balancing of the relative Q's, the output from the mixer can be made first order insensitive to both residual amplitude and frequency modulation in the source. This reduction of noise from the source makes it possible to measure small time varying frequency deviations of the crystals. Special care must be taken to operate the mixer at the ac quadrature point, which is usually slightly different from the dc zero point, in order to realize the full potential of this method. The high sensitivity of this system for detecting frequency changes of the crystals can be found from analyzing the equivalent circuit shown in Fig. 2.

The complex amplitude of the voltage appearing at the mixer input from either crystal, labeled V_{out}, is propor-

Manuscript received October 16, 1974.
The authors are with the Time and Frequency Division, National Bureau of Standards, Boulder, Colo. 80302.

Reprinted from *IEEE Trans. Instrum. Meas.*, vol. IM-24, pp. 15–20, Mar. 1975.

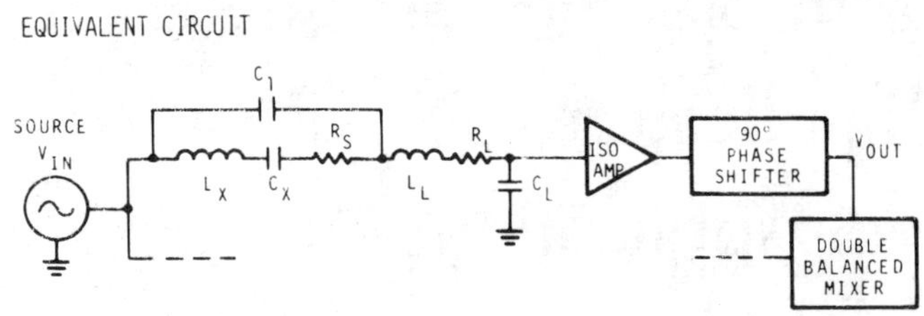

Fig. 2. Equivalent circuit.

tional to the following frequency dependent terms

$$\frac{V_{out}}{V_{in}} \simeq -\frac{i}{\omega_0 C_L (R_S + R_X)}\left(1 - i\frac{\Delta\omega}{\gamma_X}\frac{R_S}{R_S + R_L}\right)$$

$$\simeq -\frac{iK}{\omega_0}\left(1 - i\frac{\Delta\omega}{\gamma_X}\frac{R_S}{R_S + R_X}\right) \quad (1)$$

where it is assumed that

$$\frac{1}{L_X C_X} = \omega_0^2, \qquad \frac{1}{L_L C_L} \doteq \omega_0^2$$

$$\omega_{\text{source}} = \omega_0 + \Delta\omega$$

$$\frac{\Delta\omega}{\gamma_X} \ll 1, \qquad \text{where } \frac{\omega_0}{2\gamma_X} = \text{unloaded } Q \text{ of the crystal}$$

$$\gamma_L \gg \gamma_X, \qquad \frac{\omega_0}{2\gamma_L} = \text{unloaded } Q \text{ of the } L_L, C_L, R_L \text{ circuit}$$

$$\frac{1}{\omega_0 C_1} > R_S > R_L. \quad (2)$$

One notes that there is a phase shift across the circuit which is proportional to the frequency difference between the source and the natural resonant frequency of the crystal. This phase shift under the preceding assumptions is just

$$\Delta\theta = \frac{R_S}{R_S + R_L}\frac{\Delta\omega}{\gamma_X} \quad (3)$$

note that

$$\frac{R_S + R_L}{R_S}\gamma_X \equiv \gamma$$

is $\frac{1}{2}$ the loaded angular bandwidth which in most cases is very close to γ_X. Identical results are of course obtained for the other crystal. The 90° phase shifter was made from a variable length delay line. The double balance mixer is used as a phase detector and produces a dc voltage which is proportional to the instantaneous phase difference from 90° between the two signals. The mixer sensitivity was typically 3 mV per degree phase difference from 90° which, when used with (3) yields a mixer output (in volts) of

$$V = \frac{\Delta\omega}{\gamma}\frac{360}{2\pi}0.003 \text{ V} \quad (4)$$

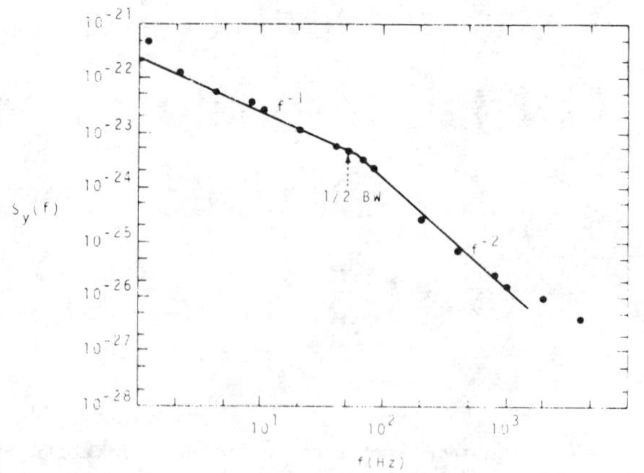

Fig. 3. Frequency domain data for a 10-MHz crystal pair with a bandwidth of 100 Hz, 10A-100, at 28°C.

for $\gamma/2\pi = 1$ Hz, the mixer output is, therefore, 170 mV per Hz frequency change in the natural resonant frequency of either quartz resonator, assuming they both have a bandwidth of 2 Hz.

The signal from the mixer is then processed for either frequency or time domain data (see Fig. 1). Spectral analysis of the mixer output yields $[S_y(f)]^{1/2}$, $S_y(f)$ being the spectral density of frequency fluctuations.

The time domain data is taken using a voltage to frequency converter and a computing counter. The average frequency measured by the computing counter during a sample time τ has a component which is proportional to the average frequency deviation between the two crystals during the sample time τ. By taking many consecutive measurements at each sample time τ, one can compute the Allan variance $\sigma_y^2(\tau)$ in the usual way [1].

The actual time domain data was taken using a second difference method which removed linear frequency drifts of the crystals. Since the object of this work was to investigate the short-term stability of crystal resonators and crystal controlled oscillators, this approach greatly reduced the requirements on the temperature stability of the enclosures without any sacrifice in data quality.

EXPERIMENTAL RESULTS

Fig. 3 shows an example of frequency domain data for a 10-MHz crystal pair. $S_y(f)$ is plotted versus Fourier frequency offset f, assuming equal contribution from each crystal. Note that the noise has a flicker of frequency character, i.e., $S_y(f) = h_{-1}f^{-1}$ for frequencies less than half-bandwidth of the crystals (50 Hz). For frequencies larger than the half-bandwidth, $S_y(f) = h_{-2}f^{-2}$ which is characteristic of a random walk frequency modulation process (the h_m's are constants). Crystal drive was approximately 200 to 250 μW.

Fig. 4 shows the time domain data for this same 10-MHz crystal pair and also the frequency domain data converted to time domain. Conversion between frequency and time domain can be easily done whenever $S_y(f)$ follows a simple power law dependence of f over an extended range [1] as in Fig. 3. The time domain stability for the

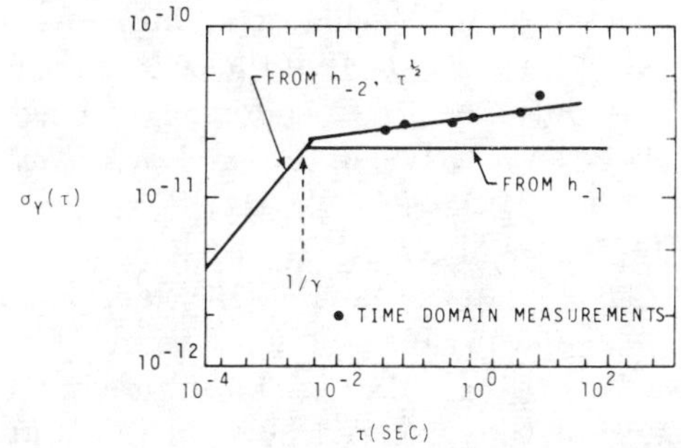

Fig. 4. Time domain data for crystal pair 10A-100 at 28°C. Also shown are the frequency domain data of Fig. 3 converted to time domain.

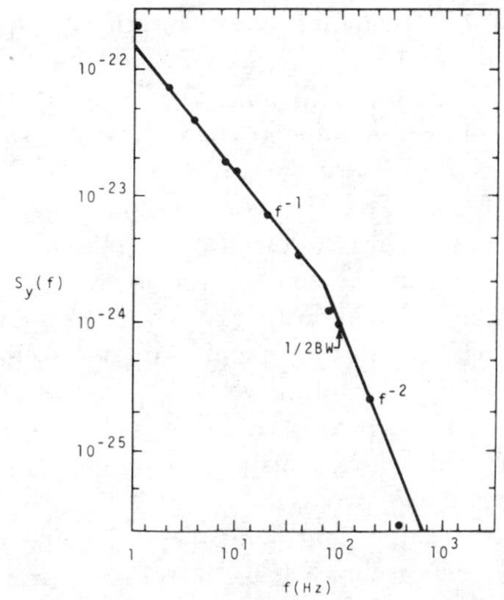

Fig. 5. Frequency domain data for a 5-MHz crystal pair with a bandwidth of 200 Hz, 5A-200, at 28°C.

flicker of frequency region is

$$\sigma_y(\tau) = (2 \ln 2 h_{-1})^{1/2} \tag{5}$$

while for the random walk frequency modulation region

$$\sigma_y(\tau) = \left(\frac{4\pi^2\tau}{6} h_{-2}\right)^{1/2}. \tag{6}$$

The most surprising result is the indication, from the frequency domain data, that the time domain stability improves as one goes to measurement times which are shorter than the inverse half angular bandwidth. The functional dependence on measurement time in this region is $\sigma_y(\tau) \sim \tau^{1/2}$.

Frequency domain measurements on a low-Q 5-MHz crystal pair are shown in Fig. 5. As in the case of the 10-MHz pair, $S_y(f)$ goes as $h_{-1}f^{-1}$ out to approximately the half-bandwidth point (100 Hz) and then goes as $h_{-2}f^{-2}$. Again, the frequency domain data indicate that $\sigma_y(\tau)$ goes as $\tau^{1/2}$ for $\tau < 1/\gamma$. Direct confirmation of this requires measurements of $\sigma_y(\tau)$ at sample times of order 10^{-4} s which could not be made with this apparatus due to dead time problems [1]. The solution to this problem is to examine the data for a high-Q crystal pair where $1/\gamma$ is of order 1 s.

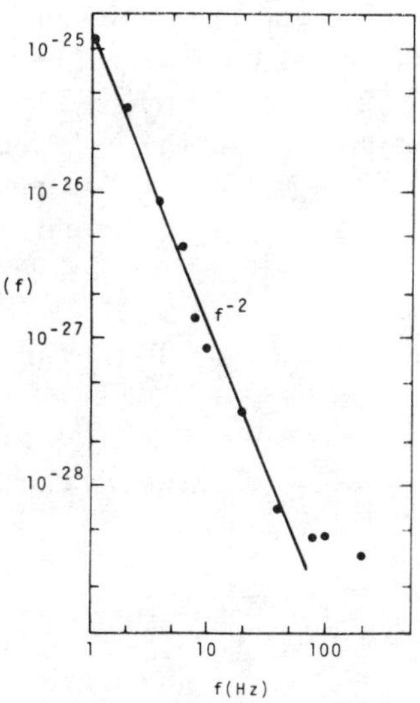

Fig. 6. Frequency domain data for a 5-MHz crystal pair with a bandwidth of 2 Hz, 5B-2, at 28°C.

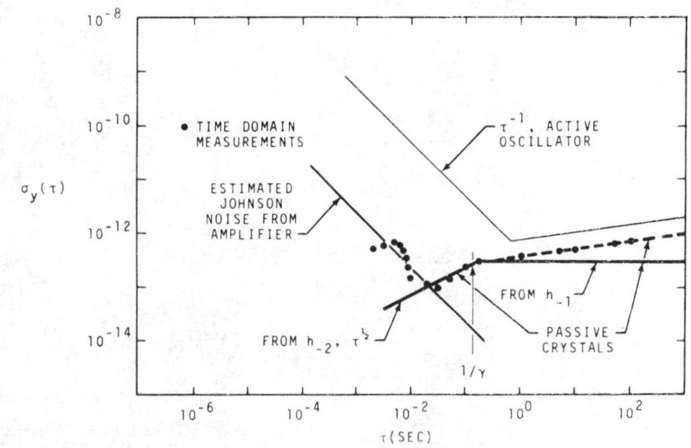

Fig. 7. Time domain data for crystal pair 5B-2 along with the frequency domain data of Fig. 6 transferred to time domain. Also shown are the time domain data for a high quality oscillator controlled by one 5B-2 crystal at 65°C.

Fig. 6 shows the frequency domain data from a 5-MHz crystal pair with a bandwidth of 2 Hz. Transition from the f^{-1} to f^{-2} behavior should occur at 1 Hz which is the lower limit of the spectrum analyzer used. Note that $S_y(f)$ goes as f^{-2} out to at least 50 Hz so that one should expect $\sigma_y(\tau)$ to go as the square root of τ from at least time of order $(2\pi 50)^{-1}$ or 3 ms or smaller. Fig. 7 shows the time domain data for this crystal pair along with the frequency domain data transferred to time domain.

The increase in stability for times less than the inverse bandwidth is confirmed by these measurements. As one goes to shorter and shorter times the stability becomes worse again due to noise in the isolation amplifier and the measurement system. This noise causes an uncertainty in the measurement of the position of the zero crossing. If the noise is white then the frequency fluctuations will have a white phase modulation character and $\sigma_y(\tau)$ will go as τ^{-1} for times larger than the inverse bandwidth of the measurement system.

The influence of the additive Johnson noise in the electronics on the measured stability can be understood using the following simplified model referred to Fig. 2. Multiplicative phase modulation in the isolation amplifiers has been reduced by the use of local negative feedback as originally suggested by [2]. Assume that the source voltage is $A_0 \cos \omega_0 t$ and that there is a white noise voltage $a(t)$ with effective bandwidth b referenced to the isolation amplifier input. Then the effective amplifier input is approximately given by

$$\nu(t) \simeq \frac{1}{\omega_0 C_L} \frac{1}{(R_S + R_L)} A_0 \cos \omega_0 t + a_C(t) \cos \omega_0 t + a_S(t) \sin \omega_0 t \quad (7)$$

where $a_C(t) \cos \omega_0 t$ and $a_S(t) \sin \omega_0 t$ are the in-phase and quadrature components of the noise, respectively, and have zero-mean value and bandwidth $b/2$. The rms values of $a_C(t)$ and $a_S(t)$ equal the rms value of $a(t)$ and there is no correlation between them, i.e.,

$$\overline{a_C^2(t)} = \overline{a_S^2(t)} = \overline{a^2(t)} \quad (8)$$

$$\overline{a_S(t) a_C(t)} = 0. \quad (9)$$

The average slope at zero crossing is

$$A'\omega_0 \equiv \omega_0 A_0 \frac{1}{\omega_0 C_L (R_S + R_L)}. \quad (10)$$

The quadrature component of the noise leads to an rms uncertainty in the zero crossing of approximately

$$\Delta t_{\text{rms}} \simeq \frac{a_S(t)_{\text{rms}}}{A'\omega_0} = \frac{a(t)_{\text{rms}}}{A'\omega_0} \quad (11)$$

for times greater than $1/2b$ since it takes about that much time for $a_S(t)$ to assume a new value. Therefore,

$$\sigma_y(\tau) \simeq \frac{\Delta t}{t} = \frac{a(t)_{\text{rms}}}{A'\omega_0 \tau}, \qquad \text{for } \tau > \frac{1}{2b}. \quad (12)$$

For times shorter than $1/2b$ it can be shown [3], [4]

$$\sigma_y(\tau) = \frac{a(t)_{\text{rms}}}{A'\omega_0} \left(\frac{b}{\tau}\right)^{1/2}, \qquad \tau < \frac{1}{2b}. \quad (13)$$

The line labeled "Johnson noise from amplifier" indicates the estimated contribution of our measurement system to $\sigma_y(\tau)$. Calculations of the Johnson noise in the series loss resistence of the crystal R_s show that this contribution to $\sigma_y(\tau)$ is very small compared to that of most active circuits.

For comparison, the stability of these same two crystals in a high performance crystal oscillator at 65°C is also indicated in Fig. 7. Note the dramatic difference in stability for times less than 1 s. This difference is just the noise contribution due to the electronics in the crystal controlled oscillators. The difference for measurement times greater than 1 s is primarily due to the decrease in the stability of this crystal pair at 65°C. Fig. 7 vividly illustrates the power of this technique for evaluating crystal controlled oscillator circuits.

The character of the oscillator stability for times less than 1 s is very similar to that of the crystal measurements below 10 ms only the level is higher. Using (12) and assuming a crystal drive of order 1 μW, $R_S = 100\ \Omega$, $a(t)_{\text{rms}} = (2\ \text{nV/Hz})(4 \times 10^4\ \text{Hz})^{1/2} = 400$ nV, yields $\sigma_y(\tau) = 10^{-12}\tau^{-1}$ as observed.

This suggests (in accordance with widely held beliefs—see, e.g., [4], [5] that Johnson noise sources in the amplifier and oscillator stages that are not filtered by the crystal are the cause of the observed noise in this and most low drive crystal oscillators. Short-term stability could be greatly improved merely by increasing crystal drive. A factor of 100 increase in crystal drive should produce a factor of 10 improvement in the short-term stability of the oscillator. Recent measurements by J. Grosslambert, G. Marianneau, M. Oliver, and J. Uebersfeld on a crystal controlled oscillator with 50 μW of crystal drive, yield a factor of 10 improvement in the short-term stability over results for the oscillator with approximately 1 μW of drive shown in Fig. 7 [6]. In both oscillators multiplicative phase modulation was reduced by local negative feedback [2]. Additional improvement can be obtained by using a high input impedance buffer amplifier and driving it with a series tuned tank from a point in the oscillator where the noise is bandwidth limited by the crystal, as in Fig. 2. This increases A without changing $a(t)$, since $a(t)$ is primarily determined by the Johnson noise in the stages following the basic oscillator.

Stability measurements on 5 crystal pairs suggest that the flicker of frequency level is roughly proportional to Q^{-1}. That is high-Q crystals generally have a correspondingly lower flicker of frequency level, however, crystals with the same Q can vary as much as a factor of 10 in stability in the flicker region. This indicates that fabrication techniques have a considerable influence on stability beyond just considerations of the obtainable Q. The measurement system described here could easily be used to examine the influence of various manufacturing processes on crystal stability. Some measurements versus temperature were also made. The 5B-2 crystal pair demonstrated a factor of 2 higher stability at 25°C than at either −75°C or +65°C. Another 5-MHz crystal pair with the same Q was a factor of 2 more stable at 65° than at 25° achieving a level about 2 times more stable than even the 5B-2 crystal pair illustrated in Fig. 7.

Since both crystal pairs were high-quality 5th overtone AT cuts, these measurements are not conclusive. It appears necessary to make stability measurement versus temperature over a very wide temperature range with many intermediate points in order to understand the temperature dependence.

APPLICATIONS FOR CRYSTAL OSCILLATORS WITH IMPROVED SHORT-TERM STABILITY

There are a number of important applications for crystal controlled oscillators with greatly improved short-term stability.

1) Use as a source for multiplication to the infrared. The linewidth could be nearly 1500 times narrower than for any previous system and nearly 10^6 times narrower than previous crystal controlled oscillator systems [7], [8].

2) Use as a short-term base in counting systems to reduce the time necessary to achieve a given level of accuracy. In some cases acquisition times could be reduced by more than a factor of 100.

3) Use as a short-term local oscillator for precision frequency standards. One of our present standards, NBS-5, is partly limited now by the stability of its local oscillator [9].

Future standards will need even better short-term stability.

MULTIPLICATION TO THE INFRARED AND THE CALCULATION OF LINEWIDTHS

There is today a growing need for frequency measurements and frequency control in the infrared to optical region of the electromagnetic spectrum. One of the problems associated with multiplying the frequency of a crystal controlled oscillator to these high frequencies is that the fast linewidth or instantaneous 3-dB linewidth of the signals becomes very large (another consideration is of course signal-to-noise). For example, a state-of-the-art commercially available quartz crystal oscillator would yield a fast linewidth W_f of nearly 1 MHz when multiplied to 1 THz—even using a perfect multiplier chain. This is to be compared with its fast linewidth of approximately 10^{-5} Hz at 5 MHz and 10^{-2} Hz at 10 GHz. This will be discussed in detail below. Analysis shows that a factor of 5 improvement in the short term stability would reduce the linewidth at 1 THz to only a few hertz. The reduction in linewidth would allow the detector bandwidth to be greatly reduced which should also improve the detected signal to noise ratio. This would improve one's ability to measure or control frequencies in the infrared and provide a platform for multiplication to the visible.

The fast linewidth W_f can be estimated from the following approximate expression

$$\int_{W_f/\pi}^{\infty} S_y(f)\,\frac{\nu^2}{f^2}\,df = 1 \text{ rad}^2 \qquad (14)$$

where ν is the frequency at which W_f is being calculated. Note that $S_y(f)\nu^2/f^2$ is just the spectral density of phase fluctuations at frequency ν.

Equation (14) integrates these phase fluctuation sa sa function of Fourier frequency from ∞ to W_f/π such that the total rms phase fluctuation is equal to 1 rad. One radian is approximately the phase fluctuation necessary to make the signal out of phase with itself.

The integral in (14) is easily broken up into parts to handle the various dependences of $S_y(f)$ on f, found in laboratory devices, e.g., see Fig. 8. For a single power law dependence of $S_y(f)$ on f, equation (14) yields values for W_f which are within a few percent of those derived by Halford in his pioneering paper [10]. The comparison for several types of commonly encountered types of noise are given in Table I. Curve a of Fig. 8 shows $S_y(f)$ calculated from the measured oscillator stability curves of Fig. 7. This was done using (5) in the flicker of frequency region and $\sigma_y(\tau) = 3bh_{-2}/(4\pi^2\tau^2)$ in the white phase modulation region. In addition, a single pole filter centered at 5 MHz with a bandwidth of +20 kHz was used to keep $S_y(f)$ finite at high frequencies. This value is also consistent with measurements of the bandwidth of the amplifier section of the oscillator. Curve b of Fig. 8 shows the measured $S_y(f)$ for our low noise 5-MHz to 9.2-GHz multiplier chains. Curve c shows $S_y(f)$ for crystal pair 5B-2 from Fig. 6 plus the estimated Johnson noise shown in Fig. 7. Again a ±20-kHz single-pole pass-band filter has been assumed.

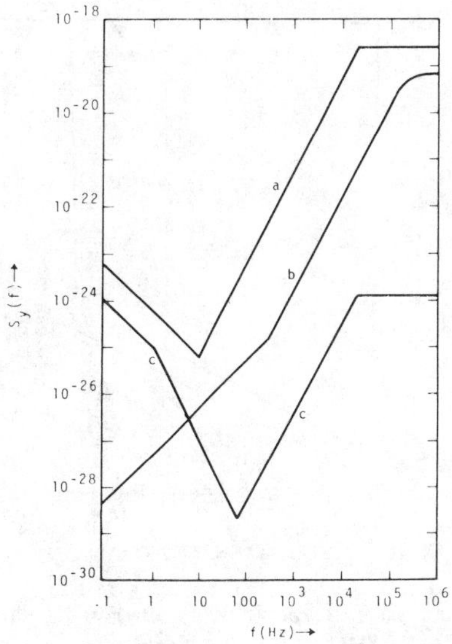

Fig. 8. Frequency domain data for: (a) high quality oscillator controlled by one 5B-2 crystal, (b) 5-MHz to 9-GHz multiplier chain, (c) 5B-2 crystal plus Johnson noise of measurement system.

TABLE I[a]

Type of Noise $S_y(f)$	Allan Variance	W_f Halford [7]	W_f equation (12)
White FM $S_y(f) = h_0 f_0$	$h_0/2\tau$	$\pi h_0 \nu^2$	$\pi h_0 \nu^2$
Flicker FM $S_y(f) = h_{-1} f^{-1}$	$2 \ln 2h_{-1}$	$\left(\frac{8\pi}{3\sqrt{3}} h_{-1}\nu^2\right)^{1/2}$	$\left(\frac{\pi^2}{2} h_{-1}\nu^2\right)^{1/2}$
Random Walk FM $S_y(f) = h_{-2} f^{-2}$	$2/3\pi^2 h_{-2}$	$(2\sqrt{2}\,\pi h_{-2}\nu^2)^{1/3}$	$\left(\frac{\pi^3}{3} h_{-2}\nu^2\right)^{1/3}$

[a] Comparison of several estimates of the fast linewidth W_f as a function of the type of noise present. $S_y(f)$ is assumed to follow a single power law dependence on Fourier frequency offset from the carrier f over the entire region of interest.

Curve a of Fig. 9 shows the linewidth calculated from curve a of Fig. 8 using (14) as a function of final output frequency ν. Although this result is very surprising and totally unanticipated from earlier calculations where $S_y(f)$ was assumed to follow a single power law dependence on f [10] it is consistent with measurements using similar oscillators and multiplier chains [7], [8], [10], [11],

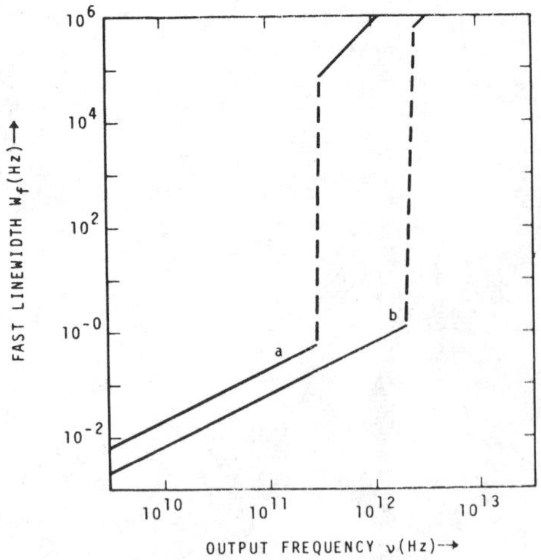

Fig. 9. Dependence of the fast linewidth on the synthesized output frequency for an input spectrum given by: (a) curve *a* in Fig. 8, (b) the sum of curves *b* and *c* in Fig. 8.

where the linewidth increased from less than 1 Hz at 10 GHz to nearly 1 MHz at 0.9 THz. The actual change in the dependence of $S_y(f)$ on f is of course more gradual than assumed in the model for the calculations. This would tend to increase the frequency span over which the linewidth step shown by curve *a* occurs. Nevertheless, the basic result of nearly 5 orders of magnitude increase in linewidth when ν is increased from approximately 130 to 500 GHz should be valid.

Curve *b* of Fig. 9 shows the resulting linewidth calculated from (14) as a function of ν, assuming that $S_y(f)$ is given by the sum of curves *b* and *c* in Fig. 8. We thus derive the fast linewidth appropriate to driving our present *X*-band multiplier chains with a crystal controlled oscillator which realizes the inherent stability of the crystal resonators shown in Fig. 7. The linewidth for frequencies above the step at approximately 2 THz is then dominated by the *X*-band multiplier chain. Further improvements in these multiplier chains would permit the linewidth step to be moved to even higher frequencies. The effect of noise on W_f in the frequency multiplication from 10 GHz to the THz region is of course reduced by 4×10^6 over the initial stages.

CONCLUSION

A new technique which allows the measurement of the inherent frequency stability of quartz crystal resonators was introduced. Measurements using this technique demonstrated that the resonators are much more stable than the corresponding crystal controlled oscillators, especially for times less than 1 s. The utility of these techniques for measuring the effects of different oscillator designs on stability, as well as the possibility of evaluation and optimizing manufacturing processes to produce crystal resonators with better short-term stability was demonstrated. A simple model of noise degradation in crystal controlled oscillators at short measurement times and suggested design changes to minimize this effect were also given. A method for estimating the fast linewidth (instantaneous 3-dB linewidth) of a signal as a function of frequency was presented. It was shown that crystal oscillators, if featuring short-term stabilities equal to those of the present crystals, would yield fast linewidths of less than 1 Hz after multiplication to 1 THz.

This is nearly a factor of 10^6 less than present crystal-controlled oscillators and a factor of 1500 less than previously obtained with any system at 1 THz [7]. Such performance would therefore greatly aid frequency metrology in this region of the spectrum as well as provide a platform for multiplication to higher frequencies. We are continuing work towards realizing this performance potential.

ACKNOWLEDGMENT

The authors acknowledge the initial suggestions of D. Halford to investigate crystals in passive circuits which stimulated this work and the many helpful discussions and suggestions of H. Hellwig, S. Jarvis Jr., D. Allan, and A. DeMarchi.

REFERENCES

[1] J. A. Barnes *et al.*, "Characterization of frequency stability," *IEEE Trans. Instrum. Meas.*, vol. IM-20, pp. 105–120, May 1971.

[2] D. Halford, A. E. Wainwright, and J. A. Barnes, "Flicker noise of phase in RF amplifiers and frequency multipliers: Characterization, cause and cure," presented at the 22th Frequency Control Symp., Atlantic City, N. J., Apr. 1968.

[3] S. Jarvis, Jr., private communication (1974). See also D. W. Allan, "Statistics of atomic frequency standards," *Proc. IEEE*, vol. 54, pp. 221–230, Feb. 1966.

[4] L. S. Cutler and C. L. Searle, "Some aspects of the theory and measurement of frequency fluctuations in frequency standards," *Proc. IEEE*, vol. 54, pp. 136–154, Feb. 1966.

[5] E. Hafner, "The effects of noise in oscillators," *Proc. IEEE*, vol. 54, pp. 179–198, Feb. 1966. See also the 1964 IEEE-NASA Symp. Short-Term Frequency Stability.

[6] J. Grosslambert, G. Marianneau, M. Oliver, and J. Uebersfeld, "Design and performance of a crystal oscillator exhibiting improved short term frequency stability," in *1974 Proc. 28th Ann. Frequency Control Symp.*, May 1974.

[7] J. S. Wells, D. G. McDonald, A. S. Risley, S. Jarvis, Jr., and J. D. Cupp, "Spectral analysis of a phase locked laser at 891 GHz, an application of Josephson Junctions in the far infrared," *Rev. Phys. Appl.*, vol. 9, pp. 285–292, Jan. 1974.

[8] K. Evenson, private commun. 1974.

[9] D. J. Glaze, H. Hellwig, D. W. Allan, S. Jarvis, Jr., and A. E. Wainwright, "Accuracy evaluation and stability of the NBS primary frequency standard," in *Conf. Rec. 1974 Precision Electromagnetic Measurements.*

[10] D. Halford, "Infrared-microwave frequency synthesis design: Some relevant conceptual noise aspects," in *Proc. Frequency Standards and Metrology Seminar*, pp. 431–466, 1971. (Available from Department of Electrical Engineering, Laval University, P. Q., Canada.)

[11] D. Glaze, private communication, 1974.

Noise Properties of PLL Systems

VENCESLAV F. KROUPA

Abstract—**This is a survey paper which begins by the derivation of the general PLL noise equation and by dividing the additive noises into the passband group and the stopband group.**

In the following paragraphs the behavior of all the major sources of additive noises is investigated and the practical numerical values of the respective power spectral densities are given.

In the terminating sections, guidelines for minimizing the additive noises in PLL systems and PLL frequency synthesizers are emphasized, and finally, phase-noise power spectral densities of several actual PLL frequency synthesizers are plotted in the normalized form.

The paper is accompanied by a copious bibliography.

I. INTRODUCTION

THE steadily increasing congestion in communications bands of the electromagnetic spectrum results in efforts for utilization of new ranges (e.g., microwave and optical frequencies) on one hand and better exploitation of the existing frequency allocations on the other hand. The latter task is met by adopting SSB modulation, telegraph multiplexing techniques, etc., and often by sharing the same communication channel by several services. In these cases the mutual interference is the major problem and it can be reduced only by increasing both the short-term and the long-term frequency stability of respective exciters and local oscillators.

Nowadays the long term stability is effectively solved with the assistance of frequency synthesis [1]-[4]. However, the situation with the short-term frequency stability is not so simple. Even in the ideal case when the frequency synthesizer is assumed to be a noiseless frequency transformer, we face an additive frequency noise [1] and often a too high phase-noise level after multiplication. The remedy to this difficulty may be a carefully designed phase-lock loop (PLL). However, this latter technique is more often used in frequency synthesizers only since it makes possible a substantial hardware simplification. And here the troubles start as nearly all PLL building blocks may add a sometimes substantial noise power to the useful signal. The problem is not yet generally understood and, in addition, is often underestimated. We shall therefore, in the following paragraphs, discuss theoretical backgrounds and summarize all the accessible experimental results to provide the leading lines for the design of low-noise PLL systems.

Manuscript received June 5, 1981; revised April 12, 1982. This paper was presented in part at the Third Symposium on Electromagnetic Compatibility, Rotterdam, The Netherlands, May 1979, and at the Conference on Precision Electromagnetic Measurements, Braunschweig, West Germany, June 1980.

The author is with the Institute of Radio Engineering and Electronics, Czechoslovak Academy of Sciences, 182 51 Prague 8, Czechoslovakia.

II. PLL NOISE EQUATION

In the following paragraphs we shall limit ourselves only to noises generated in the PLL building blocks and in a "low-noise" reference, leaving out the large class of cases with the reference frequency embedded in atmospheric or man-made noise which have been extensively dealt with in earlier works, from which many we shall mention only [5]-[7].

In Fig. 1 we have drawn a fairly general PLL arrangement with a phase detector (PD), a low-pass filter $F_L(s)$, and a voltage-controlled oscillator (VCO) in the forward path and a mixer (—), an IF filter with the effective modulation transfer function $F_M(s)$ [1, p. 234] or [5, p. 147], and a divider ($\div N$) in the feedback path. For completeness we have placed a divider ($\div Q$) between the reference generator (RG) and the phase detector and a multiplier ($\times M$) between RG and the second input to the mixer. However, we have to keep in mind that these two latter blocks, in actual PLL systems, are often replaced by more complicated frequency synthesis circuits.

Since all the noises generated or added in individual blocks in Fig. 1 are small compared with the useful signals, we have applied the rule of superposition and simply add them either at the respective inputs or outputs. Note that the subscript "n" indicates the noise signal throughout. Furthermore, the small signal theory makes it possible to use the Laplace transform approach to find the output noise of the considered PLL system or, more exactly, the respective power spectral densities.

By assuming a locked loop and by considering Fig. 1, we may write for the forward path of the loop

$$\phi_{o,n} = [(\phi_{i,n} - \phi_{o,n}')K_d + V_{\mathrm{PD},n} + V_{F,n}]F_L(s) \cdot \frac{K_0}{s} + \phi_{\mathrm{osc},n} \tag{1}$$

and for the feedback path

$$\phi_{o,n}' = (\phi_{o,n} - \phi_{m,n} + \phi_{MI,n})\frac{F_M(s)}{N} + \phi_{DN,n} \tag{2}$$

where

$$\phi_{i,n} = \frac{\phi_{r,n}}{Q} + \phi_{DQ,n} \tag{3}$$

and

$$\phi_{m,n} = M\phi_{r,n} + \phi_{MU,n}. \tag{4}$$

Reprinted from *IEEE Trans. Commun.*, vol. COM-30, pp. 2244–2252, Oct. 1982.

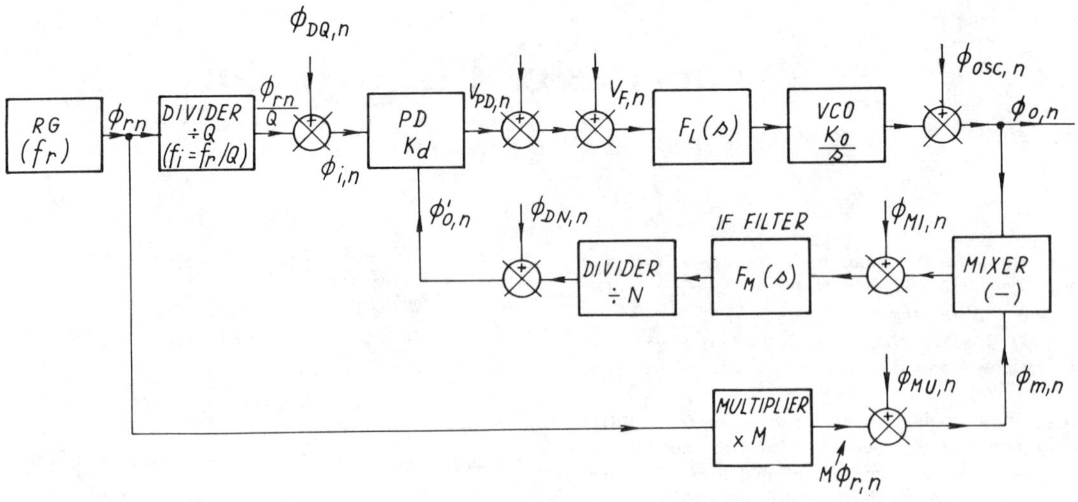

Fig. 1. Block diagram of a general PLL system with additive noise sources.

Note that all noise components $\phi \cdots, n$ and $V \cdots, n$ are Laplace transformed quantities, i.e., $\phi \cdots, n(s)$, etc.

After introducing (2) into (1) we get for the output phase noise $\phi_{o,n}$

$$\phi_{o,n} = \left[\left(\phi_{i,n} - \phi_{DN,n} + \frac{V_{\mathrm{PD},n} + V_{F,n}}{K_d}\right)\frac{N}{F_M(s)} + \phi_{m,n} - \phi_{MI,n}\right]\frac{F_M(s)F_L(s)K_dK_0/Ns}{1 + F_M(s)F_L(s)K_dK_0/Ns} + \phi_{\mathrm{osc},n}\frac{1}{1 + \dfrac{F_M(s)F_L(s)K_dK_0}{Ns}} \tag{5}$$

which can be simplified with the assistance of the effective loop transfer function $H'(s)$ [1, ch. 7] and (3) and (4) into

$$\phi_{o,n} = \left[\phi_{r,n}\left(M + \frac{N}{Q}\frac{1}{F_M(s)}\right) + \left(\phi_{DQ,n} - \phi_{DN,n} + \frac{V_{\mathrm{PD},n} + V_{F,n}}{K_d}\right)\frac{N}{F_M(s)} + \phi_{MU,n} - \phi_{MI,n}\right] \cdot H'(s) + \phi_{\mathrm{osc},n}[1 - H'(s)] \tag{6}$$

where

$$H'(s) = \frac{\dfrac{F_L(s)F_M(s)K_dK_0}{Ns}}{1 + \dfrac{F_L(s)F_M(s)K_dK}{Ns}}. \tag{7}$$

We feel that it is time to discuss the results we have arrived at until now. Let us start with the last equation.

First, we see that the loop gain $K_dK_0 = K$ is reduced in proportion to the division factor N to a new value

$$K' = K/N = K_dK_0/N \tag{8}$$

and as a consequence we face a reduced effective natural frequency ω_n' and a prolongated settling time—the remedy might be a compensating dc amplifier incorporated into the $F_L(s)$ block.

The second difficulty arrises from the IF filter in the feedback path, respectively, from its effective modulation transfer function $F_M(s)$. In many instances it may be replaced by a mere time delay function which causes a reduction of the pull-in range $\Delta\omega_P$ [8]. However, when the IF filter is a more complicated circuit we face the danger of false locks [5, p. 151] and the degradation of the loop stability.

Finally, by investigating the behavior of the PLL transfer function $H'(s)$ in the frequency domain we easily arrive at the following conclusions: for $\omega < \omega_n'$

$$|H'(s)| \approx 1 \quad \text{and} \quad |1 - H'(s)| \approx 0 \tag{9a}$$

whereas for $\omega > \omega_n'$

$$|H'(s)| \approx 0 \quad \text{and} \quad |1 - H'(s)| \approx 1. \tag{9b}$$

The consequence is that in the PLL passband the output noise is given by

$$\phi_{o,n} \approx \phi_{r,n}\left(M + \frac{N}{Q}\right) + \left(\phi_{DQ,n} - \phi_{DN,n} + \frac{V_{\mathrm{PD},n} + V_{F,n}}{K_d}\right)N + \phi_{MU,n} - \phi_{MI,n} \tag{10}$$

(note that we have neglected $F_M(s)$, the influence of which in the passband is small) and in the stopband by

$$\phi_{o,n} \approx \phi_{\mathrm{osc},n} \tag{11}$$

i.e., the PLL output noise is equal to the voltage-controlled oscillator noise. However, we shall see later that this rule of thumb is often invalidated by the improper choice of ω_n' or of the loop filter $F_L(s)$.

Now reverting to the analysis of the PLL output noise in the passband we see that the first term on the right-hand side of (10) is inevitable since it is merely a multiplied reference generator noise. The situation with the second term is not so simple; here we encounter both divider noises, the phase detector noise and the "filter noise" $V_{F,n}$, all multiplied by the division factor N. Finally, with the third term we add the multiplier and the mixer noises; however, they will be generally small compared with the second term.

As all the considered noises are random by nature and uncorrelated, we may sum the respective spectral densities.

$$S_{\varphi o,n}(f) = S_{\varphi r,n}(f)\left[M + \frac{N}{Q}\right]^2 + \left[S_{\varphi DQ,n}(f) + S_{\varphi DN,n}(f) + \frac{S_{V_{PD,n}}(f) + S_{V_{F,n}}(f)}{K_d{}^2}\right] \cdot N^2 + S_{\varphi MU,n}(f) + S_{\varphi MI,n}(f). \tag{12}$$

III. NOISES GENERATED IN INDIVIDUAL PLL BUILDING BLOCKS

In the following sections we shall investigate the quantitative and qualitative share of the individual terms, in the above equation, to the total effective "reference noise."

A. Loop Filter Noise

When the loop filter $F_L(s)$ is a passive one (see Fig. 1), i.e., a simple RC lag or lag-lead network, there are two major sources of noise, namely, some types of capacitors and resistors (carbon resistors) which can generate appreciable amounts of $1/f$ noise. As a consequence the low-noise design requires their individual selection (the use of metal-film type resistors is a necessity). The second source may be the decoupling resistor R_{dc}, separating the varactor circuit from the loop filter and the phase detector. The respective noise power density is

$$S_{V,F} = 4kT \cdot R_{dc} \approx R_{dc} \times 1.66 \times 10^{-20}\ [\mathrm{V^2/Hz}] \tag{13}$$

and for the typical value of the phase detector gain, $K_d = 0.3$, we get

$$S_{\varphi,F} = R_{dc} \times 1.84 \times 10^{-19} \qquad [\mathrm{rad^2/Hz}]. \tag{14}$$

B. Phase Noise in dc Amplifiers

In many instances we need to introduce either an active lag-lead filter [1], [5] or merely a dc amplifier.

The design of a low-noise dc amplifier is not an easy task [9] even in instances where the tested circuit diagram is used [10]-[14]. Typical equivalent input noise voltage is only several nV/$\sqrt{\mathrm{Hz}}$ with the corner-low-frequency between 10-100 Hz. Similar performance is also achieved with some modern IC operational amplifiers [9], [15].

C. Phase Noise in HF Amplifiers

Here, our investigations must start with the famous paper by Halford *et al.* [16]. They found that power spectral density of the flicker phase noise close to the carrier was approximately the same, i.e.,

$$S_\varphi(f) \approx \frac{10^{-11.2}}{f} + S_\varphi(f)_{\mathrm{white}} \tag{15}$$

for the surveyed range from 5 to 100 MHz, quite independent of the transistor type and even of the multiplication factor. Laboratory experiments proved that the intrinsic, direct phase modulation of the RF carrier by transistors was responsible for the phenomenon. The improvement—typically more than 30 dB (and up to 40 dB in some cases)—has been achieved by applying local RF negative feedback (small unbypassed resistor in the emitter—typically from 10 to 100 Ω). These findings were later supported theoretically by Healey [17] and experimentally by other authors [12], [18], [19]. Low amplifier currents and high voltages help to keep the $1/f$ noise current low. The best white noise levels $S\varphi(f)_{\mathrm{white}}$ reported [18] are of the order of 10^{-17}.

D. Phase Noise in Phase Detectors and Mixers

The experience with measurement of $S_{\varphi,n}(f)$ of low-noise crystal oscillators has taught that the best phase detectors are double-balanced mixers with Schottky barrier diodes in the ring configuration. A further improvement may be achieved by placing two diodes in each arm [11], [18]. Measurements performed by different authors [11], [13], [14], [18], [19] reveal

$$S_\varphi(f) \approx \frac{10^{-14\pm 1}}{f} + 10^{-17}. \tag{16}$$

On the other hand, there is not yet fully proved evidence that the logic circuit or digital phase detectors are much noisier [19], [20]. Šojdr [19] has measured noise properties of popular digital phase-frequency detectors in the range from 0.1 to 1 MHz and found

$$S_\varphi(f) = \frac{10^{10.6\pm 0.3}}{f}. \tag{17}$$

Furthermore, he verified the statement by Underhill *et al.* [20] that phase detectors built of ECL and CMOS logic families exhibit a better noise behavior up to about −22 dB in the flicker noise region.

E. Phase Noise in Digital Frequency Dividers

Since the frequency or phase modulation index decreases proportionally to the division factor N, the ideal noise figure is

$$F_{\mathrm{div}} = -20 \log N. \tag{18}$$

However, there is an additional noise generated in the divider itself; thus, the output phase noise is given by

$$S_{\varphi,D,n}(f) = \frac{S_{\varphi,\text{in}}(f)}{N^2} + S_{\varphi,\text{add}}(f). \tag{19}$$

To find out the properties of the additive term we have normalized the above equation with respect to the output frequency, i.e. (cf. Section III-G),

$$\frac{S_{\varphi,D,n}}{f_{\text{out}}^2} = \sum_{k=-2}^{2} h_{k,D,n}$$

$$= \sum_{k=-2}^{2} h_k f^{k-2} + \frac{h_{1,\text{add}}}{f} + h_{2,\text{add}} \tag{20}$$

and plotted the respective coefficients $h_{1,\text{add}}$ and $h_{2,\text{add}}$, computed from accessible experimental results [3], [19], [21]-[23] in Fig. 2. We see that the overall behavior is the same, namely, for large division factors (small output frequency) the data fit a straight line with the slope of 20 db/dec. Be referring to the lowest noise points we may write for the additive term

$$S_{\varphi,\text{add}} \approx \frac{10^{-14.7}}{f} + 10^{-16.5}. \tag{21}$$

On the other hand, for higher output frequencies (above 1 MHz) both coefficients $h_{1,\text{add}}$ and $h_{2,\text{add}}$ become constants which indicate that the input noise predominates.

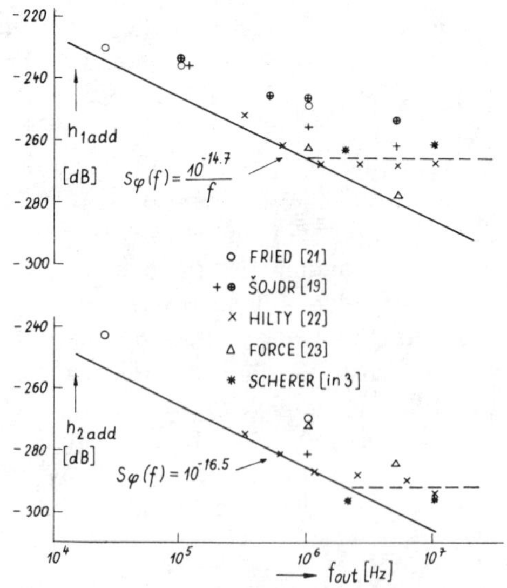

Fig. 2. Plot of the normalized additive phase-noise coefficients $h_{1,\text{add}}$ and $h_{2,\text{add}}$ in respect to the divider output frequency as measured by different authors.

F. Phase Noise in Frequency Multipliers

We have to refer first to the above mentioned paper by Halford [16] and further to that by Baugh [24] where guidelines for the design of low-noise frequency multipliers are given (RF negative feedback—see Section III-C—and steep zero crossings). To get more information we have collected published results about noise properties of transistor frequency multipliers [12], [24]-[29] and diode frequency multipliers [12], [28], [30], [31] as well and found that both noise constants $a_{-1,\text{inp}}$ and $a_{0,\text{inp}}$ are nearly the same for properly designed transistor frequency multipliers, irrespective of the frequency, i.e.,

$$S_{\varphi MU,\text{inp}}(f) \approx \frac{10^{-14}}{f} + 10^{-16.5}. \tag{22}$$

In diode frequency multipliers the flicker phase noise level is higher, typically

$$10^{-12.9}/f. \tag{23}$$

We shall see later that the flicker and white phase-noise spectral densities of the best crystal oscillators are of the same order or rather worse as the respective additive terms in frequency multipliers; thus, we can conclude that the phase noise is not appreciably deteriorated by passing the signal through a properly designed frequency multiplier. (Note that this is hardly true with the diode frequency multipliers.)

G. Phase Noise in Oscillators[1]

The theory of noise in free-running oscillators has been dealt with by many authors for nearly half a century and we feel that the best we can do is to mention only a few of them [33]-[36]. However, they do not provide the information the designer of the PLL systems is looking for, at least by the first fast reading. On the other hand, when we start from the condition of the zero phase shift around the oscillating loop [37] we easily arrive at the heuristic oscillator phase-model suggested by Leeson [38], [17].

In accordance with this model any oscillator can be simplified into a loop containing a resonator and an amplifier-limiter. As a consequence its output spectral density is given by

$$S_\varphi(\omega)_{\text{osc},n} = S_\varphi(\omega)_{A,n}[1 + (\omega_0/2Q_L\omega)^2] \tag{24}$$

where the amplifier-limiter noise is

$$S_\varphi(f)_{A,n} = a_{-1}/f + a_0. \tag{25}$$

The magnitude of the flicker noise constant a_{-1} has been found experimentally [16] in the range from 5 to 100 MHz to be

$$a_{-1} = F_{-1} \cdot 10^{-11.2} \quad [\text{rad}^2]. \tag{26}$$

The noise factor F_{-1} depends on the emitter RF feedback and may be made as small as 10^{-3}.

The white noise constant is the ratio of the noise power $\overline{\varphi_{\text{noise}}^2(t)}$ to the oscillator power P_0 reduced to 1 Hz bandwidth and multiplied by a noise factor F_0:

$$a_0 = F_0 \cdot kT/P_0 \approx 4 \times 10^{-21} F_0/P_0 \quad [\text{rad}^2/\text{Hz}]. \tag{27}$$

[1] This section is based on the paper read by the author at the EMC-79 Symposium, Rotterdam, The Netherlands [32].

By introducing (25) into (24) we arrive at a power law relation

$$S_{\varphi \text{osc},n}(f) = \left(\frac{f_0}{f}\right)^2 \cdot \frac{a_{-1}}{f \cdot 4Q_L{}^2} + \left(\frac{f_0}{f}\right)^2 \cdot \frac{a_0}{4Q_L{}^2} + \frac{a_{-1}}{f} + a_0. \tag{28}$$

To compare oscillators with different output frequencies we face the difficulty that the noise sidebands close to the carrier are proportional to the square of the resonant frequency f_0. This problem is often solved by referring to the fractional-frequency power spectral density

$$S_y(f) = (f/f_0)^2 S_\varphi(f), \tag{29}$$

i.e.,

$$S_{y,\text{osc},n}(f) = \frac{a_{-1}}{f \cdot 4Q_L{}^2} + \frac{a_0}{4Q_L{}^2} + \frac{a_{-1}}{f_0{}^2} f + \frac{a_0}{f_0} f^2 \tag{30a}$$

$$= \frac{h_{-1}}{f} + h_0 + h_1 f + h_2 f^2 \tag{30b}$$

or to the spectral density of the phase-noise time [14]

$$S_x(f) = S_y(f)/(2\pi f)^2. \tag{31}$$

The advantage of this second normalization is the close resemblance to the actual phase noise characteristic and the ease with which $S\varphi(f)$ is calculated—the proportionality factor being $(2\pi f_0)^2$. By dropping the factor $(2\pi)^2$ we arrive at a very useful simplification:

$$S_{\varphi \text{osc},n}(f)/f_0{}^2 = \frac{h_{-1}}{f^3} + \frac{h_0}{f^2} + \frac{h_1}{f} + h_2 \tag{32}$$

where

$$h_{-1} = a_{-1}/4Q_L{}^2; \qquad h_0 = a_0/4Q_L{}^2; \tag{33a}$$

$$h_1 = a_{-1}/f_0{}^2; \qquad h_2 = a_0/f_0{}^2. \tag{33b}$$

We have verified the validity of (28)–(32) by computing coefficients h_{-1} and h_0 for different types of oscillators from a wealth of published data and plotted them as functions of quoted Q_L. The results are shown in Fig. 3 and indicate a good agreement with the simplified oscillator noise theory. The mean value of the flicker noise constant a_{-1} is

$$a_{-1} \approx 10^{-11} \tag{34}$$

and is practically independent of the oscillator type in the whole frequency range from 5 MHz to 100 GHz; the same is true also for the white noise constant a_0, the mean value of which is

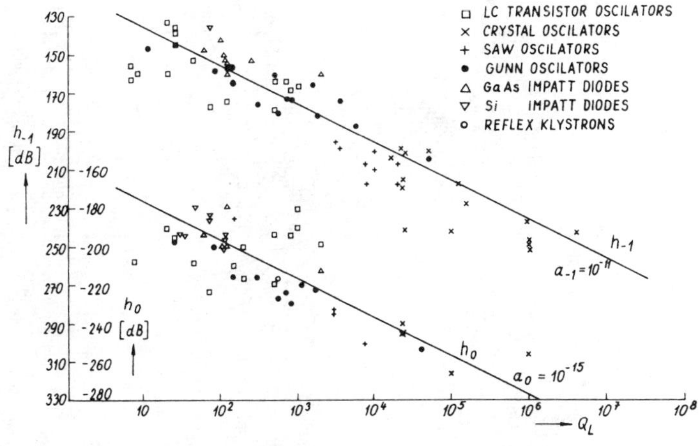

Fig. 3. Plot of the normalized phase-noise coefficients h_{-1} and h_0 in respect to the loaded Q_L of different oscillators.

$$a_0 \approx 10^{-15}. \tag{35}$$

Similar experimental verification of the dependence of h_1 and h_2 on $1/f_0{}^2$ is prevented by the lack of data for the noise power at higher Fourier frequencies. Some insight provides the investigation of noise properties of crystal oscillators discussed in the following section.

Summarizing all the above results we can write a fairly general oscillator noise equation

$$\frac{S_{\varphi \text{osc},n}(f)}{f_0{}^2} = \frac{1}{f^3} \cdot \frac{10^{-11.6}}{Q_L{}^2} + \frac{1}{f^2} \frac{10^{-15.6}}{Q_L{}^2} + \frac{1}{f} \cdot \frac{10^{-11}}{f_0{}^2} + \frac{10^{-15}}{f_0{}^2} \tag{36}$$

and plot the normalized oscillator phase-noise characteristics which consist of two sets of straight lines with parameters Q_L and f_0; see Fig. 4.

H. Phase Noise in Reference Frequency Generators

The reference generator in low-noise PLL systems (frequency synthesizers) is a spectrally pure crystal oscillator. By inspecting (36) or Fig. 4 we see that a low close-to-carrier noise requires the use of resonators with the highest possible Q. It has been found earlier [39] that for the AT-cuts the intrinsic losses in the quartz crystal material are related to the resonant frequency by

$$f_0 Q \approx 1.5 \times 10^{13}. \tag{37}$$

This product is slightly lower for the advantageous SC-cuts and nearly two times larger for BT-cuts. Consequently, the low flicker and white frequency noise, i.e., small h_{-1} and h_0 in (32) and (33), requires the lowest possible frequency f_0. However, to keep dimensions of the crystal resonators in practical limits, we hardly can go below 5 MHz (cf. [39, Fig. 2]). Very often 10 MHz crystal oscillators are used as reference frequency generators.

To find practical values of h_k coefficients in (32) we have recently investigated noise characteristics of about 60 crystal oscillators in the range from 5 to 170 MHz [40] and

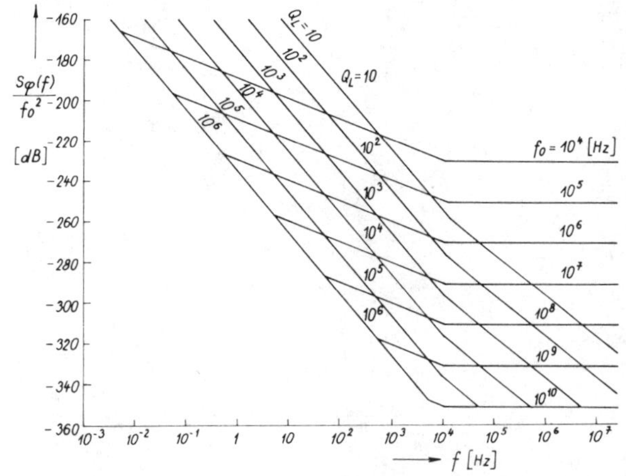

Fig. 4. Normalized phase-noise characteristics of oscillators: parameters are the loaded Q_L of the resonator and the output frequency f_0.

get for an average crystal oscillator the following noise equation.

$$\frac{S_{\varphi\mathrm{osc},n}(f)}{f_0^2} = \frac{1}{f^3}\,10^{-37.25} f_0^2 + \frac{1}{f^2} 10^{-39.4} f_0^2 + \frac{1}{f}\,\frac{10^{-12.15}}{f_0^2} + \frac{10^{-14.9}}{f_0^2}. \quad (38)$$

By investigating the dispersion of the h_k values, particularly of the best crystal oscillators we have arrived at the conclusion that in the $1/f^3$ and $1/f^2$ regions the crystal resonator noise [41], rather than the transistor noise, is the limiting factor [42].

IV. PLL LOOP FILTERS FOR LOW NOISE BEHAVIOR

The problem has been discussed earlier by the author [43] and here only the results will be summarized.

By formulating the rule of thumb in (9) in Section II we have assumed that the transfer function $H'(s)$ has a rectangular behavior. However, this is not true in real systems and a closer investigation of the second- and third-order systems gives

$$H'(s) = \frac{s\omega_n'(2\xi - \omega_n'/AK') + \omega_n'^2}{s^2 + 2\xi\omega_n' s(1 + s^2\kappa\omega_n'^2) + \omega_n'^2} \quad (39)$$

where

$$\omega_n' = \sqrt{K'/T_1}; \quad \xi = \frac{\omega_n'}{2}\,T_2 + \frac{1}{AK'};$$

$$\kappa = T_3/T_2; \quad K' = K_d K_0/N \quad (40)$$

and A is the gain of the operational amplifier used.

Since the asymptotic approximation will generally supply sufficient information for the noise behavior of the studied PLL system, we shall consider the four most important configurations.

1) Simple RC filter, i.e., $A = 1$, $\xi = \omega_n'/2K'$, and $\kappa = 0$. After introducing the normalization

$$\frac{\omega}{\omega_n} = x \quad (41)$$

we shall find

$$H(jx) \approx -1/x^2 \qquad x \gg 1 \quad (42a)$$

and

$$1 - H(jx) \approx 2j\xi x \qquad x \ll 1. \quad (42b)$$

2) Passive lag-lead filter, i.e., $A = 1$, and $\kappa = 0$.

$$H(jx) \approx -j(2\xi - \omega_n'/K')/x \qquad x \gg 1 \quad (43a)$$

$$1 - H(jx) \approx jx\omega_n'/K' \qquad x \ll 1. \quad (43b)$$

3) Active lag-lead filter, i.e., $A \to \infty$, and $\kappa = 0$.

$$H(jx) \approx -2j\xi/x \qquad x \gg 1 \quad (44a)$$

$$1 - H(jx) \approx -x^2 \qquad \omega_n'/AK' \ll x \ll 1. \quad (44b)$$

4) Active lag-lead filter with an additional RC section (it has been shown [43] that for practical applications $\kappa < 0.3$).

$$H(jx) \approx -1/\kappa x^2 \qquad x \gg 1 \quad (45a)$$

$$1 - H(jx) \approx -x^2 \qquad \omega_n'/AK' \ll x \ll 1. \quad (45b)$$

To get a better insight we shall consider the problem of phase-locking a 100 MHz crystal oscillator and a low Q LC-oscillator to a 5 MHz reference signal. All normalized phase noise characteristics are plotted in Fig. 5. By considering first the 100-to-5 MHz PLL system we find the crossover point to be approximately 200 Hz. In instances where f_n is smaller than 200 Hz we face a large amount of additive noise, the origin of which is the attenuated 100 MHz oscillator noise; the dashed lines indicate the situation with passive filters and dot-and-dash lines indicate the improvement when the active lag-lead filter is used. Similarly, we encounter an unnecessary additive noise caused by the attenuated 5 MHz oscillator noise in cases where $f_n > 200$ Hz (dashed line). A remedy can be provided by additional filtering; however, the stability of the loop deteriorates.

In the second example of phase-locking a low Q oscillator to a crystal reference, the use of the active lag-lead filter is necessary since even for f_n equal to the crossover frequency we face a large additive noise with passive filters only (see again the dashed line in Fig. 5).

V. NOISE IN PLL FREQUENCY SYNTHESIZERS

By considering the basic PLL configuration, as shown in Fig. 1, we easily arrive, with the assistance of (12) and the condition of only a 3 dB noise increase, at the conclusion that

$$S_{\varphi,r,n}(f)\left[M + \frac{N}{Q}\right]^2 \leq N^2 \cdot S_{\varphi,\mathrm{PLL}}(f) \quad (46)$$

where

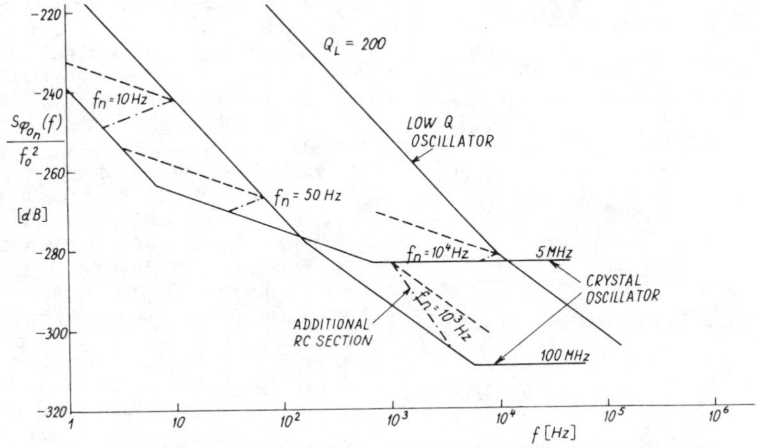

Fig. 5. Normalized phase noise of the PLL system of the 100 MHz crystal oscillator locked to a 5 MHz reference crystal oscillator and a low Q oscillator locked to the same reference oscillator.

$$S_{\varphi,\mathrm{PLL}}(f) = 2S_{\varphi,D}(f) + S_{\varphi,\mathrm{PD}}(f) + S_{\varphi,A,n}(f). \tag{47}$$

The power spectral densities of individual noise sources in the above equation have been evaluated in Section III. In a large majority of cases where active filters are used the last term on the right-hand side of (47) will dominate. With the assistance of [9, Fig. 12] and a reasonable value for K_d, $K_d = 0.3$, we find

$$S_{\varphi,\mathrm{PLL}}(f) \approx \frac{10^{-14}}{f} + 10^{-15.5}. \tag{48}$$

Since the flicker and white noise spectral densities of good reference oscillators in the 5 and 10 MHz frequency ranges are of the same order we arrive at the rule of thumb that the division factor N should not exceed the multiplication factor M, i.e.,

$$N \gtreqless M. \tag{49}$$

However, this condition can hardly be met in instances where small frequency steps are desired at the synthesizer output. One solution is provided by the application of fractional frequency dividers [2, p. 74], [44], [45]. The spurious phase modulation which is often quite large but predictable [46] is suppressed by compensation.

Another solution may be provided by a subtracting PLL-system, the principle of which will be explained with the assistance of Fig. 6. In the case where N in Fig. 1 is much larger than M, the output noise of the first digital PLL in its passband is given approximately by

$$S_{\varphi,01,n}(f) \approx N^2 \cdot S_{\varphi,\mathrm{PLL},1}(f). \tag{50}$$

With the assistance of (12) we find for the output noise (in the passband) of the subtracting loop

$$S_{\varphi,02,n}(f) \approx \frac{N^2 S_{\varphi,\mathrm{PLL},1}(f)}{P^2} + M^2 S_{\varphi r,n}(f) + S_{\varphi,\mathrm{PLL},2}(f). \tag{51}$$

Generally, the last term on the right-hand side of the above equation may be neglected and the application of the 3 dB noise-increase condition reveals the second rule of thumb, i.e.,

$$\frac{N}{P} \leq M. \tag{52}$$

By a judicious combination of both these techniques one can expect a very low additive phase noise from frequency synthesizers at frequency ranges above 100 MHz [44], [45], [47]. For generation of lower frequencies the use of dividers instead of output mixers should be preferred [21], [44], [45].

To demonstrate the state of the art we have plotted in Fig. 7 the normalized power spectral densities of the output phase noise of several commercial PLL frequency synthesizers. The advantage of the normalization is the possibility to compare the noise properties of frequency synthesizers with different output frequencies and that of the reference generator together in the same figure. The progress achieved in the last ten years is impressive.

VI. CONCLUSIONS

In this survey paper we have called the readers' attention to all major sources of additive noise in the PLL system. Furthermore, we have shown that, in the first approximation, these noises add to the input or the reference noise.

In the second we have investigated noises generated in individual PLL building blocks and tried to find numerical values for the respective power spectral densities with the assistance of the experimental findings published by different authors all over the world.

The major guideline rules for minimizing the additive noises in complicated PLL systems or PLL frequency synthesizers have been discussed in the last two sections. Finally, phase-noise power spectral densities of several commercial PLL frequency synthesizers have been plotted in normalized form in Fig. 7. On one hand, this figure demonstrates the state of the art and, on the other hand, the progress achieved in the last decade.

For readers who intend to go deeper into the problem we have collected a copious bibliography.

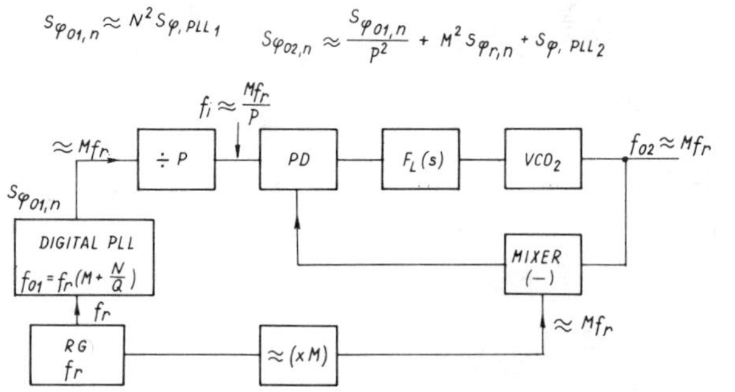

Fig. 6. A subtracting PLL system.

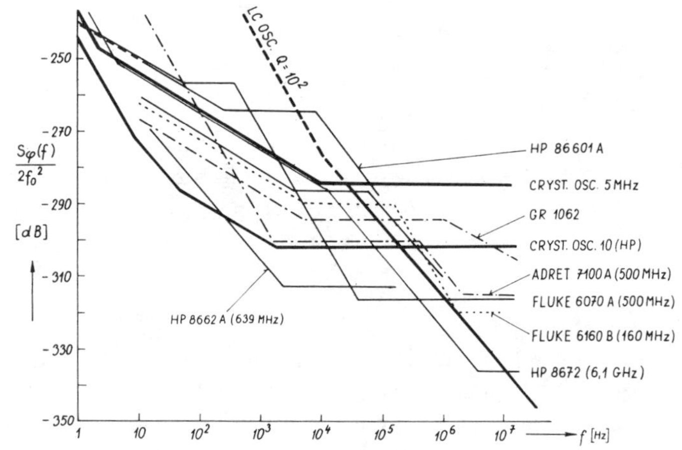

Fig. 7. Normalized phase noise of several commercial PLL frequency synthesizers, two reference crystal oscillators, and one typical LC oscillator (VCO).

REFERENCES

[1] V. F. Kroupa, *Frequency Synthesis, Theory, Design and Applications*. London, England: Griffin, 1973; New York: Wiley, 1973.

[2] J. Gorski-Popiel, *Frequency Synthesis: Techniques and Applications*. New York: IEEE Press, 1975.

[3] V. Manassewitsch, *Frequency Synthesizers, Theory and Design*. New York: Wiley, 1980.

[4] W. F. Egan, *Frequency Synthesis by Phase Lock*. New York: Wiley, 1981.

[5] F. M. Gardner, *Phaselock Techniques*, 2nd ed. New York: Wiley, 1979.

[6] W. C. Lindsey, *Synchronization Systems in Communication and Control*. Englewood Cliffs, NJ: Prentice-Hall, 1972.

[7] W. C. Lindsey and M. K. Simon, *Telecommunication Systems Engineering*. Englewood Cliffs, NJ: Prentice-Hall, 1973.

[8] J. A. Develet, Jr., "The influence of time delay on second-order phase-lock loop acquisition range," in *Proc. Int. Telemetering Conf.*, vol. 1, Sept. 23–27, 1963, pp. 432–437; see also, W. C. Lindsey and M. K. Simon, Eds., *Phase-Locked Loops and Their Application*. New York: IEEE Press, 1978.

[9] Y. Netzer, "The design of low-noise amplifiers," *Proc. IEEE*, vol. 69, pp. 728–741, June 1981.

[10] B. E. Blair, Ed., *Time and Frequency: Theory and Fundamentals*, Nat. Bur. Stand., Boulder, CO, Monog. 140, May 1974, p. 169.

[11] D. G. Meyer, "A test set for the accurate measurement of phase noise on high-quality signal sources," *IEEE Trans. Instrum. Meas.*, vol. IM-19, pp. 215–227, Nov. 1970.

[12] S. G. Andresen and J. K. Nesheim, "Phase noise of various frequency doublers," *IEEE Trans. Instrum. Meas.*, vol. IM-22, pp. 185–188, June 1973.

[13] K. Hilty, "Messtechnik zur Bestimmung des Phasenrauschspektrums," *Technische Mitteilungen PTT*, pp. 1–15, Jan. 1980.

[14] P. Kartaschoff, *Frequency and Time*. London, England: Academic, 1978.

[15] G. Erdi *et al.*, "Op amps tackle noise—And for once, noise loses," *Electron. Design*, vol. 28, pp. 65–71, Dec. 20, 1980.

[16] D. Halford, A. E. Wainright, and J. A. Barnes, "Flicker noise of phase in RF amplifiers and frequency multipliers: Characterization, cause, and cure," in *Proc. 22nd Annu. Freq. Contr. Symp.*, Atlantic City, NJ, Apr. 1968.

[17] D. J. Healey, "Flicker of frequency and phase and white frequency and phase fluctuations in frequency sources," in *Proc. 26th Annu. Symp. Freq. Contr.*, Atlantic City, NJ, June 1972, pp. 29–49.

[18] F. L. Walls, S. R. Stein, J. E. Gray, and D. J. Glaze, "Design considerations in state-of-the-art signal processing and phase noise measurement systems," in *Proc. 30th Annu. Symp. Freq. Contr.*, Atlantic City, NJ, June 1976, pp. 269–274.

[19] L. Šojdr, "Phase noise in oscillators and synthesizers" (in Czech), Dissertation, Prague, Czechoslovakia, June 1979.

[20] M. J. Uderhill, P. A. Jordan, M. A. G. Clark, and R. I. H. Scott, "A general purpose LSI frequency synthesizer system," in *Proc. 32nd Annu. Symp. Freq. Contr.*, Atlantic City, NJ, May–June 1978, pp. 365–372.

[21] R. Fried, "A wide-band non-heterodyned frequency synthesizer," John Fluke, Inc., Seattle, WA, 1974.

[22] K. Hilty, private communication at the occasion of the CPEM-80 Conf., Braunschweig, West Germany, 1980.

[23] M. Force, "Low-noise interface circuits format crystal-oscillator signals," *EDN*, vol. 25, pp. 125–129, 131, Oct. 20, 1980.

[24] R. A. Baugh, "Low noise frequency multiplication," in *Proc. 26th Annu. Symp. Freq. Contr.*, Atlantic City, NJ, June 1972, pp. 50–54

[25] D. J. Glaze, "Improvements in atomic cesium beam frequency standards at the National Bureau of Standards," *IEEE Trans. Instrum. Meas.*, vol. IM-19, pp. 156–160, Aug. 1970.

[26] D. J. Healey, "$\mathcal{L}(f)$ measurements on UHF sources comparising VHF crystal controlled oscillator followed by a frequency multiplier," in *Proc. 28th Annu. Symp. Freq. Contr.*, Atlantic City, NJ, May 1974, pp. 190–202.

[27] F. L. Walls and A. De Marchi, "RF spectrum of a signal after frequency multiplication; Measurement and comparison with a simple calculation," *IEEE Trans. Instrum. Meas.*, vol. IM-24, pp. 210–217, Sept. 1975.

[28] E. Bava, A. De Marchi, and A. Godone, "Spectral analysis of synthesized signals in the mm wavelength region," *IEEE Trans. Instrum. Meas.*, vol. IM-26, pp. 128–132, June 1977.

[29] A. L. Lance, W. D. Seal, F. G. Mendoza, and N. W. Hudson, "Automating phase noise measurements in the frequency domain," in *Proc. 31st Annu. Symp. Freq. Contr.*, Atlantic City, NJ, June 1977, pp. 347–358.

[30] E. Bava, A. De Marchi, and A. Godone, "A narrow output linewidth multiplier chain for precision frequency measurements in the 1 THz region," in *Proc. 31st Annu. Symp. Freq. Contr.*, Atlantic City, NJ, June 1977, pp. 578–582.

[31] M. B. Bloch and A. Vulcan, "Low noise measurement techniques," in *Proc. 28th Annu. Symp. Freq. Contr.*, Atlantic City, NJ, May 1974, pp. 184–189.

[32] V. F. Kroupa, "Electromagnetic compatibility of frequency synthesizers," in *Proc. 3rd Symp.*, Rotterdam, The Netherlands, May 1979, pp. 335–340.

[33] I. Berstein, "On fluctuations in the neighbourhood of periodic motion of an auto-oscillating system," *Comptes Rendus de l'Academie des Sciences de l'URSS*, vol. XX, no. 1, 11–16, 1938.

[34] W. A. Edson, "Noise in oscillators," *Proc. IRE*, vol. 48, pp. 1454–1466, Aug. 1960.

[35] K. Kurokawa, "Noise in synchronized oscillators," *IEEE Trans. Microwave Theory Tech.*, vol. MTT-16, pp. 234–240, 1968.

[36] K. F. Schünemann and K. Behm, "Nonlinear noise theory for synchronized oscillators," *IEEE Trans Microwave Theory Tech.*, vol. MTT-27, pp. 452–458, May 1979.

[37] E. J. Baghdady, R. N. Lincoln, and B. D. Nelin, "Short-term frequency stability: Characterization, theory, and measurement," *Proc. IEEE*, vol. 53, pp. 704–722, July 1965.

[38] D. B. Leeson, "A simple model of feedback oscillator noise spectrum," *Proc. IEEE*, vol. 54, pp. 329–330, Feb. 1966.

[39] A. W. Warner, "Design and performance of ultraprecise 2.5-mc quartz crystal units," *Bell Syst. Tech. J.*, vol. 39, pp. 1193–1217, Sept. 1960.

[40] V. F. Kroupa, "Noise of crystal oscillators" (in Czech), Rep. Z-1042/I, Prague, Czechoslovakia, Nov. 1979.

[41] F. L. Walls and A. E. Wainwright, "Measurement of the short-term stability of quartz crystal resonators and the implications for crystal oscillator design and applications," *IEEE Trans. Instrum. Meas.*, vol. IM-24, pp. 15–17, Mar. 1975.

[42] V. F. Kroupa, J. Pavlovec, and L. Šojdr, "Noise in standard frequency sources," to be published.

[43] V. F. Kroupa, "Spectral properties of third order phase-locked loops," in *Proc. 5th Summer Symp. Circuit Theory*, Kladno, Sept. 1977, pp. 201–215.

[44] K. L. Astrof, "Frequency synthesis in a microwave signal generator," *Hewlett-Packard J.*, vol. 29, pp. 8–15, Nov. 1977.

[45] Special Issue on PLL Frequency Synthesizer Type 8662A, *Hewlett-Packard J.*, vol. 32, Feb. 1981.

[46] V. F. Kroupa, "Spectra of pulse rate frequency synthesizers," *Proc. IEEE*, vol. 67, pp. 1680–1682, Dec. 1979.

[47] G. Mackiw and G. W. Wild, "Microwave frequency synthesis for satellite communications ground terminals," in *Proc. 30th Annu. Symp. Freq. Contr.*, Atlantic City, NJ, June 1976, pp. 420–437.

Author Index

Subject Index

Editor's Biography

Věnceslav F. Kroupa was born in Třebíč, Czechoslovakia, on September 26, 1923. He received the Ing. degree in electrical engineering from Prague Technical University, Prague, Czechoslovakia, in 1948. After attending an additional year of post-graduate study in radio engineering, he then studied theoretical physics for three years at Charles University, Prague, Czechoslovakia, and received the Ph.D. degree from the Czechoslovak Academy of Sciences, Prague, in 1967.

From 1948 to 1951 he was a Development and Research Engineer in the Transmitter Department of Tesla n.p. From 1951 to 1955 he was with the Research Institute of Physics and Electronics, Prague, Czechoslovakia. In 1955 he joined the Institute of Radio Engineering and Electronics of the Czechoslovak Academy of Sciences. For the past 30 years, his major research interests have been frequency stability, frequency synthesis, and frequency measurements, subjects on which he has presented approximately 20 papers at international conferences and published more than 40 papers, as well as two books: *Frequency Measurement* (in Czech) and *Frequency Synthesis* (in English). He has also written several book reviews.